DYNAMICS OF CLOSE BINARY SYSTEMS

ASTROPHYSICS AND SPACE SCIENCE LIBRARY

A SERIES OF BOOKS ON THE RECENT DEVELOPMENTS
OF SPACE SCIENCE AND OF GENERAL GEOPHYSICS AND ASTROPHYSICS
PUBLISHED IN CONNECTION WITH THE JOURNAL
SPACE SCIENCE REVIEWS

VOLUME 68

ZDENĚK KOPAL

University of Manchester

DYNAMICS OF CLOSE BINARY SYSTEMS

SPRINGER-SCIENCE+BUSINESS MEDIA, B.V.

Library of Congress Cataloging in Publication Data

CIP

Kopal, Zdeněk, 1914 –
Dynamics of close binary systems.

(Astrophysics and space science library; v. 68)
Bibliography: p.
Includes indexes.
1. Stars, Double. 2. Stars – Evolution.
I. Title. II. Series.
QB821.K76 523.8'41 77-20081
ISBN 978-94-009-9782-0 ISBN 978-94-009-9780-6 (eBook)
DOI 10.1007/978-94-009-9780-6

Originally published by D. Reidel Publishing Company, Dordrecht, Holland in 1978
Softcover reprint of the hardcover 1st edition 1978

TABLE OF CONTENTS

PREFACE

The aim of the present book will be to provide a comprehensive account of our present knowledge of the theory of dynamical phenomena exhibited by close binary systems; and on the basis of such phenomena as have been attested by available observations to outline probable evolutionary trends of such systems in the course of time.

The evolution of the stars – motivated by nuclear as well as gravitational energy sources – constitutes nowadays a well-established branch of stellar astronomy. No theories of such an evolution are as yet sufficently specific – let alone infallible – not to require continual tests by a confrontation of their consequences with the observed properties of actual stars at different stages of their evolution. The discriminating power of such tests depends, of course, on the range of information offered by the test objects. Single stars which move alone in space are now known to represent only a minority of objects constituting our Galaxy (cf. Chapter I-2); and are, moreover, not very revealing of their basic physical characteristics – such as their masses or absolute dimensions. If there were no binary systems in the sky, the only star whose vital statistics would be fully known to us would be our Sun.

The presence of 'visual' binary systems at known distance permits a determination of the absolute masses of their components up to a relatively small distance (of the order of 10^2 pc) from the Sun; and beyond this limit absolute masses as well as dimensions of the stars can be inferred only from the observations of 'spectroscopic' binaries which happen to be 'eclipsing variables' by virtue of a favourable orientation of their orbital planes to the line of sight. And not only are the external properties of the stars available for inspection if these stars happen to be components of such binaries; but even their internal structure may not (under certain conditions) remain concealed from us. At first sight, it may appear incredible that we could penetrate by observations into stellar interiors concealed from direct sight by enormous opacity of their material. However, even these forbidding regions are not wholly concealed from the inquisitive human mind. A gravitational field emanates from the interior, which the opacity of the overlying layers cannot appreciably modify; and radiant energy originating in the deep interior will govern (in accordance with the structure of the intervening layers) the distribution of brightness over the visible surface (cf. Chapter IV-4).

As long as a star is single, no empirical way exists for gauging (or analyzing) the 'anatomy' of its external gravitational field; or learning anything about the distribution of light on its surface. Place, however, another star in its proximity: and various properties of the combined gravitational field of both components can be deduced from observable characteristics of their motions, discussed in Chapter V. The variation of light induced by the rotation of distorted components of close eclipsing systems may permit us to deduce many additional physical properties of their constituent stars from an analysis of their observed light changes.

And what is true of a determination of the vital statistics of individual stars, is even more true of a quest for their evolutionary stage. A sample of single stars selected at random in any part of the galactic space would contain objects of very different ages as well as different masses and chemical composition – facts rendering an identification of their individual evolutionary stages impossible. A more favourable situation would be obtained if we could identify a group of stars of the same age and initial composition, which may differ from each other only in mass (or angular momentum). Such groups are indeed known to exist – to wit, stellar associations and star clusters of different types, whose colour-magnitude diagrams reflect evolutionary dispersion of equally old stars of different masses after a certain time. However – except for relative luminosities – the individual physical properties of cluster stars are not known to us any better than if such stars were solitary travellers through space.

This situation changes, however, drastically if we turn our attention to the extreme case of associations limited to *two* (or a few more) objects which we call *double* (or multiple) *stars* – and especially to the *close binary systems* which constitute gravitational liaisons durable enough to outlast a galaxy (cf. sec. I-2). The importance of such systems as a source of documentation of the entire story of stellar evolution form cradle to grave is obvious, and impossible indeed to exaggerate; for they provide a unique opportunity for tracing the effects of *differential evolution* of the stars of (virtually) the same age and initial composition, differing initially in mass. This story will be unfolded to the extent of our present knowledge in the concluding Chapter VIII of this book; though an acquaintance with the preceding chapters is largely a prerequisite for its understanding.

Since the discovery of the first double stars in the sky in the latter part of the 18th century, double-star astronomy had been the 'astronomer's astronomy' – without much direct interface with other branches of the physical sciences. This situation began, however, to change dramatically in the past ten years by the discovery, among close binary systems, of extreme end-products of the evolution of massive stars, which manifest themselves as binary X-ray sources, pulsars and (hypothetical) black holes. The possibilities of the existence of such objects have already attracted the attention they deserve on the part of the relativists and cosmologists, who have recognized their value for observational tests of theories of the behaviour of matter and space metric under extreme conditions not encountered anywhere else in the Universe.

This is only to be welcomed; and the outcome of joint work may benefit more than one branch of science. At the same time, the relativists – who until quite recently have heard but little of close binary systems in the sky – should be cautioned to get acquainted with all aspects of the problem before committing themselves to conclusions which may be premature (such as seeking evidence for 'gravitational waves' in the observed period changes of eclipsing variables of certain types, or finding 'black holes' in every single-spectrum system), and susceptible of interpretation in terms of a less exotic framework. If the present book can make this clear to the more general reader, it would have fulfilled a useful purpose – apart from a possible interest which it may hold for the specialists.

With these aims in mind, the contents of this book have been divided into eight chapters – each constituting a more or less self-contained unit. Chapter I which follows this preface will introduce to us the 'dramatis personae' – i.e., different classes of binary stars encountered in the sky, and to outline their cosmic setting – both within our Galaxy and beyond. Chapter II will be concerned with a specification of the *equilibrium form* of fluid components of close binary systems of arbitrary structure in the prevalent field of force arising from axial rotation and equilibrium tides or their interaction. In Chapter III we shall turn our attention to *dynamical tides* arising if the rotation of both components is not synchronized with their orbital revolution, and (or) the equators of the components are inclined to the invariable plane of the system. In doing so, we shall pay particular attention to the role played by dissipative forces (arising from gas as well as radiative viscosity) which were largely neglected in previous work, and whose action is responsible for many observable phenomena exhibited by close binary systems.

Chapter IV (Generalized Rotation) will contain the development of a consistent theory of rotational motion, in three dimensions, of selfgravitating configurations consisting of compressible viscous fluid. A mechanism will be investigated in detail by which the rotation of such configurations of arbitrary structure will be synchronized with orbital revolution through viscous tides; and the amount of kinetic energy determined which gets converted into heat in the course of this process. Its investigation has been presented in such detail because of the fundamental importance which this mechanism will turn out to have for tidal evolution of close binary systems (cf. Chapter VIII-4). The tidal friction produced by this process represents the most effective mechanism available (in the absence of magnetic fields) for an exchange between the rotational moments of the components and the orbital momentum of the system.

In Chapter V – the longest of this volume, which has also given it its name – we shall address ourselves to the problems of relative motions, in three dimensions, of close binary systems perturbed by the departures of its components from spherical shape – again with emphasis on phenomena arising from dissipative forces (tidal lag!) whose effects have so far received but inadequate attention, even though they are responsible for the salient processess influencing fundamentally the evolution of close binary systems – such as (1) rectification of the axes of rotation of both components to a position perpendicular to the orbital plane; (2) synchronization of their rotation and revolution; and (3) secular circularization of their orbits.

Chapter VI (The Roche Model) will contain an exhaustive discussion of geometrical properties of a double-star model consisting of a gravitational dipole of two revolving mass-points surrounded by opaque envelopes of negligible mass. The relevance of this model to the 'surfaces of zero-velocity' of the restricted problem of three bodies will be pointed out, as well as an extraordinary extent to which the geometrical properties of such a model can simulate those of actual binary systems consisting of components whose density concentrations – while not infinite – are high.

Chapter VII (Stability of Close Binary Systems) will be concerned with the problems of secular as well as ordinary (vibrational) stability of the components, and its determination by methods closely allied with those of Chapter II for secular stability, and of Chapter VI-3 for ordinary stability. Ultimately, in Chapter VIII, we shall marshal all evidence bearing on evolutionary processes in close binary systems, and attempt an interpretation of the observed facts, consistent with the essentials of their dynamics expounded in Chapters II–VII. The contents of these chapters contain the necessary – though not yet sufficient – material for such an interpretation; and while the choice of the respective subject matter (omitting as it does, for instance, a fuller discussion of the electromagnetic phenomena) does not provide a complete basis for an unambiguous interpretation of the observed facts bearing on the evolution of close binary systems, no such interpretation can claim acceptance in violation of the dynamical principles expounded in Chapters II–VII.

A glance at the contents of these chapters will disclose that not all of them cover their respective subjects to the same degree of completeness. For example, the contents of sections 1–4 of Chapter II are classical; and their results can be refined to an arbitrary precision by known methods. This, however, is not yet the case with the subject matter of sections II-4 and II-5; and especially the latter offers much scope for further exciting work. Throughout Chapter III the emphasis will be on dissipative hydrodynamics of radiating systems; and in Chapter IV on interaction between axial rotation and viscous dynamical tides, the importance of which for tidal evolution of close binary systems (section VIII-4) has already been emphasized.

The same leading motive will follow us largely through Chapter V as well. It is the dissipative phenomena, caused by the viscous friction, which will see to it that the axes of rotation of the components of close binary systems will be secularly rectified to a position normal to the invariable plane of the system (cf. sec. V-2), or the eccentricity of the relative orbit reduced asymptotically to zero by tidal lag (esc. V-3).

The contents of sections 1 and 2 of Chapter VI are again classical; but that of section VI-3 is new, and basic for problems considered in section VII-3. As regards Chapter VII, the reader will undoubtedly note that while the methods of investigation are fairly well known, specific results of their application to actual binary systems are still largely conspicuous by their absence – due to overwhelming weight of the requisite numerical work. As a result we do not know, for instance, to this date whether or not the Roche model loses secular and (or) ordinary stability at the Roche Limit – a purely static property of the model – for a given structure of the evanescent envelope; or does so already before this limit has been reached?

The concluding Chapter VIII of this book constitutes, in a sense, a culmination of our effort to understand the evolutionary processes in close binary systems on the basis of theoretical knowledge and observed facts available so far. The critical reader will, however, discern the extent to which any hypothesis advanced so far falls short of the ultimate solution of the problem. The principal theoretical desiderata for investigation – without which no real further progress can be made – are (a) a knowledge of the onset of the dynamical instability of the Roche model; and (b) the nature of the physical processes causing the loss of mass.

Each chapter of this book has been concluded by a set of Bibliographical Notes (not necessarily complete), the aim of which will be to acquaint the reader with a brief history of the respective subject and with further collateral literature. Now, when the literature on our subject is rapidly growing in unprecedented dimensions, it is all the more important to keep the bibliographical record straight, and to attempt sifting the grains from the chaff in so far as this can be done by any single writer. The literature is also rapidly becoming replete with re-discoveries – some bona fide, others assuredly not – which should be kept in check; and the future alone can judge to what extent the present writer has proved equal to this task.

It is almost 20 years since the author addressed himself to the problems of close binaries in a book (*Close Binary Systems*, 1959) which has since gone out of print as well as (in many respects) out of date. A tremendous amount of work done in this field since that time made the idea of a new edition of its original text impracticable. The present volume covers only a part of the subject matter of the 1959 book, with emphasis on dynamical phenomena in close binary systems; and in so far as it represents an original contribution by the writer, it largely summarizes his work in the past ten years (1966–1976).

No room has been devoted in this volume to the second half of the contents of the 1959 book – that concerned with an interpretation of light or radial velocity curves of close binary systems. New developments in this aspect of our general field in the past few years have also been dramatic – mainly by a transfer of the underlying problem from the time- to the frequency-domain. However, a proper account of them would really require a separate volume of a size comparable with this book, which the author hopes to complete – God willing – in the next few years.

Last but not least, this preface would not be complete without an expression of the author's sincere appreciation to Mrs. Ellen B. Carling-Finlay, who typed the entire difficult manuscript of this book for the press with the same care that she bestowed on the 1959 volume; and whose editorial assistance at all stages throughout these years has been truly invaluable.

On the Day of Epiphany, A.D. 1977 ZDENĚK KOPAL

CHAPTER I

BINARY STARS IN THE SKY

The term *binary star* was apparently used by William Herschel to designate . . . "a real double star – the union of two stars that are formed together in one system, by the laws of attraction" (Herschel, 1802). The term *double star* is, on the other hand, of much earlier origin: at least its Greek equivalent was already used by Ptolemy to describe the appearance of ν Sagittarii, two fifth-magnitude stars whose angular separation is about 14′ (i.e., a little less than the apparent radius of the Moon); and it has been used ever since to describe close pairs of stars resolvable with the aid of a telescope. Not every 'double star' defined in this sense constitutes, to be sure, a 'binary system'; for a large majority of them may be optical pairs owing the accidental proximity of the projections on the celestial sphere to the laws of chance. The first double star which we know to form a binary system was ζ Ursae Maioris (Mizar), which was discovered around 1650 by Father Giovanni Baptista Riccioli in Bologna (and whose principal component was recognized by E. C. Pickering as the first spectroscopic binary in 1889). In 1656, Christian Huyghens saw θ Orionis resolved into the principal stars of the Trapezium; and in 1664 Robert Hooke noted that γ Arietis consisted of two stars. At least two additional pairs (one of which proved to be of more than ordinary interest) were discovered before the end of the seventeenth century – namely, α Crucis, discovered in 1685 by Father Fontenay, Jesuit missionary at the Cape of Good Hope; and α Centauri, discovered by his confrère, Father Richaud, while observing a comet at Pondicherry, India, in December 1689.

All these discoveries were accidental, and made in the course of observations taken for other purposes. No suspicion seems to have been entertained by all these observers or their contemporaries that the proximity of the two stars in such pairs was due to other reason than chance; but although they were regarded as mere curiosities and no special effort made to augment their list, their number grew from decade to decade until, around 1750, several dozen of such pairs had been noted and recorded. As their number gradually became sufficiently large to lend itself to rudimentary statistical analysis, the cosmic significance of double stars appeared to conceal more than met the eye of their discoverers.

The first scientific argument in favour of the view that at least some, and probably many, double stars then known were the result of physical rather than optical association we owe to another faithful servant of Christ, the Reverend John Michell. On May 7th and 14th of the year 1767, Michell read a remarkable paper before the Royal Society in London, published subsequently in the Society's *Transactions* (Michell, 1767), in which he pointed out that the frequency-distribution of the angular separations of double stars known in his time deviated grossly from one that could be expected for chance association of stars uniformly distributed in space – there appeared to be far too many close pairs among them – and according to Michell, "the natural conclusion from hence is, that it is highly probable, and next to a certainty in general, that such double stars as appear to consist of two or more stars placed very near together, do really consist of stars placed nearly together, and under the influence of some general law . . . to whatever cause this may be owing, whether to their mutual gravitation, or to some other law or appointment of the Creator" (op. cit., p. 249).

The directness of Michell's expression left perhaps something to be desired, but the logic of his argument was unimpeachable and appears convincing to us today. Unfortunately, Michell's contemporaries did not see it in quite the same light. Consider, for example, the reaction of Michell's younger contemporary, William Herschel. In a paper entitled 'On the Parallaxes of Fixed Stars' (Herschel, 1782), Herschel expressed conviction that the components of double stars which are very unequal in brightness must be at very different distances from us and, therefore, particularly suitable for measurements of the relative parallax of their brighter (i.e., nearer) components. This prompted him to embark upon a systematic search for such paris, and the results of his efforts were gathered in the *First* and *Second Catalogues of Double Stars* published in 1782 and 1785. Michell was, to be sure, quick to point out that Herschel's new discoveries greatly strengthened his earlier probabilistic argument (Michell, 1784), but Herschel still remained unimpressed. It was not till in 1803 that the ageing astromer admitted, in a paper entitled 'Account of the Changes that have happened during the last Twenty-five Years, in the relative Situation of Double-stars: with an Investigation of the Cause to which they are owing' (Herschel, 1803) there must indeed be true binary system.

And more; for Herschel's sustained observations of several such systems (of which Castor was one) for over a third of a century demonstrated that the relative motion of the fainter component around the brighter represents (in projection) an *ellipse* on the celestial sphere, with the brighter component situated at one of its *foci*. A geometry of the orbits of the two stars around their common centre of mass is a direct consequence of the characteristics of the field of force binding them together to form a binary system. The elliptical form of orbital trajectories by itself is consistent (cf. Bertrand, 1873) with *two* distinct types of fields, in which the force varies either with the direct first, or inverse second, power of the distance between the two stars. In the former case, the centre of mass would be located at the centre of the apparent ellipse (i.e., at the intersection of its principal axes); in the latter, at its focus. The fact that Herschel's observations demonstrated the latter alternative to be true was of great cosmogonical importance in his day; for it supplied an observational proof that Newton's law of gravitation is universally valid also outside the confines of our solar system.

Since Herschel's time, double-star astronomy by visual (and, later, photographic) means has made great strides, and has been enriched by the discovery of thousands of new pairs; moreover, those binaries – less than one hundred in number – for which absolute orbits as well as parallaxes could be determined, have become an important source of our knowledge on the masses of the stars. Should we, however, be limited only to the data forthcoming from this source, our knowledge of the masses (and absolute dimensions) of the stars would be severely limited, and confined largely to those with masses smaller than that of the Sun. Fortunately, Nature provided for other sources to this end in the sky – in the form of *close binaries* – which are (mostly) beyond the means of actual optical resolution on account of the close proximity of their components, but which can be recognized as such by other means. Thus, with increasing proximity, the velocity of orbital motions is bound to increase, and may eventually become sufficiently large for its radial component to produce measurable Doppler shifts in the line spectra of such systems. Many hundreds of close binaries of this nature – we call them *spectroscopic binaries* – have been discovered in this way since the first such object was recognized by Vogel in 1889; and their observations have provided invaluable data on physical properties of a much wider class of stars.

While spectroscopic observations by their very nature are capable of providing information on absolute properties of the stars regardless of their distance, they cannot furnish actual masses or absolute dimensions unless the inclination of the orbits of such pairs of stars to the celestial sphere can be established by other methods. This can indeed be done if this inclination is large enough for the two stars to eclipse (partially or totally) each other at the times of conjunctions – thus giving rise to characteristic variation of light which earned this type of binary systems the name of *eclipsing variables.*

Eclipsing binaries represent, in fact, the oldest known close binary systems – which were also recognized as such long before Herschel demonstrated the existence of visual binaries. The Italian astronomer G. Montanari (then in Bologna) appears to be the first man on record to have noted (in 1670) that the star Algol (*El Ghūl* of the Arabs) in the constellation of Perseus did not shine always with constant light, but the latter occasionally dropped below the normal; and became so impressed by this curious phenomenon as to write a special pamphlet about it (Montanari, 1671). Twenty-five years later the variability of Algol was confirmed by Maraldi. However, neither he, nor Montanari, seems to have had any inkling of the periodicity of Algol's light changes; which was discovered by Goodricke in 1782. In his communication which has justly become famous, John Goodricke – a deaf-and-dumb astronomer then 18 years of age – not only reported the discovery of the first known short-periodic variable star in the sky (those known before were long-periodic or irregular in nature), accompanied by a remarkably close estimate of its period*; but at the end of his communication (Goodricke, 1783) we find the following sentence which had truly made it historic: ". . . if it were not perhaps too early to hazard even a conjecture on the cause of this variation, I should imagine it could hardly be accounted for otherwise than . . . by the interposition of a large body revolving around Algol. . .".

Nature denied much to young Goodricke, but certainly not the gift of a splendid imagination. For it seldom happens in the annals of science that the first conjecture of a discoverer happens to strike the nail on the head more accurately than this suggestion of a young man of 18, who "for some years amused himself with astronomy" (Banks, 1783); and who within the short life vouchsafed to him (he died at the age of 22) found time enough to discover, besides Algol, also the variability of β Lyrae (as well as δ Cephei).

At any rate, Goodricke's bold suggestion that Algol was an eclipsing binary was made too early to gain speedy acceptance. William Herschel himself was plainly non-comittal: "the idea of a small Sun revolving around a large opake body has also been mentioned in the list of such conjectures" (Herschel, 1783). He had reasons for doubt, he thought; for prior to his report, he had observed Algol repeatedly in the focus of his 7-foot telescope (of the discovery of Uranus fame) and found it "distinctly single". Twenty years later it

* Goodricke's original value for the period of Algol was 2 days, 20 hours and 45 minutes, differing from its true period by only 4 minutes. A year later (1784) Goodricke revised his period to 2 days, 20 hours, 49 minutes and 9 seconds – a result on which modern observations have but little to improve!

became Herschel's destiny to demonstrate that many visual double stars observed by him form indeed physical binary systems conforming to this definition (Herschel, 1803); but whether or not the ageing astronomer ever made up his mind about Algol we do not know. It looks very much as if the uninhibited boldness which was so characteristic of most of his other astronomical speculations somehow failed when confronted with this particular kind of stellar symbiosis – there is not a bit of evidence in all his prolific writings to suggest a rational cause for this apparent reluctance. At any rate, Goodricke's brilliant suggestion of 1783 was destined to remain in the realm of hypotheses for many more decades. It was not until 1889, when Vogel (1890) recognized Algol as a spectroscopic binary whose conjunctions coincided with the minima of light, that the binary nature of Algol and other similar eclipsing variables was at last established beyond any doubt. And it may be edifying for the reader – especially if he is a theoretician – to reflect on the historical fact that, while the first representatives of wide (i.e., visual) as well as close (i.e., eclipsing) binary systems were discovered by Riccioli (1650) and Montanari (1670) who both happened to observe in Bologna – the site of the oldest of all Universities – under the clear Italian skies, the binary nature of each group was recognized by pure reasoning by two British amateurs – Michell (1767) and Goodricke (1783) – long before the observers like Herschel (1803) or Vogel (1889) provided the compelling observational proof.

Another historical fact – borne clearly by the record – may be noted in this place: namely, that contrary to what is true of the recognition of the existence of binary systems in the sky, in subsequent interpretation of the observed facts the theoreticians have consistently trailed behind the observers for many decades. This was, to be sure, not true of the visual binary systems, where an analysis of the observations presented essentially geometrical problems that received adequate attention since the days of Bessel. But it remains a fact that while the first eclipsing variables were discovered in 1782 (Algol) and 1783 (β Lyrae), it was not till 1880 that Edward C. Pickering attempted to interpret the light changes of the former (cf. Pickering, 1880); and the research so initiated gathered momentum but very slowly because of the difficulties inherent in the problems challenged by the early investigators.

In 1912, H. N. Russell and H. Shapley developed the first general method for an analysis of the light changes of eclipsing variables (cf. Russell, 1912; Russell and Shapley, 1912) applicable to any type of eclipse, and provided adequate means for preliminary solutions for the elements on the basis of observations of moderate accuracy. In order to do so, however, several reasonable but oversimplified assumptions had to be made concerning the form of components (spheres, or similar ellipsoids) or the distribution of brightness over their apparent discs – assumptions prompted by the desire for simplicity of treatment rather than by their physical justification; but as time went on, the gradually increasing accuracy of underlying observations on the one hand, and a fuller understanding of physical properties of close binary systems on the other, have provided a sustained incentive for further efforts to understand more fully the nature of the problem.

The weakest feature of the Russell-Shapley methods* has been the fact that, in order to render a direct solution possible, a part of it (the position of fixed points!) had to be

* Referred to, in more recent years, also as Russell-Merrill methods; though nothing of real significance has been added to them since 1912.

anticipated – a fact liable to introduce bias whose effects could not be controlled, or subsequently removed from the result. In point of fact, the Russell-Shapley approach did not lead to any 'solution' of the problem in the mathematical sense at all – only to an approximate picture of the system which must be verified by more systematic work. Since the underlying problem in the time-domain is highly non-linear (in fact, transcendental) in the unknown elements to be determined, the only sound way to solve it is to seek the unknowns by successive approximations.

Systematic methods for the construction of such solutions by iteration have been initiated by Kopal (1941a), and developed subsequently by Kopal (1946a, 1948a, 1950a, 1959) and Piotrowski (1947, 1948) to a stage at which we could speak of a mathematical solution of the problem: the conditions under which iterations converge for different types of eclipses have been investigated; and error analysis furnished permitting us to express the uncertainty of the individual elements in terms of the observational errors. Yet, from the physical point of view, the models underlying Kopal and Piotrowski's nonlinear analysis were still the same as that employed by Russell and Shapley; and a greater mathematical rigour had to be paid for by an increasing load of numerical work – which only recently could be automated for solution with electronic computers (cf. Huffer and Collins, 1962; Jurkevich, 1970; Linnell and Proctor, 1970, 1971; Budding, 1973; or Söderhjelm, 1974).

The real breakthrough with the solution of this problem did not occur till still more recently, when the present writer was able to show (by a resort to Fourier techniques) that the underlying problem which is *transcendental* in the *time*-domain becomes *algebraic* in the *frequency*-domain; and thus can be solved without the need of any special functions which beclouded so much the time-domain approach in the past (cf. Kopal, 1974a, b, c, d, 1975a, b). What is more – the light changes arising from the eclipses as well as from the proximity effects (ellipticity, reflection), which are hopelessly scrambled together and whose separation (by 'rectification') turned out to be the principal stumbling block to further progress in the time-domain, proved to be neatly separable in the frequency-domain by virtue of the different nature of their frequency spectra.

It is, however, not the aim of this volume to concern ourselves with the methods of analysis of the observations of different types of close binary systems, but only with the results obtained with their aid. While the principal data which can be extracted from the light changes of eclipsing variables are the fractional radii $r_{1,2}$ of the constituent components and the inclination i of their orbital plane to the celestial sphere, spectroscopic observations provide the calibration of the masses and dimensions of the system in absolute units: it is a combination of both which provides most part of what we know of the absolute properties of close binary systems and of their components.

Since the 1780's when our subject began to engage astronomical attention, up to the early 1930's the accuracy of available observations was sufficiently low to permit us to treat the light curves of eclipsing systems known up to that time as arising from the eclipses of 'billiard-ball' spherical (or, at most, ellipsoidal) stars, by purely geometrical methods, without any need of physical considerations. The first investigator who pointed out that the components of 'typical Algol systems' must be very dissimilar in shape was K. Walter (1931); and the consequences of his pioneer suggestion were soon independently explored by S. Takeda (1934, 1937) on the basis of Chandrasekhar's 1933 work on the equilibrium figures of heterogeneous stellar models; it was then that the idea took root

– and gained gradual recognition – that the actual shape of fluid components of close binary systems must be the resultant of their axial rotation, fractional dimensions and mass-ratio – an idea whose consequences will be fully explored in Chapters II–IV of this book.

Moreover, from the purely astronomical point of view, all components of close binary systems known up to 1930 were situated, in the Hertzsprung-Russell diagram, on the Main Sequence or to the right of it. Since that time, however, a series of new and unexpected discoveries have gradually ushered our subject to its present 'golden age'. In 1933, S. Beliavsky discovered a new inconspicuous eclipsing variable – designated in due course as UX Ursae Maioris – whose single distinguishing feature was a remarkably short period (4 h and 43 min). At first his discovery failed to attract much attention; for the system was assumed to consist of a pair of dwarf stars of G- or M-type for which a period so short would be natural. In 1941, however, G. P. Kuiper (1941b) took a spectrum of this star and found it to be, not G or M, but B3! This fact renewed interest in UX UMa; and photometric work by A. P. Linnell (1950) established firmly a series of remarkable properties of this first known representative of a new class of subdwarf binary systems, of which several more have been discovered since that time (such as AE Aqr, EM Cyg, W Pup, RW Tri) – with periods ranging from 1 h and 40 min for W Pup to 16h 49m for AE Aqr. Subsequently, all but the first one of these systems proved to be eclipsing variables as well (cf., e.g., Warner, 1971: or Walker, 1971).

And more: on the heels of a realization that T CrB (recurrent Nova Coronae Borealis 1866 and 1946) is an eclipsing variable with a period of 227.6 days (Sanford, 1949; cf. also Kraft, 1958) came a discovery by M. Walker (1956) that Nova Herculis 1934 is an eclipsing variable (DQ Her), which consists of a close pair of subdwarfs (one of which is the Nova) revolving around their common centre of gravity in a period of 4 h and 39 min. Moreover, subsequent search among stars of this type led to a discovery of the binary nature of several other post-Novae – such as GK Per (Nova Persei 1901), V 603 Aql (Nova Aquilae 1918), WZ Sge (Nova Sagittae 1913 and 1946) or T Aur (Nova Aurigae 1891) – the last two of which turned out to be also eclipsing variables. The orbital periods of these systems range from 1 h 21.5 min for WZ Sge to 4 h 54 min for T Aur; only GK Per (possibly eclipsing) has a period longer than a day (1.904d). These data already on hand strongly support a contention that Nova outbursts occur only among the components of close binary systems; and that binary nature represents a necessary prerequisite for such phenomena.

The same appears, moreover, to be true not only for triggering off instabilities which produce Nova outbursts, but also the (more frequent) outbursts of light exhibited by cataclysmic variables of the SS Cyg- and U Gem-type. In point of fact, both these prototypes of their class turned out to be binaries consisting of close pairs of subdwarfs – revolving around their common centre of mass in a period of 6 h and 38 min for SS Cyg and 4 h 10.5 min for U Gem. All other known stars of this type (such as RX And, SS Aur, RU Peg, etc.) proved likewise to be binaries with orbital periods less than 9 h. Only U Gem has so far been found to eclipse (Krzeminski, 1965): while SS Cyg is a spectroscopic binary (Joy, 1956), but not an eclipsing variable (Grant, 1955).

In addition to binaries consisting of close pairs of subdwarfs, at least one eclipsing system (V 471 Tau) has been discovered (cf. Nelson and Young, 1970), in which a Main-Sequence K0 red dwarf is attended by a white dwarf. The orbital period of this system is likewise very short (12 h 30.5 min), and appears to be variable (Young and Lanning, 1975). A discovery of further systems of this type can be regarded only as a matter of time.

But it is not only among subdwarfs and stars of generally small masses that important discoveries in our field have been made in the past decades; for even greater surprises have come to light in more massive systems. In 1940, S. Gaposchkin discovered the Wolf-Rayet star HD 193576 to be an eclipsing variable (V 444 Cyg), in which the star exhibiting the Wolf-Rayet spectrum (of mass 9.8 of the Sun) constitutes its less massive component. Although the discovery light curve appeared still to be consistent with the 'billiard-ball' model of the system (cf. C. P. Gaposchkin, 1941), subsequent photoelectric work by Kron and Gordon (1943), showing a conspicuous disparity in the form of both minima of light, disclosed clearly the semi-transparent ('edgeless') nature of the expanding envelope of the Wolf-Rayet component – the first object of this type whose existence was established by the observations, but not the only one for long. In subsequent years, the Wolf-Rayet stars HD 214419 (CQ Cep), HD 168206 (CV Ser), or CX Cep proved likewise to be eclipsing variables – in addition to at least seven others which do not eclipse. If one takes account of observational selection (random distribution of orbital inclinations) it is possible – even probable – that all known Wolf-Rayet stars are components of close binary systems (cf. Underhill, 1973).

But the greatest surprises of recent years in the astronomy of close binary systems were in store for us in the nascent domain of X-ray astronomy. As is well known, the first discrete X-ray source in the sky (Sco X-1) was discovered in 1962 by Rossi and his colleagues by use of high-altitude rockets*; and when, two years later, this source was identified with its faint blue counterpart in the optical domain of the spectrum, it transpired that this object emits 10^3 times more energy in the X-ray domain than in the visible part of the spectrum. This source soon proved not to be alone in the sky; and especially after the launch of the X-ray Uhuru satellite in December 1970 the number of known X-ray sources began to grow by leaps and bounds. Most of these are (as yet) outside the scope of this chapter; but some have already proved to be of entrancing interest for double-star astronomy as well.

The spectrum of Sco X-1 source in the optical domain proved to be strikingly Nova-like; and as a binary nature of post-Novae was by then well known (cf., e.g., Kraft, 1963) the suspicion began to form that the Sco X-1 source may be the component of a close binary system. This suspicion was strengthened when in 1971, the Uhuru satellite discovered that two other known X-ray sources – Cen X-3 and Her X-1 (HZ Her) – are binaries with periods 2.087 and 1.700 days, respectively; and, moreover, exhibit intensity oscillations with a period of 4.84 s and 1.24 s. We may recall that Walker (1956) found the system DQ Her (Nova Her 1934) to exhibit optical oscillations with a period of 70 s; and a pos-

* A chance discovery if there ever was one; as the avowed aim of the experiment was to detect the X-ray bremsstrahlung from the lunar surface (cf. Rossi, 1973). The lunar X-rays proved too weak to register at the distance of the Earth (their actual discovery did not come till four years later with the Soviet Luna 12; cf. Mandelshtam *et al.*, 1968). However, at the time of Rossi's experiment the Moon happened to be in the constellation of Scorpius; and the source now known as Sco X-1 chanced to be near the Moon's limb, within the field of view of the apparatus employed. It soon proved to have nothing to do with the Moon (as it did not share in the Moon's motion), but to represent a new and previously unknown phenomenon in the sky.

sibility suggested itself that in Cen X-3 and Her X-1 we may have to do with components of close binaries similar to the post-Nova Her 1934, but (because of the short period of oscillation) of smaller size and (presumably) higher density – akin to neutron stars rather than to ordinary white dwarfs.

The Centaurus X-3 source proved to be considerably more interesting of the two (cf. Chapter VIII-5C) on account of its very large mass; its X-ray source undergoing total eclipses by its mate every 2.087 days (Krzeminski, 1973). The same is also true of the Cygnus X-1 source in the northern sky (period of 5.60 days) whose massive secondary component (of approximately 10 times the mass of the Sun) appears to be the most promising (though by no means certain) candidate known so far for a 'black hole'.

Another recent development of high astronomical as well as physical interest was the discovery in 1974 of the fact that the pulsar PSR 1913 + 16 appears to be the component of a close binary system (cf. Hulse and Taylor, 1975). The binary nature of such a system would not, of course, be detectable by any optical means; it is, however, indicated by the fact that the time interval of 0.05903 s between successive radio pulses of PSR 1913 + 16 (shorter than that of any other known pulsar except for the one in the Crab nebula) does not remain constant (or increase, as for all single objects of this type, uniformly with time), but fluctuates in a period of 0.3230 days – presumably because the distance of the object from us varies in the course of the pulsar's revolution around the centre of mass of a binary system. If so, the time interval between successive pulses would be bound to fluctuate periodically on account of the 'light equation' in the system – a subject to be discussed in section V-6D of this book. So tiny a 'clock' moving in a closed orbit would be of great interest to astronomers and relativists alike – the first of many no doubt still awaiting discovery – and its existence has forged another link between double-star astronomy and cosmology.

A fuller account of facts which bear on these and other similarly fascinating systems will be given in section VIII-5B of this book, together with a discussion of their possible meaning. In this place we wish to stress, however, that not only the stars themselves constituting close binary systems have proved invaluable sources of information for stellar astronomy, but the same has turned out to be true also of their neighbourhood. This story begins again with Algol – the first eclipsing variable discovered in 1782 – whose light (and, later, radial-velocity) changes have been under observation now for almost two centuries. Some forty years ago, D. B. McLaughlin (1934) discovered, however, that the radial velocity with which the system of Algol AB travels through space is not constant within the limits of observational errors, but oscillates in a period of 1.873 years. This phenomenon McLaughlin explained by the presence, in the neighbourhood of the close binary Algol AB, of a third star (Algol C), with which Algol AB revolves around the common centre of gravity in a period of 1.873 years.

The probability of this interpretation was strengthened by the fact that the orbital period of Algol AB exhibits fluctuations, with the same period, on account of the 'light equation' in the third orbit (cf. Chapter V-6D); and glimpses of the lines of Algol C in the visible part of the combined spectrum of this triple system were reported from time to time by several observers (cf. Pearce, 1939; Meltzer, 1957; Struve and Sahade, 1957; Fletcher, 1964; or Ebbighausen, 1970); Periodic oscillations of Algol's position in the sky have been detected also astrometrically (cf. van de Kamp *et al.*, 1951; Bachmann and Hershey, 1975). Lastly, by a resort to ingenious techniques of 'speckle interferometry',

Labeyrie and his colleagues (1974) were recently able to resolve the image of Algol C from that of Algol AB; thus confirming the correctness of previous deductions based on indirect observational data; though the techniques used by these investigators proved incapable of resolving optically Algol AB (or any other eclipsing binary) so far.

But since the commencement of this century, in addition to its other historical claims, Algol 'pioneered' also another line of investigation of the nature of close binary systems: namely, that concerning the presence of gaseous matter in close binaries which may surround the entire system. In studying the spectra of Algol secured by Schlesinger and others at Allegheny Observatory between 1907–1912, Miss Barney (1923) noted the fact that certain lines (mainly metallic) in the spectra of Algol failed to participate in the Doppler shifts of all other lines caused by the orbital motion of its components. Somewhat later, Carpenter (1930) discovered another related, and even more remarkable, fact in the spectrum of U Cephei. As is well known, the alternate minima of this eclipsing system are constantly situated half-way between each other – a fact demonstrating that the relative orbit of the two stars must be circular; and yet the radial-velocity curve was found by Carpenter to be plainly asymmetric! This fact baffled many astronomers at that time (Carpenter himself described it, in the title of his paper, an "anomalous result"); and it was not till Carpenter's results were later confirmed by Struve (1944) – and analogous results detected in several other systems (SX Cas, UX Mon, etc.) by Struve and his collaborators – that the phenomenon was realized to signify the presence of moving gas in the respective system, generally related with the evolutionary stage of its components.

The physical relation between these stars and the gas surrounding them will be discussed more fully in the concluding chapter of this book. At present we merely wish to add that, in Algol, circumstellar gas was recently found to constitute also a radio source – fluctuating in intensity in an irregular manner (cf. Hughes and Woodsworth, 1972; Hjellming, 1972; Pooley and Ryle, 1973; and many others) quite unrelated with the light changes of this eclipsing system. Moreover, recent speckle-interferometric work (Labeyrie *et al.*, 1974) disclosed that the centre of intensity of the radio emission did not coincide with the position of Algol AB, but is displaced significantly from it – a fact indicating that the observed radio emission originates in gas (or plasma) distributed asymmetrically with respect to the main mass of the system.

All these discoveries did not merely increase the diversity of known species of close binary systems and of their individual characteristics. By doing so they provided also important clues to the absolute properties of these stars – standard as well as exceptional – and to their symbiosis with the gaseous matter surrounding them. An interpretation of these clues as indicators of the evolution of these stars from cradle to grave will constitute the subject matter of the concluding Chapter VIII of this volume; while in what follows we propose to explore another clue offered by Nature for assessing the significance of close binary systems in the Universe: namely, their absolute as well as relative frequency in space.

I-1. Binary-Star Population in our Galaxy

In the introductory part of this chapter we outlined a brief history of the discoveries which led to the contemporary state of double-star astronomy, and pointed out the principal types of binary systems encountered in this quest. The aim of the present section will be to consider more quantitative aspects of a search for binary-star population in our Galaxy (and beyond), and their statistical characteristics. What proportion of binaries of different types do we encounter in our neighbourhood, or in more distant parts of the space?

Any attempt to answer this question must be preceded by a more quantitative definition of the meaning of the term 'binary star' than that given so far. If we define a binary as a pair of stars for which the probability of describing at least one orbital revolution around their (temporary) common centre of gravity is equal to (say) one-half, it has been shown by Chandrasekhar (1944) that the mean life-time τ of their dissolution by the cumulative effects of perturbations of nearby stars will be given by the expression

$$\tau \cong \frac{(m_1 + m_2)^{1/2}}{4\pi \sqrt{GmNa^{3/2}}} \tag{1.1}$$

where $m_{1,2}$ are the masses of the two components of the binary pair; m, the average mass of nearby stars; N, the number of such stars (of average mass m) per unit volume of space; a, the semi-major axis of their orbit; and G, the constant of gravitation. If, moreover, we express the masses on the r.h.s. of Equation (1.1) in solar units; N in cubic parsecs (pc^{-3}) and a in astronomical units (AU), Equation (1.1) can be rewritten as

$$\tau = 1.11 \times 10^{14} \frac{(m_1 + m_2)^{1/2}}{mNa^{3/2}} \quad \text{yr.} \tag{1.2}$$

Let us, next, identify this time of dissolution with the period

$$P^2 = \frac{4\pi^2 a^3}{G(m_1 + m_2)} \tag{1.3}$$

of the Keplerian orbit. On insertion from (1.3) in (1.2), the latter equation can be solved for a to yield

$$a_{\max} = 4.81 \times 10^4 \left(\frac{m_1 + m_2}{mN} \right)^{1/3} \text{AU.} \tag{1.4}$$

In the neighbourhood of the Sun, it is fair to set $m = 0.5\ M_\odot$, $N = 0.1\ pc^{-3}$ and, for an average binary, $m_1 + m_2 = 1\ M_\odot$. Under these conditions, Equation (1.4) yields for a maximum separation $a_{\max}$, at which the respective components would have a chance to describe at least one complete revolution around their common centre of gravity, the value equal to

$$a_{\max} = 130\,500 \text{ AU}\,; \tag{1.4}$$

or a little more than half a parsec; and the corresponding orbital period P (obtained by an insertion from (1.4) in (1.3)) would be equal to 3×10^8 yr. Beyond this distance, the kinetic energy acquired in the course of accidental encounters with the neighbouring stars would exceed the gravitational binding energy of the pair; and, as a result, such a pair would be dissolved in a time comparable with P. Therefore, all stars in our neighbourhood separated from each other by more than half a parsec should be regarded only as chance rendez-vous, and not gravitational liaisons of any permanence.

Conversely, the pairs with mutual separations between 10^4 and 10^5 AU can already be regarded to constitute real physical binaries; though their orbital periods are still too long to disclose any indications of orbital motion; their only distinguishing mark (necessary, though not sufficient) being a common proper motion of the components in the sky. It is only when we get down to separations of the order of 10^2-10^3 AU (corresponding to orbital periods of the order 10^4 yr) that indications of orbital motion can be astrometrically established.

Many such pairs have been discovered in the sky; and we shall say more about them later. For the present we wish to stress that, for close binaries with separations of the order of 1 AU or less – which can manifest themselves spectroscopically (by fluctuations of their observed radial velocity) or photometrically (eclipsing variables) – Equations (1.1) or (1.2) lead to dissolution times exceeding by several orders of magnitude the age of the entire Galaxy. As citizens of our Galaxy such close pairs are, therefore, virtually impervious to the disrupting effects of fluctuations in the gravitational field surrounding them; and (unless other phenomena intervene, on which more again will be said in later parts of this book) their union should be regarded as permanent.

What is the extent to which all these expectations are borne out by the observed facts? Starting with the widest pairs actually encountered in our neighbourhood, we note that a 10th-magnitude pair of common-proper-motion red dwarfs (spectral type M) $-32°$ 16135 A and $-31°17815$, separated by $1\overset{\circ}{.}3$ in the sky and some 7 pc distant, may represent a physical binary with a spatial separation of 33 000 AU, not far below the limit indicated by Equation (1.4); for also the radial velocities of both components appear to be the same within the limits of observational errors (cf. Kuiper, 1942); and many other pairs of this type are known in our neighbourhood whose components are separated by less than 10 000 AU in space. In point of fact, a large majority of the 17 180 individual entries in Aitken's *New General Catalogue of Double Stars* (1932) must be objects of this type (with some, no doubt, constituting mere optical pairs); though only less than one-tenth of these have so far exhibited any indications of orbital motion; while the periods and other elements of their orbits have not been established (with any reliability) for much more than one hundred pairs. These periods range from years to millenia; the shortest known established visually being that of ξ UMa (P = 1.832 yr) at a distance of 8 parsecs (cf. van den Bos, 1928); though the recently introduced methods of speckle interferometry (cf. Labeyrie *et al.*, 1974) will no doubt extend this limit to even closer pairs.

It should, moreover, be clear that such binaries can be discovered (and identified) only in the relatively close proximity from us in space – it is virtually hopeless to search for them at distances in excess of (say) a few hundred parsecs. And although distance is the most important selection factor in the discovery of 'visual' binaries, an almost equally important factor is the difference in brightness between the component stars. A faint companion is harder to detect by visual methods than one nearly equal in brightness to the primary star – especially if their angular separation is small. On the other hand, an intrinsically luminous star is more likely to have a much fainter companion than is a faint star; so that this selection factor operates also against the discovery of visual binaries among the luminous stars. A detection of white-dwarf companions of Sirius or Procyon, separated by several seconds of arc from their central luminaries, would have been easy even for small telescopes if their primaries were not so bright!

In spite of these facts, the observations of visual binary systems found to exist in our proximity furnished astronomy with many valuable data – such as the masses and luminosities of their components of known absolute orbits; as well as the form and orientation of such orbits in space. The fact that the relative orbits are ellipses, with one component located at its focus, provided also the first observational proof of the validity of Newton's law of gravitation beyond the confines of our solar system. More recent work (cf., e.g., Lippincott, 1966) has pointed out that the orbital planes of visual binaries are orientated at random with respect to the plane of galactic equator; or that (for elliptical orbits) the space directions of their semi-major axes are likewise distributed at random. Nor is there any indication of an 'equipartition of energy' between pairs of different separations (cf. Ambartsumian, 1937). In other words – *any coupling between the galactic and double-star phenomena appears to be conspicuous by its absence* – the properties of the two types of systems on so different a scale are clearly uncorrelated.

But wide binaries of this type, or their particular properties, are largely outside the scope of this book; and only a brief mention of them has been included for the sake of completeness. When we turn our attention to *close binaries* – which will hereafter remain our principal objective – we meet a completely different situation. As was already mentioned in the first part of this chapter, the binary nature of such systems can be recognized by periodic variation in radial velocity, or (for systems with suitable orientation of the orbital plane with respect to the line of sight) by periodic variation of the brightness of the system due to eclipses (and other causes).

As a result, the observational selection will facilitate (or hamper) discovery of such pairs on completely different grounds. For close binaries – and in contrast with the visual ones – the probability of discovery will increase with diminishing distance between the components; and also with an increase of their mass or luminosity. The closer (or more massive) the system, the faster the orbital motion (and, therefore, the larger the Doppler shifts which are spectroscopically measurable). And also the closer the system, the greater the probability that the components will mutually eclipse each other, thus giving rise to a characteristic variation of light – variation which may persist, for very close systems, even outside (or in the absence of) eclipses on account of the photometric 'proximity' phenomena (ellipticity, reflection).

On the other hand, there are limits beyond which a decreasing period will cease to assist discovery; for in order to do so, this period should be sufficiently long in comparison with the response-time of the equipment (spectroscopic, photometric) used to make the observations. By photometric means, it is easy to detect a light variation lasting hours or minutes (and much less by a resort to chopping techniques); though spectroscopically this is much more difficult (also on account of the fact that fast motion will widen the spectral lines and thus make their shifts more difficult to measure).

As far as selection by distance is concerned, close binaries can, of course, be discovered at much greater distance than 'visual' binaries. In the neighbourhood of our cosmic station in the Galaxy – at the outskirts of its 'Orion arm' – the binary nature of double stars can be recognized by discernible orbital motion (or, for wider pairs, by a common proper motion up to distances of the order of 10^2 pc). Beyond this limit, both these indicators of binary nature become insignificant – so that no 'visual' (or 'astrometric') binary can be discovered – at least by observational techniques available to astronomers at the present time. Spectroscopic binaries can, on the other hand, be discovered up to distances at which their luminosity does not become too low for spectrographic work, of sufficient dispersion, with reasonable exposures. These facts will again favour discovery of massive pairs over less massive ones on both counts: not only of being brighter (because of greater luminosity), but also of exhibiting greater Doppler shifts (measurable with smaller dispersion) on account of faster orbital motion. Close binaries consisting of pairs of dwarf stars may, in fact, not be detectable spectroscopically at much greater distance than visual binaries. However, massive binaries can be detected with the aid of large modern reflectors up to distances of a few thousand parsecs, and up to ten thousand and more for very massive systems.

Beyond these limits the existence of binary stars can be established if, and only if, their components are close enough – and their orbital planes do not deviate too much from the direction of the line of sight – to eclipse each other in the course of their revolution, and thus give rise to characteristic variation of light which can be measured by photometric means. As is well known, the intensity of starlight can nowadays be measured photoelectrically, with telescopes of given aperture, to limiting magnitudes which cannot be seen, but even photographed with the same aperture, after any (practicable) exposure. As a result, our present knowledge of the existence and distribution of eclipsing variables in the sky is much less hampered by their luminosity than is the case for any other group of binary stars. A selection by orbit orientation is basic, but not physically important; as it does not discriminate against any particular types of such binaries (of the same proximity of their components). In consequence of these facts, eclipsing variables can be detected in all parts of our Galaxy; and in external galaxies as well if these can be decomposed into stars. The most distant eclipsing variable known to us today is the one detected in a dwarf galaxy in Ursa Maior (cf. Kholopov, 1971); the limiting distance being of the order of 10^6 pc.

A search for close binaries among different groups of stars by spectroscopic means has led to many thought-provoking results. The first is a very high proportion of such binaries among stars on the upper part of the Main Sequence (and among absolutely bright stars in general). Already Plaskett and Pearce (1931) noted that approximately one-third of all O and B stars listed in their *Catalogue of Radial Velocities* exhibited discrepancies between individual measurements made at different times, indicative of their binary nature; and a subsequent work by Blaauw (1961) has increased an estimate of this proportion considerably – up to almost 100% for the O stars. In contrast, Harper (1937) found that the same indications of binary nature dropped to less than 5% among the A stars. Although subsequent re-discussion of these data tended somewhat to lessen this disparity, the following conclusions based on the material now available may be legitimate:

(1) Four-fifths or more of the absolutely brightest stars in the spiral arms of our Galaxy are close binaries.

(2) The percentage of close binaries diminishes, in general, as we proceed down the Main Sequence towards cooler spectral types (Wilson, 1966). It is significantly smaller among A stars (barring the Ap or Am stars, among which the incidence of binaries is

again very high); and becomes very low among stars of advanced spectral types – exceptions being again W UMa-type binary systems – the commonest type of close binaries encountered in the space around us – on which more will be said in Chapter VIII.

(3) The incidence of binaries in each group appears to be strongly correlated with the angular velocity of axial rotation encountered among the respective types of the stars: those characterized with a high incidence of binaries are also found spectroscopically to be fast rotators, and vice versa.

(4) An overall outcome of all these findings indicates that not less than a half of the stars around us constitute close binary systems; and this figure still represents probably only the lower limit (because of observational selection) of their actual abundance. Close binaries are, therefore, by no means exceptional phenomena in our Galaxy; it is the stars which managed to remain single that are in the minority in the stellar population at large!

This very high percentage of binary stars in the sky is even more amply borne out when we take into consideration all types of binaries within a given volume of space. This can of course, properly be done only in the immediate proximity of the Sun. A census of such stars (cf. Kuiper, 1942; Heintz, 1967, 1969) indicates that the percentage of binary (or multiple) systems within a distance of (say) 10 pc from us may exceed 80% of the sample – possibly more.

TABLE I-1
Eclipsing Systems nearer than 30 pc

Star	Period	Spectrum	Max. brightness	Parallax
i Boo	$0\overset{d}{.}268$	G2 + F9	$5\overset{m}{.}3$	$0\overset{''}{.}079 \pm 0\overset{''}{.}005$
YY Gem	0.814	M1 + M1	9.1	0.073 ± 0.002
δ Cap	1.023	A6 + F2	2.9	0.065 ± 0.006
VW Cep	0.278	K1 + G6	7.1	0.053 ± 0.008
α CrB	17.400	A0 + G6	2.3	0.046 ± 0.002
β Aur	3.960	A0 + A0	2.1	0.038 ± 0.004
β Per	2.867	B8 + K0	2.2	0.037 ± 0.003
V 1143 Cyg	7.641	F5 + F5	5.9	0.036 ± 0.006
RW Dor	0.286	K5 + K2	9.8	0.035 ± 0.010
R CMa	1.136	F0 + G9	5.9	0.033 ± 0.004

Not all these binaries, or even close pairs among them, happen to manifest themselves as eclipsing variables. But at least 10 such objects – listed in the accompanying Table I-1 – are known to be located within 30 pc from the Sun; and as a volume of space of this radius contains some 10 000 stars, eclipsing variables would seem to constitute about 0.1% of the total population. This fraction of the sample represents probably an underestimate – as there may be others, still undetected, objects of this type in our vicinity. However, even if we take its present estimate on its face value and apply it to our entire Galaxy with its 10^{12} individual stars, the total number of eclipsing variables in it should be of the order of 10^9. Eclipsing variables are, therefore, manifestly no exceptional or uncommon phenomena!

It is true that those discovered so far, for which at least the period and amplitude of the light changes have been established, do not add to much more than 4000 (cf. Kukarkin *et al.*, 1969). This number represents, however, but little more than a tribute to the zeal of astronomers who have been cultivating this branch of our science; for the anticipated total in our Galaxy alone renders them all quite beyond the means of individual discovery.

When we turn from our stellar vicinity to other homogenous groups of stars whose members can be enumerated – such as in galactic clusters or stellar associations – we meet, however, a somewhat different situation. A list of the clusters and associations in which binaries (mainly close) have been found is given in the accompanying Table I-2; the

TABLE I-2
Incidence of close binaries in galactic clusters and associations

Cluster	Percentage of close binaries	References
α Per	9%	Kraft (1967); Petrie and Heard (1970)
Coma	20	Kraft (1965)
Hyades	17	Kraft (1965)
NGC 6745	40?	Abt *et al.* (1970)
Pleiades	12	Abt *et al.* (1965); Anderson *et al.* (1966)
Praesepe	14	Dickens *et al.* (1968); Treanor (unpublished)
Ursa Maior	16	Geary and Abt (1970)
Lacerta 1	17	Abt and Hunter (1962)
Orion 1	17	Abt and Hunter (1962)
Perseus 2	17	Treanor (1960); Blaauw and van Albada (1963)
Scorpio-Centaurus	10	Slettebak (1968)

second column of which gives the observed percentage of the binary stars in the total population. Although the counts of binaries in these systems may still be incomplete (the same is also true of the percentage of known binaries in the general galactic substrate), it is difficult to avoid a conclusion that *the percentage of close binaries in stellar clusters or associations is noticeably smaller* – less than a half – *than that encountered in the galactic substrate*, formed by a dissolution of associations and clusters in the course of time. Whether or not this represents an age-effect is a question to which we shall return in Chapter VIII-5A; for the present we wish to note that it looks as if close binaries avoid regions where the density of stars is high.

The same situation will stare us in the face even more when we turn from the galactic to globular clusters. As is well known, most globular clusters have been thoroughly searched for variable stars; and over 2000 of them were actually discovered – mostly short-period cepheids. However, Sawyer's latest *Catalogue of Variable Stars in Globular Clusters* (1975) lists the presence of only 11 eclipsing variables in three globular clusters; and any or all of them may be foreground stars. Attempts made (cf. e.g., Batten, 1973) to explain away this disparity in terms of evolutionary effects, which may render ageing close binaries less conspicuous and more immune to observational detection, have so far proved unconvincing.

Whatever the reason (or reasons) may be, they are no doubt connected with a different upper part of the Main Sequence – which had time to evolve – the unevolved (lower) parts of the Main Sequence (which have likewise been resolved by our telescopes for several clusters) should not be affected by it. An evolutionary difference of 5000 million years for stars of the mass of the Sun (or smaller) would not make much, if any, difference to the characteristics of the eclipsing systems formed by such stars. In the spiral arms of our Galaxy many such systems (of disc-type population) are known; while, in globular clusters, they appear to be conspicuous by their absence. Rather than to camouflage this fact by artificially distributing the discrepancy among several factors that can lend themselves for this purpose, we should probably recognize that a real difference may exist between the binary-star content of Population I and II stars – even though the specific reasons why this should be so may still elude us.

Whatever the reason (or reasons) may be, they are no doubt connected with different age (and therefore, evolutionary stage) of the two stellar populations. Further evidence that evolutionary stage of stellar systems may influence also the kind of their binary-star population can be drawn from the close (eclipsing) binary content of external galaxies in our neighbourhood. To this date, 29 eclipsing variables have been identified as such in the Large Magellanic Cloud (LMC) by Shapley and McKibben; and 42 such variables in the Smaller Magellanic Cloud (SMC) by the Harvard and other investigators. Moreover, Baade, Gaposchkin and Swope discovered 65 eclipsing variables in the Andromeda nebula (M31); awaiting supplementation by new discoveries no doubt due to be made in the future.

The average maximum apparent brightness of eclipsing variables in the Andromeda nebula proved to be +21 (only 2 out of 65 turned out to be brighter than +20); which at the distance of the nebula (and with due regard to the absorption of light in our Galaxy) corresponds to the absolute magnitude of –3. This makes them at least 2–3 magnitudes fainter than the brightest known eclipsing variables in our own Galaxy, whose magnitudes are comparable with those of long-period cepheids or Novae at maximum brightness. Novae and long-period cepheids of luminosities comparable with their counterparts in our Galaxy have been found in M31 in considerable numbers; but not eclipsing variables of comparable brightness; and the possibility cannot be ruled out that (if real) this difference may be due to a somewhat different evolutionary stage of these two neighbouring galaxies.

Turning to the Magellanic clouds, we may note that the average apparent brightness of the brightest eclipsing systems in LMC is +15 (only 5 out of 29 such variables are brighter than 14) corresponding to – 4 abs. mag; while for SMC the same quantity proves to be closer to –3. These figures imply that the binary-star population in LMC behaves, in this respect, more than that of our own Galaxy; whereas that of SMC may be closer to that of the Andromeda nebula. The differences are however, not large; and any verdict on their significance must await the advent of additional observational data which are not yet at our disposal.

This, then, completes our brief introduction of the 'dramatis personae' of our narrative, whose acts in the sky will occupy us through successive Chapters II–VIII of this book.

BIBLIOGRAPHICAL NOTES

For a fuller account of the early history of double-star astronomy the best reference still remains R. G. Aitken's *The Binary Stars* (1935), recently reprinted in a paperback edition (Dover, 1964).

For a history of the discovery of Algol cf. Z. Kopal (1959), Chapter I and references quoted therein.

A fuller discussion of the gravitational dissolution of wide binaries as a result of stellar encounters can be found in V. A. Ambartsumian (1937), S. Chandrasekhar (1944), B. Takase (1953) or S. Yabushita (1966).

A comprehensive summary of observational data bearing on stellar population (including binary stars) in our neighbourhood cf., e.g., W. Gliese's *Catalogue of Nearby Stars* (1969); or the *Catalogue of Stars* within 25 parsecs of the Sun by R. v. d. R. Woolley *et al.* (1970). For a short summary of the data on stars nearer than 5 pc cf. P. v. d. Kamp (1969).

For an account of the data on visual binary systems the standard published sources remain R. G. Aitken's *New General Catalogue of Double Stars* (1932); and an *Index Catalogue of Visual Double Stars* by H. M. Jeffers, W. H. van den Bos, and F. M. Greeby (1963). A corresponding source for the data on spectroscopic binaries is the *Sixth Catalogue of the Orbital Elements of Spectroscopic Binary Systems* by A. H. Batten (1967).

Statistical discussions of the spatial distribution of the elements of visual binary systems can be found, e.g., in F. Berglund (1938), S. Arend (1950); A. Opolski (1952), P. Muller (1956), or H. B. Süer (1970). For the latest list of the pairs of nearby stars with common proper motions cf. I.N. Latyshev (1974).

The frequencies with which binary stars occur among objects of different spectral types and other properties have been the subjects of more detailed investigations by C. and M. Jaschek (1957), R. M. Petrie (1960), T. Kirillova and E. D. Pavlovskaya (1963), P. Brosche (1964), C. Jaschek and E. A. Gomez (1970), J. Dommanget (1970), and others. An anomalously high percentage of close binaries among stars of spectral types Ap or Am was pointed but by H. Abt (1961) or G. C. L. Aikman (1971). A very high spatial density of close binaries of the W UMa-type was pointed out first by H. Shapley (1948); and although subsequent work by R. W. Kraft (1965b) or O. J. Eggen (1967) tended to diminish Shapley's estimates somewhat, the difference may be due to observational selection. A good comprehensive discussion of the problem of distribution of binary systems among stars of different types may be found in Chapter 2 of the *Binary and Multiple Systems of Stars* by A H. Batten (1973).

A scarcity of close (eclipsing) binaries – other than those of the W UMa-type (cf. O. J. Eggen, 1967; O. J. Eggen and A. Sandage, 1969) – appears, however, to be significant. No eclipsing variable is known to us, for example in the Pleiades; only one (TX Cnc – a variable of W UMa-type) in Praesepe; and none in the Hyades – unless, as has been conjectured, the well-known variables RS CVn or Z Her (cf. sec. VIII-3B) are loosely associated with that group. Moreover, among the 2776 stars which P. Th. Oosterhoff (1937) regards as physical members of the twin cluster χ and h Persei, only one (No. 1021 of Oosterhoff's catalogue) is probably an eclipsing variable; and the membership of close eclipsing pairs of SZ Cam in the cluster NGC 1502 or of V448 and V453 Cyg in NGC 6871 is likewise still conjectural.

As far as the contents of close binaries in globular star clusters is concerned, the *Second Catalogue of 1421 Variable Stars in Globular Clusters* by Helen Sawyer-Hogg (1955) listed only three eclipsing variables in three clusters (NGC 3201, 5139, and 6338) which may all be foreground stars. The best known of these is, perhaps, No. 78 variable in the giant cluster ω Centauri, the known characteristics of which are consistent with the membership of the cluster (cf., E. H. Geyer, 1967; or R. F. Sistero, 1968); but the evidence is by no means conclusive; and the most recent work by Geyer (as yet unpublished) indicates that it probably is a foreground star.

For eclipsing variables in the Andromeda nebula cf., e.g., W. Baade and H. H. Swope (1965). A catalogue of eclipsing variables discovered so far in nearby external galaxies was recently compiled by J. M. Kreiner (1974).

CHAPTER II

FIGURES OF EQUILIBRIUM

The aim of the present chapter will be to provide a systematic introduction to the study of phenomena exhibited by close binary systems – with special regard to such effects as may become photometrically or spectroscopically observable. If their components were – as in visual binaries – sufficiently far apart to attract each other as as pair of mass-points, their orbits would be Keplerian ellipses – perturbed only very seldom by chance encounters with neighbouring stars or interstellar clouds. In a field of star density comparable with that obtaining in our galactic neighbourhood, cumulative effects of chance encounters may be safely discounted even on the nuclear time-scale (cf. Ambartsumian, 1937; Chandrasekhar, 1944; Yabushita, 1966; or – for encounters with gas clouds – Takase, 1953); and the dimensions of the components will in no way affect their motion or other observable manifestations. If, however, consistent with our definition of close binaries already set forth in Chapter I, the *proximity phenomena* arising from mutual interaction of the components are to be regarded as the essential cause of the problems we wish hereafter to consider, the consequences of chance encounters with external celestial objects can utterly be ignored (except, possibly, their effects on the apparent (observed) orbital periods in very dense star fields – existing, for example, in the interiors of globular clusters or elliptical galaxies). On the other hand, the proximity phenomena – observable both photometrically and spectroscopically – arising from the *deformation* of both components by their axial rotation and mutual tidal action may become noticeable, or even conspicuous. In the present chapter we wish to outline a theory of the *equilibrium* form of the components of close binary systems, as a consequence of the fact that if no element of a fluid constituting them moves relative to the adjacent elements, the surfaces of equal density ρ – of which the external shape corresponding to $\rho = 0$ constitutes the limiting case – must be *equipotentials.*

II-1. Equipotential Surfaces

In order to prove that this is the case, let us depart from the general equations of hydrodynamic motion

$$\rho \frac{D\mathbf{u}}{Dt} = \rho \operatorname{grad} \Psi - \operatorname{grad} P + \operatorname{div} \mathfrak{T} \tag{1.1}$$

in the Eulerian form, where ρ denotes the density at any internal point of the respective configuration; P, the corresponding pressure; and

$$\Psi = \Omega + V', \tag{1.2}$$

the potential arising from self-attraction (Ω) plus disturbing (V') due to rotation or tides. Moreover, $\mathbf{u}$ in (1.1) signifies the (vector) velocity of any displacement in the fluid body under consideration; $\mathfrak{T}$, the stress tensor due to dissipative forces whatever their source (viscosity, radiation), and

$$\frac{D}{Dt} \equiv \frac{\partial}{\partial t} + \mathbf{u} \text{ grad} \tag{1.3}$$

stands of the Lagrangian derivative with respect to the time.

Consistent with the assumption of equilibrium to which we shall adhere in this chapter, we shall hereafter set $\mathbf{u} = 0$ rendering also $\mathfrak{T} = 0$. If so, however, Equation (1.1) reduces then to

$$\text{grad } P = \rho \text{ grad } \Psi$$

or, in scalar form,

$$\frac{\partial P}{\partial x} = \rho \frac{\partial \Psi}{\partial x}, \quad \frac{\partial P}{\partial y} = \rho \frac{\partial \Psi}{\partial y}, \quad \frac{\partial P}{\partial z} = \rho \frac{\partial \Psi}{\partial z}. \tag{1.5}$$

On equating the mixed second derivatives $\partial^2 P/\partial x \partial y$, $\partial^2 P/\partial x \partial z$ and $\partial^2 P/\partial y \partial z$ obtained by appropriate differentiation of Equations (1.5) we find that

$$\frac{\partial \rho}{\partial x}\frac{\partial \Psi}{\partial y} = \frac{\partial \rho}{\partial y}\frac{\partial \Psi}{\partial x}, \quad \frac{\partial \rho}{\partial x}\frac{\partial \Psi}{\partial z} = \frac{\partial \rho}{\partial z}\frac{\partial \Psi}{\partial x}, \quad \frac{\partial \rho}{\partial y}\frac{\partial \Psi}{\partial z} = \frac{\partial \rho}{\partial z}\frac{\partial \Psi}{\partial y}, \tag{1.6}$$

from which it follows that

$$\frac{\dfrac{\partial \rho}{\partial x}}{\dfrac{\partial \Psi}{\partial x}} = \frac{\dfrac{\partial \rho}{\partial y}}{\dfrac{\partial \Psi}{\partial y}} = \frac{\dfrac{\partial \rho}{\partial z}}{\dfrac{\partial \Psi}{\partial z}}. \tag{1.7}$$

A structure of these equations discloses at once that the surfaces ρ = constant coincide necessarily with those of Ψ = constant. Under these circumstances, Equations (1.7) can be rewritten in the form of a single a total differential equation

$$dP = \rho \, d\Psi, \tag{1.8}$$

implying that if Ψ is a function of ρ only, so must be P – a fact which requires the existence of an 'equation of state' of the form

$$P \equiv f(\rho), \tag{1.9}$$

where $f(\rho)$ stands for an arbitrary function of the density. Any surface over which ρ and P are constant must, therefore, be an equipotential

$$\Psi = \text{constant}. \tag{1.10}$$

This statement contains, in fact, a complete specification of our problem; but in an implicit form requiring explicit development. While this task will be completed, in some detail, in subsequent sections II-2 to II-4 for deformations due to axial rotation and tides (or their combination), the aim of the present section will be to prepare the ground for such a solution by a specification of the *gravitational potential* of distorted self-gravitating bodies of arbitrary structure.

In order to do so, let us fix our attention to an arbitrary point M in the interior of our configuration, in a position specified by the coordinates

$$\begin{aligned} x &= r \cos\phi \sin\theta \\ y &= r \sin\phi \sin\theta \, , \\ z &= r \cos\theta \, , \end{aligned} \tag{1.11}$$

acted upon by the attraction of a stratum comprised between the radii $r = r_0$ and r_1. Let, moreover, $M'(r', \theta', \phi,)$ be an arbitrary point of this stratum (see Figure 2-1). If so, the *interior potential* U at M will evidently be given by the integral

$$U = G \int_{r_0}^{r_1} \frac{\mathrm{d}m'}{\Delta}, \tag{1.12}$$

where G denotes, as before, the gravitation constant; $\mathrm{d}m'$, the mass element

$$\mathrm{d}m' = \rho \, \mathrm{d}x'\mathrm{d}y'\mathrm{d}z' = \rho r'^2 \, \mathrm{d}r' \sin \mathrm{d}\theta' \, \mathrm{d}\phi';$$

and from the triangle OMM' (cf. again Figure 2-1)

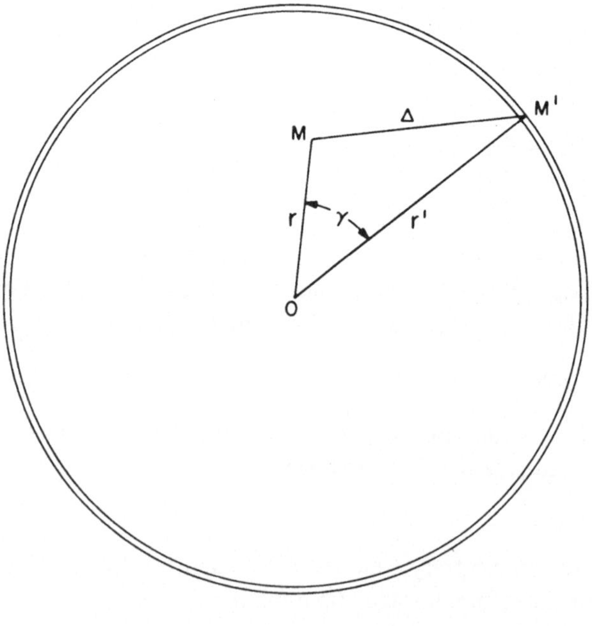

Fig. 2-1. Geometry for the determination of the potential.

$$\Delta^2 = r^2 + r'^2 - 2rr' \cos \gamma, \tag{1.13}$$

where

$$\cos \gamma = \cos \theta \cos \theta' + \sin \theta \sin \theta' \cos(\phi - \phi'). \tag{1.14}$$

Moreover, the *exterior potential* V will be given by an analogous expression of the form

$$V = G \int_0^{r_0} \frac{\mathrm{d}m'}{\Delta}, \tag{1.15}$$

similar to Equation (1.12), in which the limits of integration extend from the origin to r_0; the sum $U + V = \Omega$ constitutes the total potential of our configuration arising from its mass.

To evaluate the two constituents of Ω, let us expand Δ^{-1} in terms of the Legendre polynomials $P_n(\cos \gamma)$ of ascending integral order n in a well-known series of the form

$$\begin{aligned} \frac{1}{\Delta} &= \frac{1}{r} \sum_{n=0}^{\infty} \left(\frac{r'}{r}\right)^n P_n(\cos \gamma), \qquad r' < r, \\ &= \frac{1}{r} \sum_{n=0}^{\infty} \left(\frac{r}{r'}\right)^n P_n(\cos \gamma), \qquad r' > r, \end{aligned} \tag{1.16}$$

which on insertion in Equations (1.12) and (1.15) permits us to express U and V as

$$U = \sum_{n=0}^{\infty} r^n U_n \tag{1.17}$$

and

$$V = \sum_{n=0}^{\infty} r^{-n-1} V_n, \tag{1.18}$$

where

$$U_n = G \int_{r_0}^{r_1} \int_0^{\pi} \int_0^{2\pi} \rho(r')^{1-n} P_n(\cos \gamma) \, \mathrm{d}r' \sin \theta' \, \mathrm{d}\theta' \, \mathrm{d}\phi' \tag{1.19}$$

and

$$V_n = G \int_0^{r_0} \int_0^{\pi} \int_0^{2\pi} \rho(r')^{2+n} P_n(\cos \gamma) \, \mathrm{d}r' \sin \theta' \, d\theta' \, \mathrm{d}\phi'. \tag{1.20}$$

Now, in the foregoing expressions, let

$$r' \equiv r(a, \theta', \phi') \tag{1.21}$$

denote symbolically the parametric equation of an equipotential surface of constant density and pressure. By virtue of the uniqueness of the potential function, only one such surface can pass through any point (i.e., belong to any value of a). Since the density must remain constant over such a surface, it follows that ρ can hereafter be regarded as a function of a *single* variable a, introduced by Equation (1.21) and representing the mean radius of the respective equipotential. As such, it is bounded so that

$$0 \leqslant a \leqslant a_1, \tag{1.22}$$

where a_1 represents the (smallest) root of the equation

$$\rho(a_1) = 0. \tag{1.23}$$

The fact that ρ can thus be regarded as a function of a single variable suggests that it should be of advantage to change over from r' to a as the new variable of integration on the right-hand side of Equation (1.19) and (1.20). Inasmuch as the Jacobian J of the transformation from $(r' \theta', \phi')$ to (a, θ', ϕ') is of the form

$$J = \begin{vmatrix} \dfrac{\partial r'}{\partial a} & \dfrac{\partial r'}{\partial \theta'} & \dfrac{\partial r'}{\partial \phi'} \\ 0 & 1 & 0 \\ 0 & 0 & 1 \end{vmatrix} = \frac{\partial r'}{\partial a}, \tag{1.24}$$

a transition from r' to a can be effected simply by setting

$$\mathrm{d}r' = \frac{\partial r'}{\partial a}\, \mathrm{d}a. \tag{1.25}$$

If so, however, then evidently

$$U_n = \frac{G}{2-n} \int_{a_0}^{a_1} \rho \frac{\partial}{\partial a} \left\{ \int_0^{\pi} \int_0^{2\pi} (r')^{2-n} P_n(\cos\gamma) \sin\theta' \,\mathrm{d}\theta' \,\mathrm{d}\phi' \right\} \mathrm{d}a \tag{1.26}$$

for $n \gtrless 2$, and

$$U_2 = G \int_{a_0}^{a_1} \rho \frac{\partial}{\partial a} \left\{ \int_0^{\pi} \int_0^{2\pi} (\log r') P_2(\cos\gamma) \sin\theta' \,\mathrm{d}\theta' \,\mathrm{d}\phi' \right\} \mathrm{d}a\,; \tag{1.27}$$

while, similarly,

$$V_n = \frac{G}{n+3} \int_0^{a_0} \rho \frac{\partial}{\partial a} \left\{ \int_0^{\pi} \int_0^{2\pi} (r')^{n+3} P_n(\cos\gamma) \sin\theta' \,\mathrm{d}\theta' \,\mathrm{d}\phi' \right\} \mathrm{d}a \tag{1.28}$$

for any value of n.

If $n = 0$,

$$U_0 = 4\pi G \int_{a_0}^{a_1} \rho a \,\mathrm{d}a \tag{1.29}$$

and

$$V_0 = G \int_0^{a_0} \mathrm{d}m' = Gm(a_0), \tag{1.30}$$

where $m(a_0)$ denotes the mass of our configuration interior to a_0.

In order to proceed further, let us assume that the radius vector r' of an equipotential surface can be expanded in a series of the form

$$r' = a\left\{1 + \sum_{j=0}^{\infty} Y_j(a, \theta', \phi')\right\}, \tag{1.31}$$

where the Y_j's stand for solid harmonic functions of the respective coordiantes, satisfying the partial differential equation

$$\frac{1}{\sin\theta'}\,\frac{\partial}{\partial\theta'}\left(\sin\theta'\,\frac{\partial Y_j}{\partial\theta'}\right) + \frac{1}{\sin^2\theta'}\,\frac{\partial^2 Y_j}{\partial\phi'^2} + j(j+1)\,Y_j = 0, \tag{1.32}$$

where j is zero or a positive integer. If the solid harmonics $Y_j(a, \theta', \phi')$ can be factored in the form

$$Y_j(a, \theta', \phi') = f_j(a)P_j(\theta', \phi'), \tag{1.33}$$

where the P_j's are the same zonal harmonics which occur on the right-hand side of Equation (1.16), it is possible to assert, by a well-known orthogonality theorem, that

$$\begin{aligned} \int_0^{\pi}\int_0^{2\pi} & P_n(\cos\gamma)\, Y_j(a, \theta', \phi') \sin\theta'\, d\theta'\, d\phi' \\ &= 0 \qquad\qquad\qquad\qquad \text{if } j \neq n, \\ &= \frac{4\pi}{2j+1}\, Y_j(a, \theta, \phi) \qquad \text{if } j = n. \end{aligned} \tag{1.34}$$

With the aid of this theorem it should be possible – on insertion for r' from Equation (1.31) – to rewrite the expansions of Equations (1.17) and (1.18) for U and V in the form

$$U = 4\pi G \sum_{j=0}^{\infty} \frac{E_j(a)}{2j+1}\, r^j\, P_j(\theta, \phi) \tag{1.35}$$

and

$$V = 4\pi G \sum_{j=0}^{\infty} \frac{F_j(a)}{2j+1}\, r^{-j-1}\, P_j(\theta, \phi), \tag{1.36}$$

where $E_j(a)$ and $F_j(a)$ are appropriate functions of a only.

An explicit evaluation of these functions will be undertaken in the next section. In this place we wish to note that although the sum $\Omega = U + V$ of the foregoing expansions of Equations (1.35) and (1.36) constitutes the complete potential arising from the mass of our configuration, it does not yet represent the total potential Ψ. In order to specify the latter we must adjoin to Ω the *disturbing potential* V', whose action will cause this configuration to depart from spherical form. In point of fact, the total potential $\Psi = U + V + V'$ must satisfy the Poisson equation

$$\nabla^2\Psi + 4\pi G\rho = 2\omega^2, \tag{1.37}$$

where ω denotes the angular velocity of axial rotation that may, but need not, be constant*. The partial sum $\Omega = U + V$ represents the particular integral and V', the complementary function of this equation; the form of which remains yet to be specified.

If our configurations were nonrotating, the disturbing potential V' would represent a solution of the homogeneous (Laplace) equation

$$\nabla^2 V' = 0\,, \tag{1.38}$$

which is known to be of the form

$$V' = \sum_j \{c_j r^j + d_j r^{-j-1}\} P_j\,(\theta, \phi), \tag{1.39}$$

where c_j and d_j are arbitrary constants. The values of these constants depend, in turn, on the nature of the disturbing forces; and the regularity of V' at the origin – if required – necessitates that $d_j = 0$.

Let us first confine our attention to the case of *rotational distortion* of our configuration, arising from uniform rotation with constant angular velocity ω about one (say, z-)axis whose direction is fixed in space. In such a case, the differential equation for V' is of the form

$$\nabla^2 V' = 2\omega^2\,; \tag{1.40}$$

and its regular solution reduces to

$$V' = \frac{1}{2}\,\omega^2 r^2 \sin^2\theta = c_2 r^2\,\{1 \mp P_2(\cos\theta)\}, \tag{1.41}$$

where

$$c_2 = \frac{1}{3}\,\omega^2. \tag{1.42}$$

If Equation (1.35), (1.36) and (1.41) are combined, the total potential of self-gravitating configuration whose distortion from the centrifugal potential Equation (1.41), will assume the form

$$\begin{aligned}\Psi(r,\theta,\phi) = 4\pi G \sum_{j=0}^{\infty} \frac{r^j E_j(a) + r^{-j-1} F_j(a)}{2j+1}\, P_j\,(\cos\theta) + \\ + \frac{1}{3}\,\omega^2 r^2\,\{1 - P_2(\cos\theta)\},\end{aligned} \tag{1.43}$$

valid to any arbitrary degree of accuracy. Therefore, over an equipotential surface specified by the radius vector r',

* An assumption of constant velocity is not essential for the validity of our procedure. The latter would continue to hold good – though in more complicated form – as long as $\omega(a, \theta, \phi)$ remains expansible in terms of spherical harmonics $P_j(\theta, \phi)$.

$$\Psi(r', \theta, \phi) = 4\pi G \sum_{j=0}^{\infty} \frac{(r')^j E_j(a) + (r')^{-j-1} F_j(a)}{2j+1} P_j(\cos\theta) + \\ + \frac{1}{3}\omega^2 (r')^2 \{1 - P_2(\cos\theta)\}, \tag{1.44}$$

where r' continues to be given by Equations (1.31)–(1.33) and where, accordingly, the last term can be expanded into

$$\begin{aligned} \frac{1}{3}\omega^2 (r')^2 \{1 - P_2(\cos\theta)\} = \frac{1}{3}\omega^2 a^2 (1 - f_2) - \\ - \frac{1}{3}\omega^2 \left\{1 - 2f_2 + \frac{6}{7} f_2^2\right\} r'^2 P_2(\cos\theta) + \\ + \frac{\omega^2}{a^2}\left\{\frac{2}{3} f_4 - \frac{18}{35} f_2^2\right\} r'^4 P_4(\cos\theta) + \ldots, \end{aligned} \tag{1.45}$$

which, if we regard ω^2 itself as a small quantity of first order, represents an expansion of the centrifugal potential $V'(r')$ correctly to terms of third order.

Suppose next that we expand the right-hand side of Equation (1.44) in a Neumann series of the form

$$\Psi(r', \theta, \phi) = \sum_{j=0}^{\infty} \alpha_j(a) P_j(\theta, \phi), \tag{1.46}$$

with coefficients defined by

$$\alpha_j(a) = \frac{2j+1}{4\pi} \int_0^{\pi} \int_0^{2\pi} \Psi(r', \theta, \phi)\, P_j(\theta, \phi) \sin\theta \, d\theta \, d\phi. \tag{1.47}$$

If now – consistent with Equation (1.11) – the total potential Ψ is to remain constant over a surface of the form Equation (1.31), it follows that *all terms on the right-hand side of Equation (1.46) factored by $P_j(\theta, \phi)$ for $j > 0$ must necessarily vanish*; and this can be true only if we set

$$\alpha_j(a) = 0, \qquad j > 0; \tag{1.48}$$

leaving us with α_0 as the constant value characterizing the respective equipotential of mean radius a.

A determination of the gravitational potential of *tidally-distorted* configurations can be made to follow a closely parallel course. The only difference rests on the fact that the disturbing potential $V'(r)$ must then be identified with that arising from the presence of an external mass. As an exterior potential, V' will continue to satisfy Laplace's equation (1.38); but the explicit form of spherical harmonics $P_j(\theta, \phi)$ as well as of the coefficients c_j factoring them will depend on the magnitude and position of the disturbing body in space. Therefore, the last term on the r.h.s. of Equation (1.39) will, for tidal distortion, be of a form different from (1.42), but still expansible in a Neumann series (1.46) with coefficients defined by (1.47). In the three sections which follow we shall, accordingly, proceed to develop the explicit form of Equations (1.48) in the case of the distortion arising from pure rotation and tides, as well as from their mutual interaction.

In other words – whatever the cause of the distortion – if the distorted surface is to represent an equipotential, n equations of the form (1.48) can be set up to specify n amplitudes $f_j(a), j = 1, 2, \ldots, n$ of the expansion

$$r' = a\left\{1 + \sum_{j=0}^{\infty} f_j(a) P_j(\theta', \phi')\right\}, \tag{1.49}$$

representing Equation (1.21); and once these amplitudes have been determined, we are in a position (cf. section VII-2) to specify also the explicit forms of $\alpha_0(a)$ from the requirement that the mass of the configuration should be uninfluenced by its distortion.

II-2. Rotational Distortion

The aim of the present section will be to establish the explicit from of the amplitudes $f_j(a)$ in the expansion (1.49) for the shape of an equipotential surface of a rotating configuration distorted by centrifugal force. If this rotation takes place about a z-axes fixed in space, it is evident (on account of symmetry) that this expansion can contain only harmonics $P_j(\cos\theta)$ of even orders. Moreover, if the amplitude $f_2(a)$ of the leading term on the r.h.s. of Equation (1.49) represents a quantity of first order in surficial distortion, $f_4(a)$ will be of the order of f_2^2 or of second order; $f_6(a)$ of third order, etc.

Now let us, in what follows, set out to describe the shape of rotating configurations to quantities of *third* order in surficial distortion – a scheme within which Equation (1.49) will be restricted to the terms

$$\begin{aligned} r' &= a\,\{1 + f_0 + f_2P_2(\cos\theta) + f_4P_4(\cos\theta) + f_6P_6(\cos\theta) + \ldots\} \\ &= a\,\{1 + \Sigma\}. \end{aligned} \tag{2.1}$$

Within the scheme of our approximation,

$$\begin{aligned} (r')^{2-n} = a^{2-n}\Big\{ & 1 - (n-2)\,\Sigma + \frac{1}{2}(n-1)(n-2)\,\Sigma^2 - \\ & - \frac{1}{6}\,n(n-1)(n-2)\,\Sigma^3 + \ldots\Big\} \end{aligned} \tag{2.2}$$

for $n \gtrless 2$ and, for $n = 2$,

$$\log r' = \log a + \Sigma - \frac{1}{2}\,\Sigma^2 + \frac{1}{3}\,\Sigma^3 - \ldots \tag{2.3}$$

while

$$\begin{aligned} (r')^{n+3} = a^{n+3}\Big\{ & 1 + (n+3)\,\Sigma + \frac{1}{2}(n+3)(n+2)\,\Sigma^2 + \\ & + \frac{1}{6}(n+3)(n+2)(n+1)\,\Sigma^3 + \ldots\Big\} \end{aligned} \tag{2.4}$$

for any value of n.

Let us decompose next the powers and cross-products of Legendre polynomials occurring in different powers of Σ on the right-hand side Equations (2.2)–(2.4) into their linear combinations by use of the well-known formula which asserts that, for $m \leqslant n$,

$$P_m P_n = \sum_{j=0}^{m} \frac{A_{m-j} A_j A_{n-j}}{A_{m+n-j}} \left\{ \frac{2m + 2n + 1 - 4j}{2m + 2n + 1 - 2j} \right\} P_{m+n-2j}, \tag{2.5}$$

where $A_0 = 1$ and, for $j > 0$,

$$A_j = \frac{1.3.5 \ldots (2j - 1)}{j!} \tag{2.6}$$

The orthogonality properties of the P_n's are such that

$$\begin{aligned} \int_{-1}^{1} P_m P_n \mathrm{d} \cos\theta &= 0 \qquad \text{if } m \neq n, \\ &= \frac{2}{2n+1} \qquad \text{if } m = n. \end{aligned} \tag{2.7}$$

If so, then the assertion of (2.1)–(2.7) together with the use of the orthogonality theorem (1.34) in Equations (1.26)–(1.28) for U_n and V_n should enable us to express the latter in the forms

$$(2n + 1)\, U_n = 4\pi G E_n(a) P_n(\cos\theta), \tag{2.8}$$

$$(2n + 1)\, V_n = 4\pi G F_n(a) P_n(\cos\theta), \tag{2.9}$$

the amplitudes of which we shall now proceed to evaluate.

To begin, we note that V_0 as defined by (1.30) can be written as

$$\begin{aligned} F_0 &= \frac{1}{12\pi} \int_0^a \rho \frac{\partial}{\partial a} \left\{ \int_0^{\pi} \int_0^{2\pi} (r')^3 \sin\theta \, \mathrm{d}\theta \, \mathrm{d}\phi \right\} \mathrm{d}a \\ &= \int_0^a \rho \frac{\partial}{\partial a} \left\{ a^3 \left[\frac{1}{3} + (1 + f_0) \left(f_0 + \frac{1}{5} f_2^2 \right) + \frac{2}{105} f_2^3 + \ldots \right] \right\} \mathrm{d}a \end{aligned} \tag{2.10}$$

(in which the zero subscript of the upper limit will hereafter be dropped) and represents the mass of our configuration interior to a. Since this mass must be independent of distortion, it follows that the zero-harmonic amplitude f_0 is constrained to satisfy the equation

$$(1 + f_0) \left(f_0 + \frac{1}{5} f_2^2 \right) + \frac{2}{105} f_2^3 + \ldots = 0, \tag{2.11}$$

which yields

$$f_0 = -\frac{1}{5} f_2^2 - \frac{2}{105} f_2^3 - \ldots \tag{2.12}$$

correctly to quantities of third order.

Taking advantage of this fact, we establish – after some algebra – that, to the same order of accuracy, the coefficients

$$E_0 = \int_a^{a_1} \rho \frac{\partial}{\partial a}\left\{a^2\left[\frac{1}{2} - \frac{1}{10} f_2^2 - \frac{2}{105} f_2^3\right]\right\} da, \tag{2.13}$$

$$E_2 = \int_a^{a_1} \rho \frac{\partial}{\partial a}\left\{f_2 - \frac{1}{7} f_2^2 + \frac{12}{35} f_2^3 - \frac{2}{7} f_2 f_4\right\} da, \tag{2.14}$$

$$E_4 = \int_a^{a_1} \rho \frac{\partial}{\partial a}\left\{\frac{1}{a^2}\left[f_4 - \frac{27}{35} f_2^2 + \frac{216}{385} f_2^3 - \frac{60}{77} f_2 f_4\right]\right\} da, \tag{2.15}$$

$$E_6 = \int_a^{a_1} \rho \frac{\partial}{\partial a}\left\{\frac{1}{a^4}\left[f_6 + \frac{90}{77} f_2^3 - \frac{25}{11} f_2 f_4\right]\right\} da; \tag{2.16}$$

and, similarly,

$$F_0 = \int_0^a \rho a^2 \, da, \tag{2.17}$$

$$F_2 = \int_0^a \rho \frac{\partial}{\partial a}\left\{a^5\left[f_2 + \frac{4}{7} f_2^2 + \frac{2}{35} f_2^3 + \frac{8}{7} f_2 f_4\right]\right\} da, \tag{2.18}$$

$$F_4 = \int_0^a \rho \frac{\partial}{\partial a}\left\{a^7\left[f_4 + \frac{54}{35} f_2^2 + \frac{108}{77} f_2^3 + \frac{120}{77} f_2 f_4\right]\right\} da, \tag{2.19}$$

$$F_6 = \int_0^a \rho \frac{\partial}{\partial a}\left\{a^9\left[f_6 + \frac{24}{11} f_2^3 + \frac{40}{11} f_2 f_4\right]\right\} da. \tag{2.20}$$

As the next step of our procedure, let us establish the explicit form of Equations (1.48) for $j = 2$, 4, and 6. On evaluating the integrals on the right-hand side of Equation (1.47) by the same method, we find that, for $j = 2, 4, 6$ Equations (1.48) assume the more explicit forms

$$\frac{a^2 E_2}{5}\left\{1 + \frac{4}{7} f_2 + \frac{4}{7} f_4 + \frac{1}{35} f_2^2\right\} + \frac{8}{63} a^4 f_2 E_4 +$$

$$-\frac{F_0}{a}\left\{f_2 - \frac{2}{7} f_2^2 + \frac{29}{35} f_2^3 - \frac{4}{7} f_2 f_4\right\} + \tag{2.21}$$

$$+\frac{F_2}{5a^3}\left\{1 - \frac{6}{7} f_2 - \frac{6}{7} f_4 + \frac{111}{35} f_2^2\right\} - \frac{10}{63} \frac{f_2}{a^5} F_4 =$$

$$= \frac{\omega^2 a^2}{12\pi G}\left\{1 - \frac{10}{7} f_2 - \frac{9}{35} f_2^2 + \frac{4}{7} f_4\right\};$$

$$\frac{a^2 E_2}{5}\left\{\frac{36}{35} f_2 + \frac{40}{77} f_4 + \frac{108}{385} f_2^2\right\} + \frac{a^4 E_4}{9}\left\{1 + \frac{80}{77} f_2\right\} +$$

$$-\frac{F_0}{a}\left\{f_4 - \frac{18}{35} f_2^2 + \frac{108}{385} f_2^3 - \frac{40}{77} f_2 f_4\right\} + \tag{2.22}$$

$$-\frac{F_2}{5a^2}\left\{\frac{54}{35} f_2 + \frac{60}{77} f_4 - \frac{648}{385} f_2^2\right\} + \frac{F_4}{9a^5}\left\{1 - \frac{100}{77} f_2\right\} =$$

$$= \frac{\omega^2 a^2}{6\pi G}\left\{\frac{18}{35} f_2 - \frac{57}{77} f_4 - \frac{9}{77} f_2^2\right\};$$

$$\frac{a^2 E_2}{5}\left\{\frac{10}{11} f_4 + \frac{18}{77} f_2^2\right\} + \frac{20}{99} a^4 f_2 E_4 + \frac{1}{13} a^6 E_6 +$$

$$-\frac{F_0}{a}\left\{f_6 + \frac{18}{77} f_2^3 - \frac{10}{11} f_2 f_4\right\} - \frac{3}{11}\frac{F_2}{a^3}\left\{f_4 - \frac{36}{35} f_2^2\right\} + \tag{2.23}$$

$$-\frac{25}{99}\frac{f_2 F_4}{a^5} + \frac{F_6}{13a^7} =$$

$$= \frac{\omega^2 a^2}{6\pi G}\left\{\frac{5}{11} f_4 + \frac{9}{77} f_2^2\right\}.$$

In order to transform these equations to more symmetrical forms, let us note that, to the *first* order in small quantities, Equation (2.21) reduces to

$$\frac{a^2 E_2}{5} + \frac{F_2}{5a^3} - \frac{f_2 F_0}{a} = \frac{\omega^2 a^2}{12\pi G} \tag{2.24}$$

and, with its aid, Equations (2.21) and (2.22) yield

$$\frac{a^2 E_2}{5} + \frac{F_2}{5a^3}\left\{1 - \frac{10}{7} f_2\right\} - \frac{f_2 F_0}{a}\left\{1 - \frac{6}{7} f_2\right\} = \frac{\omega^2 a^2}{12\pi G}\left\{1 - 2f_2\right\} \tag{2.25}$$

and

$$\frac{a^4 E_4}{9} + \frac{F_4}{9a^5} - \frac{18}{35}\frac{f_2 F_2}{a^3} - \frac{F_0}{a}\left\{f_4 - \frac{54}{35} f_2^2\right\} = 0, \tag{2.26}$$

correctly to quantities of *second* order. By insertion from the foregoing Equations (2.25)–(2.26) in (2.21)–(2.23) for terms which are multiplied by small quantities, it is possible to rewrite (2.21)–(2.23) correctly to terms of third order in the following alternative forms

$$\frac{a^2E_2}{5} + \frac{F_2}{5a^3} - \frac{f_2F_0}{a} = -\frac{F_0}{a}\left\{\frac{6}{7}f_2^2 - \frac{748}{245}f_2^3 + \frac{16}{7}f_2f_4\right\} + + \frac{F_2}{5a^3}\left\{\frac{10}{7}f_2 - \frac{338}{49}f_2^2 + \frac{10}{7}f_4\right\} + + \frac{2f_2}{7a^5}F_4 + \frac{a^2}{3}\left(\frac{\omega^2}{4\pi G}\right)\left\{1 - 2f_2 + \frac{6}{7}f_2^2\right\}, \tag{2.27}$$

$$\frac{a^4E_4}{9} + \frac{F_4}{9a^5} - \frac{f_4F_0}{a} = -\frac{F_0}{a}\left\{\frac{54}{35}f_2^2 - \frac{1890}{2695}f_2^3 + \frac{160}{77}f_2f_4\right\} + + \frac{F_2}{5a^3}\left\{\frac{18}{7}f_2 - \frac{2988}{539}f_2^2 + \frac{100}{77}f_4\right\} + + \frac{20f_2}{77a^5}F_4 - \frac{\omega^2a^2}{4\pi G}\left\{\frac{2}{3}f_4 - \frac{18}{35}f_2^2\right\}, \tag{2.28}$$

and

$$\frac{a^6E_6}{13} + \frac{F_6}{13a^7} - \frac{f_6F_0}{a} = \frac{F_0}{a}\left\{\frac{216}{77}f_2^3 - \frac{40}{11}f_2f_4\right\} + + \frac{F_2}{a^3}\left\{\frac{5}{11}f_4 - \frac{90}{77}f_2^2\right\} + \frac{5f_2}{11a^5}F_4. \tag{2.29}$$

The same equations would have been obtained had we expanded the right-hand side of Equation (1.44) for $\Psi(r', \theta, \phi)$ in terms of the products $r'^jP_j(\cos\theta)$ and equated their coefficients for $j = 2, 4, 6 \ldots$ to zero.

The foregoing Equations (2.27)–(2.29) contain the unknown amplitudes f_j both in front of, and behind, the integral signs in the expressions for E_j and F_j as given by Equations (2.13)–2.20). To lure them out from behind the integral signs, multiply Equations (2.27)–(2.29) by a^j ($j = 2, 4, 6$), respectively (so as to render the coefficients of E_j on the left-hand sides constant), and differentiate with respect to a. The derivatives of E_j and F_j are merely equal to the integrands on the right-hand side of Equations (2.13)–(2.20); however, since (unlike for F_j) the independent variable a occurs in the lower limit of the definite integrals for E_j, the derivatives of E_j are equal to the integrands on the right-hand side of (2.13)–(2.16) taken with the negative sign. If, subsequently, we eliminate the terms F_n for $n \neq j$ factored by small quantities with the aid of Equations (2.24)–(2.26) valid to the accuracy of lower orders, it is possible to express the F_j's for $j = 2, 4, 6$ in the form

$$F_2(a) = a^2F_0\left\{3f_2 - af_2' + \frac{12}{7}f_2^2 - \frac{4}{7}f_2(af_2') + \frac{2}{7}(af_2')^2 + - \frac{102}{35}f_2^3 - \frac{2}{7}f_2^2(af_2') - \frac{4}{7}f_2(af_2')^2 - \frac{8}{35}(af_2')^3 + \right. \tag{2.30}$$

$$+\frac{52}{7}f_2f_4-\frac{4}{7}a(f_2f_4'+f_4f_2')+\frac{4}{7}a^2f_2'f_4'\Big\}+$$

$$+\frac{2}{3}\left(\frac{\omega^2a^5}{4\pi G}\right)\left\{af_2'+\frac{4}{7}f_2(af_2')-\frac{2}{7}(af_2')^2\right\},$$

$$F_4(a)=a^4F_0\Big\{5f_4-af_4'$$

$$+\frac{108}{35}f_2^2-\frac{72}{35}f_2(af_2')+\frac{18}{35}(af_2')^2+$$

$$+\frac{972}{385}f_2^3-\frac{612}{385}f_2^2(af_2')+\frac{324}{385}f_2(af_2')^2-\frac{108}{385}(af_2')^3+ \tag{2.31}$$

$$+\frac{520}{77}f_2f_4-\frac{80}{77}a(f_2f_4'+f_4f_2')+\frac{40}{77}a^2f_2'f_4'\Big\}+$$

$$+2\left(\frac{\omega^2a^7}{4\pi G}\right)\left\{-\frac{2}{3}f_4+\frac{1}{3}af_4'+\frac{18}{35}f_2^2+\frac{24}{35}f_2(af_2')^2-\frac{6}{35}(af_2')^2\right\},$$

and

$$F_6(a)=a^6F_0\Big\{7f_6-af_6'+\frac{270}{77}f_2^3-\frac{18}{7}f_2^2(af_2')+$$

$$+\frac{90}{77}f_2(af_2')^2-\frac{18}{77}(af_2')^3+\frac{130}{11}f_2f_4+$$

$$-\frac{30}{11}a(f_2f_4'+f_4f_2')+\frac{10}{11}a^2f_2'f_4'\Big\}, \tag{2.32}$$

where primes denote differentiation with respect to a, correctly to terms of third order.

As the last step of our analysis, let us differentiate the foregoing Equations (2.30)–(2.32) once more with respect to a, and insert for $F_j'(a)$ from (2.17)–(2.20); the outcome discloses that the amplitudes $f_j(a)$ for $j = 2, 4, 6$ should satisfy the following second-order differential equations

$$a^2f_2''+6D(af_2'+f_2)-6f_2=\frac{2}{7}\Big\{2\eta_2(\eta_2+9)-9D\eta_2(\eta_2+2)\Big\}f_2^2+$$

$$-\frac{4}{35}\Big\{(7\eta_2^3+33\eta_2^2+180\eta_2+66)+$$

$$+3D\left(2\eta_2^3-15\eta_2^2-27\eta_2+5\right)\Big\}f_2^3+$$

$$+\frac{4}{7}\Big\{2(\eta_2\eta_4+15\eta_2+8\eta_4)- \tag{2.33}$$

$$- 3D(3\eta_2\eta_4 + 3\eta_2 + 3\eta_4 - 7)\Big\} f_2 f_4 +$$

$$+ \frac{3\omega^2}{\pi G\rho}(1-D)\left\{f_2 + af_2' + \frac{6}{7}f_2(af_2') + \frac{3}{7}(af_2')^2\right\} +$$

$$+ \frac{1}{6}\left(\frac{3\omega^2}{\pi G\rho}\right)^2 (1-D)(\eta_2+1)f_2 .$$

$$\begin{aligned} a^2 f_4'' + 6D(af_4' + f_4) - 20f_4 &= \frac{18}{35}\left\{2\eta_2(\eta_2+2) - 3D(3\eta_2^2 + 6\eta_2 + 7)\right\} f_2^2 + \\ &+ \frac{36}{385}\Big\{2(3-\eta_2)(1-5\eta_2) + \\ &- 3D\left(3\eta_2^3 + 9\eta_2^2 + 12\eta_2 + 4\right)\Big\} f_2^3 + \\ &+ \frac{40}{77}\left\{(2\eta_2\eta_4 + 23\eta_2 + 9\eta_4) - 9D(\eta_2\eta_4 + \eta_4)\right\} f_2 f_4 + \\ &+ \frac{3\omega^2}{\pi G\rho}(1-D)\left\{f_4 + af_4' + \frac{9}{35}\left[7f_2^2 + 6f_2(af_2') + 3(af_2')^2\right]\right\}, \end{aligned} \tag{2.34}$$

and

$$\begin{aligned} a^2 f_6'' &+ 6D(af_6' + f_6) - 42f_6 = \\ &= \frac{18}{77}\left\{4(3-\eta_2)(\eta_2+2) - 3D(\eta_2^3 + 3\eta_2^2 + 15\eta_2 + 5)\right\} f_2^3 + \\ &+ \frac{10}{11}\left\{2(\eta_2\eta_4 + 6\eta_2 - \eta_4) - 3D(3\eta_2\eta_4 + 3\eta_2 + 3\eta_4 + 11)\right\} f_2 f_4 , \end{aligned} \tag{2.35}$$

equivalent to (2.21)–(2.23), where

$$\eta_j \equiv \frac{a}{f_j}\frac{\partial f_j}{\partial a} , \tag{2.36}$$

and where we have abbreviated

$$\bar{\rho} = \frac{3}{a^3}\int_0^a \rho a^2 \mathrm{d}a , \qquad D = \frac{\rho}{\bar{\rho}} . \tag{2.37}$$

The *boundary conditions* necessary for complete specification of the particular solutions of the foregoing equations, which are to represent the amplitudes $f_j(a)$ of the individual harmonic terms on the right-hand side of the expansion (2.1) for r', are imposed partly at the centre and partly at the boundary of our configuration. As, at the centre, all the $f_j(a)$'s are to be a minimum, the necessary condition for this to be so is that, for $a = 0$,

$$f_j'(0) = 0 \qquad \text{for } j = 2, 4, 6, \ldots . \tag{2.38}$$

On the other hand, at the boundary $a = a_1$, all $E_j(a_1)$'s as defined by Equations (2.13)–(2.16) are equal to zero and for $j > 0$, the F_j's continue to be given by (2.30)–(2.32); whereas, for $j = 0$, $4\pi F_0(a_1) = m_1$ by (2.71), where m_1 denotes the total mass of our configuration. Inserting them in Equations (2.27)–(2.29) we find that, for $a = a_1$,

$$2f_2 + af_2' + \frac{5}{3}\left(\frac{\omega^2 a_1^3}{Gm_1}\right) = \frac{2}{3}\left(\frac{\omega^2 a_1^3}{Gm_1}\right)\left\{(\eta_2+5)f_2 - \frac{1}{7}(2\eta_2^2 + 6\eta_2 + 15)f_2^2\right\} + \tag{2.39}$$

$$+ \frac{2}{35}\left\{5(\eta_2^2 + 3\eta_2 + 6)f_2^2 - (4\eta_2^3 + 30\eta_2^2 + 60\eta_2 + 76)f_2^3 + \right.$$

$$\left. + 5(2\eta_2\eta_4 + 3\eta_2 + 3\eta_4 + 26)f_2f_4\right\},$$

$$4f_4 + af_4' = \frac{2}{3}\left(\frac{\omega^2 a_1^3}{Gm_1}\right)\left\{(\eta_4+7)f_4 - \frac{9}{35}(2\eta_2^2 + 10\eta_2 + 21)f_2^2\right\} +$$

$$+ \frac{18}{35}(\eta_2^2 + 5\eta_2 + 6)f_2^2 + \tag{2.40}$$

$$- \frac{36}{385}(3\eta_2^3 + 18\eta_2^2 + 44\eta_2 + 54)f_2^3 +$$

$$+ \frac{20}{77}(2\eta_2\eta_4 + 5\eta_4 + 26)f_2f_4,$$

and

$$6f_6 + af_6' = -\frac{18}{77}(\eta_2+2)(\eta_2^2 + 6\eta_2 + 12)f_2^3 + \frac{5}{11}(2\eta_2\eta_4 + 7\eta_4 + 7\eta_4 + 26)f_2f_4. \tag{2.41}$$

A construction (numerically or otherwise) of the desired particular solutions of the simultaneous system of differential equations (2.33)–(2.35) specified by the boundary conditions (2.38) and (2.39)–(2.41) can be accomplished by successive approximations in the following manner. Within the scheme of a first-order approximation, (2.33) reduces to the well-known equation

$$a^2f_2'' + 6D(af_2' + f_2) = 6f_2, \tag{2.42}$$

which for

$$D = 1 - \lambda a^2 + \ldots \tag{2.43}$$

admits, in the proximity of the origin, of a solution varying as

$$f_2 = k_2\left(1 + \frac{3}{7}\lambda a^2 + \ldots\right), \tag{2.44}$$

where k_2 stands for an arbitrary constant. Integrating Equation (2.42) we can proceed hereafter until $a = a_1$, at which point the left-hand side of Equation (2.39) discloses that

$$2f_2 + af_2' + \frac{5}{3}\left(\frac{\omega^2 a_1^3}{Gm_1}\right) = 0; \tag{2.45}$$

and this (algebraic) equation can be used to specify the value of $\omega^2 a_1^3/Gm_1$ corresponding to the initially adopted value of k_2.

With a *first-order* approximation to $f_2(a)$ thus in our hands, we can now proceed to the *second* approximation, which consists of finding a solution of the equations

$$a^2 f_2'' + 6D(af_2' + f_2) - 6f_2 = \tfrac{2}{7}\{2\eta_2(\eta_2 + 9) - 9\eta_2(\eta_2 + 2)\} f_2^2 + \\ + (3\omega^2/\pi G\rho)(1 - D)(\eta_2 + 1) f_2 \tag{2.46}$$

and

$$a^2 f_4'' + 6D(af_4' + f_4) - 20 f_4 = \\ = \tfrac{18}{35}\{2\eta_2(\eta_2 + 2) - 3D(3\eta_2^2 + 6\eta_2 + 7)\} f_2^2, \tag{2.47}$$

subject to the boundary conditions requiring that, at the centre,

$$f_2'(0) = f_4'(0) = 0, \tag{2.48}$$

while on the surface $a = a_1$,

$$2f_2 + a_1 f_2' + \frac{5}{3}\left(\frac{\omega^2 a_1^3}{Gm_1}\right) = \tfrac{1}{7}(2\eta_2^2 + 6\eta_2 + 12) f_2^2 + \\ + \frac{2}{3}\left(\frac{\omega^2 a_1^3}{Gm_1}\right)(\eta_2 + 5) f_2^2, \tag{2.49}$$

and

$$4f_4 + a_1 f_4' = \tfrac{18}{35}(\eta_2^2 + 5\eta_2 + 6) f_2^2. \tag{2.50}$$

As Equation (2.46) is independent of f_4, its solution satisfying (2.48) can, near the origin, be expanded in a series in ascending even powers of a. If, moreover, we note that

$$\frac{3\omega^2}{\pi G\rho} = \frac{D}{\rho/\rho_c}\left(\frac{3\omega^2}{\pi G\rho_c}\right), \tag{2.51}$$

where $\rho_c \equiv \rho(0)$ and, consistent with (2.43),

$$\frac{\rho}{\rho_c} = 1 - \tfrac{5}{2}\lambda a^2 + \ldots, \tag{2.52}$$

we find that, correctly to terms of the order of the squares of surficial distortion,

$$f_2 = k_2\{1 + \tfrac{3}{7}(1 + \upsilon)\lambda a^2 + \\ + \tfrac{3}{14}(1 - \tfrac{11}{5}\upsilon)\lambda^2 a^4 - \frac{4k_2}{147}\lambda^2 a^4 + \ldots\}, \tag{2.53}$$

where

$$\upsilon = \frac{\omega^2}{2\pi G\rho_c} \tag{2.54}$$

denotes a constant.*

* Strictly speaking, one should augment the right-hand sides of (2.53) still by quantities arising from possible bi-quadratic terms on the right-hand sides of Equations (2.43) or (2.52) which are not spelled out explicity in the latter; but their inclusion may be left as an exercise for the interested reader.

Moreover, the structure of Equation (2.47) for f_4, solved in a similar way (cf. James and Kopal, 1963) discloses that, near the origin.

$$f_4 = k_4 a^2 \{1 + \tfrac{18}{55} \lambda a^2 + \ldots\} + $$
$$+ \tfrac{27}{35} k_2^2 \{1 - \tfrac{8}{99} k_2 \lambda^2 a^4 +$$
$$+ \tfrac{156}{385} k_2^{-1}(1 + v)\lambda^2 a^4 + \ldots\}, \qquad (2.55)$$

consisting of two parts: the first (factored by k_4) represents the 'complementary function' of the homogeneous version of (2.47) with its left-hand side equated to zero, while the second (factored by k_2^2) stands for a 'particular integral' arising from non-vanishing right-hand side. The constant k_4 introduced through the complementary function is new, and its value must be specified (after integration has been completed) from the boundary condition (2.50) – just as k_2 needs to be re-computed from (2.49).

With a second-order approximation to $f_2(a)$ near the origin as represented by Equation (2.53), we can proceed to evaluate the right-hand sides of Equations (2.46) and (2.47) correctly to quantities of the third (or any higher) order. In general, the structure of the differential Equations (2.46)–(2.47) governing the $f_j(a)$'s makes it evident that, near the origin, the complementary function of each f_j will vary as $k_j a^{j-2}$, while its particular integral will be factored by $k_2{}^{j/2}$. If, therefore, we set

$$f_2(0) = k_2, \qquad (2.56)$$

it follows from (2.33)–(2.35) that (for $\eta_j(0) = 0$),

$$\begin{aligned} f_4(0) &= \tfrac{27}{35} f_2^2(0) + \tfrac{108}{2695} f_2^3(0) + \ldots \\ &= \tfrac{27}{35} \{k_2^2 + \tfrac{4}{77} k_2^3 + \ldots\} \end{aligned} \qquad (2.57)$$

and

$$f_6(0) = \tfrac{5}{6} f_2(0) f_4(0) - \frac{9}{154} f_2^3(0) + \ldots = \tfrac{45}{77} k_2^3 + \ldots \qquad (2.58)$$

etc., while the lowest derivatives of the f_j's which do not vanish at the origin are given by

$$f_2''(0) = \tfrac{6}{7} k_2 (1 + v)\lambda \qquad (2.59)$$

and, for $j > 2$,

$$f_j^{(j-2)}(0) = (j-2)! k_j, \qquad j = 4, 6, \ldots \qquad (2.60)$$

The constants $k_2, k_4, k_6, \ldots,$ (not to be confused with those defined later by Equation (6.4)) constitute a set of the 'eigen-parameters' of our rotational problem; and their values for any given value of $V \equiv \omega^2/2\pi G\rho_c$ must be determined algebraically from the boundary conditions (2.39)–(2.41) at $a = a_1$. Inasmuch as the right-hand sides of Equations (2.33)–(2.35) for f_j are known algebraic functions of $f_2, f_4 \ldots f_{j-2}$, their system readily lends itself for a solution by successive approximations: first we solve (2.42) for f_2 to accuracy of first order; next (2.46) and (2.47) for f_2 and f_4 to quantities of second order; and, eventually, all three for f_2, f_4, f_6 to third order. Moreover, in case of need, the same process can evidently be extended to attain the accuracy of any order. Such an extension consistent to quantities of *fourth* order has indeed been already worked out by Kopal and Mahanta (1974). Its details are beyond the scope of this section; but can be found in the paper just quoted.

The results established in this section describe the explicit form of the amplitudes $f_j(a)$ in the expansion (2.1) for the equipotential surface of a rotating configuration in terms of the spherical harmonics of the form $P_2(\cos\theta)$. A choice of the latter for the angle-dependent part of the solid harmonics $Y_j(a, \theta, \phi)$ on the right-hand side of Equation (2.1) has behind it a tradition of more than two hundred years established by Legendre (1793) and Laplace (1825); and all subsequent investigations of the subject were content to follow in this respect in their footsteps. Yet this strategy is neither unique, nor necessary, for the treatment of the problem. A choice of the form of the harmonics on the right-hand side of the expansion (2.1) is, to be sure, intimately connected with those involved in the disturbing potential V' causing distortion. But an equally valid representation of the centrifugal potential (1.41) can be written up as expansions in terms of harmonics other than zonal – sectorial harmonics, for instance; or others – and the radius-vector r' of the respective equipotential surfaces expanded accordingly.

A consistent theory of the form of rotating configurations – equivalent to Clairaut's – in terms of sectorial harmonics $P_n^n(\cos\theta)$ as well as in terms of Hansen-Tisserand coefficients (cf., e.g., Hagihara, 1972) has recently been developed by El-Shaarawy (1974) correctly to quantities of third order in surficial distorition. To a first-order approximation, the differential equations governing the amplitudes $f_j(a)$ of these harmonics as well as the associated boundary conditions proved to be the same as when the zonal (or any other) harmonics are used; they begin to differ from them (and from each other) only through terms of higher orders, commencing with the second. El-Shaarawy established their explicit forms, for the families of harmonics adopted by him, to terms of third order. Since, however, his results turned out to be no simpler than those met with by the use of zonal harmonics earlier in this section, we shall not reproduce them in this place; and for fuller detail refer the reader to the source already quoted (El-Shaarawy, 1974).

II-3. Tidal Distortion

A determination of the gravitational potential and form of a fluid configuration of mass m_1 distorted by the attraction of a body of mass m_2 in its neighbourhood raising tides on m_1 can follow a course so parallel with the one developed already in the preceding section that only its gist needs to be given in this place.

The principal distinguishing feature of our present problem will be a different form of the disturbing potential V'. In order to specify the potential arising from tidal interaction, let the positions of the masses $m_{1,2}$ be described by the rectangular coordinates $x_{1,2}, y_{1,2}, z_{1,2}$ and let, what follows, $x_1 = y_1 = z_1 = 0$ (making the centre of mass m_1

the origin of coordinates) and $x_2 = R, y_2 = 0, z_2 = 0$ (assigning m_2 a position on the X-axis at a distance R from the origin; cf. Figure 2-2). Let, furthermore, r denote an arbitrary radius vector, of orientation specified by the direction cosines

$$\begin{aligned} \lambda &= \cos\phi \sin\theta , \\ \mu &= \sin\phi \sin\theta , \\ \nu &= \cos\theta . \end{aligned} \tag{3.1}$$

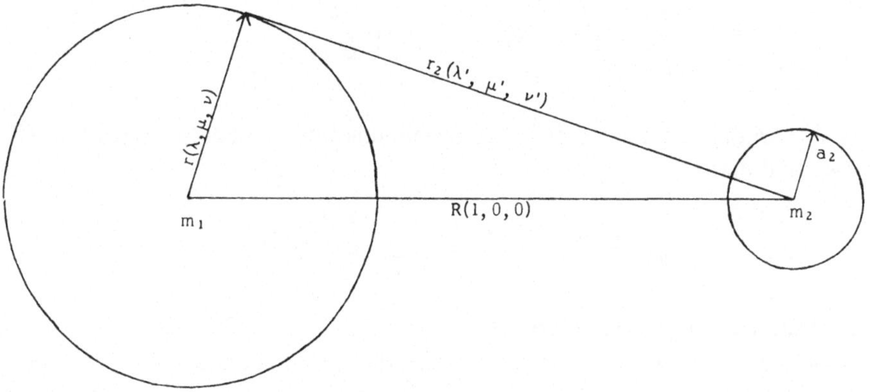

Fig. 2-2. Geometry of tidal distortion.

If, moreover, $a_{1,2}$ denote the mean radii of the two components and $(k_j)_{1,2}$, the coefficients specifying the internal structure of the two stars*, it is well known that the exterior potential $V(r)$ of mass m_1 can be expressed as

$$V(r) = \frac{Gm_1}{r}\left\{1 + \frac{m_2}{m_1}\sum_{j=2}^{\infty} 2(k_j)_1 \left(\frac{a_1}{R}\right)^{j+1}\left(\frac{a_1}{r}\right)^{j} P_j(\lambda)\right\}, \tag{3.2}$$

in conformity with the particular solution (1.39) of Laplace's Equation (1.38) for $d_j \neq 0$, where $r > a_1$ and λ represents the cosine of an angle between the radii-vectors $r(\lambda, \mu, \nu)$ and $R(1, 0, 0)$.

The exterior potential $V'(r)$ arising from the mass m_2 of the secondary component can be likewise obtained by an appropriate permutation of indices in (3.2) as

$$V'(r) = \frac{Gm_2}{r_2}\left\{1 + \frac{m_1}{m_2}\sum_{j=2}^{\infty} 2(k_j)_2 \left(\frac{a_2}{R}\right)^{j+1}\left(\frac{a_2}{r_2}\right)^{j} P_j(\lambda_2)\right\}, \tag{3.3}$$

where r_2 denotes the distance of an arbitrary element of mass m_1 from the centre of m_2, and λ_2 is the cosine of an angle between r_2 and R.

From the triangles marked on Figure 2-2 it is evident that the radii-vectors r, r_2 and R are connected with the direction cosines λ and λ_2 by the equations

* For their specification in terms of the density distribution inside the mass m_j cf. section II-6.

$$r^2 = R^2 + r_2^2 - 2Rr_2\lambda_2 , \tag{3.4}$$

$$r_2^2 = R^2 + r^2 - 2Rr\lambda. \tag{3.5}$$

Eliminating r_2 between them we find that

$$\lambda_2 = \frac{R - \lambda r}{\sqrt{R^2 + r^2 - 2Rr\lambda}} = \left\{1 - \frac{r\lambda}{R}\right\} \sum_{j=0}^{\infty} \left(\frac{r}{R}\right)^j P_j(\lambda) =$$

$$= 1 + \sum_{j=1}^{\infty} \frac{j}{2j+1} \{P_{j+1} - P_{j-1}\} \left(\frac{r}{R}\right)^{j+1}, \tag{3.6}$$

where $P_j \equiv P_j(\lambda)$. On the other hand, by eliminating λ_2 between Equations (3.4) and (3.5) we find that

$$\frac{1}{r_2} = \frac{1}{\sqrt{R^2 + r^2 - 2Rr\lambda}} = \frac{1}{R} \sum_{j=0}^{\infty} \left(\frac{r}{R}\right)^j P_j(\lambda). \tag{3.7}$$

An inspection of the structure of Equations (3.2) or (3.3) discloses that the leading terms of the summations on the right-hand sides are of the order of $a_{1,2}^5$; their squares being of the order of $a_{1,2}^{10}$, etc. Consequently, terms which are of the first order in surficial distortion correspond to $j = 2, 3, 4$; those of second order, to $j = 5, 6, 7$; etc. *Let us, in what follows, limit ourselves to consider terms in* V′ *of no higher than the second order.* If so, then by noting that, within this scheme of approximation,

$$(R/r_2)^3 = 1 + 3\,(r/R)\,P_1(\lambda) + (r/R)^2\ \{5P_2(\lambda) + 1\} + \ldots \tag{3.8}$$

and

$$P_2(\lambda_2) = 1 + (r/R)^2\ \{P_2(\lambda) - 1\} + \ldots, \tag{3.9}$$

while, for $j > 2$,

$$P_j(\lambda_2) = 1, \tag{3.10}$$

it follows from (3.4) that – correctly to quantities of second order – the disturbing potential $V'(r)$ which raises tides on the configuration of mass m_1 can be expressed as

$$V'(r) = G\frac{m_2}{R}\left\{1 + \sum_{j=2}^{7}\left(\frac{r}{R}\right)^j P_j(\lambda)\right\} +$$

$$+ G\frac{m_1}{R}(k_2)_2\left(\frac{a_2}{R}\right)^5\left\{1 + 3\left(\frac{r}{R}\right)P_1(\lambda) + 6\left(\frac{r}{R}\right)^2 P_2(\lambda) + \ldots\right\} +$$

$$+ G\frac{m_1}{R}(k_3)_2\left(\frac{a_2}{R}\right)^7\{1 + \ldots\} = \sum_{j=0}^{7} c_j r^j P_j(\lambda) \tag{3.11}$$

in accordance with the general form of Equation (1.39), where the constants

$$c_0 = G\frac{m_2}{R} + G\frac{m_1}{R}(k_2)_2\left(\frac{a_2}{R}\right)^5 + G\frac{m_1}{R}(k_3)_2\left(\frac{a_2}{R}\right)^7 + \ldots, \tag{3.12}$$

$$c_1 = \quad + 3G\frac{m_1}{R^2}(k_2)_2\left(\frac{a_2}{R}\right)^5 + \ldots, \tag{3.13}$$

$$c_2 = G\frac{m_2}{R^3} + 6G\frac{m_1}{R^3}(k_2)_2\left(\frac{a_2}{R}\right)^5 + \ldots; \tag{3.14}$$

and, for $j > 2$,

$$c_j = G\frac{m_2}{R^{j+1}}. \tag{3.15}$$

If the secondary (disturbing) component of mass m_2 could be regarded as a mass-point – either because $a_2 = 0$ (zero size) or $(k_j)_2 = 0$ (infinite density concentration) – the entire expression for $V'(r)$ on the right-hand side of Equation (3.11) would reduce to its first term; all others represent the effects of the departure of the secondary's behaviour from that of a mass-point.

The occurrence in the expression (3.11) for $V'(r)$ of harmonics $P_j(\lambda)$ for $j = 1(1)7$ to describe all effects of first and second order necessitates the radius-vector $r'(a, \theta', \phi')$ of the corresponding equipotential surface to be expressible as

$$r' = a\left\{1 + \sum_{j=0}^{7} f_j(a) P_j(\lambda')\right\}, \tag{3.16}$$

where

$$\lambda' = \cos\phi' \sin\theta', \tag{3.17}$$

correctly to quantities of second order – in contrast with the third-order Equation (2.1) of the rotational problem. The amplitudes $f_j(a)$ on the right-hand side of Equation (3.16) represent, however, now the amplitudes of the respective partial tides invoked by the disturbing potential V'; and the principal aim of the present section will be to specify them in terms of the star's structure and of the properties of the external field of force.

The products $f_j(a)P_j(\lambda')$ continue to be solid harmonics of the form $Y_j(a, \theta', \phi')$, satisfying the differential Equation (1.32) and the orthogonality conditions (1.34) of section II-1. Therefore, raising Equation (3.16) to the $(2-n)$th and $(n+3)$rd powers by use of the binomial theorem, and making use of the decomposition theorem (2.5) for Legendre polynomials we find that, correctly to quantities of second order,

$$(r')^{2-n} = a^{2-n}\left\{1 - (n-2)\sum_{j=0}^{7}[f_j - (n-1)X_j]P_j\right\} \tag{3.18}$$

and

$$(r')^{n+3} = a^{n+3}\left\{1 + (n+3)\sum_{j=0}^{7}[f_j + (n+2)X_j]P_j\right\}, \tag{3.19}$$

where

$$X_0 = \tfrac{1}{10}(f_2^2 + \tfrac{5}{7}f_3^2 + \ldots), \tag{3.20}$$

$$X_1 = \tfrac{9}{35} f_2 f_3 + \ldots, \tag{3.21}$$

$$X_2 = \tfrac{1}{7}(f_2^2 + \tfrac{2}{3} f_3^2 + 2 f_2 f_4 + \ldots), \tag{3.22}$$

$$X_3 = \tfrac{4}{15} f_2 f_3 + \ldots, \tag{3.23}$$

$$X_4 = \tfrac{9}{35}(f_2^2 + \tfrac{5}{11} f_3^2 + \tfrac{100}{99} f_2 f_4 + \ldots), \tag{3.24}$$

$$X_5 = \tfrac{10}{21} f_2 f_3 + \ldots, \tag{3.25}$$

$$X_6 = \tfrac{50}{231}(f_3^2 + \tfrac{21}{10} f_2 f_4 + \ldots), \tag{3.26}$$

and all X_j''s for $j > 6$ are identically equal to zero. Like in section II-1, an expansion of the form (3.8) continues to be valid for $n \gtrless 2$; while, for $n = 2$, the coefficient of P_2 in the expansion of $\log r'$ becomes equal to $f_2 - X_2$.

If we insert now the foregoing expansions (3.18) and (3.19) with the coefficients X_j as given by (3.20)–(3.26) in Equations (1.36) or (1.37) and (1.28) and take advantage of the orthogonality property of our harmonics as given by Equation (1.34), the interior and exterior potentials U and V of tidally-distorted configurations can again be expressed in the forms

$$U(r) = 4\pi G \sum_{j=0}^{\infty} \frac{r^j E_j(a)}{2j+1} P_j(\lambda), \tag{3.27}$$

$$V(r) = 4\pi G \sum_{j=0}^{\infty} \frac{F_j(a) P_j(\lambda)}{(2j+1) r^{j+1}}, \tag{3.28}$$

analogous to Equations (1.35) and (1.36) where (again correctly to quantities of second order)

$$E_0(a) = \int_a^{a_1} \rho \frac{\partial}{\partial a} \{a^2 [\tfrac{1}{2} + f_0 + X_0]\}\, da, \tag{3.29}$$

and

$$F_0(a) = \int_0^a \rho \frac{\partial}{\partial a} \{a^3 [\tfrac{1}{3} + f_0 + 2X_0]\}\, da; \tag{3.30}$$

while, for $j > 0$,

$$E_j(a) = \int_a^{a_1} \rho \frac{\partial}{\partial a} \{a^{2-j} [f_j + (1-j) X_j]\}\, da \tag{3.31}$$

and

$$F_j(a) = \int_a^{a_1} \rho \frac{\partial}{\partial a} \{a^{j+3} [f_j + (j+2) X_j]\}\, da, \tag{3.32}$$

respectively.

The amplitude f_0 of the zero-order harmonic on the right-hand side of Equation (3.16) can – as in the rotational case – be readily specified from the requirement that the total mass of our configuration must be independent of distortion – a requirement leading (by

Equation 2.10) to the condition

$$f_0 = -2X_0 = -\tfrac{1}{5}f_2^2 - \tfrac{1}{7}f_3^2 - \ldots, \tag{3.33}$$

analogous to Equation (2.12) of the rotational problem. Moreover, in order to determine the f_j's for $j > 0$, recourse must again be made to the fact that the level surfaces represented by Equation (3.16) are equipotentials, over which the total potential

$$\Psi(r', \lambda) = U(r', \lambda) + V(r', \lambda) + V'(r', \lambda) \tag{3.34}$$

must remain constant for the tide-generating potential V' as given by Equation (3.11).

Suppose that, as before, we expand the total potential $\Psi(r', \lambda)$ in a Neumann series of the form

$$\Psi(r', \lambda) = \sum_{j=0}^{\infty} \alpha_j(a) P_j(\lambda), \tag{3.35}$$

with coefficients defined by

$$\alpha_j(a) = \frac{2j+1}{2} \int_{-1}^{1} \Psi(r', \lambda) P_j(\lambda)\, d\lambda. \tag{3.36}$$

If so, the constancy of $\Psi(r', \lambda)$ over a level surface will evidently be ensured by setting

$$\alpha_j(a) = 0 \qquad \text{for} \qquad j > 0; \tag{3.37}$$

and j equations of this form constitute the necessary as well as sufficient conditions for a specification of the respective amplitudes f_j.

A determination of the explicit forms of Equations (3.37) is rather involved; but the results turn out to be relatively simple; for it can be shown (cf. pp. 46–48 of Kopal, 1960) that, for $j = 1(1)7$,

$$\frac{a\,E_1}{3} + \frac{F_1}{3a^2} - \frac{f_1 F_0}{a} + \frac{c_1 a}{4\pi G} = -\tfrac{9}{5} f_2 f_3 \frac{F_0}{a^2} + \tfrac{9}{35} f_3 \frac{F_2}{a^3} + \tfrac{9}{35} f_2 \frac{F_3}{a^4}, \tag{3.38}$$

$$\frac{a^2 E_2}{5} + \frac{F_2}{5a^3} - \frac{f_2 F_0}{a} + \frac{c_2 a^2}{4\pi G} = -\{\tfrac{6}{7} f_2^2 + \tfrac{16}{21} f_3^2 + \tfrac{16}{7} f_2 f_4\}\frac{F_0}{a} +$$

$$+ \{\tfrac{2}{7} f_2 + \tfrac{2}{7} f_4\}\frac{F_2}{a^3} + \tfrac{4}{21} f_3 \frac{F_3}{a^4} + \tfrac{2}{7} f_2 \frac{F_4}{a^5}, \tag{3.39}$$

$$\frac{a^3 E_3}{7} + \frac{F_3}{7a^4} - \frac{f_3 F_0}{a} + \frac{c_3 a^3}{4\pi G} = -\tfrac{9}{5} f_2 f_3 \frac{F_0}{a} + \tfrac{4}{15} f_3 \frac{F_2}{a^3} + \tfrac{4}{15} f_2 \frac{F_3}{a^4}, \tag{3.40}$$

$$\frac{a^4 E_4}{9} + \frac{F_4}{9a^5} - \frac{f_4 F_0}{a} + \frac{c_4 a^4}{4\pi G} = -\{\tfrac{54}{35} f_2^2 + \tfrac{72}{77} f_3^2 + \tfrac{80}{77} f_2 f_3\}\frac{F_0}{a} +$$

$$+ \{\tfrac{18}{35} f_2 + \tfrac{20}{77} f_4\}\frac{F_2}{a^3} + \tfrac{18}{77} f_3 \frac{F_3}{a^4} + \tfrac{20}{77} f_2 \frac{F_4}{a^5}, \tag{3.41}$$

$$\frac{a^5E_5}{11} + \frac{F_5}{11a^6} - \frac{f_5F_0}{a} + \frac{c_5a^5}{4\pi G} = -\tfrac{10}{3} f_2f_3 \frac{F_0}{a} + \tfrac{10}{21} f_3 \frac{F_2}{a^3} + \tfrac{10}{21} f_2 \frac{F_3}{a^4}, \tag{3.42}$$

$$\frac{a^6E_6}{13} + \frac{F_6}{13a^7} - \frac{f_6F_0}{a} + \frac{c_6a^6}{4\pi G} = -\left\{\tfrac{400}{231} f_3^2 + \tfrac{40}{11} f_2f_4\right\} \frac{F_0}{a} + $$

$$+ \tfrac{5}{11} f_2 \frac{F_2}{a^3} + \tfrac{100}{231} f_3 \frac{F_3}{a^4} + \tfrac{5}{11} f_2 \frac{F_4}{a^5}, \tag{3.43}$$

$$\frac{a^7E_7}{15} + \frac{F_7}{15a^8} - \frac{f_7F_0}{a} + \frac{c_7a^7}{4\pi G} = 0, \tag{3.44}$$

where the constants c_j continue to be given by Equations (3.13)–(3.15), and

$$F_0 = \int_0^a \rho a^2 \, da. \tag{3.45}$$

The foregoing Equations (3.38)–(3.44) constitute the tidal equivalents of Equations (2.27)–(2.29) relevant to the case of the rotational problem. Like in this former case, the left-hand sides of (3.38)–(3.44) consist of terms which are of both first, and second orders. Those on the right are, however, all of second order – consisting as they do of the products of various integrals $F_j (j = 0, 2, 3, 4)$ with different combinations of the amplitudes f_2, f_3, and f_4 of the most important partial tides. The latter are, moreover, known to possess leading terms of first order, satisfying the equation

$$\frac{a^j}{2j+1} \int_a^{a_1} \rho \frac{\partial}{\partial a} (a^{2-j} f_j) \, da + \frac{1}{(2j+1)\, a^{j+1}} \int_0^a \rho \frac{\partial}{\partial a} (a^{j+3} f_j) \, da +$$

$$- \frac{f_j}{a} \int_0^a \rho a^2 \, da = \frac{c_j a^j}{4\pi G} \tag{3.46}$$

for $j = 2, 3$, and 4. Divide now this equation by a^j, differentiate with respect to a, and multiply subsequently by a^{2j+1}; in doing so we find that

$$\int_0^a \rho \frac{\partial}{\partial a} (a^{j+3} f_j) \, da = a^{j+1} \left\{ \frac{j+1}{a} f_j - \frac{\partial f_j}{\partial a} \right\} \int_0^a \rho a^2 \, da; \tag{3.47}$$

or, correctly, to quantities of first order,

$$F_j(a) = a^{j+1} \left\{ \frac{j+1}{a} f_j - \frac{\partial f_j}{\partial a} \right\} F_0, \tag{3.48}$$

corresponding to Equations (2.30) and (2.32) of section II-2.

If we insert now Equation (3.48) on the right-hand sides of Equations (3.38)–(3.44) we can express the latter, more concisely, for $j = 1(1)7$ as integral equations for f_j, of the form

$$f_j \int_0^a \rho a^2 \, da - \frac{1}{(2j+1)a^j} \int_0^a \rho \frac{\partial}{\partial a} (a^{j+3} f_j) \, da +$$

$$- \frac{a^{j+1}}{2j+1} \int_a^{a_1} \rho \frac{\partial}{\partial a} (a^{2-j} f_j) \, da = \Re_j(a) + \frac{c_j a^{j+1}}{4\pi G}, \tag{3.49}$$

$$\Re_j(a) = a \frac{\partial X_j}{\partial a} \int_0^a \rho a^2 \, da + \frac{j+2}{(2j+1)a^j} \int_0^a \rho \frac{\partial}{\partial a} (a^{j+3} X_j) \, da +$$

$$- \frac{(j-1)a^{j+1}}{2j+1} \int_a^{a_1} \rho \frac{\partial}{\partial a} (a^{2-j} X_j) \, da \tag{3.50}$$

in terms of the X_j's as given by Equations (3.20)–(3.26).

Equations (3.49) define the amplitudes f_j in integral form. In order to reduce them to differential equations (in which they are more amenable to numerical solution), differentiate (3.49) with respect to a and divide by a^{2j}; then differentiate the quotient once more with respect to a and multiply by a^{j+2}. The result of these operations discloses that

$$a^2 f_j'' + 6D(af_j' + f_j) - j(j+1) f_j =$$
$$= F_0^{-1} \{a^2 \Re_j'' - j(j+1)\Re_j\} \equiv T_j(a), \tag{3.51}$$

where primes denote differentiation with respect to a;

$$D \equiv \frac{\rho}{\bar{\rho}} = \frac{\rho a^3}{3 \int_0^a \rho a^2 \, da} = \frac{\rho a^3}{3F_0}, \tag{3.52}$$

in which $\bar{\rho}$ denotes the mean density of our configuration interior to a; and

$$T_1(a) = \tfrac{18}{35} \{2(\eta_2\eta_3 + 10\eta_2 + 7\eta_3) - 3D(3\eta_2\eta_3 + 3\eta_2 + 3\eta_3 - 5)\} f_2 f_3, \tag{3.53}$$

$$T_2(a) = \tfrac{2}{7} \{2\eta_2(\eta_2 + 9) - 9D\eta_2(\eta_2 + 4)\} f_2^2 + \tfrac{4}{21} \{2\eta_3(\eta_3 + 21) +$$
$$- 9D(\eta_3^2 + 2\eta_3 - 2)\} f_3^2 + \tfrac{4}{7} \{2(\eta_2\eta_4 + 15\eta_2 + 8\eta_4) +$$
$$- 3D(3\eta_2\eta_4 + 3\eta_2 + 3\eta_4 - 7)\} f_2 f_4 \tag{3.54}$$

$$T_3(a) = \tfrac{8}{15} \{2\eta_2\eta_3 + 15\eta_2 - 9\eta_3 - 9D(\eta_2\eta_3 + \eta_2 + \eta_3)\} f_2 f_3, \tag{3.55}$$

$$T_4(a) = \tfrac{18}{35} \{2\eta_2(\eta_2 + 2) - 3D(3\eta_2^2 + 6\eta_2 + 7)\} f_2^2 +$$
$$+ \tfrac{18}{77} \{2\eta_3(\eta_3 + 14) - 3D(3\eta_3^2 + 6\eta_3 + 1)\} f_3^2 +$$
$$+ \tfrac{40}{77} \{2\eta_2\eta_4 + 23\eta_2 + 9\eta_4 - 9D(\eta_2\eta_4 + \eta_2 + \eta_4)\} f_2 f_4, \tag{3.56}$$

$$T_5(a) = \tfrac{20}{21}\{2\eta_2^3(\eta_3 + 3) - 3D(3\eta_2\eta_3 + 3\eta_2 + 3\eta_3 + 5)\} f_2 f_3, \tag{3.57}$$

$$T_6(a) = \tfrac{100}{231}\{2\eta_3^2(\eta_3 + 3) - 9D(\eta_3^2 + 2\eta_3 + 4)\} f_3^2 + \\ + \tfrac{10}{11}\{2(\eta_2\eta_4 + 6\eta_2 - \eta_4) + \\ - 3D(3\eta_2\eta_4 + 3\eta_2 + 3\eta_4 + 11)\} f_2 f_4, \tag{3.58}$$

$$T_7(a) = 0, \tag{3.59}$$

where we have abbreviated

$$\eta_j = \frac{af_j'}{f_j} \tag{3.60}$$

and taken advantage of the fact that – in accordance with (3.51) – the function η_j (j = 2, 3, 4) satisfies the differential equation

$$a\eta_j' + 6D(\eta_j + 1) + \eta_j(\eta_j - 1) = j(j + 1) \tag{3.61}$$

correctly to quantities of first order.

Equations (3.51) with their right-hand sides as defined by (3.53)–(3.59) constitute second-order analogies to Equations (2.33)–(2.35) of the rotational problem. In the present case, Equations (3.51) constitute a simultaneous system of seven (j = 1(1)7) second-order differential equations for the f_j's, subject to the boundary conditions requiring that, at the centre ($a = 0$),

$$f_j'(0) = 0 \qquad \text{for} \qquad j = 1(1)7; \tag{3.62}$$

while at the boundary $a = a_1$ (cf. Kopal, 1960; pp. 49–50)

$$f_j(a_1) = \frac{(2j + 1)\, c_j a_1^{j+1}}{G(j + \eta_j)_1 m_1} + \left\{\frac{S_j(a)}{j + \eta_j}\right\}_{a_1}, \tag{3.63}$$

where

$$S_1 = \tfrac{9}{35}\, b_1^{(2,3)} f_2 f_3, \tag{3.64}$$

$$S_2 = \tfrac{1}{7}\, b_2^{(2,2)} f_2^2 + \tfrac{2}{21}\, b_2^{(2,3)} f_3^2 + \tfrac{2}{7} b_2^{(2,4)} f_2 f_4 \tag{3.65}$$

$$S_3 = \tfrac{4}{15}\, b_3^{(2,3)} f_2 f_3, \tag{3.66}$$

$$S_4 = \tfrac{9}{35}\, b_4^{(2,2)} f_2^2 + \tfrac{9}{77}\, b_4^{(3,3)} f_3^2 + \tfrac{20}{77} b_4^{(2,4)} f_2 f_4, \tag{3.67}$$

$$S_5 = \tfrac{10}{21}\, b_5^{(2,3)} f_2 f_3, \tag{3.68}$$

$$S_6 = \tfrac{50}{231}\, b_6^{(3,3)} f_3^2 + \tfrac{5}{11}\, b_6^{(2,4)} f_2 f_4, \tag{3.69}$$

$$S_7 = 0; \tag{3.70}$$

and where we have abbreviated

$$b_j^{(i,k)} = 2\eta_i\eta_k + (j+1)(\eta_i+\eta_k) + i(i+1) + k(k+1), \tag{3.71}$$

evaluated at the boundary $a = a_1$.

For arbitrary models of the stars – i.e., for arbitrary distribution of density ρ throughout the interior – the desired particular solutions of Equations (3.51)–(3.61) subject to the boundary conditions (3.62) and (3.63)–(3.71) can be constructed only by numerical methods; and their construction can be left as an exercise for the interested reader. While a determination of the amplitudes $f_j(a_1)$'s of the respective partial tides may, therefore, become a matter of some complexity, the symmetry of a configuration deformed by the tides is simple: namely, the crest of each tide is always in the direction of the radius-vector R connecting the mass centre of the two stars (i.e., for an observer on the crest of each tide the centre of mass m_2 is always in his zenith). This is, however, true only of the equilibrium tides investigated in this chapter. When, in Chapter III, we shall turn our attention to dynamical tides in close binary systems we shall encounter a very different situation.

In conclusion of the present section, one additional consideration of some interest should be pointed out. In the case of rotational distortion treated previously in section II-2, the structure of the centrifugal potential (1.41) implied that the expansion (2.1) for the radius-vector r' contained only even harmonics of the respective angular variable $\cos\theta$; and, in such a case, the centre of mass of the rotating configuration is bound to coincide with the origin of our coordinate system. For if x, y, z denote the coordinates of the centre of gravity of mass m_1 arbitrarily placed, it follows by definition of the centre of mass that (cf. Equations (1.20) and (1.28))

$$\begin{aligned}\begin{Bmatrix} x \\ y \\ z \end{Bmatrix} m_1 &= \int_0^{r_1}\int_0^{\pi}\int_0^{2\pi} \rho r'^2 \begin{Bmatrix} r'\sin\theta'\cos\phi' \\ r'\sin\theta'\sin\phi' \\ r'\cos\theta', \end{Bmatrix} dr'\sin\theta'\,d\theta'\,d\phi' \\ &= \int_0^{a_1} \rho \frac{\partial}{\partial a}\left\{\int_0^{\pi}\int_0^{2\pi} \frac{r'^4}{4}\begin{bmatrix} \sin\theta'\cos\phi' \\ \sin\theta'\sin\phi' \\ \cos\theta' \end{bmatrix} \sin\theta'\,d\theta',d\theta'\right\} da. \end{aligned} \tag{3.72}$$

The column matrix on the right-hand side of the foregoing equation consists of the components of the solid harmonic $Y_1(a, \theta, \phi)$ of order one, known to be of the form

$$Y_1(a, \theta', \phi') = A\sin\theta'\cos\phi' + B\sin\theta'\sin\phi' + C\cos\theta', \tag{3.73}$$

where A, B, C are constants or functions of a. Therefore, unless the expansion of r'^4 itself contains a solid harmonic of the type Y_1, the orthogonality properties of spherical harmonics will compel all terms on the right-hand side of (3.72) to vanish identically; and if so, the left-hand side must similarly vanish – which it can do only if $x = y = z = 0$.

If the expansion on the right-hand side of (3.16) were to consist only of the harmonics of even orders – as was the case for purely rotational distortion – the term r'^4 did not contain any harmonic of order one; and, therefore, the centre of gravity coincided with the origin of our coordinate system regardless of the extent of the distortion. However, if – as in the case of mutual tidal distortion – the expansion on the right-hand side contains harmonics of even as well as odd orders, the expression for r'^4 will contain products of the form $Y_j Y_{j-1}$; and each of these on decomposition will furnish a non-zero term varying as Y_1. In such a case, the right-hand side of Equation (3.72) will no longer vanish; and neither can then the left-hand one – a phenomenon indicating that not all three coordinates

x, y, z of the centre of mass of m_1 can be set equal to zero in our coordinate system. In point of fact, the positions of the centres of gravity and symmetry begin gradually to drift apart with increasing degree of tidal distortion, until the initial single configuration may actually break up in two. The emergence of this tendency can be traced to terms factored by the powers and cross-products of the spherical harmonics contained in the tide-generating potential V', and introduced through it into the expansion (3.16) for r'. Whether or not a trend so initiated may, under sufficiently extreme conditions, result in a fission into two bodies of comparable mass constitutes a problem to which no satisfactory answer can so far be given. It may, however, be of interest to note that the first mathematical symptoms of a situation which may lead to this end appear with the introduction of terms which are of *second* order in surficial distortion.

II-4. Interaction between Rotation and Tides

In the preceding sections of this chapter we have investigated the rotational as well as tidal distortion of self-gravitating configurations of arbitrary structure, and established the explicit form of differential equations governing the first seven spherical-harmonic deformations ($j = 1$ to 7). Within the framework of a consistent theory including all tides with amplitudes of the order of the squares of the surficial distortion, all seven harmonics are present with non-vanishing coefficients; while in the case of rotational distortion (containing no odd harmonics as long as the configuration remains spheroidal) the use of the first three even harmonics has permitted us to attain the accuracy of third order. The leading terms of the rotational as well as tidal distortion are, however, both of first order. Therefore, no second-order theory of either distortion can be regarded as complete until the effects of the cross-terms between first-order rotational and tidal distortion have been investigated. This constitutes a problem to which we propose to address ourselves in this section.

Before we proceed to tackle this delicate task, we should clearly point out the limitations imposed by the need to maintain the equilibrium form of our configuration. In the preceding section we developed an equilibrium theory of tides as a 'two-centre' problem – with the tide-generating mass m_2 kept stationary on the X-axis. In actual binary systems both components of masses $m_{1,2}$ must, to be sure, revolve in a plane orbit around the common centre of gravity – so that the X-axis actually rotates in space, with the motion of the radius-vector R of the relative orbit of the two stars. As long as the motion is uniform, and the radius-vector constant, tidal distortion will produce no motion in the fluid mass m_1 rotating with the Keplerian angular velocity $\omega_K^2 = G(m_1 + m_2)/R^3$; and the results based on the equilibrium theory continue to be directly applicable to reality.

Suppose, however, that the radius-vector R ceases to be constant (on account of a finite eccentricity of the relative orbit): if so, the height of a j-th partial tide should vary as R^{-j-1} in the course of each orbital cycle, thus setting the fluid in motion. Moreover, if the distorted component also rotates about an axis which is not perpendicular to the orbital plane, and its angular velocity ω of rotation is different from ω_K, the fact that the crest of each tide follows (in the absence of dissipative forces) the radius-vector R will give rise to finite velocity components in the rotating system of coordinates – constituting dynamical tides.

Such tides will become the subject of appropriate discussion in the subsequent Chapter III of this book. If we wish to remain in the framework of an equilibrium theory, and to avoid the emergence of velocity components arising from differential motions within the fluid, the following conditions must be satisfied:

(1) The equator of the rotating configuration must be coplanar with the relative orbit of the disturbing body; and the orbit itself circular (i.e., R = constant).

(2) The (constant) angular velocity ω of the distorted configuration must be identical with the (constant) Keplerian angular velocity ω_K of orbital revolution – in other words, the rotation and revolution must be synchronous.

A breakdown of either one of these conditions is bound to give rise to *dynamical tides*, a study of which is being postponed for the next chapter. In the present section we shall assume that the foregoing conditions (1) and (2) are indeed fulfilled; so that the cross-terms between rotation and tides which we wish to consider will not give rise to any motion in the rotating frame of reference.

In order to investigate the nature of such terms, let the direction cosines of the vectors r and r' diagramatically shown on Figure 2-1 be given by

$$\begin{aligned} \lambda &= \cos\phi \sin\theta, \\ \mu &= \sin\phi \sin\theta, \\ \nu &= \cos\theta; \end{aligned} \tag{4.1}$$

and

$$\begin{aligned} \lambda' &= \cos\phi' \sin\theta', \\ \mu' &= \sin\phi' \sin\theta', \\ \nu' &= \cos\theta', \end{aligned} \tag{4.2}$$

respectively; so that the angle γ between them (cf. Equation (1.14)) is given by the equation

$$\cos\gamma = \lambda\lambda' + \mu\mu' + \nu\nu'. \tag{4.3}$$

Let, moreover, the direction of the tide-generating body in the same coordinates be specified by the direction cosines λ'', μ'', ν''; so that the angles Θ, Θ' between the radii-vectors r, r' and the line (of length R) joining the centres of the two mutually attracting masses $m_{1,2}$ will be given by

$$\cos\Theta = \lambda\lambda'' + \mu\mu'' + \nu\nu'' \tag{4.4}$$

and

$$\cos\Theta' = \lambda'\lambda'' + \mu'\mu'' + \nu'\nu'', \tag{4.5}$$

respectively. Within the framework of the equilibrium theory of tides, the X-axis of our rectangular system becomes identical with the orbital radius-vector, and the Z-axis with that of rotation of the respective configuration. If so, however, then

$$\lambda'' = 1 \quad \text{and} \quad \mu'' = \nu'' = 0; \tag{4.6}$$

and, accordingly,

$$\cos\Theta = \lambda \qquad \text{and} \qquad \cos\Theta' = \lambda'. \tag{4.7}$$

A development of the radius-vector r' of the distorted surface consistent to quantities of second order is bound to contain also cross-terms varying as $P_i(\nu')\,P_j(\lambda')$ for $j = 2,3$ and 4 within the scheme of second-order approximation, which must be added to the summation on the right-hand side of Equation (3.16), together with purely rotational terms investigated already in section II-2.

Suppose that we do so, and include the terms of this form for the expression for r' used to evaluate the U_j's and V_j's constituting our potential expansions in section II-1. However, we find it necessary to consider – in addition to the classical orthogonality theorem (2.7) – also more general orthogonality conditions for the *triple* products $P_i(\nu')\,P_j(\lambda')\,P_n(\cos\gamma)$ the integrals of which must be evaluated over the whole sphere in terms of the unprimed angular coordinates θ and ϕ.

Accordingly let us observe that, in the case of synchronism between rotation and revolution with co-planar equator and orbit, the addition theorem for spherical harmonics discloses that

$$P_n(\cos\phi'\sin\theta') = P_n(0)\,P_n(\cos\theta') + + 2\sum_{i=1}^{n} \frac{(n-i)!}{(n+i)!}\, P_n^i(0)\,P_n^i(\cos\theta')\cos i\phi', \tag{4.8}$$

where

$$P_{2j}(0) = (-1)^j\,\frac{1.3.5\ldots(2j-1)}{2.4.6\ldots 2j} \tag{4.9}$$

if n happens to be an even integer, and

$$P_{2j+1}(0) = 0 \tag{4.10}$$

if it is odd.

With the aid of this addition theorem and the decomposition formula (2.5) it is possible to establish that

$$\int_0^{\pi}\int_0^{2\pi} P_j(\lambda')\,P_2(\nu')\,P_n(\cos\gamma)\sin\theta'\,d\theta'\,d\phi' = \frac{4\pi}{2n+1}\,\mathfrak{P}_n^{(j)}(\theta,\phi), \tag{4.11}$$

where the only terms non-vanishing for $j = 2(1)4$ assume (in terms of the Clebsch-Gordan coefficients) the explicit forms

$$\mathfrak{P}_0^{(2)} = -\tfrac{1}{10}, \tag{4.12}$$

$$\mathfrak{P}_2^{(2)} = -\tfrac{2}{7}\,\{P_2(\lambda) + P_2(\nu)\}, \tag{4.13}$$

$$\mathfrak{P}_4^{(2)} = \tfrac{1}{10} + \tfrac{2}{7}\,\{P_2(\lambda) + P_2(\nu)\} + P_2(\lambda)\,P_2(\nu); \tag{4.14}$$

$$\mathfrak{P}_1^{(3)} = -\tfrac{9}{70}\,P_1(\lambda), \tag{4.15}$$

$$\mathfrak{P}_3^{(3)} = -\tfrac{2}{15} P_1(\lambda) - \tfrac{1}{3} P_3(\lambda) - \tfrac{2}{3} P_1(\lambda) P_2(\nu), \tag{4.16}$$

$$\mathfrak{P}_5^{(3)} = \tfrac{11}{42} P_1(\lambda) + \tfrac{1}{3} P_2(\lambda) + \tfrac{2}{3} P_1(\lambda) P_3(\nu) + P_3(\lambda) P_2(\nu); \tag{4.17}$$

and

$$\mathfrak{P}_2^{(4)} = \tfrac{1}{21} P_2(\nu) - \tfrac{5}{42} P_2(\lambda), \tag{4.18}$$

$$\mathfrak{P}_4^{(4)} = -\tfrac{20}{77} P_2(\nu) - \tfrac{27}{77} P_2(\lambda) - \tfrac{4}{11} P_4(\lambda) - \tfrac{10}{11} P_2(\lambda) P_2(\nu) - \tfrac{1}{11}, \tag{4.19}$$

$$\mathfrak{P}_6^{(4)} = \tfrac{7}{33} P_2(\nu) + \tfrac{25}{66} P_2(\lambda) + \tfrac{4}{11} P_4(\lambda) + \tfrac{10}{11} P_2(\lambda) P_2(\nu) + \tfrac{1}{11} + P_4(\lambda) P_2(\nu); \tag{4.20}$$

satisfying a general relation of the form

$$\mathfrak{P}_{j-2}^{(j)} + \mathfrak{P}_j^{(j)} + \mathfrak{P}_{j+2}^{(j)} = P_j(\lambda) P_2(\nu). \tag{4.21}$$

In consequence, a full-dress expression for the radius-vector r' of a tidally-distorted configuration rotating in synchronism with its revolution about an axis perpendicular to the orbital plane should – correctly to quantities of second order in surficial distortion – contain the terms*

$$r' = a \left\{ 1 + \ldots + \sum_{j=0}^{4} g_j(a) P_j(\lambda') + \sum_{j=0}^{4} h_j(a) P_j(\lambda') P_2(\nu') \right\} \tag{4.22}$$

additional to those arising from pure rotation or tides. The latter have already been investigated in previous parts of our work; but the explicit form of the mixed terms with amplitudes g_j and h_j remain yet to be established.

In order to do so, let – in accordance with Equations (1.17) and (1.18) – the interior and exterior potential of our distorted configuration be expressible as

$$(2n + 1)\, U = 4\pi G \sum_{n=0}^{\infty} r^n U_n \tag{4.23}$$

and

$$(2n + 1)\, V = 4\pi G \sum_{n=0}^{\infty} r^{-n-1} V_n, \tag{4.24}$$

respectively, where U_n and V_n continue to be given by Equations (1.26)–(1.28) for any value of n. Since, moreover, by use of the generalized orthogonality theorem (4.11),

$$\int_0^{\pi}\int_0^{2\pi} \left\{ \sum_{j=0}^{4} g_j P_j(\lambda') + \sum_{j=0}^{4} h_j P_j(\lambda') P_2(\nu') \right\} P_n(\cos\gamma) \sin\theta'\, d\theta'\, d\phi' =$$

$$= \frac{4\pi}{2n+1} \left\{ g_n P_n(\lambda) + \sum_{j=0}^{4} h_j \mathfrak{P}_n^{(j)}(\lambda, \nu) \right\}, \tag{4.25}$$

* The expansion of r' on right-hand side of Equation (4.22) has been carried out in terms of the zonal harmonic $P_2(\nu)$, rather than the sectorial harmonic $P_2^2(\nu)$, of the colatitude. Since, however, $P_2^2 = -$ $- 2(1 - P_2)$, the coefficients f_j and h_j of different harmonics used in our previous work (Kopal, 1960) are related with g_j and h_j used presently by $f_j = g_j + h_j$ and $h_j = -\frac{1}{2} h_j$.

the contributions to U_n and V_n arising from the *interaction* between rotation and tides (and, therefore, *additive* to those due to pure rotation and tides) will be given by the equations

$$U_0 = \int_a^{a_1} \rho \frac{\partial}{\partial a} (a^2 g_0)\, da + \mathfrak{P}_0^{(2)}(\lambda, \nu) \int_a^{a_1} \rho \frac{\partial}{\partial a} (a^2 h_2)\, da + \ldots, \tag{4.26}$$

$$3U_1 = P_1(\lambda) \int_{a_1}^{a_1} \rho \frac{\partial}{\partial a} (a g_1)\, da + \mathfrak{P}_1^{(1)}(\lambda, \nu) \int_a^{a_1} \rho \frac{\partial}{\partial a} (a h_1)\, da + $$

$$+ \mathfrak{P}_1^{(3)}(\lambda, \nu) \int_a^{a_1} \rho \frac{\partial}{\partial a} (a h_3)\, da + \ldots, \tag{4.27}$$

$$5U_2 = P_2(\lambda) \int_a^{a_1} \rho \frac{\partial}{\partial a} (g_2)\, da + \mathfrak{P}_2^{(0)}(\lambda, \nu) \int_a^{a_1} \rho \frac{\partial}{\partial a} (h_0)\, da + $$

$$+ \mathfrak{P}_2^{(2)}(\lambda, \nu) \int_a^{a_1} \rho \frac{\partial}{\partial a} (h_2)\, da + \mathfrak{P}_2^{(4)}(\lambda, \nu) \int_a^{a_1} \rho \frac{\partial}{\partial a} (h_4)\, da + \ldots, \tag{4.28}$$

$$7U_3 = P_3(\lambda) \int_a^{a_1} \rho \frac{\partial}{\partial a} \left(\frac{g_3}{a}\right) da + \mathfrak{P}_3^{(1)}(\lambda, \nu) \int_a^{a_1} \rho \frac{\partial}{\partial a} \left(\frac{h_1}{a}\right) da + $$

$$+ \mathfrak{P}_3^{(3)}(\lambda, \nu) \int_a^{a_1} \rho \frac{\partial}{\partial a} \left(\frac{h_3}{a}\right) da + \ldots, \tag{4.29}$$

$$9U_4 = P_4(\lambda) \int_a^{a_1} \rho \frac{\partial}{\partial a} \left(\frac{g_4}{a^2}\right) da + \mathfrak{P}_4^{(2)}(\lambda, \nu) \int_a^{a_1} \rho \frac{\partial}{\partial a} \left(\frac{h_2}{a^2}\right) da + $$

$$+ \mathfrak{P}_4^{(4)}(\lambda, \nu) \int_a^{a_1} \rho \frac{\partial}{\partial a} \left(\frac{h_4}{a^2}\right) da + \cdots, \tag{4.30}$$

$$11U_5 = \mathfrak{P}_5^{(3)}(\lambda, \nu) \int_a^{a_1} \rho \frac{\partial}{\partial a} \left(\frac{h_3}{a^3}\right) da + \cdots, \tag{4.31}$$

$$13U_6 = \mathfrak{P}_6^{(4)}(\lambda, \nu) \int_a^{a_1} \rho \frac{\partial}{\partial a} \left(\frac{h_4}{a^4}\right) da + \cdots; \tag{4.32}$$

and

$$V_0 = \int_0^a \rho \frac{\partial}{\partial a} (a^3 g_0)\, da + \mathfrak{P}_0^{(2)}(\lambda, \nu) \int_0^a \rho \frac{\partial}{\partial a} (a^3 h_2)\, da + \cdots, \tag{4.33}$$

$$3V_1 = P_1(\lambda) \int_0^a \rho \frac{\partial}{\partial a} (a^4 g_1)\, da + \mathfrak{P}_1^{(1)}(\lambda, \nu) \int_0^a \rho \frac{\partial}{\partial a} (a^4 g_1)\, da + $$

$$+\mathfrak{P}_1^{(3)}(\lambda,\nu)\int_0^a \rho \frac{\partial}{\partial a}(a^4 h_3)\,da + \cdots, \tag{4.34}$$

$$5V_2 = P_2(\lambda)\int_0^a \rho \frac{\partial}{\partial a}(a^5 g_2)\,da + \mathfrak{P}_2^{(0)}(\lambda,\nu)\int_0^a \rho \frac{\partial}{\partial a}(a^5 h_0)\,da +$$

$$+\mathfrak{P}_2^{(2)}(\lambda,\nu)\int_0^a \rho \frac{\partial}{\partial a}(a^5 h_2)\,da + \mathfrak{P}_2^{(4)}(\lambda,\nu)\int_0^a \rho \frac{\partial}{\partial a}(a^5 h_4)\,da + \cdots, \tag{4.35}$$

$$7V_3 = P_3(\lambda)\int_0^a \rho \frac{\partial}{\partial a}(a^6 g_3)\,da + \mathfrak{P}_3^{(1)}(\lambda,\nu)\int_0^a \rho \frac{\partial}{\partial a}(a^6 h_1)\,da +$$

$$+\mathfrak{P}_3^{(3)}(\lambda,\nu)\int_0^a \rho \frac{\partial}{\partial a}(a^6 h_3)\,da + \cdots, \tag{4.36}$$

$$9V_4 = P_4(\lambda)\int_0^a \rho \frac{\partial}{\partial a}(a^7 g_4)\,da + \mathfrak{P}_4^{(2)}(\lambda,\nu)\int_0^a \rho \frac{\partial}{\partial a}(a^7 h_4)\,da +$$

$$+\mathfrak{P}_4^{(4)}(\lambda,\nu)\int_0^a \rho \frac{\partial}{\partial a}(a^7 h_4)\,da + \cdots, \tag{4.37}$$

$$11V_5 = \mathfrak{P}_5^{(3)}(\lambda,\nu)\int_0^a \rho \frac{\partial}{\partial a}(a^8 h_3)\,da + \cdots, \tag{4.38}$$

$$13V_6 = \mathfrak{P}_6^{(4)}(\lambda,\nu)\int_0^a \rho \frac{\partial}{\partial a}(a^9 h_4)\,da + \cdots. \tag{4.39}$$

where – in addition to the $\mathfrak{P}_n^{(j)}$'s given by Equations (4.12)–(4.20) earlier in this section,

$$\mathfrak{P}_2^{(0)} = P_2(\nu) \tag{4.40}$$

and

$$\mathfrak{P}_1^{(1)} = -\tfrac{1}{5}P_1(\lambda), \tag{4.41}$$

$$\mathfrak{P}_3^{(1)} = \tfrac{1}{5}P_1(\lambda) + P_1(\lambda)P_2(\nu). \tag{4.42}$$

If we factor out common harmonics in the Expressions (4.26)–(4.39) for U_n and V_n, the latter can be rewritten more concisely in the form

$$U_n = \sum_{j,k} P_j(\nu)P_k(\lambda)\int_a^{a_1} \rho \frac{\partial}{\partial a}\{a^{2-n}\Phi_{j,k}^{(n)}\}\,da \tag{4.43}$$

and

$$V_n = \sum_{j,k} P_j(\nu)P_k(\lambda)\int_0^a \rho \frac{\partial}{\partial a}\{a^{n+3}\Phi_{j,k}^{(n)}\}\,da \tag{4.44}$$

for $j = 0, 2$ and $k = 0(1)4$, where

$$\Phi_{0,n}^{(n)} = G_n, \qquad n = 0(1)4; \tag{4.45}$$

$$\Phi_{2,0}^{(2)} = H_0, \tag{4.46}$$

$$\Phi_{2,1}^{(3)} = 5\Phi_{0,1}^{(3)} = H_1, \tag{4.47}$$

$$\Phi_{2,2}^{(4)} = \tfrac{7}{2}\Phi_{0,2}^{(4)} = \tfrac{7}{2}\Phi_{2,0}^{(4)} = 10\Phi_{0,0}^{(4)} = H_2, \tag{4.48}$$

$$\Phi_{2,3}^{(5)} = 3\Phi_{0,3}^{(5)} = \tfrac{3}{2}\Phi_{2,1}^{(5)} = \tfrac{42}{11}\Phi_{0,1}^{(5)} = H_3, \tag{4.49}$$

$$\Phi_{2,4}^{(6)} = \tfrac{11}{4}\Phi_{0,4}^{(6)} = \tfrac{11}{10}\Phi_{2,2}^{(6)} = \tfrac{66}{25}\Phi_{0,2}^{(6)} = \tfrac{33}{7}\Phi_{2,0}^{(6)} = 11\Phi_{0,0}^{(6)} = H_4, \tag{4.50}$$

in which

$$G_0 = g_0 - \tfrac{1}{10}h_2, \tag{4.51}$$

$$G_1 = g_1 - \tfrac{1}{5}h_1 - \tfrac{9}{70}h_3, \tag{4.52}$$

$$G_2 = g_2 - \tfrac{2}{7}h_2 - \tfrac{5}{42}h_4, \tag{4.53}$$

$$G_3 = g_3 - \tfrac{1}{3}h_3, \tag{4.54}$$

$$G_4 = g_4 - \tfrac{1}{11}h_4; \tag{4.55}$$

and

$$H_0 = h_0 - \tfrac{2}{7}h_2 + \tfrac{1}{21}h_4, \tag{4.56}$$

$$H_1 = h_1 - \tfrac{2}{3}h_3, \tag{4.57}$$

$$H_2 = h_2 - \tfrac{10}{11}h_4, \tag{4.58}$$

$$H_3 = h_3, \tag{4.59}$$

$$H_4 = h_4. \tag{4.60}$$

With the potential expansions for U and V now explicitly formulated by Equations (4.23)–(4.24) and (4.43)–(4.44) in terms of amplitudes g_j and h_j of the 'mixed' tides, we are in a position to set up the contraints which these amplitudes are to obey if the total potential

$$\Psi(r', \lambda, \nu) = U(r', \lambda, \nu) + V(r', \lambda, \nu) + V'(r', \lambda, \nu) \tag{4.61}$$

is to remain constant over a distorted surface of radius r'. As pointed out in section II-1, this constancy will be ensured if a sum of the coefficients of all harmonics of equal orders on the right-hand-side of the foregoing Equation (4.61) will be made to add up to zero for any value of $j > 0$; thus rendering $\Psi(r', \lambda, \nu)$ independent of angular variables θ and ϕ. The conditions so obtained for $j > 0$ should then be sufficient completely to specify the new functions g_j and h_j introduced on the right-hand side of the expansion (4.22) for r'; and our aim, in what follows, will be to establish their explicit form.

In doing so, we should observe that whereas the expected interaction terms in the first power of r' are stated on the right-hand side of (4.22), in the binomial expansion for r'^m $(m > 0)$ the complete interaction terms will be given by

$$r'^m = a^m \left\{ 1 + \cdots + m(m-1) \sum_{j=2}^{4} \mathfrak{f}_2 f_j P_2(\nu) P_j(\lambda) + \right.$$
$$\left. + m \sum_{j=0}^{4} g_j P_j(\lambda) + m \sum_{j=0}^{4} h_j P_2(\nu) P_j(\lambda) + \cdots \right\} \tag{4.62}$$

correctly to quantities of second order, where $\mathfrak{f}_2$ and f_j represent the first order-effects of rotational and tidal distortion, respectively.

With the aid of the foregoing expansion, we are now in a position to proceed with the formulation of the necessary conditions on g_j and h_j to ensure the constancy of the total potential $\Psi(r', \lambda, \nu)$ over a level surface. In order to do so, let us expand $\Psi(r' \lambda, \nu)$ in a series of the form

$$\Psi(r', \lambda, \nu) = \sum_{i,j} \alpha_{i,j}(a) \; P_j^{(i)} (\lambda, \nu), \tag{4.63}$$

and seek to obtain the explicit form of the equations

$$\alpha_{i,j}(a) = 0, \tag{4.64}$$

analogous to (1.48), for all types of surface harmonics $Y_j^{(i)} (\theta, \phi)$ involved in r'.

Turning last to the disturbing potential $V' (r')$, we note that the structure of the one of tidal origin – as given by Equation (3.11) – is such that its terms varying as $r'^j P_j(\lambda)$ contain mixed terms of the form $jc_j a^j \mathfrak{f}_2 P_2(\nu) P_j(\lambda)$ for j = 2, 3, and 4 of second order. In addition, the centrifugal potential arising from axial rotation – as represented by Equation (1.41) – will contribute mixed terms of the form $\frac{2}{3}\omega^2 a^2 f_j P_j(\lambda)$ and $\frac{2}{3}\omega^2 a^2 f_j P_2(\nu)$ $P_j(\lambda)$, likewise for j = 2, 3 and 4. Therefore, the total disturbing potential of the rotational as well as tidal origin will contain interaction terms of the form

$$V_t^{(2)}(r', \lambda, \nu) = \tfrac{2}{3}\omega^2 a^2 \sum_{j=2}^{4} f_j P_j(\lambda) +$$
$$+ \sum_{j=2}^{4} \{ jc_j a^j \mathfrak{f}_2 - \tfrac{2}{3} \omega^2 a^2 f_j \} P_2(\nu) P_j(\lambda); \tag{4.65}$$

and if so, the conditions (4.64) required for the constancy of the total potential $\Psi(r', \theta, \phi)$ over a level surface $r'(a, \theta, \phi)$ will assume the form

$$G_j \int_0^a \rho a^2 \, da - \frac{1}{(2j+1)a^j} \int_0^a \rho \frac{\partial}{\partial a} (a^{j+3} G_j) \, da +$$
$$- \frac{a^{j+1}}{2j+1} \int_a^{a_1} \rho \frac{\partial}{\partial a} (a^{2-j} G_j) \, da = \mathfrak{G}_j + \frac{\omega^2 a^2}{6\pi G} f_j, \tag{4.66}$$

$$H_j\int_0^a \rho a^2\,\mathrm{d}a\;\frac{1}{(2j+5)a^{j+2}}\int_a^{a_1}\rho\,\frac{\partial}{\partial a}\,(a^{j+5}H_j)\,\mathrm{d}a\;+$$

$$-\frac{a^{j+3}}{2j+5}\int_a^{a_1}\rho\,\frac{\partial}{\partial a}\,(a^{-j}H_j)\,\mathrm{d}a=\mathfrak{H}_j+\frac{jc_ja^{j+1}}{4\pi G}\,\mathfrak{f}_j-\frac{\omega^2a^3}{6\pi G}\,f_j \tag{4.67}$$

for $j=0(1)4$, where

$$\mathfrak{G}_j=\frac{2(2j+1)}{j+2}\,\mathfrak{g}_j\int_0^{a_1}\rho a^2\,\mathrm{d}a+\frac{1}{a^j}\int_0^{a}\rho\,\frac{\partial}{\partial a}\,(a^{j+3}\,\mathfrak{g}_j)\,\mathrm{d}a\;-$$

$$-\frac{j-1}{j+2}\,a^{j+1}\int_a^{a_1}\rho\,\frac{\partial}{\partial a}\,(a^{2-j}\,\mathfrak{g}_j)\,\mathrm{d}a \tag{4.68}$$

and

$$\mathfrak{H}_j=\frac{2(2+5)}{j+4}\,\mathfrak{h}_i\int_0^{a}\rho a^2\,\mathrm{d}a+\frac{1}{a^{j+2}}\int_0^{a}\rho\,\frac{\partial}{\partial a}\,(a^{j+5}\,\mathfrak{h}_i)\,\mathrm{d}a\;-$$

$$-\frac{j+1}{j+4}\,a^{j+3}\int_a^{a_1}\rho\,\frac{\partial}{\partial a}\,(a^{-j}\,\mathfrak{h}_j)\,\mathrm{d}a, \tag{4.69}$$

where

$$\mathfrak{g}_0=-\tfrac{1}{5}\,\mathfrak{f}_2f_2, \tag{4.70}$$

$$\mathfrak{g}_1=-\tfrac{9}{70}\,\mathfrak{f}_2f_3, \tag{4.71}$$

$$\mathfrak{g}_2=-\tfrac{8}{35}\,\mathfrak{f}_2f_2-\tfrac{2}{21}\,\mathfrak{f}_2f_4, \tag{4.72}$$

$$\mathfrak{g}_3=-\tfrac{5}{21}\,\mathfrak{f}_2f_3, \tag{4.73}$$

$$\mathfrak{g}_4=-\tfrac{8}{33}\,\mathfrak{f}_2f_4; \tag{4.74}$$

and

$$\mathfrak{h}_0=-\tfrac{8}{35}\,\mathfrak{f}_2f_2+\tfrac{4}{105}\,\mathfrak{f}_2f_4, \tag{4.75}$$

$$\mathfrak{h}_1=-\tfrac{10}{21}\,\mathfrak{f}_2f_3, \tag{4.76}$$

$$\mathfrak{h}_2=\tfrac{2}{3}\,\mathfrak{f}_2f_2-\tfrac{20}{33}\,\mathfrak{f}_2f_4, \tag{4.77}$$

$$\mathfrak{h}_3=+\tfrac{7}{11}\,\mathfrak{f}_2f_3, \tag{4.78}$$

$$\mathfrak{h}_4=+\tfrac{8}{13}\,\mathfrak{f}_2f_4. \tag{4.79}$$

It may also be noted that the tidal amplitudes f_j on the right-hand side of Equations (4.66) and (4.67) for $j=0$ and 1 are (unlike those for $j=2$, 3 and 4) quantities of second order – a fact which justifies the neglect of their products with ω^2 in our present approximation. The same is, moreover, true of c_1 as given by Equation (3.13) – which justifies again a disregard, on the right-hand side of Equations (4.66) and (4.67), of its products with ϕ_2.

The foregoing Equations (4.66) and (4.67) will play the same role in the study of the effects produced by interaction between rotation and tides in close binary systems as did Equations (3.49) for purely tidal distortion; and Equations (4.68) with (4.69) are then analogous to (3.50). Their subsequent treatment can, consequently, follow similar lines. In order to do so, multiply both sides of Equations (4.66) by a^j, and differentiate with respect to a; the result will be the relation

$$\frac{\partial}{\partial a}(a^j G_j)\int_0^a \rho a^2\, da - a^{2j}\int_a^{a_1} \rho \frac{\partial}{\partial a}(a^{2-j}G_j)\, da =$$

$$= \frac{\partial}{\partial a}(a^j \mathfrak{G}_j) + \frac{\omega^2}{6\pi G}\frac{\partial}{\partial a}(a^{j+3}f_j); \tag{4.80}$$

while if (4.67) is multiplied by a^{j+2} and differentiated with respect to a, we find that

$$\frac{\partial}{\partial a}(a^{j+2}H_j)\int_0^a \rho a^2\, da - a^{2(j+2)}\int_a^{a_1} \rho \frac{\partial}{\partial a}(a^{-j}H_j)\, da =$$

$$= \frac{\partial}{\partial a}(a^{j+2}\mathfrak{H}_j) + \frac{jc_j}{4\pi G}\frac{\partial}{\partial a}(a^{2j+3}\phi_2) +$$

$$- \frac{\omega^2}{6\pi G}\frac{\partial}{\partial a}(a^{j+5}f_j). \tag{4.81}$$

Next, divide the foregoing Equations (4.80) and (4.81) by a^{2j} and $a^{2(j+2)}$, respectively; differentiate once more with respect to a; and divide subsequently by $a^{-(j+2)}F_0$ and $a^{-(j+4)}F_0$. The outcome of these operations will disclose that

$$a^2 G_j'' + 6D(aG_j' + G_j) - j(j+1)G_j = \mathfrak{A}_j + \frac{3\omega^2}{\pi G\rho}(1-D)(f_j + af_j') \tag{4.82}$$

and

$$a^2 H_j'' + 6D(aH_j' + H_j) - (j+2)(j+3)H_j =$$

$$= \mathfrak{B}_j + \frac{jc_j a^{j+1}}{Gm(a)}\{2(j+1)a\phi_2' + (j^2 - 3j - 6)\phi_2 - 6D(a\phi_2 + \phi_2)\} \tag{4.83}$$

by use of Equation (3.47) which is bound to be satisfied by both f_j and ϕ_2; where primes continue to denote differentiation with respect to a; $m(a)$ represents the mass of the distorted configuration interior to a; and $D \equiv \rho/\bar{\rho}$ continues to be given by Equation (3.52). Moreover, the quantities

$$\mathfrak{A}_j \int_0^a \rho a^2\, da = a^2 \mathfrak{g}_j'' - j(j+1)\,\mathfrak{g}_j \tag{4.84}$$

and

$$\mathfrak{B}_j \int_0^a \rho a^2 \, da = a^2 \mathfrak{H}_j'' - (j+2)(j+3)\, \mathfrak{H}_j \tag{4.85}$$

occurring on the right-hand side of Equations (4.66) and (4.67) can be obtained by repeated differentiation of the expressions (4.68) and (4.69) in the form

$$\begin{aligned} \mathfrak{A}_j(\mathfrak{g}_j) = & \frac{2(2j+1)}{j+2} \{a^2 \mathfrak{g}_j'' - j(j+1)\, \mathfrak{g}_j\} + \\ & + \frac{3(2j+1)}{j+2} \{a^2 \mathfrak{g}_j'' + 7a\, \mathfrak{g}_j' - (j^2 + j + 8)\, \mathfrak{g}_i\} D + \\ & + \frac{3(2j+1)}{j+2} \{a\, \mathfrak{g}_j' + 6\, \mathfrak{g}_j\} (aD' + 3D^2) = \mathfrak{B}_{j-2}(\mathfrak{h}_{j-2}), \end{aligned} \tag{4.86}$$

in which advantage has been taken of the relation

$$a^2 \rho' / \bar{\rho} = aD' + 3D(D-1) \tag{4.87}$$

obtained by differentiation of (3.52).

The functions $\mathfrak{G}_j$ and $\mathfrak{H}_j$ as given by Equations (4.70) –(4.79) are linear combinations of the products $\mathfrak{f}_2 f_j$. Successive differentiation of the latter discloses, moreover, that

$$a \frac{\partial}{\partial a} (\mathfrak{f}_2 f_j) = (\mathring{\eta}_2 + \eta_j)\, \mathfrak{f}_2 f_j \tag{4.88}$$

and

$$a^2 \frac{\partial^2}{\partial a^2} (\mathfrak{f}_2 f_j) = \{2\mathring{\eta}_2 \eta_j - 6D(\mathring{\eta}_2 + \eta_2 + 2) + j(j+1) + 6\}\, \mathfrak{f}_2 f_j \tag{4.89}$$

in terms of the logarithmic derivatives

$$\mathring{\eta}_2 = a\, \mathfrak{f}'/\mathfrak{f} \quad \text{and} \quad \eta_j = a f_j'/f_j \tag{4.90}$$

if advantage is again taken of the fact that, correctly to quantities of first order, $\mathring{\eta}_2$ as well as η_j satisfy the differential Equation (3.61).

With the functions $\mathfrak{A}_j$ and $\mathfrak{B}_j$ thus explicitly formulated, Equations (4.82) and (4.83) represent second-order nonhomogeneous linear differential equations for G_j and H_j – analogous to Equation (3.51) for the purely tidal problem. In comparing the latter with (4.82)–(4.83) we may note that, whereas its nonhomogeneous terms $T_j(a)$ as given by Equations (3.53)–(3.59) depend on the internal structure of the respective configuration only through the first powers of the ratio $D \equiv \rho/\bar{\rho}$, those on the right-hand sides of Equations (4.82) and (4.83) governing the interaction between rotation and tides involve terms factored not only by D, but also D^2 and aD'.

Like in the case of purely tidal distortion discussed in section 3, the particular solutions of Equations (4.82) and (4.83) representing the amplitudes of 'mixed' tides are subject to specific boundary conditions imposed at both ends of the interval of integration. At the origin ($a = 0$), both $G_j(a)$ as well as $H_j(a)$ should be minimum – the necessary conditions of which require that

$$G'(0) = H'(0) = 0 \quad \text{for} \quad j = 0(1)4\,. \tag{4.91}$$

At the other end – on the surface $a = a_1$ defined by $\rho(a_1) = 0$ – Equations (4.80) and (4.81) reduce to

$$\left\{\frac{\partial}{\partial a}(a^j G_j)\right\}_{a_1} = \frac{4\pi}{m_1}\left\{\frac{\partial}{\partial a}\left(a^j\,\mathfrak{G}_j + \frac{\omega^2 f_j}{6\pi G}\,a^{j+3}\right)\right\}_{a_1} \tag{4.92}$$

and

$$\left\{\frac{\partial}{\partial a}(a^{j+2} H_j)\right\}_{a_1} = \frac{4\pi}{m_1}\left\{\frac{\partial}{\partial a}\left(a^{j+2}\,\mathfrak{H}_j - \frac{\omega^2 f_j}{6\pi G}\,a^{j+5} + \frac{jc_j\,\mathfrak{f}_2}{4\pi G}\,a^{2j+3}\right)\right\}_{a_1}, \tag{4.93}$$

where m_1 denotes the total mass of the distorted configuration, and where

$$\left\{\frac{\partial}{\partial a}[a\mathfrak{G}_j(\mathfrak{g}_j)]\right\}_{a_1} = \frac{(2j+1)m_1}{2(j+2)\pi}\{a\,\mathfrak{g}_j' + \mathfrak{g}_j\}_{a_1} +$$

$$-\frac{j-1}{a_1^j}\int_0^{a_1}\rho\,\frac{\partial}{\partial a}(a^{j+3}\,\mathfrak{g}_j)\,da = \left\{\frac{\partial}{\partial a}[a\,\mathfrak{H}_{j-2}(\mathfrak{h}_{j-2})]\right\}_{a_1}. \tag{4.94}$$

It may be noted that whereas the conditions (4.91) at $a = 0$ are homogeneous in the dependent variables, those given by Equations (4.92) and (4.93) representing linear relations between G_j and G_j' or H_j and H_j' at $a = a_1$ are nonhomogeneous and, as such, impose specific amplitudes on the respective tides.

Moreover, once the particular solutions of (4.82) and (4.83) subject to the boundary conditions (4.91) and (4.92)–(4.94) have been constructed (by numerical or other methods), the amplitudes g_j and h_j of the 'mixed' equilibrium terms arising from interaction between rotation and tides will– by an inversion of Equations (4.51)–(4.60) – be given by the relations

$$g_0 = G_0 + \tfrac{1}{10}H_2 + \tfrac{1}{11}H_4, \tag{4.95}$$

$$g_1 = G_1 + \tfrac{1}{5}H_1 + \tfrac{11}{42}H_3, \tag{4.96}$$

$$g_2 = G_2 + \tfrac{2}{7}H_2 + \tfrac{25}{66}H_4, \tag{4.97}$$

$$g_3 = G_3 + \tfrac{1}{3}H_3, \tag{4.98}$$

$$g_4 = G_4 + \tfrac{4}{11}H_4; \tag{4.99}$$

and

$$h_0 = H_0 + \tfrac{2}{7}H_2 + \tfrac{7}{33}H_4, \tag{4.100}$$

$$h_1 = H_1 + \tfrac{2}{3} H_3, \tag{4.101}$$

$$h_2 = H_2 + \tfrac{10}{11} H_4, \tag{4.102}$$

$$h_3 = H_3, \tag{4.103}$$

$$h_4 = H_4. \tag{4.104}$$

In conclusion of the present section on the interaction effects between rotation and tides, it should be noted that the total mass m_1 of our distorted configuration, defined by Equation (1.28) for $n = 0$ as

$$m(a_1) = \frac{1}{3} \int_0^{a_1} \rho \frac{\partial}{\partial a} \left\{ \int_0^{\pi} \int_0^{2\pi} (r')^3 \sin\theta' \, d\theta' \, d\phi' \right\} da \tag{4.105}$$

requires for its constancy (i.e., independence of distortion) that the coefficients of zero-order harmonics should, within the scheme of our approximation, be related with those of higher harmonics by the equation

$$\begin{aligned} f_0 + \mathfrak{f}_2 + g_0 = & -\tfrac{1}{5} f_2^2 - \tfrac{1}{7} f_2^3 - \cdots \\ & -\tfrac{1}{5} \mathfrak{f}_2^2 + \tfrac{1}{5} \mathfrak{f}_2 f_2 + \cdots \\ & + \tfrac{1}{10} h_2 + \cdots, \end{aligned} \tag{4.106}$$

where the terms of purely tidal origin on the right-hand side have already established by Equation (3.33); while the 'mixed' terms go back to that factored by $\mathfrak{P}_0^{(2)}$ on the right-hand side of Equation (4.33).

II-5. Effects of Internal Structure

In the preceding sections 2–4 of this chapter we established that the form of the rotationally and (or) tidally distorted self-gravitating fluids can be described by expansions of the form (2.1), (3.16) or (4.22), in which the amplitudes $f_j(a)$ or $g_j(a)$ and $h_j(a)$ of the respective harmonic are defined as particular solutions of Equations (2.33)–(2.35), (3.51)–(3.61), or (4.82)–(4.87) subject to appropriate boundary conditions. Correctly to quantities of first order in surficial distortion, all amplitudes f_j have been found to satisfy the linear second-order differential equation

$$a^2 \frac{\partial^2 f_j}{\partial a^2} + 6 \frac{\rho}{\bar{\rho}} \left(a \frac{\partial f_j}{\partial a} + f_j \right) - j(j+1) f_j = 0\,, \tag{5.1}$$

where

$$\bar{\rho} = \frac{3}{a^3} \int_0^a \rho a^2 \, da \tag{5.2}$$

denotes the mean density of the respective configuration interior to the radius a. *It is through this equation that the equilibrium form (or potential) of a distorted body depends on its internal structure.*

Moreover, the form of the boundary conditions at $a = a_1$ discloses *that the surface values of $f_j(a_1)$ depend on the internal structure of the distorted configuration only through the logarithmic derivative*

$$\frac{a}{f_j}\frac{\mathrm{d}f_j}{\mathrm{d}a} \equiv \eta_j(a)\,, \tag{5.3}$$

introduced already through Equations (2.36), (3.60) or (4.90) to simplify our previous results. If we insert (5.3) in (5.1) we find – in accordance with Equations (3.52) and (3.61) – that Equation (5.1) transforms into

$$a\frac{\mathrm{d}\eta_j}{\mathrm{d}a} + 6D(\eta_j + 1) + \eta_j(\eta_j - 1) = j(j + 1), \tag{5.4}$$

which is of first order, but second degree, in the dependent variable.

In the contemporary literature Equation (5.1) is usually referred to as *Clairaut's equation* – in honour of the distinguished 18th-century French mathematician who first deduced its form for the case of $j = 2$; while Equation (5.4) often carries the name of *Radau's equation.* As is frequently the case in the annals of science, such summary labels are to be regarded as matters of convenience rather than historical truth. For Equation (5.1) with arbitrary j did not appear till in the works of Laplace some 80 years after Clairaut's time; while Radau's equation (5.4) was already known to Clairaut for $j = 2$. As it came, however, to play – very much later – a central part in Radau's investigations (Radau, 1885a, b) in the second half of the 19th century, it seems appropriate to associate (5.4) henceforth with his name – if alone to distinguish it by name from its second-order form (5.1).

The initial condition which a solution of Radau's equation (5.4) must satisfy at the origin – and which by itself will specify uniquely the requisite particular solution for any $\rho(r)$ – is easy to establish. For finite values of $\rho(0)$ it follows that $\rho/\bar{\rho} \equiv D \to 1$ as $a \to 0$; and if so, the structure of Equation (5.4) makes it evident that

$$\eta_j(0) = j - 2. \tag{5.5}$$

Moreover, if our configuration were homogeneous (ρ = constant) and, accordingly, $D = 1$ for $0 \leqslant a \leqslant a_1$, we find that $\eta_j(a)$ remains constant and equal to

$$\eta_j(a) = j - 2 \tag{5.6}$$

throughout the interior.

If, on the other hand, the entire mass of our configuration were condensed at its centre – i.e., if $D = \infty$ for $a = 0$ and zero for $a > 0$ – the solution of (5.4) becomes

$$\eta_j(a) = j + 1, \qquad a > 0. \tag{5.7}$$

If, conversely, the whole mass of our configuration were confined to an infitesimally thin shell – so that $D = 0$ for $0 \leqslant a \leqslant a_1$, and ∞ at a_1 – Equation (5.4) yields

$$\eta_j(a) = -1. \tag{5.8}$$

Therefore – if we presume no knowledge of the density distribution in the interior of our configuration – the absolute limits of the solution of (5.4) at $a = a_1$ are

$$-1 \leqslant \eta_j(a_1) \leqslant j + 1\,; \tag{5.9}$$

while if the internal density $\rho(a)$ is supposed nowhere to increase outwards (i.e., for $\rho'(a) \leqslant 0$), the foregoing inequality (5.9) becomes restricted to

$$j - 2 \leqslant \eta_j(a_1) \leqslant j + 1. \tag{5.10}$$

It may also be noted that, whatever the structure,

$$\eta_j(a) < \eta_{j+1}(a). \tag{5.11}$$

The existence of the inequality (5.10) has first been pointed out (for $j = 2$) by Clairaut (1743); and that of (5.9), by Kopal (1941b). The latter discloses that, for any given external field of force, *the surficial distortion of a configuration whose mass is confined to an infinitesimally thin surface shell would – to the first order in small quantities – be twice as large as if the configuration were homogeneous, and five times as large as that appropriate for a mass-point model.*

It may be added that, for a continuous distribution of density in the interior, the inequality (5.10) may be refined further in the following manner. In order to do so, let us return to Equation (3.47) which, on integration by parts of its left-hand side, assumes the form

$$a^j f_j \left\{ [j + 1 - \eta_j(a)] \int_0^a \rho a^2 \,\mathrm{d}a - \rho a^3 \right\} = -\int_0^a \frac{\mathrm{d}\rho}{\mathrm{d}a} f_j\, a^{j+3}\, \mathrm{d}a. \tag{5.12}$$

Since, however, it follows from (5.2) and (3.52) that

$$\int_0^a \rho a^2 \,\mathrm{d}a = \tfrac{1}{3} a^3 \bar{\rho} \qquad \text{and} \qquad \rho = D\bar{\rho}\,. \tag{5.13}$$

on insertion of these expressions in the left-hand side of (5.12) this latter equation can be rewritten as

$$a^{j+3} f_j \,\{j + 1 - 3D - \eta_j(a)\}\, \bar{\rho} = 3 \int_0^a \left(-\frac{\mathrm{d}\rho}{\mathrm{d}a} \right) f_j\, a^{j+3}\, \mathrm{d}a\,. \tag{5.14}$$

Now let us assume next that $\rho(a)$ is a diminishing function of a. If so, however, both sides of (5.14) must be non-negative*; and this can obviously be true only if

$$\eta_j(a) \leqslant j + 1 - 3D(a)\,, \tag{5.15}$$

where the quantity $D(a)$ introduced by Equation (3.52) is, in turn, constrained to obey the inequality

$$0 \leqslant D(a) \leqslant 1. \tag{5.16}$$

The left-hand side of (5.15) is necessarily true if $d\bar{\rho}/da$ is negative; the right-hand side, because ρ is positive. Therefore, any variation of $\eta_j(a)$ throughout the interior must be comprised within the range

$$j - 2 \leqslant \eta_j(a) \leqslant j + 1 - 3D(a), \tag{5.17}$$

which constitutes a refinement of our previous inequality (5.10).

How does the function $\eta_j(a)$ behave within this permitted interval? In considering a manifold of the solutions of Radau's equation (5.4) in the $(a - \eta)$-plane, we may note the Equation (5.4) defines the first derivative η_j' in terms of and η_j uniquely everywhere except at the origin. Along the axis $a = 0$ the function $\eta_j'(a)$ ceases to be holomorphic, and (5.4) can be satisfied by η_j assuming the alternative values of $j - 2$ or $-j - 3$. At the points $(0, j - 2)$ and $(0, -j - 3)$ the integral curves obeying Equation (5.4) are thus not uniquely defined. It can, however, be shown (cf. Poincaré, 1903) that, of the total manifold of such curves passing through an arbitrary point of the $(a - \eta)$-plane, only *one* will pass through $(0, j - 2)$; all others passing through $(0, -j - 3)$. Therefore, the initial condition (5.5) is sufficient to ensure the uniqueness of the particular solution of the (nonlinear) Equation (5.4) which is of interest to us in this connection.

Let us inquire next as to the behaviour of such solutions in the $(\eta - \eta')$-plane. For the limiting model of a homogeneous configuration $(D = 1)$, Equation (5.4) reduces to the parabola

$$a\eta' = j(j + 1) - (\eta + 2)(\eta + 3); \tag{5.18}$$

while for a centrally-condensed model ($D = 0$ for $a > 0$), Radau's equation reduces likewise to

$$a\eta' = j(j + 1) - \eta(\eta - 1)\,. \tag{5.19}$$

On the other hand, for values of $D(a)$ for which (5.15) becomes an equality,

$$a\eta' = (j + 1)(j - 2) - (2j - 1)\eta + \eta^2. \tag{5.20}$$

Any point of the solution of Equation (5.4) for a configuration intermediate in structure between a homogeneous and a mass-point model must, in the $(\eta - \eta')$-plane, lie in the region limited between the foregoing parabolae; in fact, for such models as render the inte-

* It may be noted that the positivity of both sides of (5.14) – and therefore, the validity of (5.15) – does not necessarily require $d\rho/da$ to remain negative throughout the interior. This latter condition is sufficient, but not necessary. The validity of (5.15) would not, for instance, be impaired by an occurrence of positive density gradients in the interior – provided that these are not sufficient to alter the sign of the integral on the right-hand side of (5.14).

gral on the right-hand side of Equation (5.14) positive, the permitted region is comprised between the parabolae (5.18) and (5.19) intersecting at the points $j-2$ and $j+2$. These curves have been plotted on the accompanying Figure 2–3; and delimit an area which $\eta'_j(a)$ can be both positive and negative.

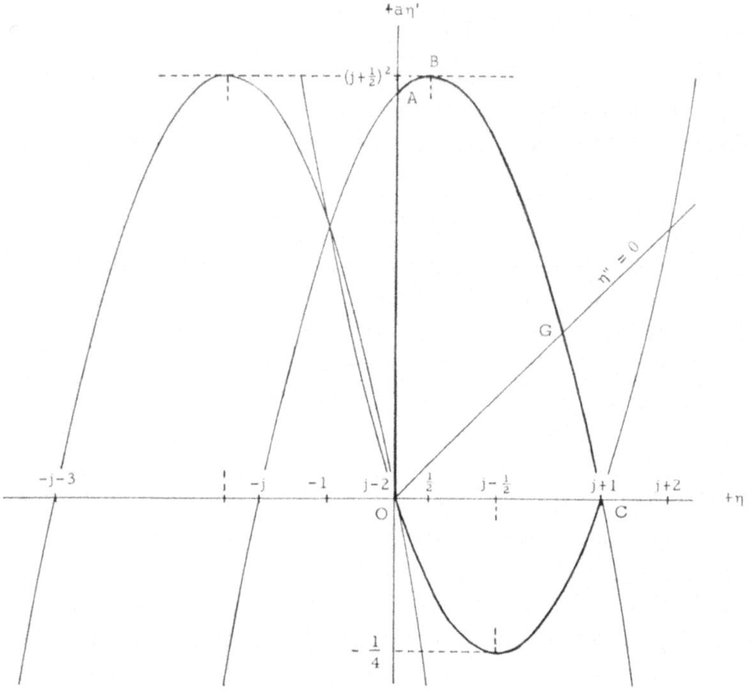

Fig. 2-3. Topology of Radau's equation.

In order to investigate further a possible behaviour of $\eta_j(a)$ in this area, let us differentiate Equation (5.4) with respect to a, to find that

$$a\eta'' + \frac{d}{da}\{\eta(\eta+6) - 6(1-D)(1+\eta)\} = 0. \tag{5.21}$$

Accordingly, $\eta''(a) = 0$ for $a \neq 0$ only if

$$\eta(\eta+6) - 6(1-D)(1+\eta) = \textit{constant}\ ; \tag{5.22}$$

and the value of this constant be obtained from the initial conditions $D(0) = 1$ and $\eta(0) = j-2$ to be equal to $(j-2)(j+4)$. If so, however, an insertion of (5.22) in (5.4) discloses that

$$a\eta' = \eta - j + 2. \tag{5.23}$$

The only solution of (5.23) consistent with (5.4) is $\eta = \text{const}$ and $D = \text{const}$, requiring $\eta' = D' = 0$. Therefore, the locus $\eta'' = \eta' = 0$ of inflection points will be represented on Figure 2–3 by a straight line passing through the point $\eta = j-2$, and inclined by 45° to the locus $\eta' = 0$ of the extrema which coincides with the η-axis.

Apart from the limiting cases (5.6) or (5.7) representing the solutions of (5.4) for the homogeneous or mass-point models, Clairaut's or Radau's equations admit also of other closed solutions for certain less extreme internal density distributions – adopted, however, for the sake of mathematical tractability rather than their physical reasonableness; and for an account of them the reader is referred, e.g., to sections II-2 of Kopal (1959, 1960), or to Lanzano (1973). Some – particularly those corresponding to weak internal density concentration – are of geophysical (or planetological) rather than astrophysical interest, and an account of their more detailed properties need not detain us in this place. The density profiles of the stars (with the exception of white dwarfs) are characterized by a generally high degree of central condensation; and a construction of the corresponding particular solutions of Radau's equation (5.4) subject to the initial condition (5.5) can, in general, proceed only by numerical integration. Extensive integrations carried out for the polytropic family of models by Brooker and Olle (1955) are now of mainly historical significance – except, perhaps, for the case of polytropic index $n = 1.5$ which should approximate closely the structure of (non-relativistic) white dwarfs. The most comprehensive integrations for a wide range of actual stellar models corresponding to different evolutionary stages we owe to Petty (1973); and their outcome bears out indeed the general properties of the η-functions discussed earlier in this section.

For relatively *high degree* of *central condensation* of the distorted configuration – i.e., if the ratio $\rho/\bar{\rho} \equiv D(a) \ll 1$ throughout most part of the range $0 \leqslant a \leqslant a_1$ – approximate solutions of Clairaut's equation (5.1) can be constructed in the following manner. Let $y(a)$ define a new function related with the amplitudes f_j by the equation

$$y(a) = \tfrac{1}{3} a^3 \bar{\rho} f_j(a) , \tag{5.24}$$

where $\bar{\rho}$ continues to be defined by Equation (5.2), so that

$$\eta_j(a) = \frac{a}{y} \frac{dy}{da} - 3 \frac{\rho}{\bar{\rho}} . \tag{5.25}$$

Let, moreover, the surface value of

$$\eta_j(a_1) = \lambda a_1 , \tag{5.26}$$

where λ represents the desired value of a parameter specifying the effects of internal structure on the external shape of the respective configuration. Its determination is obviously tantamount (cf. Kopal, 1953) to a search for the characteristic parameters λ of a Sturm-Liouville problem, consisting of the equation

$$L[y] = \alpha(a)\, y, \tag{5.27}$$

subject to the homogeneous boundary conditions

$$y(0) = 0 \qquad \text{and} \qquad y'(a_1) = \lambda\, y(a_1) , \tag{5.28}$$

where

$$\alpha(a) \equiv \frac{3}{a\bar{\rho}} \frac{d\rho}{da} = \frac{3}{a^2 \bar{\rho}} \frac{d}{da}\left(a^2 \frac{d\bar{\rho}}{da}\right) , \tag{5.29}$$

and L stands for the operator

$$L \equiv \frac{d^2}{da^2} - \frac{j(j+1)}{a^2} \quad . \tag{5.30}$$

The operator L in (5.30) is evidently self-adjoint. Its Green's function appropriate for the given boundary conditions will, therefore, be symmetrical and of the form

$$G(a, \alpha) = -\frac{a^{j+1}}{2j+1} \{\alpha^{-j} + C_j \alpha^{j+1}\}, \quad \alpha \geqslant a \,, \tag{5.31}$$

in which we have abbreviated

$$a_1^{2j+1} C_j = \frac{j + \lambda a_1}{j - \lambda a_1 + 1} . \tag{5.32}$$

Our foregoing boundary-value problem can, accordingly, be rewritten in the form of the single integral equation

$$\begin{aligned}(2j+1)\, y(a) = & -a^{j+1} C_j \int_0^{a_1} \mathrm{a}(\alpha)\, y(\alpha) \alpha^{j+1}\, d\alpha \\ & - a^{-j} \int_0^{a} \mathrm{a}(\alpha)\, y(\alpha)\, \alpha^{j+1}\, d\alpha \\ & - a^{j+1} \int_a^{a_1} \mathrm{a}(\alpha)\, y(\alpha) \alpha^{-j}\, d\alpha \,.\end{aligned} \tag{5.33}$$

Now in the preceding sections of this chapter we found that, for homogeneous configurations $(\rho/\bar{\rho} = D = 1)$, $\eta(a) = j - 2$; while for a mass-point model ($D = 0$ for $a > 0$), $\eta(a) = j + 1$. In the former case, therefore, $\lambda a_1 = j - 2$ and, in the latter, $\lambda a_1 = j + 1$ rendering the constant C_j as defined by the Equation (5.33) infinite. In actual stars this limit can never be attained; for the central condensation of no existing configuration can be infinite. If it, however, happens to be high (such as is likely to be in most stars), the constants C_j may become very large. The second and third term on the right-hand side of our integral Equation (5.33) will then be small in comparison with the first; for this latter alone is magnified by a multiplication with C_j. To the order of accuracy to which their disparity is such as to render the second and third terms ignorable, Equation (5.33) should become essentially equivalent to

$$(2j+1)\, y(a) = -a^{j+1} C_j \int_0^{a_1} \mathrm{a}(\alpha)\, y(\alpha) \alpha^{j+1}\, d\alpha \,; \tag{5.34}$$

and since the limits of the integral on the right-hand side now are constant, it implies that

$$y(a) = A a^{j+1} , \tag{5.35}$$

where A ist a constant – or, by (3.38)–(3.44)

$$f_j(a) = \frac{c_j\, a^{j+1}}{4\pi G F_0} \quad , \tag{5.36}$$

where $4\pi F_0 \equiv m(a)$ continues to be given by Equations (2.17) or (3.45), and the constants c_j are given by Equations (1.42) or (3.12)–(3.15), depending on whether the distortion is caused by the rotation or tides.

The reader may, furthermore, note from Equations (2.27)–(2.29) or (3.38)–(3.44) that – to be first order in small quantities – we have

$$F_0 f_j(a) = \frac{c_j a^{j+1}}{4\pi G} + \frac{a^{j+1} E_j + a^{-j} F_j}{2j+1}, \tag{5.37}$$

where $j = 2, 3, 4$, for configurations of *any* structure. If, therefore, in accordance with Equation (5.35), the first term on the right-hand side of (5.36) represents a good approximation to the actual solution for a pronounced degree of central condensation, it follows that the second term on the right-hand side of (5.37) should generally be small. In such a case, however, the idea suggests itself to construct a solution of Equation (5.37) by an *iterative process*, which can generate successive approximations to the solution of our problem in a systematic manner.

Let, in what follows, Y_n stand for the n-th approximation to the desired function $Y \equiv f_j(a)$. In order to generate such approximations, let us replace Equation (5.37) symbolically by a *quadrature* formula of the form

$$Y_n = \frac{c_j a^{j+1}}{Gm(a)} + J[Y_{n-1}], \tag{5.38}$$

where (for sufficiently high central condensation) the second term on the right-hand side should be small in comparison with the first to justify the use in it of an approximation of lower order.

Let, moreover, the absolute upper bound of Y be denoted by M. If so, then

$$|J[Y]| < MJ[1]; \tag{5.39}$$

and as (by the Mean Value Theorem)

$$J[Y] < \frac{3}{2j+1}, \tag{5.40}$$

it follows that

$$|J[Y]| < \frac{3M}{2j+1}. \tag{5.41}$$

Now let

$$|J[Y_n - Y_{n-1}]| < \frac{3M_n}{2j+1}, \tag{5.42}$$

where M_n stands for the upper bound of the individual Y_n's in the interval $(0, a_1)$. If so, however, then Equation (5.42) yields

$$|Y_{n+1} - Y_n| < \frac{3M_n}{2j+1} < \left(\frac{3}{2j+1}\right)^n M_1 \tag{5.43}$$

and, therefore, the sequence

$$x = Y_0 + (Y_1 - Y_0) + (Y_2 - Y_1) + \cdots \tag{5.44}$$

converges absolutely and uniformly throughout the interval $(0, a_1)$ to a definite limit x.

Does this limit actually satisfy Clairaut's equation? In order to demonstrate that this is indeed the case, let us rewrite Equation (5.45) as

$$x - J[x] - \frac{ca^{j+1}}{\int_0^a \rho a^2 da} = x - Y_{n+1} - J[x - Y_n] \tag{5.45}$$

Now if ϵ_n denotes the absolute upper bound of the differences $x - Y_n$ in the interval $(0, a_1)$, Equation (5.45) reveals that

$$|J[x - Y_n] < \frac{3\,\epsilon_n}{2j+1}\,, \tag{5.46}$$

so that

$$x - J[x] - \frac{c_j a^{j+1}}{Gm(a)} < \epsilon_{n+1} + \frac{3\,\epsilon_n}{2j+1} \tag{5.47}$$

which can be true only if the left-hand side of this inequality vanishes – i.e., if $x \equiv Y$.

The sequence of the Y_n's in terms of which we have defined x depends on the form of the adopted starting function Y_0. However, the limit x is clearly independent of it; for if an arbitrary function X_0 is used to generate another sequence of the X_n's through the mill of Equation (5.47) and if – analogously with Equation (5.4),

$$|Y_n - X_n| < \left\{\frac{3}{2j+1}\right\}^n N \tag{5.48}$$

where N represents the absolute upper bound of $J[X]$, it follows that (as $3/(2j + 1)$ is less than one for $j > 1$) the difference $Y_n - X_n$ can be diminished arbitrarily by choosing a sufficiently large value of n – no matter how large N may be.

In order to demonstrate this procedure in a more concrete case, let us return to Equation (5.27) for y, and replace the latter by its logarithmic derivative

$$\mu z = \frac{1}{y}\frac{dy}{da}\,, \tag{5.49}$$

where

$$\mu^2 = j(j+1)\,. \tag{5.50}$$

If so, (5.27) transforms into a non-homogeneous Riccati equation

$$\frac{1}{\mu}\frac{dz}{da} + z^2 = \frac{1}{a^2} + \frac{\mathfrak{a}(a)}{\mu^2} \tag{5.51}$$

of first order but second degree, where $\mathfrak{a}(a)$ continues to be defined by Equation (5.29).

Since the constant parameter $\mu \gg 1$, the possibility suggests itself (cf. Kopal and Lanzano, 1973) to seek the solution of (5.51) for $a > 0$ in the form of an expansion

$$z = \sum_{j=0}^{\infty} z_j(a)\,\mu^{-1} \tag{5.52}$$

in inverse powers of μ. If we insert this expansion in (5.51) and set the coefficients of successive powers of μ equal to zero, we find that

$$z_0 = a^{-1}, \tag{5.53}$$

$$z_0' + 2z_0 z_1 = 0\,, \tag{5.54}$$

$$z_1' + 2z_0 z_2 + z_1^2 = \mathfrak{a}\,(a)\,; \tag{5.55}$$

while, for $j > 2$,

$$z_{j-1}' + 2z_0 z_j + \sum_{k=1}^{j-1} z_k z_{j-k} = 0. \tag{5.56}$$

The foregoing equations constitute a set of recursion relations by which individual terms z_j on the right-hand side of Equation (5.52) can be successively generated. The only condition for the validity of such a procedure is the requirement that the function $\mathfrak{a}(a)$ be analytic within $0 < a \leqslant a_1$. Moreover, once this has been done, the function $\eta_j(a)$ can by a combination of Equations (5.24), (5.25) and (5.49) be expressed as

$$\eta_j(a) = a\mu z - 3(\rho/\bar{\rho}) = \sum_{j=0}^{\infty} \frac{a z_j(a)}{\mu^{j-1}} - 3\frac{\rho}{\bar{\rho}} \tag{5.57}$$

reducing, on the surface ($a = a_1$), to

$$\eta_j(a_1) = a_1\mu \sum_{j=0}^{\infty} z_j(a_1)\,\mu^{-j} =$$

$$= \mu + \frac{1}{2} + \frac{1}{8\mu} + \frac{3}{2\mu}\left(\frac{a\rho'}{\rho}\right)_{a_1} + a_1\mu \sum_{j=3}^{\infty} z_j(a)\,\mu^{-j} \tag{5.58}$$

by (5.53)–(5.55); while, for $j > 3$, the z_j's can be generated by use of (5.56).

For a centrally-condensed model $\rho'(a_1) = 0$. The reader is invited to verify that, in such a case, the first three terms on the right-hand side of (5.58), approximating the exact expression $\eta_j(a_1) = j + 1$, err by excess of only 52, 19 and 9 units of the fifth decimal place – a quality of approximation which is unlikely to change much for models of finite density concentration as long as the latter remains high. The structure of the expansion on the right-hand side of Equation (5.58) demonstrates convincingly the extent to which *the values of $\eta_j(a_1)$ depend primarily on the surface density gradient $\rho'(a_1)$ of the respective configuration.*

II-6. Gravity, Density Distribution, and Moments of Inertia

Before concluding the present chapter we wish first to return to one task left unfinished in section II-3: namely, a specification of the constants k_j in the expansion (3.2) for the exterior potential $V(r)$ in terms of the internal structure of the respective configuration. In order to do so, let us note that the potential is of the form (3.28), where, by (3.48),

$$F_j(a_1) = a_1^j f_j(a_1)\ \{j + 1 - \eta_j(a_1)\}\ F_0(a_1)\ . \tag{6.1}$$

Since, however, $F_0(a_1) = m_1/4\pi$ and, by (3.63),

$$f_j(a_1) = \frac{(2j + 1)\, a_1^{j+1} c_j}{G(j + \eta_j)\, m_1} \tag{6.2}$$

to the first order in small quantities, it follows that

$$F_j(a_1) = \frac{(2j + 1)\,(j + 1 - \eta_j)\, c_j}{4\pi G\,(j + \eta_j)}\ a_1^{2j+1}\ . \tag{6.3}$$

If, lastly, we insert for c_j from (3.15), we find (3.2) to become indeed an expansion of the form (3.28) if we set

$$k_j = \frac{j + 1 - \eta_j(a_1)}{2[j + \eta_j(a_1)]}\ . \tag{6.4}$$

The foregoing coefficients define the constants k_j in the expansion of the exterior potential of a distorted configuration in terms of its internal structure; and as such they are going to play an important role in subsequent chapters concerned with the perturbations of the orbital elements in close binary systems. For the present we may note that if the internal density concentration were infinite – and, in accordance with (5.7), $\eta_j(a) = j + 1$ – all k_j's will become zero, and the potential expansion (3.28) will reduce (as it should) to its leading term appropriate for a mass point.

For a finite degree of mass concentration, the surface values of $\eta_j(a_1)$ for any particular stellar model must in general be obtained by a numerical integration of Equations (5.1)–(5.4). If, however, this concentration is high enough to enable us to approximate the function $y(a)$ by (5.35), the fact that the integral equation (5.34) is homogeneous in y discloses that – within this scheme of approximation – the value of C_j involved in it should be given by the equation

$$C_j^{-1} = 2a_1^{2j+1}\, k_j = -\frac{3}{2j + 1} \int_0^{a_1} \left(\frac{1}{\bar{\rho}}\, \frac{d\rho}{da}\right) a^{2j+1} da. \tag{6.5}$$

A partial integration of the expression on its right-hand side reveals that

$$\begin{aligned} \int_0^{a_1} \frac{1}{\bar{\rho}}\, \frac{d\rho}{da}\ a^{2j+1} da &= -\int_0^{a_1} \rho\, \frac{d}{da}\left(\frac{a^{2j+1}}{\bar{\rho}}\right) da \\ &= -\int_0^{a_1} \frac{\rho}{\bar{\rho}} \left\{2j + 1 - \frac{a}{\bar{\rho}}\, \frac{d\bar{\rho}}{da}\right\} a^{2j} da\ ; \end{aligned} \tag{6.6}$$

and as, by (5.2)

$$\frac{1}{\bar{\rho}}\frac{d\bar{\rho}}{da} = \frac{3}{a}\left\{\frac{\rho}{\bar{\rho}} - 1\right\}, \tag{6.7}$$

it follows that

$$\int_0^{a_1} \frac{1}{\bar{\rho}}\frac{d\rho}{da}\, a^{2j+1}da = 3\int_0^{a_1} D^2 a^{2j} da - 2(j+1)\int_0^{a_1} Da^{2j}da, \tag{6.8}$$

where, as before, $D \equiv \rho/\bar{\rho}$. For centrally-condensed configurations $D \ll 1$ and, consequently, $D^2 \ll D$. In such cases, the first integral on the right-hand side of the foregoing equation is likely to be very small in comparison with the second and may, therefore, be neglected. Doing so we arrive at the equation

$$a_1^{2j+1}\, k_j = \frac{3(j+2)}{2j+1}\int_0^{a_1} Da^{2j}da, \tag{6.9}$$

permitting us to approximate closely the numerical values of the constants k_j for centrally-condensed configurations by simple quadratures of the products $a^{2j}D$ throughout the interior.

Therefore, outside the distorted body ($r > a_1$), where the interior potential $U(a)$ becomes identically zero, the potential arising from the mass will – to the first order in surficial distortion – reduce to

$$\Omega \equiv V(r) = \frac{Gm_1}{r}\left\{1 + \sum_{j=2}^{4} \frac{j+1-\eta_j}{2j+1}\,\frac{a_1^j}{r^{j+1}}\, Y_j(a_1)\right\} \tag{6.10}$$

where, by Equation (3.47) particularized for $a = a_1$, it follows that

$$Y_j(a_1) = \frac{2j+1}{j+\eta_j(a_1)}\,\frac{a_1^{j+1}}{Gm_1}\, c_j P_j(\theta, \phi) \tag{6.11}$$

in agreement with (6.2). Inserting (6.11) in (6.10), we find that

$$\begin{aligned} V(r) &= \frac{Gm_1}{r} + \sum_{j=2}^{4} \frac{j+1-\eta_j}{j+\eta_j}\,\frac{a_1^{2j+1}}{r^{j+1}}\, c_j P_j(\theta,\phi) \\ &= \frac{Gm_1}{r} + \sum_{j=2}^{4} 2k_j\,\frac{a_1^{2j+1}}{r^{j+1}}\, c_j P_j, \end{aligned} \tag{6.12}$$

in agreement with Equation (3.2), where the constants k_i continue to be given (6.4).

Let us inquire next about the *gravitational acceleration* g inside, and up to the surface, of distorted configurations constituting the components of close binary systems. As is well known, the square of this acceleration is defined as a sum of the squares of the components of the gradient of the total potential $\Psi \equiv \Omega + V' \equiv U + V + V'$ of the potential Ω arising from the mass of the distorted configuration and V' arising from the disturbing force. By an introduction of the coordinates

$$\left.\begin{aligned} x &= r'(a, \theta, \phi) \cos\phi \sin\theta \ , \\ y &= r'(a, \theta, \phi) \sin\phi \sin\theta \ , \\ z &= r'(a, \theta, \phi) \cos\theta \ , \end{aligned}\right\} \tag{6.13}$$

where $r'(a, \theta, \phi)$ represents the radius-vector of an equipotential surface as defined by Equations (1.31), we succeeded in expressing the total potential Ψ in terms of a Neumann series of the form

$$\Psi = \sum_{j=0}^{\infty} \alpha_j(a) P_j(\theta, \phi) \ , \tag{6.14}$$

with coefficients $\alpha_j(a)$ defined by Equation (1.47). The constancy of (6.14) over a level surface requires, moreover, that (in accordance with Equation (1.48), $\alpha_j(a) = 0$ for $j > 0$; leaving us with

$$\Psi(a) \equiv \alpha_0(a) \tag{6.15}$$

as the expression for the value of the total potential $\Psi(a)$ over a level surface of mean radius a and density $\rho_0(a)$.

Unlike the polar coordinates r', θ, ϕ which constitute a rectilinear orthogonal set, the trio a, θ, ϕ constitute a set of *curvilinear non-orthogonal coordinates*. related with the rectangular system x, y, z by the differential transformation

$$\begin{aligned} (\mathrm{d}x)^2 + (\mathrm{d}y)^2 + (\mathrm{d}z)^2 &= g_{11}(\mathrm{d}a)^2 + g_{22}(\mathrm{d}\theta)^2 + g_{33}(\mathrm{d}\phi)^2 + \\ &+ 2g_{12}\mathrm{d}a\mathrm{d}\theta + 2g_{13}\mathrm{d}a\mathrm{d}\phi + 2g_{23}\mathrm{d}\theta\mathrm{d}\phi, \end{aligned} \tag{6.16}$$

whose metric coefficients are given by

$$g_{11} = (r'_a)^2 , \tag{6.17}$$

$$g_{22} = r'^2 + (r'_\theta)^2 , \tag{6.18}$$

$$g_{33} = r'^2 \sin^2\theta + (r'_\phi)^2 ; \tag{6.19}$$

and

$$g_{12} = g_{21} = r'_a r'_\theta , \tag{6.20}$$

$$g_{13} = g_{31} = r'_a r'_\phi , \tag{6.21}$$

$$g_{23} = g_{32} = r'_\theta r'_\phi , \tag{6.22}$$

where the radius-vector r' continues to be defined by Equation (1.31), and the subscripts denote partial differentiation by the respective variables.

Now if x^i stands for any one of the coordinates x, y, z of the original rectangular system and q^i for those of the curvilinear system a, θ, ϕ, the components of the gradient in the latter coordinates are known to be given by

$$\nabla \equiv \mathbf{e}_2 \times \mathbf{e}_3 \left(\frac{g_{22}g_{33}}{g}\right)^{1/2} \frac{\partial}{\partial q^1} +$$

$$+\mathbf{e}_3 \times \mathbf{e}_1 \left(\frac{g_{11}g_{33}}{g}\right)^{1/2} \frac{\partial}{\partial q^2} + \mathbf{e}_1 \times \mathbf{e}_3 \left(\frac{g_{11}g_{22}}{g}\right)^{1/2} \frac{\partial}{\partial q^3}, \tag{6.23}$$

where $\mathbf{e}_i$ are unit vectors parallel with the intersections of the coordinate surfaces, and

$$g \equiv \det \|g_{ij}\| = (r'^2 r'_a \sin\theta)^2 . \tag{6.24}$$

Since, moreover, the potential $\Psi(a) \equiv \alpha_0(a)$ depends only on one variable $q^1 \equiv a$, it follows that the gravitational acceleration g of the respective configuration should be given by

$$\mathrm{g} = -\left(\frac{g_{22}g_{33}}{g}\right)^{1/2} \frac{\partial \alpha_0}{\partial a} . \tag{6.25}$$

In order to evaluate this acceleration, we may note that the partial derivatives of r' with respect to the angular variables are small quantities of first order in terms of the amount of distortion. Therefore, to the first order in small quantities, it follows from Equations (6.18), (6.19) and (6.24) that

$$\sqrt{g_{22}} = r', \qquad \sqrt{g_{33}} = r' \sin\theta , \tag{6.26}$$

and

$$\sqrt{g} = r'^2 r'_a \sin\theta , \tag{6.27}$$

rendering

$$\left(\frac{g_{22}g_{33}}{g}\right)^{1/2} = \frac{1}{r'_a} . \tag{6.28}$$

Since, moreover, by (1.31)

$$r_a = 1 + \sum_j \left\{ Y_j + a \frac{\partial Y_j}{\partial a} \right\} = 1 + \sum_j (1 + \eta_j) f_j(a) P_j(\theta, \phi), \tag{6.29}$$

while

$$\alpha_0(a) = U_0 + \frac{V_0}{a} = 4\pi G \left\{ \int_a^{a_1} \rho a \, da + \frac{1}{a} \int_0^a \rho a^2 \, da \right\} \tag{6.30}$$

by (1.29) and (1.30); so that

$$\frac{\partial \alpha_0}{\partial a} = -\frac{V_0}{a^2} = -\frac{Gm(a)}{a^2} . \tag{6.31}$$

Therefore,

$$\mathrm{g} = \frac{Gm}{a^2}\left\{1 - \sum_j (1+\eta_j) f_i(a) P_j(\theta,\phi)\right\} \tag{6.32}$$

or

$$\frac{\mathrm{g} - \mathrm{g}_0}{\mathrm{g}_0} = -\sum_j (1+\eta_j) f_j P_j, \tag{6.33}$$

an equation which we shall put to use in section IV-4 in connection with stellar gravity-darkening, and where

$$\mathrm{g}_0 = \frac{Gm(a)}{a^2} \tag{6.34}$$

stands for the mean value of the gravitational acceleration over a level surface characterized by the mean radius a. Equations (6.32) or (6.33) as they stand hold good, to be sure, only to quantities of first order in the amplitudes $f_j(a)$ of the respective distortion; but their generalization to accuracy of any order can be carried by the above process at a price no greater than that of additional algebraic work.

Next, let us consider the effects of distortion the *density distribution* in the interior of the respective configuration; for these too can be deduced form perturbations of the total potential Ψ in the following manner. Let

$$\rho(a,\theta,\phi) \equiv \rho_0(a) + \rho'(a,\theta,\phi) \tag{6.35}$$

represent the actual density at any point of the distorted configuration; and $\rho_0(a)$ the density which the configuration would possess in its undistorted state. If so, the relation between ρ and Ψ is provided by Poisson's equation

$$\nabla^2\Psi = -4\pi G\rho\ ; \tag{6.36}$$

so that once we know the potential the actual distribution of density $\rho(a,\theta,\phi)$ can be extracted from it by operating on Ψ by the Laplacian ∇^2.

In order to be able to do so, however, we have first to express the Laplacian operator in terms of the curvilinear coordinates a, θ, ϕ. As is well known, the general expression for such an operator in non-orthogonal curvilinear coordinates can be written in the form

$$\nabla^2 \equiv \frac{1}{\sqrt{g}}\frac{\partial}{\partial q^i}\left(\sqrt{g}\, g^{ij}\frac{\partial}{\partial q^j}\right), \tag{6.37}$$

where the g^{ij}'s are co-factors of the corresponding g_{ij}'s as given by Equations (6.17)–(6.22) in the determinant

$$g \equiv \begin{vmatrix} g_{11} & g_{12} & g_{13} \\ g_{21} & g_{22} & g_{23} \\ g_{31} & g_{32} & g_{33} \end{vmatrix}, \tag{6.38}$$

divided by the value of g itself (as given by Equation 6.24 above). Since, however, $\Psi =$

$\alpha_0(a)$ in (6.36) is by definition a function of $q^1 \equiv a$ alone, it follows from (6.15), (6.26) and (6.36) that

$$4\pi G\rho = -\frac{1}{\sqrt{g}}\frac{\partial}{\partial q^i}\left(\sqrt{g}\,g^{i1}\frac{\partial\alpha_0}{\partial a}\right) , \tag{6.39}$$

which relates the actual density ρ inside a distorted configuration with the first term $\alpha_0(a)$ on the right-hand side of the potential expansion (1.46).

In order to illustrate this procedure on a practical example, consider the case of distortion caused by *axial rotation* with an angular velocity ω, in which the expansion for the radius-vector r' of the equipotential surface distorted by centrifugal force is given Equation (2.1). If so, then correctly to the *third* order in small quantities,

$$\begin{aligned}\alpha_0(a) = E_0 + \tfrac{2}{25}a^2 (f_2 + \tfrac{1}{7}f_2^2)E_2 + \\ + \frac{1}{a}(1 + \tfrac{2}{5}f_2^2 - \tfrac{4}{105}f_2^3)F_0 - \tfrac{3}{25}\frac{f_2}{a^3}F_2 + \\ + \tfrac{1}{3}\omega^2 a^2(1 - \tfrac{2}{5}f_2 - \tfrac{1}{5}f_2^2) ,\end{aligned} \tag{6.40}$$

where the coefficients $E_{0,2}(a)$, $F_{0,2}(a)$ as well as the amplitudes $f_{2,4}(a)$ are the same as previously defined in section II-2.

Moreover, since in such a case r' is a function of a and θ only, it follows that

$$\sqrt{g}\,g^{11} = [(r'^2 + r_\theta'^2)/r_a'] \sin\theta , \tag{6.41}$$

$$\sqrt{g}\,g^{21} = -r_\theta' \sin\theta , \tag{6.42}$$

$$\sqrt{g}\,g^{31} = 0 ; \tag{6.43}$$

and, accordingly, Poisson's equation in the rotational case assumes the form

$$\begin{aligned}4\pi G\rho = -\frac{1}{r'^2 r_a' \sin\theta}\left\{\frac{\partial}{\partial a}\left(\frac{r'^2 + r_\theta'^2}{r_a'}\sin\theta\,\frac{\partial\alpha_0}{\partial a}\right) - \right. \\ \left. - \frac{\partial}{\partial\theta}(r_\theta' \sin\theta)\frac{\partial\alpha_0}{\partial a}\right\} + 2\omega^2 ;\end{aligned} \tag{6.44}$$

where, to the third order in small quantities,

$$r_a' = 1 + \sum_{j=1}^{3}(1 + \eta_j) f_{2j} P_{2j}(\cos\theta) \tag{6.45}$$

and

$$r_\theta' = a\sum_{j=1}^{3} f_{2j}\frac{\partial P_{2j}}{\partial\theta} . \tag{6.46}$$

An explicit evaluation of the expressions on the right-hand side of Equation (6.44) to this order of accuracy constitutes an arduous piece of work, which has recently been carried out by Lanzano (1975); with the result disclosing that

$$\begin{aligned}\rho(a,\theta) = \rho_0(a) &+ q\,\rho_1(a,\theta)\\ &+ q^2\rho_2(a,\theta)\\ &+ q^3\rho_3(a,\theta) + \cdots,\end{aligned} \tag{6.47}$$

where

$$\tfrac{1}{2}\rho_1(a,\theta) \equiv -f_{21}\bar{\rho}_0 P_2, \tag{6.48}$$

$$\begin{aligned}\tfrac{1}{2}\rho_2(a,\theta) \equiv{}& \tfrac{1}{5}(3-\eta_{21})f_{21}^2\bar{\rho}_0 - \tfrac{3}{5}f_{21}^2\rho_0 + \\ &+ [2f_{21}\bar{\rho}_0(a_1) + (\tfrac{6}{7}f_{21}^2 - f_{22})\bar{\rho}_0 - \tfrac{3}{7}f_{21}^2\rho_0]P_2 + \\ &+ \{\tfrac{36}{35}f_{21}^2\rho_0 + [\tfrac{6}{35}(9+7\eta_{21})f_{21}^2 - \tfrac{10}{3}f_{42}]\bar{\rho}_0\}P_4,\end{aligned} \tag{6.49}$$

$$\begin{aligned}\tfrac{1}{2}\rho_3(a,\theta) \equiv{}& \tfrac{2}{5}\eta_{21}f_{21}^2\bar{\rho}_0(a_1) + \tfrac{6}{35}(f_{21}^3 - 7f_{21}f_{22})\rho_0 + \\ &+ [\tfrac{1}{5}(6-\eta_{21}-\eta_{22})f_{21}f_{22} + \tfrac{2}{35}(\eta_{21}^2+\eta_{21}-3)f_{21}^3]\bar{\rho}_0 \\ &+ \{2(f_{22} - \tfrac{3}{7}f_{21}^2)\rho_0(a_1) - \tfrac{2}{7}(3f_{21}f_{22} + 10f_{21}f_{42} - 3f_{21}^3)\rho_0 - \\ &- [f_{23} - \tfrac{12}{7}f_{21}f_{22} - \tfrac{2}{21}(39-7\eta_{42})f_{21}f_{42} - \\ &- \tfrac{2}{35}(6\eta_{21}^2 - 5\eta_{21} - 23)f_{21}^3]\bar{\rho}_0\}P_2 + \\ &+ \{[\tfrac{20}{3}f_{42} - \tfrac{12}{35}(15+7\eta_{21})f_{21}^2]\bar{\rho}_0(a_1) + \\ &+ [\tfrac{72}{35}f_{21}f_{22} - \tfrac{36}{385}f_{21}^3 - \tfrac{60}{77}f_{21}f_{42}]\rho_0 - \\ &- [\tfrac{10}{3}f_{43} - \tfrac{6}{35}(7\eta_{21} + 7\eta_{22} + 18)f_{21}f_{22} - \\ &- \tfrac{20}{231}(39+7\eta_{21})f_{21}f_{42} + \\ &+ \tfrac{12}{385}(11\eta_{21}^2 + 41\eta_{21} + 57)f_{21}^3]\bar{\rho}_0\}P_4 + \\ &+ \{(\tfrac{40}{11}f_{21}f_{42} - \tfrac{72}{77}f_{21}^3)\rho_0 - \\ &- [7f_{63} - \tfrac{5}{33}(39 + 18\eta_{21} + 11\eta_{42})f_{21}f_{42} + \\ &+ \tfrac{6}{77}(11\eta_{21}^3 + 32\eta_{21} + 30)f_{21}^3]\bar{\rho}_0\}P_6,\end{aligned} \tag{6.50}$$

in which $\rho_0(a)$ stands (in accordance with Equation (5.2)) for the mean density of the respective configuration interior to a; $\rho_0(a_1) \equiv \rho_m$, for the mean density of the entire configuration; and where we have abbreviated

$$q \equiv \frac{\omega^2}{4\pi G\rho_m}\;; \tag{6.51}$$

the successive powers of which are *not*, however, to be confused with the generalized coordinates q^i introduced in Equations (6.23) and (6.37).

Lastly, the double-subscript amplitudes $f_{ij}(a)$ are related with the single-subscript $f_j(a)$'s by

$$f_0 \equiv \qquad + f_{02}\, q^2 + f_{03}\, q^3 + \cdots, \tag{6.52}$$

$$f_2 \equiv f_{21}\, q + f_{22}\, q^2 + f_{23}\, q^3 + \cdots, \tag{6.53}$$

$$f_4 \equiv \qquad + f_{42}\, q^2 + f_{43}\, q^3 + \cdots, \tag{6.54}$$

$$f_6 \equiv \qquad\qquad + f_{63}\, q^3 + \cdots; \tag{6.55}$$

and the double-subscript η_{ij}'s represent logarithmic derivatives of the respective f_{ij}'s.

It should be particularly noted that, in order to evaluate the effects of rotation on the internal structure of a spinning star, it is not necessary (as has been previously attempted by many writers) to solve the equations governing its internal structure for all variables of state (such as the density, pressure, or gravitational potential) simultaneously. By a recourse to Clairaut's artifice expounded in section II-1 it is possible to eliminate the pressure altogether from our considerations, and to express the density ρ in terms of the potential Ψ from Poisson's equation (6.36) for an arbitrary equation of state of the form (1.9).

In conclusion of the present section let us formulate in terms of Clairaut's theory also explicit expressions for the *moments of inertia* of distorted configurations about their principal axes, which we shall need in later parts of this book. In order to do so, let us identify these axes with our xyz-rectangular system as defined by Equations (6.13); and let, furthermore, the symbols A, B, C denote the moments of inertia of our configurations about the axes x, y, z, respectively. If so, it follows by definition that

$$A = \int (y^2 + z^2)\, dm', \tag{6.56}$$

$$B = \int (x^2 + z^2)\, dm', \tag{6.57}$$

$$C = \int (x^2 + y^2)\, dm', \tag{6.58}$$

where

$$\int \equiv \int_0^{r_1} \int_0^{\pi} \int_0^{2\pi} \tag{6.59}$$

and the mass element dm continues to be given by

$$dm' = \rho r'^2\, dr' \sin\theta'\, d\theta'\, d\phi', \tag{6.60}$$

in accordance with previous usage in section II-1.

In order to evaluate the foregoing integrals, let us resort again to Clairaut's artifice (1.25) which enables us to rewrite (6.56)–(6.58) alternatively as

$$A = \frac{1}{5} \int_0^{a_1} \rho \frac{\partial}{\partial a} \left\{ \int_0^{\pi} \int_0^{2\pi} r'^5 (\sin^2\theta' \sin^2\phi' + \cos^2\theta') \sin\theta' d\theta' d\phi' \right\} da, \tag{6.61}$$

$$B = \frac{1}{5} \int_0^{a_1} \rho \frac{\partial}{\partial a} \left\{ \int_0^{\pi} \int_0^{2\pi} r'^5 (\sin^2\theta' \cos^2\phi' + \cos^2\theta') \sin\theta' d\theta' d\phi' \right\} da, \tag{6.62}$$

and

$$C = \frac{1}{5} \int_0^{a_1} \rho \frac{\partial}{\partial a} \left\{ \int_0^{\pi} \int_0^{2\pi} r'^5 \sin^3\theta' d\theta' d\phi' \right\} da \tag{6.63}$$

respectively, where

$$r' = a \{1 + \Sigma f_j P_j\} \tag{6.64}$$

continues to be given by (2.1) or (3.16).

In order to progress further, let us consider first the case of *rotational distortion*, and to this end identify the z-axis with the axis of rotation of our configuration (while xy coincides with its equatorial plane). If so, then obviously $P_j \equiv P_j(\cos\theta')$; and a term-by-term integration of the expressions in curly brackets on the right-hand sides of Equations (6.61)–(6.63) reveals that, for a configuration of constant mass, the principal moments A', B', C' of inertia assume the more explicit forms

$$A' = B' = \frac{2}{5}\pi \int_0^{a_1} \rho \frac{\partial}{\partial a} \left\{ a^5 \left[\frac{4}{3} + \frac{2}{3} f_2 + \frac{12}{7} f_2^2 + \ldots \right] \right\} da \tag{6.65}$$

and

$$C' = \frac{2}{5}\pi \int_0^{a_1} \rho \frac{\partial}{\partial a} \left\{ a^5 \left[\frac{4}{3} - \frac{4}{3} f_2 + \frac{4}{7} f_2^2 + \ldots \right] \right\} da , \tag{6.66}$$

correctly to quantities of second order.

In consequence,

$$A' - C' = \frac{4}{5}\pi \int_0^{a_1} \rho \frac{\partial}{\partial a} \left\{ a^5 \left[f_2 + \frac{4}{7} f_2^2 + \ldots \right] \right\} da ; \tag{6.67}$$

and if, moreover,

$$I = \frac{2}{5}\pi \int_0^{a_1} \rho \frac{\partial}{\partial a} \left\{ \frac{4}{3} a^5 \right\} da = \frac{8}{3}\pi \int_0^{a_1} \rho a^4 da \tag{6.68}$$

denotes the moment of inertia of our configuration about its centre of mass, it follows from the foregoing results that

$$A' + B' + C' - 3I = \frac{8}{5}\pi \int_0^{a_1} \rho \frac{\partial}{\partial a} (a^5 f_2^2) \, da , \tag{6.69}$$

where the function $f_2(a)$ continues to be given by Equations (2.46) and (2.48)–(2.49), or their simplified version (2.42). The reader may note that, within the scheme of our approximation, all three moments of inertia involve solely f_2 and are independent of f_4; moreover, the quantity $A' + B' + C' - 3I$ turns out to be one of second order.

Turning now to the effects of *tidal distortion*, let us identify our x-axis with the semi-major axis of the prolate configuration (coinciding in direction with the line joining the centres of the distorted and disturbing body), so that now $P_j \equiv P_j(\cos\phi' \sin\theta')$ in (3.16). If we raise this latter expansion to the fifth power and retain terms consistently to the quantities of second order, a term-by-term integration of (6.61)–(6.63) reveals that the principal momens A'', B'', C'' of inertia of a tidally-distorted configuration of constant mass assume the more explicit forms

$$A'' = \frac{2}{5}\pi \int_0^{a_1} \rho \frac{\partial}{\partial a}\left\{a^5\left[\frac{4}{3} + \frac{5}{3}f_1 - \frac{4}{3}f_2 - \frac{5}{3}f_3 + \frac{65}{96}f_5 - \frac{13}{32}f_7 + \frac{4}{7}f_2^2 + \frac{4}{9}f_3^2 + \frac{25}{12}f_2f_3 - \frac{32}{21}f_2f_4\right]\right\}da \tag{6.70}$$

and

$$B'' = C'' = \frac{2}{5}\pi \int_0^{a_1} \rho \frac{\partial}{\partial a}\left\{a^5\left[\frac{4}{3} + \frac{15}{4}f_1 + \frac{2}{3}f_2 - \frac{5}{12}f_3 + \frac{55}{192}f_5 - \frac{3}{16}f_7 + \frac{12}{7}f_2^2 + \frac{76}{63}f_3^2 + \frac{95}{24}f_2f_3 + \frac{16}{21}f_2f_4\right]\right\}da\ ; \tag{6.71}$$

so that

$$C'' - A'' = \frac{2}{5}\pi \int_0^{a_1} \rho \frac{\partial}{\partial a}\left\{a^5\left[\frac{5}{4}f_1 + 2f_2 + \frac{5}{4}f_3 - \frac{25}{64}f_5 + \frac{7}{32}f_7 + \frac{8}{7}f_2^2 + \frac{16}{21}f_3^2 + \frac{15}{8}f_2f_3 + \frac{16}{7}f_2f_4\right]\right\}da \tag{6.72}$$

and

$$A'' + B'' + C'' - 3I = 4\pi \int_0^{a_1} \rho \frac{\partial}{\partial a}\left\{a^5\left[f_1 - \frac{1}{4}f_3 + \frac{1}{8}f_5 - \frac{5}{64}f_7 + \frac{2}{5}f_2^2 + \frac{2}{7}f_3^2 + f_2f_3\right]\right\}da\ , \tag{6.73}$$

where the individual f_j's have already been formulated, to the requisite degree of accuracy, in he preceding sections II-2 and 3. The reader may, moreover, note that – unlike (6.69) – the right-hand side of the foregoing Equation (6.73) proves to be a quantity of first order, due to the presence of the coefficient f_3 of the third-harmonic tidal distortion; all other quantities in the square brackets being again of second order.

BIBLIOGRAPHICAL NOTES

II-1: The method followed in this section for a specification of the gravitational potential of distorted configurations has its source in a book, by Alexis Claude Clairaut (1713–1765), entitled *Théorie de la figure de la terre, tirée des principles de l'hydrostatique* (Paris, 1743); but its general version (for any order of spherical-harmonic distortion) was not developed till by André-Marie Legendre (1752–1833) in his *Recherches sur la figure des planètes* (published in *Mémoires de mathématique par divers savants pour 1789*; but not actually published in Paris till 1793). A comprehensive summary of this work can be found in the fifth volume of *Mécanique céleste* (Paris, 1825) by P. S. Laplace (1749–1827); and, more recently, in volume 2 of F. Tisserand's *Traité de la mécanique céleste* (Paris, 1891) or H. Poincaré's *Leçons sur les figure d'équilibre* (Paris, 1903), Chapter 4.

Throughout the 18th and 19th centuries a development of this subject remained almost exclusively in French hands. In the present century shorter summaries of it appeared in H. Jeffreys, *The Earth* (Cambridge, 1924), Chapter 13; L. Lichtenstein, *Gleichgewichtsfiguren rotierender Flüssigkeiten* (Berlin, 1933); W. S. Jardetzky, *Theories of Figures of Celestial Bodies* (New York, 1958); or Chapter 2 of *Close Binary Systems* (New York and London, 1958) by Z. Kopal. Cf. also N.R. Lebovitz (1970).

Kopal's *Figures of Equilibrium of Celestial Bodies* (1960) contains the most comprehensive treatment of our subject in the English language prior to the appearance of this book. Lichtenstein's as well as Jardetzky's books contain, moreover, a number of further references to purely mathematical investigations concerning Clairaut's equation; among which a fundamental memoir by A. M. Liapounov (1904) should, in particular, be listed.

II-2: A theory of the rotational distortion of self-gravitating fluid configurations of arbitrary structure, correct to quantities of first order in surficial distortion, goes back to A.C. Clairaut (1743); but more than a century elapsed before anyone set out to extend Clairaut's theory to quantities of higher orders; no doubt because the axial rotation of the Earth – the prime object of applications in the minds of the 19th-century investigators – is so slow as to make its quadratic or higher effects barely significant. It was the much more rapid rotation of Jupiter or Saturn which provided the first motive for exploration of the quadratic terms by the planetologists; with astrophysicists following rather belatedly in their wake.

The first investigator who addressed himself to the task of expanding Clairaut's theory to quantities of *second* order was O. Callandreau (1889), followed by G. H. Darwin (1900), A. Véronnet (1912) and W. de Sitter (1924); and their results were later applied to geophysical and planetological problems by E. C. Bullard (1948), H. Jeffreys (1953), W. C. de Marcus (1958) or R. James and Z. Kopal (1963); the only worker with astrophysical interest was L. Evrard (1951). All these investigators limited, however, themselves to a derivation (or use) of second-order Clairaut's *integral* equations of the form (2.27)–(2.29) of the present section – rather than of the *differential* equations (2.33)–(2.35). The latter were derived first by Z. Kopal (1960), and applied to the Earth and the major planets by R. James and Z. Kopal (1963).

The first investigator who set out to extend this whole theory to *third*-order terms was P. Lanzano (1962) who, however, left again the corresponding Clairaut's equation in its integral form. A complete treatment of the subject (largely followed in this section) was not given till several years later by Z. Kopal (1973); while Lanzano soon thereafter confirmed Kopal's results by a somewhat different method (Lanzano, 1974).

A further extension of the entire theory to terms of *fourth* order was subsequently carried out by Z. Kopal and M. Kamala Mahanta (1974); while a development of a parallel theory based on the expansion of the radius-vector in terms of the sectorial (rather than zonal) harmonics, as well as of the Hansen-Tisserand coefficients was carried out by M. B. El-Shaarawy (1974).

For a relativistic treatment of the underlying problem cf. J.B. Hartle (1967) or P. Lanzano (1968).

II-3: A first-order treatment of the subject of this section is severely classical, and little of significance has been added to it since the time of Laplace. However, an extension of this theory to quantities of second order – essential for astrophysical applications – was not given till by Z. Kopal (1960); and its present version follows largely Kopal (1974).

II-4: The subject of this section – constituting an integral part of a second-order theory of stellar distortion – was first treated by Z. Kopal (1960); and its present (abridged) version follows Kopal (1974).

II-5: For solutions of the Clairaut or Radau equations in the case of limiting homogeneous or mass-point models cf., e.g., F. Tisserand (1891); though that corresponding to a shell-model does not seem to have been noted before Kopal (1941). The inequality (5.10) is likewise classical, and implicit (for $j = 2$) already in Clairaut's work of 1743; though its rigorous proof had to await Poincaré (1903). The inequality (5.9) is due to Kopal (1941b); while that represented by Equation (5.17) was (for $j = 2$) proved already by Callandreau (1888); cf. also Véronnet (1912).

A discussion of the topology of the solutions of Radau's equation in the $(a - \eta)$- and $(\eta - \eta')$-spaces, as given in this section, represents a generalization (for an arbitrary value of j) of a discussion previously given (for $j = 2$) by Poincaré (1903); and for any j, by Kopal (1959).

For particular models of density distribution which render Clairaut's equation solvable in terms of hypergeometric series cf. M. Lipschitz (1863), F. Tisserand (1884), or M. Levy (1888); and for solutions expressible in terms of Bessel functions (obtaining in the case of an internal density distribution corresponding to a polytropic gas sphere specified by the index $n = 1$) cf. Z. Kopal (1938). P. Lanzano (1973) proved subsequently that these constitute the only models for which Clairaut's equation is solvable in terms of such functions.

A list of references to numerical solutions of Radau's equation for different types of stellar models cf., e.g., the Bibliographical Notes to section II-2 of Kopal (1959).

The formulation (5.27)–(5.30) of our problem as one of the Sturm-Liouville type is due to Kopal (1953); later, Kopal and Lanzano (1973) discovered other substitutions leading to equations of self-adjoint form. The convergence arguments for an iterative solution of Equation (5.38) is due to A. M. Liapounov (1904); and for subsequent examples of its application to stellar problems cf. section II-4 of Kopal (1960), or to Kopal and Lanzano (1973).

II-6: The argument leading to a derivation of the formula (6.9), approximating the internal-structure constants k_j by the j-th order moments of the respective density-distribution, was first given by Kopal (1953).

For a determination of the internal density distribution in distorted configurations in terms of their total potential Ψ to quantities of second order in the amount of distortion cf. section III-7 of Kopal (1960). This method has subsequently been extended to quantities of third order by P. Lanzano (1975), whose treatment of the subject has been largely followed in this section. It may be added that, by a recourse to a fourth-order theory of the rotational distortion developed recently by Kopal and Mahanta (1974) it should be possible to extend the work of this section explicity to quantities of *fifth* order.

A discussion of the moments of inertia as given in this section follows that developed in section III-8 of Kopal (1960).

CHAPTER III

DYNAMICAL TIDES

In the preceding chapter of this book we gave an outline of a theory of equilibrium figures of the components in close binary systems, which should permit us to predict their shape correctly to quantities of second order in surficial distortion caused by the tides mutually raised by the two stars on each other, and to the third order in polar flattening caused by axial rotation.

A rigorous applicability of such a theory to actual binary systems encountered in the sky is, however, limited by the three requirements already set forth at the commencement of section II-4: namely, that (1) the relative orbit of the two stars is circular; (2) the axial rotation has been synchronized with that of revolution; and (3) the equatorial plane of each component is coplanar with that of their orbit. A departure from any one of these three conditions will make the tides raised by each component on its mate to cause *fluid motion* in the interiors of both stars in the *rotating* system of coordinates – i.e., the individual fluid elements will be compelled to *move* relative to their neighbours in astrocentric longitude as well as latitude; and the ensemble of these motions is known as the *dynamical tides.*

While in the subsequent two chapters of this book we shall have a repeated opportunity to mention that the bulk of the observed distortion of the components of close binary systems are describable in terms of the equilibrium theory – that is why so much attention was devoted to it in Chapter II – dynamical tides are bound to make the actual surfaces of these components depart from their equilibrium form, to the extent to which any one of the above-mentioned conditions (1) – (3) may not be fulfilled in the respective system. Moreover, many phenomena or processes observed in close binaries could not be accounted for otherwise than by the operation of dynamical tides – such as the tidal lag, or tidal evolution of such a system due to dissipation of energy through viscous friction – phenomena to which particular attention will be paid in this and subsequent chapters of this book.

In order to do so, however, we must outline first the mathematical foundations which are to serve as a basis for such investigations. In the section which follows these introductory remarks the equations will accordingly be set up, which govern dynamical tides in close binary systems – with particular attention to the effects of *dissipative* forces (viscosity) and to the interaction of mass with radiation. These equations will then be *linearized* in section III-2 to a form applicable for studies of *small oscillations* of tidal (or other) origin about a figure of equilibrium which we shall assume to be a sphere. Moreover, such oscillations will subsequently be restricted to those *harmonic* in type and *spheroidal* in shape.

The motions governed by such equations will next be particularized to the requisite type of tides by an introduction of the appropriate type of *disturbing potential* (sec. III-3); and systems of equations set up which govern phenomena arising from viscous tides in systems whose components possess a structure exhibiting a pronounced degree of central condensation. Lastly, in the concluding section III-4 we shall ascertain the rate at which the kinetic energy of a binary system is being dissipated (into heat) through continuous operation of viscous tides – as a prerequisite for the tasks awaiting discussion in Chapter VIII.

III-1. Equations of the Problem

The fundamental equations governing the deformations – due to tides or any other cause – of the fluid components of close binary systems represent the consequences of the three great 'principles of conservation' of the physical sciences – viz. that of mass, energy, and momentum. A conservation of momentum is safeguarded by the 'equation of motion'; that of mass, by the 'equations of continuity'; while a conservation of energy may be accomplished by any one of the many 'equations of state' whose specific form depends on the mode of energy transport throughout the interior of our configuration.

A. DISSIPATIVE SYSTEMS

The equations of motion – which, in what follows, we shall write out in their Eulerian form – can be expressed in tensor form (cf. e.g., Landau and Lifschitz, 1959) as

$$\rho\left\{\frac{\partial v_i}{\partial t} + v_k \frac{\partial v_i}{\partial x_k}\right\} = \rho \frac{\partial \Psi}{\partial x_i} - \frac{\partial P}{\partial x_i} + \frac{\partial}{\partial x_k}\left\{\mu_1\left(\frac{\partial v_i}{\partial x_k} + \frac{\partial v_k}{\partial x_i} - \frac{2}{3}\delta_{ik}\frac{\partial v_l}{\partial x_l}\right)\right\} + \frac{\partial}{\partial x_i}\left\{\mu_2 \frac{\partial v_l}{\partial x_l}\right\}, \tag{1.1}$$

where x_i denote the rectangular coordinates in an inertial system and t, the time. Moreover, $v_i \equiv \dot{x}_i$ are the velocity components in the direction of increasing x_i, and δ_{ik} stands for the Dirac δ-function (i.e., 0 if $i \neq k$ and 1 for $i = k$). Since $x_i \equiv x_i(t)$, the ordinary (total) and partial derivatives of the coordinates x_i with respect to the time t are identical. Moreover, as in Chapter II, the symbols P and ρ will continue to denote the pressure and density of the material constituting our configuration; Ψ, the potential of all forces (due to self-attraction, or external field if any) acting upon our fluid; while $\mu_{1,2}$ will represent the 'first' and 'second' coefficient of viscosity: the former being a measure of the rate at which macroscopic motion (specified by the velocity components v_i) is being converted by our physical system into molecular motion (i.e., heat); while μ_2 stands for the rate at which a change in pressure P brings about an appropriate change in density ρ (or vice versa) in accordance with the prevalent equation of state.

Let us write out first the tensor equation (1.1) in terms of the rectangular Cartesian coordinates

$$x_1 \equiv x, \qquad x_2 \equiv y, \qquad x_3 \equiv z, \tag{1.2}$$

with the origin at the centre of mass of our configuration, and such that

$$\dot{x} \equiv u, \qquad \dot{y} \equiv v, \qquad \dot{z} \equiv w \tag{1.3}$$

are in the direction of increasing x, y and z. If so, Equation (1.1) can be rewritten as a system of three simultaneous scalar equations of the form

$$\rho \frac{Du}{Dt} = \rho \frac{\partial \Psi}{\partial x} - \frac{\partial P}{\partial x} + \frac{\partial \sigma_{xx}}{\partial x} + \frac{\partial \sigma_{xy}}{\partial y} + \frac{\partial \sigma_{xz}}{\partial z} + \frac{\partial}{\partial x}(\mu_2 \Delta), \tag{1.4}$$

$$\rho \frac{Dv}{Dt} = \rho \frac{\partial \Psi}{\partial x} - \frac{\partial P}{\partial y} + \frac{\partial \sigma_{yx}}{\partial x} + \frac{\partial \sigma_{yy}}{\partial y} + \frac{\partial \sigma_{yz}}{\partial z} + \frac{\partial}{\partial y}(\mu_2 \Delta), \tag{1.5}$$

$$\rho \frac{Dw}{Dt} = \rho \frac{\partial \Psi}{\partial z} - \frac{\partial P}{\partial z} + \frac{\partial \sigma_{zx}}{\partial x} + \frac{\partial \sigma_{zy}}{\partial y} + \frac{\partial \sigma_{zz}}{\partial z} + \frac{\partial}{\partial z}(\mu_2 \Delta), \tag{1.6}$$

where

$$\frac{D}{Dt} \equiv \frac{\partial}{\partial t} + u \frac{\partial}{\partial x} + v \frac{\partial}{\partial y} + w \frac{\partial}{\partial z} \tag{1.7}$$

denotes the Lagrangian time-derivative (following the motion);

$$\Delta \equiv \frac{\partial u}{\partial x} + \frac{\partial v}{\partial y} + \frac{\partial w}{\partial z}, \tag{1.8}$$

stands for the divergence of the velocity vector; and

$$\sigma_{zz} = \frac{2}{3} \mu_1 \left\{ 3 \frac{\partial w}{\partial z} - \Delta \right\}, \tag{1.9}$$

$$\sigma_{yy} = \frac{2}{3} \mu_1 \left\{ 3 \frac{\partial v}{\partial y} - \Delta \right\}, \tag{1.10}$$

$$\sigma_{xx} = \frac{2}{3} \mu_1 \left\{ 3 \frac{\partial u}{\partial x} - \Delta \right\}, \tag{1.11}$$

$$\sigma_{xy} = \mu_1 \left\{ \frac{\partial v}{\partial x} + \frac{\partial u}{\partial y} \right\} = \sigma_{yx}, \tag{1.12}$$

$$\sigma_{zx} = \mu_1 \left\{ \frac{\partial u}{\partial z} + \frac{\partial w}{\partial x} \right\} = \sigma_{xz}, \tag{1.13}$$

$$\sigma_{yz} = \mu_1 \left\{ \frac{\partial w}{\partial y} + \frac{\partial v}{\partial z} \right\} = \sigma_{zy}, \tag{1.14}$$

constitute the respective components of the viscous stress tensor.

The conservation of mass will be safeguarded next if the changes in density ρ are related with the velocity components u, v, w of fluid flow by the 'equation of continuity' of the form

$$\frac{D\rho}{Dt} + \rho\Delta = 0\,, \tag{1.15}$$

while the total potential Ψ satisfies Poisson's equation

$$\nabla^2\Psi = -4\pi G\rho\,, \tag{1.16}$$

where

$$\nabla^2 \equiv \frac{\partial^2}{\partial x^2} + \frac{\partial^2}{\partial y^2} + \frac{\partial^2}{\partial z^2} \tag{1.17}$$

stands for the Laplacian operator, and G denotes the constant of gravitation.

Equations (1.4)–(1.6), (1.15) and (1.16) constitute a simultaneous system of five differential equations for six dependent variables (i.e., u, v, w; P, ρ, and Ψ) as functions of four independent variables (i.e., x, y, z, and t); in which the coefficients $\mu_{1,2}$ of viscosity may be treated as constants or variable quantities. In order to render this system determinate, one more equation must be adjoined to those we already have; and that will take the form of an 'equation of state' safeguarding the conservation of *energy*.

In order to formulate this condition in its most general form, let us suppose that the fluid medium constituting our stars contains both matter and radiation (i.e., plasma and photon gas) contributing to the total pressure P. Let, in what follows, κ denote the coefficient of thermal conduction of the stellar material; and ϵ, the amount of energy generated by unit mass of such a material per unit time. Let, moreover, $\mathbf{F}$ represent the (vector) flux of radiant energy arising from this source. If so, the most general form of the 'equation of state' in stellar interiors can be expressed as

$$\rho\frac{DE}{Dt} - \frac{P + aT^4}{\rho}\frac{D\rho}{Dt} = \text{div}\,(\kappa\ \text{grad}\ T) + \rho\epsilon - \text{div}\,\mathbf{F} + \Phi\,, \tag{1.18}$$

where

$$E \equiv \tfrac{1}{2}\,\mathrm{v}^2 + C_v T + aT^4/\rho \tag{1.19}$$

stands for the total energy (kinetic plus thermal) per unit mass of stellar material moving with a (vector) velocity $\mathbf{v}$, maintained at a temperature T, and characterized by a specific heat C_V at constant volume (for an appropriate gas-radiation mixture). Furthermore,

$$P = R\rho T + \tfrac{1}{3}\,aT^4 \tag{1.20}$$

constitutes the (scalar) total pressure arising from (ideal) gas as well as radiation (with R representing the universal gas constant and a, the Stefan constant). Lastly, the symbol Φ

on the right-hand side of Equation (1.18) stands for the viscous dissipation function

$$\Phi \equiv \frac{1}{2}\mu_1\left(\frac{\partial v_i}{\partial x_k}+\frac{\partial v_k}{\partial x_i}-\frac{2}{3}\,\delta_{ik}\,\frac{\partial v_l}{\partial x_l}\right)^2+\mu_2\left(\frac{\partial v_l}{\partial x_l}\right)^2 , \tag{1.21}$$

which is quadratic (positive definite) in the velocity components v_i, and where $\mu_{1,2}$ denote – as before – the 'first' and 'second' coefficient of viscosity.

An inspection of the relative importance of the different terms on the right-hand side of Equation (1.18) discloses that the conductive term div(κ grad T) is likely to be negligible in all stars except (possibly) white dwarfs, and will hereafter be ignored. Moreover, equating terms on both sides of (1.18) which are quadratic in the velocity components v_i we find that, in Cartesian coordinates,

$$\begin{aligned}\frac{1}{2}\frac{D}{Dt}(u^2+v^2+w^2) = \mu_1\Bigg\{2\left[\left(\frac{\partial u}{\partial x}\right)^2+\left(\frac{\partial v}{\partial y}\right)^2+\left(\frac{\partial w}{\partial z}\right)^2\right]+ \\ +\left(\frac{\partial w}{\partial y}+\frac{\partial v}{\partial z}\right)^2+\left(\frac{\partial u}{\partial z}+\frac{\partial w}{\partial x}\right)^2+\left(\frac{\partial v}{\partial x}+\frac{\partial u}{\partial y}\right)^2\Bigg\}+ \\ +\mu_2\left(\frac{\partial u}{\partial x}+\frac{\partial v}{\partial y}+\frac{\partial w}{\partial z}\right)^2\end{aligned} \tag{1.22}$$

Lastly, for a regime in *radiative equilibrium* it should be legitimate to satisfy Equation (1.18) by setting

$$\operatorname{div}\mathbf{F} = \rho\epsilon, \tag{1.23}$$

while the changes in the variables of state P, ρ, T are then given by

$$\rho\,\frac{D}{Dt}\left(C_V T+\frac{aT^4}{\rho}\right) = \left(\frac{P+aT^4}{\rho}\right)\frac{D\rho}{Dt}. \tag{1.24}$$

Equation (1.22) controlling the dissipation of motion by viscous friction will constitute the basis for much of our discussion in subsequent sections of this book, and (1.23) represents the equation of radiative transfer familiar from a theory in the internal structure of the stars. As regards (1.24), its significance becomes clear if we regard P to represent the gas pressure alone, and treat C_V as a constant. An appeal to (1.20), where the gas constant

$$R = C_P - C_V \tag{1.25}$$

can be identified with a difference between the specific heats of an ideal gas at constant pressure and volume permits us to eliminate T from the left-hand side of (1.24) in favour of P. Under these circumstances we find Equation (1.24) to be equivalent to

$$\frac{DP}{Dt} = a^2 \frac{D\rho}{Dt} , \tag{1.26}$$

where

$$a^2 = \frac{C_P}{C_V}\left(\frac{P}{\rho}\right) \equiv \gamma \frac{P}{\rho} \tag{1.27}$$

stands for the square of the local velocity of sound.

Equation (1.26) discloses that the assumptions at the basis of its derivation render the corresponding changes of state *adiabatic* – an approximation to which we shall adhere in all our subsequent work. Inasmuch as Equation (1.26) contains no new dependent variables other than those already encountered in (1.4)–(1.6), (1.15) and (1.16), the latter together with (1.26) constitute a simultaneous system of six equations for six dependent variables $u, v, w; P, \rho, \Psi$; and can be solved for them uniquely (subject, of course, to a specification of the boundary condition requisite for this purpose).

While the scalar form of the equations of our problem retains maximum symmetry if Equation (1.1) is written out in terms of the Cartesian coordinates, this is manifestly not true of their requisite boundary conditions. Indeed, the very fact that the form of self-gravitating fluid configurations is known to be spherical in the absence of any disturbing force field suggests that the boundary conditions of our problem should become simplest if we change over from the rectangular coordinates x, y, z to *spherical polar coordinates* r, θ, ϕ, related with the former by the equations

$$\begin{aligned} x &= r \cos\phi \sin\theta , \\ y &= r \sin\phi \sin\theta , \\ z &= r \cos\theta . \end{aligned} \tag{1 28}$$

Let, further,

$$U = \dot{r}, \qquad V \equiv r\dot{\theta}, \qquad W = (r \sin\theta)\,\dot{\phi} \tag{1.29}$$

stand for the velocity components in the direction of increasing r, θ, ϕ. these components can be shown to be related with the rectangular velocity components u, v, w by means of the transformation

$$\begin{Bmatrix} U \\ V \\ W \end{Bmatrix} = \begin{Bmatrix} \cos\phi\sin\theta & \sin\phi\sin\theta & \cos\theta \\ \cos\phi\cos\theta & \sin\phi\cos\theta & -\sin\phi \\ \sin\theta & \cos\phi & 0 \end{Bmatrix} \begin{Bmatrix} u \\ v \\ w \end{Bmatrix} \tag{1.30}$$

If, moreover, we apply the same transformation to (1.4)–(1.6), we find our equations of motion in spherical polar coordinates assume the forms

$$\rho \frac{DU}{Dt} - \rho \frac{V^2 + W^2}{r} = \rho \frac{\partial\Psi}{\partial r} - \frac{\partial P}{\partial r} + \mu \left[\nabla^2 U + \frac{1}{3}\frac{\partial\Delta}{\partial r} - \frac{2}{r^2}\frac{\partial V}{\partial\theta} - \frac{2}{r^2 \sin\theta}\frac{\partial W}{\partial\phi} - \frac{2U}{r^2} - \frac{2V\cot\theta}{r^2}\right] + 2\frac{\partial\mu}{\partial r}\left[\frac{\partial U}{\partial r} - \frac{\Delta}{3}\right] \tag{1.31}$$

$$+\frac{1}{r}\frac{\partial\mu}{\partial\theta}\left[\frac{1}{r}\frac{\partial U}{\partial\theta}+\frac{\partial V}{\partial r}\right]-\frac{V}{r}+\frac{1}{r\sin\theta}\frac{\partial\mu}{\partial\phi}\left[\frac{1}{r\sin\theta}\frac{\partial U}{\partial\phi}+\right.$$

$$\left.+\frac{\partial W}{\partial r}-\frac{W}{r}\right],$$

$$\rho\frac{DV}{Dt}+\rho\frac{UV}{r}-\rho\frac{W^2\cot\theta}{r}=\frac{1}{r}\left[\rho\frac{\partial\Psi}{\partial\theta}-\frac{\partial P}{\partial\theta}\right]+\mu\left[\nabla^2V+\frac{1}{3r}\frac{\partial\Delta}{\partial\theta}\right.$$

$$\left.+\frac{2}{r^2}\frac{\partial U}{\partial\theta}-\frac{2\cos\theta}{\sin^2\theta}\frac{\partial W}{\partial\phi}-\frac{V}{r^2\sin^2\theta}\right]$$

$$+\frac{\partial\mu}{\partial r}\left[\frac{\partial V}{\partial r}+\frac{1}{r}\frac{\partial U}{\partial\theta}-\frac{V}{r}\right]+\frac{2}{r}\frac{\partial\mu}{\partial\theta}\left[\frac{1}{r}\frac{\partial V}{\partial\theta}+\frac{U}{r}-\frac{\Delta}{3}\right]$$

$$+\frac{1}{r\sin\theta}\frac{\partial\mu}{\partial\phi}\left[\frac{1}{r\sin\theta}\frac{\partial V}{\partial\phi}+\frac{1}{r}\frac{\partial W}{\partial\theta}-\frac{W\cot\theta}{r}\right], \tag{1.32}$$

$$\rho\frac{DW}{Dt}+\rho\frac{W(U+V\cot\theta)}{r}=\frac{1}{r\sin\theta}\left[\rho\frac{\partial\Psi}{\partial\phi}-\frac{\partial P}{\partial\phi}\right]+\mu\left[\nabla^2W+\frac{1}{3r\sin\theta}\frac{\partial\Delta}{\partial\phi}\right.$$

$$\left.+\frac{2}{r^2\sin\theta}\frac{\partial U}{\partial\phi}+\frac{2\cos\theta}{r^2\sin^2\theta}\frac{\partial V}{\partial\phi}-\frac{W}{r^2\sin^2\theta}\right]$$

$$+\frac{\partial\mu}{\partial r}\left[\frac{\partial W}{\partial r}+\frac{1}{r\sin\theta}\frac{\partial U}{\partial\phi}-\frac{W}{r}\right]+\frac{1}{r}\frac{\partial\mu}{\partial\theta}\left[\frac{1}{r}\frac{\partial W}{\partial\theta}+\frac{1}{r\sin\theta}\frac{\partial V}{\partial\phi}-\frac{W\cot\theta}{r}\right]$$

$$+\frac{2}{r\sin\theta}\frac{\partial\mu}{\partial\phi}\left[\frac{1}{r\sin\theta}\frac{\partial W}{\partial\phi}+\frac{U+V\cot\theta}{r}-\frac{\Delta}{3}\right], \tag{1.33}$$

where now the Lagrangian time-derivative

$$\frac{D}{Dt}\equiv\frac{\partial}{\partial t}+U\frac{\partial}{\partial r}+\frac{V}{r}\frac{\partial}{\partial\theta}+\frac{W}{r\sin\theta}\frac{\partial}{\partial\phi}\,; \tag{1.34}$$

the divergence

$$\Delta=\frac{1}{r^2}\frac{\partial}{\partial r}(r^2U)+\frac{1}{r\sin\theta}\left\{\frac{\partial}{\partial\theta}(V\sin\theta)+\frac{\partial W}{\partial\phi}\right\}; \tag{1.35}$$

and the Laplacian operator

$$\nabla^2\equiv\frac{1}{r^2}\frac{\partial}{\partial r}\left(r^2\frac{\partial}{\partial r}\right)+\frac{1}{r^2\sin\theta}\frac{\partial}{\partial\theta}\left(\cos\theta\frac{\partial}{\partial\theta}\right)+\frac{1}{r^2\sin^2\theta}\frac{\partial^2}{\partial\phi^2}. \tag{1.36}$$

Moreover, the viscous dissipation function (1.21) in terms of spherical polar coordinates assumes the form

$$\Phi = \frac{1}{2}\mu_1\left\{\sigma_{rr}^2 + \sigma_{\theta\theta}^2 + \sigma_{\phi\phi}^2 + 2(\sigma_{r\theta}^2 + \sigma_{\theta\phi}^2 + \sigma_{\phi r}^2)\right\} +$$

$$+\left(\mu_2 - \frac{2}{3}\mu_1\right)(\Delta)^2, \tag{1.37}$$

where

$$\sigma_{rr} = 2\frac{\partial U}{\partial r}, \tag{1.38}$$

$$\sigma_{\theta\theta} = 2\left(\frac{1}{r}\frac{\partial V}{\partial\theta} + \frac{U}{r}\right), \tag{1.39}$$

$$\sigma_{\phi\phi} = 2\left(\frac{1}{r\sin\theta}\frac{\partial W}{\partial\phi} + \frac{U}{r} + \frac{V\cot\theta}{r}\right); \tag{1.40}$$

and

$$\sigma_{r\theta} = \frac{1}{r}\frac{\partial U}{\partial\theta} + \frac{\partial V}{\partial r} - \frac{V}{r} = \sigma_{\theta r}, \tag{1.41}$$

$$\sigma_{r\phi} = \frac{1}{r\sin\theta}\frac{\partial U}{\partial\phi} + \frac{\partial W}{\partial r} - \frac{W}{r} = \sigma_{\phi r}, \tag{1.42}$$

$$\sigma_{\theta\phi} = \frac{1}{r\sin\theta}\frac{\partial V}{\partial\phi} + \frac{1}{r}\frac{\partial W}{\partial\theta} - \frac{W\cot\theta}{r} = \sigma_{\phi\theta} \tag{1.43}$$

stand for the respective components of the viscous stress tensor, and the divergence Δ continues to be given by (1.35).

Equations (1.31)–(1.33) in spherical polar coordinates have certainly lost much of the symmetry they possessed in the Cartesian coordinates. This loss is, however, more apparent than real; and full benefits of this shift in strategy will become transparent in the next section. For the present, we wish to proceed with a more general discussion of our subject by pointing out the existence of further dissipative forces affecting our equations, which arise from the presence of *radiation.*

B. RADIATING SYSTEMS

In non-radiating systems, the dissipative forces are limited to those factored in our equations of motion of the form (1.1) by the two viscosity coefficients $\mu_{1,2}$. Their significance and numerical magnitudes in stellar interiors will be discussed, in some detail, in the last section of this chapter. In formulating appropriate consequences of the conservation of energy earlier in this section we noted already – through Equation (1.23) – the role played by radiation in the energy transfer in stellar interiors. However, radiation carries not only energy, but also momentum – a fact which suggests that not only the Equation (1.18) of energy, but also Equations (1.4)–(1.6) or (1.31)–(1.33) of motion should be influenced by the radiation field, and its interaction with matter. Radiation transports, to be sure, less mass than ordinary hydrodynamical mass flow. It carries, however; this mass much faster; so that the amounts of monentum transferred per unit time both ways need not be too unequal. The decisive factor is the mean free path; radiation having in general a longer free-path than the material carriers of momentum – regardless of whether these are molecules, neutral atoms, or plasma particles of either charge.

As was pointed out first by Jeans (1926), and subsequently elaborated in greater detail by Kippenhahn (1954), Ledoux (1958) or Hazlehurst and Sargent (1959), the equations safeguarding the conservation of momentum must, in the presence of radiation, be augmented by the respective components of the divergence of the energy-momentum tensor $\mathfrak{P}$, the elements P_j^i of which can (cf. Hazlehurst and Sargent, 1959; p. 283) be expressed as

$$P_j^i = \frac{4\,p_R}{5\,k\rho c}\left(\frac{\partial v_i}{\partial x_k} + \frac{\partial v_k}{\partial x_i} + \frac{1}{3}\,\delta_{ij}\,\frac{\partial v_l}{\partial x_l}\right) +$$

$$+ \frac{1}{k\rho c}\left(v_i\,\frac{\partial}{\partial x_k} + v_k\,\frac{\partial}{\partial x_i} + \delta_{ik}\,\frac{D}{Dt}\right) p_R \tag{1.44}$$

$$\equiv \overset{(1)}{P}{}_j^i + \frac{1}{k\rho c}\,\overset{(2)}{P}{}_j^i\,,$$

where, in accordance with (1.35)

$$\frac{D}{Dt} \equiv \frac{\partial}{\partial t} + v_l\,\frac{\partial}{\partial x_l}\,; \tag{1.45}$$

and

$$p_R = \frac{1}{3}\,aT^4 \tag{1.46}$$

stands for the scalar part of the radiation pressure and k denotes the absorption coefficient of stellar material per unit mass, while c is the velocity of light.

The reader may note that the first term $\overset{(1)}{P}{}_j^i$ on the right-hand side of Equation (1.44)* is identical with the respective element of stress tensor on the right-hand side of Equation (1.1) arising from gas viscosity, with

$$\mu_1 = \mu_2 = \frac{4\,p_R}{5\,k\rho c} = \frac{4\,aT^4}{15\,k\rho c}\,, \tag{1.47}$$

which we shall hereafter refer to as the 'radiative viscosity' μ_R of photon gas. The latter should, therefore, be added to those appropriate for gas (plasma) viscosity to represent the sum total $\mu_G + \mu_R$ of dissipative effects caused by combined particle-photon viscosity.

The second term on the right-hand side of (1.44) represents effects of somewhat different nature: namely, those arising from a systematic difference which exists in stellar

* The factor 1/3 should replace unity in the Hazlehurst and Sargent result if the symbol P on the right-hand side of Equation (1.1) is to represent the total pressure arising from both gas and radiation.

interiors between the average motions of photons and gas (or plasma) particles. In particular, if we abbreviate

$$\left\{v_i \frac{\partial}{\partial x_k} + v_k \frac{\partial}{\partial x_i} + \delta_{ik} \frac{D}{Dt}\right\} p_R \equiv \overset{(2)}{P}{}^i_k , \tag{1.48}$$

then in spherical polar coordinates – in which

$$v_1 \equiv U, \qquad v_2 \equiv V, \qquad v_3 \equiv W \tag{1.49}$$

and

$$\frac{\partial}{\partial x_1} \equiv \frac{\partial}{\partial r}, \quad \frac{\partial}{\partial x_2} \equiv \frac{1}{r}\frac{\partial}{\partial \theta}, \quad \frac{\partial}{\partial x_3} = \frac{1}{r \sin\theta} \frac{\partial}{\partial \phi}, \tag{1.50}$$

the diagonal components $\overset{(2)}{P}{}^i_i \, (i = 1, 2, 3)$ assume the forms

$$\overset{(2)}{P}{}^1_1 = \frac{Dp_R}{Dt} + 2U \frac{\partial p_R}{\partial r}, \tag{1.51}$$

$$\overset{(2)}{P}{}^2_2 = \frac{Dp_R}{Dt} + 2\frac{V}{r}\frac{\partial p_R}{\partial \theta}, \tag{1.52}$$

$$\overset{(2)}{P}{}^3_3 = \frac{Dp_R}{Dt} + \frac{2W}{r \sin\theta} \frac{\partial p_R}{\partial \phi} ; \tag{1.53}$$

while

$$\overset{(2)}{P}{}^1_2 = V \frac{\partial p_R}{\partial r} + \frac{U}{r}\frac{\partial p_R}{\partial \theta} = \overset{(2)}{P}{}^2_1 \tag{1.54}$$

$$\overset{(2)}{P}{}^1_3 = W \frac{\partial p_R}{\partial r} + \frac{U}{r \sin\theta}\frac{\partial p_R}{\partial \phi} = \overset{(2)}{P}{}^3_1 , \tag{1.55}$$

$$\overset{(2)}{P}{}^2_3 = \frac{W}{r}\frac{\partial p_R}{\partial \theta} + \frac{V}{r \sin\theta}\frac{\partial p_R}{\partial \phi} = \overset{(2)}{P}{}^3_2 . \tag{1.56}$$

On the other hand, in the state of radiative equilibrium the transfer equation

$$\frac{1}{c^2}\frac{D\mathbf{F}}{Dt} + \operatorname{grad} p_R + \frac{k\rho}{c}\mathbf{F} = 0 \tag{1.57}$$

for the flux **F** yields

$$\frac{1}{k\rho c}\frac{\partial p_R}{\partial x_j} = -\frac{1}{c^2}\left\{F_j + \frac{1}{k\rho c}\frac{DF_j}{Dt}\right\}, \tag{1.58}$$

where F_j denotes the components of the flux **F** for $j = 1, 2, 3$.

Since the mean-free-path $(k\rho)^{-1}$ of the photon gas is small in comparison with ct, the second term on the right-hand side of Equation (1.58) is likely to be small in comparison with the first. Ignoring it we conclude that, to a sufficient approximation,

$$\frac{1}{k\rho c}\frac{\partial p_R}{\partial x_j} = -\frac{F_j}{c^2}\,. \tag{1.59}$$

If, moreover,

$$F_1 \equiv F_r, \qquad F_2 \equiv F_\theta, \qquad F_3 \equiv F_\phi \tag{1.60}$$

denote the respective flux component in the direction of increasing r, θ, ϕ, Equation (1.59) permits us to rewrite the components $\overset{(2)}{P}{}^i_j$ in terms of those of the flux **F**. It may be noted that, in his treatment of the subject, Kippenhahn (1954) omitted the term, on the left-hand side of (1.48), factored by the Dirac δ_{ij}-function; but its existence was later brought to light by Ledoux (1958) and Hazlehurst and Sargent (1959).

Next – by analogy with our previous treatment of the viscous part of the energy stress tensor – let us introduce a set of nine new stress components σ'_{ij}, defined by

$$(k\rho c)^{-1}\,\overset{(2)}{P}{}^i_j \equiv -c^{-2}\,\sigma'_{ij} \tag{1.61}$$

By use of (1.59) – (1.61) we find that, in *steady* state (when $\partial p_R/\partial t = 0$),

$$\sigma'_{rr} = 3UF_r + VF_\theta + WF_\phi, \tag{1.62}$$

$$\sigma'_{\theta\theta} = UF_r + 3VF_\theta + WF_\phi, \tag{1.63}$$

$$\sigma'_{\phi\phi} = UF_r + VF_\theta + 3WF_\phi; \tag{1.64}$$

and

$$\sigma'_{r\theta} = VF_r + UF_\theta = \sigma'_{\theta r}\,, \tag{1.65}$$

$$\sigma'_{r\phi} = WF_r + UF_\phi = \sigma'_{\phi r}\,, \tag{1.66}$$

$$\sigma'_{\theta\phi} = WF_\theta + VF_\phi = \sigma'_{\phi\theta}\,. \tag{1.67}$$

The last three equations agree with those previously deduced by Kippenhahn (1954); but the diagonal elements (1.62)–(1.64) differ from his because of Kippenhahn's neglect of the sum $UF_r + VF_\theta + WF_\phi$ in $\overset{(2)}{P}{}^i_j$.

In order to derive an explicit form of the dynamical effects of interaction between mass motions characterized by the velocity components U, V, W and the radiative flux of components F_r, F_θ, F_ϕ, we take the divergence of the components $\overset{(2)}{P}{}^i_j$ of the radiative stress tensor, and augment the right-hand sides of Equations (1.31) – (1.33) by the terms

$$-c^2\mathfrak{R} = \frac{1}{r^2}\frac{\partial}{\partial r}(r^2\sigma'_{rr}) + \frac{1}{r\sin\theta}\left\{\frac{\partial}{\partial\theta}(\sigma'_{r\theta}\sin\theta) + \right.$$

$$\left. + \frac{\partial\sigma'_{r\phi}}{\partial\phi} - (\sigma'_{\theta\theta} + \sigma'_{\phi\phi})\sin\theta\right\}, \tag{1.68}$$

$$-c^2\mathfrak{S} = \frac{1}{r^2}\frac{\partial}{\partial r}(r^2\sigma'_{\theta r}) + \frac{1}{r\sin\theta}\left\{\frac{\partial}{\partial\theta}(\sigma'_{\theta\theta}\sin\theta) + \right.$$

$$\left. + \frac{\partial\sigma'_{\theta\phi}}{\partial\phi} + \sigma'_{\theta r}\sin\theta - \sigma'_{\phi\phi}\cos\theta\right\}, \tag{1.69}$$

$$-c^2\mathfrak{T} = \frac{1}{r^2}\frac{\partial}{\partial r}(r^2\sigma'_{\phi r}) + \frac{1}{r\sin\theta}\left\{\frac{\partial}{\partial\theta}(\sigma'_{\phi\theta}\sin\theta) + \right.$$

$$\left. + \frac{\partial\sigma'_{\phi\phi}}{\partial\phi} + \sigma'_{\phi r}\sin\theta + \sigma'_{\phi\theta}\cos\theta\right\}. \tag{1.70}$$

On insertion in (1.68)–(1.70) for the stress components from (1.62)–(1.67) we find that

$$\mathfrak{R} = -\frac{1}{c^2}\left\{\frac{\partial}{\partial r}(3UF_r + VF_\theta + WF_\phi) + \frac{2}{r}(2UF_r - VF_\theta - WF_\phi)\right.$$

$$+ \frac{1}{r\sin\theta}\frac{\partial}{\partial\theta}[(UF_\theta - VF_r)\sin\theta]$$

$$\left. + \frac{1}{r\sin\theta}\frac{\partial}{\partial\phi}[UF_\phi + WF_r]\right\}, \tag{1.71}$$

$$\mathfrak{S} = -\frac{1}{c^2}\left\{\left[\frac{\partial}{\partial r} + \frac{3}{r}\right](UF_\theta + VF_r) + \frac{1}{r}\frac{\partial}{\partial\theta}(UF_r + 3VF_\theta + WF_\phi) + \right.$$

$$\left. + \frac{2\cot\theta}{r}(VF_\theta - WF_\phi) + \frac{1}{r\sin\theta}\frac{\partial}{\partial\phi}(VF_\phi + WF_\theta)\right\}, \tag{1.72}$$

and

$$\mathfrak{T} = -\frac{1}{c^2}\left\{\left[\frac{\partial}{\partial r} + \frac{3}{r}\right](UF_\phi + WF_r) + \right.$$

$$+ \frac{1}{r}\left[\frac{\partial}{\partial\theta} + 2\cot\theta\right](VF_\phi + WF_\theta) + \tag{1.73}$$

$$+ \frac{1}{r\sin\theta}\frac{\partial}{\partial\phi}[UF_r + VF_\theta + 3WF_\phi] \; .$$

Should the angular flux components F_θ and F_ϕ be negligible in comparison with F_r for a configuration which may not depart much from spherical symmetry, the foregoing Equations (1.71) – (1.73) reduce to

$$\mathfrak{R} = -\frac{1}{c^2}\left\{\frac{2}{r^2}\frac{\partial}{\partial r}(r^2 UF_r) + \frac{\partial}{\partial r}(UF_r) + \right.$$

$$\left. + \frac{1}{r\sin\theta}\left[\frac{\partial}{\partial r}(VF_r\sin\theta) + \frac{\partial}{\partial\phi}(WF_r)\right]\right\} , \tag{1.74}$$

$$\mathfrak{S} = -\frac{1}{c^2}\left\{\frac{1}{r^2}\frac{\partial}{\partial r}(r^2 VF_r) + \frac{1}{r}\frac{\partial}{\partial \theta}(UF_r) + \frac{VF_r}{r}\right\} \tag{1.75}$$

$$\mathfrak{T} = -\frac{1}{c^2}\left\{\frac{1}{r^2}\frac{\partial}{\partial r}(r^2 WF_r) + \frac{1}{r\sin\theta}\frac{\partial}{\partial \phi}(UF_r) + \frac{WF_r}{r}\right\}. \tag{1.76}$$

Lastly, if in the far interior (where tides can become of importance) the energy generation ϵ per unit mass becomes negligible, it is legitimate to set

$$F_r = \frac{L}{4\pi r^2}, \tag{1.77}$$

where L denotes the (constant) luminosity of the respective star. In such a case, Equations (1.74)–(1.76) reduce further to

$$\mathfrak{R} = -\frac{L}{4\pi c^2 r^2}\left\{2r^2\frac{\partial}{\partial r}\left(\frac{U}{r^2}\right) + \Delta\right\}, \tag{1.78}$$

$$\mathfrak{S} = -\frac{L}{4\pi c^2 r^2}\left\{\frac{1}{r}\frac{\partial U}{\partial r} + \frac{1}{r}\frac{\partial}{\partial r}(rV)\right\}, \tag{1.79}$$

$$\mathfrak{T} = -\frac{L}{4\pi c^2 r^2}\left\{\frac{1}{r\sin\theta}\frac{\partial U}{\partial \phi} + \frac{1}{r}\frac{\partial}{\partial r}(rW)\right\}, \tag{1.80}$$

expression which should enable us at least to estimate the order of magnitude of these terms.

It should, in conclusion, be noted that an interaction between particles (plasma) and light (photons) in radiating dynamical systems is bound to affect not only the equations of motion safeguarding the conservation of momentum, but also the energy equation (1.18); and this will occur in two ways. First, the coefficients $\mu_{1,2}$ of gas viscosity on the right-hand side of Equations (1.21) or (1.37) for Φ should be augmented by the coefficients $\mu_{1,2}$ of radiative viscosity as given by Equation (1.47).

Secondly, the right-hand side of Equation (1.21) for the dissipation function Φ should be augmented (cf., e.g., Chandrasekhar, 1961; pp. 14–15) by the term $-c^2\sigma_{ij}\sigma'_{ii}$, where the σ_{ij}'s stand for the components of the viscous stress tensor as given by Equations (1.38) – (1.43), while the σ'_{ij}'s are given by Equations (1.62) – (1.67).

Using these and neglecting the flux components F_θ and F_ϕ as small while F_r continues to be approximated by (1.77), we find that the right-hand side of the expression (1.37) for Φ should be augmented by the term

$$\begin{aligned} -c^{-2}\{&\sigma_{rr}\sigma'_{rr} + \sigma_{\theta\theta}\sigma'_{\theta\theta} + \sigma_{\phi\phi}\sigma'_{\phi\phi} + \\ &+ 2(\sigma_{r\theta}\sigma'_{r\theta} + \sigma_{r\phi}\sigma'_{r\phi} + \sigma_{\theta\phi}\sigma'_{\theta\phi})\} = \\ &= \frac{1}{2\pi r^2 c^2}\{U\Delta + U\sigma_{rr} + V\sigma_{r\theta} + W\sigma_{r\phi}\}, \end{aligned} \tag{1.81}$$

where the divergence Δ continues to be given by Equation (1.35), and which plays the same role for the conservation of energy as do the terms $\mathfrak{R}, \mathfrak{S}, \mathfrak{T}$ deduced earlier for the conservation of the momentum.

III-2. Linearized Equations

The equations of our problem developed in the preceding section hold good regardless of the magnitude of the displacements brought about by the velocity components v_i. In order to study the nature of such displacements in the neighbourhood of a *spherical* figure of equilibrium, let us hereafter assume that *all three velocity-components U, V, W are small enough for their squares and cross-products to be negligible.* Let, moreover, the pressure P, density ρ, and potential Ψ be expressible as

$$P = P_0(r) + P'(r, \theta, \phi; t), \tag{2.1}$$

$$\rho = \rho_0(r) + \rho'(r, \theta, \phi; t), \tag{2.2}$$

$$\Psi = \Psi_0(r) + \Psi'(r, \theta, \phi; t), \tag{2.3}$$

where P_0, ρ_0 and Ψ_0 refer to the respective variables characterizing our configuration in its equilibrium (spherical) form; and P', ρ', Ψ' stand for their *changes* brought about by motion characterized by the velocity components U, V, W.

In the state of equilibrium (when $U = V = W = 0$), Equation (1.31) discloses at once that

$$\frac{\partial P_0}{\partial r} = \rho_0 \frac{\partial \Psi_0}{\partial r} = -g\rho_0 , \tag{2.4}$$

where the gravitational acceleration

$$g = G \frac{m(r)}{r^2} = \frac{4\pi G}{r^2} \int_0^r \rho_0 r^2 \, \mathrm{d}r. \tag{2.5}$$

If, moreover, we regard the coefficients $\mu_{1,2}$ of viscosity to be functions of r only and realize that the primed functions P', ρ', Ψ' are – like the velocity components giving rise to them – small enough for their squares and cross-products to be negligible, our *linearized* equations of motion (1.31) – (1.33) will assume the forms

$$\begin{aligned} \rho \frac{\partial U}{\partial t} = \rho \frac{\partial \Psi'}{\partial r} - \frac{\partial P'}{\partial r} - g\rho' + \frac{\mu_1}{r} \left[\nabla^2(rU) - 2\Delta + \frac{r}{3} \frac{\partial \Delta}{\partial r} \right] + \\ + 2 \frac{\partial \mu_1}{\partial r} \left[\frac{\partial U}{\partial r} - \frac{\Delta}{3} \right] + \frac{\partial}{\partial r} (\mu_2 \Delta) - \\ - \frac{L}{4\pi c^2 r^2} \left[2r^2 \frac{\partial}{\partial r} \left(\frac{U}{r^2} \right) + \Delta \right] , \end{aligned} \tag{2.6}$$

$$\rho \frac{\partial V}{\partial t} = \frac{1}{r}\left[\rho \frac{\partial \Psi'}{\partial \theta} - \frac{\partial P'}{\partial \theta}\right] + \frac{\mu_1}{r^2}\left[r^2 \nabla^2(V) + 2\frac{\partial U}{\partial \theta} - \frac{V}{\sin^2\theta} - \right.$$

$$\left. - \frac{2\cos\theta}{\sin^2\theta}\frac{\partial W}{\partial \phi} + \frac{r}{3}\frac{\partial \Delta}{\partial \theta}\right] + \frac{\partial \mu_1}{\partial r}\left[\frac{\partial V}{\partial r} + \frac{1}{r}\frac{\partial V}{\partial \theta} - \frac{V}{r}\right] +$$

$$+ \frac{\mu_2}{r}\frac{\partial \Delta}{\partial \theta} - \frac{L}{4\pi c^2 r^2}\left[\frac{1}{r}\frac{\partial U}{\partial \theta} + \frac{1}{r}\frac{\partial}{\partial r}\right](rV) , \tag{2.7}$$

$$\rho \frac{\partial W}{\partial t} = \frac{1}{r\sin\theta}\left[\rho \frac{\partial \Psi'}{\partial \phi} - \frac{\partial P'}{\partial \phi}\right] + \frac{\mu_1}{r^2\sin^2\theta}\left[(r^2\sin^2\theta)\nabla^2 W + 2\frac{\partial U}{\partial \phi} + \right.$$

$$\left. + 2\cos\theta \frac{\partial V}{\partial \phi} - W + \frac{1}{3}(r\sin\theta)\frac{\partial \Delta}{\partial \phi}\right] +$$

$$+ \frac{\partial \mu_1}{\partial r}\left[\frac{\partial W}{\partial \phi} + \frac{1}{r\sin\theta}\frac{\partial U}{\partial \phi} - \frac{W}{r}\right] +$$

$$+ \frac{\mu_2}{r\sin\theta}\frac{\partial \Delta}{\partial \phi} - \frac{L}{4\pi c^2 r^2}\left[\frac{1}{r\sin\theta}\frac{\partial U}{\partial \phi} + \frac{1}{r}\frac{\partial}{\partial r}\right](rW) . \tag{2.8}$$

where it is understood that, in radiating systems,

$$\mu_{1,2} \equiv (\mu_{1,2})_{\text{gas}} + (\mu_{1,2})_{\text{rad}} , \tag{2.9}$$

and where we inserted for the 'braking' terms $\mathfrak{R}$, $\mathfrak{S}$, $\mathfrak{T}$ from (1.78)–(1.80). The reader may note that, inasmuch as the viscous terms on the right-hand sides of the exact Equations (1.31) – (1.33) are linear, they went through our foregoing process of linearization without change.

Moreover, the Equation (1.15) of continuity and Poisson's equation (1.16) can be similarly linearized to yield

$$\frac{\partial \rho'}{\partial t} + U\frac{\partial \rho}{\partial r} + \rho\Delta = 0 \tag{2.10}$$

and

$$\nabla^2 \Psi' = -4\pi G\rho' ; \tag{2.11}$$

while the linearized Equation (1.26) of state similarly yields

$$\frac{\partial P'}{\partial t} + U\frac{\partial P_0}{\partial r} = a_0^2\left\{\frac{\partial \rho'}{\partial t} + U\frac{\partial \rho_0}{\partial r}\right\}, \tag{2.12}$$

which by use of (2.4) and (2.10) can be rewritten as

$$\frac{\partial P'}{\partial t} = \rho_0(gU - a_0^2\Delta) , \tag{2.13}$$

where

$$a_0^2 = \gamma \frac{P_0}{\rho_0} . \tag{2.14}$$

denotes the square of the local velocity of sound inside our configurations in its equilibrium state.

A. VISCOUS SPHEROIDAL DEFORMATIONS

In order to proceed further, let us hereafter assume that the *deformation* of this configuration caused by the velocity components U, V, W is spheroidal which implies that the velocity components U, V, W are constrained to be of the form

$$U = u(r, t)\, Y_j^i(\theta, \phi), \tag{2.15}$$

$$V = v(r, t) \frac{\partial Y_j^i}{\partial \theta} , \tag{2.16}$$

$$W = \frac{v(r, t)}{\sin\theta} \frac{\partial Y_j^i}{\partial \phi} , \tag{2.17}$$

where $u(r, t)$ as well as $v(r, t)$ are functions of r and t only, while the $Y_j^i(\theta, \phi)$'s surface harmonics of index i and order j, obeying the differential equation

$$\frac{1}{\sin^2\theta} \frac{\partial^2 Y}{\partial \phi^2} + \frac{1}{\sin\theta} \frac{\partial}{\partial \theta} (\sin\theta \frac{\partial Y}{\partial \theta}) + j\,(j+1)\,Y = 0. \tag{2.18}$$

If, furthermore, we abbreviate

$$\frac{1}{r^2} \frac{\partial}{\partial r} (r^2 u) - j\,(j+1)\frac{v}{r} = y \tag{2.19}$$

and

$$\frac{1}{r} \frac{\partial}{\partial r} (rv) - \frac{u}{r} = z, \tag{2.20}$$

if follows by insertion of (2.15) – (2.17) in (1.35) that

$$\Delta = y Y_j^i ; \tag{2.21}$$

while the linearized equation (2.10) of continuity will assume the form

$$\frac{\partial \rho'}{\partial t} = -\left\{ \frac{1}{r^2} \frac{\partial}{\partial r} (\rho r^2 u) - j(j+1) \frac{\rho v}{r} \right\} Y_j^i = -f Y_j^i ; \tag{2.22}$$

and the energy equation (2.13) transforms into

$$\frac{\partial P'}{\partial t} = \rho\,(gu - a^2 y)\,Y_j^i = -a^2 h Y_j^i\,, \tag{2.23}$$

where zero subscripts on P and ρ will hereafter be omitted. As a result

$$\frac{\partial}{\partial t}\left\{\frac{\partial P'}{\partial r} + g\rho'\right\} = -\left\{gf + \frac{\partial}{\partial r}\,(a^2 h)\right\} Y_j^i \tag{2.24}$$

and

$$\frac{\partial}{\partial t}\left(\frac{\partial P'}{\partial \theta}\right) = -a^2 h\,\frac{\partial Y}{\partial \theta}\,, \qquad \frac{\partial}{\partial t}\left(\frac{\partial P'}{\partial \phi}\right) = -a^2 h\,\frac{\partial Y}{\partial \phi}\,. \tag{2.25}$$

On the other hand, an insertion of (2.15) – (2.17) on the right-hand sides of Equations (2.6) – (2.8) reveals that the viscous terms transform into

$$\frac{\mu_1}{r}\left\{\nabla^2(rU) - 2\Delta + \frac{r}{3}\,\frac{\partial \Delta}{\partial r}\right\} + 2\,\frac{\partial \mu_1}{\partial r}\left\{\frac{\partial U}{\partial r} - \frac{\Delta}{3}\right\}$$

$$+ \frac{\partial}{\partial r}\,(\mu_2 \Delta) - \frac{L}{4\pi c^2 r^2}\left\{2r^2\,\frac{\partial}{\partial r}\left(\frac{U}{r^2}\right) + \Delta\right\} =$$

$$= \left\{2\left[\mu_1\,\frac{\partial y}{\partial r} + \frac{\partial \mu_1}{\partial r}\,\frac{\partial u}{\partial r}\right] + \frac{j(j+1)}{r}\,\mu_1 z -\right.$$

$$\left. - \frac{2}{3}\,\frac{\partial}{\partial r}\,(\mu_1 y) + \frac{\partial}{\partial r}\,(\mu_2 y) - \frac{L_1}{4\pi c^2 r^2}\left[y + 2\,\frac{\partial u}{\partial r} - \frac{4u}{r}\right]\right\} Y_j^i$$

$$\equiv \mathfrak{F}\,(r, t)\; Y_j^i\,(\theta, \phi), \tag{2.26}$$

while

$$\mu_1\left\{\nabla^2 V + \frac{2}{r^2}\,\frac{\partial U}{\partial \theta} - \frac{V}{r^2 \sin^2\theta} - \frac{2\cos\theta}{r^2\sin^2\theta}\,\frac{\partial W}{\partial \phi} + \frac{1}{3r}\,\frac{\partial \Delta}{\partial \theta}\right\}$$

$$+ \frac{\partial \mu_1}{\partial r}\left\{\frac{\partial V}{\partial r} + \frac{1}{r}\,\frac{\partial U}{\partial \theta} - \frac{V}{r}\right\} + \frac{1}{r}\,\frac{\partial}{\partial \theta}\,(\mu_2 \Delta) -$$

$$- \frac{L}{4\pi c^2 r^2}\left\{\frac{1}{r}\,\frac{\partial U}{\partial \theta} + \frac{1}{r}\,\frac{\partial}{\partial r}\,(rV)\right\} = \tag{2.27}$$

$$= \left\{\frac{1}{r}\,\frac{\partial}{\partial r}\,(r\mu_1 z) + \frac{4\mu_1}{3r}\,y + \frac{2}{r}\,\frac{\partial \mu_1}{\partial r}\,(u - v) +\right.$$

$$\left. + \mu_2\,\frac{y}{r} - \frac{L}{4\pi c^2 r^2}\left[z + \frac{2u}{r}\right]\right\}\frac{\partial Y}{\partial \theta} \equiv \frac{\mathfrak{G}(r, t)}{r}\,\frac{\partial Y}{\partial \theta}$$

and

$$\mu_1 \left\{ \nabla^2 W + \frac{2}{r^2 \sin^2\theta} \frac{\partial U}{\partial \phi} + \frac{2\cos\theta}{r^2 \sin^2\theta} \frac{\partial V}{\partial \phi} - \frac{W}{r^2 \sin^2\theta} + \right.$$

$$\left. + \frac{1}{3r\sin\theta} \frac{\partial \Delta}{\partial \phi} \right\} + \frac{\partial \mu_1}{\partial r} \left\{ \frac{\partial W}{\partial r} + \frac{1}{r\sin\theta} \frac{\partial U}{\partial \phi} - \frac{W}{r} \right\}$$

$$+ \frac{1}{r\sin\theta} \frac{\partial}{\partial \phi} (\mu_2 \Delta) - \frac{L}{4\pi c^2 r^2} \left\{ \frac{1}{r\sin\theta} \frac{\partial U}{\partial \phi} + \frac{1}{r} \frac{\partial}{\partial r} (rW) \right\}$$

$$\equiv \frac{\mathfrak{G}(r,t)}{r\sin\theta} \frac{\partial Y}{\partial \phi}. \tag{2.28}$$

If, lastly, we set

$$\frac{\partial \Psi'}{\partial t} = R(r,t)\, Y_j^i(\theta, \phi) \tag{2.29}$$

a differentiation of the first Equation (2.6) of motion with respect to the time makes the latter to assume the form

$$\rho \frac{\partial^2 u}{\partial t^2} = \rho \frac{\partial R}{\partial r} + \frac{\partial}{\partial r} (a^2 h) + gf + \frac{\partial \mathfrak{F}}{\partial t}; \tag{2.30}$$

while Equation (2.7) and (2.8) similarly treated both reduce to

$$\rho r \frac{\partial^2 v}{\partial t^2} = \rho R + a^2 h + \frac{\partial \mathfrak{G}}{\partial t}, \tag{2.31}$$

where quantities f and h are defined by Equations (2.22) and (2.23); the dissipative terms $\mathfrak{F}$ and $\mathfrak{G}$ by (2.26)–(2.28); and R, by (2.29). The latter can, in turn, be obtained from the linearized version (2.11) of Poisson's equation, which after differentiation with respect to t and insertion from the continuity equation (2.10) is found to satisfy the relation

$$\frac{\partial^2 R}{\partial r^2} + \frac{2}{r} \frac{\partial R}{\partial r} - j(j+1) \frac{R}{r^2} = -4\pi G \frac{\partial \rho'}{\partial t} = 4\pi G f. \tag{2.32}$$

The foregoing Equations (2.30)–(2.32) constitute a set of simultaneous partial differential equations for $u(r,t)$, $v(r,t)$ and $R(r,t)$. The latter plays, however, in the present section only an auxiliary role – as our interest centres on u and v. In order to solve for them it is, therefore, desirable to eliminate R between (2.30)–(2.32); and this can best be done in the following manner. First divide both sides of Equations (2.31) by ρ, differentiate with respect to r, and then eliminate $\partial R/\partial r$ between the outcome and (2.30); in doing so we find that

$$\frac{\partial^2}{\partial t^2} (rz) + \frac{\partial}{\partial t} \left\{ \frac{\mathfrak{F}}{\rho} - \frac{\partial}{\partial r} \left(\frac{\mathfrak{G}}{\rho} \right) \right\} + a^2 A y = 0, \tag{2.33}$$

where y and z continue to be given in terms of u and v by (2.19)–(2.20), and where we have abbreviated

$$A = \frac{1}{\rho}\frac{\partial \rho}{\partial r} - \frac{1}{\gamma P}\frac{\partial P}{\partial r}, \tag{2.34}$$

so that, by (2.4) and (2.14)

$$a^2 A = g + \frac{a^2}{\rho}\frac{\partial \rho}{\partial r}. \tag{2.35}$$

Next, multiply Equation (2.30) by r^2/ρ and differentiate with respect to r: the result will be

$$\frac{\partial}{\partial r}\left(r^2 \frac{\partial R}{\partial r}\right) = \frac{\partial^2}{\partial t^2}\left\{\frac{\partial}{\partial r}(r^2 u)\right\} - \frac{\partial}{\partial r}\left\{\frac{r^2}{\rho}\left[\frac{\partial}{\partial r}(a^2 h + gf)\right]\right\} - \frac{\partial}{\partial t}\left\{\frac{\partial}{\partial r}\left(\frac{r^2 \mathfrak{F}}{\rho}\right)\right\},$$

which, inserted in (2.32) together with (2.31) discloses that (2.36)

$$\begin{aligned}\frac{\partial^2 y}{\partial t^2} = 4\pi G f &+ \frac{1}{r^2}\frac{\partial}{\partial r}\left\{\frac{r^2}{\rho}\frac{\partial}{\partial r}(a^2 h)\right\} + \frac{1}{r^2}\frac{\partial}{\partial r}\left\{\frac{r^2 gf}{\rho}\right\} - \\ &- \frac{j(j+1)}{r^2}\frac{a^2 h}{\rho} + \frac{1}{r^2}\frac{\partial}{\partial t}\left\{\frac{\partial}{\partial r}\left(\frac{r^2\mathfrak{F}}{\rho}\right) - j(j+1)\frac{\mathfrak{G}}{\rho}\right\},\end{aligned} \tag{2.37}$$

where, it may be remembered,

$$4\pi G\rho = \frac{1}{r^2}\frac{\partial}{\partial r}(r^2 g) \tag{2.38}$$

by (2.5).

If, eventually, we insert for $\mathfrak{F}$ and $\mathfrak{G}$ from (2.26) and (2.27) or (2.28) our fundamental Equations (2.33) and (2.37) for u and v will assume the more explicit forms

$$\begin{aligned}\frac{\partial^2}{\partial t^2}(rz) + a^2 A y = \frac{1}{\rho}\frac{\partial}{\partial t}\Bigg\{&\left[\rho\frac{\partial}{\partial t}\left(\frac{1}{\rho}\frac{\partial}{\partial r}\right) - \frac{j(j+1)}{r^2}\right](\mu_1 rz) + \\ &+ 2\left(y - \frac{\partial u}{\partial r}\right)\frac{\partial \mu_1}{\partial r} + \\ &+ 2\rho\frac{\partial}{\partial r}\left(\frac{u-v}{\rho}\frac{\partial \mu_1}{\partial r}\right) - \\ &- \left(\frac{1}{\rho}\frac{\partial \rho}{\partial r}\right)\left(\frac{4}{3}\mu_1 + \mu_2\right)y + \\ &+ \frac{L}{4\pi c^2}\left[\frac{y}{r^2} - \frac{\partial}{\partial r}\left(\frac{z}{r}\right) + \frac{1}{\rho}\frac{\partial \rho}{\partial r}\left(\frac{z}{r} + \frac{2u}{r^2}\right)\right]\Bigg\}\end{aligned} \tag{2.39}$$

and

$$\frac{\partial^2}{\partial t^2}(r^2 y) - 4\pi G r^2 f - \frac{\partial}{\partial r}\left\{\frac{r^2}{\rho}\left[\frac{\partial}{\partial r}(a^2 h) + gf\right] + \right.$$

$$\left. + j(j+1)\frac{a^2 h}{\rho}\right\} = \frac{\partial}{\partial t}\left\{\left[\frac{\partial}{\partial r}\left(\frac{r^2}{\rho}\frac{\partial}{\partial r}\right) - \frac{j(j+1)}{\rho}\right]\left(\frac{4}{3}\mu_1 + \mu_2\right) y - \right.$$

$$-2\frac{\partial}{\partial r}\left[\frac{r^2}{\rho}\left(y - \frac{\partial u}{\partial r}\right)\frac{\partial \mu_1}{\partial r}\right] -$$

$$-\frac{j(j+1)}{\rho}\left[2(u-v)\frac{\partial \mu_1}{\partial r} + \frac{1}{\rho}\frac{\partial \rho}{\partial r}(\mu_1 r z)\right] -$$

$$\left. -\frac{L}{4\pi c^2}\left[\frac{\partial}{\partial r}\left(\frac{1}{\rho}\left[y + 2\frac{\partial u}{\partial r} - \frac{4u}{r}\right]\right) - \frac{j(j+1)}{\rho r^2}(rz + 2u)\right)\right]\right\}. \tag{2.40}$$

The foregoing system (2.39)–(2.40) of two simultaneous partial differential equations for u and v is evidently one of fourth order in t, and sixth order in r. In the absence of viscosity, the right-hand sides of both (2.39) and (2.40) would vanish identically (except for the 'braking' terms due to interaction between matter and radiation); and their left-hand sides equated to zero would constitute equations which would continue to be of second order in t. However, Equation (2.39) so truncated would reduce to a first-order differential equation with respect to r; while (2.40) would remain of third order in the radial variable. The emergence of viscosity would raise the order of (2.40) from one to three in r even if the coefficients $\mu_{1,2}$ were constant; but the order of (2.40) would remain unaltered.

A further elimination of u and v between (2.39) and (2.40) becomes impracticable. In order to proceed with their solution let us, in what follows, confine our attention to motions – governed by (2.39) and (2.40) – which are *harmonic* in time. If so, it should be legitimate to replace the time-operators in (2.39) and (2.40) by

$$\frac{\partial}{\partial t} \equiv i\sigma \quad \text{and} \quad \frac{\partial^2}{\partial t^2} \equiv -\sigma^2, \tag{2.41}$$

where the constants σ represent frequencies that can be real or complex.

An insertion of (2.41) and (2.39) and (2.40) converts the latter at once form partial to ordinary differential equations, with r as the sole independent variable. If, moreover, the dissipative terms on the right-hand sides of these equations were negligible (i.e., $\mathfrak{F} = \mathfrak{G} = 0$), it is possible (cf. Lebovitz, 1965) to replace the latter by an equivalent system of four first-order differential equations of the form

$$\begin{Bmatrix} \dfrac{\partial x_1}{\partial r} \\[2ex] \dfrac{\partial x_2}{\partial r} \\[2ex] \dfrac{\partial x_3}{\partial r} \\[2ex] \dfrac{\partial x_4}{\partial r} \end{Bmatrix} = \begin{Bmatrix} A - \dfrac{1}{r} & \dfrac{j(j+1)}{r} - \dfrac{\sigma^2 r}{a^2} & -\dfrac{\rho r}{a^2} & 0 \\[2ex] \dfrac{gA}{\sigma^2 r} & -\dfrac{g}{a^2} & -\dfrac{\rho A}{\sigma^2} & 0 \\[2ex] 0 & 0 & 0 & \dfrac{1}{r} \\[2ex] 4\pi G A & -\dfrac{4\pi G r \sigma^2}{a^2} & -\dfrac{4\pi G \rho r}{a^2} + \dfrac{j(j+1)}{r} & -\dfrac{1}{r} \end{Bmatrix} \begin{Bmatrix} x_1 \\[2ex] x_2 \\[2ex] x_3 \\[2ex] x_4 \end{Bmatrix} \tag{2.42}$$

in terms of the dependent variables

$$\rho r u \equiv x_1, \quad \rho r v \equiv x_2, \quad R \equiv x_3, \quad r\frac{\partial R}{\partial r} \equiv x_4 . \tag{2.43}$$

In particular, the first equation of the foregoing matrix (for $\partial x_1/\partial r$) is identical with Equation (2.31); and the second (for $\partial x_2/\partial r$), with (2.34) of the inviscid case, if account is taken of the fact that, by virtue of Equation (2.31) Equation (2.19) can be written as

$$a^2 y = gu - \sigma^2 r v - R. \tag{2.44}$$

The third one of our matrix equations (2.43) is an obvious identity; while the fourth becomes identical with (2.33) in which we set

$$f \equiv \rho\left\{y + \frac{u}{\rho}\frac{\partial \rho}{\partial r}\right\} = A\rho u - \frac{\sigma^2 \rho r}{a^2} v - \frac{\rho}{a^2} R . \tag{2.45}$$

In the case of purely *radial* motion – when $j = 0$ and $v = 0$ – the first and second of our first-order matrix equations (2.43) for x_1 and x_2 do not hold, as the former was derived from (2.31) now nonexistent, and must be replaced by direct use of (2.30) which is of second order in u. The derivative $\partial R/\partial r$ occurring in this latter equation can be solved for from the last one of the matrix equations (2.43), of the form

$$\frac{\partial}{\partial r}\left(r\frac{\partial R}{\partial r}\right) = 4\pi G A(\rho r u) - \frac{4\pi G \rho r}{a^2} R - \frac{\partial R}{\partial r}, \tag{2.46}$$

i.e.,

$$\frac{\partial}{\partial r}\left(r\frac{\partial R}{\partial r}\right) = 4\pi G \rho r^2\left\{Au - \frac{R}{a^2}\right\} = 4\pi G\frac{\partial}{\partial r}(\rho r^2 u) , \tag{2.47}$$

which readily integrates into

$$\frac{\partial R}{\partial r} = 4\pi G\rho u \; ; \tag{2.48}$$

and on insertion in (2.30) furnishes the well-known equation for radial pulsations of second order.

For nonradial oscillations ($j \neq 0$) the matrix equations (2.43) constitute a simultaneous system of fourth order, first derived (though not in this form) by Pekeris (1938). The validity of these equations implies no restriction on the properties of the equilibrium model to which they are applicable. However, if the density concentration in such a model is high enough to render R an ignorable quantity (and a fuller discussion of the conditions under which this may be legitimate will be given in the next part of this section), the third and fourth equation of the system (2.43) vanish identically, while the first and second reduce to a system of the following two first-order differential equations

$$\frac{\partial}{\partial r}(\rho r u) = (rA - 1)\,\rho u + \{j(j+1) - \sigma^2 r^2 a^{-2}\}\,\rho v \, , \tag{2.49}$$

$$\frac{\partial}{\partial r}(\rho r v) = (g\sigma^{-2}A + 1)\,\rho u - g\rho r a^{-2} v, \tag{2.50}$$

(in a somewhat different form) considered first by Cowling (1941).

B. BOUNDARY CONDITIONS

Before we proceed, in the next section, to construct the actual solutions of these and other equations, it remains for us still to specify the *boundary conditions* which such solutions will be called upon to satisfy. Since our system (2.40)–(2.41) has proved to be generally (i.e., for viscous fluid, and no restriction on internal structure) one of sixth order in r, six boundary conditions are necessary for the complete specification of any desired particular solution; and in what follows we shall enumerate these in turn.

First, the obvious requirement that there be no displacement at the centre necessitates that

$$u(0,t) = v(0,t) = 0 \tag{2.51}$$

at all times. Next, we shall require that there be no variation in pressure over the free surface of a distorted configuration – a condition which is automatically satisfied by (2.23) provided that $u(r_1, t)$ remains bounded at $r = r_1$.

The fourth condition (and sufficient for a complete specification in the case of inviscid fluid) requires that the potential Ψ and its normal derivative (i.e., acceleration) be continuous across the boundary of radius r_1. In order to establish the explicit form of this condition, let us return to the second-order Equation (2.32) for R and set out to construct its complete primitive. The two linearly-independent solutions which would satisfy (2.32) if $f = 0$ are obviously $C_j r^j$ and $D_j r^{-j-1}$, with C_j and D_j denoting integration constants; and the sum $C_j r^j + D_j r^{-j-1}$ represents the 'complementary function' of Equation (2.32). The complete primitive of (2.32) is then obtained if we augment this complementary function by the 'particular integral' arising from a presence of the term $4\pi Gf$ on the right-hand-side of (2.32); and can be shown (on insertion for f from (2.22)) to assume the form

$$R = C_j r^j + D_j r^{-j-1} + \frac{4\pi G}{2j+1}\left\{\frac{1}{r^{j+1}}\int_0^r \rho\left[\frac{\partial}{\partial r}(r^{j+2}u) - \right.\right.$$

$$\left.\left. - yr^{j+2}\right] dr + r^j \int_r^{r_1} \rho\left[\frac{\partial}{\partial r}(r^{1-j}u) - yr^{1-j}\right] dr\right\}, \tag{2.52}$$

where r_1 denotes the radius of our configuration.

In order to ensure the continuity of the potential and its radial derivative on the surface $r = r_1$, let us differentiate Equation (2.53) to form the expression

$$\frac{\partial R}{\partial r} + \frac{j+1}{r} R = (2j+1)\,C_j r^{j-1} + 4\pi G\rho u +$$
$$+ \frac{4(j+2)}{2j+1}\pi G r^j \int_r^{r_1} \rho\left[\frac{\partial}{\partial r}(r^{1-j}u) - yr^{1-j}\right] dr, \tag{2.53}$$

which at the boundary $r = r_1$, where $\rho(r_1) = 0$, reduces to

$$\left\{\frac{\partial R}{\partial r} + \frac{j+1}{r} R\right\}_{r_1} = (2j+1)\,C_j r_1^{j-1}, \tag{2.54}$$

representing the fourth required boundary condition of our problem.

Should, moreover, our configuration be allowed to oscillate freely in space – in the absence of any external field of force – then also $C_j = 0$; and Equation (2.54) reduces to a homogeneous requirement that

$$\left(\frac{\partial R}{\partial r}\right)_{r_1} = -\frac{j+1}{r_1} R_1 . \tag{2.55}$$

In the presence of an external field – such as represented, in close binary systems, by the attraction of a companion – $C_j \neq 0$; and oscillations so constrained represent then the 'dynamical tides', of which more will be said in the next section of this chapter.

A specification of the values of C_j in such a case is being postponed till section III-3C. For the present we wish to note that the four boundary conditions represented by Equations (2.51) and (2.54) or (2.55) together with the requirement that $u(r_1, t)$ and $v(r_1, t)$ be bounded on the surface are sufficient for a complete specification of the desired particular solutions of our systems of linearized equations regardless of whether oscillations represented by them are free or forced.

This is, however, the case only if the configuration in question were to consist of an inviscid fluid. We have seen, however, before that the introduction of viscosity on the right-hand side of Equation (2.40) has raised the order of our differential system (2.40)–(2.41) from four to six – a fact which necessitates an imposition of two additional boundary conditions for a complete specification of our problem; and these are obtained from the radial viscous stress components

$$\sigma_{rr} = \sigma_{r\theta} = \sigma_{r\phi} = 0 \tag{2.56}$$

at $r = r_1$.

The explicit forms of all six components σ_{ij} of viscous stresses in spherical polar coordinates have already been given by Equations (1.38)–(1.42). If we particularize these now for the spheroidal velocity components U, V, W as given by Equations (2.15)–(2.17) we find that

$$\sigma_{rr} = \mu \frac{\partial U}{\partial r} = \mu\left(\frac{\partial u}{\partial r}\right) Y_j^i, \tag{2.57}$$

$$\sigma_{r\theta} = \mu\left\{\frac{1}{r}\frac{\partial U}{\partial \theta} - \frac{\partial V}{\partial r} - \frac{V}{r}\right\} = \mu\left\{\frac{\partial v}{\partial r} + \frac{u-v}{r}\right\}\frac{\partial Y}{\partial \theta}, \tag{2.58}$$

and

$$\begin{aligned}\sigma_{r\phi} &= \mu\left\{\frac{1}{r\sin\theta}\frac{\partial U}{\partial \phi} + \frac{\partial W}{\partial r} - \frac{W}{r}\right\} = \\ &= \mu\left\{\frac{\partial v}{\partial r} + \frac{u-v}{r}\right\}\frac{1}{\sin\theta}\frac{\partial Y}{\partial \phi}.\end{aligned} \tag{2.59}$$

In consequence, all three radial components of viscous stresses will vanish on the surface provided that

$$\frac{\partial u}{\partial r} = 0 \quad \text{and} \quad \frac{\partial v}{\partial r} = \frac{v-u}{r} \tag{2.60}$$

for $r = r_1$. These two conditions are to be added to those previously established to render the specification of our problem complete; and in the next section we shall turn to put them to practical use.

C. HOMOGENEOUS MODEL

Before we do so, however, we wish to illustrate the use of equations derived in this section by applying them to free oscillations of a particular model consisting of *inviscid, homogeneous but compressible* fluid – not because such a model could readily be applied to any situation of astrophysical significance, but because – in such a case – our problem admits of a solution in a closed form.

In order to do so, let ρ_0 denote the constant density of the respective configuration of radius r_1; and let τ, x stand for the normalized time and radial distance, defined by the equations

$$\tau = (2\pi G\rho_0)^{1/2}\tau \quad \text{and} \quad r = r_1 x \tag{2.61}$$

such that

$$0 \leqslant x \leqslant 1 \tag{2.62}$$

between the centre and the surface of our configuration. If so, then for constant ρ_0 Equation (2.4) can be readily integrated to

$$P_0 = \frac{2}{3}\pi G\rho_0^2 r_1^2(1-x^2), \tag{2.63}$$

while (2.5) yields

$$g = \frac{4}{3}\pi G\rho_0 r_1 x \tag{2.64}$$

and from (2.14) combined with (2.63)

$$a_0^2 = \frac{2}{3}\pi G\rho_0 r_1 \gamma(1-x^2). \tag{2.65}$$

Moreover, in accordance with (2.34),

$$A = \frac{2x}{\gamma r_1(1-x^2)}, \tag{2.66}$$

so that (2.35) assumes the form

$$a_0^2 A = \frac{4}{3}\pi G\rho_0 r_1 x \tag{2.67}$$

and, lastly, by (2.22) and (2.23),

$$gf = \frac{4}{3}\pi G\rho_0^2 x\tilde{y} \tag{2.68}$$

and

$$a_0^2 h = \frac{2}{3}\pi G\rho_0^2 r_1\,\{\gamma(1-x^2)\,\tilde{y} - 2xu\}, \tag{2.69}$$

where we have abbreviated

$$\tilde{y} = r_1 y \qquad \text{and} \qquad \tilde{z} = r_1 z. \tag{2.70}$$

If so, the system (2.39)–(2.40) of our fundamental equations of motion can, for constant μ, be rewritten as

$$\frac{3}{2}\frac{\partial^2\tilde{y}}{\partial\tau^2} = 3\tilde{y} + \frac{1}{x^2}\frac{\partial}{\partial x}(x^3\tilde{y}) + \left[\frac{\partial^2}{\partial x^2} + \frac{2}{x}\frac{\partial}{\partial x} - \frac{j(j+1)}{x^2}\right]\left[\frac{\gamma}{2}(1-x^2)\,\tilde{y} - xu\right]$$
$$+\,\tilde{\mu}\frac{\partial}{\partial\tau}\left[\frac{\partial^2}{\partial x^2} + \frac{2}{x}\frac{\partial}{\partial x} - \frac{j(j+1)}{x^2}\right]\tilde{y}, \tag{2.71}$$

and

$$\frac{3}{2}\frac{\partial^2 z}{\partial\tau^2} + \tilde{y} = \frac{3}{4}\tilde{\mu}\frac{\partial}{\partial\tau}\left[\frac{\partial^2}{\partial x^2} + \frac{2}{x}\frac{\partial}{\partial x} - \frac{j(j+1)}{x^2}\right]\tilde{z}, \tag{2.72}$$

where

$$\tilde{\mu} = \frac{2\mu}{\rho_0 r_1^2\sqrt{2\pi G\rho_0}} \tag{2.73}$$

stands for a nondimensional constant parameter proportional to the viscosity.

Let us disregard at first the terms factored by μ in the foregoing Equations (2.71)–(2.72) and – anticipating the motion governed by them to be harmonic – set

$$\frac{\partial^2}{\partial \tau^2} \equiv -\tilde{\nu}^2, \tag{2.74}$$

where $\tilde{\nu}$ stands for the (normalized) frequency of the respective motion. If so, the system (2.71)–(2.72) obviously reduces to

$$\left\{\frac{\partial^2}{\partial x^2} + \frac{2}{x}\frac{\partial}{\partial x} - \frac{j(j+1)}{x^2}\right\}\left[\frac{\gamma}{2}(1-x^2)\tilde{y} - xu\right] + $$

$$+ x\frac{\partial \tilde{y}}{\partial x} + (\tilde{\nu}^2 + 6)\tilde{y} = 0 \tag{2.75}$$

and

$$\tilde{y} = \tilde{\nu}^2 \tilde{z}, \tag{2.76}$$

respectively.

The foregoing Equations (2.75) and (2.76) constitute a fourth-order simultaneous system for the velocity components u and v. However, in this particular case of a homogeneous configuration, their integration can be split up in two stages. First, let us note that, by virtue of Equations (2.19)–(2.20) and (2.76).

$$\left\{\frac{\partial^2}{\partial x^2} + \frac{2}{x}\frac{\partial}{\partial x} - \frac{j(j+1)}{x^2}\right\}(xu) = \frac{1}{x}\frac{\partial}{\partial x}(x^2\tilde{y}) + \frac{j(j+1)}{\tilde{\nu}^2}\tilde{y}, \tag{2.77}$$

which on insertion in (2.75) yields

$$x^2(1-x^2)\frac{\partial^2 \tilde{y}}{\partial x^2} + 2x(1-3x^2)\frac{\partial \tilde{y}}{\partial x} + [Kx^2 - j(j+1)]\tilde{y} = 0 \tag{2.78}$$

where we have abbreviated

$$K = \frac{2}{\gamma}\left\{\tilde{\nu}^2 + 4 - \frac{j(j+1)}{\tilde{\nu}^2}\right\} + (j-2)(j+3). \tag{2.79}$$

The foregoing relation (2.77) constitutes a second-order differential equation for $\tilde{y}$ which can be integrated as it stands; and once its solution has been obtained, the velocity components u and v can be solved for from the equations

$$\tilde{y} = \frac{\partial u}{\partial x} + \frac{2u}{x} - j(j+1)\frac{v}{x} = \tilde{\nu}^2\left\{\frac{\partial v}{\partial x} + \frac{v-u}{x}\right\} \tag{2.80}$$

resulting from (2.70) and (2.76).

Let us, however, return now to Equation (2.78). If we substitute $x^2 \equiv \xi$, the latter can be rewritten as

$$\xi^2(\xi - 1)\frac{\partial^2 \tilde{y}}{\partial \xi^2} + \frac{7\xi - 3}{2}\xi\frac{\partial \tilde{y}}{\partial \xi} + \frac{1}{4}\{j(j+1) - K\xi\}\tilde{y} = 0, \tag{2.81}$$

and its complete primitive can be expressed as a linear combination of two hypergeometric series of the form

$$\tilde{y} = Ax^j F(a, b, c, x^2) + Bx^{-j-1} F(a - c + 1, b - c + 1, 2 - c, x^2), \tag{2.82}$$

where A, B are arbitrary integration constants, and

$$\left.\begin{aligned} a &= \tfrac{1}{4}[2j + 5 \pm \sqrt{25 + 4K}], \\ b &= \tfrac{1}{4}[2j + 5 \mp \sqrt{25 + 4K}], \\ c &= j + \frac{3}{2} \quad . \end{aligned}\right\} \tag{2.83}$$

The finiteness of $\tilde{y}$ as given by Equation (2.82) at the origin obviously requires that $B = 0$. Moreover, inasmuch as, by (2.83),

$$a + b - c = 1, \tag{2.84}$$

an application of standard tests for the convergence of hypergeometric series discloses that both series on the right-hand side of (2.82) diverge for $x = 1$. The solution (2.83) for $\tilde{y}$ expressed in their terms can, therefore, remain finite at the boundary only if the respective hypergeometric series are made to *terminate* by setting (say)

$$\tfrac{1}{4}\{2j + 5 \mp \sqrt{25 + 4K}\} = -k, \tag{2.85}$$

where k stands for an arbitrary positive integer. Since, however, then

$$a = j + k + \frac{5}{2} \qquad \text{and} \qquad b = -k, \tag{2.86}$$

the particular solution of (2.81) which remain finite for $0 \leqslant x \leqslant 1$ assumes the form

$$\begin{aligned} \tilde{y} &= A_{j,k} x^j F(j + k + \frac{5}{2}, -k, j + \frac{3}{2}, x^2) \\ &= A_{j,k} x^j G_k (j + \frac{5}{2}, j + \frac{3}{2}, x^2), \end{aligned} \tag{2.87}$$

where G_k denotes the corresponding Jacobi polynomial of degree k.

Moreover, Equation (2.85) implies that

$$(2j + 4k + 5)^2 = 25 + 4K, \tag{2.88}$$

which on insertion for K from (2.79) and after some rearrangement of terms assumes the form

$$\tilde{\nu}^2 + 4 - \frac{j(j+1)}{\tilde{\nu}^2} = \gamma(k + 1)(2j + 2k + 3); \tag{2.89}$$

and the frequency of the corresponding oscillation will be given by

$$\tilde{\nu}^2 = \omega \pm \sqrt{\omega^2 + j(j + 1)}, \tag{2.90}$$

where we have abbreviated

$$\omega = \gamma(k+1)(j+k+\frac{3}{2}) - 2. \tag{2.91}$$

For $j > 0$, one of the conjugate roots of (2.89) is bound to be negative – implying dynamical instability. If ω (i.e., k) is large, the conjugate roots (2.90) will be led by the terms

$$\tilde{\nu}^2 = 2\omega + \frac{j(j+1)}{2\omega} + \ldots \quad \text{or} \quad -\frac{j(j+1)}{2\omega} + \ldots . \tag{2.92}$$

Thus, for any given j, the requirement that $\tilde{y}$ be finite throughout the interval $0 \leqslant x \leqslant 1$ leads to two types of the spectra: one consisting of positive eigenvalues tending towards infinity as k increases; the other of negative eigenvalues tending towards zero.

With the explicit form of y as given by (2.87), the Equations (2.80) für u and ν assume the explicit form

$$\begin{aligned} \frac{1}{x^2}\left\{\frac{\partial}{\partial x}(x^2 u) - j(j+1)x\nu\right\} &= \frac{\tilde{\nu}^2}{x}\left\{\frac{\partial}{\partial x}(x\nu) - u\right\} \\ &= A_{j,k}x^j F(a,b,c,x^2), \end{aligned} \tag{2.93}$$

where, as before,

$$\left.\begin{aligned} a &= j + \frac{5}{2} + k, \\ b &= \quad -k, \\ c &= j + \frac{3}{2}; \end{aligned}\right\} \tag{2.94}$$

and can be integrated to furnish polynomial solutions for u and ν in the following manner.

First, we note that an elimination of $x\nu$ between the two parts of the Equation (2.93) yields a relation of the form

$$\frac{\partial^2}{\partial x^2}(x^2 u) - j(j+1)u = A_{j,k}\left\{\frac{\partial}{\partial x}(x^{j+2}F) + \frac{j(j+1)}{\tilde{\nu}^2}\, x^{j+1}F\right\}, \tag{2.95}$$

with $F \equiv F(a, b, c, x^2)$; and its particular integral which remains finite at the origin becomes

$$\begin{aligned} u_{j,k}(x) = \frac{A_{j,k}}{2j+1}\Bigg\{ & j\left[1 + \frac{j+1}{\tilde{\nu}^2}\right] x^{j-1}\int_0^x xF\mathrm{d}x \\ & + (j+1)\left[1 - \frac{j}{\tilde{\nu}^2}\right] x^{-j-2}\int_0^x x^{2(j+1)}F\mathrm{d}x\Bigg\}. \end{aligned} \tag{2.96}$$

Since, moreover,

$$xF(a,b,c,x^2) = \frac{c-1}{2(a-1)(b-1)}\,\frac{\partial}{\partial x}F(a-1,b-1,c-1,x^2) \tag{2.97}$$

and

$$(2j+3)x^{2(j+1)}F(a,b,c,x^2) = \frac{\partial}{\partial x}\left\{x^{2j+3}F(a,b,c+1,x^2)\right\} \tag{2.98}$$

the integrals on the right-hand side of (2.96) can be evaluated in a closed form to yield

$$u_{j,k}(x) = \frac{A_{j,k}x^{j-1}}{4(a-1)(b-1)\tilde{\nu}^2}\{(j+1)(j-\tilde{\nu}^2)F(a-1,b-1,c,x^2)$$

$$+ 2(c-1)\tilde{\nu}^2F(a-1,b-1,c-1,x^2) - j(j+\tilde{\nu}^2+1)\}, \tag{2.99}$$

where advantage has been taken of the identity

$$x^2F(a,b,c+1,x^2) = \frac{c(c-1)}{(a-1)(b-1)}\{F(a-1,b-1,c-1,x^2) -$$

$$-F(a-1,b-1,c,x^2)\}. \tag{2.100}$$

The arguments $b - 1 = -(k+1)$ in both hypergeometric series on the right-hand side of (2.99) are negative integers; hence, both series are terminating and can be expressed in terms of the Jacobi polynomials

$$F(a-1,b-1,c,x^2) \equiv G_{k+1}(j+\frac{1}{2}, j+\frac{3}{2}, x^2) \tag{2.101}$$

and

$$F(a-1,b-1,c-1,x^2) \equiv G_{k+1}(j+\frac{1}{2}, j+\frac{1}{2}, x^2) \tag{2.102}$$

of degrees $2(k+1)$ in x. Moreover, inasmuch as

$$(j+1)(j-\tilde{\nu}^2) + 2(c-1) = j(j+\tilde{\nu}^2+1), \tag{2.103}$$

the leading term of the polynomial expression (2.99) will be

$$u_{j,0}(x) = A_{j,0}\left\{\frac{j(j+1)+(j+2)\tilde{\nu}^2}{2(2j+3)\tilde{\nu}^2}\right\}x^{j+1}. \tag{2.104}$$

With a polynomial solution for u thus established, that for v follows algebraically from the equality of the first and third term of the Equation (2.93), disclosing that

$$j(j+1)v = \frac{1}{x}\frac{\partial}{\partial x}(x^2u) - x^{j+1}F(a,b,c,x^2). \tag{2.105}$$

Inserting in (2.105) for u form (2 99) we find that

$$v_{j,k}(x) = \frac{A_{j,k}x^{j-1}}{4(a-1)(b-1)\tilde{\nu}^2}\{2(c-1)F(a-1,b-1,c-1,x^2)$$

$$+ (\tilde{\nu}^2 - j)F(a-1,b-1,c,x^2) - (j+\tilde{\nu}^2+1)]\} \tag{2.106}$$

which for $k = 0$ reduces again to

$$v_{j,0}(x) = A_{j,0}\left\{\frac{j+\tilde{\nu}^2+3}{2(2j+3)\tilde{\nu}^2}\right\} x^{j+1}. \tag{2.107}$$

The foregoing Equations (2.99) and (2.106) represent closed polynomial solutions of Equations (2.75) and (2.76) for free non-radial oscillations of order j and mode k of a homogeneous configuration of compressible inviscid fluid, with characteristic frequencies $\tilde{\nu}$ as given by Equations (2.90). Such oscillations were previously investigated by Pekeris (1938), who noted their instability, and who was the first to express the divergence of motion in terms of a second-order differential equation (2.71). He failed, however, to notice (and take advantage of) the hypergeometric character of this equation; nor did he carry out the explicit solution far enough to specify the actual velocity components u and v – a task first performed by Kopal (1968).

It may be added that, for purely radial oscillations ($j = 0$), Equation (2.99) reduces to

$$u_{0,k}(x) = \frac{A_{0,k}}{2(k+1)(2k+3)x}\left\{G_{k+1}\left(\frac{1}{2}, \frac{1}{2}, x^2\right) - G_{k+1}\left(\frac{1}{2}, \frac{3}{2}, x^2\right)\right\}; \tag{2.108}$$

and Equation (2.89) for the frequency, to

$$\tilde{\nu}^2 = 2\left\{\gamma(k+1)\left(k+\frac{3}{2}\right) - 2\right\}. \tag{2.109}$$

The polynomial nature of (2.108) for a homogeneous compressible model was noted already by Sterne (1937); but its hypergeometric character was not recognized till by Kopal (1948b, 1950b).

In conclusion, let us point out the way in which the foregoing results simplify further if our homogeneous configuration becomes *incompressible*. Incompressibility implies the vanishing of the divergence Δ of the velocity vector as defined by (1.8) or (1.35). If so, however, then $y = 0$ by (2.21); and $z = 0$ by (2.76). Equations (2.19) and (2.20) can then be solved for u and v to yield

$$u = jv = a_1 x^{j-1}, \tag{2.110}$$

where a_1 is a constant.

For constant ρ_0, the vanishing of y implies – by (2.22) – that $f = 0$. On the other hand, incompressibility implies also that the ratio γ of specific heats of the respective fluid becomes infinite – but in such a way that the product $\gamma\tilde{y}$ remains a finite quantity, governed by the differential equation

$$\left\{(1-x^2)\frac{\partial^2}{\partial x^2} + 2\,\frac{1-3x^2}{x}\,\frac{\partial}{\partial x} + (j-2)(j+3) - \frac{j(j+1)}{x^2}\right\}(\gamma\tilde{y}) = 0, \tag{2.111}$$

which can be satisfied by

$$\gamma\tilde{y} = \frac{x^j}{1-x^2} \, . \tag{2.112}$$

Accordingly, by (2.23),

$$a_0^2 h = -\frac{2}{3}\pi G\rho_0^2 r_1 a_1 x^j \, . \tag{2.113}$$

Lastly, on insertion of $\tilde{y} = 0$ from (2.112) and of the expressions (2.110) for u and v on the right-hand side of Equation (2.52) in which we set (for free oscillations) $C_j = D_j = 0$, the particular integral of (2.32) assumes the form

$$R = \frac{4\pi G\rho_0}{2j+1} \, a_1 r_1 x^j . \tag{2.114}$$

An insertion of all the foregoing results in the inviscid ($\mathfrak{F} = 0$) form of Equation (2.30) discloses that

$$a_1 r_1 \tilde{\nu}^2 x^j = 2\pi G\rho_0 r_1 \left\{\frac{2}{2j+1} - \frac{1}{3}\right\} a_1 x^j \, ; \tag{2.115}$$

while Equation (2.31) for $\mathfrak{G} = 0$ yields

$$j a_1 r_1 \tilde{\nu}^2 x^{j-1} = 2\pi G\rho_0 r_1 \left\{\frac{2j}{2j+1} - \frac{j}{3}\right\} a_1 x^{j-1} . \tag{2.116}$$

Equation (2.116) proves merely to be a radial derivative of (2.115); and both are satisfied – for any x – by

$$\frac{\nu^2}{2\pi G\rho_0} \equiv \tilde{\nu}^2 = \frac{4j(j-1)}{3(2j+1)} \, ; \tag{2.117}$$

a result obtained first by Lord Kelvin (Thompson, 1863). It discloses that – unlike in the compressible case – an incompressible configuration is incapable of performing radial oscillations (corresponding to $j = 0$); its harmonic oscillations with finite periods are possible only for $j > 1$.

The emergence of effects arising from viscosity will not alter this general picture; but the relation (2.117) between the frequency $\tilde{\nu}$ and the mode of oscillations becomes very much more involved; for fuller details cf. secs. 7 and 8 of Kopal (1968a).

III-3. Forced Oscillations: Dynamical Tides

In the preceding sections of this chapter the general equations have been set up which govern small spheroidal oscillations of self-gravitating viscous fluids, and boundary conditions established to specify free oscillations of such configurations in the absence of any external field of force. Moreover, in the concluding part of section III-2 we particularized these equations for the case of inviscid homogeneous (though compressible), fluid, for which alone the respective equations can be solved in a closed form.

In spite of a mathematical interest which may attach to such a case, its astrophysical interest is severely limited by the fact that, for no known class of binary systems (except, possibly, those formed by a pair of white dwarfs, or neutron stars) can a homogeneous model offer any tolerable approximation. On the contrary, the actual density concentrations in the interiors of the components of most binary systems are likely to be so high that a much better approximation to their dynamical behaviour should be provided by a mass-point model – in which the entire mass of the star is confined to an infinitesimally small region at the centre, surrounded by an envelope which, though opaque, is so tenuous that its total mass can be neglected in comparison with that of the central mass-point. The gravitational potential of such a model reduces then to that of a point-mass; and, as such, becomes independent of the time. This latter fact will simplify the analysis of section III-2 far-reachingly; and the aim of the present section will be to take advantage of it.

A mass-point model will be shown (cf. Chapter VI) to approximate certain properties of close binaries – such as, for instance, the shape of its components – so well that refinements arising from finite density concentration are scarcely necessary for an interpretation of (say) their observed light (or radial velocity) changes as are available to us at the present time. However, close binary systems exhibit also other phenomena, of considerable interest, which do depend on the actual degree of internal density concentration of the actual components in a more sensitive manner. Such are the rates of apsidal motion in binary systems whose components revolve in eccentric orbits (cf. sec. V-3); or the periods of oscillations of the constituent stars which will be subject of the present section (cf. also sec. VI-3); or the rate of dissipation of their kinetic energy into internal heat through viscous friction – a process which can secularly alter the angular momenta of the components, and thus the periods of their orbits (cf. sec. VIII-4). The cumulative effects of such period changes are sensitively observable over long intervals of time (cf. sec. V-4); and for their interpretation a mass-model is obviously inadequate.

In order to provide a basis for appropriate studies of these phenomena, the aim of part A of the present section will be to particularize the general theory of spheroidal oscillations of the stars, as developed in sections III-1 and 2, to a mass-point model; while in part B we shall take up the case in which the density concentration – though high – remains finite. As we shall see, the mathematical process by which this can be done will enable us to retain many of the simplifying features of the mass-point model; but by not ignoring completely the self-attraction of opaque layers surrounding the massive core we shall lay down a more realistic basis for investigations of phenomena associated with tidal friction. In the concluding part C of this section we shall specify the explicit form of the coefficients which govern the amplitudes of the dynamical tides in close binary systems in terms of the elements of their orbits. Applications of these results to an investigation of the extent to which such tides produce an irreversible dissipation of the kinetic energy into heat will be postponed to the concluding section IV of this chapter.

A. MASS-POINT MODEL

In order to outline the way in which the theory developed in sections 1 and 2 of this chapter become simplified in the case when the gravitational potential of the distorted stars can be replaced by that of a mass point, let us return to Equation (2.53) fo the time-derivative R of the gravitational potential. The regularity of this expression at the centre requires that the constants D_j are all equal to zero; so that

$$R = \frac{4\pi G}{2j+1}\left\{\frac{1}{r^{j+1}}\int_0^r \rho\left[\frac{\partial}{\partial r}(r^{j+2}u) - yr^{j+2}\right]\mathrm{d}r\right.$$

$$\left. + r^j \int_0^{r_0}\left[\rho\frac{\partial}{\partial r}(r^{1-j}u) - yr^{1-j}\right]\right\}\mathrm{d}r + r^j\frac{\partial C_{k,j}}{\partial t}, \tag{3.1}$$

where the two integrals in the curly brackets represent the time-derivatives of the external (V) and internal (U) potential arising from self-attraction of our star; and the coefficients $C_{k,j}(t)$ are those of the harmonic expansion of the tide-generating potential

$$V_T = \sum_{j,k} C_{k,j}(t)\, r^j Y_j^k(\theta,\phi)\,. \tag{3.2}$$

The foregoing expression for R is (within the scheme of our linearized theory) exact for the configurations of any structure. However, in the limiting case of a mass-point model, both integrals on the right-hand side of (1) must clearly vanish*, leaving us with

$$R = r^j\frac{\partial C_{kj}}{\partial t}, \tag{3.3}$$

which in the absence of external forces (or in a constant field of force) reduces likewise to zero.

In the more general case, however, the fundamental equations of motion (2.30)–(2.31) fr the velocity components u, v can be restated as

$$\rho\frac{\partial^2 u}{\partial t^2} = \frac{\partial}{\partial r}(a^2 h) + gf + \frac{\partial}{\partial t}(\mathfrak{F} + j\rho r^{j-1}C_{k,j}) \tag{3.4}$$

$$\rho r\frac{\partial^2 v}{\partial t^2} = a^2 h + \frac{\partial}{\partial t}(\mathfrak{G} + \rho r^j C_{i,j})\,, \tag{3.5}$$

where, as before,

$$f = \rho y + u\frac{\partial p}{\partial r}, \tag{3.6}$$

$$a^2 h = \rho(a^2 y - gu), \tag{3.7}$$

in which, for a mass-point model, the gravitational acceleration

$$g = G\frac{m}{r^2}, \tag{3.8}$$

* Since the potential Gm/r of a mass-point is independent of the time.

where m denotes the total mass of the distorted configuration and G, the gravitational constant. Moreover, by Equations (2.26)–(2.28), the viscous terms $\mathfrak{F}$ and $\mathfrak{G}$ are defined by

$$\mathfrak{F} = 2\left\{\mu\frac{\partial y}{\partial r} + \frac{\partial \mu}{\partial r}\frac{\partial u}{\partial r}\right\} + \frac{j(j+1)}{r}\mu z - \frac{2}{3}\frac{\partial}{\partial r}(\mu y) \tag{3.9}$$

and

$$\mathfrak{G} = 2(u-v)\frac{\partial \mu}{\partial r} + \frac{\partial}{\partial r}(\mu r z) + \frac{4}{3}\mu y; \tag{3.10}$$

while the $C_{kj}(t)$'s are the coefficients of the expansion (3.2) of the disturbing potential V_T.

Let us, moreover, assume hereafter that this latter potential consists of a series of periodic functions of the time, of generally complex frequency σ. If so, it is legitimate to set in the foregoing equations

$$\frac{\partial}{\partial t} \equiv i\sigma, \quad \frac{\partial^2}{\partial t^2} \equiv -\sigma^2, \quad i \equiv \sqrt{-1}; \tag{3.11}$$

as a result of which Equations (3.4) and (3.5) can be written as

$$\frac{\partial}{\partial r}\{\rho(a^2 y - gu)\} + g\left\{\rho y + u\frac{\partial \rho}{\partial r}\right\} + \rho\sigma^2 u + i\sigma\,\mathfrak{F} + j\rho^{j-1}\dot{C}_{k,j} = 0, \tag{3.12}$$

$$a^2 y - gu + r\sigma^2 v + i\sigma\rho^{-1}\,\mathfrak{G} + r^j\dot{C}_{k,j} = 0, \tag{3.13}$$

where $\dot{C}_{k,j}$ stand for the time-derivatives of the respective coefficients; and where, it may be remembered,

$$y = \frac{\partial u}{\partial r} + \frac{2u}{r} - j(j+1)\frac{v}{r}. \tag{3.14}$$

Equations (3.12)–(3.14) constitute a simultaneous system for the (complex) velocity components u and v of spheroidal deformation of a self-gravitating configuration of compressible viscous fluid, built up in accordance with a mass-point model. The system (3.12)–(3.13) can be diminished further by the following strategy. An elimination of the system of Equations (2.39)–(2.40),valid for configurations of any structure); and non-homogeneous in the dependent variables unless $C_{k,j} = 0$. Moreover, the boundary conditions which the solutions of (3.12)–(3.14) must fulfil to represent spheroidal deformation of self-gravitating globes have already been investigated in the preceding section; and necessitate that, at the centre,

$$u(0,t) = v(0,t) = 0, \tag{3.15}$$

while, on the surface $r = r_1$,

$$\frac{\partial u}{\partial r} = 0 \quad \text{and} \quad \frac{\partial v}{\partial r} = \frac{v-u}{r}. \tag{3.16}$$

In the *inviscid* case $\mu = 0$, in which $\mathfrak{F} \equiv \mathfrak{G} \equiv 0$, the order of the system (3.4)–(3.5) or (3.12)–(3.13) can be diminished further by the following strategy. An elimination of the pressure-oscillations, as represented by the term a^2h, between Equations 3.4) and (3.5) leads to the simple relation

$$\frac{\partial^2}{\partial t^2}(rz) + a^2 Ay = 0, \tag{3.17}$$

where the functions y and z, as well as a^2A continue to defined by Equations (2.19)–(2.20) and (2.35). Inserting from them we find that, for harmonic motion, the above Equation (16) can be rewritten in terms of the velocity components u, v as

$$\frac{\partial}{\partial r}(rv) - u = \frac{a^2 A}{a^2}\left\{\frac{\partial u}{\partial r} + \frac{2u}{r} - j(j+1)\frac{v}{r}\right\}. \tag{3.18}$$

In order to eliminate v from this latter equation, let us note that an insertion of (3.7) in (3.5) reveals that

$$rv = \varphi\left\{\frac{\partial u}{\partial r} + \frac{2u}{r} - \frac{gu}{a^2} + \frac{r^j}{a^2}\right\}\dot{C}_{k,j} \tag{3.19}$$

where we have abbreviated

$$\frac{1}{\varphi} = \frac{j(j+1)}{r^2} - \frac{\sigma^2}{a^2}\ ; \tag{3.20}$$

and an elimination of v between (3.18) and (3.19) leads to a second-order differential equation for u of the form

$$\frac{\partial^2 u}{\partial r^2} + \left\{\frac{\partial}{\partial r}(\log r^2\rho\varphi\right\}\frac{\partial u}{\partial r} -$$

$$-\left\{\frac{2}{r^2} + \frac{1}{\varphi} + \frac{\partial}{\partial r}\left(\frac{g}{a^2}\right) - \frac{2A}{r} + \frac{j(j+1)g}{\sigma^2 r^2}A + \left(\frac{g}{a^2} - \frac{2}{r}\right)\left(\frac{1}{\varphi}\frac{\partial\varphi}{\partial r}\right)\right\}u$$

$$= -\frac{1}{a^2}\left\{\frac{\partial}{\partial r}\left(\log\frac{\varphi r^j}{a^2}\right) + \frac{j(j+1)a^2A}{\sigma^2 r^2}\right\}r^j\dot{C}_{k,j}, \tag{3.21}$$

where g continues to be given by Equation (3.8); and once (3.21) has been solved for u, the second velocity component v follows readily from (3.19).

If our configuration were incompressible – i.e., if

$$\Delta = y = 0 \quad \text{and} \quad a^2 = \infty, \tag{3.22}$$

but so that the product a^2y remains a finite quantity, it follows from (2.35) and (3.20) that

$$\frac{1}{\varphi} = \frac{j(j+1)}{r^2}, \qquad A = \frac{1}{\rho}\left(\frac{\partial\rho}{\partial r}\right), \tag{3.23}$$

in which case the second-order differential Equation (3.21) for u reduces to

$$\frac{\partial^2 u}{\partial r^2} + \left\{\frac{\partial}{\partial r} \log (\rho r^4)\right\} \frac{\partial u}{\partial r}$$

$$+\left\{\frac{2}{r^2} + \left[\frac{2}{r} - \frac{g}{\sigma^2 \varphi}\right] A - \frac{1}{\varphi}\right\} u + j(j+1) A r^{j-2} \dot{C}_{k,j} = 0, \tag{3.24}$$

while the v-component follows from (3.19) – or from (3.14) for $y = 0$ – to be given by

$$j(j+1) rv = \frac{\partial}{\partial r} (r^2 u). \tag{3.25}$$

Returning to the more general compressible case, suppose that we wish to obtain an explicit differential equation for the pressure-variations $a^2 h$ – i.e., to retain $a^2 h$ and u as the dependent variables of our problem, in place of u and v. This can be easily accomplished if we remember that – by Equations (2.22)–(2.23) and (2.35)–the adiabatic fluctuations in density and pressure produced by motion are related by the equation

$$f = h + u\rho A, \tag{3.26}$$

with the aid of which we can rewrite the Equation (3) as

$$\frac{\partial}{\partial r} (a^2 h) + \frac{g}{a^2} (a^2 h) + \frac{\rho}{r^2} \{\sigma^2 + gA\} (r^2 u) = -j\rho r^{j-1} \dot{C}_{k,j}. \tag{3.27}$$

On the other hand, a combination of (3.7) and (3.14) with (3.5) in the inviscid case ($\mathfrak{F} = \mathfrak{G} = 0$) permits us to express the outcome in the symmetrical form

$$\frac{\partial}{\partial r} (r^2 u) - \frac{1}{\varphi}\left\{\frac{r^2}{a^2} - \frac{j(j+1)}{\sigma^2}\right\} (a^2 h) = -j(j+1) r^j \dot{C}_{k,j}, \tag{3.28}$$

which together with (3.27) constitutes a simultaneous second-order system for $a^2 h$ and $r^2 u$. Its homogeneous form (obtaining in the absence of external forces, when $C_{k,j} = 0$) becomes identical with Equations (2.49)–(2.50).

If the matter constituting such configurations were incompressible (i.e., if $a^2 = \infty$). the foregoing Equations (3.27) and (3.28) would reduce to

$$\frac{\partial}{\partial r} (a^2 h) + \frac{\rho}{r^2} (\sigma^2 + gA) (r^2 u) + j r^{j-1} \dot{C}_{k,j} = 0. \tag{3.29}$$

and

$$\frac{\partial}{\partial r} (r^2 u) + \frac{j(j+1)}{\rho \sigma^2} (a^2 h) + j(j+1) r^j \dot{C}_{k,j} = 0, \tag{3.30}$$

where A continues to be given by (3.23). An elimination of $r^2 u$ between them leads to a second-order equation for $a^2 h$ of the form

$$\frac{\partial}{\partial r}\left\{\frac{r^2}{\rho(\sigma^2 + gA)} \frac{\partial}{\partial r} (a^2 h)\right\} - \frac{j(j+1)}{\rho \sigma^2} (a^2 h)$$

$$= \frac{1}{\sigma^2(\sigma^2 + gA)}\left\{(j+1) gA + \frac{\sigma^2 r}{\sigma^2 + gA} \frac{\partial}{\partial r} (gA)\right\} j r^j \dot{C}_{k,j}, \tag{3.31}$$

the homogeneous form of which (for $C_{k,j} = 0$) was given first by Rosseland (1932); while a similar elimination of $a^2 h$ between (3.29) and (3.30) reproduces (3.24).

B. CENTRALLY-CONDENSED MODEL

In order to investigate next the case in which the density concentration of our configuration – while high – is allowed to remain finite, let us return to Equation (3.1) defining the time derivative R of the changes in gravitational potential due to deformation, which in the case of a mass-point model could be replaced by (3.3). Suppose now that the internal density distribution $\rho(r)$ – though continuous in the entire interval $0 \leqslant r \leqslant r_1$ – is characterized by a steep peak in the neighbourhood of the origin, so that ρ in the interval $(0, r)$ is much larger than in the interval (r, r_1). If so, however, the second integral in the curly brackets on the right-hand side of Equation (3.1) should be much smaller than the first; and, consequently, the equation

$$R = \frac{4\pi G}{(2j+1)r^{j+1}} \int_0^r \rho \left[\frac{\partial}{\partial r}(r^{j+2}u) - yr^{j+2} \right] dr + r^j \frac{\partial C_{kj}}{\partial t} \tag{3.32}$$

should provide a closer approximation to the exact Equation (3.1) than (3.3). In other words, for configurations which exhibit strong internal concentration of density near the origin the forces acting upon any point derive predominantly from the mass interior to it; the potential of the exterior shell being relatively unimportant.

In such a case, a multiplication of both sides of Equation (3.32) and subsequent differentiation with respect to r discloses that, within the framework of our approximation

$$\frac{\partial R}{\partial t} + \frac{j+1}{r} R = \frac{4\pi G \rho r^{j+1}}{2j+1} \left\{ \frac{\partial}{\partial r}(r^{j+2}u) - yr^{j+2} \right\} + (2j+1)r^{j-1} \frac{\partial C_{kj}}{\partial t}, \tag{3.33}$$

where, by an insertion from (3.14),

$$\frac{\partial}{\partial r}(r^{j+2}u) - yr^{j+2} = jr^{j+1}[u + (j+1)v]. \tag{3.34}$$

The reader may note that, for $r = r_2$ such that $\rho(r_1) = 0$, the foregoing Equation (3.33) becomes identical with the boundary condition (2.54); in the present section we shall consider it to be approximately valid also through most part of the interior provided that the central condensation of the respective configuration is sufficiently high.

In order to take advantage of such a situation, let us depart from the exact equations

$$R = r \frac{\partial^2 v}{\partial t^2} - \frac{a^2 h}{\rho} - \frac{1}{\rho} \frac{\partial \mathfrak{G}}{\partial t}, \tag{3.35}$$

$$\frac{\partial R}{\partial r} = \frac{\partial^2 u}{\partial t^2} - \frac{1}{\rho} \frac{\partial}{\partial r}(a^2 h) - \frac{gf}{\rho} - \frac{1}{\rho} \frac{\partial \mathfrak{F}}{\partial t}, \tag{3.36}$$

of Section III-2A, and insert them in the left-hand side of (3.33); the result will assume the form

$$\left\{\frac{\partial^2}{\partial t^2} - \frac{4\pi Gj}{2j+1}\right\}[u + (j+1)v] - a^2 Ay - \left\{\frac{\partial}{\partial r} + \frac{j+1}{r}\right\}(a^2 y - gu)$$

$$= \frac{1}{\rho}\frac{\partial}{\partial t}\left\{\mathfrak{F} + \frac{j+1}{r}\mathfrak{G}\right\} + (2j+1)\, r^{j-1}\frac{\partial C_{kj}}{\partial t}, \tag{3.37}$$

where, by (3.9) and (3.10)

$$\mathfrak{F} + \frac{j+1}{r}\mathfrak{G} = \left\{\frac{\partial}{\partial r} + \frac{j+1}{r}\right\}\{(j+1)\mu z + \tfrac{4}{3}\mu y\} + \frac{2(j-1)}{r}[u + (j+1)v]\frac{\partial \mu}{\partial r}, \tag{3.38}$$

and where, for harmonic motion, the time-derivatives $\partial/\partial t$ and $\partial^2/\partial t^2$ can be replaced by σ by means of (3.11).

The foregoing Equation (3.37)) is one of the second order in u and v – in viscous as well as inviscid case. The second independent relation between these variables, resulting from an elimination of R between (3.35) and (3.36) by differentiation will, as before, be of the form

$$\frac{\partial^2}{\partial t^2}(rz) + a^2 Ay + \frac{1}{\rho}\frac{\partial}{\partial t}\left\{\mathfrak{F} - \rho\frac{\partial}{\partial r}\left(\frac{\mathfrak{G}}{\rho}\right)\right\} = 0, \tag{3.39}$$

where the explicit form of the viscous term can be found on the right-hand side of Equation (2.40).

Equation (3.39) above is of first order in u and v in the inviscid case, and of third order when the effects of viscosity are considered. Therefore the simultaneous system of (3.37) and (3.39) represents one of third order in the inviscid case, and of fifth order with the effects of viscosity included – i.e., in either case of one order more than for the mass-point model cònsidered in section 3A, but of one order less than the exact case discussed in Section 2.

C. DISTURBING POTENTIAL

Throughout this chapter devoted to an outline of the theory of dynamical tides in close binary systems we have described them so far only in terms of a set of the coefficients $C_{k,j}(t)$ factoring the non-homogeneous parts of the equations which govern their amplitudes. These coefficients must, however, derive from those of an expansion of the tide-generating potential V_T as given by Equation (3.2). The aim of this concluding part of the present section will, therefore, be to establish the explicit form of the connection between these two sets of the coefficients in terms of the elements of the respective system.

In order to do so, let in what follows m' represent the external tide-raising mass at a distance Δ from the centre of gravity of the distorted configuration, and let

$$\left.\begin{aligned} \lambda' &= \cos\phi\sin\theta\,, \\ \mu' &= \sin\phi\sin\theta\,, \\ \nu' &= \cos\theta\,, \end{aligned}\right\} \tag{3.40}$$

denote the direction cosines of an arbitrary radius-vector r in the rotating body axes; and

$$\left.\begin{aligned} \lambda'' &= \cos u \cos \Omega - \sin u \sin \Omega \cos I , \\ \mu'' &= \cos u \sin \Omega + \sin u \cos \Omega \cos I , \\ \nu'' &= \sin u \sin I , \end{aligned}\right\} \tag{3.41}$$

be the direction cosines, in the space axes, of the radius-vector Δ joining the centres of mass of the two bodies, where I denotes the inclination of the orbital plane of the disturbing body to the equatorial plane $x'y'$ of the distorted configuration; and u, the true longitude of the mass m' from the longitude Ω of the ascending node in which the equatorial and orbital planes intersect.

If so then, as is well known, the attractive force of mass m' on our configuration will derive from the disturbing potential

$$V_T = \frac{Gm'r^2}{\Delta^3} P_2(\cos \Theta) + \cdots , \tag{3.42}$$

where

$$\cos \Theta = \lambda'\lambda'' + \mu'\mu'' + \nu'\nu'' ; \tag{3.43}$$

so that, by the addition theorem for spherical harmonics,

$$P_j(\cos \Theta) = P_j(\nu') P_j(\nu'') + 2 \sum_{k=1}^{j} \frac{(j-k)!}{(j+k)!} P_j^k(\nu') P_j^k(\nu'') \cos k(\phi - p), \tag{3.44}$$

where

$$p = \Omega + \tan^{-1} (\cos I \tan u). \tag{3.45}$$

In consequence, by rewriting the circular functions involved on the right-hand side of (3.44) in terms of imaginary exponentials, we find that, for $j = 2$,

$$\begin{aligned} P_2(\cos \Theta) = &-\tfrac{1}{2} Y_2^0(\theta) \, \{P_2^0(q) + \tfrac{1}{4} P_2^2(q) \, e^{\pm 2iu} \} \\ &- \frac{i}{8} Y_2^1(\theta, \phi) \, \{2q e^{\pm i\Omega} - (1+q) \, e^{\pm i(\Omega + 2u)} + (1-q) \, e^{\pm i(\Omega - 2u)} \} \sqrt{1-q^2} \\ &+ \tfrac{1}{96} Y_2^2(\theta, \phi) \, \{2(1-q^2) \, e^{\pm 2i\Omega} + (1+q)^2 \, e^{\pm 2i(\Omega+u)} + (1-q)^2 \, e^{\pm 2i(\Omega-u)} \}, \end{aligned} \tag{3.46}$$

where

$$\left.\begin{aligned} Y_2^0(\theta) &= P_2^0(\nu) , \\ Y_2^1(\theta, \phi) &= e^{\mp i\phi} P_2^1(\nu) , \\ Y_2^2(\theta, \phi) &= e^{\mp 2i\phi} P_2^2(\nu) ; \end{aligned}\right\} \tag{3.47}$$

$$\left.\begin{aligned} P_2^0(\nu) &= \tfrac{1}{2}(3\nu^2 - 1), \\ P_2^1(\nu) &= -3\nu\sqrt{1-\nu^2}, \\ P_2^2(\nu) &= 3(1-\nu^2); \end{aligned}\right\} \tag{3.48}$$

$i \equiv \sqrt{-1}$ denotes the imaginary unit; and

$$q = \cos I. \tag{3.49}$$

It is, furthermore, obvious that expressions analogous to (3.46) can be constructed for any solid harmonic $P_j(\cos\Theta)$ that may occur on the right-hand side of Equation (3.42) of any order; but the requisite algebra becomes progessively more complicated and may be left as an exercise for he interested reader.

On insertion of (3.47) in (3.42) the latter can evidently be identified with an expansion of the form

$$V_T = \sum_{k=0}^{2} C_{k,2}(t) r^2 Y_2^k(\theta, \phi), \tag{3.50}$$

where

$$C_{0,2}(t) = -\frac{Gm'}{2\Delta^3}\{1 - \tfrac{3}{2}(1 - e^{\pm 2iu})\sin^2 I\}, \tag{3.51}$$

$$C_{1,2}(t) = -i\frac{Gm'}{4\Delta^3}\{[e^{\pm i\Omega} - e^{\pm i(\Omega+2u)}]\cos^2 \tfrac{1}{2}I + [e^{\pm i\Omega} - e^{\pm i(\Omega-2u)}]\sin^2 \tfrac{1}{2}I\}\sin I, \tag{3.52}$$

$$C_{2,2}(t) = \frac{Gm'}{24\Delta^3}\{e^{\pm 2i(\Omega+u)}\cos^4 \tfrac{1}{2}I + + 2e^{\pm 2i\Omega}\sin^2 \tfrac{1}{2}I\cos^2 \tfrac{1}{2}I + e^{\pm 2i(\Omega-u)}\sin^4 \tfrac{1}{2}I\}. \tag{3.53}$$

On the other hand, the reciprocal of the radius-vector Δ of the relative orbit of both components can be expressed as

$$\frac{1}{\Delta} = \frac{1 + e\cos\upsilon}{A(1-e^2)}, \tag{3.54}$$

where A denotes the semi-major axis of the elliptical orbit; e, its eccentricity; and υ, the true anomaly of the secondary (disturbing) component measured from the apsidal line in the plane of the orbit. In what follows we shall find it convenient to rewrite the right-hand side of (3.54) in terms of imaginary exponentials. In order to do so, let η denote an auxiliary quantity related with the orbital eccentricity e by means of the equation

$$e = \frac{2\eta}{1+\eta^2} \tag{3.55}$$

If so, Equation (3.54) can be obviously rewritten symmetrically as

$$\frac{A}{\Delta} = \frac{1+\eta^2}{(1-\eta^2)^2}(1+\eta e^{iv})(1+\eta e^{-iv}); \tag{3.56}$$

and, hence

$$\frac{A^3}{\Delta^3} = \left\{\frac{1+\eta^2}{(1-\eta^2)^2}\right\}\{(1+9\eta^2+9\eta^4+\eta^6)+3\eta(1+3\eta^2+\eta^4)\,e^{\pm iv} + 3\eta^2(1+\eta^2)\,e^{\pm 2iv}+\eta^3 e^{\pm 3iv}\} \tag{3.57}$$

where, as before, the expression $\exp(\pm ix)$ stands for the sum $\exp(ix) + \exp(-ix)$.

Since, moreover, the true longitudes u and v of m' are related by

$$u = v + \omega, \tag{3.58}$$

where ω denotes the longitude of the periastron (i.e., the angle between the nodal and apsidal line in the plane of the orbit), an insertion of (3.57), in (3.50)–(3.52) permits us to rewrite the latter as

$$C_{0,2} = \frac{Gm'}{4}\left\{\frac{1+\eta^2}{A(1-\eta^2)^2}\right\}^3 \{(1-3q^2)\,[E_0+\eta E_1 e^{\pm iv} + \eta^2 E_2 e^{\pm 2iv}+\eta^3 E_3 e^{\pm 3iv}] + 3(1-q^2)\times \times [E_0 e^{\pm 2iu}+\eta E_1 e^{\pm 2iu\pm iv}+\eta^2 E_2 e^{\pm 2iu\pm 2iv}+\eta^3 E_3 e^{\pm 2iu\pm 3iv}]\}; \tag{3.59}$$

$$iC_{1,2} = -\frac{Gm'}{16}\left\{\frac{1+\eta^2}{A(1-\eta^2)^2}\right\}^3\sqrt{1-q^2}\;\{(1+q)\,[E_0 e^{\pm i(\Omega+2u)} + \eta E_1 e^{\pm i(\Omega+2u)\pm iv}+\eta^2 E_2 e^{\pm i(\Omega+2u)+2iv}+\eta^3 E_3 e^{\pm i(\Omega+2u)+3iv}] - 2q\,[E_0 e^{\pm i\Omega}+\eta E_1 e^{\pm i\Omega\pm iv}+\eta^2 E_2 e^{\pm i\Omega\pm 2iv}+\eta^3 E_3 e^{\pm i\Omega\pm 3iv}] - (1-q)\,[E_0 e^{\pm i(\Omega-2u)}+\eta E_1 e^{\pm i(\Omega-2u)\pm iv} + \eta^2 E_2 e^{\pm i(\Omega-2u)\pm 2iv}+\eta^3 E_3 e^{\pm i(\Omega-2u)\pm 3iv}]\}, \tag{3.60}$$

and

$$C_{2,2} = \frac{Gm'}{96}\left\{\frac{1+\eta^2}{A(1-\eta^2)^2}\right\}^3\{(1+q)^2\,[E_0 e^{\pm 2i(\Omega+u)}+\eta E_1 e^{\pm 2i(\Omega+u)\pm iv} + \eta^2 E_2 e^{\pm 2i(\Omega+u)\pm 2iv}+\eta^3 E_3 e^{\pm 2i(\Omega+u)\pm 3iv}] + 2(1-q)^2\,[E_0 e^{\pm 2i\Omega}+\eta E_1 e^{\pm 2i\Omega\pm iv}+\eta^2 E_2 e^{\pm 2i\Omega\pm 2iv}+\eta^3 E_3 e^{\pm 2i\Omega\pm 3iv}] + (1-q)^2\,[E_0 e^{\pm 2i(\Omega-u)}+\eta E_1 e^{\pm 2i(\Omega-u)\pm iv} +$$

$$+ \eta^2 E_2 e^{\pm 2i(\Omega - u) \pm 2iv} + \eta^3 E_3 e^{\pm 2i(\Omega - u) \pm 3iv}] \}, \tag{3.61}$$

where u and v are related by (3.58) and where we have abbreviated

$$\left. \begin{aligned} E_0 &= 1 + 9\eta^2 + 9\eta^4 + \eta^6, \\ E_1 &= 3(1 + 3\eta^2 + \eta^4), \\ E_2 &= 3(1 + \eta^2), \\ E_3 &= 1. \end{aligned} \right\} \tag{3.62}$$

If the equator and the orbit were coplanar (i.e. if $q = 1$), the foregoing Equations (3.59)–(3.61) would reduce to

$$C_{0,2} = -\frac{Gm'}{2} \left\{ \frac{1 + \eta^2}{A(1 - \eta^2)^2} \right\}^3 \{ E_0 + \eta E_1 e^{\pm iv} + \eta^2 E_2 e^{\pm 2iv} + \eta^3 E_3 e^{\pm 3iv} \}, \tag{3.63}$$

$$C_{1,2} = 0, \tag{3.64}$$

$$C_{2,2} = \frac{Gm'}{24} \left\{ \frac{1 + \eta^2}{A(1 - \eta^2)^2} \right\}^3 \{ E_0 e^{\pm 2i(\Omega + u)} + \eta E_1 e^{\pm 2i(\Omega + u) \pm iv} + \eta^2 E_2 e^{\pm 2i(\Omega + u) \pm 2iv} + \eta^3 E_3 e^{\pm 2i(\Omega + u) \pm 3iv} \}; \tag{3.65}$$

and if, lastly, the relative orbit were to become circular (i.e., $\eta = 0$), these equations would reduce further to

$$C_{0,2} = -\frac{Gm'}{2A^3} \tag{3.66}$$

$$C_{1,2} = 0, \tag{3.67}$$

$$C_{1,2} = \frac{Gm'}{24A^3} e^{\pm 2i(\Omega + u)}. \tag{3.68}$$

In the general case, however, the $C_{i,2}$'s are multi-periodic functions of the time, with latent periods as given (with the aid of (3.58)), in the following Table II-1.

TABLE II-1

Latent periods

Coefficient	Argument of periodic terms	
	$i = 0$	
(1): $1-3\cos^2 I$:	kv	$k = 0, \pm 1, \pm 2, \pm 3.$
(2): $3\sin^2 I$:	$(kv + 2\omega)$	$k = 0, \pm 1, 2, 3, 4, 5$
	$i = 1$	
(3): $2\sin I \sin^2 \frac{1}{2}I$:	$(\Omega - kv - \omega)$,	$k = 0, \pm 1, 2, 3, 4, 5$
(4): $2\sin I \cos^2 \frac{1}{2}I$:	$(\Omega + kv + \omega)$,	$k = 0, \pm 1, 2, 3, 4, 5$
(5): $2\sin I \cos I$:	$(\Omega + kv)$,	$k = 0, \pm 1, \pm 2, \pm 3.$

	$i = 2$	
(6): $4\sin^4\frac{1}{2}I$:	$(2\Omega - kv - \omega)$	$k = 0, \pm 1, 2, 3, 4, 5$
(7): $4\cos^4\frac{1}{2}I$:	$(2\Omega + kv + \omega)$	$k = 0, \pm 1, 2, 3, 4, 5$
(8): $2\sin^2 I$:	$(2\Omega \pm kv)$	$k = 0, \pm 1, \pm 2, \pm 3.$

The foregoing tabulation contains, on the whole, 52 distinct and non-vanishing periodic arguments, all of which are present in the leading term of the expansion (41) for the disturbing potential V_T: 10 are associated with its zonal harmonic $i = 0, j = 2$; 21 of them with the tesseral harmonic $i = 1, j = 2$; and another 21 with the sectional harmonic $i = j = 2$. Each of these terms corresponds to a partial tide, the sum of which represents the total distortion. If the orbits were circular ($e = 0$), only those in (5) and (8) of the above tabulation characterized by $k = 0$, and those in (2), (3), (4), (6) and (7) characterized by $k = 2$ would survive. On the other hand, for coplanar orbits ($q = 1$) the only periodic terms are those in (1) and (7); of which only $2(\Omega \pm v)$ alone would survive if $e = 0$ and $q = 1$ at the same time.

The surface harmonics Y_j^k as introduced by Equations (3.47) are, in general (for $i > 0$), complex; and so are their coefficients $C_{k,j}$ on the right-hand side of the potential expansion (3.50). This complex formulation was of advantage in our analysis of the perturbing function for latent periodicities, and for its decomposition in individual harmonic terms. Needless to stress, however, the entire right-hand side of Equation (3.50) must be real; and the imaginary terms in the coefficients $C_{k,j}$ were introduced to offset those in (3.47). If, alternatively, we expand the first term of V_T as defined by Equations (3.42) –(3.45) in a series of the form

$$V_T = \sum_{k=0}^{2} \{C_2^k \cos k\phi + S_2^k \sin k\phi\} r^2 P_2^k(\cos\theta), \tag{3.69}$$

it follows (on re-arrangement of terms on the right-hand side of (3.50)) that

$$C_2^0 = \frac{Gm'}{4\Delta^3}\{(1 - 3q^2) - 3(1 - q^2)\cos 2u\}; \tag{3.70}$$

$$C_2^1 = -\frac{Gm'}{2\Delta^3}\{\cos\Omega\sin 2u - q\sin\Omega(1 - \cos 2u)\}\sqrt{1 - q^2}; \tag{3.71}$$

$$S_2^1 = -\frac{Gm'}{2\Delta^3}\{\sin\Omega\sin 2u + q\cos\Omega(1 - \cos 2u)\}\sqrt{1 - q^2}; \tag{3.72}$$

$$C_2^2 = \frac{Gm'}{24\Delta^3}\{(1 - q^2)\cos 2\Omega + (1 + q^2)\cos 2\Omega\cos 2u - 2q\sin 2\Omega\sin 2u\} \tag{3.73}$$

and

$$S_2^2 = \frac{Gm'}{24\Delta^3}\{(1 - q^2)\sin 2\Omega + (1 + q^2)\sin 2\Omega\cos 2u + 2q\cos 2\Omega\sin 2u\}, \tag{3.74}$$

all of which are real and differentiable periodic functions of the time. For coplanar case ($q = 1$) these reduce to

$$C_2^0 = -\frac{Gm'}{2\Delta^3}, \tag{3.75}$$

$$C_2^2 = \frac{Gm'}{12\Delta^3}\cos 2(\Omega + u), \tag{3.76}$$

$$S_2^2 = \frac{Gm'}{12\Delta}\sin 2(\Omega + u); \tag{3.77}$$

and in the case of circular orbits (i.e., for constant Δ) the first of these becomes independent of the time.

In principle, all three quantities ω, Ω and υ involved in the periodic arguments of Table II-1 are functions of the time. Thus, the longitude ω of periastron in close binary systems consisting of distorted components are known to undergo both secular and periodic perturbations (cf. section V-3); but its rate of secular advance as well as the amplitudes of its periodic perturbations are so small that the time-derivative $\dot{\omega}$ can hereafter be regarded as ignorable.* If so, however, the number of distinct periodic terms in Table II-1 corresponding to second-harmonic tides reduces at once from 52 to 35 (i.e., 7 for $i = 0$, and 14 for $i = 1$ and 2 each).

However, an interpretation of the meaning of the longitude Ω of the node in Equations (3.41) and the sequel requires some care. Inasmuch as the direction cosines (3.40) refer to the *body* axes x', y' rotating about the z-axis of the direction cosines (3.41) referred to the fixed *space* axes, it follows that if we set (without any loss of generality) $\theta = 0°$ and let the component under consideration rotate about the fixed $z \equiv z'$ axis with a uniform angular velocity ω_z,

$$\Omega = \mp \omega_z t; \tag{3.78}$$

the upper and lower sign referring to the case of direct and retrograde rotation, respectively. Moreover, in accordance with Kepler's second law of elliptic motion, the true anomaly υ in all preceding equations of this section can be expanded in a Fourier series of the time t in the form

$$\upsilon = nt + 2e\sin nt + \tfrac{5}{4}e^2 \sin 2nt + \cdots, \tag{3.79}$$

where n denotes the mean daily motion of the disturbing body.

In particular, if the relative orbit of the two components were circular (i.e., $\eta = 0$), the second-harmonic term $P_2(\cos\Theta)$ in V_T will alone give rise to the following seven pairs of frequencies in the disturbing function:

$$\dot{\Omega} = \mp \omega_z,$$

$$2\dot{\Omega} = \mp 2\omega_z,$$

$$2\dot{\upsilon} = \pm n,$$

* The same is also true of the time-derivatives of the Eulerian angles θ and ϕ in the tesseral harmonics $Y_j^k(\theta, \phi)$, which would depart from zero only on account of the precession and nutation of the components in space (cf. sec. V-2).

$$\dot{\Omega} \pm 2\dot{v} = \mp \omega_z \pm 2n,$$
$$2(\dot{\Omega} \pm \dot{v}) = 2(\mp \omega_z \pm n).$$

If the rotation is direct (i.e., occurs in the same direction as the orbital motion of the disturbing body), it follows from (3.78) that

$$\dot{\Omega} = -\omega_z; \tag{3.80}$$

moreover, should the rotation and revolution be synchronous,

$$\dot{\Omega} + n = n - \omega_z = 0. \tag{3.81}$$

In such a case all the foregoing frequencies would reduce to $\pm n$ *and* $\pm 2n$. On the other hand, if the orbit were eccentric but coplanar, the only remaining terms in the disturbing function would be characterized by the frequencies

$$i = 0: \quad k\dot{v}, \qquad k = 0, 1, 2, 3,$$
$$i = 2: \quad 2\dot{\Omega} + k\dot{v}, \qquad k = 0, \pm 1, 2, 3, 4, 5;$$

which for circular orbits would reduce to

$$i = 2: \quad 2(\dot{\Omega} + \dot{v}) = 2(\omega_z - n)$$

for direct rotation, vanishing if the rotation is synchronous. In such a case, all $C_{k,j}$'s cease to depend on the time altogether and, as a result, our configuration will no longer exhibit any *dynamical* tides. Time-independent distortions described by *constant* values of $C_{k,j}$ represent then the corresponding *equilibrium* tides causing permanent deformation of the body in question, but no motion of any of its mass elements in the rotating system of coordinates.

With a complete specification of all requisite tide-generating coefficients $C_{k,j}(t)$ our task of setting up the differential equations (together with their associated boundary conditions) which govern dynamical tides in close binary systems has thus been accomplished, and we are in a position to proceed to the construction of particular solutions of our equations of motion for the amplitudes $u(r, t)$ and $v(r, t)$ corresponding to each respective partial tide. Since some of the $C_{i,j}$'s proved to be imaginary – and so are the terms involving viscosity in our equations of motion – the solutions of these equations must be sought in terms of *complex* variables

$$u = u_1 + i u_2 \quad \text{and} \quad v = v_1 + i v_2 \tag{3.82}$$

where $u_{1,2}$ and $v_{1,2}$ are real functions of r and t.

If we substitute the forms (3.82) in Equations (3.37) and (3.39) for the centrally condensed model, or Equations (3.12)–(3.13) for the mass-point model, and equate the real and imaginary terms in these equations, each will split up in two and the order of the respective simultaneous system will be doubled – thus furnishing the exact number of equations necessary to specify the four function $u_{1,2}$ and $v_{1,2}$. Moreover, once these have been solved (numerically or otherwise), the real parts of the respective velocity components will be given by the *moduli*

$$|u| = \sqrt{u_1^2 + u_2^2} \quad \text{and} \quad |v| = \sqrt{v_1^2 + v_2^2} \tag{3.83}$$

of the complex functions (81); while their *phase-lags* will follow as the amplitudes of the same functions from the equations

$$u = |u| (\cos \epsilon_u + i \sin \epsilon_u) \tag{3.84}$$

and

$$v = |v| (\cos \epsilon_v + i \sin \epsilon_v) \tag{3.85}$$

as

$$\tan \epsilon_u = \frac{u_2}{u_1} \tag{3.86}$$

and

$$\tan \epsilon_v = \frac{v_2}{v_1}, \tag{3.87}$$

respectively.

Moreover, in the absence of more detailed integrations it is possible (cf. Kopal, 1968d) at least to estimate from the structure of our equations that

$$\tan \epsilon_u = \frac{u_2}{u_1} \sim \frac{\sigma\mu}{g\rho r^2 A}, \tag{3.88}$$

where $gr^2 = Gm$, and A denotes the adiabatic constant defined by Equation (2.35); with similar results for the ratio v_2/v_1.

The magnitude of the phase-lag that follows from the outcome of the present investigation appears in general to be very small. If we adopt $\sigma \sim 10^{-5}$ s^{-1}, $r \sim 10^{10}$ cm and $g \sim 10^3$ cm s^{-2} as typical frequency, scale-length, and gravitational acceleration in close binary systems, these would correspond by (3.86)–(3.87) to phase-lags of the order of

$$\epsilon_u \sim \epsilon_v \sim \frac{10^{-18}\mu}{\rho(rA)}, \tag{3.89}$$

a result revealing that the term $\mu/\rho rA$ would have to be of the order of 10^{16} to invoke phase-lags of the order of one degree of arc. The kinematic viscosity μ/ρ of hydrogen plasma can scarcely be expected to exceed $10^7 - 10^8$ cm^2 s^{-1} in those parts of stellar interiors where tidal phenomena are of importance – which means that the angles ϵ_u or ϵ_v should remain utterly negligible unless the adiabatic term A is extremely close to zero (corresponding to adiabatic equilibrium); or unless very much larger viscosity than that of hydrogen plasma is produced by turbulence – two phenomena which may, in fact, be coupled.

It is only if turbulent viscosity (characterized by Reynolds numbers of the order of 10^5 or greater) occurs in sub-surface layers under near-adiabatic conditions that tidal lags in close binary systems attain observable magnitudes. Therefore, *the significance of observable tide-lags in close binary systems is primarily that of an indicator of high degree of turbulence in sub-surface stellar layers.* As is well known, in early-type Main Sequence stars turbulent regions are relegated to their deep interiors, where tides are of very minor importance. Sub-surface turbulent viscosity sufficient to produce observationally significant tide-lags can, however, arise among late-type Main Sequence systems, or those with subgiant components which had evolved away from the Main Sequence. If and when eclipsing systems are identified whose observed characteristics (such as asymmetry of the light changes between minima) require the existence of tidal lag for their explanation, then the theory developed in this section should provide a bridge between the observed facts and the degree of internal viscosity necessary to account for them.

Further numerical applications of these results are being postponed to the next section. In conclusion of the present one we wish to stress that, throughout this section, we have not concerned ourselves with the effects of axial rotation on the shape of tidally-distorted components; nor with the effects of Coriolis forces produced by the revolution of our binary. This should indeed be legitimate as long as the axial rotation (which may, but need not, be synchronized with the Keplerian angular velocity of orbital revolution) is sufficiently slow for the polar flattening produced by it to represent a small quantity of first order. For, within the scheme of our approximation, both rotation and tides produce first-order effects which are independant of each other; and the interaction between them would (cf. sec. II-4) constitute second-order effects negligible within the framework of a linearized theory developed in this chapter.

III-4. Dissipation of Energy by Dynamical Tides

In the preceding sections of this chapter differential equations have been set up together with their associated boundary conditions, which govern the dynamical tides in close binary systems consisting of the components of any structure; and the properties of such tides have been related explicitly with the disturbing function which arises from mutual attraction as well as axial rotation with arbitrary angular velocity. The resulting time-dependent deformations have been found to consist of a considerable number of discrete partial tides possessing certain spherical-harmonic symmetry, each of which sweeps around the respective component with a constant amplitude, velocity, and phase. Their motions relative to the centre of gravity of a fluid mass of given viscosity is, however, bound to bring about also a dissipation of the kinetic energy of motion into heat through viscous friction – an irreversible process which, in turn, is bound to bring about *secular changes* in the internal structure of the components and the elements of their orbit.

The aim of the present section will be to investigate quantitatively the *rate* at which such dissipation is operative in close binary systems. This rate has already been specified in general terms by Equation (1.21) of this chapter*; and in spherical polar coordinates assumes the form represented by Equation (1.37) in terms of the stress-components $\sigma_{i,j}$ ($i, j = r, \theta, \phi$) as given by Equations (1.38)–(1.43), which for spherical deformations characterized by the velocity components (2.15)–(2.17) assume the particular forms

$$\sigma_{rr} = 2\,\frac{\partial u}{\partial r}\,Y, \tag{4.1}$$

$$\sigma_{\theta\theta} = \frac{2}{r}\left(v\,\frac{\partial^2 Y}{\partial\theta^2} + uY\right), \tag{4.2}$$

* For radiating systems Equation (1.21) should be augmented by terms arising from interaction between mass and radiation as given by Equation (1.81).

$$\sigma_{\phi\phi} = \frac{2}{r}\left(uY - j(j+1)\,vY - v\frac{\partial^2 Y}{\partial\theta^2}\right), \tag{4.3}$$

$$\sigma_{r\theta} = \left(\frac{\partial v}{\partial r} + \frac{u-v}{r}\right)\frac{\partial Y}{\partial\theta}\ , \tag{4.4}$$

$$\sigma_{r\phi} = \left(\frac{\partial v}{\partial r} + \frac{u-v}{r}\right)\frac{1}{\sin\theta}\frac{\partial Y}{\partial\phi}\ , \tag{4.5}$$

$$\sigma_{\theta\phi} = \frac{2v}{r\sin\theta}\frac{\partial}{\partial\phi}\left(\frac{\partial Y}{\partial\theta} - Y\cot\theta\right), \tag{4.6}$$

and

$$\Delta = \left\{\frac{\partial u}{\partial r} + \frac{2u}{r} - j(j+1)\frac{v}{r}\right\} Y \equiv yY \tag{4.7}$$

in accordance with (2.21), where $Y \equiv Y_j^i(\theta, \phi)$ are spherical harmonics satisfying Equation (2.18).

The total rate of dissipation of the energy E throughout the oscillating configuration is then given by the volume integral

$$\frac{\mathrm{d}E}{\mathrm{d}t} \equiv \dot{E} = \Phi \mathrm{d}V = \int_0^{a_1}\int_0^{\pi}\int_0^{2\pi} \Phi r^2 \mathrm{d}r \sin\theta\, \mathrm{d}\theta\, \mathrm{d}\phi, \tag{4.8}$$

where a_1 denotes the mean radius of the respective star. If we insert the foregoing stress-components (4.1) – (4.6) and (4.7) on the right-hand side of Equation (1.37) for Φ, the volume integral of Φ as given by Equation (1.37) can be evaluated term by term. Because of the fact that the dissipation function Φ is quadratic in the velocity components of viscous flow, this task proves to be one of considerable complexity. However, with the aid of auxiliary integrals of the squares and cross-products of the spherical harmonics Y_j^i established by Higgins and Kopal (1968) it is possible to perform the integrations with respect to the angular variables θ and ϕ in a closed form; and to express Equation (4.8) as

$$\dot{E} = \sum_{i,j} \dot{E}_{i,j}\ , \tag{4.9}$$

where, for $i = 0$,

$$\begin{aligned} E_{0,j} = \frac{4\pi}{2j+1}\int_0^{a_1} \mu\Bigg\{2\left(\frac{\partial u}{\partial r}\right)^2 &+ \left[\frac{2u}{r} - j(j+1)\frac{v}{r}\right]^2 \\ &+ j(j+1)\left[\frac{\partial v}{\partial r} + \frac{u-v}{r}\right]^2 \\ &+ (j-1)j(j+1)(j+2)\frac{v^2}{r^2} - \frac{2}{3}y^2\Bigg\} r^2 \mathrm{d}r; \end{aligned} \tag{4.10}$$

while, for $i > 0$,

$$\dot{E}_{i,j} = \frac{(j+i)!\,\dot{E}_{0,j}}{(j-i)!\;2}\,. \tag{4.11}$$

In any applications of this result it should be kept in mind that the functions $u(r, t)$ and $v(r, t)$ in the integrand on the right-hand side of Equation (4.10) are different for each value of j; and, moreover, that the expressions $E_{i,j}(t)$ represent the energy dissipation contributed, per unit time, by each single partial tide characterized by the i, j-th spherical harmonic symmetry. The total amount of energy dissipated by each tide per each orbital cycle of period P should then be given by

$$\dot{E}_{i,j} = \int_0^P \dot{E}_{i,j}(t)\,\mathrm{d}t.. \tag{4.12}$$

In order to estimate the rate of the anticipated dissipation of energy by the tides in actual binary systems on this basis, let us attempt to evaluate the radial integral (4.10) in so far as this can be done in general terms. The exact evaluation requires, of course, a prior knowledge of the functions $u(r, t)$ and $v(r, t)$ for the appropriate structure of the star. The only closed solutions which we were able to construct in section 2C pertained to a homogeneous model; and these led us to expect that the leading terms of both u and v will vary with the radius as r^{j+1}, with appropriate coefficients.

In what follows, two limiting cases will be considered: one corresponding to

$$\text{(A)}\qquad \begin{aligned} u &= j(j+1)\,c_{i,j}r^{j+1} \\ v &= (j+3)\,c_{i,j}r^{j+1} \end{aligned}$$

where $c_{i,j}$ are appropriate constants (or functions of the time) which specify the nature of the oscillation; the other to

$$\text{(B)}\qquad \begin{aligned} u &= (j+2)\,c_{i,j}r^{j+1}, \\ v &= c_{i,j}r^{j+1} \end{aligned}$$

The reader may note that these forms constitute the limiting cases of Equations (2.104) or (2.107) if, in these equations, the frequency $\tilde{\nu}$ is allowed to tend to (A) zero, or (B) infinity; moreover, an insertion of the above-adopted forms for u and v in Equations (2.19) or (2.20) reveals that, in these particular cases,

$$\text{(A)}: y = 0, \tag{4.13}$$

or

$$\text{(B)}: z = 0, \tag{4.14}$$

respectively. The former condition implies *poloidal* oscillations, and will always be fulfilled in the case of incompressibility; while the latter implies u, v to be *gradients* of the same functions.

Inserting successively (A) and (B) on the right-hand side of our present Equation (21) and performing the requisite integrations we find that, in the case (A),

$$\dot{E}_{0j} = \frac{[8\pi j(j+1)(2j+3)-3]}{2j+1}\;[j(j+2)(2j+1)-3]\,c^2_{0,j}\int_0^{a_1} \mu r^{2j+2}\mathrm{d}r; \tag{4.15}$$

while in the (B) case,

$$\dot{E}_{0,j} = \frac{8\pi j(2j+3)}{3(2j+1)} [6j^2 + 9j + 1] c_{0,j}^2 \int_0^{a_1} \mu r^{2j+2} dr; \tag{4.16}$$

and the corresponding $E_{i,j}$'s can then be obtained from (4.11).

In order to estimate the rate of the anticipated dissipation of energy on this basis, let us confine our attention to the effects produced by the principal second-harmonic tides (corresponding to $j = 2$). From the discussion of section 3C of this chapter we deduce that the most important zonal-harmonic tide arises from the eccentricity of the relative orbit, with the height of the tide varying ('breathing') in inverse proportion of the cube of the instantaneous radius-vector. If, as before, Δ denotes the radius-vector of the relative orbit; e, its eccentricity; m'/m, the ratio of the masses of the disturbing (m') and the disturbed (m) component; n, their mean daily motion; and a_1, the mean radius of the distorted star, the most important partial tide corresponding to $j = 0, j = 2$ will be factored by the cofficient

$$c_{0,2} = \frac{1}{2}\left(\frac{m'}{m}\right)\frac{na_1}{\Delta^3} e \sin nt, \tag{4.17}$$

where t denotes the time.

Partial tides of the type $i = 1, j = 2$ arise in connection with a finite inclination of the plane of the equator of a rotating distorted star to that of the orbit. If ω denotes the angular velocity of axial rotation of the component of mass m, and I, the angle between the equator and the orbit, this latter inclination will give rise to a tesseral-harmonic tide led by the coefficient

$$c_{1,2} = \frac{1}{2}\left(\frac{m'}{m}\right)\frac{\sin I}{\Delta^3} \{[\omega a_1] \sin \omega t + [(2n \mp \omega)a_1] \sin (2u \mp \omega t)\}, \tag{4.18}$$

where u stands for the true longitude of the component of mass m' in its relative orbit, measured from the line of the nodes; and the $\mp$ sign of ω corresponds to the case of direct ($-$) or retrograde ($+$) rotation. The first time-dependent term in curly brackets on the right-hand side of Equation (4.18) represents the periodic motion of the crest of the respective tide in (astrocentric) latitude; the second, in longitude. Should the axial rotation be synchronized with the revolution (i.e. $\omega = n$) and $u = nt$, the right-hand side of (4.18) becomes independent of the longitude and both terms in curly brackets are identical.

The most important tide associated with the sectorial harmonic $i = j = 2$ is characterized by the coefficient

$$c_{2,2} = \frac{1}{6}\left(\frac{m'}{m}\right)\frac{(n \mp \omega) a_1}{\Delta^3} \cos 2(n \mp \omega)\, t, \tag{4.19}$$

and represents a wave sweeping around the equator of the star of mass m with an angular velocity $n \mp \omega$, equal to a difference between those of rotation and revolution.

Partial tides of the type $i = 0$, $j = 2$ vanish whenever $e = 0$ and $I = 0$; but persist if either one of these conditions fails to be met. Tides characterized by $i = 1, j = 2$, arise only if $I > 0$ regardless of the eccentricity; while the sectorial tides ($i = j = 2$) vanish in the case of synchronism between rotation and revolution only if $I = 0$; otherwise other tides of this type arise with non-vanishing coefficients. A full account of such tides can be found in section 3C. In the most general case (i.e. $e > 0, I > 0, \omega \neq n$) altogether 14 partial tides of different frequencies are associated with the zonal harmonics ($i = 0$, $j = 2$) of the disturbing function; 19 with the tesseral harmonics $i = 1, j = 2$; and 21 with the sectorial harmonics $i = j = 2$. The damping of each in a viscous medium will contribute to the total dissipation of kinetic energy into heat; and their individual contributions would have to be summed up in exact work to obtain the total effect. This we do not propose to carry out in the present section, since we are concerned merely with estimates of the expected magnitude of the total effect; and to this end the results obtained so far should be sufficient.

A. VISCOUS FRICTION

In order to complete our task, it remains for us to evaluate the integrals on the right-hand sides of Equations (4.15) or (4.16); and to this end we must specify the viscosity of stellar matter which gives rise to friction. Apart from turbulent viscosity (which we shall postpone for subsequent discussion) its principal constituents in stellar interiors are the gas (i.e. plasma) viscosity μ_G and the radiative viscosity μ_R, the sum of which will hereafter be identified with our μ. As is well known (cf. e.g., Chapman 1954; Oster, 1957), the coefficient of viscosity μ_G of stellar plasma (consisting essentially of hydrogen) is sensibly equal to

$$\mu_G = 0.96\,\mu_i, \qquad \mu_i = \frac{5\sqrt{m_H}(kT)^{5/2}}{4\sqrt{\pi}\,\epsilon^4 A_2(\xi)}, \tag{4.20}$$

where T denotes the local temperature; k, the Boltzmann constant; m_H, the mass of a proton; and where

$$A_2(\xi) = \log(1 + \xi^2) - \frac{\xi^2}{1 + \xi^2}, \tag{4.21}$$

with

$$\xi = \frac{4kT}{\epsilon^2}\left(\frac{m_H}{\rho}\right)^{1/3}; \tag{4.22}$$

ρ denoting the density and ϵ, the electronic charge.

If in the preceding formulae we insert $k = 1.379 \times 10^{-16}$ erg deg^{-1}, $m_H = 1.672 \times$ $\times\ 10^{-24}$ g, and $\epsilon = 4.802 \times 10^{-10}$ e.s.u., we find that

$$\mu_G = \frac{3.68 \times 10^{-15}\,T^{5/2}}{\log(1 + \xi^2) - \dfrac{\xi^2}{1 + \xi^2}}\ \frac{\text{g}}{\text{cm sec}}, \tag{4.23}$$

where

$$\xi = 2.84 \times 10^{-5}\,T\rho^{-1/3}. \tag{4.24}$$

On the other hand, the radiative viscosity μ_R is known from Equation (1.47) to be given by

$$\mu_R = \frac{4aT^4}{15\, c\kappa\rho} \quad \frac{\text{g}}{\text{cm sec}}, \tag{4.25}$$

where $a = 7.55 \times 10^{-15}$ erg/cm^{-3} deg^{-4} denotes the Stefan constant; $c = 2.998 \times 10^{10}$ cm s^{-1}, the velocity of light; and κ, the absorption coefficient of stellar matter per unit mass.

If we combine Equations (4.23) and (4.25) for the plasma and radiative viscosity, the integrals on the right-hand sides of Equations (19) and (21) for $j = 2$ can be expressed as

$$\int_0^{a_1} (\mu_G + \mu_R) r^6 \, dr = a_1^7 \, \{(\mu_G)_c I_G + (\mu_R)_c I_R\}, \tag{4.26}$$

where $(\mu_G)_c$, $(\mu_R)_c$ stand for the central values of gas or radiative viscosity, and

$$I_G = \int_0^1 \frac{T}{T_c}^{5/2} \frac{A_2(\xi_c)}{A_2(\xi)} x^6 \, dx, \tag{4.27}$$

$$I_R = \int_0^1 \frac{(\kappa\rho)_c}{\kappa\rho} \left(\frac{T}{T_c}\right)^4 x^6 \, dx, \tag{4.28}$$

are non-dimensional quantities, the values of which can be ascertained by numerical integration for any desired model of a star.

It what follows, we wish to evaluate these parameters for six models of typical stars, published by Schwarzschild (1958) on pp. 254–259 of his book *Structure and Evolution of the Stars.* All these models pertain to Main-Sequence stars of different masses and evolutionary stages; and their principal characteristics have been compiled in Table II-2, the columns of which indicate, successively, the star's mass m (in units of $\odot = 1.985 \times \times 10^{33}$ g); radius a_1 (expressed in solar units $\odot = 0.695 \times 10^{11}$ cm); the luminosity L (in terms of $\odot = 3.78 \times 10^{33}$ erg s^{-1}); the corresponding spectral class; decimal logarithm of the central temperature T_c (in degrees K); logarithm of the central density ρ_c (in g cm^{-3}); the ratio ρ_c/ρ_m of the central to the mean density of the respective configuration; and h, its fractional radius of gyration.

TABLE II-2
Physical properties of stellar models

Model No.	m (in $\odot$)	a_1 (in $\odot$)	L (in $\odot$)	Spectrum	$\log T_c$ (in deg)	$\log \rho_c$ (in g cm^{-3})	ρ_c/ρ_m	h
I	10	3.65	3000	O8	7.442	0.892	26.9	0.29
II	2.5	1.59	21.2	A2	7.297	1.684	55.0	1.24
III	1.0	1.021	0.578	K1	6.906	1.813	51.3	0.25
IV	0.603	0.644	0.565	F8	6.906	1.813	20.5	0.32
V	10	6.09	5220	B0	7.545	1.075	191	0.18
VI	1	1	1	G2	7.165	2.128	95.2	0.22

Models I – IV represent initial Main-Sequence stars of different masses and luminosities (Model III approximating the properties of the initial Sun; and Model IV, the initial state of the components of the eclipsing variable Castor C); while Models V – VI correspond to evolved Main-Sequence stars (model VI to that of the present Sun).

The viscous properties of these models are listed in the accompanying Table II-3, the contents of the columns of which are self-explanatory. The central values of μ_G and μ_R (in g cm^{-1} s^{-1}) have been computed from Equations (4.23) and (4.25) with the aid of the data compiled in Table II-2; and those for I_G and I_R obtained from (4.27) and (4.28)

TABLE II-3
Viscous properties of stellar models

Model No.	$\log(\mu_G)_c$ (in g cm s)	$\log(\mu_R)_c$ (in g cm^{-1} s^{-1})	$10^3 I_G$	$10^3 I_R$	$(\mu_G)_c I_G + (\mu_R)_c I_R$ (in g cm^{-1}s^{-1})	
I	3.47	3.97	1.70	3.87	5.0	36.3
II	3.20	2.35	0.91	2.58	1.45	0.58
III	2.77	1.16	1.41	2.31	0.83	0.03
IV	2.35	0.22	4.86	3.26	1.10	0.01
V	3.72	4.31	0.23	2.03	1.2	41.7
VII	2.95	1.33	1.14	2.78	1.02	0.06

by numerical integration; the last column then gives the effective viscosity (in g gm^{-1} s^{-1}) of the respective configuration as a whole, as it occurs in curly brackets on the right-hand side of Equation (4.26).

An inspection of the individual columns of the foregoing tables discloses several noteworthy facts. First, it reveals that the central viscosity of all stellar models under consideration proves to be remarkably high – of the order of $10^2 - 10^4$ g cm^{-1} s^{-1} – thus bearing out an earlier surmise by Eddington (1926) that ". . . For hydrodynamical purposes, one must think of the star as thick oily liquid. This applies even to the regions of low density. . . . I suppose that even the photosphere will be rather sticky" (op. cit., p. 281).

Secondly, our present computations disclose that, whereas in stars of masses comparable with (or smaller than) that of the Sun radiative viscosity remains unimportant in comparison in comparison with plasma viscosity, in massive stars ($m > 5\odot$) radiative viscosity becomes dominant not only near the centre, but throughout the interior (as is borne out by the fact that $I_R \gg I_G$), because the ratio $T^4/\kappa\rho$ diminishes outwards less rapidly than $T^{3/2}$.

B. APPLICATION TO BINARY SYSTEMS

With the aid of the numerical results listed in Tables II-2 and II-3 we are now in a position to evaluate the total rate of the energy dissipation, through viscous tides, in close binary systems consisting of the components for which our Models I–VI can be regarded as representative.

In order to do so, let us return to Equations (4.8) – (4.11), which for the dominant second-harmonic tides (i.e., with $j = 2$) assume the forms

$$\left.\begin{aligned} E_{0,2} &= \frac{12432\pi}{5} c_{0,2}^2 \int_0^{a_1} \mu r^6 \, \mathrm{d}r \ldots \text{(A)} \\ &= \frac{4816\pi}{15} c_{0,2}^2 \int_0^{a_1} \mu r^6 \, \mathrm{d}r \ldots \text{(B)} \end{aligned}\right\} \tag{4.29}$$

and, in either case,

$$\dot{E}_{i,2} = \frac{(2+i)!}{(2-i)!} \frac{\dot{E}_{0,2}}{2} \tag{4.30}$$

by (4.11).

Now for $j = 0$ (i.e. the zonal-harmonic tides), a combination of Equations (4.17) and (4.26) discloses that

$$c_{0,2}^2 \int_0^{a_1} \mu r^6 \, \mathrm{d}r = a_1 \left\{ \frac{1}{2} \frac{m'}{m} \left(\frac{a_1}{\Delta} \right)^3 n a_1 e \sin nt \right\}^2 \{(\mu_{\mathrm{G}})_c J_{\mathrm{G}} + (\mu_{\mathrm{R}})_c J_R\} . \tag{4.31}$$

where the second expression in curly brackets on the right-hand side has been tabulated for our models in the ultimate column of Table II-3. Inserting (4.31) in (4.29) and adopting, typically,

$$\frac{m'}{m} \sim 1 \quad \text{and} \quad \frac{a_1}{\Delta} \sim 0.3 \tag{4.32}$$

we find that, in this case.

$$\dot{E} \sim a_1 (n a_1 e)^2 \{(\mu_{\mathrm{G}})_c J_{\mathrm{G}} + (\mu_{\mathrm{R}})_c J_{\mathrm{R}}\}, \tag{4.33}$$

where the constant of proportionality – 1.424 in Case (A) and 0.092 in Case (B) – will generally be of the order of 0.1. If, lastly, we adopt for a_1 and na_1 the values of 10^{11} cm (i.e., 1.4 ☉) and 10^6 cm s^{-1}, respectively (which should well represent typical cases), we find that, ultimately,

$$E = (10^{22} - 10^{23})\, e^2 \sin^2 nt \text{ erg s}^{-1}, \tag{4.34}$$

corresponding to a secular energy loss of

$$\dot{E} = (10^{27} - 10^{28})\, e^2 \text{ erg per cycle.} \tag{4.35}$$

In the case of partial tides corresponding to $i = 1$, the result proves to be identical with the preceding one, provided that the angular velocity m of axial rotation is made to replace the mean daily motion n of the binary pair, and that $\sin I$ replaces e as the factor whose non-zero value is responsible for the tide. As regards the sectorial tide corresponding to $i = 2$, the same continues to be true provided that n is replaced by $n \mp \omega$ and the result multiplied by $\frac{1}{3}(n \mp \omega)$; but it contains no other small factor. A combination $n \mp \omega$, (the sign $\mp$ corresponding to direct and retrograde rotation, respectively) need not, moreover, be small; but can be of the same order of magnitude as n or even larger if $\omega \gg$

n. In such a case, a contribution to E arising from sectorial second-harmonic tide due to non-synchronism between rotation and revolution can, in effect, be 10–100 times as large as that arising from the orbital eccentricity or equatorial inclination, and attain the order of magnitude of $10^{23} - 10^{24}$ erg s^{-1} or $10^{28} - 10^{29}$ erg per cycle.

Let us compare these rates with the total amount of kinetic energy possessed by typical components of close binary systems, and with their energy loss due to radiation. As the latter are mostly of the order of 10^{33} erg s^{-1}, it is immediately obvious that a production of heat by viscous friction at a rate of 10^{23} or 10^{24} erg s^{-1} is utterly negligible in comparison with the rate of nuclear energy production in stellar interiors – or even with the rate of gravitational energy release during the stages of contraction. *The heat produced by tidal interaction between components in close binary systems cannot*, therefore, *affect the internal structure or evolution of such stars to any appreciable extent* – at least as far as the effects of plasma or radiative viscosity are concerned – and the corresponding terms can be safely ignored in the equations for the energy-balance. This does not necessarily mean yet that the evolution of the individual components of close binary systems will be unaffected by their symbiosis, and proceed as if they were single; for the effects of their mutual attraction on their mechanical equilibrium are very much larger; but the interaction effects on the energy balance are clearly negligible.

Let us compare next the kinetic energy – both orbital and rotational – of the components in typical binary systems with the rate of its loss by tidal interaction. If possible orbital eccentricity is ignored, the kinetic energy $\mathfrak{T}_0$ of orbital motion of the system can be expressed by

$$\mathfrak{T}_0 = \frac{Gmm'}{2\Delta}, \tag{4.36}$$

where $G = 6.68 \times 10^{-8}$ cm^3 g^{-1} s^{-2} denotes the constant of gravitation; m, m', the masses of the constituent components; and Δ, their mutual separation. For typical values of $m \sim m' \sim 10^{33}$ g and $\Delta \sim 10^{12}$ cm, the orbital kinetic energy $_0$ of the corresponding pair proves to be a quantity of the order of 10^{46} erg. On the other hand, the kinetic energy $\mathfrak{T}_0$ of axial rotation of the component of mass m, radius a_1, and angular velocity ω, is known to be given by

$$\mathfrak{T}_\omega = \tfrac{4}{3}\pi\omega^2 \int_0^{a*} \rho r^4 dr = \tfrac{1}{2} m\,(a_1 h\omega)^2, \tag{4.37}$$

where h, the fractional radius of gyration of the respective star, has been tabulated for our six models in the ultimate column of Table II-2; and its average value proves to be close to 0.2. Adopting again $m \sim 10^{33}$g, $a_1 \sim 10^{11}$ cm, and $\omega \sim 10^{-5}$ s^{-1}, we find that $\mathfrak{T}_\omega \sim 10^{44}$ erg – i.e., a quantity some 100 times smaller than the orbital kinetic energy of the pair.

If, now – in accordance with the results stated earlier in this section – each component is to lose $10^{22} - 10^{23}$ erg of kinetic energy by dissipative action of viscous tides (mainly those corresponding to the case of $i = j = 2$), it follows that *the kinetic energy of axial rotation would be affected by them appreciably* – say, within 10 % of the actual value – *only after time-intervals of the order of* 10^{20} s *or* 10^{13} y – *i.e., on a slow nuclear rather, than gravitational* (Kelvin) *time-scale;* while during time-span of the order of 10^6 or 10^7 y the axial rotation would be thoroughly uninfluenced by tides.

This appears incontrovertibly so as long as the dissipative action is due solely to plasma or radiative viscosity. However, should *turbulent viscosity* appear, it may be a very different story. Proper quantitative treatment of turbulent viscosity in stellar interiors still encounters difficulties which force us to postpone its more specific discussion for the future. However, it is known that turbulent zones of stellar interiors are characterized by macroscopic viscosity of the order of $\Re \mu_G$, where $\Re$ – the Reynolds number – must be a quantity of the order of 10^5 or 10^6 for turbulence to occur at all. In other words, in turbulent regimes the macroscopic viscosity becomes at least $10^5 - 10^6$ times as large as the corresponding plasma viscosity μ_G; and so will be the corresponding rate of energy dissipation.

On the Main Sequence, turbulent zones are known to develop only in central parts of massive stars, where tidal effects are minor. However, as soon as the star begins to evolve away from the Main Sequence, turbulent zones begin to develop on the outside – where tides are relatively of greatest importance – and extend rapidly into the interior for stars which frequently occur as *secondary* components in close binary system so such stars exhibit spectroscopic characteristics of subgiants; and it is in systems possessing such components where dissipative effects of viscosity should be primarily anticipated; for *turbulent viscosity of the order of* 10^6 μ_G *could make the respective component lose (say)* 10 % *or its rotational kinetic energy in* 10^6 yr *or even shorter intervals of time – and thus produce dissipative effects on the contractional rather than nuclear time-scale.*

As is well known, the most sensitive detector of such effects are the period changes which should occur in binary systems as a result of the degradation of kinetic energy into heat. It may indeed be recalled (cf. sec. VIII-3) that orbital periods of eclipsing binaries with both components on the Main Sequence are generally stable; while systems with one (or both) components evolved away from the Main Sequence exhibit as a rule complicated period fluctuation. This fact has in recent years been mostly attributed to mass loss, or exchange of mass between evolved components, stimulated by the coincidence of the surfaces with their 'Roche limits' (cf. sec. VI-2). However, this latter characteristic alone would not explain why similar fluctuations in periods are observed also in systems containing 'undersize subgiants'. In the light of the results of the present investigation, dissipative phenomena of viscous tides should be likewise considered as possible causes of the observed period fluctuations (the magnitude of which could, in fact, disclose to us the extent – and the Reynolds numbers – of the turbulent zones in the respective components); but a closer analysis must be postponed for future investigations.

One last remark may be added in this connection, and this concerns the role of viscous tides in close binaries which may consist of a pair of white dwarfs. At present no such system is known with certainty to exist. Since, however, their discovery may be only a matter of time, it may be of interest to consider the role which the tides raised in such systems would play in the conversion of mechanical energy into heat.

In doing so we should first recall that the viscosity of a degenerate fermion gas (cf. e.g., Nishimura and Mori, 1961) is by several orders of magnitude *larger* than that of a non-degenerate plasma; and should exceed that expected in turbulent zones of subgiant stars. Secondly, the typical values of a or na to be expected in close binaries consisting of white dwarfs (cf. Kopal, 1957a) should be of the order 10^9 cm and 10^8 cm s^{-1}, respectively – which together with the fermion-gas viscosity should lead to an energy dissipation at a rate of 10^{30} erg s^{-1}. Since, moreover, the total luminosity output of such objects is of the order of $10^{30} - 10^{31}$ erg s^{-1}, it follows that *a dissipation of the kinetic energy into heat through viscous tides in fermion-gas systems could, by itself, be able to provide a major part of the source of the luminosity of such objects.*

Secondly, since the kinetic energy of a rotating white dwarf (of the radius of gyration $h = 0.453$, corresponding to that of a polytrope $n = 1.5$ (Motz, 1852)) is about 10^{49} erg, it follows that a depletion of so large a store even at a rate of 10^{30} erg s^{-1} could maintain a steady source of heat arising from the dissipation of viscous tides in white-dwarf systems for astronomically long intervals of time.

BIBLIOGRAPHICAL NOTES

III-1: The equations of motion for non-radiating fluid systems are classical (cf., e.g., L. D. Landau and E. M. Lifschitz, *Fluid Mechanics,* London 1959); for radiating systems cf. an excellent summary of them by P. Ledoux in vol. **51** (pp. 432–452) of the *Handbuch der Physik* (Ledoux, 1958).

The first investigator to note that interaction between moving matter and radiation will give rise to effects simulating those of viscosity was J. H. Jeans (1925 and 1926). For subsequent investigations of the interaction between radiation field and moving matter cf. S. Rosseland (1926), H. Vogt (1928), E. A. Milne (1929 and 1930), and L. H. Thomas (1930) who gave the first relativistic treatment of the subject – a treatment further elaborated by J. L. Synge (1957).

Most of these investigators concerned themselves with the energy transfer in moving media by radiation. Of the investigations concerned with a transfer of momentum by radiation the prime references to literature in more recent years are R. Kippenhahn (1954), P. Ledoux (1958) and J. Hazlehurst and W. L. W. Sargent (1958). This latter paper constitutes the most complete treatment of its subject available to date; the work of Ledoux is somewhat less general (in so far as he employed the steady-state equation of radiative energy transfer); while Kippenhahn confined his attention only to the terms involving the products of the velocity components of moving gas with those of the radiation flux in the respective directions.

In his pioneer work of 1926, Jeans underestimated, however, the coefficient of 'radiative viscosity' by a factor 2 (a slip repeated later by Milne); and dropped also a factor 4 in his energy equation. This latter slip was subsequently rectified by Vogt; and the former (with some others) by Thomas. This was not, however, the end of the role which mathematical gremlins tried their best to impede the growth of this subject. Thus Ledoux (1958) dropped the term U_{R0} in the last term on the right-hand side of his Equation (49.13) for p_R; while Hazlehurst and Sargent (1958) included in their Equations (38), (40), (42), (45) and (47) a superfluous factor 15 (besides, the subscripts t on the right-hand side of their Equation (47) should in reality be τ). When, however, these minor slips have been corrected, the results of all investigators can be brought in mutual satisfactory agreement. Several consequences of their work are, however, presented in this section for the first time.

III-2: The linearization of the equations of motion, carried out in this section and leading to Equations (2.39) – (2.40) governing spheroidal oscillations of self-gravitating viscous fluids, follows the previous work by Z. Kopal (1968a) in which these equations made their first appearance.

In the inviscid case, a system equivalent to (2.39) – (2.40) but consisting of four first-order differential equations (2.42) was derived first by N. R. Lebovitz (1965). An equivalent form of the system (2.49) – (2.50) obtaining when perturbations of the gravitational potential are neglected was obtained first by T. G. Cowling (1941).

An explicit treatment of the exact (i.e., 4-th order) system of equations for non-radial oscillations of self-gravitating configurations consisting of homogeneous compressible fluid goes back to C. L. Pekeris (1938); but the hypergeometric nature of these equations reproduced in this section was not recognized till 20 years later (cf. Kopal, 1968a).

III-3: An analytical treatment of the dynamical tides in close binary systems, regarded as forced oscillations in a prevalent field of force which are governed by linearized equations of the preceding section, follows closely a previous investigation of this subject by Z. Kopal (1968b) with only minor modifications.

For previous work on this subject cf. J.-P. Zahn (1966, 1970); a G. R. Roach (1968).

III-4: A treatment of the subject given in this section follows closely Z. Kopal (1968c, d).

CHAPTER IV

GENERALIZED ROTATION

A treatment of various phenomena associated with dynamical tides, as contained in the preceding chapter, is still insufficient for direct application to phenomena observed in close binary systems in one essential respect: namely, we have not yet taken account of the fact that the components constituting such systems *rotate* about one (or more) axes of the rectangular coordinates x, y, z with respect to which our fundamental equations of motion have been referred. In Chapter II the effects of rotation about *one* axis have been explored to quantities of third order in centrifugal force (sec. II-2). However, throughout Chapter III we regarded our fundamental set of coordinates x, y, z as representing an *inertial* system, devoid of systematic motion other than uniform translation; and the only allowance made for axial rotation in section III-4 was through Equation (3.78).

Yet all stars – or stellar systems – observed in the sky are found to rotate – no doubt as a consequence of the vorticity of pre-existing gas (or plasma) from which such formations originated – and the angular momentum associated with rotation constitutes as fundamental a property of celestial bodies as their mass or chemical composition.

In particular, the components of all binary systems observed by the requisite means do exhibit clear evidence of *axial rotation* – restricted, however, to motions of certain type. Thus in all systems for which spectroscopic observations of the 'rotational effect' within eclipses are available (cf., e.g., Kopal, 1959; sec. V-2) the rotation is found to be *direct* – i.e., takes place in the same direction as the orbital motion of the respective components; though the angular velocity ω of the axial rotation does not seem to bear any unique relation to the Keplerian angular velocity ω_K of orbital motion. For a large majority of known systems the angular velocities ω and ω_K are comparable in magnitude; and for a good many of them (especially in systems with nearly circular orbits) the differences $|\omega - \omega_K|$ appear to be insignificant. However, for systems characterized by eccentric orbits we almost invariably find that $\omega > \omega_K$; and in some – e.g., α CrB (for which $e = 0.36$) the primary component rotates as much as a hundred times faster than it revolves (cf., McLaughlin, 1933). The converse case of $\omega < \omega_K$ appears to be much scarcer; though at least one is known – namely, the primary (visible) component of β Lyrae – where this is indubitably encountered (cf. Struve, 1952; Böhm-Vitense, 1954; Mitchell, 1954).

Another question – less easy to answer from available evidence – concerns the direction of the principal axis of rotation with respect to the orbital plane of the respective binary system. A deviation of this direction from perpendicularity would produce, e.g., an asymmetry of the spectroscopic rotational effect with respect to the moments of conjunctions. A less direct, but more sensitive, method for the detection of such inclination is through apparent fluctuations of the orbital period – a method which will be developed more fully later (cf. see V-4) in the present volume. A full range of the dynamical phenomena to be expected if the equatorial planes of one (or both) components fail to coincide with the plane of the orbit will, however, be taken up already in section V-2.

But even if these planes happen to be coincident, and the axis of rotation remains permanently in a position normal to the orbital plane, the angular velocity ω of rotation about that axis itself cannot remain constant in binary systems whose components consist of viscous fluid. The reason is the fact that, in such a case, dynamical tides arising when $\omega \gtrless \omega_K$ are bound to slow down (or speed up) axial rotation, in a manner tending to establish *synchronism* between rotation and revolution as a result of secularly acting *viscous friction.* The action of such a friction constitutes, in turn, a motive power for *tidal evolution* of close binary systems, which we shall discuss in some detail in the concluding chapter of this book.

All these phenomena make it desirable that we prepare the ground for their discussion by considering first the general problem of *three-axial rotation* of deformable self-gravitating bodies, in which the velocity components u, v, w in the direction of increasing x, y, z-coordinates will be replaced by angular velocity components ω_x, ω_y, ω_z of rotation about the respective axes. This problem attracted many investigators in the past – mainly in connection with the precession and nutation of the Earth or other planetary globes of incomplete rigidity – but our treatment of it will be carried out by a different method than that followed by most previous investigators. Whereas those of the past – from Liouville (1858) to Poincaré (1910) – did so by departing from the Lagrangian equations of rational mechanics, we shall – consistent with our treatment of the subject in the preceding Chapter III – depart from the Eulerian hydrodynamical equations of viscous flow represented by Equations (1.4)–(1.6) of that chapter.

In section IV-1 which follows these introductory remarks, we shall set out systematically to replace, in these equations, the linear by angular velocity components – without resort to any linearization in terms of the dependent variables, or to other simplification. In particular, no limit will be imposed on the magnitude of the viscosity coefficients $\mu_{1,2}$ of our fluid, nor on the way in which $\mu_{1,2}$ (x, y, z) can vary inside our configuration. Only one simplifying hypothesis will be made which will essentially restrict the generality of our work: namely, an assumption that *all three angular velocity components* $\omega_{x,y,z}$ *do not depend on relative position*, and are functions only of the time. In invoking such a hypothesis we restrict, in effect, our configuration to rotate at any moment as a rigid body, with a velocity which may vary with the time; but shall allow for a time dependent deformability of its shape.

A converse case in which ω will be allowed to depend on the position x,y,z but not on the time t – corresponding to non-rigid steady-state rotation – will be taken up in section IV-4, in connection with von Zeipel's theorem and its ramifications. However, the most general case in which $\omega \equiv \omega(x, y, z; t)$ is not yet amenable to analytic treatment; and numerical integration of the corresponding set of non-linear partial differential equations in four independent variables – a task apt to discourage most contemporary analysts – remains the only avenue of approach.

IV-1. Equations of Motion for Deformable Bodies

In order to apply the system (1.4)–(1.16) of equations set up in the preceding chapter for a study of the motions of a self-gravitating fluid body about its centre of mass, consider a transformation of rectangular coordinates between an *inertial* (fixed) system of *space* axes x, y, z, and a *rotating* system of *body* (primed) axes x', y', z', possessing the same origin, but with the primed axes rotated with respect to the space axes by the Eulerian angles ϕ, θ, ψ, in accordance with a scheme illustrated on the accompanying Figure 4-1.

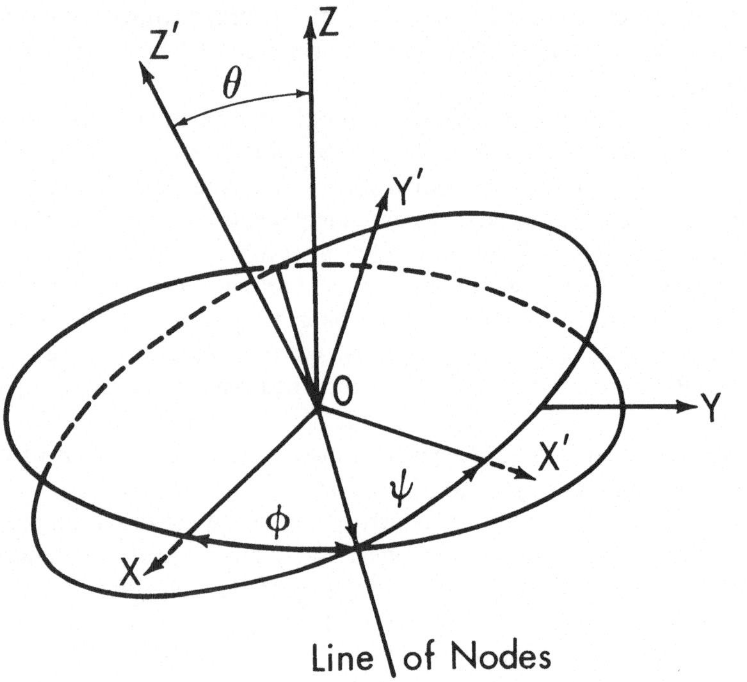

Fig. 4-1. Definition of Eulerian angles.

As is well known, the transformation of coordinates from the space to the body axes is governed by the matrix equation

$$\begin{Bmatrix} x \\ y \\ z \end{Bmatrix} = \begin{Bmatrix} a_{11} & a_{12} & a_{13} \\ a_{21} & a_{22} & a_{23} \\ a_{31} & a_{32} & a_{33} \end{Bmatrix} \begin{Bmatrix} x' \\ y' \\ z' \end{Bmatrix} , \tag{1.1}$$

where the coefficients a_{ik}, expressed in terms of the Eulerian angles assume the explicit forms

$$\left.\begin{aligned} a_{11} &= \cos\psi\cos\phi - \cos\theta\sin\phi\sin\psi, \\ a_{12} &= -\sin\psi\cos\phi - \cos\theta\sin\phi\cos\psi, \\ a_{13} &= \sin\theta\sin\phi \quad ; \end{aligned}\right\} \tag{1.2}$$

$$\left.\begin{aligned} a_{21} &= \cos\psi \sin\phi + \cos\theta \cos\phi \sin\psi , \\ a_{22} &= -\sin\psi \sin\phi + \cos\theta \cos\phi \cos\psi , \\ a_{23} &= \qquad\qquad -\sin\theta \cos\phi \qquad ; \end{aligned}\right\} \tag{1.3}$$

$$\left.\begin{aligned} a_{31} &= \sin\psi \sin\theta , \\ a_{32} &= \cos\psi \sin\theta , \\ a_{33} &= \qquad \cos\theta . \end{aligned}\right\} \tag{1.4}$$

In order to obtain the corresponding *space* velocity-components u, v, w let us differentiate equations (1.1) with respect to the time. With dots denoting hereafter ordinary (total) derivatives with respect to t, we find that

$$\dot{x} = u = \dot{a}_{11}x' + \dot{a}_{12}y' + \dot{a}_{13}z' + a_{11}\dot{x}' + a_{12}\dot{y}' + a_{13}\dot{z}', \tag{1.5}$$

$$\dot{y} = v = \dot{a}_{21}x' + \dot{a}_{22}y' + \dot{a}_{23}z' + a_{21}\dot{x}' + a_{22}\dot{y}' + a_{23}\dot{z}', \tag{1.6}$$

$$\dot{z} = w = \dot{a}_{31}x' + \dot{a}_{32}y' + \dot{a}_{33}z' + a_{31}\dot{x}' + a_{32}\dot{y}' + a_{33}\dot{z}'; \tag{1.7}$$

whereas the *body* velocity-components u', v', w' follow from

$$\begin{aligned} \dot{x}' = u' &= \dot{a}_{11}x + \dot{a}_{21}y + \dot{a}_{31}z \\ &\quad + a_{11}\dot{x} + a_{21}\dot{y} + a_{31}\dot{z}, \end{aligned} \tag{1.8}$$

$$\begin{aligned} \dot{y}' = v' &= \dot{a}_{12}x + \dot{a}_{22}y + \dot{a}_{32}z \\ &\quad + a_{12}\dot{x} + a_{22}\dot{y} + a_{32}\dot{z}, \end{aligned} \tag{1.9}$$

$$\begin{aligned} \dot{z}' = w' &- \dot{a}_{13}x + \dot{a}_{23}y + \dot{a}_{33}z \\ &\quad + a_{13}\dot{x} + a_{23}\dot{y} + a_{33}\dot{z}; \end{aligned} \tag{1.10}$$

where

$$\left.\begin{aligned} \dot{a}_{11} &= a_{12}\dot{\psi} - a_{21}\dot{\phi} + a_{31}\dot{\theta}\sin\phi = a_{31}\omega_y - a_{21}\omega_z \\ &= a_{12}\omega_{z'} - a_{13}\omega_{y'}, \end{aligned}\right\} \tag{1.11}$$

$$\left.\begin{aligned} \dot{a}_{12} &= -a_{11}\dot{\psi} - a_{22}\dot{\phi} + a_{32}\dot{\theta}\sin\phi = a_{32}\omega_y - a_{22}\omega_z \\ &= a_{13}\omega_{x'} - a_{11}\omega_{z'}, \end{aligned}\right\} \tag{1.12}$$

$$\left.\begin{aligned} \dot{a}_{13} &= \qquad -a_{23}\dot{\phi} + a_{33}\dot{\theta}\sin\phi = a_{33}\omega_y - a_{23}\omega_z \\ &= a_{11}\omega_{y'} - a_{12}\omega_{x'}; \end{aligned}\right\} \tag{1.13}$$

$$\left.\begin{aligned} \dot{a}_{21} &= a_{22}\dot{\psi} + a_{11}\dot{\phi} - a_{31}\dot{\theta}\cos\phi = a_{11}\omega_z - a_{31}\omega_x \\ &= a_{22}\omega_{z'} - a_{23}\omega_{y'}, \end{aligned}\right\} \tag{1.14}$$

$$\left.\begin{aligned} \dot{a}_{22} &= -a_{21}\dot{\psi} + a_{12}\dot{\phi} - a_{32}\dot{\theta}\cos\phi = a_{12}\omega_z - a_{32}\omega_x \\ &= a_{23}\omega_{x'} - a_{21}\omega_{z'}, \end{aligned}\right\} \tag{1.15}$$

$$\left.\begin{aligned} \dot{a}_{23} &= \qquad + a_{13}\dot{\phi} - a_{33}\dot{\theta}\cos\phi = a_{13}\omega_z - a_{33}\omega_x \\ &= a_{21}\omega_{y'} - a_{22}\omega_{x'}; \end{aligned}\right\} \tag{1.16}$$

$$\left.\begin{aligned} \dot{a}_{31} &= a_{32}\dot{\psi} + \dot{\theta}\sin\psi\cos\theta = a_{21}\omega_x - a_{11}\omega_y \\ &= a_{32}\omega_{z'} - a_{33}\omega_{y'}, \end{aligned}\right\} \tag{1.17}$$

$$\left.\begin{aligned} \dot{a}_{32} &= -a_{31}\dot{\psi} + \dot{\theta}\cos\psi\cos\theta = a_{22}\omega_x - a_{12}\omega_y \\ &= a_{33}\omega_{x'} - a_{31}\omega_{z'}, \end{aligned}\right\} \tag{1.18}$$

$$\left.\begin{aligned} \dot{a}_{33} &= \qquad -\dot{\theta}\sin\theta \qquad = a_{23}\omega_x - a_{13}\omega_y \\ &= a_{31}\omega_{y'} - a_{32}\omega_{x'}; \end{aligned}\right\} \tag{1.19}$$

where, taking advantage of the fact that

$$\left.\begin{aligned} a_{11}\dot{a}_{11} + a_{12}\dot{a}_{12} + a_{13}\dot{a}_{13} &= 0, \\ a_{21}\dot{a}_{21} + a_{22}\dot{a}_{22} + a_{23}\dot{a}_{23} &= 0, \\ a_{31}\dot{a}_{31} + a_{32}\dot{a}_{32} + a_{33}\dot{a}_{33} &= 0, \end{aligned}\right\} \tag{1.20}$$

and

$$\left.\begin{aligned} a_{11}\dot{a}_{11} + a_{21}\dot{a}_{21} + a_{31}\dot{a}_{31} &= 0, \\ a_{12}\dot{a}_{12} + a_{22}\dot{a}_{22} + a_{32}\dot{a}_{32} &= 0, \\ a_{13}\dot{a}_{13} + a_{23}\dot{a}_{23} + a_{33}\dot{a}_{33} &= 0, \end{aligned}\right\} \tag{1.21}$$

the respective angular velocities of rotation are given by

$$\left.\begin{aligned} \omega_x &= +(a_{21}\dot{a}_{31} + a_{22}\dot{a}_{32} + a_{23}\dot{a}_{33}) \\ &= -(a_{31}\dot{a}_{21} + a_{32}a_{22} + a_{33}\dot{a}_{23}), \end{aligned}\right\} \tag{1.22}$$

$$\left.\begin{aligned} \omega_y &= +(a_{31}\dot{a}_{11} + a_{32}\dot{a}_{12} + a_{33}\dot{a}_{13}) \\ &= -(a_{11}\dot{a}_{31} + a_{12}\dot{a}_{32} + a_{13}\dot{a}_{33}), \end{aligned}\right\} \tag{1.23}$$

$$\left.\begin{aligned} \omega_z &= +(a_{11}\dot{a}_{21} + a_{12}\dot{a}_{22} + a_{13}\dot{a}_{23}) \\ &= -(a_{21}\dot{a}_{11} + a_{22}\dot{a}_{12} + a_{23}\dot{a}_{13}), \end{aligned}\right\} \tag{1.24}$$

with respect to the space axes; or

$$\left.\begin{aligned} \omega_{x'} &= +(a_{13}\dot{a}_{12} + a_{23}\dot{a}_{22} + a_{33}\dot{a}_{32}) \\ &= -(a_{12}\dot{a}_{13} + a_{22}\dot{a}_{23} + a_{32}\dot{a}_{33}), \end{aligned}\right\} \tag{1.25}$$

$$\left.\begin{aligned} \omega_{y'} &= +(a_{11}\dot{a}_{13} + a_{21}\dot{a}_{23} + a_{31}\dot{a}_{33}) \\ &= -(a_{13}\dot{a}_{11} + a_{23}\dot{a}_{21} + a_{33}\dot{a}_{31}), \end{aligned}\right\} \tag{1.26}$$

$$\left.\begin{aligned}\omega_{z'} &= +(a_{12}\dot{a}_{11} + a_{22}\dot{a}_{21} + a_{32}\dot{a}_{31}) \\ &= -(a_{11}\dot{a}_{12} + a_{21}\dot{a}_{22} + a_{31}\dot{a}_{32}),\end{aligned}\right\} \tag{1.27}$$

with respect to the body axes; the pairs of alternative equations arising from the fact that, by a time-differentiation of the relations $a_{ij}a_{ik} = \delta_{jk}$ it follows that $a_{ij}\dot{a}_{ik} + a_{ik}\dot{a}_{ij} = 0$.

Inserting in the Equations (1.22)–(1.27) from (1.11)–(1.19) we find that, in terms of the Eulerian angles,

$$\omega_x = \dot{\theta}\cos\phi + \dot{\psi}\sin\theta\sin\phi, \tag{1.28}$$

$$\omega_y = \dot{\theta}\sin\phi - \dot{\psi}\sin\theta\cos\phi, \tag{1.29}$$

$$\omega_z = + \dot{\psi}\cos\theta + \dot{\phi}; \tag{1.30}$$

while

$$\omega_{x'} = \dot{\phi}\sin\theta\sin\psi + \dot{\theta}\cos\psi, \tag{1.31}$$

$$\omega_{y'} = \dot{\phi}\sin\theta\cos\psi - \dot{\theta}\sin\psi, \tag{1.32}$$

$$\omega_{z'} = \dot{\phi}\cos\theta + \dot{\psi}, \tag{1.33}$$

as could be also directly verified by an application of the inverse of the transformation (1.1), in accordance with which

$$\left.\begin{aligned}\omega_{x'} &= a_{11}\omega_x + a_{21}\omega_y + a_{31}\omega_z, \\ \omega_{y'} &= a_{12}\omega_x + a_{22}\omega_y + a_{32}\omega_z, \\ \omega_{z'} &= a_{13}\omega_x + a_{23}\omega_y + a_{33}\omega_z.\end{aligned}\right\} \tag{1.34}$$

With the aid of the preceding results the Equations (1.5)–(1.7) or (1.8)–(1.10) for the velocity-components with respect to the space or body axes can be reduced to the forms

$$u = z\omega_y - y\omega_z + u_0', \tag{1.35}$$

$$v = x\omega_z - z\omega_x + v_0', \tag{1.36}$$

$$w = y\omega_x - x\omega_y + w_0', \tag{1.37}$$

or

$$u' = -z'\omega_{y'} + y'\omega_{z'} + u_0, \tag{1.38}$$

$$v' = -x'\omega_{z'} + z'\omega_{x'} + v_0, \tag{1.39}$$

$$w' = -y'\omega_{x'} + x'\omega_{y'} + w_0, \tag{1.40}$$

where

$$\left.\begin{aligned}u_0 &= a_{11}u + a_{21}v + a_{31}w, \\ v_0 &= a_{12}u + a_{22}v + a_{32}w, \\ w_0 &= a_{13}u + a_{23}v + a_{33}w,\end{aligned}\right\} \tag{1.41}$$

are the *space* velocity components in the direction of the *rotating* axes x', y', z'; and

$$\left.\begin{aligned} u_0' &= a_{11}u' + a_{12}v' + a_{13}w', \\ v_0' &= a_{21}u' + a_{22}v' + a_{23}w', \\ w_0' &= a_{31}u' + a_{32}v' + a_{33}w', \end{aligned}\right\} \tag{1.42}$$

are the *body* velocity components in the direction of the *fixed* axes x, y, z.

In order to specify the appropriate forms of the components of *acceleration*, let us differentiate the foregoing expressions (1.35)–(1.40) for the velocity components with respect to the time. Doing so we find that those with respect to the *space* axes assume the forms

$$\dot{u} = w\omega_y + z\dot{\omega}_y - v\omega_z - y\dot{\omega}_z + \dot{u}_0', \tag{1.43}$$

$$\dot{v} = u\omega_z + x\dot{\omega}_z - w\omega_x - z\dot{\omega}_x + \dot{v}_0', \tag{1.44}$$

$$\dot{w} = v\omega_x + y\dot{\omega}_x - u\omega_y - x\dot{\omega}_y + \dot{w}_0', \tag{1.45}$$

where the velocity components u, v, w have already been given by equations (1.35)-(1.37); and where, by differentiation of (1.42),

$$\dot{u}_0' = a_{11}\dot{u}' + a_{12}\dot{v}' + a_{13}w' + \dot{a}_{11}u' + \dot{a}_{12}v' + \dot{a}_{13}w', \tag{1.46}$$

$$\dot{v}_0' = a_{21}\dot{u}' + a_{22}\dot{v}' + a_{23}w' + \dot{a}_{21}u' + \dot{a}_{22}v' + \dot{a}_{23}w', \tag{1.47}$$

$$\dot{w}_0' = a_{31}\dot{u}' + a_{32}\dot{v}' + a_{33}w' + \dot{a}_{31}u' + \dot{a}_{32}v' + \dot{a}_{33}w'. \tag{1.48}$$

The first three terms in each of these expressions represent obviously the body accelerations with respect to the space axes; and we shall abbreviate them as

$$\left.\begin{aligned} a_{11}\dot{u}' + a_{12}\dot{v}' + a_{13}\dot{w}' &= (\dot{u})_0', \\ a_{21}\dot{u}' + a_{22}\dot{v}' + a_{23}\dot{w}' &= (\dot{u})_0', \\ a_{31}\dot{u}' + a_{32}\dot{v}' + a_{33}\dot{w}' &= (\dot{w})_0'. \end{aligned}\right\} \tag{1.49}$$

Since, moreover, by insertion from (1.11)–(1.13) and (1.42),

$$\begin{aligned} a_{11}u' + a_{12}v' + \dot{a}_{13}w' &= (a_{31}\omega_y - a_{21}\omega_z)\,u' \\ &\quad + (a_{32}\omega_y - a_{22}\omega_z)\,v' \\ &\quad + (a_{33}\omega_y - a_{23}\omega_z)\,w' \\ &= \omega_y(a_{31}u' + a_{32}v' + a_{33}w') - \\ &\quad - \omega_z(a_{21}u' + a_{22}v' + a_{23}w') = \\ &= \omega_y w_0' - \omega_z v_0'; \end{aligned} \tag{1.50}$$

and, similarly,

$$\dot{a}_{21}u' + \dot{a}_{22}v' + \dot{a}_{23}w' = \omega_z u_0' - \omega_x w_0' \tag{1.51}$$

while

$$\dot{a}_{31}u' + \dot{a}_{32}v' + \dot{a}_{33}w' = \omega_x v_0' - \omega_y u_0', \tag{1.52}$$

Equations (1.43)–(1.45) can be rewritten in a more explicit form

$$\dot{u} = -x(\omega_y^2 + \omega_z^2) + y(\omega_x\omega_y - \dot{\omega}_z) + z(\omega_x\omega_z + \dot{\omega}_y) + (\dot{u})_0' + 2(w_0'\omega_y - v_0'\omega_z), \tag{1.53}$$

$$\dot{v} = -y(\omega_z^2 + \omega_x^2) + z(\omega_y\omega_z - \dot{\omega}_x) + x(\omega_x\omega_y + \dot{\omega}_z) + (\dot{v})_0' + 2(u_0'\omega_z - w_0'\omega_x), \tag{1.54}$$

and

$$\dot{w} = -z(\omega_x^2 + \omega_y^2) + x(\omega_x\omega_z - \dot{\omega}_y) + y(\omega_y\omega_z + \dot{\omega}_x) + (\dot{w})_0' + 2(v_0'\omega_x - u_0'\omega_y). \tag{1.55}$$

The foregoing equations refer to accelerations with respect to the inertial system of spaces axes. Those with respect to the (rotating) *body* axes can be obtained by an analogous process from the equations

$$\dot{u}' = -w'\omega_{y'} - z'\dot{\omega}_{y'} + v'\omega_{z'} + y'\omega_{z'} + \dot{u}_0, \tag{1.56}$$

$$\dot{v}' = -u'\omega_{z'} - x'\dot{\omega}_{z'} + w'\omega_{x'} + z'\dot{\omega}_{x'} + \dot{v}_0, \tag{1.57}$$

$$\dot{w}' = -v'\omega_{x'} - y'\dot{\omega}_{x'} + u'\omega_{y'} + x'\dot{\omega}_{y'} + \dot{w}_0, \tag{1.58}$$

equivalent to (1.43)–(1.45); which on being treated in the same way as the latter can eventually be reduced to the form

$$\dot{u}' = -x'(\omega_{y'}^2 + \omega_{z'}^2) + y'(\omega_{x'}\omega_{y'} + \dot{\omega}_{z'}) + z'(\omega_{x'}\omega_{z'} - \dot{\omega}_{y'}) + (\dot{u})_0 - 2(w_0\omega_{y'} - v_0\omega_{z'}), \tag{1.59}$$

$$\dot{v}' = -y'(\omega_{z'}^2 + \omega_{x'}^2) + z'(\omega_{y'}\omega_{z'} + \dot{\omega}_{x'}) + x'(\omega_{x'}\omega_{y'} - \dot{\omega}_{z'}) + (\dot{v})_0 - 2(u_0\omega_{z'} - w_0\omega_{x'}), \tag{1.60}$$

$$\dot{w}' = -z'(\omega_{x'}^2 + \omega_{y'}^2) + x'(\omega_{x'}\omega_{z'} + \dot{\omega}_{y'}) + y'(\omega_{y'}\omega_{z'} - \dot{\omega}_{x'}) + (\dot{w})_0 - 2(v_0\omega_{x'} - u_0\omega_{y'}), \tag{1.61}$$

where the space velocity components u_0, y_0, w_0 in the direction of increasing x', y', z' continue to be given by Equations (1.41); while the corresponding components of the accelerations are given by

$$\left.\begin{aligned} (\dot{u})_0 &= a_{11}\dot{u} + a_{21}\dot{v} + a_{31}\dot{w}, \\ (\dot{v})_0 &= a_{12}\dot{u} + a_{22}\dot{v} + a_{32}\dot{w}, \\ (\dot{w})_0 &= a_{13}\dot{u} + a_{23}\dot{v} + a_{33}\dot{w}. \end{aligned}\right\} \tag{1.62}$$

If, in particular, we consider the restricted case of a rotation about the z-axis only (so that $\omega_x = \omega_y = 0$), Equations (1.53)–(1.55) will reduce to the system

$$\left.\begin{aligned}\boxed{\dot{u}} &= (\dot{u})'_0 - 2v'_0\omega_z + x\omega_z^2 - y\dot{\omega}_z,\\ \boxed{\dot{v}} &= (\dot{v})'_0 + 2u'_0\omega_z + y\omega_z^2 + x\omega_z,\\ \boxed{\dot{w}} &= (\dot{w})'_0\ ;\end{aligned}\right\}\tag{1.63}$$

while Equations (1.59)–(1.61) will likewise reduce to

$$\left.\begin{aligned}\dot{u}' &= \boxed{(\dot{u})_0} + 2v_0\omega_{z'} + x'\omega_{z'}^2 + y'\omega_{z'},\\ \dot{v}' &= \boxed{(\dot{v})_0} - 2u_0\omega_{z'} + y'\omega_{z'}^2 + x'\omega_{z'},\\ \dot{w}' &= \boxed{(\dot{w})_0}\ .\end{aligned}\right\}\tag{1.64}$$

It is the acceleration in the cartouches of the two systems – referred as they are to the inertial space axes – which should be identified with the Lagrangian time-derivatives DV/Dt on the left-hand sides of the Equations (1.2)–(1.3) of motion if these are referred to the inertial or rotating axes of coordinates.

A closing note concerning the time differentiation of the coordinates or velocities should be added in this place. As

$$x \equiv x(t), \qquad y \equiv y(t), \qquad z \equiv z(t),\tag{1.65}$$

it follows that

$$\left.\begin{aligned}\dot{x} \equiv u &= \frac{dx}{dt} = \frac{\partial x}{\partial t},\\ \dot{y} \equiv v &= \frac{dy}{dt} = \frac{\partial y}{\partial t},\\ \dot{z} \equiv w &= \frac{dz}{dt} = \frac{\partial z}{\partial t},\end{aligned}\right\}\tag{1.66}$$

i.e., the ordinary (total) and partial derivatives of the coordinates with respect to the time are obviously identical. This is, however, no longer true of the time-differentiation of the velocities – whether linear or angular. As

$$u \equiv u(x, y, z; t), \qquad v \equiv v(x, y, z; t), \qquad w \equiv w(x, y, z; t);\tag{1.67}$$

or

$$\omega_{x,y,z} \equiv \omega_{x,y,z}(x, y, z; t)\tag{1.68}$$

where the coordinates (1.65) are themselves functions of the time. In consequence,

$$\begin{aligned}u = \frac{du}{dt} &= \frac{\partial u}{\partial t} + \frac{\partial u}{\partial x}\frac{\partial x}{\partial t} + \frac{\partial u}{\partial y}\frac{\partial y}{\partial t} + \frac{\partial u}{\partial z}\frac{\partial z}{\partial t}\\ &= \frac{\partial u}{\partial t} + u\frac{\partial u}{\partial x} + v\frac{\partial u}{\partial y} + w\frac{\partial u}{\partial z}\end{aligned}\tag{1.69}$$

by virtue of (1.66); and similarly for $\dot{v}$ and $\dot{w}$. Likewise,

$$\omega \equiv \frac{d\omega}{dt} = \frac{\partial\omega}{\partial t} + u\,\frac{\partial\omega}{\partial x} + v\,\frac{\partial\omega}{\partial y} + w\,\frac{\partial\omega}{\partial z} \tag{1.70}$$

for $\omega \equiv \omega_{x,y,z}$.

For coordinate systems referred to the rotating body axes similar relations hold good; care being merely needed to replace the unprimed coordinates or velocity components by the primed ones.

A. EULERIAN EQUATIONS

In the preceding part of this section we set up the general form of the equations governing the motion of compressible viscous fluids in rectangular coordinates, and expressed its velocity components in terms of arbitrary rotations about three rectangular axes. The aim of the present section will be to combine the fundamental Equations (1.4)–(1.7) of Chapter III, rewritten in terms of the angular variables $\omega_{x,y,z}$ in a form suitable for their solution.

In order to embark on this task, let us multiply Equations (1.53)–(1.55) by x, y, z and form their following differences:

$$\begin{aligned} y\dot{w} - z\dot{v} = {} & (y^2 + z^2)\,\dot{\omega}_x + (y^2 - z^2)\,\omega_y\omega_z \\ & - xy(\dot{\omega}_y + \omega_x\omega_z) - xz(\dot{\omega}_z + \omega_x\omega_y) - yz(\omega_y^2 - \omega_z^2) \\ & + \{y(\dot{w})'_0 - z(\dot{v})'_0\} + 2y\,\{v'_0\omega_x - u'_0\omega_y\} \\ & - 2z\,\{u'_0\omega_z - w'_0\omega_y\}\,, \end{aligned} \tag{1.71}$$

$$\begin{aligned} z\dot{u} - x\dot{w} = {} & (z^2 + x^2)\,\dot{\omega}_y + (z^2 - x^2)\,\omega_x\omega_z \\ & - yz(\dot{\omega}_z - \omega_y\omega_x) - yx(\dot{\omega}_x + \omega_y\omega_z) - zx(\omega_z^2 - \omega_x^2) \\ & + \{z(\dot{u})'_0 - x(\dot{w})'_0\} + 2z\,\{w'_0\omega_y - v'_0\omega_z\} \\ & - 2x\,\{v'_0\omega_x - u'_0\omega_y\}, \end{aligned} \tag{1.72}$$

$$\begin{aligned} x\dot{v} - y\dot{u} = {} & (x^2 + y^2)\,\dot{\omega}_2 + (x^2 - y^2)\,\omega_x\omega_y \\ & - zx(\dot{\omega}_x - \omega_y\omega_z) - zy(\dot{\omega}_y + \omega_y + \omega_x\omega_z) - xy(\omega_x^2 - \omega_y^2), \\ & + \{x(\dot{v})'_0 - y(\dot{u})'_0\} + 2x\,\{u'_0\omega_z - w'_0\omega_x\} \\ & - 2y\,\{w'_0\omega_y - v'_0\omega_z\}\,. \end{aligned} \tag{1.73}$$

If so, however, Equations (1.4)–(1.7) of Chapter III can be combined accordingly to yield

$$y\dot{w} - z\dot{v} + \frac{1}{\rho}\,\{y\,\frac{\partial}{\partial z} - z\,\frac{\partial}{\partial y}\}P - \{y\frac{\partial}{\partial z} - z\,\frac{\partial}{\partial y}\}\,\Omega = y\,\mathfrak{H} - z\mathfrak{G}, \tag{1.74}$$

$$z\dot{u} - x\dot{w} + \frac{1}{\rho}\{z\frac{\partial}{\partial z} - x\frac{\partial}{\partial z}\}P - \{z\frac{\partial}{\partial x} - x\frac{\partial}{\partial z}\}\Omega = z\mathfrak{F} - x\mathfrak{H}, \tag{1.75}$$

$$x\dot{v} - y\dot{u} + \frac{1}{\rho}\{x\frac{\partial}{\partial y} - y\frac{\partial}{\partial x}\}P - \{x\frac{\partial}{\partial y} - y\frac{\partial}{\partial x}\}\Omega = x\mathfrak{G} - y\mathfrak{F}, \tag{1.76}$$

where

$$\rho\mathfrak{F} = \frac{\partial\sigma_{xx}}{\partial x} + \frac{\partial\sigma_{xy}}{\partial y} + \frac{\partial\sigma_{xz}}{\partial z}, \tag{1.77}$$

$$\rho\mathfrak{G} = \frac{\partial\sigma_{yx}}{\partial x} + \frac{\partial\sigma_{yy}}{\partial y} + \frac{\partial\sigma_{yz}}{\partial z}, \tag{1.78}$$

$$\rho\mathfrak{H} = \frac{\partial\sigma_{zx}}{\partial x} + \frac{\partial\sigma_{zy}}{\partial y} + \frac{\partial\sigma_{zz}}{\partial z}, \tag{1.79}$$

represent the effects of viscosity.

In order to proceed further, let us rewrite the foregoing expressions in terms of the respective velocity components. Inserting for the components σ_{ij} of the viscous stress tensor from (1.9)–(1.14) of Chapter III we find the expressions on the right-hand sides of Equations (1.77)–(1.79) to assume the more explicit forms*

$$\begin{aligned}\frac{\partial\sigma_{xx}}{\partial x} + \frac{\partial\sigma_{xy}}{\partial y} + \frac{\partial\sigma_{xz}}{\partial z} = {} & \mu\nabla^2 u + \frac{\mu}{3}\frac{\partial\Delta}{\partial x} \\ & + 2\left\{\frac{\partial u}{\partial x} - \frac{\Delta}{3}\right\}\frac{\partial\mu}{\partial x} \\ & + \left\{\frac{\partial v}{\partial x} + \frac{\partial u}{\partial y}\right\}\frac{\partial\mu}{\partial y} \\ & + \left\{\frac{\partial u}{\partial z} + \frac{\partial w}{\partial x}\right\}\frac{\partial\mu}{\partial z},\end{aligned} \tag{1.80}$$

$$\begin{aligned}\frac{\partial\sigma_{yx}}{\partial x} + \frac{\partial\sigma_{yy}}{\partial y} + \frac{\partial\sigma_{yz}}{\partial z} = {} & \mu\nabla^2 w + \frac{\mu}{3}\frac{\partial\Delta}{\partial z} \\ & + 2\left\{\frac{\partial w}{\partial z} - \frac{\Delta}{3}\right\}\frac{\partial\mu}{\partial z} \\ & + \left\{\frac{\partial u}{\partial z} + \frac{\partial w}{\partial x}\right\}\frac{\partial\mu}{\partial x} \\ & + \left\{\frac{\partial v}{\partial x} + \frac{\partial u}{\partial y}\right\}\frac{\partial\mu}{\partial x},\end{aligned} \tag{1.81}$$

* In these and subsequent equations, the symbol μ stands hereafter for the 'first' coefficient of viscosity $\mu_1 = \mu_G + \mu_R$ of section II-1, where μ_G and μ_R denote the gas (plasma) and radiative (photon) viscosity, respectively. The 'second' coefficient μ_2 of viscosity is negligible for problems considered in this chapter.

and

$$\frac{\partial \sigma_{zx}}{\partial x} + \frac{\partial \sigma_{zy}}{\partial y} + \frac{\partial \sigma_{zz}}{\partial z} = \mu \nabla^2 w + \frac{\mu}{3} \frac{\partial \Delta}{\partial z} + 2\left\{\frac{\partial w}{\partial z} - \frac{\Delta}{3}\right\} \frac{\partial \mu}{\partial z} + \left\{\frac{\partial u}{\partial z} + \frac{\partial w}{\partial x}\right\} \frac{\partial \mu}{\partial x} + \left\{\frac{\partial w}{\partial y} + \frac{\partial v}{\partial z}\right\} \frac{\partial \mu}{\partial y}, \tag{1.82}$$

where Δ denotes, as before, the divergence of the velocity vector; and ∇^2 stands for the Laplacean operator.

Next. let us insert for the velocity components u, v, w from (1.35)–(1.37); by doing so we find that

$$\nabla^2 u = 2\nabla^2 \omega_y - y\nabla^2 \omega_z + \nabla^2 u_0' + 2\left\{\frac{\partial \omega_y}{\partial z} - \frac{\partial \omega_z}{\partial y}\right\}, \tag{1.83}$$

$$\nabla^2 v = x\nabla^2 \omega_z - z\nabla^2 \omega_x + \nabla^2 v_0' + 2\left\{\frac{\partial \omega_z}{\partial x} - \frac{\partial \omega_x}{\partial z}\right\}, \tag{1.84}$$

$$\nabla^2 w = y\nabla^2 \omega_x - x\nabla^2 \omega_y + \nabla^2 w_0' + 2\left\{\frac{\partial \omega_x}{\partial y} - \frac{\partial \omega_y}{\partial x}\right\}, \tag{1.85}$$

and

$$\Delta = \left\{y \frac{\partial}{\partial z} - z \frac{\partial}{\partial y}\right\} \omega_x + \left\{z \frac{\partial}{\partial x} - x \frac{\partial}{\partial z}\right\} \omega_y + \left\{x \frac{\partial}{\partial y} - y \frac{\partial}{\partial x}\right\} \omega_z + \frac{\partial u_0'}{\partial x} + \frac{\partial v_0'}{\partial y} + \frac{\partial w_0'}{\partial z}. \tag{1.86}$$

Before proceeding further, one feature of basic importance should be brought out which we by-passed without closer discussion at an earlier stage: namely, when by virtue of Equations (1.35)–(1.37) or (1.38)–(1.40) we replaced the *three* dependent variables u, v, w or u', v', w' on their left-hand sides by *six* new variables $\omega_x, \omega_y, \omega_z$ and u_0', v_0', w_0' or $\omega_{x'}, \omega_{y'}, \omega_{z'}$ and u_0, y_0, w_0 on their right-hand sides. This deliberately created redundancy permits us to impose without the loss of generality additional constraints on these variables, not embodied in the fundamental equations; and this we shall use at the present time. We propose, in particular, to assume that the primed axes $x'y'z'$ obtained by a rotation of the inertial system xyz, about a fixed origin, in accordance with the transformation (1.1) remain rectangular – an assumption which implies, in effect, that *the Eulerian angles* ϑ, ϕ, ψ involved in the direction cosines a_{ik} and, therefore, in the angular velocity components $\omega_{x,y,z}$ or $\omega_{x',y',z'}$ as defined by Equations (1.28)–(1.30) or (1.31)–(1.33) *are functions of the time t alone* (for should they depend, in addition, on the spatial coordinates x, y, z, a rotation as represented by Equations (1.1) would result in a curvilinear coordinate system).

This assumption will neatly separate the physical meaning of the two groups of variables: for while the angular velocity components $\omega_{x,y,z}$ will describe a *rigid-body rotation* of our dynamical system (during which the position of each particle remains unchanged in the primed coordinates), the remaining velocity components u_0', v_0', w_0' will represent *deformation* of our body, in the primed system, in the course of time. It is, therefore, the latter which will be of particular interest for the main problem which we have in mind; and in what follows, we propose to investigate the extent to which their occurrence may modify the structure of our equations.

In order to do so we notice first that, inasmuch as the angular velocity components are hereafter to be regarded as functions of t alone it follows from (1.83)–(1.85) that

$$\left.\begin{aligned} \nabla^2 u &= \nabla^2 u_0', \\ \nabla^2 v &= \nabla^2 v_0', \\ \nabla^2 w &= \nabla^2 w_0'; \end{aligned}\right\} \tag{1.87}$$

and, similarly, the divergence (1.86) of the velocity vector will reduce to

$$\Delta_0' = \frac{\partial u_0'}{\partial x} + \frac{\partial v_0'}{\partial y} + \frac{\partial w_0'}{\partial z}. \tag{1.88}$$

In consequence, the corresponding expressions on the right-hand sides of Equations (1.80) –(1.82) are obtained if the velocity components u, v, w present there are replaced by u_0', v_0', w_0'; and Δ by Δ_0'.

Therefore,

$$\begin{aligned} \rho\{y\mathfrak{H} - \mathfrak{G}\} = \mu\,&\{y\nabla^2 w_0' - z\nabla^2 v_0' + \frac{1}{3} D_1 \Delta_0'\} \\ &+ \frac{\partial\mu}{\partial x}\{D_1 u_0' + \frac{\partial}{\partial x}(yw_0' - zv_0')\} \\ &+ \frac{2}{3}\frac{\partial\mu}{\partial y}\{2D_1 v_0' + D_4 w_0'\} \\ &+ \frac{2}{3}\frac{\partial\mu}{\partial z}\{2D_1 w_0' - D_4 v_0'\} \\ &- \frac{2}{3}\frac{\partial\mu_0'}{\partial x} D_1\mu + \frac{1}{3}\xi D_4\mu, \end{aligned} \tag{1.89}$$

$$\begin{aligned} \rho\{z\mathfrak{F} - x\mathfrak{H}\} = \mu&\{z\nabla^2\mu_0' - x\nabla^2 w_0' + \frac{1}{3} D_2\Delta_0'\} \\ &+ \frac{2}{3}\frac{\partial\mu}{\partial x}\{2D_2 u_0' - D_5 w_0'\} \end{aligned}$$

$$+\frac{\partial \mu}{\partial y}\{D_2 v_0' + \frac{\partial}{\partial y}(zu_0' - xw_0')\}$$

$$+\frac{2}{3}\frac{\partial \mu}{\partial z}\{2D_2 w_0' + D_5 u_0'\}$$

$$-\frac{2}{3}\frac{\partial v_0'}{\partial y} D_2 \mu + \frac{1}{3}\eta D_5 \mu, \tag{1.90}$$

and

$$\rho\{x\mathfrak{G} - y\mathfrak{F}\} = \mu\{x\nabla^2 v_0' - y\nabla^2 u_0' + \frac{1}{3} D_3 \Delta_0'\}$$

$$+\frac{2}{3}\frac{\partial \mu}{\partial x}\{2D_3 u_0' + D_6 v_0'\}$$

$$+\frac{2}{3}\frac{\partial \mu}{\partial y}\{2D_3 v_0' - D_6 u_0'\}$$

$$+\frac{\partial \mu}{\partial z}\{D_3 w_0' + \frac{\partial}{\partial z}(xv_0' - yu_0')\}$$

$$-\frac{2}{3}\frac{\partial w_0'}{\partial z} D_3 \mu + \frac{1}{3}\zeta D_6 \mu, \tag{1.91}$$

where the symbols D_j ($j = 1, \ldots 6$) stand for the following operators

$$D_1 \equiv y\frac{\partial}{\partial z} - z\frac{\partial}{\partial y}, \tag{1.92}$$

$$D_2 \equiv z\frac{\partial}{\partial x} - x\frac{\partial}{\partial z}, \tag{1.93}$$

$$D_3 \equiv x\frac{\partial}{\partial y} - y\frac{\partial}{\partial x}, \tag{1.94}$$

$$D_4 \equiv z\frac{\partial}{\partial z} + y\frac{\partial}{\partial y}, \tag{1.95}$$

$$D_5 \equiv x\frac{\partial}{\partial x} + z\frac{\partial}{\partial z}, \tag{1.96}$$

$$D_6 \equiv x\frac{\partial}{\partial x} + y\frac{\partial}{\partial y}; \tag{1.97}$$

and where

$$\xi = \frac{\partial w_0'}{\partial y} - \frac{\partial v_0'}{\partial z}, \tag{1.98}$$

$$\eta = \frac{\partial u_0'}{\partial z} - \frac{\partial w_0'}{\partial x} , \tag{1.99}$$

$$\zeta = \frac{\partial v_0'}{\partial x} - \frac{\partial u_0'}{\partial y} , \tag{1.100}$$

denote the components of vorticity of the deformation vector.

As the next step of our analysis, let us integrate both sides of the Equations (1.74)–(1.76) over the entire mass of our configuration with respect to the mass element

$$\mathrm{d}m = \rho \, \mathrm{d}V = \rho \, \mathrm{d}x\mathrm{d}y\mathrm{d}z. \tag{1.101}$$

If, as usual,

$$A = \int (y^2 + z^2) \, \mathrm{d}m, \tag{1.102}$$

$$B = \int (x^2 + z^2) \, \mathrm{d}m, \tag{1.103}$$

$$C = \int (x^2 + y^2) \, \mathrm{d}m \tag{1.104}$$

denote the *moments of inertia* of our configuration with respect to the axes x, y, z; and

$$D = \int yz\mathrm{d}m, \tag{1.105}$$

$$E = \int xz\mathrm{d}m, \tag{1.106}$$

$$F = \int xy\mathrm{d}m, \tag{1.107}$$

stand for the respective *products of inertia*, the mass integrals of the Equations (1.74)–(1.76) combined with (1.71)–(1.73) will assume the forms

$$\begin{aligned} A\dot{\omega}_x &+ (C-B)\,\omega_y\omega_z - D(\omega_y^2 - \omega_z^2) - E(\dot{\omega}_z + \omega_x\omega_y) - F(\dot{\omega}_y - \omega_x\omega_z) + \\ &+ 2\omega_x \int (yv_0' + zw_0')\,\mathrm{d}m - 2\omega_y \int yu_0'\mathrm{d}m - 2\omega_z \int zu_0'\mathrm{d}m + \\ &+ \int D_1 P\,\mathrm{d}V - \int D_1\psi\,\mathrm{d}m = \int \{z(\dot{v})_0' - y(\dot{w})_0'\}\,\mathrm{d}m + \\ &+ \int \rho\{y\mathfrak{H} - z\mathfrak{G}\}\,\mathrm{d}V , \end{aligned} \tag{1.108}$$

$$\begin{aligned} B\dot{\omega}_y &+ (A-C)\,\omega_x\omega_z - D(\dot{\omega}_z - \omega_x\omega_y) - E(\omega_z^2 - \omega_x^2) - F(\dot{\omega}_x + \omega_y\omega_z) + \\ &+ 2\omega_y \int (xu_0' + zw_0')\,\mathrm{d}m - 2\omega_z \int zv_0'\mathrm{d}m - 2\omega_x \int xv_0'\mathrm{d}m + \\ &+ \int D_2 P\,\mathrm{d}V - \int D_2\psi\,\mathrm{d}m = \int \{x(\dot{w})_0' - z(\dot{u})_0'\}\,\mathrm{d}m + \\ &+ \int \rho\{z\mathfrak{F} - x\mathfrak{H}\}\,\mathrm{d}V, \end{aligned} \tag{1.109}$$

and

$$C\dot{\omega}_z + (B-A)\omega_x\omega_y - D(\dot{\omega}_y + \omega_x\omega_z) - E(\dot{\omega}_x - \omega_y\omega_z) - F(\omega_x^2 - \omega_y^2) +$$

$$+ 2\omega_z \int (yv_0' + xu_0')\,\mathrm{d}m - 2\omega_x \int xw_0'\,\mathrm{d}m - 2\omega_y \int yw_0'\,\mathrm{d}m +$$

$$+ \int D_3 P\,\mathrm{d}V - \int D_3\psi\,\mathrm{d}m = \int \{y(\dot{u})_0' - x(\dot{v})_0'\}\,\mathrm{d}m +$$

$$+ \int \rho\{x\mathfrak{G} - y\mathfrak{F}\}\,\mathrm{d}V, \tag{1.110}$$

The preceding three equations represent the exact form of the generalized Eulerian equations governing the axial rotation as well as precession and nutation of self-gravitating configurations which consist of a viscous fluid. They constitute a system of three ordinary differential equations for $\omega_{x,y,z}$ considered as functions of the time t alone. If the body in question were rigid (non-deformable) – or, if deformable, it were subject to no time-dependent deformation – all three velocity components u', v', w' relative to the rotating frame of reference (and thus, by (1.42), u_0', v_0', w_0') would be identically zero. In such a case, Equations (1.108)–(1.110) would reduce to their more familiar form

$$A\dot{\omega}_x + (C-B)\omega_y\omega_z - D(\omega_y^2 - \omega_z^2) - E(\dot{\omega}_z + \omega_x\omega_y)$$

$$- F(\dot{\omega}_y - \omega_x\omega_z) + \int D_1 P\,\mathrm{d}V - \int D_1\Omega\,\mathrm{d}m = \int D_1 V'\,\mathrm{d}m, \tag{1.111}$$

$$B\dot{\omega}_y + (A-C)\,\omega_x\omega_z - D(\dot{\omega}_z - \omega_x\omega_y) - E(\omega_z^2 - \omega_x^2)$$

$$- F(\omega_x + \omega_y\omega_z) + \int D_2 P\,\mathrm{d}V - \int D_2\Omega\,\mathrm{d}m = \int D_2 V'\,\mathrm{d}m, \tag{1.112}$$

and

$$C\dot{\omega}_z + (B-A)\omega_x\omega_y - D(\dot{\omega}_y + \omega_x\omega_z) - E(\dot{\omega}_x - \omega_y\omega_z)$$

$$- F(\omega_x^2 - \omega_y^2) + \int D_3 P\,\mathrm{d}V - \int D_3\Omega\,\mathrm{d}m = \int D_3 V'\,\mathrm{d}m, \tag{1.113}$$

where we have decomposed the total gravitational potential

$$\Psi = \Omega + V' \tag{1.114}$$

into its part arising from the mass of the respective body (Ω) and that arising from external disturbing forces (V') if any.

In the case of non-deformable bodies, the existence of hydrostatic equilibrium would require that

$$\int D_i P\,\mathrm{d}V = \int D_i\,\Omega\,\mathrm{d}m \tag{1.115}$$

for $i = 1, 2, 3$. If, moreover, the system of coordinates were to coincide with the principal axes of inertia of our configuration, all three products of inertia D, E and F can be made to vanish. In such a case, Equations (1.111)–(1.113) will reduce further to

$$A\dot{\omega}_x + (C-B)\omega_y\omega_z = \int D_1 V'\,\mathrm{d}m, \tag{1.116}$$

$$B\dot{\omega}_y + (A - C)\omega_z\omega_x = \int D_2 V' \mathrm{d}m, \tag{1.117}$$

$$C\dot{\omega}_z + (B - A)\omega_x\omega_y = \int D_3 V' \mathrm{d}m, \tag{1.118}$$

which constitute the familiar form of the Eulerian equations for the rotation of rigid bodies, well known in celestial mechanics (cf., e.g., Tisserand, 1892; Chapters 22–30).

If, however, our rotating configuration is fluid and subject to distortion by external forces, Equations (1.116)–(1.118) cease to be exact and must be replaced by the simultaneous system (1.108)–(1.110), the solutions of which specify generalized rotation of deformable configurations consisting of viscous fluid. In this chapter we shall be concerned with the effects of viscous tides on uni-axial rotation of deformable bodies; while a discussion of the effects on three-axial rotation (including the precession and nutation) is being postponed till the second section of Chapter V.

IV-2. Rotation of Deformable Bodies

In order to embark on the task set forth in the preceding section, let us return to Equations (1.108)–(1.110) which are fundamental for the solution of our problem; and set out to make their form more explicit by a specification of the velocity components u'_0, v'_0, w'_0 arising from the deformation of the rotating components of close binary systems due to their mutual tidal distortion (dynamical tides). In order to do so, let δx, δy, δz denote the changes in position of any mass element of our configuration, relative to the inertial axes, which are produced by dynamical tides. The body velocity components u'_0, v'_0, w'_0, produced by tidal deformation and referred to the same axes, can then be *defined* by

$$u'_0 \equiv \frac{D}{Dt}\,\delta x, \qquad v'_0 \equiv \frac{D}{Dt}\,\delta y, \quad w'_0 \equiv \frac{D}{Dt}\,\delta z, \tag{2.1}$$

where D/Dt signifies total derivatives with respect to the time, operating on the (time-dependent) displacements.

In order to specify these displacements, let us avail ourselves of three sets of rectangular coordinates possessing the same origin (which, for the sake of convenience, we shall identify with the centre of mass of the rotating star we shall regard as the 'primary' component of the pair – regardless of whether its mass is larger or smaller than that of its companion), but differing in the orientation of their axes in space. The xyz-axes (without primes) will constitute the inertial set of coordinates whose directions remain invariant in space. The singly-primed axes $x'y'z'$ will rotate with primary component (the 'body axes' in a way defined by the Eulerian angles θ, ϕ, ψ in accordance with a scheme illustrated on Figure 4-1 of the preceding section; while the doubly-primed axes $x''y''z''$ will revolve with the secondary component in such a way that the x''-axis remains constantly coincident with the radius-vector joining the centres of the two components.

Accordingly, a transformation of coordinates from the inertial to the rotating or revolving aces is governed by the matrix equation

$$\begin{Bmatrix} x \\ y \\ z \end{Bmatrix} = \begin{Bmatrix} a_{11} & a_{12} & a_{13} \\ a_{21} & a_{22} & a_{23} \\ a_{31} & a_{32} & a_{33} \end{Bmatrix} \begin{Bmatrix} x' \\ y' \\ z' \end{Bmatrix} \tag{2.2}$$

$$= \begin{Bmatrix} a''_{11} & a''_{12} & a''_{13} \\ a''_{21} & a''_{22} & a''_{23} \\ a''_{31} & a''_{32} & a''_{33} \end{Bmatrix} \begin{Bmatrix} x'' \\ y'' \\ z'' \end{Bmatrix} ,$$

where the direction cosines a_{ij} continue to be given by Equations (1.2)–(1.4), and the a''_{ij}'s by

$$\left.\begin{aligned} a''_{11} &= \cos u \cos \Omega - \sin u \sin \Omega \cos i , \\ a''_{12} &= -\sin u \cos \Omega - \cos u \sin \Omega \cos i , \\ a''_{13} &= + \sin \Omega \sin i ; \end{aligned}\right\} \tag{2.3}$$

$$\left.\begin{aligned} a''_{21} &= \cos u \sin \Omega + \sin u \cos \Omega \cos i , \\ a''_{22} &= -\sin u \sin \Omega + \cos u \cos \Omega \cos i , \\ a''_{23} &= - \cos \Omega \sin i ; \end{aligned}\right\} \tag{2.4}$$

$$\left.\begin{aligned} a''_{31} &= \sin u \sin i , \\ a''_{32} &= \cos u \sin i , \\ a''_{33} &= \cos i . \end{aligned}\right\} \tag{2.5}$$

where Ω, i and u denote, successively, the longitude of the node (from the x-axis), the inclination of the orbital plane from to that of $z = 0$, and the longitude of the component of mass m_2 in its orbital plane as measured from the line of the nodes.

Moreover, a transformation from the rotating to the revolving axes obeys the matrix equation

$$\begin{Bmatrix} x' \\ y' \\ z' \end{Bmatrix} = \begin{Bmatrix} \lambda''_1 & \lambda''_2 & \lambda''_3 \\ \mu''_1 & \mu''_2 & \mu''_3 \\ \nu''_1 & \nu''_2 & \nu''_3 \end{Bmatrix} \begin{Bmatrix} x'' \\ y'' \\ z'' \end{Bmatrix} , \tag{2.6}$$

where

$$\begin{Bmatrix} \lambda''_j \\ \mu''_j \\ \nu''_j \end{Bmatrix} = \begin{Bmatrix} a_{11} & a_{21} & a_{31} \\ a_{12} & a_{22} & a_{32} \\ a_{13} & a_{23} & a_{33} \end{Bmatrix} \begin{Bmatrix} a''_{1j} \\ a''_{2j} \\ a''_{3j} \end{Bmatrix} \tag{2.7}$$

for $j = 1, 2, 3$.

A. EFFECTS OF DEFORMATION

Let, moreover, the distorted surface of the primary component in the rotating (primed) system of coordinates be represented parametrically by the equations

$$\left.\begin{aligned} X' &= x' + \delta x' = x'(1+f), \\ Y' &= y' + \delta y' = y'(1+f), \\ Z' &= z' + \delta z' = z'(1+f), \end{aligned}\right\} \tag{2.8}$$

where f represents the fractional height of the tides. The case of f dependent only on position would correspond to tides of the equilibrium type; and, in such a case, the coordinates x', y', z' as well as X', Y', Z' are independent of the time. However, a dependence of f on the time – characteristic of dynamical tides – will render also x', y', z' time-dependent; because the shape of the rotating body will undergo time-dependent deformations with respect to the body axes.

The body velocity components u_0', v_0', w_0' in the direction of the fixed (inertial) axes x,y,z should then (in accordance with Equations (1.42)) be given by

$$\begin{Bmatrix} u_0' \\ v_0' \\ w_0' \end{Bmatrix} = \begin{Bmatrix} a_{11} & a_{12} & a_{13} \\ a_{21} & a_{22} & a_{23} \\ a_{31} & a_{32} & a_{33} \end{Bmatrix} \begin{Bmatrix} \delta\dot{x}' \\ \delta\dot{y}' \\ \delta\dot{z}' \end{Bmatrix} \tag{2.9}$$

where, by a differentiation of (3.4) and use of (2.5),

$$\left.\begin{aligned} \delta\dot{x}' &= \dot{x}'f + x'\dot{f} = u'f + x'\dot{f}, \\ \delta\dot{y}' &= \dot{y}'f + y'\dot{f} = v'f + y'\dot{f}, \\ \delta\dot{z}' &= \dot{z}'f + z'\dot{f} = w'f + z'\dot{f}, \end{aligned}\right\} \tag{2.10}$$

in which

$$u' = \dot{x}', \qquad v' = \dot{y}', \qquad w' = \dot{z}' \tag{2.11}$$

stand for the velocity components with which the tides move relative to the rotating system, and

$$\dot{f} = \frac{\partial f}{\partial t} + u'\frac{\partial f}{\partial x'} + v'\frac{\partial f}{\partial y'} + w'\frac{\partial f}{\partial z'} \cdot \tag{2.12}$$

In order to specify the velocity components u', v', w' let us return to the transformation (2.3) and observe that – by the deliberate choice of the direction of the x''-axis – the phenomena described as dynamical tides in the rotating system of singly-primed coordinates are bound to reduce to equilibrium tides in the revolving (doubly-primed) coordinates, so that

$$\dot{x}'' = \dot{y}'' = \dot{z}'' = 0. \tag{2.13}$$

Accordingly, a time-differentiation and repeated use of the transformation (2.3) discloses that

$$\begin{Bmatrix} u' \\ v' \\ w' \end{Bmatrix} = \begin{Bmatrix} \dot{\lambda}_1'' & \dot{\lambda}_2'' & \dot{\lambda}_3'' \\ \dot{\mu}_1'' & \dot{\mu}_2'' & \dot{\mu}_3'' \\ \dot{\nu}_1'' & \nu_2'' & \nu_3'' \end{Bmatrix} \begin{Bmatrix} x'' \\ y'' \\ z'' \end{Bmatrix} =$$

$$= \begin{Bmatrix} \dot{\lambda}_1'' & \dot{\lambda}_2'' & \dot{\lambda}_3'' \\ \dot{\mu}_1'' & \dot{\mu}_2'' & \dot{\mu}_3'' \\ \dot{\nu}_1'' & \dot{\nu}_2'' & \dot{\nu}_3'' \end{Bmatrix} \begin{Bmatrix} \lambda_1'' & \mu_1'' & \nu_1'' \\ \lambda_2'' & \mu_2'' & \nu_2'' \\ \lambda_3'' & \mu_3'' & \nu_3'' \end{Bmatrix} \begin{Bmatrix} x' \\ y' \\ z' \end{Bmatrix} \quad ; \tag{2.14}$$

or, more explicitly,

$$\begin{aligned} u' = {} & (\lambda_1''\dot{\lambda}_1'' + \lambda_2''\dot{\lambda}_2'' + \lambda_3''\dot{\lambda}_3'')\, x' \\ & + (\mu_1''\dot{\lambda}_1'' + \mu_2''\dot{\lambda}_2'' + \mu_3''\dot{\lambda}_3'')\, y' \\ & + (\nu_1''\dot{\lambda}_1'' + \nu_2''\dot{\lambda}_2'' + \nu_3''\dot{\lambda}_3'')\, z', \end{aligned} \tag{2.15}$$

$$\begin{aligned} v' = {} & (\lambda_1''\dot{\mu}_1'' + \lambda_2''\dot{\mu}_2'' + \lambda_3''\dot{\mu}_3'')\, x' \\ & + (\mu_1''\dot{\mu}_1'' + \mu_2''\dot{\mu}_2'' + \mu_3''\dot{\mu}_3'')\, y' \\ & + (\nu_1''\mu_1'' + \nu_2''\dot{\mu}_2'' + \nu_3''\dot{\mu}_3'')\, z', \end{aligned} \tag{2.16}$$

$$\begin{aligned} w' = {} & (\lambda_1''\dot{\nu}_1'' + \lambda_2''\dot{\nu}_2'' + \lambda_3''\dot{\nu}_3'')\, x' \\ & + (\mu_1''\dot{\nu}_1'' + \mu_2''\dot{\nu}_2'' + \mu_3''\dot{\nu}_3'')\, y' \\ & + (\nu_1''\dot{\nu}_1'' + \nu_2''\dot{\nu}_2'' + \nu_3''\dot{\nu}_3'')\, z'. \end{aligned} \tag{2.17}$$

In order to simplify this latter equation let us observe, first, that

$$\left.\begin{aligned} & \lambda_1''\dot{\lambda}_1'' + \lambda_2''\dot{\lambda}_2'' + \lambda_3''\dot{\lambda}_3'' = 0, \\ & \mu_1''\dot{\mu}_1'' + \mu_2''\dot{\mu}_2'' + \mu_3''\dot{\mu}_3'' = 0, \\ & \nu_1''\dot{\nu}_1'' + \nu_2''\dot{\nu}_2'' + \nu_3''\dot{\nu}_3'' + 0. \end{aligned}\right\} \tag{2.18}$$

Next, in differentiating the direction cosines λ_j'', μ_j'', ν_j'' ($j = 1, 2, 3$) as given by Equation (2.7) with respect to the time, let us assume that the time-dependence of the direction cosines a_{ij}'' as given by Equations (2.2)–(2.5) arises solely from the presence in them of periodic functions of the time anomaly u of the secondary component in its relative orbit, measured from the node Ω. If we regard, accordingly, the longitude Ω of the node as well as the inclination i of the orbital plane to invariable plane $Z = 0$ of the system as constant, we readily find that

$$\left.\begin{aligned} \dot{a}_{i1}'' &= +\dot{u}a_{i2}'' , \\ \dot{a}_{i2}'' &= -\dot{u}a_{i1}'' , \\ \dot{a}_{i3}'' &= \ \ 0 , \end{aligned}\right\} \tag{2.19}$$

$i = 1, 2, 3$; where, for circular orbit, $\dot{u} \equiv n$ denotes the mean daily motion of the revolving system with respect to the space axes.

Remembering, furthermore, that the time-derivatives of the direction cosines a_{ij} can be expressed in terms of the angular velocity components $\omega_{x,y,z}$ by Equations (1.11)–(1.19) we find that

$$\left.\begin{aligned} \lambda_1''\dot{\mu}_1'' + \lambda_2''\dot{\mu}_2'' + \lambda_3''\dot{\mu}_3'' &= -\,\omega_z' + n\nu_3'',\\ \lambda_1''\dot{\nu}_1'' + \lambda_2''\dot{\nu}_2'' + \lambda_3''\dot{\nu}_3'' &= +\,\omega_y' - n\mu_3''; \end{aligned}\right\} \tag{2.20}$$

$$\left.\begin{aligned} \mu_1''\dot{\lambda}_1'' + \mu_2''\dot{\lambda}_2'' + \mu_3''\dot{\lambda}_3'' &= +\,\omega_z' - n\lambda_3'',\\ \mu_1''\dot{\nu}_1'' + \mu_2''\dot{\nu}_2'' + \mu_3''\dot{\nu}_3'' &= -\,\omega_x' + n\lambda_3''; \end{aligned}\right\} \tag{2.21}$$

$$\left.\begin{aligned} \nu_1''\dot{\lambda}_1'' + \nu_2''\dot{\lambda}_2'' + \nu_3''\dot{\lambda}_3'' &= -\,\omega_y' + n\mu_3'',\\ \nu_1''\dot{\mu}_1'' + \nu_2''\dot{\mu}_2'' + \nu_3''\dot{\mu}_3'' &= +\,\omega_x' - n\lambda_3''; \end{aligned}\right\} \tag{2.22}$$

where

$$\begin{Bmatrix} \omega_x' \\ \omega_y' \\ \omega_z' \end{Bmatrix} = \begin{Bmatrix} a_{11} & a_{21} & a_{31} \\ a_{12} & a_{22} & a_{32} \\ a_{13} & a_{23} & a_{33} \end{Bmatrix} \begin{Bmatrix} \omega_x \\ \omega_y \\ \omega_z \end{Bmatrix} \tag{2.23}$$

are the angular velocity-components about the rotating system of body-axes $x'y'z'$.

If we insert now Equations (2.18)–(2.22) in (2.15)–(2.17) we find that the body velocity components u', v', w' in the direction of the rotating axes of coordinates $x',y'z'$ should be given by

$$u' = \;\;(\omega_z' - n\nu_3'')y' - (\omega_y' - n\mu_3'')z', \tag{2.24}$$

$$v' = -(\omega_z' - n\nu_3'')x' + (\omega_x' - n\lambda_3'')z', \tag{2.25}$$

$$w' = \;\;(\omega_y' - n\mu_3'')x' - (\omega_x' - n\lambda_3'')y'; \tag{2.26}$$

and, therefore,

$$\begin{aligned} u'\frac{\partial f}{\partial x'} + v'\frac{\partial f}{\partial y'} + w'\frac{\partial f}{\partial z'} = -\,&\{(\omega_z' - n\nu_3'')D_3' + \\ &+ (\omega_y' - n\mu_3'')D_2' + (\omega_x' - n\lambda_3'')D_1'\}f, \end{aligned} \tag{2.27}$$

where the D_j''s stand for the cyclic differential operators

$$D_1' \equiv y'\frac{\partial}{\partial z'} - z'\frac{\partial}{\partial y'}\,, \tag{2.28}$$

$$D_2' \equiv z'\frac{\partial}{\partial x'} - x'\frac{\partial}{\partial z'}\,, \tag{2.29}$$

$$D_3' \equiv x'\frac{\partial}{\partial y'} - y'\frac{\partial}{\partial z'}\,. \tag{2.30}$$

Equation (2.27) can be transformed in the system of inertial coordinates xyz by remembering (cf. Equations (1.92)–(1.94)) that

$$\begin{Bmatrix} D_1' \\ D_2' \\ D_3' \end{Bmatrix} = \begin{Bmatrix} a_{11} & a_{21} & a_{31} \\ a_{12} & a_{22} & a_{32} \\ a_{13} & a_{23} & a_{33} \end{Bmatrix} \begin{Bmatrix} D_1 \\ D_2 \\ D_3 \end{Bmatrix} . \tag{2.31}$$

where the D_j's denote the same operators in the inertial coordinates as the D_j''s do in the rotating system. Making use of (2.23) with (2.32), and abbreviating

$$\begin{Bmatrix} a_{11} & a_{12} & a_{13} \\ a_{21} & a_{22} & a_{23} \\ a_{31} & a_{32} & a_{33} \end{Bmatrix} \begin{Bmatrix} u' \\ v' \\ w' \end{Bmatrix} = \begin{Bmatrix} u \\ v \\ w \end{Bmatrix} , \tag{2.32}$$

we find that Equation (2.27) can be alternatively rewritten as

$$\begin{aligned} u'f_{x'} + v'f_{y'} + w'f_z &= uf_x + vf_y + wf_z \\ &= -(p_1D_1 + p_2D_2 + p_3D_3)f , \end{aligned} \tag{2.33}$$

where

$$\left.\begin{aligned} u &= p_3y - p_2z, \\ v &= p_1z - p_3x, \\ w &= p_2x - p_1y ; \end{aligned}\right\} \tag{2.34}$$

and

$$\left.\begin{aligned} p_1 &= \omega_x - n \cos\gamma_1 , \\ p_2 &= \omega_y - n \cos\gamma_2 , \\ p_3 &= \omega_z - n \cos\gamma_3 , \end{aligned}\right\} \tag{2.35}$$

in which we have abbreviated

$$\cos\gamma_j = a_{j1}\lambda_3'' + a_{j2}\mu_3'' + a_{j3}\,\nu_3'' = a_{j3}'' ; \tag{2.36}$$

with a_{j3}'' as given by Equations (2.3)–(2.5).

Equation (2.33) enables us to express (2.12) in the form

$$\dot{f} = f_t - (p_1D_1 + p_2D_2 + p_3D_3)f, \tag{2.37}$$

which is exact for any values of f or p_j; and of which frequent use will be made in the sequel. Moreover, from (2.9)–(2.10) and (2.24)–(2.25) it follows now that, likewise exactly,

$$u_0' = (p_3y - p_2z)f + x\dot{f}, \tag{2.38}$$

$$v_0' = (p_1z - p_3x)f + y\dot{f}, \tag{2.39}$$

$$w_0' = (p_2x - p_1y)f + z\dot{f} ; \tag{2.40}$$

and, in accordance with Equations (1.49)

$$\begin{Bmatrix} (\dot{u})_0' \\ (\dot{v})_0' \\ (\dot{w})_0' \end{Bmatrix} = \begin{Bmatrix} a_{11} & a_{12} & a_{13} \\ a_{21} & a_{22} & a_{23} \\ a_{31} & a_{32} & a_{33} \end{Bmatrix} \begin{Bmatrix} \dot{u}'f + 2u'\dot{f} + x'\ddot{f} \\ \dot{v}'f + 2v'\dot{f} + y'\ddot{f} \\ \dot{w}'f + 2w'\dot{f} + z'\ddot{f} \end{Bmatrix}, \tag{2.41}$$

where u', v', w' continue to be given by Equations (2.24)–(2.26).

Differentiating the latter with respect to the time, we find that

$$\begin{aligned} \dot{u}' &= (\omega_{z'} - n\nu_3'')\,v' - (\omega_{y'} - n\mu_3'')\,w' + \\ &+ (\dot{\omega}_{z'} - \dot{n}\nu_3'' - n\dot{\nu}_3'')\,y' - (\dot{\omega}_{y'} - \dot{n}\mu_3'' - n\dot{\mu}_3'')z', \end{aligned} \tag{2.42}$$

$$\begin{aligned} \dot{v}' &= (\omega_{y'} - n\lambda_3'')w' - (\omega_{z'} - n\nu_3'')u' \\ &+ (\dot{\omega}_{x'} - \dot{n}\lambda_3'' - n\dot{\lambda}_3'')z' - (\dot{\omega}_{z'} - \dot{n}\nu_3'' - n\dot{\nu}_3'')x', \end{aligned} \tag{2.43}$$

$$\begin{aligned} \dot{w}' &= (\omega_{y'} - n\mu_3'')u' - (\omega_{x'} - n\lambda_3'')v' \\ &+ (\dot{\omega}_{x'} - \dot{n}\mu_3'' - n\dot{\mu}_3'')x' - (\dot{\omega}_{x'} - \dot{n}\lambda_3'' - n\dot{\lambda}_3'')y'; \end{aligned} \tag{2.44}$$

so that

$$a_{11}\dot{u}' + a_{12}\dot{v}' + a_{13}\dot{w}' = \dot{p}_3 y - p_2 z + \dot{p}_3 v - p_2 w, \tag{2.45}$$

$$a_{21}\dot{u}' + a_{22}\dot{v}' + a_{23}\dot{w}' = \dot{p}_1 z - p_3 x + \dot{p}_1 w - p_3 u, \tag{2.46}$$

$$a_{21}\dot{u}' + a_{32}\dot{v}' + a_{33}\dot{w}' = \dot{p}_2 x - p_1 y + \dot{p}_2 u - p_1 v. \tag{2.47}$$

where u, v, w and $p_{1,2,3}$ continue to be given by Equation (2.34) and(2.35). Accordingly, from Equation (2.41) it follows that

$$\begin{aligned} y(\dot{w})_0' - z(\dot{v})_0' &= 2[(p_1 x + p_2 y + p_3 z)x - p_1 r^2]\dot{f} \\ &+ [(\dot{p}_2 + p_1 p_3)xy + (\dot{p}_3 - p_1 p_2)xz + (p_3^2 - p_2^2)\,yz \\ &- (\dot{p}_1 - p_2 p_3)y^2 - (\dot{p}_1 + p_2 p_3)z^2]f, \end{aligned} \tag{2.48}$$

$$\begin{aligned} z(\dot{u})_0' - x(\dot{w})_0' &= 2[(p_1 x + p_2 y + p_3 z)y - p_2 r^2]\dot{f} \\ &+ [(\dot{p}_3 + p_2 p_1)yz + (\dot{p}_1 - p_2 p_3)yx + (p_1^2 - p_3^2)xz \\ &- (\dot{p}_2 - p_3 p_1)z^2 - (\dot{p}_2 + p_3 p_1)x^2]f, \end{aligned} \tag{2.49}$$

$$\begin{aligned} x(\dot{v})_0' - y(\dot{u})_0' &= 2[(p_1 x + p_2 y + p_3 z)z - p_3 r^2]\dot{f} \\ &+ [(\dot{p}_1 + p_3 p_2)zx + (\dot{p}_2 - p_3 p_1)zy + (p_2^2 - p_1^2)\,xy \\ &- (\dot{p}_3 - p_1 p_2)\,x^2 - (\dot{p}_3 + p_1 p_2)y^2]f, \end{aligned} \tag{2.50}$$

again without any restriction on the extent of the tidal distortion f or of the magnitudes of the coefficients $p_{1,2,3}$.

B. MOMENTS AND PRODUCTS OF INERTIA

As is well-known, the moments and products of inertia occurring on the left-hand sides of our fundamental Equations (1.108)–(1.110) are defined (cf. sec. II-6) by the expressions

$$A = \int (Y^2 + Z^2)\, dm, \tag{2.51}$$

$$B = \int (X^2 + Z^2)\, dm, \tag{2.52}$$

$$C = \int (X^2 + Y^2)\, dm; \tag{2.53}$$

$$D = \int YZ\, dm, \tag{2.54}$$

$$E = \int XZ\, dm, \tag{2.55}$$

$$F = \int XY\, dm, \tag{2.56}$$

where

$$X = x(1+f), \qquad Y = y(1+f), \qquad Z = z(1+f), \tag{2.57}$$

so that

$$R = r(1+f); \tag{2.58}$$

and the mass-element

$$dm = \rho R^2 dR \sin\theta\, d\theta\, d\phi, \tag{2.59}$$

where r, θ, ϕ denote now the spherical polar coordinates in inertial xyz-system; and the limits of integration in (2.51)–(2.56) are to be extended over entire mass of the respective configuration of density ρ which may vary in its interior in an appropriate manner.

If we regard now f to be small enough for its squares and higher powers to be ignorable – an assumption to which we shall adhere throughout this section – and change over from R to r by the transformation

$$dR = \frac{\partial R}{\partial r}\, dr = \left\{1 + \frac{\partial}{\partial r}(rf)\right\} dr, \tag{2.60}$$

it follows that

$$\begin{aligned} dm &= \rho\left\{r^2 + \frac{\partial}{\partial r}(r^3 f)\right\} dr \sin\theta\, d\theta\, d\phi \\ &= \rho\left\{1 + \frac{1}{r^2}\frac{\partial}{\partial r}(r^3 f)\right\} dV; \end{aligned} \tag{2.61}$$

the evaluation of the moments and products of inertia as given by the integrals on the right-hand sides of Equations (2.51)–(2.56) offers no difficulty.

In order to do so, we find it expedient to change over from the inertial to doubly-primed coordinates, because the tides are symmetrical with respect to the x''-axis coincident with the radius-vector joining the centres of the two bodies. Therefore, the limits of integration on the right-hand sides of Equations (2.51)–(2.56), which would vary with the phase in the (inertial) xyz-coordinates, should become constant in the $x''y''z''$ revolving frame of reference* – a fact which greatly simplifies their evaluation. Since, moreover, the volume elements

$$dx\,dy\,dz = dx''\,dy''\,dz'', \tag{2.62}$$

the mass-element dm obviously remains of the same form in both sets of coordinates; and the only modification necessary in (2.61) is to replace the spherical polar coordinates r, θ, ϕ of the inertial system, defined by

$$\left.\begin{aligned} x &= r\cos\phi\sin\theta &&= r\lambda, \\ y &= r\sin\phi\sin\theta &&= r\mu, \\ z &= r\cos\theta &&= r\nu, \end{aligned}\right\} \tag{2.63}$$

by their doubly-primed versions

$$\left.\begin{aligned} x'' &= r\cos\phi''\sin\theta'' &&= r\lambda'', \\ y'' &= r\sin\phi''\sin\theta'' &&= r\mu'', \\ z'' &= r\cos\theta &&= r\nu'', \end{aligned}\right\} \tag{2.64}$$

and to rewrite the inertial coordinates in the integrands of (2.51)–(2.56) in terms of the doubly-primed revolving system by the second one of the transformations (2.2). The limits of integration of the expressions (2.51)–(2.56) in the doubly-primed coordinates then assume the form

$$\int \ldots dm = \int_0^{R_1}\int_0^{\pi}\int_0^{2\pi} \ldots \rho r^2 dr \sin\theta''\,d\theta''\,d\phi'', \tag{2.65}$$

where R_1 denotes the mean radius of our configuration.

The actual integration between these limits need not be pursued here in any detail, as the method is already known to us from Chapter II; so that only the essential points of the underlying operations need to be mentioned in this place. First, inasmuch as the weight functions

$$\begin{aligned} y^2 + z^2 &= r^2\{1 - (a''_{11}\lambda'' + a''_{12}\mu'' + a''_{13}\nu'')^2\}, \\ x^2 + z^2 &= r^2\{1 - (a''_{21}\lambda'' + a''_{22}\mu'' + a''_{23}\nu'')^2\}, \\ x^2 + y^2 &= r^2\{1 - (a''_{31}\lambda'' + a''_{32}\mu'' + a''_{33}\nu'')^2\}, \end{aligned} \tag{2.66}$$

* Except for the phenomenon of 'tidal lag', which would cause a displacement of the axes of tidal symmetry with respect to the $x''y''z''$-axes by a constant angle. This angle should, however, be generally small enough for its squares and higher powers to be negligible (cf. sec. III-3) which should maintain the symmetry; and even if it were not, a simple rotation to new axes of symmetry of dynamical tides should restore the constancy of the corresponding limits of integration.

of the three moments of inertia (as well as of the products of inertia) can be expresssed in terms of the tesseral harmonics in θ and ϕ of second order ($j = 2$), the integral operator (2.65) will annihilate all terms in f except those associated with the second harmonics. These alone can survive in our results with non-vanishing coefficients; all others will become victims of the orthogonality properties of the respective harmonics.

Secondly, from the theory of tides developed in Chapters II and III we can expect that, for configurations of pronounced density concentration, the relative heights of the partial tides associated with the j-th harmonic $P_j(\lambda'')$ should (to the first order in small quantities) be given by Equation (6.2) of Chapter II, and expressible as

$$f_j = (1 + 2k_j)\frac{m_2}{m_1}\left(\frac{r}{\Delta}\right)^{j+1} P_j(\lambda'') \equiv K_j r^{j+1} P_j(\lambda'') \tag{2.67}$$

for $j = 2, 3, 4$, where $m_{1,2}$ denote the masses of the distorted (m_1) and disturbing (m_2) component, respectively; $P_j(\lambda'')$, the Legendre polynomial in λ'' of j-th order; and

$$\Delta = \frac{A(1-e^2)}{1 + e \cos v}\ , \tag{2.68}$$

the radius-vector of the relative orbit of the two stars, of semi-major axis A and eccentricity e, representing their mutual separation for the true anomaly v; while the k_j's are constants depending on the internal structure of the configuration of mass m_1, which tend rapidly to values defined by Equation (6.4) of Chapter II.

If we adopt the foregoing expression (2.67) for the fractional height of partial tides which distort our configuration, on performing the requisite integrations we find that the contributions to the moments of inertia arising from the tides will be of the form

$$A'' = -\frac{64}{15}\pi K_2 P_2 (a''_{11}) \int_0^{R_1} \rho r^7 \mathrm{d}r, \tag{2.69}$$

$$B'' = -\frac{64}{15}\pi K_2 P_2 (a''_{21}) \int_0^{R} \rho r^7 \mathrm{d}r, \tag{2.70}$$

$$C'' = -\frac{64}{15}\pi K_2 P_2 (a''_{31}) \int_0^{R} \rho r^7 \mathrm{d}r; \tag{2.71}$$

while the corresponding products of inertia are similarly found to be

$$D'' = \frac{32}{5}\pi a''_{21} a''_{31} K_2 \int_0^{R} \rho r^7 \mathrm{d}r, \tag{2.72}$$

$$E'' = \frac{32}{5}\pi a''_{11} a''_{31} K_2 \int_0^{R} \rho r^7 \mathrm{d}r, \tag{2.73}$$

$$F'' = \frac{32}{5}\pi a''_{11} a''_{21} K_2 \int_0^{R} \rho r^7 \mathrm{d}r\ . \tag{2.74}$$

In addition to tidal distortion, our configuration of mass m_1 must also suffer flattening at the poles on account of its rotation about the z'-axis with an angular velocity $\omega_{z'}$.

The rotational distortion arising from this cause is of the same order of magnitude as that arising from the tides, and is describable by the function

$$g = -\frac{\omega_z^{2\prime} r^3}{3\,Gm_1} P_2(\nu'), \tag{2.75}$$

where G denotes the constant of gravitation, and

$$\nu' = a_{13}\lambda + a_{23}\mu + a_{33}\nu = \cos\theta', \tag{2.76}$$

expressing the effects of centrifugal force of axial rotation as the function f represented the effects of tidal attraction.

If we replace f behind the integral signs in (2.51)–(2.56) by g as given by Equation (2.75), and change over to the volume element

$$\mathrm{d}x'\,\mathrm{d}y'\,\mathrm{d}z' = r^2 \sin\theta'\,\mathrm{d}r\,\mathrm{d}\theta'\,\mathrm{d}\phi' \tag{2.77}$$

in the coordinate system rotating with the body, the integration limits on the right-hand side of (2.65) remain unaltered in the new variables. On performing the requisite integrations we find that the contributions of the polar flattening to the moments and products of inertia of the rotating mass m_1 will assume the forms

$$A' = \frac{64\,\omega_z^{2\prime}}{45\,Gm_1} P_2(a_{13}) \int_0^{R_1} \rho r^7\,\mathrm{d}r, \tag{2.78}$$

$$B' = \frac{64\,\omega_z^{2\prime}}{45\,Gm_1} P_2(a_{23}) \int_0^{R_1} \rho r^7\,\mathrm{d}r, \tag{2.79}$$

$$C' = \frac{64\,\omega_z^{2\prime}}{45\,Gm_1} P_2(a_{33}) \int_0^{R_1} \rho r^7\,\mathrm{d}r, \tag{2.80}$$

$$D' = -\frac{64\,\omega_z^{2\prime}}{15\,Gm_1} a_{23}a_{33} \int_0^{R_1} \rho r^7\,\mathrm{d}r, \tag{2.81}$$

$$E' = -\frac{64\,\omega_z^{2\prime}}{15\,Gm_1} a_{13}a_{33} \int_0^{R_1} \rho r^7\,\mathrm{d}r, \tag{2.82}$$

$$F' = -\frac{64\,\omega_z^{2\prime}}{15\,Gm_1} a_{13}a_{23} \int_0^{R_1} \rho r^7\,\mathrm{d}r, \tag{2.83}$$

In the absence of a rotational or tidal distortion,

$$A_0 = B_0 = C_0 = \frac{8}{3}\pi \int_0^{R_1} \rho r^4\,\mathrm{d}r = m_1 h_1^2, \tag{2.84}$$

where h_1 denotes the radius of gyration of the respective configuration, and

$$D_0 = E_0 = F_0 = 0. \tag{2.85}$$

As, moreover, for configurations of pronounced central condensation it is possible (with the aid of Equation (6.9) of Chapter II, in which we insert $D \cong 4\pi\rho a^3/3m_1$) to approximate the remaining integrals on the right-hand sides of Equations (2.69)–(2.74) or (2.78) – (2.83) by

$$\int_0^{R_1} \rho a^{2j+3}\, da = \frac{(2j+1)k_j}{4\pi(j+2)} R_1^{2j+1} m_1 \,, \tag{2.86}$$

a combination of (2.69)–(2.74) with (2.78)–(2.83) eventually yields

$$A = m_1 h_1^2 - \frac{2}{3} k_2 R_1^5 \left\{ \frac{m_2}{\Delta^3} P_2(a_{11}'') - \frac{\omega_z^{2\prime}}{3G} \right\} P_2(a_{13}) \,, \tag{2.87}$$

$$B = m_1 h_1^2 - \frac{2}{3} k_2 R_1^5 \left\{ \frac{m_2}{\Delta^3} P_2(a_{21}'') - \frac{\omega_z^{2\prime}}{3G} \right\} P_2(a_{23}) \,, \tag{2.88}$$

$$C = m_1 h_1^2 - \frac{2}{3} k_2 R_1^5 \left\{ \frac{m_2}{\Delta^3} P_2(a_{31}'') - \frac{\omega_z^{2\prime}}{3G} \right\} P_2(a_{33}) \,, \tag{2.89}$$

and

$$D = k_2 R_1^5 \left\{ \frac{m_2}{\Delta^3} a_{21}'' a_{31}'' - \frac{\omega_z^{2\prime}}{3G} a_{23} a_{33} \right\} , \tag{2.90}$$

$$E = k_2 R_1^5 \left\{ \frac{m_2}{\Delta^3} a_{11}'' a_{31}'' - \frac{\omega_z^{2\prime}}{+G} a_{13} a_{33} \right\} , \tag{2.91}$$

$$F = k_2 R_1^5 \left\{ \frac{m_2}{\Delta^3} a_{11}'' a_{21}'' - \frac{\omega_z^{2\prime}}{3G} a_{13} a_{23} \right\} ; \tag{2.92}$$

to the first order in surficial distortion, but without any restriction on the magnitude of the direction cosines a_{ij} or a_{ij}''.

C. COEFFICIENTS OF DEFORMATION

The next task confronting us in our problem is to formulate explicitly the gyroscopic terms in the fundamental equations (1.108)–(1.110), consisting of the products of the angular velocity components $\omega_{x,y,z}$ with the 'coefficients of deformation' arising from dynamical tides. All such coefficients prove expressible as mass-integrals of functions which are homogeneous in the velocity components u_0', v_0', w_0' as given by Equations (2.38)–(2.40); and the method of their integration can follow closely the one adopted in the preceding section for an evaluation of the moments and products of inertia.

In particular, because of the constancy of the limits of integration in the doubly-primed coordinates, we propose to integrate again with respect to the volume element $dx''dy''dz''$; and to this end the inertial coordinates x, y, z involved in u_0', v_0', w_0' must first be rewritten in terms of x'', y'', z'' by means of the linear transformation (2.2). Moreover, for f_j of the form (2.68), only terms associated with the second harmonic $P_2(\lambda'')$

will survive again the process of integration, between the limits given by (2.65), with respect to the surface element $dS = \sin\theta''\, d\theta''\, d\phi''$; and these are found to reduce to

$$\int \lambda''^2 P_2(\lambda'')\, dS = \frac{8}{15}\pi \tag{2.93}$$

and

$$\int \mu''^2 P_2(\lambda'')\, dS = \int \nu''^2 P_2(\lambda'')\, dS = -\frac{4}{15}\pi . \tag{2.94}$$

The operators $D_{1,2,3}$ in (2.37) transform into the doubly-primed coordinates in accordance with the scheme

$$\begin{Bmatrix} D_1 \\ D_2 \\ D_3 \end{Bmatrix} = \begin{Bmatrix} a''_{11} & a''_{12} & a''_{13} \\ a''_{21} & a''_{22} & a''_{23} \\ a''_{31} & a''_{32} & a''_{33} \end{Bmatrix} \begin{Bmatrix} D''_1 \\ D''_2 \\ D''_3 \end{Bmatrix} , \tag{2.95}$$

where

$$D''_1 \equiv y'' \frac{\partial}{\partial z''} - z'' \frac{\partial}{\partial y''} \equiv \qquad + \mu'' \frac{\partial}{\partial \nu''} , \tag{2.96}$$

$$D''_2 \equiv z'' \frac{\partial}{\partial x''} - x'' \frac{\partial}{\partial z''} \equiv \nu'' \frac{\partial}{\partial \lambda''} - \lambda'' \frac{\partial}{\partial \nu''} , \tag{2.97}$$

$$D''_3 \equiv x'' \frac{\partial}{\partial y''} - y'' \frac{\partial}{\partial z''} \equiv - \mu'' \frac{\partial}{\partial \lambda''} , \tag{2.98}$$

by (2.64); so that

$$D''_1 f_2 = 0, \tag{2.99}$$

$$D''_2 f_2 = 3 K_2 r^3 \lambda'' \nu'', \tag{2.100}$$

$$D''_3 f_2 = 3 K_2 r^3 \lambda'' \mu'' ; \tag{2.101}$$

and the only quadratic terms in x, y, z which are not annihilated by integration with respect of $d\phi''$ between 0 and 2π are

$$\int x'' z'' D''_2 f_2\, dS = \frac{4}{5}\pi K_2 r^5 \tag{2.102}$$

and

$$\int x'' y'' D''_3 f_2\, dS = \frac{4}{5}\pi K_2 r^5 . \tag{2.103}$$

By use of (2.93)–(2.94) and (2.102)–(2.103), and by resorting again to (2.86) for an evaluation of the radial integrals of ρr^5 with respect to the spherical volume element $r^2 dr$, we find that

$$\int x v'_0\, dm = \alpha \left\{ \frac{2}{3} H_2 P_2(a''_{11}) + K_2 a''_{11} (p_2 a''_{31} - p_3 a''_{21}) \right\} , \tag{2.104}$$

$$\int yu_0' \, dm = \alpha\left\{H_2 a_{11}'' a_{21}'' + K_2\left[\frac{2}{3} p_3 p_2(a_{11}'') - p_1 a_{11}'' a_{31}''\right]\right\}, \tag{2.105}$$

$$\int zu_0' \, dm = \alpha\left\{H_2 a_{11}'' a_{31}'' - K_2\left[\frac{2}{3} p_2 p_2(a_{11}'') - p_1 a_{11}'' a_{21}''\right]\right\}; \tag{2.106}$$

$$\int xv_0' \, dm = \alpha\left\{H_2 a_{11}'' a_{21}'' - K_2\left[\frac{2}{3} p_3 p_2(a_{21}'') - p_2 a_{21}'' a_{31}''\right]\right\}, \tag{2.107}$$

$$\int yv_0' \, dm = \alpha\left\{\frac{2}{3} H_2 P_2(a_{21}'') + K_2 a_{21}''(p_3 a_{11}'' - p_1 a_{31}'')\right\}, \tag{2.108}$$

$$\int zv_0' \, dm = \alpha\left\{H_2 a_{21}'' a_{31}'' + K_2\left[\frac{2}{3} p_1 p_2(a_{21}'') - p_2 a_{11}'' a_{21}''\right]\right\}; \tag{2.109}$$

$$\int xw_0' dm = \alpha\left\{H_2 a_{11}'' a_{31}'' + K_2\left[\frac{2}{3} p_2 p_2(a_{31}'') - p_3 a_{21}'' a_{31}''\right]\right\}, \tag{2.110}$$

$$\int yw_0' dm = \alpha\left\{H_2 a_{21}'' a_{31}'' - K_2\left[\frac{2}{3} p_1 p_2(a_{31}'') - p_3 a_{11}'' a_{31}''\right]\right\}, \tag{2.111}$$

$$\int zw_0' dm = \alpha\left\{\frac{2}{3} H_2 P_2(a_{31}'') + K_2 a_{31}''(p_1 a_{21}'' - p_2 a_{11}'')\right\}; \tag{2.112}$$

where we have abbreviated

$$\alpha = \tfrac{1}{4} k_2 m_1 R_1^5 , \tag{2.113}$$

and where H_2 stands for the coefficient in the equation

$$f_t = H_j r^{j+1} P_j(\lambda''). \tag{2.114}$$

The only factor in f_j (as given by Equation (2.67)) which renders it explicitly dependent on the time is the radius-vector Δ; and a partial differentiation with respect to t discloses that, for $j = 2$,

$$H_2 = -3K_2 \frac{\dot{\Delta}}{\Delta} = -3\frac{K_2}{\Delta^3}\left\{\frac{e \sin v}{1 + e \cos v}\right\} \dot{v}. \tag{2.115}$$

Equations (2.104)–(2.112) make it, moreover, evident that

$$\int (yv_0' + zw_0') dm = \alpha\left\{K_2 a_{11}''(p_3 a_{21}'' - p_2 a_{31}'') - \frac{2}{3} H_2 P_2(a_{11}'')\right\}, \tag{2.116}$$

$$\int (xu_0' + zw_0') dm = \alpha\left\{K_2 a_{21}''(p_1 a_{31}'' - p_3 a_{11}'') - \frac{2}{3} H_2 P_2(a_{21}'')\right\}, \tag{2.117}$$

$$\int (yv_0' + xu_0') dm = \alpha\left\{K_2 a_{31}''(p_2 a_{11}'' - p_1 a_{21}'') - \frac{2}{3} H_2 P_2(a_{31}'')\right\}. \tag{2.118}$$

As the next step of our analysis, it remains for us to evaluate similarly the mass integrals of the differences in body accelerations with respect to the space axes, as given by

Equations (2.48)–(2.50). Proceeding exactly in the same way as we did in evaluating the coefficients of deformation (2.104)–(2.112) we establish, after some algebra, that

$$\int \{y(\dot{w})'_0 - z(\dot{v})'_0\}\, dm = 2\alpha\, \{a''_{11}\omega_{x''} - \tfrac{1}{3}\, p_1\}\, H_2 + \tag{2.119}$$

$$+\, \alpha\{a''_{11}(\omega_{x''} - n\omega_{y''}) + a''_{12}(\omega_{z''} - n)\omega_{x''} - a''_{13}\omega_{x''}\omega_{y''} - \tfrac{1}{3}\, p_1\}\, K_2,$$

$$\int \{z(u)'_0 - x(w)'_0\}\, dm = 2\alpha\, \{a''_{21}\omega_{x''} - \tfrac{1}{3}\, p_2\}\, H_2 +$$

$$+\, \alpha\, \{a''_{21}(\dot{\omega}_{x''} - n\omega_{y''}) + a''_{22}(\omega_{z''} - n)\omega_{x''} - a''_{23}\omega_{x''}\omega_{y''} - \tfrac{1}{3}\, \dot{p}_2\}\, K_2, \tag{2.120}$$

$$\int \{x(\dot{v})'_0\}\, dm = 2\alpha\, \{a''_{31}\omega_{x''} - \tfrac{1}{3}\, p_3\}\, H_2 +$$

$$+\, \alpha\, \{a''_{31}(\dot{\omega}_{x''} - n\omega_{y''}) + a''_{32}(\omega_{z''} - n)\omega_{x''} - a''_{33}\omega_{x''}\omega_{y''} - \tfrac{1}{3}\, \dot{p}_3\}\, K_2, \tag{2.121}$$

where

$$\begin{Bmatrix} \omega_{x''} \\ \omega_{y''} \\ \omega_{z''} \end{Bmatrix} = \begin{Bmatrix} a''_{11} & a''_{21} & a''_{31} \\ a''_{12} & a''_{22} & a''_{32} \\ a''_{13} & a''_{23} & a''_{33} \end{Bmatrix} \begin{Bmatrix} \omega_x \\ \omega_y \\ \omega_z \end{Bmatrix} = \begin{Bmatrix} \lambda''_1 & \mu''_1 & \nu''_1 \\ \lambda''_2 & \mu''_2 & \nu''_2 \\ \lambda''_3 & \mu''_3 & \nu''_3 \end{Bmatrix} \begin{Bmatrix} \omega_{x'} \\ \omega_{y'} \\ \omega_{z'} \end{Bmatrix} \tag{2.122}$$

stand for the angular velocity components with respect to the revolving (doubly-primed) axes of coordinates.

It may also be noted that, in the coefficients of K_2 on the right-hand sides of Equations (2.116)–(2.118),

$$\left.\begin{aligned} p_3 a''_{21} - p_2 a''_{31} &= a''_{13}\,\omega_{y''} - a''_{12}\,(\omega_{z''} - n)\,, \\ p_1 a''_{31} - p_3 a''_{11} &= a''_{23}\,\omega_{y''} - a''_{22}(\omega_{z''} - n)\,, \\ p_2 a''_{11} - p_1 a''_{21} &= a''_{33}\omega_{y'} - a_{32}(\omega_{z''} - n)\,. \end{aligned}\right\} \tag{2.123}$$

All transformations used, and results obtained, in the present section so far continue to be exact, and hold good without any restriction on the fractional height of the dynamical tides f, on the magnitude of the coefficients $p_{1,2,3}$; or on the values of the direction cosines a''_{ij}.

Lastly, let us turn our attention to the integrals of the form

$$\int D_i\, \Psi\, dm \quad \text{and} \int D_i\, P dV, \qquad i = 1, 2, 3, \tag{2.124}$$

encountered on the left-hand sides of Equations (1.108)–(1.110), where Ψ denotes the total potential of all forces acting upon our configuration, and P, its internal pressure. Let us identify Ψ first with that part of the total potential which arises from self-attraction (i.e., Ω), and set

$$\Omega = \Omega_0(r) + \Omega'(r, \theta, \phi; t), \tag{2.125}$$

$$P = P_0(r) + P'(r, \theta, \phi; t), \tag{2.126}$$

where Ω_0, P_0 are the respective functions in the non-rotating (i.e., spherical) state, and Ω', P' represent the effects of rotation. The operators $D_{1,2,3}$, being purely angular, readily annihilate the radial parts Ω_0 as well as P_0 of the potential or pressure; thus leaving us with an evaluation of

$$\int D_i\Omega' \, dm \quad \text{and} \quad \int D_i P' \, dV .$$

It is, moreover, known (cf. Chapter II) that Ω' is generally expansible in terms of spherical harmonics which, for tidal distortion considered in this paper, are of the form $P_j(\lambda'')$. On the other hand, the D_i's are expressible (by (2.95)) as linear combinations of the operators D_i'' in doubly-primed coordinates. Of the latter, D_1'' readily annihilates $P_j(\lambda'')$; while operations with $D_{2,3}''$ lead to expressions of the form

$$\nu'' \frac{\partial P_j}{\partial \lambda''} \quad \text{and} \quad \mu \frac{\partial P_j}{\partial \lambda''}$$

or, for $j = 2$, of $3\lambda''\nu''$ and $3\lambda''\mu''$ which vanish both when integrated over the whole sphere. Therefore,

$$\int D_{1,2,3}\, \Omega \, dm = 0. \tag{2.127}$$

If, moreover, the perturbations P' in pressure are likewise expansible (as they should) in terms of the same spherical harmonics of the type $P_j(\lambda'')$, it follows that

$$\int D_{1,2,3}\, P \, dV = 0 \tag{2.128}$$

as well. The reason is the fact that if the changes Ω' in potential Ω are expansible in terms of $r^{j+1}\, P_j(\lambda'')$, then from Poisson's equation

$$\nabla^2 \Omega' = -4\pi G\rho', \tag{2.129}$$

it follows that

$$r^2 \nabla^2 \Omega' = 2(j+1)\Omega' = -4\pi G r^2 \rho' . \tag{2.130}$$

Therefore, if Ω' is expansible in terms of $P_j(\lambda'')$, so is ρ'; and as $P' = a^2 \rho'$ (a = local velocity of sound) from the energy equation for adiabatic changes of state, P' must be the same function of angular variables as ρ' or Ω' – Q.E.D.

The only part of the total potential of forces acting upon our configuration whose mass-integral need not vanish is that arising from the disturbing forces – such as centrifugal force due to axial rotation, or tidal attraction of the companion. The latter is, however, bound to act always in the radial direction, and its effects are, therefore, annihilated by the angular operators $D_{1,2,3}$. On the other hand, the terms $D_i\Omega_c$ deriving from the centrifugal potential Ω_c can give rise to nonvanishing contributions to the prevalent field of force, after integration over the whole mass of our rotating configuration, provided that the equator of this configuration is inclined to the orbital plane of its companion.

As is well known, the corresponding terms in the equations (1.108)–(1.110) of motion represent the effects of the companion's attraction on the equatorial bulge of the rotating configuration. In the inertial system of coordinates, the explicit form of such terms is found to be

$$(1+2k_2)\int D_1\Omega_c\,dm = 8\alpha K_2\,(\qquad + a_{33}a_{21}'' - a_{23}a_{31}'')\,\nu_1''\omega_{z'}^2, \quad (2.131)$$

$$(1+2k_2)\int D_2\Omega_c\,dm = 8\alpha K_2\,(-a_{33}a_{11}'' \qquad + a_{13}a_{31}'')\,\nu_1''\omega_{z'}^2, \quad (2.132)$$

$$(1+2k_2)\int D_3\Omega_c\,dm = 8\alpha K_2\,(+a_{23}a_{11}'' - a_{13}a_{21}'' \qquad)\,\nu_1''\omega_{z'}^2, \quad (2.133)$$

where $\omega_{z'}$ denotes the angular velocity of axial rotation of our configuration. In the rotating frame of coordinates, the foregoing expressions (2.131)–(2.133) transform readily into the components of applied force; but their derivation can be left as an exercise for the interested reader.

IV-3. Effects of Viscosity

In the preceding section of this chapter we have evaluated the explicit forms of all coefficients of the fundamental equations (1.108)–(1.110) of our problem in terms of the velocity components u_0', v_0', w_0' arising from dynamical tides – except for the last terms on the right-hand sides of these equations which arise from the effects of viscosity. In order to complete our task, our aim will be to evaluate these terms to the same degree of approximation to which the deformation coefficients were formulated in the preceding parts of this section.

A. EQUATIONS OF THE PROBLEM

In order to do so, let us depart from the functions

$$\begin{aligned}\rho\{y\mathfrak{H} - z\mathfrak{G}\} &= \mu\{y\nabla^2 w_0' - z\nabla^2 v_0' + \tfrac{1}{3}D_1\Delta_0'\}\\ &+ \frac{\partial u}{\partial x}\{D_1 u_0' + \frac{\partial}{\partial x}(yw_0' - zv_0')\} +\\ &+ \frac{2}{3}\frac{\partial\mu}{\partial y}\{2D_1 v_0' + D_4 w_0'\}\\ &+ \frac{2}{3}\frac{\partial\mu}{\partial z}\{2D_1 w_0' + D_4 v_0'\}\\ &- \frac{2}{3}\frac{\partial u_0'}{\partial x}D_1\mu + \tfrac{1}{3}\xi D_4\mu,\end{aligned} \qquad (3.1)$$

$$\begin{aligned}\rho\{z\mathfrak{F}-x\mathfrak{H}\} &= \mu\{z\nabla^2 u_0' - x\nabla^2 w_0' + \tfrac{1}{3}D_2\Delta_0'\}\\ &+\frac{2}{3}\frac{\partial\mu}{\partial x}\{2D_2u_0' - D_5w_0'\}\\ &+\frac{2}{3}\frac{\partial\mu}{\partial y}\{D_2v_0' + \frac{\partial}{\partial y}(zu_0' - xw_0')\}\\ &+\frac{2}{3}\frac{\partial\mu}{\partial z}\{2D_2w_0' + D_5u_0'\}\\ &-\frac{2}{3}\frac{\partial v_0'}{\partial y}D_2\mu + \tfrac{1}{3}\eta D_5\mu,\end{aligned} \tag{3.2}$$

and

$$\begin{aligned}\rho\{x\mathfrak{G}-y\mathfrak{H}\} &= \mu\{x\nabla^2 v_0' - y\nabla^2 u_0' + \tfrac{1}{3}D_3\Delta_0'\}\\ &+\frac{2}{3}\frac{\partial\mu}{\partial x}\{2D_3u_0' + D_6v_0'\}\\ &+\frac{2}{3}\frac{\partial\mu}{\partial y}\{2D_3v_0' - D_6u_0'\}\\ &+\frac{\partial\mu}{\partial z}\{D_3w_0' + \frac{\partial}{\partial z}(xv_0' - yu_0')\}\\ &-\frac{2}{3}\frac{\partial w_0'}{\partial z}D_3\mu + \tfrac{1}{3}\zeta D_6\mu,\end{aligned} \tag{3.3}$$

constituting the integrands of the terms in question, and defined already by Equations (1.89)–(1.91). The first group of terms on the right-hand sides of these equations, factored by the viscosity coefficient μ, involves the Laplacians of the different velocity-components as well as their divergence (representing the effects of compressibility). The actual velocity components u_0', v_0', w_0' of motion produced by dynamical tides and referred to the inertial space-axes are given by Equations (2.38)–(2.40). An application of the Laplacian operator ∇^2 in the inertial coordinates discloses that

$$\nabla^2 u_0' = (p_3y - p_2z)\,\nabla^2 f + 2p_3f_y - 2p_2f_z + x\,\nabla^2\dot{f} + 2f\dot{x}, \tag{3.4}$$

$$\nabla^2 v_0' = (p_1z - p_3x)\,\nabla^2 f + 2p_1f_z - 2p_3f_x + y\,\nabla^2\dot{f} + 2f\dot{y}, \tag{3.5}$$

$$\nabla^2 w_0' = (p_2x - p_1y)\,\nabla^2 f + 2p_2f_x - 2p_1f_y + z\,\nabla^2\dot{f} + 2f\dot{z}\,; \tag{3.6}$$

so that

$$\begin{aligned}y\nabla^2 w_0' - z\nabla^2 v_0' &= 2D_1f - 2p_1D_4f + 2(p_2y + p_3z)f_x\\ &- [p_1(y^2+z^2) - (p_2y + p_3z)x]\,\nabla^2 f,\end{aligned} \tag{3.7}$$

$$z\nabla^2 u_0' - x\nabla^2 w_0' = 2D_2 f - 2p_2 D_5 f + 2(p_1 x + p_3 z) f_y - [p_2(z^2 + x^2) - (p_1 x + p_3 z)y]\,\nabla^2 f, \tag{3.8}$$

$$x\nabla^2 v_0' - y\nabla^2 u_0' = 2D_3 f - 2p_3 D_6 f + 2(p_1 x + p_2 y) f_z - [p_3(x^2 + y^2) - (p_1 x + p_2 y)z]\,\nabla^2 f, \tag{3.9}$$

where the operators $D_{1,2,3}$ continue to be given by Equations (1.92)–(1.94);

$$D_4 \equiv z\frac{\partial}{\partial z} + y\frac{\partial}{\partial y} \equiv r\frac{\partial}{\partial r} - x\frac{\partial}{\partial x}, \tag{3.10}$$

$$D_5 \equiv x\frac{\partial}{\partial x} + z\frac{\partial}{\partial z} \equiv r\frac{\partial}{\partial r} - y\frac{\partial}{\partial y}, \tag{3.11}$$

$$D_6 \equiv x\frac{\partial}{\partial x} + y\frac{\partial}{\partial y} \equiv r\frac{\partial}{\partial r} - z\frac{\partial}{\partial z}; \tag{3.12}$$

and the coefficients $p_{1,2,3}$ continue to be given by (2.35).

The divergence

$$\Delta_0' \equiv \frac{\partial u_0'}{\partial x} + \frac{\partial v_0'}{\partial y} + \frac{\partial w_0'}{\partial z}, \tag{3.13}$$

arising from compressibility, can be expressed in terms of the tidal deformation f with equal ease. Since a differentiation of (2.38)–(2.40) yields

$$\frac{\partial u_0'}{\partial x} = (p_3 y - p_2 z) f_x + \dot{f} + x\dot{f}_x, \tag{3.14}$$

$$\frac{\partial v_0'}{\partial y} = (p_1 z - p_3 x) f_y + \dot{f} + y\dot{f}_y, \tag{3.15}$$

$$\frac{\partial w_0'}{\partial z} = (p_2 x - p_1 y) f_z + \dot{f} + z\dot{f}_z; \tag{3.16}$$

an insertion of the preceding equations in (3.13) discloses that

$$\begin{aligned}\Delta_0' &= r\dot{f}_r + 3\dot{f} - (p_1 D_1 + p_2 D_2 + p_3 D_3) f \\ &= (r f_r + 3f)_t - (p_1 D_1 + p_2 D_2 + p_3 D_3)(r f_r + 4f),\end{aligned} \tag{3.17}$$

by virtue of (2.37).

Let us next turn our attention to terms on the right-hand sides of Equations (3.1)–(3.3) which are factored by the derivatives of μ. An application of the operators $D_{1,2,\ldots,6}$ to the expressions (2.38)–(2.40) for u_0', v_0', w_0' discloses that

$$D_1 u_0' = (-p_2 y - p_3 z) f + (p_3 y - p_2 z) D_1 f + x D_1 \dot{f}, \tag{3.18}$$

$$D_2 u_0' = (+p_2 x \qquad) f + (p_3 y - p_2 z) D_2 f + x D_2 \dot{f} + z\dot{f}, \tag{3.19}$$

$$D_3u_0' = (\qquad +p_3x)f + (p_3y - p_2z)D_3f + xD_3\dot{f} - y\dot{f}; \tag{3.20}$$

$$D_4u_0' = (-p_2z - p_3y)f + (p_3y - p_2z)D_4f + xD_4\dot{f}, \tag{3.21}$$

$$D_5u_0' = (-p_2z \qquad)f + (p_3y - p_2z)D_5f + xD_5\dot{f} + x\dot{f}, \tag{3.22}$$

$$D_6u_0' = (\qquad +p_3y)f + (p_3y - p_2z)D_6f + xD_6\dot{f} + x\dot{f}; \tag{3.23}$$

$$D_1v_0' = (+p_1y \qquad f + (p_1z - p_3x)D_1f + yD_1\dot{f} - z\dot{f}, \tag{3.24}$$

$$D_2v_0' = (-p_1x - p_3z)f + (p_1z - p_3x)D_2f + yD_2\dot{f}, \tag{3.25}$$

$$D_3v_0' = (\qquad +p_3y)f + (p_1z - p_3x)D_3f + yD_3\dot{f} + x\dot{f}; \tag{3.26}$$

$$D_4v_0' = (+p_1z \qquad f + (p_1z - p_3x)D_4f + yD_4\dot{f} + y\dot{f}, \tag{3.27}$$

$$D_5v_0' = (+p_1z - p_3x)f + (p_1z - p_3x)D_5f + yD_5\dot{f}, \tag{3.28}$$

$$D_6v_0' = (\qquad -p_3x)f + (p_1z - p_3x)D_6f + yD_6\dot{f} + y\dot{f}; \tag{3.29}$$

$$D_1w_0' = (+p_1z \qquad)f + (p_2x - p_1y)D_1f + zD_1\dot{f} + y\dot{f}, \tag{3.30}$$

$$D_2w_0' = (\qquad +p_2z)f + (p_2x - p_1y)D_2f + zD_2\dot{f} - x\dot{f}, \tag{3.31}$$

$$D_3w_0' = (-p_1x - p_2y)f + (p_2x - p_1y)D_3f + zD_3\dot{f}, \tag{3.32}$$

$$D_4w_0' = (-p_1y \qquad)f + (p_2x - p_1y)D_4f + zD_4\dot{f} + z\dot{f}, \tag{3.33}$$

$$D_5w_0' = (\qquad +p_2z)f + (p_2x - p_1y)D_5f + zD_5\dot{f} + z\dot{f}, \tag{3.34}$$

$$D_6w_0' = (-p_1x + p_2y)f + (p_2x - p_1y)D_6f + zD_6\dot{f}. \tag{3.35}$$

Accordingly,

$$\frac{\partial\mu}{\partial x}D_1u_0' + \frac{\partial\mu}{\partial y}D_1v_0' + \frac{\partial\mu}{\partial z}D_1w_0' = -(p_2y + p_3z)f\frac{\partial\mu}{\partial x} + p_1fD_4\mu -$$

$$-(p_1D_1\mu + p_2D_2\mu + p_3D_3\mu)D_1f + r\frac{\partial\mu}{\partial r}D_1\dot{f} + \dot{f}D_1\mu\,, \tag{3.36}$$

$$\frac{\partial\mu}{\partial x}D_2u_0' + \frac{\partial\mu}{\partial y}D_2v_0' + \frac{\partial\mu}{\partial z}D_2w_0' = -(p_1x + p_3z)f\frac{\partial\mu}{\partial y} + p_2fD_5\mu -$$

$$-(p_1D_1\mu + p_2D_2\mu + p_3D_3\mu)D_2f + r\frac{\partial\mu}{\partial r}D_2\dot{f} + \dot{f}D_2\mu, \tag{3.37}$$

$$\frac{\partial\mu}{\partial x}D_3u_0' + \frac{\partial\mu}{\partial y}D_3v_0' + \frac{\partial\mu}{\partial z}D_3w_0' = -(p_1x + p_2y)f\frac{\partial\mu}{\partial z} + p_1fD_6\mu -$$

$$-(p_1D_1\mu+p_2D_2\mu+p_3D_3\mu)D_3f+r\frac{\partial\mu}{\partial r}D_3\dot{f}+\dot{f}D_3\mu; \tag{3.38}$$

and

$$\frac{\partial\mu}{\partial y}D_4w_0'-\frac{\partial\mu}{\partial z}D_4v_0' = -p_1(f+D_4f)D_4\mu+ +x\left(p_2\frac{\partial\mu}{\partial y}+p_3\frac{\partial\mu}{\partial z}\right)D_4f- -(\dot{f}+D_4\dot{f})D_1\mu, \tag{3.39}$$

$$\frac{\partial\mu}{\partial x}D_5w_0'-\frac{\partial\mu}{\partial z}D_5u_0' = +p_2(f+D_5f)D_5\mu- -y\left(p_1\frac{\partial\mu}{\partial x}+p_3\frac{\partial\mu}{\partial z}\right)D_5f+ +(\dot{f}+D_5\dot{f})D_2\mu, \tag{3.40}$$

$$\frac{\partial\mu}{\partial x}D_6v_0'-\frac{\partial\mu}{\partial y}D_6u_0' = -p_3(\dot{f}+D_6\dot{f})D_6\mu+ +z\left(p_1\frac{\partial\mu}{\partial x}+p_2\frac{\partial\mu}{\partial y}\right)D_6f- -(\dot{f}+D_6\dot{f})D_3\mu. \tag{3.41}$$

Moreover, the curl-components

$$\xi=\frac{\partial w_0'}{\partial y}-\frac{\partial v_0'}{\partial z},\quad \eta=\frac{\partial u_0'}{\partial z}-\frac{\partial w_0'}{\partial x},\quad \zeta=\frac{\partial v_0'}{\partial x}-\frac{\partial u_0'}{\partial y}, \tag{3.42}$$

follow on differentiation of (2.38)–(2.40) as

$$\xi=-p_1[2f+D_4f]+x(p_2f_y+p_3f_z)-D_1\dot{f}, \tag{3.43}$$

$$\eta=-p_2[2f+D_5f]+y(p_1f_x+p_3f_z)-D_2\dot{f}, \tag{3.44}$$

$$\zeta=-p_3[2f+D_6f]+z(p_1f_x+p_2f_y)-D_3\dot{f}, \tag{3.45}$$

and, similarly,

$$\frac{\partial}{\partial x}(yw_0'-zv_0') = (p_2y+p_3z)(f+xf_x)-p_1(y^2+z^2)f_x, \tag{3.46}$$

$$\frac{\partial}{\partial y}(zu_0'-xw_0') = (p_3z+p_1x)(f+yf_y)-p_2(x^2+z^2)f_y, \tag{3.47}$$

$$\frac{\partial}{\partial z}(xv_0'-yu_0') = (p_1x+p_2y)(f+zf_z)-p_3(x^2+y^2)f_z. \tag{3.48}$$

If we insert the results represented by Equations (3.4)–(3.48) in (3.1)–(3.3), the latter assume the more explicit forms

$$\rho\{y\mathfrak{H}-z\mathfrak{G}\} = \mu\{2[D_1\dot{f}-p_1D_4f+(p_2y+p_3z)f_x]-$$
$$-[p_1r^2-x(p_1x+p_2y+p_3z)]\nabla^2f+\frac{1}{3}D_1\Delta_0'\}-$$
$$-p_1\left\{A(y^2+z^2)+\frac{1}{3}D_1fD_1\mu\right\}+$$
$$+p_2\left\{A\,xy-\frac{2}{3}Bz-\frac{1}{3}D_1fD_2\mu\right\}+$$
$$+p_3\left\{A\,xz+\frac{2}{3}By-\frac{1}{3}D_1fD_3\mu\right\}+$$
$$+\left(r\frac{\partial\mu}{\partial r}\right)D_1\dot{f}-\frac{2}{3}\left(r\frac{\partial\dot{f}}{\partial r}\right)D_1\mu, \qquad (3.49)$$

$$\rho\{z\mathfrak{F}-x\mathfrak{H}\}=\mu\{2[D_2\dot{f}-p_2D_5\dot{f}+(p_1x+p_3z)f_y]-$$
$$-[p_2r^2-y(p_1x+p_2y+p_3z)\nabla^2f+\frac{1}{3}D_2\Delta_0']\}+$$
$$+p_1\{A\,xy+\frac{2}{3}Bz-\frac{1}{3}D_2fD_1\mu\}-$$
$$-p_2\{A(x^2+z^2)+\frac{1}{3}D_2fD_2\mu\}+$$
$$+p_3\{A\,yz-\frac{2}{3}Bx-\frac{1}{3}D_2fD_3\mu\}+$$
$$+\left(r\frac{\partial\mu}{\partial r}\right)D_2\dot{f}-\frac{2}{3}\left(r\frac{\partial\dot{f}}{\partial r}\right)D_2\mu, \qquad (3.50)$$

and

$$\rho\{x\mathfrak{G}-y\mathfrak{F}\}=\mu\{2[D_3\dot{f}-p_3D_6f+(p_1x+p_2y)f_z-$$
$$-[p_3r^2-z(p_1x+p_2y+p_3z)]\nabla^2f+\frac{1}{3}D_3\Delta_0'\}+$$
$$+p_1\{A\,xz-\frac{2}{3}By-\frac{1}{3}D_3fD_1\mu\}+$$
$$+p_2\{A\,yz+\frac{2}{3}Bx-\frac{1}{3}D_3fD_2\mu\}-$$
$$-p_3\{A(x^2+y^2)+\frac{1}{3}D_3fD_3\mu\}+$$
$$+\left(r\frac{\partial\mu}{\partial r}\right)D_3\dot{f}-\frac{2}{3}\left(r\frac{\partial\dot{f}}{\partial r}\right)D_3\mu, \qquad (3.51)$$

where we have abbreviated

$$A = \frac{\partial\mu}{\partial x}\frac{\partial f}{\partial x} + \frac{\partial\mu}{\partial y}\frac{\partial f}{\partial y} + \frac{\partial\mu}{\partial z}\frac{\partial f}{\partial z} \tag{3.52}$$

and

$$B = \frac{\partial\mu}{\partial x} D_1 f + \frac{\partial\mu}{\partial y} D_2 f + \frac{\partial\mu}{\partial z} D_3 f. \tag{3.53}$$

The foregoing expressions (3.49)–(3.51) are correct to quantities of first order in the tidal distortion f, but without any restriction on the magnitude of the coefficients $p_{1,2,3}$ as defined by Equations (2.35), or on the behaviour of the viscosity $\mu(x, y, z)$ throughout the interior of the distorted configuration.

The last task remaining to complete the explicit formulation of the fundamental equations (1.108)–(1.110) of our problem is to integrate the expressions (3.49)–(3.51) over the entire volume of the distorted configuration. In order to do so we propose, however, to avail ourselves of the following additional simplifications:

(1) The heights f_j of the j-th harmonic partial tides will (consistent with (2.67)) be regarded as functions of r and λ only; and accordingly, $f_\nu = 0$. If so, then

$$\left.\begin{aligned} f_x &= \lambda f_r + [(1-\lambda^2)/r] f_\lambda, \\ f_y &= \mu f_r - (\lambda\mu/r) f_\lambda, \\ f_z &= \nu f_r - (\lambda\nu/r) f_\lambda, \end{aligned}\right\} \tag{3.54}$$

where the direction cosines λ, μ, ν are defined by Equations (2.63).

(2) The viscosity $\mu(x, y, z)$ of the material will hereafter be regarded to depend essentially or r only; and so weakly on angular variables that the products of angular derivatives of μ with f or its derivatives can be neglected. If so, however, it follows that throughout Equations (3.49)–(3.51) we are entitled to reduce (3.52) and (3.53) to

$$A = \frac{\partial\mu}{\partial r}\frac{\partial f}{\partial r} \quad \text{and} \quad B = 0; \tag{3.55}$$

while

$$D_1\mu = D_2\mu = D_3\mu = 0; \tag{3.56}$$

but

$$\left.\begin{aligned} r D_4\mu &= (y^2 + z^2)\frac{\partial\mu}{\partial r}, \\ r D_5\mu &= (x^2 + z^2)\frac{\partial\mu}{\partial r}, \\ r D_6\mu &= (x^2 + y^2)\frac{\partial\mu}{\partial r}. \end{aligned}\right\} \tag{3.57}$$

(3) The inclinations θ as well as i of the equatorial and orbital planes to the invariable plane ($z = 0$) of our binary system are – if not zero – small enough for their cross-products with f to be negligible. Within this scheme of approximation we are, therefore, entitled to set in (3.49)–(3.51)

$$p_1 = p_2 = 0 \quad \text{and} \quad p_3 = \omega - n, \tag{3.58}$$

which considerably simplifies the results.

In order to prepare the remaining terms on the right-hand sides of Equations (3.49)–(3.51) for integration between the limits (2.65), the integrands must again be rewritten in terms of the revolving coordinates $x''y''z''$. Consistent with the approximation outlined by the point (3) above, we are entitled to set $a''_{13} = a''_{23} = 0, a''_{31} = a''_{32} = 0$ and $a''_{33} = 1$ in any terms involving f – which means that (to this approximation) $D_3 f \equiv \boldsymbol{D}''_3 f$. Secondly, if follows from (2.67) that, for any j-th harmonic tide, $r(f_j)_r = (j+1) f_j$; which inserted in (3.17) enables us to simplify the latter into

$$\Delta'_0 = (j+4) f_t - (j+5) p_3 D''_3 f. \tag{3.59}$$

Therefore, by use of (2.95) and (2.122),

$$D_1 \Delta'_0 = a''_{12} \nu'' \{(j+4) f_{\lambda'' t} + (j+5) p_3 \mu'' f_{\lambda''\lambda''}\} - a''_{11} (j+5) p_3 \nu'' f_{\lambda''}, \tag{3.60}$$

$$D_2 \Delta'_0 = a''_{22} \nu'' \{(j+4) f_{\lambda'' t} + (j+5) p_3 \mu'' f_{\lambda''\lambda''}\} - a''_{21} (j+5) p_3 \nu'' f_{\lambda''}, \tag{3.61}$$

$$D_3 \Delta'_0 = -(j+4) \mu'' f_{\lambda'' t} - (j+5) p_3 (\mu''^2 f_{\lambda''\lambda''} - \lambda'' f_{\lambda''}). \tag{3.62}$$

However, all terms of the preceding equations vanish identically when integrated over the whole sphere; so that

$$\int \mu D_1 \Delta'_0 \mathrm{d}V = \int \mu D_2 \Delta'_0 \mathrm{d}V = \int \mu D_3 \Delta'_0 \, \mathrm{d}V = 0. \tag{3.63}$$

Turning to the volume integrals of the terms involving $\nabla^2 f$, we note that the Laplacian operator is invariant to a transformation from the inertial to the rotating or revolving coordinates (i.e., $\nabla^2 \equiv \nabla''^2$). Moreover, if we extend the integration over the whole sphere, the only term which will survive this operation in non-vanishing amount will be

$$\int \mu (x^2 + y^2) \nabla^2 f \, \mathrm{d}V = \frac{8}{5} \pi K_2 \int_0^{R_1} \mu r^5 \, \mathrm{d}r\,; \tag{3.64}$$

for all other j's the respective terms vanish identically. The same is true of the volume integrals

$$\int \mu D_1 \dot{f} \, \mathrm{d}V = \int \mu D_2 \dot{f} \, \mathrm{d}V = \int \mu D_3 \dot{f} \, \mathrm{d}V = 0 \tag{3.65}$$

$$\int \mu z f_x \, \mathrm{d}V = \int \mu z f_y \, \mathrm{d}V = 0, \tag{3.66}$$

involving the derivatives $\dot{f}$ or $\dot{f}_x$, the only nonvanishing term of this latter type being

$$\int \mu D_6 f \, \mathrm{d}V = \frac{8}{5} \pi K_2 \int_0^{R_1} \mu r^5 \, \mathrm{d}r. \tag{3.67}$$

If we turn next to the evaluation of volume integrals in (3.49)–(3.51) involving derivatives of the viscosity μ, we note that those containing $D_{1,2,3} f$ are all zero when integrated with respect to the angular variables over the entire sphere; and the same is true of

$r\dot{f}_r$. Since, moreover, the coefficient B is zero by virtue of the condition (2) on p. 176 the only nonvanishing terms may arise from A. In point of fact, only one of them happens to be different from zero: namely,

$$\int (x^2 + y^2) A \, dV = (j+1) \int r \frac{\partial \mu}{\partial r} (1 - \nu''^2) f \, dV$$

$$= \frac{4}{5} \pi K_2 \int_0^{R_1} \frac{\partial \mu}{\partial r} r^6 \, dr \qquad (3.68)$$

for $j = 2$; all others contribute nothing to the result.

Therefore, to the first order in tidal distortion f, and within restrictions (1)–(3) specified on p. 176, the volume integrals of Equations (3.49–(3.50) become

$$\int \rho \, \{y\mathfrak{H} - z\mathfrak{G}\} \, dV = 0, \qquad (3.69)$$

and

$$\int \rho \{z\mathfrak{F} - x\mathfrak{H}\} \, dV = 0; \qquad (3.70)$$

while Equation (3.51) yields

$$\int \rho \{x\mathfrak{G} - y\mathfrak{F}\} \, dV = -\frac{24}{5} \pi p_3 K_2 \int_0^{R_1} \mu r^5 \, dr - \qquad (3.71)$$

$$- \frac{4}{5} \pi p_3 K_2 \int_0^{R_1} \frac{\partial \mu}{\partial r} r^6 \, dr.$$

B. TIDAL FRICTION: SPHERICAL CONFIGURATIONS

With the results established in the preceding parts of this section we are now in a position to fulfill one main task of expressing the fundamental equations (1.108)–(1.110) of our problem explicity in terms of the effects produced by dynamical tides, and proceed with their solution. The aim of the present part of this section will, however, be to restrict this task to the third equation (1.110) governing axial rotation about the z-axis; and forego a simultaneous treatment of Equations (1.108) and (1.109) governing the precession and nutation of this axis in the inertial space – a task postponed for subsequent section VI-2.

Within the scheme of the present approximation we set – for the time being – $\omega_x = \omega_y = 0$ in Equations (1.108)–(1.110) and drop subscript of ω_z which becomes now the sole dependent variable of our problem. Let us, further, identify the equatorial plane of the rotating configuration and the orbital plane of its companion with the inertial plane $z = 0$ – so that the angles θ and i in the direction cosines a_{ij} and a''_{ij} can be set equal to zero. If, moreover, the eccentricity of the relative orbit of the secondary component of mass m_2 is sufficiently small for the product of H_2 with $k_2 K_2 R_1^3$ to be negligible, Equations (2.119)–(2.120), (2.127)–(2.128) and (3.69) – (3.70) make it evident that both sides of the fundamental equations (1.108)–(1.109) vanish identically; while Equation (1.110) on insertion from (3.71) reduces to

$$C\frac{d\omega}{dt} + \frac{4}{5}\pi K_2 \left\{ 6\int_0^{R_1} \mu r^5\, dr + \int_0^{R_1} \frac{\partial\mu}{\partial r} r^6 dr \right\} (\omega - n) = 0, \tag{3.72}$$

where the moment of inertia C about the z-axis continues to be given by (2.89).

The foregoing equation governs (within the scheme of our approximation) the rate $\dot{\omega}$ at which the angular velocity of rotation of the configuration of mass m_1 will vary as a result of tidal friction represented by the second term on the left-hand side of (3.72). Since, moreover, by partial integration

$$\int_0^{R_1} \frac{\partial\mu}{\partial r} r^6 dr = [\mu r^6]_0^{R_1} - 6\int_0^{R_1} \mu r^5\, dr, \tag{3.73}$$

the expression in the curly brackets on the left-hand side of (3.72) reduces simply to the integrated part $[\mu r^6]_0^{R_1}$ which will obviously vanish at the centre – thus leaving us with

$$C\frac{d\omega}{dt} + \frac{4}{5}\pi K_2 \mu(R_1) R_1^6 (\omega - n) = 0. \tag{3.74}$$

In this latter equation, the viscosity enters no longer through the whole march of its distribution within the interior, but only through a constant equal to the surface of $\mu(R_1)$.

It should be stressed, however, that the partial integration represented by (3.73) is legitimate only for a closed domain in which the viscosity $\mu(r)$ remains a continuous function of its argument.

For gaseous configurations of stellar mass, the value which should be assigned to $\mu(R_1)$ depends on the type of the dominant physical process responsible for the effective viscosity of stellar material. Should the latter be due to predominantly plasma (Chapman, 1954) or radiative (Hazlehurst and Sargent, 1959) viscosity, $\mu(r)$ should depend on r through the density $\rho(r)$ and temperature $T(r)$ which are both continuous functions of r for $0 \geqslant r \geqslant R_1$ vanishing on the boundary. In such a case – which should generally apply to stellar configurations in radiative equilibrium – the numerical value of $\mu(R_1)$ could be expected to be theoretically zero, and practically very small.

A very different case may, however, arise if the configuration in question possesses a convective zone near the centre, or below the surface. The actual motions of gas in such zones are almost necessarily turbulent; and turbulent viscosity μ_T associated with them is likely to exceed that due to plasma or radiation by several orders of magnitude. Suppose – for the sake of argument – that the configuration in question possesses a convective core of radius $R_* \geqslant R_1$* and turbulent viscosity $\mu_T(r)$; while the gas and radiative viscosity in the surrounding zone ($R_* < r \geqslant R_1$) can be ignored. If so, the integrated part in (3.73) should be represented by

$$[\mu r^6]_0^{R_1} = \mu_T(R_*) R_*^6 \, ; \tag{3.75}$$

while if the core is radiative and R_* marks the inner radius of a convective shell,

* The equality sign should apply to wholly convective configurations during the initial contraction to the Main Sequence.

$$[\mu r^6]_0^{R_1} = \mu_T(R_1)R_1^6 - \mu_T(R_*)R_*^6. \tag{3.76}$$

Whatever the case may be, however, Equation (3.72) should be generally approximable by a nonhomogeneous first-order differential equation for ω, of the form

$$\frac{d\omega}{dt} + k(\omega - n) = 0, \tag{3.77}$$

where

$$k = \frac{4\pi K_2}{5C}[\mu r^6]_0^{R_1} = \frac{4\pi(1+2k_2)m_2}{5h_1^2 m_1^2 \Delta^3}[\mu r^6]_0^{R_1} \tag{3.78}$$

can be regarded (for orbits of small eccentricity) as very nearly constant. Within the scheme of such an approximation, we readily find the solution of (3.77) to be of the form

$$\omega(t) = n + (\omega_0 - n)e^{-kt}, \tag{3.79}$$

where $\omega_0 \equiv \omega(0)$. Equation (3.79) makes it evident that, whatever the value of ω may have been for $t = 0$, it is bound to approach n asymptotically as $t \to \infty$ – in other words, *no matter what the initial disparity* $\omega \gtreqless n$ may have been in magnitude or sign, *viscous forces are bound eventually to synchronize them in the course of time.*

The most significant limitation of a theory of tidal friction as developed in this section so far is, however, its underlying assumption that the tide-generating component of mass m_2 revolves around the mass m_1 in a relative orbit which is circular. Should this not be the case, the mean daily motion n of the secondary component will of course cease to be constant, and vary as

$$n = \frac{\sqrt{G(1-e^2)(m_1+m_2)A}}{\Delta^2} \tag{3.80}$$

in accordance with Kepler's law of the areas, where Δ denotes the radius-vector (2.68) of the relative orbit of the two stars, of semi-major axis A and eccentricity e. Since n can obviously be identified with the time-derivative of the true anomaly υ, it follows that

$$\dot{n} \equiv \ddot{\upsilon} = -2\left(\frac{\dot{\Delta}}{\Delta}\right)\dot{\upsilon} = -2\left(\frac{\dot{\Delta}}{\Delta}\right)n \tag{3.81}$$

and, therefore,

$$nH_2 = \frac{3}{2}K_2\dot{n} \tag{3.82}$$

by (2.105) and (3.80).

Since, in the elliptic case, both Δ and n are known functions of the time as given by Equations (2.68) and (3.80), the complete primitive of (3.72) should be expressible as

$$\omega(t) = e^{-q\int_0^t \Delta^{-3}dt}\left\{\omega_0 + q\int_0^t n\Delta^{-3}e^{q\int_0^t \Delta^{-3}dt}\,dt\right\}, \tag{3.83}$$

where

$$q = \frac{4\pi(1 + 2k_2)m_2}{5C\, m_1} \left[\mu r^6\right]_0^{R_1} \tag{3.84}$$

denotes a constant. By known integrals of elliptic motion both Δ as well as n are given in closed form as functions of the true anomaly v. Changing over from t to v as the variable of integration by

$$\mathrm{d}t = \frac{\Delta^2 \mathrm{d}v}{\omega_K A^2 \sqrt{1 - e^2}}\ , \tag{3.85}$$

where

$$\omega_K^2 = \frac{G(m_1 + m_2)}{A^3} \tag{3.86}$$

denotes the Keplerian angular velocity of orbital revolution, we find that

$$\int_0^t \Delta^{-3} \mathrm{d}t = \frac{v + e \sin v}{\omega_K A^3 (1 - e^2)^{3/2}} \tag{3.87}$$

if time is measured from the moment of periastron passage; and, consequently,

$$\omega(t) = e^{-\frac{q(v + e \sin v)}{\omega_K A^3 (1 - e^2)^{3/2}}} \left\{ \omega_0 + \frac{q}{A^3 (1 - e^2)^{3/2}} \int_0^v (1 + e \cos v)^3 \times \right.$$
$$\left. \times\ e^{\frac{q(v + e \sin v)}{\omega_K A^3 (1 - e^2)^{3/2}}} \mathrm{d}v \right\} . \tag{3.88}$$

The remaining integral on the right-hand side of the preceding equation can no longer be evaluated in a closed form. The terms factored by the eccentricity e represent the effects, on axial rotation, of 'tidal breathing' of the configuration of mass m_1, caused by the periodic oscillations in the height of the tides f which vary with inverse cube of the instantaneous radius-vector. This variation whould cause $\omega(t)$ to oscillate even if the Keplerian angular velocity ω_K were identical with the mean value of $\omega(t)$ in the course of each cycle.

It should, however, be added that if n and Δ are allowed to vary with the time as a result of orbital eccentricity, Equation (3.72) should be augmented by inclusion of the terms arising from (2.119) and (2.121). As can be easily verified, for $a''_{31} = a''_{32} = 0$, $a''_{33} = 1$; and $p_1 = p_2 = 0$ while $p_3 = \omega - n$, Equation (2.118) readily reduces to

$$\int (xu'_0 + yv'_0)\, \mathrm{d}m = \frac{1}{12}\, k_2 m_1 R_1^5 H_2 , \tag{3.89}$$

while (2.121) can similarly be reduced to

$$\int \{x(\dot{v})'_0 - y(\dot{u})'_0\}, \mathrm{d}m = \frac{1}{12} k_2 m_1 R_1^5\ \{K_2 \dot{p}_3 - 2H_2 p_3\}\ ; \tag{3.90}$$

which disclose that

$$2\omega_z \int (xu_0' + yv_0')\, dm + \int \{x(\dot{v})_0' - y(\dot{u})_0'\}\, dm =$$

$$= \frac{1}{12} k_2 m_1 R_1^5 (2H_2 n - K_2 \dot{p}_3)$$

$$= \frac{1}{3} k_2 m_1 R_1^5 K_2 \dot{n} - \frac{1}{12} k_2 m_1 R_1^5 K_2 \dot{\omega} \qquad (3.91)$$

by (3.82). If we incorporate the coefficient of $\dot{\omega}$ on the right-hand side of the preceding equation into the angular momentum C, we find that – even in the complete absence of viscosity – the orbital eccentricity alone would render (3.72) a differential equation of the form

$$C\omega + \frac{1}{3} k_2 m_1 K_2 R_1^5 \dot{n} = 0, \qquad (3.92)$$

that must be fulfilled for the sake of the preservation of the momentum, for time-varying bodily tides, in the course of an orbital cycle.

So simplified an equation can be readily integrated; for changing over from t to v as our independent variable, we find by (3.81) and (3.85) that

$$\begin{aligned} \dot{n}\, dt &= -2 \frac{\dot{\Delta}}{\Delta} dv \\ &= -\frac{2\omega_K e}{\sqrt{(1-e^2)^3}} (1 + e \cos v) \sin v\, dv, \end{aligned} \qquad (3.93)$$

which (after insertion for K_2) enables us to solve (3.92) by quadratures in the closed form of

$$\omega = \frac{ek_2(1 + 2k_2)\,\omega_K}{3(1-e^2)^{9/2}} \left(\frac{m_2}{m_1}\right)\left(\frac{A}{h_1}\right)^2 \left(\frac{R_1}{A}\right)^5 \int (1 + e \cos v)^4 \sin v\, dv, \qquad (3.94)$$

leading to

$$\frac{\omega(v) - \omega(\pi/2)}{\omega_K} = \frac{2}{15} k_2(1 + 2k_2)\sqrt{1-e^2}\, \frac{m_2}{m_1} \left(\frac{A}{h_1}\right)^2 \left(\frac{R_1}{\Delta}\right)^5 \{(1 + e \cos v)^{-5} - 1\} \qquad (3.95)$$

if C has been approximated by $m_1 h_1^2$.

A periodic oscillation of ω as represented by the foregoing equation would become necessary for the conservation of momentum only if one rotating configuration were inviscid, and no interaction were possible witht the orbital momentum. In the presence of viscosity (or of dissipative forces of other kind) the rotational momenta of both components are coupled with those of their orbital motions through viscous interaction terms in Equations (1.108)–(1.110), the explicit form of which we investigated in this section. In reality, therefore, the second term on the left-hand side of Equation (3.89) should be added to the left-hand side of (3.72) before a more general solution could be obtained; and, through this term, a study of the variation of ω will be coupled with the variational equations for the orbital elements of the system. Needless to say, the solution of so generalized a problem cannot be obtained in a closed form; and any effort to obtain it must be postponed for subsequent investigations.

C. TIDAL FRICTION: SPHEROIDAL CONFIGURATIONS

In addition to the restrictions mentioned on p. 176, the results obtained in the foregoing parts of this section are subject to one additional limitation: namely, a neglect of the cross-products between rotational and tidal distortion. Throughout all preceding developments we have considered, in effect, the tides f to be raised by the attraction of mass m_2 over a spherical globe of mass m_1. This configuration was, on the other hand, allowed to be in the state of three-axial rotation; and the effect of this rotation on the moments of inertia was duly taken into account in earlier parts of this section. However, the effects of rotation on dynamical tides raised on the configuration of mass m_1 by its companion of mass m_2 have been neglected so far – a policy which can be justified only if the effects of both rotation and tides on the configuration of mass m_1 are small enough for their cross-products to be negligible.

That the fractional height f of partial tides can be regarded as a small quantity of first order constitutes an approximation to which we propose to adhere throughout our work; for to do otherwise would entail far-reaching complications. However, the axial rotation need not necessarily be slow; and if so, the products ωf need not remain negligible because – for fast-rotating configuration – the angular velocity ω of axial rotation may become a quantity of zero-order. The aim of the present section will, therefore, be to generalize the results of our previous work to fast-rotating configurations and investigate, in particular, the extent to which the terms nonlinear in ω will influence the rate of viscous friction through dynamical tides.

In order to do so, we shall revert once more to our previous approximations of regarding the relative orbit of the companion of mass m_2 as circular and its plane coinciding with the equator of the rotating mass m_1; and set out to generalize Equation (3.72) governing the rate of viscous friction to more rapidly-rotating configurations. The main consideration to be kept in mind in this connection is the fact that, in the case of rapid rotation, the radius-vector r in all the volume-integrals in the fundamental equations (1.108)–(1.110) should be replaced by $R \equiv r\,(1 + g)$, where g stands for the extent of rotational distortion as defined already by Equation (2.75) – in which, for co-planar orbits, Equation (2.122) reduces (2.76) to $\nu' \equiv \nu \equiv \nu''$. This change is bound to effect all previous developments of this section in two ways: by altering appropriately all respective terms on the right-hand sides of Equations (3.49)–(3.59), as well as the volume element $\mathrm{d}V$ with respect to which these terms are to be subsequently integrated.

The requisite change in the volume element follows readily form (2.61) in which we replace f by g; doing so we find that, in the present ellipsodial case,

$$\mathrm{d}V = r^2\left\{1 - \frac{2\omega^2 r^3}{Gm_1} P_2(\nu'')\right\}\mathrm{d}r \sin\theta''\,\mathrm{d}\theta''\,\mathrm{d}\phi''. \tag{3.96}$$

As regards the requisite modifications of the integrand, let us note that – inasmuch as g depends only on θ'' and not on ϕ'' – all terms on the right-hand sides of (3.49)–(3.51) which are annihilated by integration with respect to ϕ'' between 0 and 2π will continue to be zero. Therefore, only integration with respect to θ'' needs hereafter to be considered; and for coplanar orbits ($\theta = i = 0$), these reduce to the terms previously given by (3.64), (3.67) and (3.68).

Let us divide, in what follows, the contributions which stem from the spheroidal shape of the rotating configuration in two groups: (a) those arising on the left-hand sides of Equations (3.64), (3.67) and (3.68) from the g-term in the volume element (3.96); and (b), those arising in the integrands from the ellipticity of figure. The former are easily obtained if we multiply the 'spherical' integrands by $6g$ and integrate between the limits in (3.65). In doing so, we find that the terms factored by the viscosity μ will call for an integration of the expression

$$\begin{aligned}&-(j+4)r^2 g\{1-\nu^2)r^2\nabla^2 f + 2D_6 f\}\mu = \\ &= \frac{2}{3}(j+4)\left(\frac{\omega^2 r^5}{Gm_1}\right)\mu\{(j+1)(1-\nu''^2)K_j r^{j+1}P_j(\lambda'') + D_6 f_j\}P_2(\nu'');\end{aligned} \tag{3.97}$$

while those factored by the gradient of viscosity will contribute the integral of

$$\begin{aligned}&-g(1-\nu''^2)(rf_r)(r\mu_r) = \\ &= (j+1)(j+4)\frac{\omega^2 r^5}{3Gm_1}K_j r^{j+1}\left(r\frac{\partial\mu}{\partial r}\right)(1-\nu''^2)P_2(\nu'')P_j(\lambda'').\end{aligned} \tag{3.98}$$

Performing the integration over the entire volume we find that the category (a) terms will contribute to the right-hand side of Equation (3.51) the following nonlinear terms

$$\begin{aligned}\text{(a):}\quad \int \rho\{x\mathfrak{G} - y\mathfrak{F}\}\mathrm{d}V = &-\frac{144}{35}\pi p_3\frac{\omega^2 K_2}{Gm_1}\int_0^{R_1}\mu r^8\,\mathrm{d}r \\ &-\frac{32}{7}\pi p_3\frac{\omega^2 K_4}{Gm_1}\int_0^{R_1}\mu r^{10}\,\mathrm{d}r \\ &-\frac{8}{7}\pi p_3\frac{\omega^2 K_2}{Gm_1}\int_0^{R_1}\frac{\partial\mu}{\partial r}r^9\,\mathrm{d}r \\ &-\frac{16}{21}\pi p_3\frac{\omega^2 K_4}{Gm_1}\int_0^{R_1}\frac{\partial\mu}{\partial r}r^{11}\,\mathrm{d}r\end{aligned} \tag{3.99}$$

arising from the ellipticity.

Turning now to the terms of group (b) factored by μ, we note that – if X, Y and R continue to be given by Equations (2.57)–(2.58) in which we now replace f by g.

$$\begin{aligned}(X^2+Y^2)\nabla^2 f_j(R) &= (1-\nu''^2)R^2\nabla^2 f_1(R) = 2(j+1)(1-\nu''^2)f_j(R) \\ &= 2(j+1)(1-\nu''^2)K_j r^{j+1}\left\{1-\frac{j+1}{3}\frac{\omega^2 r^3}{Gm_1}P_2(\nu'')\right\}P_j(\lambda'')\end{aligned} \tag{3.100}$$

and

$$
\begin{aligned}
D_6 f_j(R) &= D_6 f(r) - \frac{\omega^2 r^3}{2Gm_1}(1-\nu''^2)^2\, r f_r \\
&= D_6 f - (j+1)\frac{\omega^2 r_j}{Gm_1}\, r^{j+4}(1-\nu''^2)^2 P_j(\lambda'')\,;
\end{aligned}
\tag{3.101}
$$

while the form involving the radial derivative of μ will now be of the form

$$
\begin{aligned}
(X^2+Y^2)A &= (1-2rg_r)(rf_r)(r\mu_r)(1-\nu''^2) \\
&= (j+1)K_j r^{j+2}\frac{\partial\mu}{\partial r}\left\{1+\frac{2\omega^2 r^3}{Gm_1}P_2(\nu'')\right\}(1-\nu''^2)P_j(\lambda'').
\end{aligned}
\tag{3.102}
$$

Within the scheme of our approximation, all three expressions (3.100)–(3.101) are to be integrated with respect to the 'spherical' volume element $r^2 \sin\theta''\, dr\, d\theta''\, d\phi''$. In doing so we find the group (b) terms arising from the ellipticity to be of the form

$$
\begin{aligned}
\text{(b):}\quad \int \rho\{x\mathfrak{G} - y\mathfrak{F}\}\,dV = &-\frac{8}{35}\pi(\omega-n)\frac{\omega^2K_2}{Gm_1}\int_0^{R_1}\mu r^8\,dr - \\
&-\frac{16}{21}\pi(\omega-n)\frac{\omega^2K_4}{Gm_1}\int_0^{R_1}\mu r^{10}\,dr + \\
&+\frac{8}{7}\pi(\omega-n)\frac{\omega^2K_2}{Gm_1}\int_0^{R_1}\frac{\partial\mu}{\partial r}r^9\,dr + \\
&+\frac{4}{7}\pi(\omega-n)\frac{\omega^2K_4}{Gm_1}\int_0^{R_1}\frac{\partial\mu}{\partial r}r^{11}\,dr.
\end{aligned}
\tag{3.103}
$$

Summing up the terms on the right-hand sides of Equations (3.99) and (3.103) we find, moreover, that the total effect of the ellipticity on the viscous terms represented by (3.51) will be of the form

$$
\begin{aligned}
\text{(a)+(b):}\quad \rho\{x\mathfrak{G} - y\mathfrak{H}\}\,dV = &-\frac{152}{35}\pi p_3\frac{\omega^2K_2}{Gm_1}\int_0^{R_1}\mu r^8\,dr - \\
&-\frac{16}{3}\pi p_3\frac{\omega^2K_4}{Gm_1}\int_0^{R_1}\mu r^{10}\,dr - \\
&-\frac{4}{21}\pi p_3\frac{\omega^2K_4}{Gm_1}\int_0^{R_1}\frac{\partial\mu}{\partial r}r^{11}\,dr\,.
\end{aligned}
\tag{3.104}
$$

If we add these terms on the right-hand side of (3.71) and insert in Equation (1.110), the latter will assume the form

$$
C\frac{d\omega}{dt} + \frac{4}{5}\pi K_2\left\{6\int_0^{R_1}\mu r^5\,dr + \int_0^{R_1}\frac{\partial\mu}{\partial r}r^6\,dr\right\}(\omega-n) +
$$

$$+ \frac{152}{35} \pi (\omega - n) \frac{\omega^2 K_2}{G m_1} \int_0^{R_1} \mu r^8 \, dr +$$

$$+ \frac{16}{3} \pi (\omega - n) \frac{\omega^2 K_4}{G m_1} \int_0^{R_1} \mu r^{10} \, dr + \tag{3.105}$$

$$+ \frac{4}{21} \pi (\omega - n) \frac{\omega^2 K_4}{G m_1} \int_0^{R_1} \frac{\partial \mu}{\partial r} r^{11} dr = 0 ,$$

which represents a generalization of the Equation (3.72) of viscous friction of spheroidal configurations. Moreover, if we assume again the viscosity function $\mu(r)$ to be continuous throughout the closed interval $(0, R_1)$ – so that it is legitimate to set

$$\int_0^{R_1} \frac{\partial \mu}{\partial r} r^{11} = [\mu r^{11}]_0^{R_1} - 11 \int_0^{R_1} \mu r^{10} \, dr, \tag{3.106}$$

in addition to the partial integration represented by (3.73), Equation (3.105) can be re-written as

$$C \frac{d\omega}{dt} + \frac{4}{5} \pi (\omega - n)(1 + 2k_2) \frac{m_2}{m_1} \frac{[\mu r^6]_0^{R_1}}{\Delta^3} +$$

$$+ \frac{4}{105} \pi (\omega - n) \frac{\omega^2 m_2}{G m_1^2} \left\{ 114 \frac{1 + 2k_2}{\Delta^3} \int_0^{R_1} \mu r^8 \, dr + \right.$$

$$+ 85 \frac{1 + 2k_2}{\Delta^5} \int_0^{R_1} \mu r^{10} \, dr \tag{3.107}$$

$$\left. + 5 \frac{1 + 2k_4}{\Delta^5} [\mu r^{11}]_0^{R_1} \right\} = 0,$$

in which C continues to be given by Equation (2.89), k_j by Equation (6.4) of Chapter II, and which represents the final result of our analysis.

This equation continues to be one of first order in ω; but its generalization to spheroidal configuration has raised its degree to three; and, for eccentric orbits, its coefficients will depend on the time. As such, this equation can no longer be integrated in a closed form; and numerical integrations offer the only avenue of approach to a construction of its solution. The reader may note, moreover, that even if $\mu(R_1) = 0$ – so that both integrated parts on the left-hand side of (3.107) are identically zero – the angular velocity ω of axial rotation will not be constant, but (for spheroidal configurations) vary as a solution of the equation

$$C \frac{d\omega}{dt} + \frac{152}{35} \pi (\omega - n) \frac{\omega^2 m_2}{G m_1^2} \frac{1 + 2k_2}{\Delta^3} \int_0^{R_1} \mu r^8 \, dr + \ldots = 0, \tag{3.108}$$

which involves expressions depending on the whole march of internal viscosity distribution throughout the interior.

D. DISSIPATION OF ENERGY BY TIDAL FRICTION

In conclusion of this section, the last task we wish to take up is to investigate the rate at which the kinetic energy of dynamical tides will undergo irreversible dissipation into heat through the effects of viscosity – a process which will tend secularly to diminish the mechanical energy of the system.

As is well known, the viscous dissipation function Φ of compressible fluid flow, characterized by the rectangular velocity components u, v, w, is – in accordance with equation (1.21) of Chapter III – given by

$$\Phi = \mu\left\{2\left(\frac{\partial u}{\partial x}\right)^2 + 2\left(\frac{\partial v}{\partial y}\right)^2 + 2\left(\frac{\partial w}{\partial z}\right)^2 + \left(\frac{\partial w}{\partial y} + \frac{\partial v}{\partial z}\right)^2 + \left(\frac{\partial u}{\partial z} + \frac{\partial w}{\partial x}\right)^2 + \left(\frac{\partial v}{\partial x} + \frac{\partial u}{\partial y}\right)^2\right\} - \frac{2}{3}\mu(\Delta_0')^2, \tag{3.109}$$

where we have ignored the second coefficient of viscosity μ_2 to set $\mu_1 \equiv \mu$; and where the divergence Δ_0' continues to be given by Equation (3.13).

For mass motions caused by dynamical tides, the rectangular velocity components u, v, w in (3.109) should, moreover, be identified with u_0', v_0', w' as given by Equations (2.38)–(2.40). Inserting the latter in Equation (3.109) we find (after some algebra) that

$$\begin{aligned}
\mu^{-1}\Phi &= 6\dot{f}^2 + 2r(\dot{f}^2)_r + (r\dot{f}_r)^2 + (r\dot{f}_n)^2 + \\
&\quad - 4\dot{f}(p_1D_1f + p_2D_2f + p_3D_3f) \\
&\quad - 2(r\dot{f}_r)(p_1D_1\dot{f} + p_2D_2\dot{f} + p_3D_3\dot{f})^2 \\
&\quad + \qquad (p_1D_1f + p_2D_2f + p_3D_3f)^2 \\
&\quad + (f_n)^2\{r^2(p_1^2 + p_2^2 + p_3^2) - (p_1x + p_2y + p_3z)^2\} - \frac{2}{3}(\Delta_0')^2 \qquad (3.110) \\
&= \frac{1}{3}(r\dot{f}_r)^2 + \frac{1}{3}(p_1D_1f + p_2D_2f + p_3D_3f)^2 \\
&\quad + (r\dot{f}_n)^2 + \frac{4}{3}(r\dot{f}_r)(p_1D_1 + p_2D_2 + p_3D_3)f \\
&\qquad - 2(rf_r)(p_1D_1 + p_2D_2 + p_3D_3)\dot{f} \\
&\quad + (f_n)^2\{r^2(p_1^2 + p_2^2 + p_3^2) - (p_1x + p_2y + p_3z)^2\}
\end{aligned}$$

on insertion for Δ_0' from (3.13), and where we have abbreviated

$$(f_n)^2 = (f_x)^2 + (f_y)^2 + (f_z)^2. \tag{3.111}$$

Equation (3.110) represents the viscous dissipation functions for dynamical tides of any type. In particular, the terms on its right-hand side which arise from the 'tidal breathing' due to orbital eccentricity will be factored by the f_t-component of $\dot{f}$ (cf. Equation

(2.37); while those arising from asynchronous rotation (which produces the relative motion of the tides in longitude) will be factored by p_3. In order to determine the relative importance of tidal friction arising from these two causes, let us consider separately the following two cases.

(a) equator lying in the plane of a circular orbit (i.e., $p_1 = p_2 = 0$ and $e = 0$), but rotation and revolution asynchronous (i.e., $p_3 \neq 0$, though constant);

(b) equator and orbit coplanar (i.e., $p_1 = p_2 = 0$), but $e > 0$ and, as result, $p_3 \equiv \omega - n$ becomes a fluctuating function of the time (as $n \equiv \dot{v}$!).

In the first case, it follows at once (for $f_t = 0$) that

$$\dot{f} = -p_3 D_3 f_2 = -p_3 D_3'' f_2 = 3p_3 K_2 r^3 \lambda'' \mu'', \tag{3.112}$$

so that

$$r\dot{f}_r = 9\, p_3 K_2 r^3 \lambda'' \mu'' , \tag{3.113}$$

$$(r\dot{f}_r)^2 = 81\, p_3^2 K_2^2 r^6 \lambda''^2 \mu''^2 , \tag{3.114}$$

and

$$\begin{aligned}(r\dot{f}_n)^2 &= (r\dot{f}_r)^2 + (1-\lambda''^2)(\dot{f}_{\lambda''})^2 + (1-\nu''^2)(\dot{f}_{\nu''})^2 - 2\lambda''\nu'' \dot{f}_{\lambda''} \dot{f}_{\nu''} \\ &= 45\, p_3^2 K_2^2 r^6 \lambda''^2 \mu''^2 + 9\, p_3^2 K_2^2 r^6 (1-\nu''^2) .\end{aligned} \tag{3.115}$$

Moreover, by (3.114),

$$(r\dot{f}_r)\, p_3 D_3 f = 4\, p_3^2 \,(D_3 f_2)^2 = -27\, p_3^2 K_2^2 r^6 \lambda''^2 \mu''^2 \tag{3.116}$$

and

$$D_3 \dot{f} = D_3'' \dot{f}_2 = 3\, p_3 K_2 r^3 (\lambda''^2 - \mu''^2); \tag{3.117}$$

which leads to

$$-2\,(r\dot{f}_r)\, p_3 D_3 \dot{f} = -18\, p_3^2 K_2^2 r^6 (\lambda''^2 - \mu''^2) P_2(\lambda'') \tag{3.118}$$

and, lastly,

$$(\dot{f}_n)^2 = 9\, K_2^2 r^4 \{[P_2(\lambda'')]^2 + (1-\lambda''^2)\lambda''^2\} ; \tag{3.119}$$

so that

$$p_3^2 (r^2 - z^2)(\dot{f}_n)^2 = 9\, p_3^2 K_2^2 r^6 \{[P_2(\lambda'')]^2 + (1-\lambda''^2)\lambda''^2\}(1-\nu''^2) . \tag{3.120}$$

Inserting now from (3.112)–(3.116), (3.118) and (3.120) in (3.110) we find that in case (a),

$$\begin{aligned}\mu^{-1}\Phi &= 75\, p_3^2 K_2^2 r^6 \lambda''^2 \mu''^2 \\ &+ \frac{9}{4} p_3^2 K_2^2 r^6 (1-\lambda''^2)\{5(1-\lambda''^2)(1-\nu''^2) + 4(\lambda''^2 - \mu''^2)\} .\end{aligned} \tag{3.121}$$

On integrating this expression over the entire volume of the respective configuration between the limits given by (2.65), we find that the rate of energy dissipation E due to longitudinal viscous tides should be given by

$$\frac{dE}{dt} = \frac{1}{2}\int \Phi \, dV = \frac{506}{35}\,\pi p_3^2 K_2^2 \int_0^{R_1} \mu r^8 \, dr \quad \text{erg s}^{-1}, \tag{3.122}$$

corresponding (on insertion for K_2 and p_3) to an energy loss per orbital cycle equal to

$$E = \frac{1012}{35n}\left\{\pi(\omega - n)\,\frac{1 + 2k_2 m_2}{\Delta^3 m_1}\right\}^2 \int_0^{R_1} \mu r^8 \, dr \quad \text{erg}, \tag{3.123}$$

where, for circular orbits, n represents the Keplerian angular velocity $2\pi/P$ of orbital revolution in the period P; and ω stands, as before, for the angular velocity of axial rotation of the tidally-distorted star.

If, however, the orbit of binary ceases to be circular, terms additional to those included in (3.121) will arise from 'tidal breathing', and the angular velocity n of orbital motion ceases to be constant. The part of viscous dissipation function due to 'tidal breathing' will stem from the term f_t in f, and assume the form

$$\begin{aligned}\mu^{-1}\Phi_e &= \frac{1}{3}(rf_{tr})^2 + (rf_{tn})^2 \\ &\quad + \frac{4}{3}(rf_{tr})\,p_3 D_3 f - 2p_3 (rf_r) D_3 f_t\,,\end{aligned} \tag{3.124}$$

which on insertion from (2.114) assumes the more explicit form

$$\begin{aligned}\mu^{-1}\Phi_e &= 3H_2^2 r^6 \{(3\lambda''^2 - 1)^2 + 3\lambda''^2(1 - \lambda''^2)\} \\ &\quad + 6\,H_2 K_2 p_3 r^6 \,\lambda''\mu''\, P_2(\lambda'')\,,\end{aligned} \tag{3.125}$$

leading to a rate of dissipation

$$\frac{dE_e}{dt} = \frac{1}{2}\int \Phi_e \, dV = \frac{36}{5}\,\pi H_2^2 \int_0^{R_1} \mu r^8 dr \quad \text{erg s}^{-1}. \tag{3.126}$$

The total dissipation of kinetic energy per cycle should now be given by

$$E = \int_0^P (\dot{E} + \dot{E}_e)\, dt\,, \tag{3.127}$$

where $\dot{E}$ and $\dot{E}_e$ are given by Equations (3.122) and (3.126), respectively. In evaluating the right-hand side of (3.127) we must keep in mind that, for $e > 0$, the angular velocity n of orbital motion will cease to be constant, and becomes a function of the time as given by Equation (3.80). If we insert for H_2 and K_2 from Equations (2.115) and (2.67) we find that

$$E = \frac{2\pi}{35}\left\{(1 + 2k_2)\frac{m_2}{m_1}\right\}^2 \int_0^{R_1} \mu r^8 \, dr \left\{253 \int_0^P \left(\frac{\omega - n}{\Delta^3}\right)^2 dt + \right. \tag{3.128}$$

$$+ 1134 \int_0^P \left(\frac{\dot{\Delta}}{\Delta}\right)^2 \frac{dt}{\Delta^6}\Bigg\} .$$

If, moreover, we note that by the laws of elliptic motion,

$$\frac{\dot{\Delta}}{\Delta} = \frac{\omega_K e \sin v}{\sqrt{(1-e^2)^3}} (1 + e \cos v) , \tag{3.129}$$

where ω_K denotes the Keplerian angular velocity (3.86), then changing over from the time t to the true anomaly v as the variable of integration by (3.85), we establish that

$$\int_0^P \left(\frac{\dot{\Delta}}{\Delta}\right)^2 \frac{dt}{\Delta^6} = \frac{e^2 \omega_K}{A^6(1-e^2)^{15/2}} \int_0^{2\pi} (1 + e \cos v)^6 \sin^2 v \, dv \tag{3.130}$$

$$= \frac{\pi e^2 \omega_K}{A^6(1-e^2)^{15/2}} \left\{1 + \frac{15}{4} e^2 + \frac{15}{8} e^4 + \frac{5}{64} e^6\right\} ;$$

while if we set (for simplicity) $\omega = \omega_K$ and, therefore,

$$\omega - n = \omega_K \left\{1 - \frac{(1 + e \cos v)^2}{(1-e^2)^{3/2}}\right\}, \tag{3.131}$$

it follows that

$$\int_0^P \left(\frac{\omega - n}{\Delta^3}\right)^2 dt = \frac{\omega_K}{A^6(1-e^2)^{15/2}} \int_0^{2\pi} \{1-e^2)^{3/2} - (1 + e \cos v)^2\}^2 (1 + e \cos v)^6 \, dv$$

$$= \frac{\pi \omega_K}{A^6(1-e^2)^{9/2}} \left\{1 + 15 e^2 + \frac{45}{4} e^4 + \frac{5}{8} e^6\right\}$$

$$- \frac{2\pi \omega_K}{A^6(1-e^2)^6} \left\{1 + 28 e^2 + \frac{105}{2} e^4 + \frac{35}{2} e^6 + \frac{35}{64} e^8\right\}$$

$$+ \frac{\pi . \omega_K}{A^6(1-e^2)^{15/2}} \left\{1 + 45 e^2 + \frac{315}{2} e^4 + \frac{525}{4} e^6 + \frac{1575}{64} e^8 + \frac{63}{128} e^{10}\right\}$$

$$= \frac{\pi \omega_K}{A^6} \left\{4 e^2 + 135 e^4 + \ldots\right\} . \tag{3.132}$$

Inserting (3.130) and (3.132) in (3.128) and limiting ourselves to powers of the eccentricity e not exceeding the fourth, we find that

$$E = \frac{1}{35} \pi^2 \omega_K \left\{\frac{1 + 2k_2}{A^3} \frac{m_2}{m_1}\right\}^2 \int_0^{R_1} \mu r^8 \, dr \, \{4292 e^2 + 93825 e^4 + \ldots\} \tag{3.133}$$

as the final result for the energy dissipation, through viscous tides, arising from orbital eccentricity if $\omega = \omega_K$.

Turning back to the exact Equation (3.128) which holds good for any value of orbital eccentricity, we may note that the first term in curly brackets on its right-hand side represents the effects of the 'tidal breathing' (i.e., radial tides) proper; while the second part accounts for those of the 'periodic lag' of dynamical tides arising from the periodic changes in the angular velocity n of orbital revolution. As a result of such changes, the tidal bulge following in phase the secondary component will be sometimes ahead, and at other times behind the position of an observer situated on the surface to the rotating stars – the corresponding phase difference varying with the inverse square of the radius-vector of the relative orbit; while the height of the tide varies (for second-harmonic tides) with inverse cube of this distance; and the effects of both must be taken into account in any more comprehensive treatment of the subject.

It should also, be added that the cases treated in some detail in this section do not exhaust the full range of tidal phenomena which can occur in close binary systems. For should the equator of the rotating configuration be inclined to the orbital plane of its companion (i.e., if $p_1 \neq 0$ and $p_2 \neq 0$), dynamical tides will shift also in latitude and thus give rise to motion which has not been considered yet in any detail. Equation (3.110) is, to be sure, exact for any values of $p_{1,2,3}$; but a detailed treatment of the most general case in which all p's are different from zero must again await future investigations.

IV-4. Nonuniform Rotation

In the preceding two sections of this chapter we have explored the effects of viscosity on axial rotation of the components in close binary systems which are subject to dynamical tides; and in doing so we regarded the angular velocity component ω_z about the z-axis to be only a function of the time. In the present section we wish to explore first some of the consequences of a converse situation, which arises if we allow ω_z to depend on the spatial coordinates, but not on the time. In other words, what we propose is to exchange an assumption of rigid-body rotation for assumed existence of *steady-state* rotation, in which neither the pressure P, density ρ, or potential ψ; nor the angular velocity ω about one (say, z-)axis depends on the time.

A. STEADY-STATE ROTATION; GRAVITY-DARKENING

Such a case is treated best in terms of the spherical polar coordinates r, θ, ϕ. On the assumptions mentioned above, the polar velocity components U, V, W as defined by Equation (1.29) of Chapter III reduce to

$$U = V = 0 \quad \text{and} \quad W = r\omega \sin\theta \; ; \tag{4.1}$$

and, moreover, the partial derivatives of all quantities with respect to the azimuthal angle ϕ vanish identically. If so, and if we insert (4.1) in the Equations (1.31)(1.33) of motion as given in Chapter III, the latter assume the forms

$$\frac{1}{\rho}\frac{\partial P}{\partial r} = \frac{\partial \Psi}{\partial r} + r\omega^2 \sin^2\theta \ , \tag{4.2}$$

$$\frac{1}{\rho}\frac{\partial P}{\partial \theta} = \frac{\partial \Psi}{\partial \theta} + r^2\omega^2 \sin\theta \cos\theta, \tag{4.3}$$

$$0 = \frac{1}{r^2}\frac{\partial}{\partial r}\left\{r^4\left[\mu\frac{\partial\omega}{\partial r} - \frac{F_r}{c^2}\,\omega\right]\right\} + $$
$$+ \frac{1}{\sin^3\theta}\frac{\partial}{\partial\theta}\left\{\left[\mu\frac{\partial\omega}{\partial\theta} - \frac{F_\theta}{c^2}\,r\omega\right]\sin^3\theta\right\}, \tag{4.4}$$

where μ on the right-hand side of Equation (4.4) standards for the sum of the gas plus radiative viscosity, and the terms involving the radiation flux components F_r and F_θ stem from the expression for $\mathfrak{T}$ as given by Equation (1.73) of Chapter III.

Equations (4.2) and (4.3), inserted in Poisson's Equation (1.16) in Chapter III lead to

$$\operatorname{div}\left\{\frac{\operatorname{grad} P}{\rho}\right\} = -4\pi G\rho + 2\omega^2 + \sin^2\theta\left\{r\frac{\partial}{\partial r} + \cot\theta\frac{\partial}{\partial\theta}\right\}\omega^2. \tag{4.5}$$

On the other hand, Equation (1.18) of Chapter III for the conservation of energy reduces in steady state to

$$\operatorname{div}\mathbf{F} = \rho\epsilon + \Phi, \tag{4.6}$$

where, in Equation (1.37) of Chapter III for the dissipation function Φ in spherical polar coordinates, the only nonvanishing components of the viscous stress tensor are

$$\sigma_{\phi r} = r\sin\theta\,\frac{\partial\omega}{\partial r} \quad \text{and} \quad \sigma_{\phi\theta} = \sin\theta\,\frac{\partial\omega}{\partial\theta}\ ; \tag{4.7}$$

to which Equation (1.81) of the same chapter contributes the term

$$-\frac{2\,Fr}{c^2}\,W\sigma_{\phi r} \equiv -\frac{Fr}{c^2}\,(r\sin\theta)^2\,\frac{\partial\omega^2}{\partial r}\ . \tag{4.8}$$

In consequence, in the rotational problem,

$$\Phi = (r\sin\theta)^2\left\{\mu(\operatorname{grad}\omega)^2 - \frac{Fr}{c^2}\frac{\partial\omega^2}{\partial r}\right\}, \tag{4.9}$$

where, in most cases, the radial component of the flux F_r can again be approximated by Equation (1.77) of Chapter III.

Now, in steady-state radiative equilibrium, consistent with Equation (1.57) of Chapter III, the vector flux

$$\mathbf{F} = -(c/k\rho)\operatorname{grad} p_{\mathrm{R}}, \tag{4.10}$$

where, in local thermodynamic equilibrium at temperature T, the scalar radiation pressure

$$p_R = \frac{1}{3} aT^4, \tag{4.11}$$

in accordance with Equation (1.46) of Chapter III. Accordingly, Equation (4.10) can be rewritten as

$$\mathbf{F} = -\kappa_R \text{ grad } T, \tag{4.12}$$

where

$$\kappa_R = \frac{c}{k\rho}\frac{\mathrm{d}p_R}{\mathrm{d}T} = \frac{4acT^3}{3k\rho} \tag{4.13}$$

represents the coefficient of radiative heat conduction.

Equation (4.12) can obviously be rewritten as

$$\mathbf{F} = -\kappa_R \frac{\mathrm{d}T}{\mathrm{d}P}\frac{\mathrm{d}P}{\mathrm{d}n}, \tag{4.14}$$

where $\mathrm{d}/\mathrm{d}n$ stands for the gradient operator. An elimination of **F** between (4.6) and (4.14) then discloses that

$$\rho\epsilon = -\operatorname{div}\left\{\kappa_R \frac{\mathrm{d}T}{\mathrm{d}P}\frac{\mathrm{d}P}{\mathrm{d}n}\right\} - \Phi . \tag{4.15}$$

The reader may note that, of the expressions in curly brackets on the right-hand side of the foregoing equation, $\kappa_R(\mathrm{d}T/\mathrm{d}P)$ represents a scalar quantity, while the gradient $\mathrm{d}P/\mathrm{d}n$ is a vector. Therefore, by an appeal to the well-known rule

$$\operatorname{div}(u\,\mathbf{v}) = u \operatorname{div}\mathbf{v} + \mathbf{v} \operatorname{grad} u \tag{4.16}$$

of vector calculus we find that

$$\operatorname{div}\left\{\kappa_R \frac{\mathrm{d}T}{\mathrm{d}P}\frac{\mathrm{d}P}{\mathrm{d}n}\right\} = \rho\kappa_R \frac{\mathrm{d}T}{\mathrm{d}P} \operatorname{div}\left\{\frac{1}{\rho}\frac{\mathrm{d}P}{\mathrm{d}n}\right\} + \tag{4.17}$$

$$+ \frac{1}{\rho}\frac{\mathrm{d}P}{\mathrm{d}n}\frac{\mathrm{d}}{\mathrm{d}n}\left\{\rho\kappa_R \frac{\mathrm{d}T}{\mathrm{d}P}\right\},$$

the last term of which can be rewritten as

$$\frac{\mathrm{d}}{\mathrm{d}n}\left\{\rho\kappa_R \frac{\mathrm{d}T}{\mathrm{d}P}\right\} = \frac{\mathrm{d}}{\mathrm{d}P}\left\{\rho\kappa_R \frac{\mathrm{d}T}{\mathrm{d}P}\right\}\frac{\mathrm{d}P}{\mathrm{d}n} . \tag{4.18}$$

By a combination of Equations (4.17) and (4.18) with (4.15), the latter can be rewritten as

$$\rho\epsilon + \rho\kappa_R \frac{\mathrm{d}T}{\mathrm{d}P} \operatorname{div}\left\{\frac{\operatorname{grad} P}{\rho}\right\} + \tag{4.19}$$

$$+ \frac{1}{\rho}\frac{\mathrm{d}}{\mathrm{d}P}\left\{\rho\kappa_R \frac{\mathrm{d}T}{\mathrm{d}P}\right\}\left(\frac{\mathrm{d}P}{\mathrm{d}n}\right)^2 + \Phi = 0 ;$$

and an insertion for $\operatorname{div}(\operatorname{grad} P/\rho)$ from Equation (4.5) yields

$$\rho\epsilon = \rho\kappa_R \frac{dT}{dP}\left\{4\pi G\rho - 2\omega^2 - \sin^2\theta\left[r\frac{\partial}{\partial r} + \cot\theta\frac{\partial}{\partial\theta}\right]\omega^2\right\}$$

$$-\frac{1}{\rho}\frac{d}{dP}\left\{\rho\kappa_R\frac{dT}{dP}\right\}\left(\frac{dP}{dn}\right)^2 - \Phi . \tag{4.20}$$

If ω were constant, Φ as given by (4.9) vanishes identically; and Equation (4.20) would reduce to

$$\frac{1}{\rho}\frac{d}{dP}\left\{\rho\kappa_R\frac{dT}{dP}\right\}\left(\frac{dP}{dn}\right)^2 + \rho\epsilon = \rho\kappa_R\frac{dT}{dP}\left\{4\pi G\rho - 2\omega^2\right\}. \tag{4.21}$$

Since, for ω = constant, Equations (4.2) and (4.3) can be expressed vectorially as

$$\text{grad}\, P = \rho\, \text{grad}\, \Psi, \tag{4.22}$$

where (cf. consistent with Equation (1.41) of Chapter II)

$$\Psi = \Omega + \frac{1}{2}(\omega r \sin\theta)^2 , \tag{4.23}$$

it follows that, in hydrostatic equilibrium, the quantities P, ρ, T or κ_R in Equation (4.21) must be constant over equipotential surfaces ψ = constant. The pressure gradient dP/dn cannot, however, remain constant over a non-spherical (distorted) equipotential – as the element dn cannot become zero (except locally), as it varies between the equator and the pole of a rotational spheroid.

Under such conditions, the only way to render both sides of Equation (4.21) constant over an equipotential Ψ = constant is, therefore, to set

$$\rho\kappa_R\frac{dT}{dP} = C, \tag{4.24}$$

where C denotes a constant; and, in such a case, Equation (4.21) then reduces to

$$\epsilon = 4\pi GC\left(1 - \frac{\omega^2}{2\pi G\rho}\right) ; \tag{4.25}$$

which constitutes the celebrated theorem discovered by H. von Zeipel in 1924.

Since its discovery the theorem was, to be sure, regarded largely of the nature of a paradox; for it is unthinkable on physical grounds that the generation ϵ of energy in the stars should depend on the speed of their rotation; and considerable thought has been given to escape the tenets of its validity. Thus Eddington (1926) and his followers set out to accomplish this by relaxing the condition $U = V = 0$ in the equations (4.2)–(4.4) of motion, and thereby generating a system of 'meridional circulation' inside a star, which is indeed sufficient to invalidate the foundation on which the paradoxical result (4.25) is based.

It is not our aim in this chapter to describe the mechanism of this meridional circulation, which represents one way of escape from the validity of (4.25); for a fuller discussion of this aspect of our problem cf., e.g., Mestel (1965). What we wish to stress is that von Zeipel's theorem makes it merely impossible for a light-emitting star in radiative equilibrium to rotate like a rigid body. Moreover, although the meridional currents postulated by Eddington and others constitute a sufficient condition for the way out of the difficulty, this condition is not in fact necessary; for the same end can be accomplished – with the retention of $U = V = 0$ – by the presence of the dissipation function Φ on the right-hand side of (4.20).

It is indeed difficult to see that the validity of von Zeipel's theorem could ever have been taken seriously in the face of the fact that non-zero value of ϵ – whether of nuclear or gravitational (contraction) origin – is bound to entail a presence of radiation; and in radiating systems the right-hand sides of equations of motion of the form (1.1) would be incomplete without addition of terms given by Equation (1.44) of Chapter III. It may, incidentally, be noted that the existence of some of these terms was pointed out by Jeans (1926) only two years after von Zeipel originally called attention to paradoxical consequences of his theorem; though their bearing on the problem was not realized at that time.

A complete expression for the dissipation function Φ in our present case has already been given by Equation (4.9); and it is to be noted that it represents a scalar quantity. Therefore, in order to satisfy Equation (4.19), it is necessary that Equation (4.24) continues to hold good; while (4.25) should be replaced by

$$\rho\epsilon = C\left\{4\pi G\rho - 2\omega^2 - \sin^2\theta\left[r\frac{\partial}{\partial r} + \cot\theta\,\frac{\partial}{\partial\theta}\right]\omega^2\right\} - (r\sin\,\theta)^2\left\{\mu\,(\text{grad}\;\omega)^2 - \frac{F_r}{c^2}\,\frac{\partial\omega^2}{dr}\right\}, \tag{4.26}$$

which represents an implicit relation between ϵ, ρ, F_r and ω^2, determining none of them uniquely in terms of any other.

In the outer parts of a star, where ϵ is effectively zero and F_r can be replaced by $L/4\pi r^2$ (L representing the star's luminosity), the foregoing Equation (4.26) reduces to a partial differential equation for $\omega(r, \theta)$, of second degree, which can be solved (numerically or otherwise) for any adopted pair of functions ρ and μ characteristic of the respective layers; but no solutions of this nature have been performed so far.

It should also be added that Equation (4.26) so simplified is not the only one which bears on the specification of the function $\omega(r, \theta)$. The other such relation is represented by the differential equation (4.4), which is of second order in ω and which we have not so far used. It contains, however, the angular flux component F_θ as an additional dependent variable; and thus does not represent an alternative form of (4.26)

Next, we wish to return once more to Equation (4.24), in order to deduce from it further consequences of interest for the students of close binary systems. A combination of (4.24) with (4.14) discloses that

$$\mathbf{F} = -\frac{C}{\rho}\,\frac{dP}{dn}\,; \tag{4.27}$$

which on insertion for grad P from (4.22) discloses further that

$$\mathbf{F} = -C\,\text{grad}\,\Psi = C\,\mathbf{g} \tag{4.28}$$

demonstrating that the flux F or radiation emergent from the surface of a star in radiative equilibrium should vary proportionally to the local gravity g. In other words, such stars distorted from spherical form should be expected to exhibit a 'gravity-darkening' which (together with the limb-darkening due to semi-transparency of the star's outer fringe) should govern the distribution of apparent brightness over the surfaces of distorted stars.

The physical meaning of the theorem (4.28) at the basis of the phenomenon of gravity-darkening – the mathematical proof of which was likewise given by von Zeipel in 1924 – can be described in the following terms. In accordance with a discussion already given in section II-1, the internal structure of a star in hydrostatic equlibrium, under the influence of its own attraction and of any disturbing potential (centrifugal, tidal) must be such that the total pressure P and the density are functions of the potential Ψ alone. Hence, the temperature T as well as all other functions depending upon ρ and T (such as opacity or viscosity) must also be functions of Ψ; and so the outward flux F of radiation across an equipotential is proportional to the temperature gradient. Since, however, equipotential surfaces are isothermal, the temperature gradient over such a surface is proportional to the gradient of Ψ – and the latter is merely equal to the local gravity g. Thus the outward flow of radiation over any equipotential surface – and, therefore, also over the visible boundary of a star – should be proportional to g.

The consequences of this theorem for an interpretation of the light changes exhibited by close binary systems between eclipses as well as within minima are profound; and their existence has been amply attested by photometric observations (cf. Kopal, 1941c; or, more recently, Kopal, 1968a,b). This is, to be sure, strictly speaking true so far only of the photometric effects of tidal distortion producing the 'ellipticity effect' between minima. Whether or not the rotational distortion gives rise to a similar gravity-darkening remains yet unsettled by purely empirical evidence; for polar flattening invokes light changes out of eclipse only if the axis of rotation is inclined to the orbital plane; and those which minima are so interlocked with consequences of tidal distortion that it is still premature to feel sure of their presence even in the best light curves now available.

Another consequence of Equation (4.24) should enable us to estimate the magnitude of the coefficient μ_R of radiative viscosity from the observed properties of the respective star in radiative equilibrium. The actual value of μ_R in terms of ρ and T has already been given by Equation (1.47) of Chapter III; while the coefficient κ_R of radiative heat conduction was given by Equation (4.13) of the present section. Their comparison discloses that

$$\kappa_R = (5c^2/T)\,\mu_R\,; \tag{4.29}$$

which by use of (4.24) can be solved for μ_R to yield

$$\mu_R = \frac{C}{5c^2}\left\{\frac{T}{\rho}\frac{dP}{dT}\right\}. \tag{4.30}$$

For a perfect gas, in the absence of radiation we can write (in accordance with Equation (1.20) of Chapter III)

$$P = R\rho T; \tag{4.31}$$

while from (4.28) it follows that (to a sufficient approximation)

$$C = \frac{F_r}{g} = \frac{L}{4\pi Gm}\,, \tag{4.32}$$

where L and m denote the luminosity and mass of our star. In consequence, Equation (4.30) can be rewritten as

$$\begin{aligned} \mu_R &= \frac{RC}{5c^2}\left\{\frac{T}{\rho}\frac{\mathrm{d}}{\mathrm{d}T}(\rho T)\right\} \\ &= \frac{RLT}{20\pi c^2 gm}\left\{1+\frac{\mathrm{d}\log\rho}{\mathrm{d}\log T}\right\}. \end{aligned} \tag{4.33}$$

If, lastly, we assume the existence of an adiabatic equation of state of the form $P\rho^\gamma =$ = constant, and remember that (in accordance with Equation (1.25) of Chapter III) $R = C_P - C_V = (\gamma - 1)\,C_v$, Equation (4.33) can be reduced further to

$$\mu_R = \frac{\gamma-1}{\gamma+1}\left(\frac{\gamma C_v L}{20\pi c^2 Gm}\right) T, \tag{4.34}$$

which could represent a tolerable approximation to radiative viscosity in the far (though not deep) interior of a star.

B. NON-STEADY ROTATION

In conclusion of the present section, let us return once more to Equation (4.4), which we shall rewrite in a form valid for *non-steady state* (i.e., allowing ω to be a function of the time t as well as of the polar coordinates r, θ), but in which the components $F_{r,\theta}$ of radiative flux will be neglected. Under these conditions, Equation (4.4) should be replaced by

$$\rho\frac{\partial\omega}{\partial t} = \frac{1}{r^4}\frac{\partial}{\partial r}\left(r^4\mu\frac{\partial\omega}{\partial r}\right) + \frac{1}{r^2}\frac{\partial}{\partial l}\left\{(1-l^2)\,\mu\frac{\partial\omega}{\partial l}\right\} - \frac{2\mu l}{r^2}\frac{\partial\omega}{\partial l}, \tag{4.35}$$

where we have abbreviated

$$l \equiv \cos\theta\ ; \tag{4.36}$$

and which for $\partial\mu/\partial l = 0$ becomes identical with one previously considered by Kippenhahn (1955).

Let us attempt now to separate the variables by setting

$$\omega(r, l; t) \equiv w(r)\,\Theta(l)\,T(t) \tag{4.37}$$

on the assumption that the viscosity μ as well as density ρ are functions of r only (a supposition which imposes a limit on the magnitude of ω, for which a dependence of μ and ρ on l through rotational flattening would represent a second-order effect). If so, an insertion of (4.37) in (4.35) and subsequent division by $\rho\omega$ transforms the latter into

$$\frac{1}{T}\frac{\partial T}{\partial t} = \frac{1}{\rho r^4 w}\frac{\partial}{\partial r}\left(r^4\mu\frac{\partial w}{\partial r}\right) + \frac{\mu}{\rho r^2\Theta}\frac{\partial}{\partial l}\left[(1-l^2)\frac{\partial\Theta}{\partial l}\right] - \frac{2\mu l}{\rho r^2\Theta}\frac{\partial\Theta}{\partial l} \tag{4.38}$$

The left-hand side of this equation depends, by definition, on t only; while the right-hand side depends on r and l. Therefore, if (4.38) is to hold good, both sides of it must obviously be *constant* and equal to (say) λ. If so, however, the equation for $T(t)$ immediately becomes of the form

$$\frac{\partial T}{\partial t} + \lambda T = 0, \tag{4.39}$$

and integrates into

$$T(t) = T_0 e^{-\lambda t}, \tag{4.40}$$

where the constants λ and T_0 can ben identified with those defined previously by Equations (3.78) and (3.79).

By the same argument, it follows that

$$\frac{1}{\rho r^4 w}\frac{\partial}{\partial r}\left(r^4\mu\frac{\partial w}{\partial r}\right) + \frac{\mu}{\rho r^2\Theta}\left\{\frac{\partial}{\partial l}\left[(1-l^2)\frac{\partial\Theta}{\partial l}\right] - 2l\frac{\partial\Theta}{\partial l}\right\} = -\lambda\,, \tag{4.41}$$

or

$$\frac{1}{\mu r^2 w}\frac{\partial}{\partial r}\left(r^4\mu\frac{\partial w}{\partial r}\right) + \frac{\lambda\rho r^2}{\mu} + \frac{1}{\Theta}\left\{(1-l^2)\frac{\partial^2\Theta}{\partial l^2} - 4l\frac{\partial\Theta}{\partial l}\right\} = 0. \tag{4.42}$$

The first two terms on the left-hand side of the foregoing equation depend, by definition, on r alone; and the third, on l. The validity of (4.42) requires then that each group be equal to a constant (say, k) of opposite algebraic sign; and this permits us to express the equations for w and Θ separately in the form

$$\frac{\partial}{\partial r}\left(r^4\mu\frac{\partial w}{\partial r}\right) + (\lambda\rho r^2 - k\mu)\,r^2 w = 0 \tag{4.43}$$

and

$$(1-l^2)\frac{\partial^2\Theta}{\partial l^2} - 4l\frac{\partial\Theta}{\partial l} + k\Theta = 0. \tag{4.44}$$

This latter equation turns out to be a particular case of the hypergeometric type, and its solution which remains regular for $l = 1$ can be expressed as

$$\Theta_n(l) = a_n F(-n, n+3, 2; \tfrac{1}{2}(1-l))\,, \tag{4.45}$$

where $n(n + 3) = k$ and a_n is a constant of integration. The series on the right-hand side of (4.45) diverges, however, for $l = -1$; and its sum will remain finite only if n happens to be zero or a positive integer – in which case the right-hand side of (4.45) will reduce to the Jacobi polynomials G_n of the form

$$G_n\left(3, 2, \frac{1-l}{2}\right) \equiv \frac{2}{(n+1)(n+2)} \frac{\partial}{\partial l} P_{n+1}(l)\,, \tag{4.46}$$

where $P_{n+1}(l)$ stands for the Legendre polynomials of the respective degree, orthogonalized between ±1.

Accordingly, Equation (4.43) now assumes the form

$$\frac{\partial}{\partial r}\left(r^4 \mu \frac{\partial w_n}{\partial r}\right) + \{\lambda \rho r^2 - n(n+3)\mu\} r^2 w_n = 0; \tag{4.47}$$

and the corresponding solution of (4.35) written in the form (4.37) can be expressed as

$$\omega(r, \theta; t) = a_0 + \sum_{n=1}^{\infty} a_n e^{-\lambda_n t} w_n(r)\, G_n\left(3, 2, \sin^2 \tfrac{1}{2}\theta\right), \tag{4.48}$$

where the a_n's are arbitrary constants.

The boundary conditions which should single out admissible solutions of (4.35) can be specified as follows. If the surface $r = R$ is determined by the conditions $\rho(R) = \mu(R) = 0$, the second term of (4.47) becomes identically zero; and if so, the equation itself will be satisfied at $r = R$ only if

$$\left[\frac{\partial}{\partial r}\left(r^4 \mu \frac{\partial w_n}{\partial r}\right)\right]_R = 0\,; \tag{4.49}$$

necessitating that

$$\left(\frac{\partial w_n}{\partial r}\right)_R = 0 \quad \text{for} \quad \left(\frac{\partial \mu}{\partial r}\right)_R \neq 0. \tag{4.50}$$

Next, let us require again that – as in section III-2 – the radial components of the viscous stress tensor should vanish over the free surface. For the velocity components the conditions (4.1)

$$\sigma_{rr} = \sigma_{r\theta} = 0 \tag{4.51}$$

are satisfied automatically; while the additional requirement that

$$\sigma_{r\phi} = 0 \tag{4.52}$$

implies that, on the boundary,

$$\left(\frac{\partial \omega}{\partial r}\right)_R = 0. \tag{4.53}$$

This condition is satisfied already by Equation (4.50) for an arbitrary set of constants a_j, which can be specified by additional boundary conditions imposed on the problem. This is, however, not the place to pursue such possibilities in further detail; and their development must be postponed for subsequent investigations.

BIBLIOGRAPHICAL NOTES

The problem of non-uniform rotation of the stars has attracted much attention in the past, and has become the centre of an extensive literature. Most of this literature is, however, of only marginal interest for the real subject of this chapter – which was to prepare the ground for investigations of the effects, on the rotation of the components of close binary systems, arising from the dissipation of the kinetic energy of generalized rotation through dynamical tides. The essential emphasis has, therefore, been on the effects caused by the *deformability* of the stars on their generalized rotation, in three dimensions, in an external field of force.

A mathematical theory of the rotation of deformable bodies dissipating energy through the action of dynamical tides has constituted a classical problem, whose investigation advanced but slowly (cf. J. Liouville, 1858; H. Gyldén, 1873; G. H. Darwin, 1879; S. Oppenheimer, 1885; H. Poincaré, 1910) in the past hundred years – mainly in response to geophysical, rather than astronomical, situations (such as the precession and nutation, or tidal friction, of the Earth and the planets. Of more recent work along similar lines cf. W. M. Kaula (1963, 1964), G. J. F. MacDonald (1964), P. Goldreich (1966), P. Melchior (1966); and others.

In contrast with all this work, section IV-1 of this chapter contains a development (the first of its kind) of a mathematical theory of such motions based on the Eulerian (rather than Lagrangian) equations of viscous flow – more appropriate for astrophysical situations, where viscous terms are of smaller magnitude (though not any lesser importance) than in the planets. Such a theory was developed by the writer several years ago (Kopal, 1968a); and applied to obtain a quantitative solution for the retardation of axial rotation of a fluid star of arbitrary structure caused by lagging tides – a phénomenon at the basis of the 'tidal evolution' of close binary systems (to be more fully discussed in Chapter VIII of this book).

In view of the fundamental importance of this process for the cosmogony of such systems, its foundations were developed in some detail in Sections IV-2 and IV-3 of this chapter, following Kopal (1972a). Subsequently, this theory has been translated into the elegant formalism of vector calculus by J. N. Tokis (1973, 1974). Since, however, a scalar form of these equations is prerequisite for any kind of analytical or numerical work to be based upon them, it was thought desirable to retain it in the text.

Throughout sections IV-2 and IV-3 we have been concerned with rigid-body, non-steady rotation of the components of close binary systems about one axis. In the concluding section IV-2 we switched our attention to the case of steady-state, but non-rigid, uni-axial rotation, in order to bring out some astrophysical consequences of such a situation – mainly those concerning the phenomenon of 'gravity-darkening' which is of fundamental importance for an interpretation of the light changes of distorted eclipsing systems.

The historically important papers in this connection are those by H. von Zeipel (1924a,b), A. S. Eddington (1926) or S. Chandrasekhar (1933). It is of historical interest to note that the first investigator who conjectured that the surface brightness of a rotating star of very large mass should vary proportionally to the local gravity was J. H. Jeans (1919). He insisted, however, that this conjecture was applicable only to stars having no sources of energy other than gravitational contraction; and he abandoned the conjecture (for reasons not stated) in a subsequent paper (Jeans, 1925).

The first investigator to deduce the photometric consequences of gravity darkening in close binary systems was S. Takeda (1934), followed by H. N. Russell (1939) and, more generally, by Z. Kopal (1941c, 1942) or T. E. Sterne (1941) who extended its theory to all terms of first order in surficial distortion due to rotation as well as tides – a theory extended later to terms of second order (cf. Z. Kopal and M. Kitamura, 1968) for observations not only in total light (as was done by Takeda or Russell), but in any particular domain of the spectrum.

The first comparison of such a theory with the observations – demonstrating the existence of the gravity darkening – was carried out by Kopal (1941c); and for the latest results of such a comparison cf. Kopal (1968d, f).

Throughout section IV-4 we have been concerned with the phenomenon of gravity-darkening only for stars in purely radiative equilibrium. For the effects of convection on gravity-darkening (not treated explicity in this chapter) cf. L. B. Lucy (1967), R. C. Smith and R. Worley (1974); and others.

CHAPTER V

DYNAMICS OF CLOSE BINARIES

In the second chapter of this volume we have investigated the ways in which axial rotation, or the proximity of another star, will influence the equilibrium figures of the components of close binary systems; and in subsequent two chapters we extended our investigation to consider the effects of dynamical tides, or of nonuniform rotation. The aim of the present chapter will be to investigate the ways in which systems consisting of the components distorted by rotation or tides behave dynamically as a whole – i.e., the three-axial motions of the two components about their respective centres of mass, as well as the motions of such centres around the common centre of gravity of the entire system.

In doing so we shall consider first (until section V-5) the total masses $m_{1,2}$ of the components of a close binary to be constant and independent of the time. Let us, moreover, assume that our binary constitutes an isolated dynamical system in space, undisturbed by any external force. Such an assumption can be amply justified for our stellar neighbourhood in the Galaxy, where even cumulative effects of chance encounters with neighbouring stars entail but negligible dynamical consequences (cf. again Ambartsumian, 1937; Chandrasekhar, 1944; Takase, 1953; or Yabushita, 1966). Under such conditions, the centre of mass of our binary system will move uniformly through space, with a constant velocity, over intervals of time which are long in comparison with the time-scale of our observations (except, possibly, in very dense clusters).

Such a picture would, however, no longer be adequate if our close binary system happens to be attended by a third star near enough to constitute with it a physical triple system. Triple systems consisting of a close pair attended by a third body are relatively frequent – about one-tenth of all known close binaries seem to possess distant companions whose dynamical (as distinct from photometric, or spectroscopic) effects may be observationally significant. An investigation of such effects will, however, be postponed till the concluding section V-6 of this chapter. Throughout sections V-1–5 we shall restrict ourselves to considering our problem strictly within the framework of the problem of two bodies, subject to perturbations arising from mutual distortion of the two stars.

In section V-1 which follows these introductory remarks we shall consider the differential equations governing our problem, and shall detail the form of the equations which govern the pertubations of the orbital elements. In section V-2 we shall digress back to Chapter IV, in order to solve the equations derived there for the *precession* and *nutation* of the equatorial planes of our components, which are inextricably coupled with perturbations of the orbital plane of our binary; and in section V-3 we shall address ourselves to an investigation of the perturbation of the elements which govern the relative motion in the orbital plane of our binary. Section V-4 will then be devoted to a discussion of the *period changes* of such pairs due – both directly and indirectly – to such perturbations; while in the remaining two sections we propose to explore the dynamical consequences of a relaxation of the assumed constancy of the two masses $m_{1,2}$

(sec. V-5); while in section V-6 we shall explore similar consequences of the presence of a third mass m_3 in the proximity of the binary $m_{1,2}$.

V-1. Equations of the Problem

Consider an isolated dynamical system consisting of two components of constant masses $m_{1,2}$, and revolving around the common centre of gravity of the system. In the absence of any external mass, we shall confine our attention to a *relative* orbit of one star (say, of mass m_2) around the centre of gravity of mass m_1 which will be identified with the origin of an inertial system of rectangular coordinates x, y, z. Accordingly, if (0, 0, 0) denote the coordinates, in this system, of the centre of gravity of mass m_1, the coordinates x, y, z of the position of the secondary's centre of gravity in its relative orbit around m_1 at a time t are known to be governed by the system of differential equations of the form

$$\frac{d^2x}{dt^2} + G(m_1 + m_2)\frac{x}{r^3} = \frac{\partial \Re_{12}}{\partial x}, \tag{1.1}$$

$$\frac{d^2y}{dt^2} + G(m_1 + m_2)\frac{y}{r^3} = \frac{\partial \Re_{12}}{\partial y}, \tag{1.2}$$

$$\frac{d^2z}{dt^2} + G(m_1 + m_2)\frac{z}{r^3} = \frac{\partial \Re_{12}}{\partial z}, \tag{1.3}$$

where

$$r^2 = x^2 + y^2 + z^2 \tag{1.4}$$

stands for the radius vector of the relative orbit; and $\Re_{12}$ represents the 'disturbing function' due to the departure of the mass distribution in both components from spherical symmetry (because of axial rotation and mutual tidal action), and related with the potential energy W_{12} of the system by

$$W_{12} = \frac{Gm_1m_2}{r} + \frac{m_1m_2}{m_1 + m_2}\Re_{12}. \tag{1.5}$$

Equations (1.1)–(1.3) constitute a simultaneous differential system of 6-th order. For $\Re_{12} = 0$ this reduces to the well-known Keplerian problem which can be integrated in a closed form in terms of plane conic curves entailing six integration constants. Two of these – the longitude Ω of the ascending node and inclination i of the orbit to the invariable plane of the system – specify the position of the orbital plane in space; while additional four: namely,

A = semi-major axis of the relative orbit;
e = orbital eccentricity;
ω = longitude of periastron measured from Ω;
ϵ = difference between the true and mean anomaly of the periastron passage;

describe the respective properties of the plane orbits.

These orbits are, moreover, known to be described in terms of these constants by the radius-vector

$$r = \frac{A(1-e^2)}{1+e\cos v} = A(1-e\cos E), \tag{1.6}$$

where v and E denote, respectively, the true and eccentric anomaly of m_2 in its relative orbit; the latter being related with the time t by Kepler's well-known equation

$$E - e\sin E = n(t-T), \tag{1.7}$$

where T denotes the time of the periastron passage, and

$$n^2 = \frac{G(m_1+m_2)}{A^3} \tag{1.8}$$

the mean daily motion of the two bodies; while the true anomaly v is related with E by the equation

$$\tan\tfrac{1}{2}v = \sqrt{\frac{1+e}{1-e}}\tan\tfrac{1}{2}E. \tag{1.9}$$

The reader may note that, under these conditions, the constant $\epsilon = \Omega + \omega - nT$.

A. PERTURBATION EQUATIONS

The foregoing Equations (1.6)–(1.9) constitute an exact solution of Equations (1.1)–(1.3) only for an unperturbed two-body problem, obtained when $\mathfrak{R}_{12} = 0$. Should this not be the case, one can continue to seek a solution for disturbed motion in the form (1.6)–(1.9), provided that the six orbital parameters involved in (1.6)–(1.9) are allowed to *vary* with the time in accordance with the 'perturbation equations' of the form

$$\frac{1}{An}\frac{d\Omega}{dt} = \frac{1}{\sqrt{1-e^2}\sin i}\frac{\partial\mathfrak{R}_{12}}{\partial i}, \tag{1.10}$$

$$\frac{1}{An}\frac{di}{dt} = -\frac{1}{\sqrt{1-e^2}\sin i}\frac{\partial\mathfrak{R}_{12}}{\partial\Omega} - \frac{\tan\frac{1}{2}i}{\sqrt{1-e^2}}\left\{\frac{\partial\mathfrak{R}_{12}}{\partial\omega} + \frac{\partial\mathfrak{R}_{12}}{\partial\epsilon}\right\}, \tag{1.11}$$

$$\frac{1}{An}\frac{dA}{dt} = 2A\frac{\partial\mathfrak{R}_{12}}{\partial\epsilon}, \tag{1.12}$$

$$\frac{1}{An}\frac{de}{dt} = -\sqrt{1-e^2}\left\{\frac{1-\sqrt{1-e^2}}{e}\frac{\partial\mathfrak{R}_{12}}{\partial\epsilon} + \frac{1}{e}\frac{\partial\mathfrak{R}_{12}}{\partial\omega}\right\}, \tag{1.13}$$

$$\frac{1}{An}\frac{d\tilde{\omega}}{dt} = \frac{\tan\frac{1}{2}i}{\sqrt{1-e^2}}\frac{\partial\mathfrak{R}_{12}}{\partial i} + \frac{\sqrt{1-e^2}}{e}\frac{\partial\mathfrak{R}_{12}}{\partial e}, \tag{1.14}$$

$$\frac{1}{An}\frac{d\epsilon}{dt} = \frac{\tan\frac{1}{2}i}{\sqrt{1-e^2}}\frac{\partial\mathfrak{R}_{12}}{\partial i} + \sqrt{1-e^2}\,\frac{1-\sqrt{1-e^2}}{e}\frac{\partial\mathfrak{R}_{12}}{\partial e} - 2A\frac{\partial\mathfrak{R}_{12}}{\partial A}, \tag{1.15}$$

where we have abbreviated

$$\bar{\omega} = \Omega + \omega. \tag{1.16}$$

Let, in what follows, x stand for any one of the six elements Ω, i, A, e, ω and ϵ of the orbit. Differentiating (1.6) with respect to x we find that

$$\frac{\partial r}{\partial x} = \frac{r}{A}\frac{\partial A}{\partial x} - A\cos E\frac{\partial e}{\partial x} + Ae\sin E\frac{\partial E}{\partial x}, \tag{1.17}$$

and a similar differentiation of (1.7) yields

$$\frac{\partial E}{\partial x} = \frac{A}{r}\left\{t\frac{\partial n}{\partial x} + \frac{\partial \epsilon}{\partial x} - \frac{\partial \bar{\omega}}{\partial x} + \sin E\frac{\partial e}{\partial x}\right\}. \tag{1.18}$$

Eliminating $\partial E/\partial x$ from (1.17) with the aid of (1.18) and making use of the fact that, from (1.6),

$$\cos E = \frac{A - r}{eA} = \frac{e + \cos v}{1 + e\cos v} \tag{1.19}$$

and, accordingly,

$$\sin E = \frac{r\sin v}{A\sqrt{1 - e^2}}, \tag{1.20}$$

we find that

$$\frac{\partial r}{\partial x} = \frac{r}{A}\frac{\partial A}{\partial x} - A\cos E\frac{\partial e}{\partial x} +$$

$$+ \frac{Ae\sin v}{\sqrt{1 - e^2}}\left\{t\frac{\partial n}{\partial x} + \frac{\partial \epsilon}{\partial x} - \frac{\partial \bar{\omega}}{\partial x} + \sin E\frac{\partial e}{\partial x}\right\}. \tag{1.21}$$

By differentiating similarly Equation (1.9) we find that

$$\frac{1}{\sin v}\frac{\partial v}{\partial x} = \frac{1}{\sin E}\frac{\partial E}{\partial x} + \frac{1}{1 - e^2}\frac{\partial e}{\partial x} =$$

$$= \left\{\frac{1}{1 - e^2} + \frac{A}{r}\right\}\frac{\partial e}{\partial x} + \frac{A}{r\sin E}\left\{t\frac{\partial n}{\partial x} + \frac{\partial \epsilon}{\partial x} - \frac{\partial \bar{\omega}}{\partial x}\right\} \tag{1.22}$$

by use of (1.18). If, moreover, we eliminate $\sin E$ form the right-hand side of Equation (1.21) by a recourse to (1.20) we obtain

$$\frac{\partial v}{\partial x} = \frac{A^2\sqrt{1 - e^2}}{r^2}\left\{t\frac{\partial n}{\partial x} + \frac{\partial \epsilon}{\partial x} - \frac{\partial \omega}{\partial x}\right\} + \sin v\left\{\frac{1}{1 - e^2} + \frac{A}{r}\frac{\partial e}{\partial x}\right\}. \tag{1.23}$$

In order to reduce the right-hand sides of Equations (1.10)–(1.15) to more tractable forms, we find it convenient to abbreviate by **R**, **S**, **W** the rectangular components of disturbing accelerations, defined so that **R** represents the component acting in the direction of the radius-vector; **S**, that acting in the plane of the orbit in the direction perpendicular to the radius-vector (positive in the direction of motion); while **W** stands for the component normal to the plane of the orbit (positive towards the north pole). If so, it can be shown (cf., e.g., Tisserand, 1889; Chapter 27) that the perturbation equations (1.10)–(1.115) can be rewritten alternatively as

$$\frac{\mathrm{d}\Omega}{\mathrm{d}t} = \frac{\mathbf{W}\, r \sin u}{nA^2\sqrt{1-e^2}\sin i}\,, \tag{1.24}$$

$$\frac{\mathrm{d}i}{\mathrm{d}t} = \frac{\mathbf{W}\, r \cos u}{nA^2\sqrt{1-e^2}}\,, \tag{1.25}$$

$$\frac{\mathrm{d}A}{\mathrm{d}t} = \frac{2}{n\sqrt{1-e^2}}\left\{\mathbf{R}\, e \sin v + \frac{A(1-e^2)}{r}\,\mathbf{S}\right\}, \tag{1.26}$$

$$\frac{\mathrm{d}e}{\mathrm{d}t} = \frac{\sqrt{1-e^2}}{nA}\left\{\mathbf{R}\sin v + (\cos v + \cos E)\,\mathbf{S}\right\}, \tag{1.27}$$

$$\frac{\mathrm{d}\bar{\omega}}{\mathrm{d}t} = \frac{\sqrt{1-e^2}}{nAe}-\left\{\mathbf{R}\cos v + \left[1+\frac{r}{A(1-e^2)}\right]\mathbf{S}\sin v\right\} + \frac{\mathbf{W}\, r\sin u}{nA^2\sqrt{1-e^2}}\tan\tfrac{1}{2}i, \tag{1.28}$$

$$\frac{\mathrm{d}\epsilon}{\mathrm{d}t} = -\frac{1}{nA}\left[\frac{2r}{A} - \frac{1-e^2}{e}\cos v\right]\mathbf{R} - \frac{1-e^2}{nAe}\left[1+\frac{r}{A(1-e^2)}\right]\mathbf{S}\sin v, \tag{1.29}$$

where

$$u = \omega + v \tag{1.30}$$

denotes the true anomaly measured from the node.

Let, moreover, $x''y''z''$ hereafter denote another system of rectangular coordinates, revolving in space in such a way that the X''-axis continues to coincide with the radius-vector r joining the centres of the two stars, while Y'' and Z'' remain perpendicular to it. If a''_{ij} denote the direction cosines of the doubly-primed axes with respect to the inertial system XYZ, the transformation between the two is governed by the matrix equation

$$\begin{Bmatrix} x \\ y \\ z \end{Bmatrix} = \begin{Bmatrix} a''_{11} & a''_{12} & a''_{13} \\ a''_{21} & a''_{22} & a''_{23} \\ a''_{31} & a''_{32} & a''_{33} \end{Bmatrix} \begin{Bmatrix} x'' \\ y'' \\ z'' \end{Bmatrix} \tag{1.31}$$

identical with Equations (2.2) of Chapter IV, in which the direction cosines $a''_{i,j}$ continue to be given by Equations (2.3) – (2.5) of that chapter.

Consequently,

$$\begin{Bmatrix} \dfrac{\partial}{\partial x} \\[2ex] \dfrac{\partial}{\partial y} \\[2ex] \dfrac{\partial}{\partial z} \end{Bmatrix} = \begin{bmatrix} a''_{11} & a''_{12} & a''_{13} \\[2ex] a''_{21} & a''_{22} & a''_{23} \\[2ex] a''_{31} & a''_{32} & a''_{33} \end{bmatrix} \begin{Bmatrix} \dfrac{\partial}{\partial x''} \\[2ex] \dfrac{\partial}{\partial y''} \\[2ex] \dfrac{\partial}{\partial z''} \end{Bmatrix} \tag{1.32}$$

and the components **R, S, W** of the disturbing accelerations introduced in Equations (1.24) – (1.29) will be given the partial derivatives

$$\mathbf{R} = \left(\frac{\partial \mathfrak{R}_{12}}{\partial x''}\right)_{x'',0,0}, \tag{1.33}$$

$$\mathbf{S} = \left(\frac{\partial \mathfrak{R}_{12}}{\partial y''}\right)_{x'',0,0}, \tag{1.34}$$

$$\mathbf{W} = \left(\frac{\partial \mathfrak{R}_{12}}{\partial z''}\right)_{x'',0,0}, \tag{1.35}$$

evaluated at the centre of secondary (disturbing) component of mass m_2, the position of which is described by the coordinates $x'' = r$ and $y'' = z'' = 0$.

In order to ascertain the forms of these derivatives, we find it expedient to change over from the rectangular coordinates x'', y'', z'' to polar coordinates r, θ'', ϕ'' in the revolving system by means of the transformation

$$\begin{Bmatrix} \dfrac{\partial}{\partial x''} \\[2ex] \dfrac{\partial}{\partial y''} \\[2ex] \dfrac{\partial}{\partial z''} \end{Bmatrix} = \begin{bmatrix} \cos\phi''\sin\theta'' & \cos\phi''\cos\theta'' & -\sin\phi'' \\[2ex] \sin\phi''\sin\theta'' & \sin\phi''\cos\theta'' & \cos\phi'' \\[2ex] \cos\theta'' & -\sin\theta'' & 0 \end{bmatrix} \begin{Bmatrix} \dfrac{\partial}{\partial r} \\[2ex] \dfrac{1}{r}\dfrac{\partial}{\partial \theta''} \\[2ex] \dfrac{1}{r\sin\theta''}\dfrac{\partial}{\partial \phi''} \end{Bmatrix} \tag{1.36}$$

Since, at the centre of the tide-generating body, $\theta'' = \frac{1}{2}\pi$ and $\phi'' = 0$, it follows that

$$\begin{Bmatrix} \mathbf{R} \\[2ex] \mathbf{S} \\[2ex] \mathbf{W} \end{Bmatrix} = \begin{bmatrix} 1 & 0 & 0 \\[2ex] 0 & 0 & 1 \\[2ex] 0 & -1 & 0 \end{bmatrix} \begin{Bmatrix} \dfrac{\partial \mathfrak{R}_{12}}{\partial r} \\[2ex] \dfrac{1}{r}\dfrac{\partial \mathfrak{R}_{12}}{\partial \theta''} \\[2ex] \dfrac{1}{r}\dfrac{\partial \mathfrak{R}_{12}}{\partial \phi''} \end{Bmatrix}_{r,90^\circ,0^\circ}, \tag{1.37}$$

i.e.,

$$\mathbf{R} = \left(\frac{\partial \mathfrak{R}_{12}}{\partial r}\right)_{r, 90^\circ, 0^\circ}, \tag{1.38}$$

$$\mathbf{S} = \left(\frac{1}{r}\frac{\partial \mathfrak{R}_{12}}{\partial \phi''}\right)_{r, 90^\circ, 0^\circ}, \tag{1.39}$$

$$\mathbf{W} = -\left(\frac{1}{r}\frac{\partial \mathfrak{R}_{12}}{\partial \theta''}\right)_{r, 90^\circ, 0^\circ}. \tag{1.40}$$

As long as the disturbing action of one component upon the other can be regarded as that of a mass-point, the disturbing function $\mathfrak{R}_{12}$ can be expressed as

$$\mathfrak{R}_{12} = \mathfrak{R}_1 + \mathfrak{R}_2, \tag{1.41}$$

where

$$\begin{aligned} \mathfrak{R}_i &= -\frac{2\mu\omega_i^2 R_i^5 (k_2)_i}{3Gm_i r^3} P_2(\nu_i') \\ &+ \mu \frac{m_{3-i}}{m_i} \sum_{j=2}^{4} \frac{(k_j)_i R_i^{2j+1}}{r^{j+1} r_e^{j+1}} P_j(\lambda_i''), \qquad i = 1, 2, \end{aligned} \tag{1.42}$$

and in which we have abbreviated $\mu \equiv G(m_1 + m_2)$; while ω_i denotes the angular velocity of axial rotation of the respective component; R_i, its mean radius; and $(k_j)_i$ the constants associated with the j-th spherical harmonic distortion and depending on the internal structure of the respective star as defined by Equation (6.4) of Chapter II.

The first term on the right-hand side of Equation (1.42), factored by ω_i^2, represents the effect of the rotational distortion of the respective star; while the second one stands for the contribution of the first three partial tides. Let, moreover, in what follows

$$\left.\begin{aligned} \lambda'' &= \cos\phi'' \sin\theta'', \\ \mu'' &= \sin\phi'' \sin\theta', \\ \nu'' &= \cos\theta'', \end{aligned}\right\} \tag{1.43}$$

denote the direction cosines of an arbitrary radius-vector in the doubly-primed system of revolving coordinates; and $\nu''_{1,2,3}$ be the direction cosines of the axis of rotation of the respective component in the same system, given by the equations

$$\nu_1'' = \mathfrak{A}_i \sin u + \mathfrak{B}_i \cos u, \tag{1.44}$$

$$\nu_2'' = \mathfrak{A}_i \cos u - \mathfrak{B}_i \sin u, \tag{1.45}$$

where

$$\mathfrak{A}_i = \cos\theta_i \sin i - \sin\theta_i \cos(\phi_i - \Omega)\cos i, \tag{1.46}$$

$$\mathfrak{B}_i = \qquad + \sin\theta_i \sin(\phi_i - \Omega); \tag{1.47}$$

and

$$\nu_3'' = \cos\theta_i \cos i + \sin\theta_i \cos(\phi_i - \Omega)\sin i, \tag{1.48}$$

in which the angles Ω and i represent, as before, the longitude of the node and inclination of the orbital plane; and the angles ϕ_i denote the longitude of the nodes at which the equator of the respective component intersects the invariable plane $Z = 0$ of our dynamical system, while θ_i stands for that between the axis of rotation of the respective configuration and the Z-axis.

If so, however, the argument ν_i' of the rotational harmonic $P_2(\nu_i')$ on the right-hand side of Equation (1.42) – representing as it does a cosine of the angle between an arbitrary radius-vector in the doubly-primed coordinates and the axis of rotation – should be given by

$$\nu_i' = \lambda'' \nu_1'' + \mu'' \nu_2'' + \nu'' \nu_3'' \tag{1.49}$$

irrespective of any tidal lag.

Furthermore, let – in what follows – the angle ϵ_i represent the amount of tidal lag (cf. sec. III-3) of the respective component in astrocentric longitude (i.e. in the $X''Y''$-plane), and η_i that in latitude (i.e., in the meridional plane). For stars consisting of viscous fluid, the former will represent a nonvanishing quantity if their angular velocity of axial rotation differs from that of orbital revolution, which remains constant if the relative orbit of the two bodies is circular, but oscillating otherwise in time. On the other hand, η_i can be non-zero only if the equators of the rotating configurations are inclined to the orbital plane, and is bound to oscillate in the course of a revolution even if the latter is synchronous with the rotation.

Accordingly, the direction cosines l_i, m_i, n_i, in the doubly-primed coordinates, of the actual axis of symmetry of the respective tidally-distorted configuration should be given by

$$\left.\begin{aligned} l_i &= \cos \epsilon_i \cos \eta_i \,, \\ m_i &= \sin \epsilon_i \cos \eta_i \,, \\ n_i &= \sin \eta_i \,; \end{aligned}\right\} \tag{1.50}$$

and the cosine λ_i'' of the angle between an arbitrary radius-vector in the revolving system of coordinates and the axis of symmetry of the lagging tidal bulge should be given by

$$\begin{aligned} \lambda_i'' &= \lambda'' l_i + \mu'' m_i + \nu'' \eta_i \\ &= \cos \theta'' \sin \eta_i + \sin \theta'' \cos \eta_i \cos (\phi'' - \epsilon_i) \,, \end{aligned} \tag{1.51}$$

which specifies the argument of the tidal harmonics $P_j(\lambda_i'')$ in the second term on the right-hand side of Equation (1.42). Lastly, in the same term,

$$r_\epsilon = \frac{A(1-e^2)}{1 + e \cos (\upsilon - \epsilon)} \tag{1.52}$$

represents a radius-vector in the direction of the lagging tide.

B. EVALUATION OF THE COEFFICIENTS; TIDAL LAG

Having established (within the framework of our approximation) the explicit form of the disturbing function $\mathfrak{R}_{12}$ in terms of the spherical polar coordinates r, θ'', ϕ'' of the doubly-primed revolving system, we are ready to proceed with the evaluation of the coefficients **R**, **S** and **W**, occurring on the right-hand sides of the fundamental equations (1.24) – (1.29) from Equations (1.38) – (1.40). In order to do so, let us rewrite (1.42) in the form

$$\mathfrak{R}_i = (\mathfrak{R}_i)_{\text{rot}} + (\mathfrak{R}_i)_{\text{tidal}}\,, \tag{1.53}$$

where, accordingly,

$$(\mathfrak{R}_i)_{\text{rot}} = -\frac{2\mu\omega_i^2 R_i^5}{3Gm_i r^3}\,(k_2)_i P_2(\nu_i') \tag{1.54}$$

and

$$(\mathfrak{R}_i)_{\text{tidal}} = \mu\,\frac{m_{3-i}}{m_i}\sum_{j=2}^{4}\frac{(k_j)_i R_i^{2j+1}}{r^{j+1}\,r_\epsilon^{j+1}}\,P_j(\lambda_i'') \tag{1.55}$$

with ν_i' and λ_i'' as given by Equations (1.49) and (1.51).

If we insert the preceding Equations (1.54) and (1.55) in (1.38) – (1.40) and perform the requisite differentiations, we find that in the case of the *rotational* distortion,

$$\mathbf{R} = \frac{2\mu}{G}\sum_{i=1}^{2}(k_2)_i\,\frac{\omega_i^2 R_i^5}{m_i r^4}\,P_2(\nu_1''), \tag{1.56}$$

$$\mathbf{S} = -\frac{2\mu}{G}\sum_{i=1}^{2}(k_2)_i\,\frac{\omega_i^2 R_i^5}{m_i r^4}\left(\nu_i'\,\frac{\partial\nu_i'}{\partial\phi''}\right)_{r,\,90^\circ,0^\circ}, \tag{1.57}$$

$$\mathbf{W} = \frac{2\mu}{G}\sum_{i=1}^{2}(k_2)_i\,\frac{\omega_i^2 R_i^5}{m_i r^4}\left(\nu_i'\,\frac{\partial\nu_i'}{\partial\theta''}\right)_{r,\,90^\circ,0^\circ}, \tag{1.58}$$

where ν_1'' continues to be given by Equation (1.44), while

$$\left(\nu_i'\,\frac{\partial\nu_i'}{\partial\phi''}\right)_{r,90^\circ,0^\circ} = \nu_1''\nu_2'' = \frac{1}{2}\,\{(\mathfrak{A}_i^2 - \mathfrak{B}_i^2)\sin 2u + 2\mathfrak{A}_i\,\mathfrak{B}_i\cos 2u\} \tag{1.59}$$

$$-\left(\nu_i'\,\frac{\partial\nu_i'}{\partial\phi''}\right)_{r,90^\circ,0^\circ} = \nu_1''\nu_3'' = (\mathfrak{A}_i\sin u + \mathfrak{B}_i\cos u)\,\nu_3'' \tag{1.60}$$

by (1.44) and (1.45).

On the other hand, in the case of *tidal* distortion, at the centre of the disturbing component (where $\theta'' = 90^\circ$ and $\phi'' = 0^\circ$),

$$P_2(\lambda_i'') = P_2(\cos\epsilon_i\cos\eta_i)\,, \tag{1.61}$$

$$\frac{\partial P_2}{\partial \theta''} = -\frac{3}{2} \sin 2\eta_i \cos \epsilon_i, \tag{1.62}$$

$$\frac{\partial P_2}{\partial \phi''} = \frac{3}{2} \sin 2\epsilon_i \cos^2 \eta_i ; \tag{1.63}$$

$$P_3(\lambda_i'') = P_3(\cos \epsilon_i \cos \eta_i) , \tag{1.64}$$

$$\begin{aligned} \frac{\partial P_3}{\partial \theta''} &= -\frac{15}{16} (\sin 3\eta_i + \sin \eta_i) \cos 2\epsilon_i - \\ &\quad - \frac{3}{16} (5 \sin 3\eta_i - 3 \sin \eta_i) , \end{aligned} \tag{1.65}$$

$$\begin{aligned} \frac{\partial P_3}{\partial \phi''} &= \frac{15}{32} (\sin 3\epsilon_i + \sin \epsilon_i) \cos 3\eta_i + \\ &\quad + \frac{3}{32} (15 \sin 3\epsilon_i - \sin \epsilon_i) \cos \eta_i ; \end{aligned} \tag{1.66}$$

$$P_4(\lambda_i'') = P_4(\cos \epsilon_i \cos \eta_i) , \tag{1.67}$$

$$\begin{aligned} \frac{\partial P_4}{\partial \theta''} &= -\frac{35}{64} (\cos 3\epsilon_i + 3 \cos \epsilon_i) \sin 4\eta_i - \\ &\quad - \frac{5}{32} (7 \cos 3\epsilon_i - 3 \cos \epsilon_i) \sin 2\eta_i , \end{aligned} \tag{1.68}$$

$$\begin{aligned} \frac{\partial P_4}{\partial \phi''} &= \frac{35}{128} (\cos 4\eta_i + 4 \cos 2\eta_i + 3) \sin 4\epsilon_i + \\ &\quad + \frac{5}{64} (7 \cos 4\eta_i + 4 \cos 2\eta_i - 3) \sin 2\epsilon_i . \end{aligned} \tag{1.69}$$

Hence, it follows from (1.38) – (1.40) and (1.61) – (1.69) that

$$\begin{aligned} \mathbf{R} &= -2 \frac{\mu}{r} \sum_{i=1}^{2} \frac{m_{3-i}}{m_i} \sum_{i=2}^{4} (j+1) (k_j)_i \times \\ &\quad \times \left\{ 1 - \frac{1}{2} \frac{r}{f} \frac{\partial f}{\partial r} \right\} \frac{R_i^{2j+1}}{(rr_\epsilon)^{j+1}} P_j (\cos \epsilon_i \cos \eta_i), \end{aligned} \tag{1.70}$$

$$\mathbf{S} = \frac{\mu}{4} \sum_{i=1}^{2} \frac{m_{3-i}}{m_i} \; 3(k_2)_i \frac{R_i^5}{r^4 r_\epsilon^3} (\cos 2\eta_i + 1) \sin 2\epsilon_i +$$

$$+ \frac{3}{8}(k_3)_i \frac{R_i^7}{r^5 r_\epsilon^4} [5(\cos 3\eta_i + 3\cos\eta_i)\cos 3\epsilon_i +$$
$$+ (5\cos 3\eta_i - \cos\eta_i)\sin\epsilon_i] +$$
$$+ \frac{5}{32}(k_4)_i \frac{R_i^9}{r^6 r_\epsilon^5} [7(\cos 4\eta_i + 4\cos 2\eta_i + 3)\sin 4\epsilon_i +$$
$$+ 2(7\cos 4\eta_i + 4\cos 2\eta_i - 3)\sin 2\epsilon_i] + \cdots \Big\}, \tag{1.71}$$

and

$$W = \frac{\mu}{2}\sum_{i=1}^{2}\frac{m_{3-i}}{m_i}\Big\{3(k_2)_i \frac{R_i^5}{r^4 r_\epsilon^3}\cos\epsilon_i \sin 2\eta_i +$$
$$+ \frac{3}{8}(k_3)_i \frac{R_i^7}{r^5 r_\epsilon^4} [10\cos^3\epsilon_i \sin 3\eta_i +$$
$$+ (5\cos 2\epsilon_i - 3)\sin\eta_i] +$$
$$+ \frac{5}{16}(k_4)_i \frac{R_i^9}{r^6 r_\epsilon^5} [14\cos^3\epsilon_i \sin 4\eta_i +$$
$$+ (7\cos 3\epsilon_i - 3\cos\epsilon_i)\sin 2\eta_i] + \cdots \Big\}, \tag{1.72}$$

where ϵ_i, η_i denote, as before, the angles of tidal lag of the respective component in longitude and latitude; and from (1.42) expanded in ascending powers of ϵ

$$\frac{r}{r_\epsilon} = \frac{1 + e\cos(v - \epsilon)}{1 + e\cos v} = 1 + \frac{\epsilon\, er\sin v}{A(1-e^2)} -$$
$$- \frac{\epsilon^2\, er\cos v}{2A(1-e^2)} + \cdots \equiv f. \tag{1.73}$$

The presence of the factor r_ϵ as distinct from r signifies that, on account of tidal lag, the extent of tidal distortion will not be maximum when the components are closest to each other, but at a somewhat later time (because of a finite speed of progression of the radial tidal waves in viscous media).

V-2. Perturbations of the Orbital Plane; Precession and Nutation

In order to study the perturbations of the angles Ω and i specifying the position in space of the plane of the orbit, it is necessary to investigate simultaneously the motion of the equatorial planes of the two components rotating about an axis which may, but need not, be perpendicular to that of the orbit. If not, the precession and nutation of these axes in space is inextricably coupled with perturbations of Ω and i; and must be investigated before Equations (1.10)–(1.11) or (1.24)–(1.25) can be solved for Ω and i. The aim of the present section will be to do so on the basis of the Eulerian equations (1.108)–

(1.110), the explicit form of which was established already in the preceding chapter. Of these three, Equation (1.110) alone was used so far to study the effects of viscous forces on the rotation about a *fixed* axis in sections IV-2 and IV-3. What we wish to do now is to 'unfreeze' the position of this axis, and to investigate its motion in space consistent with the Eulerian equations (1.108) and (1.109) of Chapter IV, on the assumption that the solution of (1.110) is known from sections IV-2 and 3.

A. EFFECTS OF VISCOSITY

In order to do so, we must complete an explicit formulation of the terms on the right-hand sides of Equations (1.108)–(1.109) of Chapter IV arising from viscosity. The integrands are already known to us from Equations (3.49)–(3.50) of that chapter, and their integrals can be evaluated term-by-term by the method followed previously. For the fractional height f of the j-th harmonic dynamical tides is given by Equation (2.67) of the preceding chapter as

$$f_j = K_j r^{j+1} P_j(\lambda''); \tag{2.1}$$

and if we regard the viscosity μ to be again a function of the radial coordinate r only, we find that

$$\int \mu x^2 \nabla^2 f \, dV = \frac{16}{5} \pi K_2 P_2(a''_{11}) \int_0^{R_1} \mu r^5 \, dr, \tag{2.2}$$

$$\int \mu y^2 \nabla^2 f \, dV = \frac{16}{5} \pi K_2 P_2(a''_{21}) \int_0^{R_1} \mu r^5 \, dr, \tag{2.3}$$

$$\int \mu z^2 \nabla^2 f \, dV = \frac{16}{5} \pi K_2 P_2(a''_{31}) \int_0^{R_1} \mu r^5 \, dr; \tag{2.4}$$

$$\int \mu xy \nabla^2 f \, dV = \frac{24}{5} \pi K_2 a''_{11} a''_{21} \int_0^{R_1} \mu r^5 \, dr, \tag{2.5}$$

$$\int \mu xz \nabla^2 f \, dV = \frac{24}{5} \pi K_2 a''_{21} a''_{31} \int_0^{R_1} \mu r^5 \, dr, \tag{2.6}$$

$$\int \mu yz \nabla^2 f \, dV = \frac{24}{5} \pi K_2 a''_{21} a''_{31} \int_0^{R_1} \mu r^5 \, dr; \tag{2.7}$$

so that

$$\int \mu(x^2 + y^2) \nabla^2 f \, dV = -\frac{16}{5} \pi K_2 P_2(a''_{31}) \int_0^{R_1} \mu r^5 \, dr, \tag{2.8}$$

$$\int \mu(x^2 + z^2) \nabla^2 f \, dV = -\frac{16}{5} \pi K_2 P_2(a''_{21}) \int_0^{R_1} \mu r^5 \, dr, \tag{2.9}$$

and

$$\int \mu(y^2 + z^2)\,\nabla^2 f\,\mathrm{d}V = -\frac{16}{5}\,\pi K_2 P_2(a''_{11}) \int_0^{R_1} \mu r^5\,\mathrm{d}r, \tag{2.10}$$

where the constant K_2 continues to be defined by Equation (2.67) of Chapter IV.

Moreover, by a similar process we establish that

$$\int \mu D_4 f\,\mathrm{d}V = -\frac{16}{5}\,\pi K_2 P_2(a''_{11}) \int_0^{R_1} \mu r^5\,\mathrm{d}r, \tag{2.11}$$

$$\int \mu D_5 f\,\mathrm{d}V = -\frac{16}{5}\,\pi K_2 P_2(a''_{21}) \int_0^{R_1} \mu r^5\,\mathrm{d}r, \tag{2.12}$$

$$\int \mu D_6 f\,\mathrm{d}V = -\frac{16}{5}\,\pi K_2 P_2(a''_{31}) \int_0^{R_1} \mu r^5\,\mathrm{d}r; \tag{2.13}$$

and, turning to the integrals of terms factored by the radial derivative of μ, we find that

$$\int \mu_r x^2 f_r\,\mathrm{d}V = \frac{8}{5}\,\pi K_2 P_2(a''_{11}) \int_0^{R_1} \frac{\partial\mu}{\partial r}\, r^6\,\mathrm{d}r, \tag{2.14}$$

$$\int \mu_r y^2 f_r\,\mathrm{d}V = \frac{8}{5}\,\pi K_2 P_2(a''_{21}) \int_0^{R_1} \frac{\partial\mu}{\partial r}\, r^6\,\mathrm{d}r, \tag{2.15}$$

$$\mu_r z^2 f_r\,\mathrm{d}V = \frac{8}{5}\,\pi K_2 P_2(a''_{31}) \int_0^{R_1} \frac{\partial\mu}{\partial r}\, r^6\,\mathrm{d}r, \tag{2.16}$$

and

$$\int \mu_r xy\, f_r\,\mathrm{d}V = \frac{12}{5}\,\pi K_2 a''_{11} a''_{21} \int_0^{R_1} \frac{\partial\mu}{\partial r}\, r^6\,\mathrm{d}r, \tag{2.17}$$

$$\int \mu_r xz\, f_r\,\mathrm{d}V = \frac{12}{5}\,\pi K_2 a''_{11} a''_{21} \int_0^{R_1} \frac{\partial\mu}{\partial r}\, r^6\,\mathrm{d}r, \tag{2.18}$$

$$\int \mu_r yz\, f_r\,\mathrm{d}V = \frac{12}{5}\,\pi K_2 a''_{21} a''_{31} \int_0^{R_1} \frac{\partial\mu}{\partial r}\, r^6\,\mathrm{d}r. \tag{2.19}$$

These are the only terms which make nonvanishing contributions to the viscous terms on the right-hand sides of Equations (1.108)–(1.109) of Chapter IV. For, consistent with our assumption that $\mu \equiv \mu(r)$,

$$D_1(\mu) = D_2(\mu) = D_3(\mu) = 0; \tag{2.20}$$

and, of the terms factored by μ,

$$\int \mu D_{1,2,3} f\,\mathrm{d}V = 0, \tag{2.21}$$

$$\int \mu(x, y, z) f_{x,y,z} \, \mathrm{d}V = 0 \tag{2.22}$$

for any combination of x,y, and z.

If we insert all the foregoing partial results on the right-hand sides of Equations (3.1) to (3.3) of Chapter IV, we find that

$$\begin{aligned}\int \rho \{y \mathfrak{H} - z \mathfrak{G}\} \, \mathrm{d}V = &\frac{24}{5} \pi K_2 \{2p_1 p_2(a''_{11}) + p_2 a''_{11} a''_{21} + \\ &+ p_3 a''_{11} a''_{31}\} \int_0^{R_1} \mu r^5 \, \mathrm{d}r + \\ &+ \frac{12}{5} \pi K_2 \left\{\frac{2}{3} p_1 P_2(a''_{11}) + p_2 a''_{11} a''_{21} + \right. \\ &\left. + p_3 a''_{11} a''_{31}\right\} \int_0^{R_1} \frac{\partial \mu}{\partial r} r^6 \mathrm{d}r ,\end{aligned} \tag{2.23}$$

$$\begin{aligned}\int \rho \{z \mathfrak{F} - x \mathfrak{H}\} \, \mathrm{d}V = &\frac{24}{5} \pi K_2 \{2p_2 P_2(a''_{21}) + p_3 a''_{21} a''_{31} + \\ &+ p_1 a''_{11} a''_{31}\} \int_0^{R_1} \mu r^5 \, \mathrm{d}r + \\ &+ \frac{12}{5} \pi K_2 \left\{\frac{2}{3} p_2 P_2(a''_{21}) + p_3 a''_{21} a''_{31} + \right. \\ &\left. + p_1 a''_{11} a''_{21}\right\} \int_0^{R_1} \frac{\partial \mu}{\partial r} r^6 \mathrm{d}r ,\end{aligned} \tag{2.24}$$

$$\begin{aligned}\int \rho \{x \mathfrak{H} - y \mathfrak{G}\} \, \mathrm{d}V = &\frac{24}{5} \pi K_2 \{2p_3 P_2(a''_{31}) + p_1 a''_{31} a''_{11} + \\ &+ p_2 a''_{21} a''_{31}\} \int_0^{R_1} \mu r^5 \, \mathrm{d}r + \\ &+ \frac{12}{5} \pi K_2 \left\{\frac{2}{3} p_3 P_2(a''_{31}) + p_1 a''_{31} a''_{11} + \right. \\ &\left. + p_2 a''_{21} a''_{31}\right\} \int_0^{R_1} \frac{\partial \mu}{\partial r} r^6 \mathrm{d}r ;\end{aligned} \tag{2.25}$$

or, on partial integration,

$$\begin{aligned}\int \rho \{y \mathfrak{F} - z \mathfrak{H}\} \, \mathrm{d}V = &\frac{12}{5} \pi K_2 \left\{\left[\mu r^6\right]_0^{R_1} - 4 \int_0^{R_1} \mu r^5 \, \mathrm{d}r\right\} a''_{11} \omega''_x \\ &- \frac{4}{5} \pi K_2 \left\{\left[\mu r^6\right]_0^{R_1} - 12 a''^{\,2}_{11} \int_0^{R_1} \mu r^5 \, \mathrm{d}r\right\} p_1 ,\end{aligned} \tag{2.26}$$

$$\int \rho \{z\mathfrak{F} - x\mathfrak{H}\}\,dV = \frac{12}{5}\pi K_2\left\{\left[\mu r^6\right]_0^{R_1} - 4\int_0^{R1} \mu r^5\,dr\right\} a''_{21}\omega''_x$$

$$- \frac{4}{5}\pi K_2\left\{\left[\mu r^6\right]_0^{R_1} - 12 a''^{\,2}_{21} \int_0^{R_1} \mu r^5\,dr\right\} p_2, \qquad (2.27)$$

$$\int \rho \{x\mathfrak{G} - y\mathfrak{F}\}\,dV = \frac{12}{5}\pi K_2\left\{\left[\mu r^6\right]_0^{R_1} - 4\int_0^{R_1} \mu r^5\,dr\right\} a''_{31}\omega''_x$$

$$- \frac{4}{5}\pi K_2\left\{\left[\mu r^6\right]_0^{R_1} - 12 a''^{\,2}_{31} \int_0^{R_1} \mu r^5\,dr\right\} p_3; \qquad (2.28)$$

where the coefficients $p_{1,2,3}$ continue to be given by Equations (2.35) of Chapter IV; and ω''_x by Equation (2.122) of the same chapter.

Should the orbit and the equator of the rotating configuration be coplanar, $p_1 = p_2 = 0$, $a''_{31} = 0$ and $\omega''_x = 0$ on the right-hand sides of the foregoing Equations (2.26)–(2.27), rendering them identical with Equations (3.19) and (3.20) of Chapter IV; while Equations (3.21) becomes identical with (2.28) above.

With the explicit expressions established now for the integrals on the right-hand sides of the fundamental equations (1.108)–(1.110) of Chapter IV, we can do the same with their left-hand sides by an appeal to the Equations (2.104)–(2.112), (2.116)–(2.118) and (2.119)–(2.121) established in Chapter IV for an arbitrary geometry of the system; and the stage has been set for the integration of the Equations (1.108) and (1.109) of motion in the same way as we did in Chapter IV with (1.110). This sets the task which will confront us in the next part of this section.

B. LINEARIZED CASE

In order to embark upon such a task with any hope of obtaining an analytic solution of our problem, we find it necessary to restrict the range of the dependent variables in such a way that none will appear in our equations of motion in higher than the first powers. This we shall obtain by regarding not only the fractional height f of the tides (or, more specifically, their amplitudes αK_2) as small enough for their squares and higher powers to be negligible – an assumption inherent in all developments of the preceding Chapter IV – but also the angles Θ and i of inclination of the equator of the rotating cofiguration, and of its orbit, to the invariable plane of the system to be likewise small quantities of first order; the same should be true of the eccentricity e of their orbit. In consequence of these assumptions, the tides make the moments of inertia A, B, C differ from their equilibrium values by small quantities of first order; and of the products of inertia, F'' will likewise be one of first order while D'' and E'' are of second order. Moreover, while the angular velocity ω_z of axial rotation will hereafter be regarded as a quantity of zero order, ω_x and ω_y will be of first order. Lastly, we anticipate that the (secular) rates ϕ and θ of precession and nutation are so small in comparison with ω_z that their ratios can likewise be regarded as small quantities of first order; while ω_z is of third order – a fact permitting us thereafter to treat now ω_z as constant.

Within the scheme of such an approximation, the dominant terms in Equations (1.108)–(1.109) of Chapter IV are found to be of second order; and those in (2.3), of third order. Stripped to their dominant terms, Equations (1.108) and (1.109) will reduce to

$$\begin{aligned} A\,\omega_x &+ (C-B)\,\omega_y\omega_z + D\omega_z^2 + F\omega_x\omega_z \\ &+ 2\alpha K_2\left\{p_3 a''_{11}a''_{31}\,\omega_x - \frac{2}{3}p_3 P_2(a''_{11})\,\omega_y + \right. \\ &\left. + \left[\frac{2}{3}p_2P_2(a''_{11}) - p_1 a''_{11}a''_{21}\right]\omega_z + \frac{1}{2}p_3 a''_{12}\omega_{x''}\right\} \\ &+ 8\alpha K_2\, a_{33}a''_{21}\, v''_1\omega_z^2 \\ &= \frac{12}{5}\pi K_2\left\{\left[\mu r^6\right]_0^{R_1} - 4\int_0^{R_1}\mu r^5\,\mathrm{d}r\right\}a''_{11}\omega_{x''} \\ &- \frac{4}{5}\pi K_2\left\{\left[\mu r^6\right]_0^{R_1} - 12a''^{\,2}_{11}\int_0^{R_1}\mu r^5\,\mathrm{d}r\right\}p_1, \end{aligned} \tag{2.29}$$

and

$$\begin{aligned} B\,\omega_y &+ (A-C)\,\omega_x\omega_z - E\omega_z^2 - F\omega_y\omega_z \\ &- 2\alpha K_2\left\{p_3 a''_{11}a''_{21}\omega_y - \frac{2}{3}p_3P_2(a''_{21})\,\omega_x\right. \\ &\left. + \left[\frac{2}{3}p_1P_2(a''_{21}) - p_2 a''_{11}a''_{21}\right]\omega_z - \frac{1}{2}p_3 a''_{22}\omega_{x''}\right\} \\ &- 8\alpha K_2\, a_{33}a''_{11}\, v''_1\omega_z^2 \\ &= \frac{12}{5}\pi K_2\left\{\left[\mu r^6\right]_0^{R_1} - 4\int_0^{R_1}\mu r^5\,\mathrm{d}r\right\} \\ &- \frac{4}{5}\pi K_2\left\{\left[\mu r^6\right]_0^{R_1} - 12a''^{\,2}_{21}\int_0^{R_1}\mu r^5\,\mathrm{d}r\right\}, \end{aligned} \tag{2.30}$$

where, by definition,

$$\omega_x = \dot\theta\cos\phi + \dot\psi\sin\theta\sin\phi, \tag{2.31}$$

$$\omega_y = \dot\theta\sin\phi - \dot\psi\sin\theta\cos\phi, \tag{2.32}$$

$$\omega_z = \qquad + \dot\psi\cos\theta \qquad + \dot\phi, \tag{2.33}$$

are the three angular velocity components expressed in terms of the three Eulerian angles $\theta, \phi, \dot\psi$; and (by Equation (2.122) of Chapter IV)

$$\omega_{x''} = a''_{11}\omega_x + a''_{21}\omega_y + a''_{31}\omega_z \,. \tag{2.34}$$

The foregoing Equations (2.31)–(2.34) are exact for any values of the Eulerian angles or their derivatives. Within the scheme of our approximation, however, in terms associated with the rotational distortion we can set

$$\omega_x = a_{13}\omega_z \qquad \text{and} \qquad \omega_y = a_{23}\omega_z \,; \tag{2.35}$$

while the corresponding moments and products of inertia are given by Equations (2.78) –(2.83) of Chapter IV. Inserting then in the left-hand sides of Equations (2.29) and (2.30) we find that

$$(C' - B')\,\omega_y\omega_z + D'\omega_z^2 + F'\omega_x\omega_z = 0 \tag{2.36}$$

and

$$(A' - C')\,\omega_x\omega_z - E'\omega_z^2 - F\,\omega_y\omega_z = 0. \tag{2.37}$$

Similarly, in the case of tidal distortion (whose axis of symmetry is 'locked-in' with the radius vector of direction cosines $a''_{11}, a''_{21}, a''_{31}$), we have

$$\omega_x = na''_{13}, \omega_y = na''_{23}, \qquad \omega_z = n, \tag{2.38}$$

where n denotes the mean daily motion of the disturbing body. Moreover, the moments and products of inertia of tidally-distorted configurations are given by Equations (2.69) – (2.74) of Chapter IV and their insertion in (2.29)–(2.30) discloses that

$$(C'' - B'')\,\omega_y\omega_z + D''\omega_z^2 + F''\omega_x\omega_z = 0 \tag{2.39}$$

and

$$(A'' - C'')\,\omega_x\omega_z - E''\omega_z^2 - F''\omega_y\omega_z = 0 \tag{2.40}$$

as well.

Let us multiply Equation (2.29) by $\cos\phi$, (2.30) by $\sin\phi$, and add; afterwards multiply (2.29) by $\sin\phi$, (2.30) by $\cos\phi$, and subtract. Since, within the scheme of our approximation,

$$a_{13} = \sin\phi\sin\theta, \quad a_{23} = -\cos\phi\sin\theta, \quad a_{33} = 1; \tag{2.41}$$

and

$$\left.\begin{aligned} &a''_{13} = \sin\Omega\sin i, \quad a''_{23} = -\cos\Omega\sin i, \quad a''_{33} = 1; \\ &a''_{31} = \sin u\sin i, \quad a''_{32} = \cos u\sin i, \end{aligned}\right\} \tag{2.42}$$

while

$$\left.\begin{aligned} &a''_{11} = \cos(u + \Omega), \quad a''_{21} = \sin(u + \Omega), \\ &a''_{12} = -\sin(u + \Omega), a''_{22} = \cos(u + \Omega), \end{aligned}\right\} \tag{2.43}$$

it follows that, correctly to quantities of second order,

$$A\dot{\omega}_x\cos\phi + B\dot{\omega}_y\sin\phi = A_0\omega_z\dot{\phi}\sin\theta, \tag{2.44}$$

$$A\dot{\omega}_x\sin\phi - B\dot{\omega}_y\cos\phi = A_0\omega_z\dot{\theta}\cos\theta\,; \tag{2.45}$$

while

$$2a''_{11}a''_{21}(\omega_x \cos\phi - \omega_y \sin\phi) = \omega_z \sin\theta \sin 2\phi \sin 2(u+\Omega), \tag{2.46}$$

$$2a''_{11}a''_{21}(\omega_x \sin\phi + \omega_y \cos\phi) = -\omega_z \sin\theta \cos 2\phi \sin 2(u+\Omega); \tag{2.47}$$

$$\left\{\frac{2}{3}p_2P_2(a''_{11}) - p_1a''_{11}a''_{21}\right\}\cos\phi - \left\{\frac{2}{3}p_1P_2(a''_{21}) - p_2a''_{11}a''_{21}\right\}\sin\phi =$$
$$= -\frac{2}{3}\sin\theta\, P_2\{\cos(u+\Omega-\phi)\} + \tag{2.48}$$
$$+ n\sin i\left\{\cos u\cos(u+\Omega-\phi) - \frac{1}{3}\cos(\Omega-\phi)\right\};$$

$$\left\{\frac{2}{3}p_2P_2(a''_{11}) - p_1a''_{11}a''_{21}\right\}\sin\phi + \left\{\frac{2}{3}p_1P_2(a''_{21}) - p_2a''_{11}a''_{21}\right\}\cos\phi$$
$$= \frac{1}{2}\omega_z \sin\theta \sin 2(u+\Omega-\phi) - \tag{2.49}$$
$$- n\sin i\cos u\sin(u+\Omega-\phi) + \frac{1}{3}n\sin i\sin(\Omega-\phi);$$

$$-\omega_yP_2(a''_{11})\cos\phi + \omega_xP_2(a''_{21})\sin\phi =$$
$$= \omega_z\sin\theta\left\{P_2[\cos(u+\Omega-\phi)] - \frac{3}{4}\sin 2\phi\sin 2(u+\Omega)\right\}; \tag{2.50}$$

$$-\omega_yP_2(a''_{11})\sin\phi - \omega_xP_2(a''_{21})\cos\phi =$$
$$= \sin\theta\sin 2\phi\cos 2(u+\Omega); \tag{2.51}$$

and

$$\omega_{x''}(a''_{12}\cos\phi + a''_{22}\sin\phi) =$$
$$= -\omega_z\{\sin u\sin i - \sin\theta\sin(u+\Omega-\phi)\}\sin(u+\Omega-\phi), \tag{2.52}$$

$$\omega_{x''}(a''_{12}\sin\phi - a''_{22}\cos\phi) =$$
$$= -\omega_z\{\sin u\sin i - \sin\theta\sin(u+\Omega-\phi)\}\cos(u+\Omega-\phi); \tag{2.53}$$

it follows that

$$2\alpha K_2\omega_z\left\{\frac{2}{3}\omega_z(p_3-\omega_z)\sin\theta\, P_2[\cos(u+\Omega-\phi)]\right.$$
$$\left. + n\sin i[\cos u\cos(u+\Omega-\phi) - \frac{1}{3}\cos(\Omega-\phi)]\right\} \tag{2.54}$$
$$= -\alpha K_2\omega_z n\left\{\frac{\partial^2 v''^2_2}{\partial\theta} + \frac{1}{3}\frac{\partial v''^2_3}{\partial\theta}\right\},$$

and

$$\alpha K_2\omega_z\left\{(\omega_z - p_3)\sin\theta\sin 2(u+\Omega-\phi) - \right.$$
$$\left. - 2n\sin i\cos u\sin(u+\Omega-\phi) + \frac{2}{3}n\sin i\sin(\Omega-\phi)\right\} = \tag{2.55}$$

$$= \alpha K_2\omega_z n\left\{\frac{1}{\sin\theta}\frac{\partial v''^2_2}{\partial\phi} + \frac{1}{3\sin\theta}\frac{\partial v''^2_3}{\partial\phi}\right\},$$

where the direction cosines ν_j'' continue to be given exactly by Equation (2.7) of Chapter IV and, within the scheme of our first-order approximation reduce to

$$\nu_1'' = \sin i \sin u - \sin\theta \sin(u + \Omega - \phi)\,, \tag{2.56}$$

$$\nu_2'' = \sin i \cos u - \sin\theta \cos(u + \Omega - \phi)\,; \tag{2.57}$$

so that

$$\sin(u + \Omega - \phi) = -\frac{1}{\sin\theta}\frac{\partial\nu_2''}{\partial\phi} = -\frac{\partial\nu_1''}{\partial\theta}\,, \tag{2.58}$$

$$\cos(u + \Omega - \phi) = +\frac{1}{\sin\theta}\frac{\partial\nu_1''}{\partial\phi} = -\frac{\partial\nu_2''}{\partial\theta}\,; \tag{2.59}$$

while from the exact equation

$$\nu_3'' = \cos\theta\cos i + \sin\theta\sin i\cos(\Omega - \phi) \tag{2.60}$$

it follows that, approximately,

$$\frac{\partial\nu_3''}{\partial\theta} = -\sin\theta + \sin i\cos(\Omega - \phi) \tag{2.61}$$

and

$$\frac{1}{\sin\theta}\frac{\partial\nu_3''}{\partial\phi} = \sin i \sin(\Omega - \phi)\,. \tag{2.62}$$

The foregoing results complete the evaluation of the 'gyroscopic' terms in the equations of motion, factored by different deformation coefficients arising from dynamical tides. The 'forcing terms' (cf. Equations (2.131)–(2.133) of Chapter IV), arising from the attraction of the external mass on the equatorial bulge of the rotating configuration, will be of the form

$$\begin{aligned}\cos\phi \int D_1\Omega_c dm + \sin\phi \int D_2\Omega_c\, dm &= \\ = 2k_2R_1^5m_2\,\Delta^{-3}\,\omega_z^2\,\nu_1''\,\sin(u + \Omega - \phi) & \\ = -\{k_2R_1^5m_2\,\Delta^{-3}\,\omega_z^2\}\frac{\partial\nu_1''^2}{\partial\theta} &\end{aligned} \tag{2.63}$$

and

$$\begin{aligned}\sin\phi \int D_1\Omega_c dm - \cos\phi \int D_2\Omega_c dm &= \\ = 2k_2R_1^5m_2\,\Delta^{-3}\,\omega_z^2\,\nu_1''\cos(u + \Omega - \phi) &= \\ = \left\{k_2R_1^5m_2\,\Delta^{-3}\,\omega_z^2\right\}\frac{1}{\sin\theta}\frac{\partial\nu_1''^2}{\partial\phi} &\end{aligned} \tag{2.64}$$

by (2.58) and (2.59).

Lastly, turning our attention to the viscous terms, we shall hereafter assume that $\mu(R_1) = 0$ which makes the integrated part $[\mu r^6]_0^{R_1}$ on the right-hand side of Equations (2.29) and (2.30) to vanish. Since, moreover, to the first order in small quantities

$$p_1 a''_{11} - \omega_{x''} = \omega_z \{\sin\theta \cos\phi \sin(u + \Omega) - \sin i \sin u\} - \\ - n \sin i \sin\Omega \cos(u + \Omega) \tag{2.65}$$

and

$$p_2 a''_{21} - \omega_{x''} = -\omega_z \{\sin\theta \sin\phi \cos(u + \Omega) + \sin i \sin u\} + \\ + n \sin i \cos\Omega \sin(u + \Omega)\,; \tag{2.66}$$

then abbreviating

$$\frac{48}{5}\pi K_2 (a''_{11} p_1 - \omega_{x''}) a''_{11} \int \mu r^5 \, dr \equiv G\,, \tag{2.67}$$

$$\frac{48}{5}\pi K_2 (a''_{21} p_2 - \omega_{x''}) a''_{21} \int \mu r^5 \, dr \equiv H\,, \tag{2.68}$$

we find that

$$G\cos\phi + H\sin\phi = \frac{48}{5}\pi K_2 \{-p_3 \sin i \sin u \cos(u + \Omega - \phi) + \\ + \frac{1}{2}\sin 2(u + \Omega)[\omega_z \sin\theta \cos 2\phi - n \sin i \cos(\Omega - \phi)]\} \int_0^{R_1} \mu r^5 \, dr, \tag{2.69}$$

and

$$G\sin\theta + H\cos\phi = \frac{48}{5}\pi K_2 \{p_3 \sin i \sin u \sin(u + \Omega - \phi) + \\ + \frac{1}{2}\sin 2(u + \Omega)[\omega_z \sin\theta \sin 2\phi + n \sin i \sin(\Omega + \phi)]\} \int_0^{R_1} \mu r^5 \, dr. \tag{2.70}$$

If we collect all results obtained so far in this section, the linearized Equations (2.29) –(2.30) can ultimately be rewritten as

$$A_0 \omega_z \dot{\phi} \sin\theta + \frac{\alpha K_2 \omega_z}{2} \frac{\partial}{\partial\theta} \left\{\omega_z v''^2_1 - n v''^2_2 + \frac{1}{3} n v''^2_3\right\} = \\ = -\left\{\omega_z^2 k_2 R_1^5 \frac{m_2}{\Delta^3}\right\} \frac{\partial v''^2_1}{\partial\theta} + \text{viscous terms} \tag{2.71}$$

and

$$A_0 \omega_z \dot{\theta} \cos\theta - \frac{\alpha K_2 \omega_z}{2} \frac{1}{\sin\theta} \frac{\partial}{\partial\phi} \left\{\omega_z v''^2_1 - n v''^2_2 + \frac{1}{3} n v''^2_3\right\} = \\ = \left\{\omega_z^2 k_2 R_1^5 \frac{m_2}{\Delta^3}\right\} \frac{1}{\sin\theta} \frac{\partial v''^2_1}{\partial\phi} + \text{viscous terms}, \tag{2.72}$$

where, in accordance with Equation (2.84) of Chapter IV,

$$A_0 = m_1 h_1^2; \tag{2.73}$$

h_1 denoting the radius of gyration of the respective configuration. The terms in curly brackets on the left-hand sides of the foregoing Equations (2.71) and (2.72) represent the effects of deformation arising from dynamical tides; while the first terms on the right-hand sides arise from the attraction of the companion on the equatorial bulge of the rotating configuration.

If $n = 0$ (corresponding to a 'two-centre' problem, in which the rotating configuration is attracted by a fixed mass-point), Equations (2.71) and (2.72) will reduce to

$$A_0 \dot{\phi} \sin\theta + \frac{\alpha K_2 \omega_z}{2} \frac{\partial v_1''^2}{\partial\theta} = -\left\{\omega_z k_2 R_1^5 \frac{m_2}{\Delta^3}\right\} \frac{\partial v_1''^2}{\partial\theta} + \text{viscous terms} \tag{2.74}$$

and

$$A_0 \dot{\theta} \cos\theta - \frac{\alpha K_2 \omega_z}{2} \frac{1}{\sin\theta} \frac{\partial v_1''^2}{\partial\phi} = \left\{\omega_z k_2 R_1^5 \frac{m_2}{\Delta^3}\right\} \frac{1}{\sin\theta} \frac{\partial v_1''^2}{\partial\phi} + \text{viscous terms.} \tag{2.75}$$

If, moreover, the configuration were rigid, the deformation terms on the left-hand side of Equations (2.74) and (2.75) as well as all viscous terms would be identically zero; and their system would reduce to

$$A_0 \dot{\phi} \sin\theta = -\left\{\omega_z k_2 R_1^5 \frac{m_2}{\Delta^3}\right\} \frac{\partial v_1''^2}{\partial\theta}, \tag{2.76}$$

$$A_0 \dot{\theta} \sin\theta \cos\theta = \left\{\omega_z k_2 R_1^5 \frac{m_2}{\Delta^3}\right\} \frac{\partial v_1''^2}{\partial\phi}, \tag{2.77}$$

in which form the problem was treated, e.g., by Brouwer (1946).

If, on the other hand, $\omega_z = 0$ but $n \neq 0$, Equations (2.71) and (2.72) would reduce to

$$A_0 \dot{\phi} \sin\theta - \frac{\alpha K_2 n}{2} \frac{\partial}{\partial\theta} \left\{v_2''^2 - \frac{1}{3} v_3''^2\right\} = \text{viscous terms,} \tag{2.78}$$

$$A_0 \dot{\theta} \cos\theta + \frac{\alpha K_2 n}{2} \frac{1}{\sin\theta} \frac{\partial}{\partial\phi} \left\{v_2''^2 - \frac{1}{3} v_3''^2\right\} = \text{viscous terms,} \tag{2.79}$$

the solution of which represents the 'stabilizing' effects of dynamical tides on a non-rotating configuration susceptible of deformation. If, lastly, both ω_z and n are set equal to zero, the case corresponds to a neutral equilibrium of a stationary body, for which the Eulerian angles θ, ϕ, ψ may assume arbitrary constant values.

C. SECULAR AND LONG-PERIODIC MOTION

Among the assumptions committed to linearize our fundamental Equations (1.108) and (1.109) of Chapter IV was the expectation that the time-derivatives θ and ϕ of the Eulerian angles are so small that their ratios to $\psi - \omega_z$ can be regarded as small quantities of first order. If so, however, all terms in (2.71) and (2.72) containing the true anomaly u of the disturbing components can obviously be *averaged* with respect to u over the orbital cycle, and values of their different trigonometric functions replaced by their averages taken over the interval $(0, 2\pi)$. As, in accordance with the exact Equation (2.7) of Chapter IV,

$$\nu_1'' = \mathfrak{A} \sin u + \mathfrak{B} \cos u \, , \tag{2.80}$$

$$\nu_2'' = \mathfrak{A} \cos u - \mathfrak{B} \sin u \, , \tag{2.81}$$

where

$$\mathfrak{A} = \sin i \cos \theta - \cos i \sin \theta \cos(\Omega - \phi) \, , \tag{2.82}$$

$$\mathfrak{B} = \qquad - \qquad \sin \theta \sin (\Omega - \phi) \, , \tag{2.83}$$

the average values over a cycle (denoted hereafter by the square brackets [...]) of $\nu_{1,2}''^2$ are obviously given by

$$[\nu_1''^2] = \frac{1}{2} (\mathfrak{A}^2 + \mathfrak{B}^2) = [\nu_2''^2] . \tag{2.84}$$

Since, moreover,

$$\nu_1''^2 + \nu_2''^2 + \nu_3''^2 = 1 \tag{2.85}$$

and, by (2.60),

$$[\nu_3''] = \nu_3'' \quad \text{and} \quad [\nu_3''^2] = \nu_3''^2 ; \tag{2.86}$$

so that

$$[\nu_1''^2] = [\nu_2''^2] = \tfrac{1}{2} (1 - \nu_3''^2), \tag{2.87}$$

it follows that

$$[\omega_z \nu_1''^2 - n\nu_2''^2 + \tfrac{1}{3} \nu_3''^2] = \tfrac{1}{2} p_3 (1 - \nu_3''^2) + \tfrac{1}{3} n\nu_3''^2 ; \tag{2.88}$$

and, likewise, from (2.69) and (2.70) that

$$[G \cos \phi + H \sin \phi] = \frac{24}{5} \pi K_2 p_3 \sin i \sin (\Omega - \phi) \int_0^{R_1} \mu r^5 \, dr \tag{2.89}$$

and

$$[G \sin \phi - H \cos \phi] = \frac{24}{5} \pi K_2 p_3 \sin i \cos(\Omega - \phi) \int_0^{R_1} \mu r^5 \, dr . \tag{2.90}$$

Accordingly, on insertion for $p_3 = \omega_z - n$, $\alpha = \frac{1}{4} k_2 m_1 R_1^5$, $A_0 = m_1 h_1^2$ and $\kappa_2 = (m_2/m_1) \Delta^{-3}$, in the secular part of the linearized Equations (2.71) and (2.72) can be expressed in a more concise form as

$$\dot{\phi} \sin \theta = -\frac{\kappa}{2} \frac{\partial \nu''^2_3}{\partial \theta} + \lambda \sin i \sin(\Omega - \phi) , \tag{2.91}$$

$$\dot{\theta} \sin \theta = +\frac{\kappa}{2} \frac{\partial \nu''^2_3}{\partial \phi} + \lambda \sin i \cos(\Omega - \phi) \sin \theta , \tag{2.92}$$

where we have abbreviated

$$\kappa = \frac{27\, \omega_z - 5n}{24}\, k_2 \left(\frac{m_2}{m_1}\right)\left(\frac{R_1}{\Delta}\right)^3 \left(\frac{R_1}{h_1}\right)^2 \tag{2.93}$$

and

$$\lambda = \frac{24\pi}{5\, m_1 h_1^2 \Delta^3} \frac{m_2}{m_1} \frac{\omega_z - n}{\omega_z} \int_0^{R_1} \mu r^5 \, dr . \tag{2.94}$$

In order to proceed further, let us set

$$\left.\begin{aligned} p &= \sin \theta \sin \phi , \\ q &= \sin \theta \cos \phi , \end{aligned}\right\} \tag{2.95}$$

$$\left.\begin{aligned} r &= \sin i \sin \Omega , \\ s &= \sin i \cos \Omega ; \end{aligned}\right\} \tag{2.96}$$

and adopt these direction cosines, rather than the constituent angles θ, ϕ or i, Ω, as the dependent variables of our problem – such that

$$\nu''_3 = pr + qs + \sqrt{(1 - p^2 - q^2)(1 - r^2 - s^2)} . \tag{2.97}$$

By a differentiation of (2.95) it readily follows that

$$\dot{\theta} \sin \theta = p\dot{p} + q\dot{q} , \tag{2.98}$$

$$\dot{\phi} \sin^2 \theta = q\dot{p} - p\dot{q} , \tag{2.99}$$

and, moreover, the operators

$$\sin \theta \frac{\partial}{\partial \theta} = p \frac{\partial}{\partial p} + q \frac{\partial}{\partial q} , \tag{2.100}$$

$$\frac{\partial}{\partial \phi} = q \frac{\partial}{\partial p} - p \frac{\partial}{\partial q} . \tag{2.101}$$

Accordingly, the system of Equations (2.91)–(2.92) can be alternatively rewritten in the more symmetrical form

$$\dot{p} = \frac{\kappa}{2} \frac{\partial \nu''^2_3}{\partial q} + \lambda r , \tag{2.102}$$

$$\dot{q} = -\frac{\kappa}{2} \frac{\partial \nu''^2_3}{\partial p} + \lambda s , \tag{2.103}$$

By a differentiation of Equaton (2.97) we find that

$$\frac{\partial v_3''^2}{\partial p} = 2v_3'' \left\{ r - \frac{p(1-r^2-s^2)}{v_3''-pr-qs} \right\}, \tag{2.104}$$

$$\frac{\partial v_3''^2}{\partial q} = 2v_3'' \left\{ s - \frac{q(1-r^2-s^2)}{v_3''-pr-qs} \right\}; \tag{2.105}$$

but if we remember that the direction cosines p, q, r, s constitute small quantities of first order whose squares and higher powers can be ignored, the foregoing expressions reduce to

$$\frac{\partial v_3''}{\partial p} = r - p, \tag{2.106}$$

$$\frac{\partial v_3''}{\partial q} = s - q. \tag{2.107}$$

On insertion of (2.106) and (2.107) in (2.102)–(2.103) the latter pair of equations assume the neat forms

$$\dot{p} = \kappa(s-q) + \lambda r, \tag{2.108}$$

$$\dot{q} = \kappa(p-r) + \lambda s, \tag{2.109}$$

where κ and λ continues to be given by Equations (2.93) and (2.94).

The foregoing Equations (2.108)–(2.109) do not define yet the variables p and q uniquely; for they involve also the direction cosines r and s. Therefore, in order to render the problem determinate, an additional two equations must be adjoined to specify r and s. These are the variational equations for the angular variables Ω and i, known to us already from section V-1; and their linearized form can be reduced to

$$\dot{r} = \gamma(q-s), \tag{2.110}$$

and

$$\dot{s} = \gamma(r-p), \tag{2.111}$$

where we have abbreviated

$$\gamma = \frac{k_2 R_1^5 \omega_z^2 n}{G m_1 \Delta^2}. \tag{2.112}$$

The simultaneous differential equations for p, q, r, s of the linearized system governing the precession and nutation of a rotating deformable configuration attracted by an external mass are, therefore, of the form

$$\left.\begin{aligned} \dot{p} &= \kappa(s-q) + \lambda r, \\ \dot{q} &= \kappa(p-r) + \lambda s, \\ \dot{r} &= \gamma(q-s), \\ \dot{s} &= \gamma(r-p), \end{aligned}\right\} \tag{2.113}$$

where the dependent variables p, q, r, and s have already been defined by Equations (2.108)–(2.111); while κ, λ, and γ are constants by Equations (2.93), (2.94) and (2.112).

D. SOLUTION OF EQUATIONS

In order to construct a solution of the foregoing simultaneous system (2.113), let us assume that

$$p = A_1 e^{\alpha_1 t}, \quad q = A_2 e^{\alpha_2 t}, \quad r = A_3 e^{\alpha_3 t}, \quad s = A_4 e^{\alpha_4 t} \tag{2.114}$$

where the A_j's and α_j's are constants satisfying the system of linear algebraic equations

$$\left.\begin{aligned} A_1\alpha + A_2\kappa - A_3\lambda - A_4\kappa &= 0 , \\ -A_1\kappa + A_2\alpha + A_3\kappa - A_4\lambda &= 0 , \\ -A_2\gamma + A_3\alpha + A_4\gamma &= 0 , \\ A_1\gamma \qquad - A_3\gamma + A_4\alpha &= 0 . \end{aligned}\right\} \tag{2.115}$$

This system can obviously possess a nontrivial solution only if the determinant

$$\begin{vmatrix} \alpha & \kappa & -\lambda & -\kappa \\ -\kappa & \alpha & \kappa & -\lambda \\ 0 & -\gamma & \alpha & \gamma \\ \gamma & 0 & -\gamma & \alpha \end{vmatrix} = 0 ; \tag{2.116}$$

this latter equation being equivalent to

$$\alpha^4 + (\sigma\alpha + \rho)^2 = 0 , \tag{2.117}$$

where we have abbreviated

$$\sigma = \kappa + \gamma \tag{2.118}$$

and

$$\rho = \gamma\lambda . \tag{2.119}$$

If we take square-root of Equation (2.117), the latter reduces to the quadratic

$$\alpha^2 \pm i(\sigma\alpha + \rho) = 0 \tag{2.120}$$

with real and imaginary coefficients and, therefore, complex roots of the form

$$\alpha = x + iy . \tag{2.121}$$

Inserting (2.121) in (2.120) and splitting up the latter in its real and imaginary parts we find that the quantities x and y on the right-hand side of (2.121) should satisfy the simultaneous equations

$$\left.\begin{aligned} x^2 - y^2 \mp \sigma y &= 0\,, \\ 2\,x\,y \pm (\sigma x + \rho) &= 0; \end{aligned}\right\} \tag{2.122}$$

or, on separation of variables,

$$y = \mp \frac{\sigma x + \rho}{2x}\,, \tag{2.123}$$

where x is a root of the equation

$$\frac{x^2}{2} = \left(\frac{\sigma}{4}\right)^2 \{\sqrt{1 + (4\rho/\sigma^2)^2} - 1\}\,. \tag{2.124}$$

If $\rho = 0$ (i.e., no viscous dissipation) the real part x of the complex root (2.121) disappears, but in such a way that

$$\lim_{\rho \to 0} (\rho/x) = \sigma\,. \tag{2.125}$$

In consequence, the imaginary part of the root (2.121) will become equal (from 2.123) to $\mp\sigma$ and, therefore, $\alpha = \pm\, i\sigma$. In such a case, the solutions (2.114) of the system (2.113) will consist of terms which are purely periodic in time.

If, however, $\rho^2 > 0$*, the real part x of the complex root (2.121) follows from (2.124) in the form

$$x = \pm \frac{\sigma}{2\sqrt{2}} \left\{\sqrt{1 + \left(\frac{4\rho}{\sigma^2}\right)^2} - 1\right\}^{1/2} = \pm \frac{|\rho|}{\sigma} + \cdots\,, \tag{2.126}$$

and, from (2.123),

$$\begin{aligned} y &= \mp \left\{\frac{\sigma}{2} \mp \frac{\rho}{\dfrac{\sigma}{\sqrt{2}}\left[\sqrt{1 + \left(\dfrac{4\rho}{\sigma^2}\right)^2} - 1\right]^{1/2}}\right\} \\ &= \mp\,\sigma \left\{1 + \frac{2\,\rho^2}{\sigma^4} + \cdots\right\}; \end{aligned} \tag{2.127}$$

but the upper algebraic signs in (2.126) and (2.127) are ruled out by physical considerations** rendering x and y single-valued.

Moreover, for values of α which satisfy the determinantal equation (2.116) the ratios of the coefficients $A_{2,3,4}/A_1$ can be determined from the system (2.115) in the form

$$\frac{A_2}{A_1} = \frac{\alpha\kappa - \rho}{\alpha(\alpha \mp i\gamma)}\,, \tag{2.128}$$

* The value of ρ itself (i.e., those of $\lambda_{1,2}$) can be positive or negative, depending on whether $\omega_z \gtrless n$.

** I.e., the need to render the XY-plane of our inertial coordinates the invariable plane of the binary system.

$$\frac{A_3}{A_1} = \mp \frac{i\gamma}{\alpha \mp i\gamma} \quad , \tag{2.129}$$

$$\frac{A_4}{A_1} = - \frac{\gamma}{\alpha \mp i\gamma} \quad ; \tag{2.130}$$

which on insertion from (2.121) yield

$$\begin{aligned} \frac{A_2}{A_1} = & \frac{\kappa x(x^2 + y^2) - \rho(x^2 - y^2 + \gamma y)}{(x^2 + y^2)\,[x^2 + (y - \gamma)^2]} - \\ & - i \left\{ \frac{\kappa(y - \gamma)\,(x^2 + y^2) - \rho(^2 y - \gamma)}{(x^2 + y^2)\,[x^2 + (y - \gamma)^2]} \right\} \quad , \end{aligned} \tag{2.131}$$

$$\frac{A_3}{A_1} = - \frac{\gamma(y - \gamma)}{x^2 + (y - \gamma)^2} - \frac{i\,\gamma x}{x^2 + (y - \gamma)^2} \quad . \tag{2.132}$$

and

$$\frac{A_4}{A_1} = - \frac{\gamma x}{x^2 + (y - \gamma)^2} + \frac{i\,\gamma(y - \gamma)}{x^2 + (y - \gamma)^2} \quad . \tag{2.133}$$

In the inviscid case, when

$$\rho = x = 0 \tag{2.134}$$

and

$$\alpha = i\sigma = i\,(\kappa + \gamma)\,, \tag{2.135}$$

Equations (2.131)–(2.133) reduce to

$$\frac{A_2}{A_1} = -i\,, \quad \frac{A_3}{A_1} = - \frac{\gamma}{\kappa}\,, \quad \frac{A_4}{A_1} = i\,\frac{\gamma}{\kappa} \quad ; \tag{2.136}$$

and, accordingly,

$$\begin{aligned} p &= A_1\,(\cos \sigma t + i \sin \sigma t)\,, \\ q &= - i A_1\,(\cos \sigma t + i \sin \sigma t), \end{aligned} \tag{2.137}$$

$$\begin{aligned} r &= -(\gamma/\kappa) A_1\,(\cos \sigma t + i \sin \sigma t)\,, \\ s &= i\,(\gamma/\kappa) A_1\,(\cos \sigma t + i \sin \sigma t)\,. \end{aligned} \tag{2.138}$$

When $\rho^2 > 0$, we note that

$$p^2 + q^2 = \sin^2\theta = \tilde{A}^2 e^{-2xt} \tag{2.139}$$

and

$$r^2 + s^2 = \sin^2 i = (\gamma/\kappa)^2 \tilde{A}^2 e^{-2xt}, \tag{2.140}$$

so that, at all times,

$$\sin i = (\gamma/\kappa) \sin \theta \ . \tag{2.141}$$

Moreover, by use of the definitions (2.96)–(2.97) and of Equations (2.137)–(2.138) we find that

$$\sin \phi = -\sin \Omega \qquad \text{and} \qquad \cos \phi = -\cos \Omega \ , \tag{2.142}$$

disclosing that

$$\Omega - \phi = \pm \pi \tag{2.143}$$

and that, moreover,

$$\phi(t) = \phi(0) - yt \ , \tag{2.144}$$

$$\Omega(t) = \Omega(0) - yt \ . \tag{2.145}$$

Therefore, the nodal lines of the equatorial plane of the rotating configuration, and of the orbital plane of its companion, recede secularly at the same uniform rate; and complete their regression in a period U which bears to the orbital period the ratio

$$\frac{P}{U} = \frac{y}{\omega_K} , \tag{2.146}$$

where ω_K denotes the Keplerian angular velocity (mean daily motion) of the relative orbit. Moreover – unlike in the inviscid case, the angle θ of inclination of the axis of rotation to the invariable plane of the system diminishes secularly in accordance with the equation

$$\sin \theta = \tilde{A} e^{-xt}. \quad ; \tag{2.147}$$

while

$$\sin i = (\gamma/\kappa) \tilde{A} e^{-xt}. \tag{2.148}$$

As, moreover, the ratio γ/κ is of the order of the square of the fractional radius of gyration $(h_1/R_1)^2$, it follows that, in general,

$$\theta > i ; \tag{2.149}$$

the foregoing inequality being the stronger, the higher the degree of central condensation of the respective configuration.

All results obtained so far are relevant to the simple case of a binary system in which the secondary (disturbing) component can be regarded as a mass-point, without finite dimensions or rotational momentum of its own. Should, however, the secondary be a finite celestial body of its own right, our entire analysis as given so far in this paper continues to hold good, provided only that the indices of different quantities referring to respective components be appropriately interchanged. If, moreover, both components of a close pair are regarded as configurations of finite size and angular momentum, their precession and nutation become mutually interlocked; and an analysis of their combined motions can be performed as follows.

Let the subscripts $j = 1$ and 2 refer hereafter to the values of θ_j, $\dot{\phi}_j$ as well as to the constants κ_j and λ_j appropriate for each individual components. If so, the system (2.113) of the fundamental equations of our problem should be generalized to a simultaneous system

$$\left.\begin{aligned} \dot{p}_1 &= \kappa_1(s - q_1) + \lambda_1 r \,, \\ \dot{p}_2 &= \kappa_2(s - q_2) + \lambda_2 r \,, \\ \dot{q}_1 &= \kappa_1(p_1 - r) + \lambda_1 s \,, \\ \dot{q}_2 &= \kappa_2(p_2 - r) + \lambda_2 s \,, \\ \dot{r} &= \gamma_1(q_1 - s) + \gamma_2(q_2 - s) \,, \\ \dot{s} &= \gamma_1(r - p_1) + \gamma_2(r - p_2) \,, \end{aligned}\right\} \tag{2.150}$$

of sixth order which (because of its linearity) can again be expected to admit of exponential solutions of the form

$$\left.\begin{aligned} p_1 &= A_1 e^{\alpha t} \,, & p_2 &= A_2 e^{\alpha t} \,, \\ q_1 &= A_3 e^{\alpha t} \,, & q_2 &= A_4 e^{\alpha t} \,, \\ r &= A_5 e^{\alpha t} \,, & s &= A_5 e^{\alpha t} \end{aligned}\right\} \tag{2.151}$$

where the constants $A_{1,2,\ldots,6}$ are constrained to obey the linear system of homogeneous equations

$$\left.\begin{array}{llllllll} +\alpha A_1 & & +\kappa_1 A_3 & & -\lambda_1 A_5 & -\kappa_1 A_6 & = 0 \,, \\ & +\alpha A_2 & & +\kappa_2 A_4 & -\lambda_2 A_5 & -\kappa_2 A_6 & = 0 \,, \\ -\kappa_1 A_1 & & +\alpha A_3 & & -\kappa_1 A_5 & -\lambda_1 A_6 & = 0 \,, \\ & -\kappa_2 A_2 & & +\alpha A_4 & +\kappa_2 A_5 & -\lambda_2 A_6 & = 0 \,, \\ & & -\gamma_1 A_3 & -\gamma_2 A_4 & +\alpha A_5 & +(\gamma_1 + \gamma_2) A_6 & = 0 \,, \\ +\gamma_1 A_1 & +\gamma_2 A_2 & & & -(\gamma_1 + \gamma_2) A_5 & +\alpha A_6 & = 0 \,, \end{array}\right\} \tag{2.152}$$

Its non-trivial solution requires that the determinant

$$\begin{vmatrix} \alpha & 0 & \kappa_1 & 0 & -\lambda_1 & -\kappa_1 \\ 0 & \alpha & 0 & \kappa_2 & -\lambda_2 & -\kappa_2 \\ -\kappa_1 & 0 & \alpha & 0 & \kappa_1 & -\lambda_1 \\ 0 & -\kappa_2 & 0 & \alpha & \kappa_2 & -\lambda_2 \\ 0 & 0 & -\gamma_1 & -\gamma_2 & \alpha & \gamma_1 + \gamma_2 \\ \gamma_1 & \gamma_2 & 0 & 0 & -\gamma_1 - \gamma_2 & \alpha \end{vmatrix} = 0 \,; \tag{2.153}$$

or, which is equivalent, that

$$\{\alpha^3 - (\sigma_1\sigma_2 - \gamma_1\gamma_2)\,\alpha + (\gamma_1\kappa_2\lambda_1 + \gamma_2\kappa_1\lambda_2)^2\}^2 + \\ + \alpha^2\,\{\sigma_1 + \sigma_2)\alpha - (\gamma_1\lambda_1 + \gamma_2\lambda_2)\}^2 = 0\,. \tag{2.154}$$

Taking a square-root of the foregoing equation we find that the exponents α in Equations (2.151) must be roots of the cubic equation

$$\alpha^3 \mp i\,(\sigma_1 + \sigma_2)\,\alpha^2 - [(\sigma_1\sigma_2 - \gamma_1\gamma_2) \mp i\,(\gamma_1\lambda_1 + \gamma_2\lambda_2)]\,\alpha \\ + (\gamma_1\kappa_2\lambda_1 + \gamma_2\kappa_1\lambda_2) = 0\,, \tag{2.155}$$

where again the upper sign of the pair $\mp$ alone is admissible for the same reasons as before. The complex nature of some of the coefficients of this equation discloses that its roots must again be expected to be complex and of the form (2.121) which inserted in (2.155) splits up the latter into a pair of simultaneous equations

$$x^3 - (3y^2 + 2Ay + B)\,x + D \quad = Cy\,, \tag{2.156}$$

$$y^3 - (3x^2 \qquad - B)\,y + Ay^2 = Cx - Ax^2\,, \tag{2.157}$$

where we have abbreviated

$$\left.\begin{aligned} A &= \sigma_1 + \sigma_2\,, \\ B &= \sigma_1\sigma_2 - \gamma_1\gamma_2\,, \\ C &= \gamma_1\lambda_1 + \gamma_2\lambda_2\,, \\ D &= \gamma_1\lambda_1\kappa_2 + \gamma_2\lambda_2\kappa_1\,. \end{aligned}\right\} \tag{2.158}$$

In the inviscid case – if $\lambda_1 = \lambda_2 = 0$ – both $C = D = 0$; and so in the real part of the complex root (2.121). In such a case, Equation (2.156) is satisfied identically; while (2.157) reduces to

$$y^2 + Ay + B = 0 \tag{2.159}$$

which will possess a pair of real and distinct roots $y_{1,2}$, since

$$(y_1 - y_2)^2 = (\sigma_1 - \sigma_2)^2 + 4\,\gamma_1\gamma_2\,. \tag{2.160}$$

In such a case, by setting

$$p_j = \tilde{A}_i \sin\, y_1\, t + \tilde{B}_j \sin\, y_2 t\,, \tag{2.161}$$

$$q_j = \tilde{A}_i \cos y_1\, t + \tilde{B}_j \cos y_2 t\,, \tag{2.162}$$

$j = 1, 2$; and

$$\begin{aligned} r &= -\,a_1 p_1 - a_2 p_2 \\ &= -\,\{a_1\tilde{A}_1 + a_2\tilde{A}_2\}\sin y_1 t - \{a_1\tilde{B}_1 + a_2\tilde{B}_2\}\sin y_2 t, \end{aligned} \tag{2.163}$$

$$\begin{aligned} s &= -\,a_1 q_1 - a_2 q_2 \\ &= -\,\{a_1\tilde{A}_1 + a_2\tilde{A}_2\}\cos y_1 t - \{a_1\tilde{B}_1 + a_2\tilde{B}_2\}\cos y_2 t \end{aligned} \tag{2.164}$$

where we have abbreviated

$$a_j \equiv \frac{\gamma_j}{\kappa_j}; \tag{2.165}$$

and where the integration constants $\tilde{A}_j, \tilde{B}_j$ bear to each other the ratios

$$\frac{\tilde{A}_2}{\tilde{A}_1} = \frac{\sigma_2 - \sigma_1 + y_2 - y_1}{2a_2\kappa_1} = \frac{2a_1\kappa_2}{\sigma_1 - \sigma_2 + y_2 - y_1} \tag{2.166}$$

and

$$\frac{\tilde{B}_2}{\tilde{B}_1} = \frac{\sigma_2 - \sigma_1 + y_1 - y_2}{2a_2\kappa_1} = \frac{2a_1\kappa_2}{\sigma_1 - \sigma_2 + y_1 - y_2} . \tag{2.167}$$

In consequence, by virtue of Equations (2.96) and (2.97) we find that

$$\sin^2\theta_j = \tilde{A}_j^2 + \tilde{B}_j^2 + 2\tilde{A}_j\tilde{B}_j \cos(y_1 - y_2)t \tag{2.168}$$

and

$$\begin{aligned}\sin^2 i = {} & (a_1\tilde{A}_1 + a_2\tilde{A}_2)^2 + (a_1\tilde{B}_1 + a_2\tilde{B}_2)^2 \\ & + 2(a_1\tilde{A}_1 + a_2\tilde{A}_2)(a_1\tilde{B}_1 + a_2\tilde{B}_2)\cos(y_1 - y_2)t .\end{aligned} \tag{2.169}$$

The physical meaning of the constants $\tilde{A}_j$ and $\tilde{B}_j$ becomes clear when we consider their role on the right-hand sides of Equations (2.168) and (2.169). Equation (2.160) makes it evident that the difference $y_1 - y_2 > 0$. As moreover, $\cos(y_1 - y_2)t$ oscillates between ± 1, $\sin\theta_j$ is bound to oscillate between $\tilde{A}_j \pm \tilde{B}_j$. Therefore, the constants $\tilde{A}_{1,2}$ are seen to specify the *mean* values of the *inclinations* if the equatorial planes of the two components to the invariable plane of the binary system; while the $\tilde{B}_{1,2}$'s denote the amplitudes of their *nutations*, in the period U' given by the equations

$$\frac{P}{U'} = \frac{|y_1 - y_2|}{\omega_z} . \tag{2.170}$$

Furthermore, the angle i of inclination between the orbital and invariable planes of the system will (in accordance with Equation (2.169)) oscillate about its mean position specified by the value of $a_1\tilde{A}_1 + a_2\tilde{A}_2$ in the period U' as given by (2.170) and with an amplitude $a_1\tilde{B}_1 + a_2\tilde{B}_2$ bearing a fixed ratio to that of nutation, but (since $a_{1,2} \ll 1$) generally much smaller.

The nodal line of the system continues to recede – though no longer uniformly – in period U given now by the equation

$$\frac{2}{U} = \frac{1}{U_1} + \frac{1}{U_2} + \frac{1}{U'} , \tag{2.171}$$

representing the harmonic mean of the periods $U_{1,2}$ of nodal regression due to the action of each component separately (as given by our previous Equation (2.146)), and augmented by that of nutation U'. In consequence, in every system consisting of two components of finite size and angular momentum,

$$U' > U \tag{2.172}$$

i.e., the period of nutation will always be longer than that of nodal revolution or procession. Lastly, the difference $\phi(t) - \Omega(t)$ will no longer be equal to π, but will oscillate around this value in the period U' and its submultiples.

All this is strictly true only in the absence of dissipative forces. Should, however, either (or both) constants $\lambda_{1,2}$ be different from zero, the coefficients C and D in Equations (2.156) and (2.157) will cease to vanish and, as a result, the variables x and y in these equations can no longer be conveniently separated. Since, moreover, each equation is one of third degree, their simultaneous solution can proceed only numerically, for specific values of the constants A, B, C, and D. The non-zero values of x will again give rise to real exponential terms in the solutions of the form (2.147) and (2.418), causing a secular decrease in the values of θ_i and i; and will likewise affect (through second-order terms) the periods of precession and nutation. An investigation of the magnitude of these effects requires, however, numerical solutions of the system (2.156)–(2.157) for specific values of its coefficients – a task which, at this stage, still must be left as an exercise for the reader.

V-3. Perturbations in the Orbital Plane

In the preceding section we have developed a procedure for systematic investigation of the perturbations of the elements Ω and i – which specify the orientation *of* the orbital plane of a close binary system in space – caused by the axial rotation of its components and their mutual tidal distortion. Our next task should be to investigate, in a similar manner, perturbations of the elements A, e and ω which describe the size, shape, and orientation of the orbit of the components of our binary *in* their (instantaneous) orbital plane; and this we shall proceed to do now on the basis of the equations set forth in section V-1.

In doing so, we shall focus our attention primarily on *secular* perturbations of the respective elements; and within the scheme of a first-order theory of such effects we shall investigate those arising from the rotational and tidal distortion of both components in turn.

A. SECULAR PERTURBATIONS: ROTATIONAL DISTORTION

The expressions (1.33)–(1.35) for **R**, **S** and **W**, established in section V-1 for the disturbing components of accelerations which arise from axial rotation about axes inclined to the orbital plane, permit us to evaluate from Equations (1.26)–(1.28) the corresponding perturbations of the Keplerian elements in the plane of the orbit. In what follows, we shall confine our interest primarily on *secular* perturbations of the respective elements, obtained by a construction of particular solutions of the variational Equations (1.26)–(1.28) in *which all coefficients oscillating periodically in the course of an orbital cycle will be replaced by their time-averages over a complete revolution.*

The short-period variation on the right-hand sides of Equations (1.26)–(1.28) are caused by terms involving periodic functions of the true anomaly v and (in eccentric orbits) of the radius-vector r. In order to obtain their time-averages over an orbital cycle of period P, we note (cf. e.g., Tisserand, 1889; Chapter 15) that

$$\int_0^P (r/A)^n \sin mv \, dt = 0 \tag{3.1}$$

for all positive integral values of m and any integral values of n (be these positive or negative); while

$$\frac{1}{P}\int_0^P \left(\frac{r}{A}\right)^n \cos mv \, dt = \frac{(1-e^2)^{n+\frac{3}{2}}}{2\pi} \int_0^{2\pi} \frac{\cos mv \, dv}{(1+e\cos v)^{n+2}} \equiv X_0^{n,m}(e), \tag{3.2}$$

where

$$X_0^{n,m}(e) = \binom{-n-2}{m}\left(\frac{e}{2}\right)^m (1-e^2)^{n+\frac{3}{2}} \times F\left\{\frac{m+n+2}{2}, \frac{m+n+3}{2}, m+1; e^2\right\} \tag{3.3}$$

are the zero-order Hansen coefficients, well-known from the expansions of elliptic motion, and expressible in terms of (terminating) hypergeometric series of the type ${}_2F_1$. If $m+n+2 \leqslant 0, X_0^{n,m} > 0$; but if $m+n+2 > 0, X_0^{n,m} = 0$.

If, in what follows, square brackets [...] are used to denote the mean values of the respective short-periodic coefficients averaged over the orbital cycle, Equation (1.26) governing the secular (or long-periodic) variation of the semi-major axis A of the relative orbit then assumes the form

$$\begin{aligned}\frac{dA}{dt} &= \frac{2}{n\sqrt{1-e^2}}\{e[\mathbf{R}\sin v] + A(1-e^2)[\mathbf{S}r^{-1}]\} \\ &= \frac{4\mu A}{n\sqrt{1-e^2}} \sum_{i=1}^{2} (k_2)_2 \frac{\omega_i^2 a_i^5}{Gm_i}\left\{\frac{3}{8}[(\mathfrak{A}_i^2 - \mathfrak{B}_i^2)\sin 2\omega + 2\mathfrak{A}_i\mathfrak{B}_i\cos 2\omega] X_0^{-4,1} \right. \\ &\left. - \frac{1}{2}[(\mathfrak{A}_i^2 - \mathfrak{B}_i^2)\sin 2\omega + 2\mathfrak{A}_i\mathfrak{B}_i\cos 2\omega](1-e^2) X_0^{-5,2}\right\},\end{aligned} \tag{3.4}$$

where we have abbreviated

$$a_i = \frac{R_i}{A} \tag{3.5}$$

and where, as before, $\mu \equiv G(m_1 + m_2)$; while $n^2 = \mu/A^3$. Since, moreover,

$$X_0^{-4,1}(e) = e(1-e^2)^{-5/2} \; X_0^{-5,2}(e) = \frac{3}{4}e^2(1-e^2)^{-7/2}, \tag{3.6}$$

it follows on insertion of (3.6) in (3.4) that

$$\frac{dA}{dt} = 0\,; \tag{3.7}$$

disclosing that *inclined axes of rotation do not produce any secular (or long-periodic) perturbation of the semi-major axis of the relative orbit*.

Equation (1.27) for secular changes in orbital eccentricity e can be developed in a similar manner. Since, in elliptic orbits, the eccentric anomaly E is known to satisfy the equation

$$\cos E = \frac{e + \cos v}{1 + e \cos v} = \frac{r(e + \cos v)}{A(1 - e^2)} \tag{3.8}$$

the secular form of Equation (1.27) can be rewritten as

$$\frac{\mathrm{d}e}{\mathrm{d}t} = \frac{\sqrt{1 - e^2}}{nA}\left\{[\mathbf{R} \sin v] + \frac{e[\mathbf{S}r] + [\mathbf{S}r \cos v]}{A(1 - e^2)} + [\mathbf{S} \cos v]\right\} =$$

$$= \frac{\mu}{4n\sqrt{1 - e^2}} \sum_{i=1}^{2} (k_2)_i \frac{\omega_i^2 a_i^5}{Gm_i} \{[(1 - e^2) X_0^{-4,1} - 2X_0^{-3,1}] (\mathfrak{A}_i^2 - \mathfrak{B}_i^2) \sin 2\omega\} . \tag{3.9}$$

Since, moreover, by (3.3)

$$X_0^{-3,1}(e) = \frac{1}{2} e (1 - e^2)^{-3/2} \tag{3.10}$$

and $X_0^{-4,1}(e)$ is given by (3.5) Equation (3.9) readily reduces to

$$\frac{\mathrm{d}e}{\mathrm{d}t} = 0 ; \tag{3.11}$$

implying that *inclined axes of rotation produce no secular or long-periodic perturbations in orbital eccentricity* as well.

Lastly, Equation (1.28) governing the secular advance of the longitude $\bar{\omega}$ of the periastron, caused by rotational distortion, assumes the more specific form

$$\frac{\mathrm{d}\bar{\omega}}{\mathrm{d}t} = \frac{\sqrt{1 - e^2}}{nAe}\left\{- [\mathbf{R} \cos v] + [\mathbf{S} \sin v] + p^{-1} [\mathbf{S}r \sin v] + e\frac{[\mathbf{W}r \sin u]}{A(1 - e^2)} \tan \tfrac{1}{2} i\right\}$$

$$= -\frac{\mu\sqrt{1 - e^2}}{2ne} \sum_{i=1}^{2} (k_2)_i \frac{\omega_i^2 a_i^5}{Gm_i} \{[3(\mathfrak{A}_i^2 + \mathfrak{B}_i^2) - 2] X_0^{-4,1} +$$

$$+ 2e(1 - e^2)^{-1} X_0^{-3,0}\, \mathfrak{A}_i\, \nu_3'' \tan \tfrac{1}{2} i\} . \tag{3.12}$$

Moreover, by (2.26)–(2.28),

$$\mathfrak{A}_i^2 + \mathfrak{B}_i^2 = 1 - \nu_3''^2 \tag{3.13}$$

and

$$\mathfrak{A}_i = -\frac{\partial \nu_3''}{\partial i} . \tag{3.14}$$

If so, and if

$$X_0^{-3,0}(e) = (1 - e^2)^{-3/2} \tag{3.15}$$

while $X_0^{-4,1}(e)$ continues to be given by (3.5), Equation (3.12) can be reduced to

$$\frac{d\bar{\omega}}{dt} = \frac{3n}{2(1-e^2)^2} \sum_{m=1}^{2} \frac{\omega_m^2}{2\pi G\bar{\rho}_m} (k_2)_m a_m^2 \left\{ P_2(\nu_3'') + \right.$$

$$\left. + \frac{1}{2}\left(\frac{\partial \nu_3''^2}{\partial i}\right) \tan\tfrac{1}{2}i \right\}, \tag{3.16}$$

where $\bar{\rho}_m$ denotes the mean density of the respective configuration.

Within the scheme of approximation to which we can set $\phi_m - \Omega = \pi$ in accordance with Equation (2.143), it follows that

$$\nu_3'' = \cos(\theta_m + i), \quad \frac{\partial \nu_3''^2}{\partial i} = -\sin 2(\theta_m + i), \tag{3.17}$$

and Equation (3.16) reduces then to

$$\frac{d\bar{\omega}}{dt} = \frac{3n}{2(1-e^2)^2} \sum_{m=1}^{2} \frac{\omega_m^2 (k_2)_m}{2\pi G\bar{\rho}_m} a_m^2 \left\{ 1 - \frac{3}{2}\sin^2(\theta_m + i) - \right.$$

$$\left. - \frac{1}{2}\sin 2(\theta_m + i)\tan\tfrac{1}{2}i \right\}, \tag{3.18}$$

deduced previously (by a different method) by Kopal (1959, p. 79). If, lastly, the equatorial planes of both components happen to coincide with that of their orbit (in which case $\theta_m = i = 0$), the second and third terms in curly brackets of the foregoing equation vanish; and its right-hand side reduces to one (the first) term only – a result first obtained by Sterne (1939). As this term is *positive* (regardless of whether the rotation is direct or retrograde), it is bound to cause the apsidal line to *advance* – rendering the longitude $\bar{\omega}$ of periastron the only element of the orbit in the plane which undergoes secular perturbations even if the axes of rotation of both components of our binary system coincide with their orbital plane.

On the other hand, the negative sign of the second and third terms in curly brackets on the right-hand side of Equation (3.18) imply that if $\theta_m \gg i > 0$, the rate of apsidal advance would *diminish* with increasing inclination. We have reasons to expect (cf. sec. V-4) that, in general, the angles θ_m as well as i will be small. Should, however, this not be the case and the angles θ_m and i be unrestricted – subject only to the requirement that they bear to each other the ratio given by Equation (2.141) – the apsidal advance may get *arrested* if

$$3\sin^2(\theta_m + i) + \sin 2(\theta_m + i)\tan\tfrac{1}{2}i = 2 \tag{3.19}$$

for $m = 1, 2$; and *change direction* (i.e., recede) if the left-hand side of the foregoing equation becomes greater than 2. Whether or not such a case could actually occur in nature depends not only on the value of i, but also on the internal structure of both components (which specifies the ratios θ_m/i). The possibility is intriguing; but again this is not the place to follow it in further detail.

B. SECULAR PERTURBATIONS: TIDAL DISTORTION

In close binary systems consisting of components whose dimensions represent an appreciable fraction of their separation, axial rotation – the effects of which were considered in the preceding part of this section – is not the only source of perturbations influencing orbital motion. As equally important source of perturbations will arise from tidal distortion caused by mutual attraction of the two stellar or planetary (satellite) configurations; and their analysis will be the principal objective of this section.

If the binary configuration consists of fluid that could be regarded as inviscid, the form of the components would be controlled by (and vary with) the instantaneous field of force; and, moreover, the axis of symmetry of each partial tide would remain the radius-vector joining the centres of the two bodies. Under these conditions,

$$\epsilon_i = \eta_i = 0 \qquad \text{and} \qquad r_\epsilon \equiv r \tag{3.20}$$

in Equations (1.70)–(1.72); so that, accordingly, both **S** and **W** as given by (1.71) and (1.72) become identically zero. The only nonvanishing component of secular acceleration of tidal origin, acting in the plane of the orbit, derives from the radial component **R**, and will be of the form

$$[\mathbf{R}\cos v] = -2\mu \sum_{i=1}^{2} \frac{m_{3-i}}{m_i} \sum_{j=2}^{4} (j+1)(k_j)_i X_0^{-(2j+3),1} \frac{R_i^{2j+1}}{A^{2j+3}}, \tag{3.21}$$

correctly to terms of the first order in surficial distortion, where $\mu \equiv G(m_1 + m_2)$ and – in accordance with Equation (3.3),

$$X_0^{-7,1}(e) = \frac{5}{2} e(1-e^2)^{-11/2} \left(1 + \frac{3}{2} e^2 + \frac{1}{8} e^4\right), \tag{3.22}$$

$$X_0^{-9,1}(e) = \frac{7}{2} e(1-e^2)^{-15/2} \left(1 + \frac{15}{4} e^2 + \frac{15}{8} e^4 + \frac{5}{64} e^6\right), \tag{3.23}$$

$$X_0^{-11,1}(e) = \frac{9}{2} e(1-e^2)^{-19/2} \left(1 + 7e^2 + \frac{35}{4} e^4 + \frac{35}{16} e^6 + \frac{7}{128} e^7\right). \tag{3.24}$$

However, it should be stressed again that inviscid fluid represents only an abstract mathematical concept; and that the material of real stars must be characterized by a finite degree of viscosity (plasma, radiation, turbulent). In consequence (cf. section III-3), dynamical tides are bound, in general, to *lag* in phase with respect to the direction of the external force; and, as a result, the radius-vector connecting the centres of the two stars will cease to represent the axis of symmetry of the respective binary configuration.

In embarking on an investigation of dynamical consequences of such lag, let us disregard, in what follows, the tidal lag in latitude (i.e., set $\eta_i = 0$ for $\iota = 1,2,$), and confine our attention to that in longitude (i.e., $\epsilon_i \gtrless 0$). The lag in latitude can arise only if the equator of the respective component is inclined to its orbital plane; and this inclination can generally be expected to be small (cf. sec. III-3). A lag in longitude will, however, arise from asynchronism between rotation and revolution even if the equators and orbit are coplanar and can, generally, be expressed as

$$\epsilon_i = c_i(\omega_i - \dot{v}) = c_i[\omega_i - n(A/r)^2\sqrt{1-e^2}\,], \tag{3.25}$$

where the constants c_i can be deduced, for configurations of given structure, by the methods developed in Chapters III and IV, and will hereafter be regarded as known.

Let us, moreover, assume in all what follows that the angles ϵ_i of tidal lag in longitude are small enough for their squares and higher powers to be neglected; and that the tidal lag in longitude $\eta_i = 0$. If so, however, it follows immediately from (1.72) that **W** = 0; while the nonvanishing components **R** and **S** of disturbing accelerations continue to be given by Equations (1.70) and (1.71).

Let us set out now to evaluate the former in more explicit form. By logarithmic differentiation of (1.73) we readily find that

$$\frac{r}{f}\frac{\partial f}{\partial r} = \frac{\epsilon_i e r \sin v}{A(1-e^2)}\left\{1 + \frac{r}{\epsilon_i}\frac{\partial \epsilon_i}{\partial r}\right\} + O(\epsilon_i^2) \tag{3.26}$$

where, by (3.25),

$$\frac{r}{\epsilon_i}\frac{\partial \epsilon_i}{\partial r} = \frac{2nA^2\sqrt{1-e^2}}{\omega_i r^2 - nA^2\sqrt{1-e^2}}. \tag{3.27}$$

Accordingly,

$$\begin{aligned}\frac{r}{f}\frac{\partial f}{\partial r} &= c_i\left\{\omega_i - \frac{nA^2\sqrt{1-e^2}}{r^2}\right\}\frac{er\sin v}{A(1-e^2)}\left\{\frac{\omega_i r^2 + nA^2\sqrt{1-e^2}}{\omega_i r^2 - nA^2\sqrt{1-e^2}}\right\}\\ &= \frac{c_i e\sin v}{Ar(1-e^2)}\left\{\omega_i r^2 + nA^2\sqrt{1-e^2}\right\}\\ &= \frac{c_i\omega_i e}{1-e^2}\left[\left(\frac{r}{A}\right)\sin v\right] + \frac{c_i ne}{\sqrt{1-e^2}}\left[\left(\frac{A}{r}\right)\sin v\right];\end{aligned} \tag{3.28}$$

which on insertion in (1.70) yields

$$\begin{aligned}\mathbf{R} = &-\frac{2\mu}{A^2}\sum_{i=1}^{2}\frac{m_{3-i}}{m_i}\left\{1 - \frac{c_i\omega_i e}{2(1-e^2)}\left[\left(\frac{r}{A}\right)\sin v\right] - \frac{c_i ne}{2\sqrt{1-e^2}}\left[\left(\frac{A}{r}\right)\sin v\right]\right\}\times\\ &\times\sum_{j=2}^{4}(j+1)(k_j)_i a_i^{2j+1}\left(\frac{A}{r}\right)^{2j+3} -\\ &-\frac{2\mu e\sin v}{A^2(1-e^2)}\sum_{i=1}^{2}\frac{m_{3-i}}{m_i}\epsilon_i\sum_{j=2}^{4}(j+1)^2(k_j)_i a_i^{2j+1}\left(\frac{A}{r}\right)^{2(j+1)},\end{aligned} \tag{3.29}$$

where ϵ_i continues to be given by (3.25).

Moreover, Equation (1.71) then discloses that, within the scheme of our present approximation,

$$\mathbf{S} = \frac{\mu}{2} \sum_{i=1}^{2} \frac{m_{3-i}}{m_i} \epsilon_i \sum_{j=2}^{4} j\,(j+1)\, \frac{R_i^{2j+1}}{r^{2j+3}} \;. \tag{3.30}$$

Next, let us average the foregoing expressions with respect to the time, in the manner developed in the preceding parts of this section. The expression [**R** $\cos v$] has already been given by Equation (3.21) above; and terms arising from the logarithmic derivative of f do not contribute anything to it, as all those multiplied by $\sin v \cos v = \frac{1}{2} \sin 2v$ average out to zero. However, terms multiplied by $\sin^2 v = \frac{1}{2}(1 - \cos 2v)$ do not vanish; and after some simple algebra we find that

$$\begin{aligned}[\mathbf{R} \sin v] = &- \frac{\mu e}{2A^2(1-e^2)} \sum_{i=1}^{2} c_i\,\omega_i \frac{m_{3-i}}{m_i} \sum_{j=2}^{4} (j+1)(2j+1) \times \\ &\times (k_j)_i\, a_i^{2j+1}\, [X_0^{-2(j+1),0} - X_0^{-2(j+1),2}] + \\ &+ \frac{\mu e}{2A^2\sqrt{1-e^2}} \sum_{i=1}^{2} c_i n \frac{m_{3-i}}{m_i} \sum_{j=2}^{4} (j+1)(2j+3)(k_j)_i\, a_i^{2j+1} \times \\ &\times [X_0^{-2(j+2),0} - X_0^{-2(j+2),2}]\,. \end{aligned} \tag{3.31}$$

When we turn to evaluate similarly the appropriate time-averages of various products of the tangential acceleration component S, we find that

$$\begin{aligned}[\mathbf{S}/r] = \frac{\mu}{2A^3} \sum_{i=1}^{2} c_i \frac{m_{3-i}}{m_i} \sum_{j=2}^{4} &j(j+1)(k_j)_i\, a_i^{2j+1}\, \{\omega_i X_0^{-2(j+2),0} - \\ &- n\sqrt{1-e^2}\; X_0^{-2(j+3),0}\}\;; \end{aligned} \tag{3.32}$$

with [**S**r] being given by the same expression as (3.32) if the superscript n of $X_0^{n,m}$ has been increased by 2, and the factor A^3 replaced by A in the denominator on the right-hand side.

Furthermore, it is easy to establish that, for lagging tides, the time-averages

$$[\mathbf{S} \sin v] = [\mathbf{S}\, r \sin v] = 0 \tag{3.33}$$

as well; while

$$\begin{aligned}[\mathbf{S} \cos v] = \frac{\mu}{2A^2} \sum_{i=1}^{2} c_i \frac{m_{3-i}}{m_i} \sum_{i=2}^{4} &j(j+1)(k_j)_i\, a_i^{2j+1}\, \{\omega_i X_0^{-(2j+3),1} - \\ &- n\sqrt{1-e^2}\; X_0^{-(2j+5),1}\}\;; \end{aligned} \tag{3.34}$$

[S r cos υ] being likewise of the same form if the first superscript on $X^{n,m}$ in (3.34) is raised by 1, and A^2 replaced by A in the denominator on the right-hand side; and the $X_j^{n,m}$'s are the Hansen coefficients (i.e., functions of the orbital eccentricity e) as defined by Equation (3.3).

At this stage, we are ready to make use of the variational Equations (1.26)–(1.28) in their averaged form, which disclose that the secular (or long-periodic) perturbations of the orbital plane, arising from the lag of the second-harmonic dynamical tides will assume the explicit forms

$$\frac{dA}{dt} = \frac{2}{n\sqrt{1-e^2}}\left\{-\frac{3\mu e}{A^2}\sum_{i=1}^{2} c_i \frac{m_{3-i}}{m_i}(k_2)_i a_i^5 [\omega_i X_0^{-7,1} - n\sqrt{1-e^2}\, X_0^{-9,1}] + \frac{3\mu(1-e^2)}{A^2}\sum_{i=1}^{2} c_i \frac{m_{3-i}}{m_i}(k_2)_i a_i^5 [\omega_i X_0^{-8,0} - n\sqrt{1-e^2}\, X_0^{-10,0}]\right., \tag{3.35}$$

which since

$$\left.\begin{aligned}(1-e^2)X_0^{-8,0} - eX_0^{-7,1} &= X_0^{-7,0},\\ (1-e^2)X_0^{-10.0} - eX_0^{-9,1} &= X_0^{-9,0},\end{aligned}\right\} \tag{3.36}$$

can further be reduced to

$$\frac{dA}{dt} = \frac{6\,nA}{\sqrt{1-e^2}}\sum_{i=1}^{2} c_i \frac{m_{3-i}}{m_i}(k_2)_i a_i^5 \;\{\omega_i X_0^{-7,0}(e) - n\sqrt{1-e^2}\, X_0^{-9,0}(e)\}. \tag{3.37}$$

Similarly, the secular perturbations of the orbital eccentricity are found to be governed by the equations

$$\frac{de}{dt} = \frac{3\mu}{nA^3\sqrt{1-e^2}}\sum_{i=1}^{2} c_i \frac{m_{3-i}}{m_i}(k_2)_i a_i^5 \;\{\omega_i\,(eX_0^{-6,0} + X_0^{-6,1}) - n\sqrt{1-e^2}\,(eX_0^{-8,0} + X_0^{-8,1})\}, \tag{3.38}$$

which for

$$\left.\begin{aligned}eX_0^{-6,0} + X_0^{-6,1} &= \frac{6}{5}(1-e^2)X_0^{-7,1},\\ eX^{-8,0} + X_0^{-8,1} &= \frac{8}{7}(1-e^2)X_0^{-9,1},\end{aligned}\right\} \tag{3.39}$$

simplifies into

$$\frac{de}{dt} = 6n\sqrt{1-e^2}\sum_{i=1}^{2} c_i \frac{m_{3-i}}{m_i}(k_2)_i a_i^5 \times \left\{\frac{3}{5}\,\omega_i X_0^{-7,1}(e) - \frac{4}{7}\,n\sqrt{1-e^2}\,X_0^{-9,1}(e)\right\} \tag{3.40}$$

The semi-latus rectum

$$p = A(1-e^2) \tag{3.41}$$

will then undergo tidal perturbations governed by the equation

$$\frac{dP}{dt} = (1-e^2)\frac{dA}{dt} - A\frac{de^2}{dt} = \frac{2\sqrt{1-e^2}}{nA}\quad [r\,\mathrm{S}]$$

$$= \frac{6\mu\sqrt{1-e^2}}{nA^2}\sum_{i=1}^{2}(k_2)_i c_i a_i^5 \frac{m_{3-i}}{m_i}\{\omega_i X_0^{-6,0} - n\sqrt{1-e^2}\,X_0^{-8,0}\}$$

$$= \frac{6\mu}{n(1-e^2)^6A^2}\sum_{i=1}^{2}(k_2)_i\ c_i a_i^5 \frac{m_{3-i}}{m_i}\Big\{(1-e^2)^2\left(1+3e^2+\frac{3}{8}e^4\right)\omega_i -$$

$$-n\sqrt{1-e^2}\left(1+\frac{15}{2}e^2+\frac{45}{8}e^4+\frac{5}{16}e^6\right)\Big\}. \tag{3.42}$$

A generalization of the foregoing results to include the effects of third- and fourth-harmonic tides by use of Equations (3.29)–(3.34) constitutes a straightforward task which can be left as an exercise for the interested reader.

Let us consider now the form to which the preceding Equations (3.37), (3.40) or (3.42) would reduce if tidal lag were absent. In accordance with Equation (3.25), the azimuthal angle ϵ_i of tidal lag will vanish if either $c_i = 0$ (signifying absence of viscosity) or $\omega_i = n$ and $e = 0$ (corresponding to tides of equilibrium type). Should the viscosity of stellar material be negligible (so that $c_i = 0$), the right-hand sides of Equations (3.37) and (3.40) vanish identically; and the equations themselves reduce to

$$\frac{dA}{dt} = 0 \quad \text{and} \quad \frac{de}{dt} = 0\,, \tag{3.43}$$

demonstrating that, *in the absence of dissipative processes arising from viscosity, the semi-major axis A and the eccentricity e of the orbit remain secularly constant.*

Consider next the case of equilibrium tides in viscous media. If (as was shown in section IV-3) the secular action of viscosity is to equalize the angular velocities ω_i of axial rotation of both components with their mean daily motion n of their orbital revolution, Equation (3.40) will, on insertion for $X_0^{-7,1}(e)$ and $X_0^{-9,1}(e)$ from (3.22) and (3.23), reduce to

$$\frac{de}{dt} = \frac{3ne^2}{(1-e^2)^7}\sum_{i=1}^{2} c_i \frac{m_{3-i}}{m_i}(k_2)_i a_i^5 \times$$

$$\times\left\{3(1-e^2)^2 F\left(-\frac{3}{2},-2,2,e^2\right) - 4F\left(-\frac{5}{2},-3,2,e^2\right)\right\}$$

$$= -\frac{3n^2e}{(1-e^2)^7}\sum_{i=1}^{2} c_i \frac{m_{3-i}}{m_i}(k_2)_i a_i^5 \times \tag{3.44}$$

$$\times\left\{1+\frac{33}{2}e^2+\frac{105}{8}e^4-\frac{55}{16}e^6-\frac{3}{8}e^8\right\}.$$

The right-hand side of this equation is *negative* for all values of $e > 0$ corresponding to closed orbits; therefore, the value of e governed by (3.44) is bound to *diminish,* and *secularly tend to zero* in the course of time. Therefore, *a synchronization of rotation and revolution by tidal friction is also bound to effect a secular circularization of their orbits.*

Moreover, as $e \to 0$, Equation (3.37) for the semi-major axis reduces to

$$\begin{aligned}\frac{dA}{dt} &= 6\,nA \sum_{i=1}^{2} c_i \frac{m_{3-i}}{m_i} (k_2)_i a_i^5 (\omega_i - n) \\ &= 6\,nA \sum_{i=1}^{2} \frac{m_{3-i}}{m_i} (k_2)_i a_i^5 \epsilon_i\end{aligned} \tag{3.45}$$

by (3.25), which for $\omega_i \to n$ (i.e., $\epsilon_i \to 0$) reduces again to (3.43), rendering A constant. Therefore, while in the absence of dissipative processes operative in the system the semi-major axis A of the orbit, and its eccentricity e, can secularly maintain any constant values, the emergence of viscous friction tending to synchronize rotation with revolution will cause the size of the orbit to tend likewise asymptotically to a constant, while its eccentricity tends exponentially to zero. Consequently, in the limiting stage of tidal evolution in which rotation and revolution become synchronized, the orbit becomes a circle or radius determined by the total momentum of the system.

Lastly, as the sum of all quantities of tidal origin on the right-hand side of Equation (1.28) turns out to be positive, the longitude ω of periastron in the orbital plane will *advance* in accordance with the equation

$$\begin{aligned}\frac{d\bar{\omega}}{dt} &= \frac{2\mu\sqrt{1-e^2}}{nA^3 e} \sum_{i=1}^{2} \frac{m_{3-i}}{m_i} \sum_{j=2}^{4} (j+1)(k_j)_i X_0^{-(2j+3),1} a_i^{2j+1} \\ &= n \sum_{i=1}^{2} \frac{m_{3-i}}{m_i} \Big\{ 15(1-e^2)^{-5}\left(1 + \frac{3}{2}e^2 + \frac{1}{8}e^4\right)(k_2)_i a_i^5 + \\ &\quad + 28(1-e^2)^{-7}\left(1 + \frac{15}{4}e^2 + \frac{15}{8}e^4 + \frac{5}{64}e^6\right)(k_3)_i a_i^7 \\ &\quad + 45(1-e^2)^{-9}\left(1 + 7e^2 + \frac{35}{4}e^4 + \frac{35}{16}e^6 + \frac{7}{128}e^8\right)(k_4)_i a_i^9 + \cdots \Big\},\end{aligned} \tag{3.46}$$

exact for the first three partial tides of spherical-harmonic distortion.

The *total* rate of apsidal advance will, of course, be the resultant of the contributions arising from the rotational as well as tidal distortion; and within the framework of first-order theory their contributions are *additive.* Therefore, the apsidal advance $\delta\omega$ per cycle of orbital period P should be given by

$$\frac{\Delta\bar{\omega}}{2\pi} = \frac{P}{U} = \frac{1}{n}\left\{\left(\frac{d\bar{\omega}}{dt}\right)_{\rm rot} + \left(\frac{d\bar{\omega}}{dt}\right)_{\rm tidal}\right\}, \tag{3.47}$$

where $(d\omega/dt)_{rot}$ and $(d\omega/dt)_{tidal}$ are given by Equations (3.18) and (3.46), respectively, and U denotes the period of the revolution of the apsidal line.

The rate of apsidal advance is governed mainly by mutual tidal distortion of both components (dominated by second-harmonic tides); with the rotational distortion making but a moderate contribution (unless the spin of one or both components is much faster than the angular rate of their revolution). Moreover, Equation (3.18) has made it manifest that the contribution of rotational distortion to the total apsidal advance should be maximum if the equatorial planes of both components coincide with that of their orbit, and diminish with increasing angles between them.

Should the orbital eccentricity e be small enough for its squares and higher powers to be ignorable, Equation (3.47) would assume (correctly to the second-harmonic distortion) the more simplified form

$$\frac{P}{U} = \frac{3}{2} \sum_{i=1}^{2} (k_i)_2 \left\{ \frac{\omega_i^2}{2\pi G\rho_i} \left(\frac{R_i}{A}\right)^2 + 10 \frac{m_{3-i}}{m_i} \left(\frac{R_i}{A}\right)^5 \right\} \tag{3.48}$$

Should, moreover, both components rotate with the Keplerian angular velocity of their orbit, so that

$$\omega_i^2 = \omega_2^2 = \frac{G(m_1 + m_2)}{A^3}, \tag{3.49}$$

the foregoing Equation (3.48) would reduce to the well-known result

$$\frac{P}{U} = k_{12} \left(1 + 16 \frac{m_2}{m_1}\right)\left(\frac{R_i}{A}\right)^5 + k_{22} \left(1 + 16 \frac{m_1}{m_2}\right)\left(\frac{R_2}{A}\right)^5 , \tag{3.50}$$

derived first (by a different method) by Cowling (1938).

We may also wish to add, in this place, that – regardless of the perturbations arising from the distortion of the components of close binary systems due to axial rotation and mutual tidal action – advance of the apsides is also bound to arise from the relativistic effects connected with the curvature of space in the proximity of our gravitational dipole. In point of fact, the equations of motion of a two-body problem in general relativity lead to a secular advance of the apsides, whose period U' of revolution bears to the orbital period P the ratio (cf., e.g., Levi-Civita, 1937) approximable by the expression

$$\begin{aligned} \frac{P}{U'} &= \frac{3\,Gm_\odot}{c^2 R_\odot (1-e^2)} \frac{m_1 + m_2}{m_\odot} \frac{R_\odot}{A} \\ &= 6.35 \times 10^{-6} \frac{m_1 + m_2}{A(1-e^2)} \end{aligned} \tag{3.51}$$

if the dimensions of the semimajor axis A of the relative orbit of a binary with components of masses $m_{1,2}$ are expressed in solar units.

This relativistic advance of the apsides is, of course, independent of (and supplementary to) that produced by rotational or tidal distortion within the framework of Newtonian mechanics; therefore, in any comparison with the observations the right-hand side of Equation (3.51) should be *added* to those, of Equations (3.48) or (3.50) before the latter are used to compute the apsidal-motion constants k_j. In actual practice, however, this relativistic advance turns out to be so slow for eclipsing systems of all evolutionary stages (except, possibly, systems consisting of white dwarfs or neutron stars) as to make a scarcely perceptible contribution to the motions actually observed; it is only for extremely close or very eccentric systems (characterized by very small values of A) that the relativistic ratios P/U' may become at all appreciable.

C. APSIDAL MOTIONS: COMPARISON WITH OBSERVATIONS

The longitude of the apsidal line is not the only orbital element of close binary systems which is subject to secular perturbations; for in the preceding section V-2 we found that the longitude Ω of the node likewise regresses secularly at a rate depending on the internal structure of the components. The effects of such a regression are, however, observable only indirectly (cf. sec. V-4); while the orientation of the apsidal line in eccentric eclipsing systems can manifest itself directly in three different ways, namely:

(1) by producing a displacement of the secondary minima of light with respect to the adjacent primary ones;

(2) by rendering the alternate minima of unequal duration;

(3) by giving rise to an asymmetry of the light curves at each minimum.

For no orientation of the apsidal line can all three of these effects vanish simultaneously; but neither are all equally conspicuous: the relative displacements of the minima (specified by the product $e \cos \omega$) are the easiest to establish from the observations; and the unequal durations of the minima (specified by $e \sin \omega$) the second next; while an asymmetry of the descending and ascending branches of the same minima become detectable only in very eccentric systems.

A closer discussion of the methods by which the instantaneous values of ω can be extraced from the photometric and (or) spectrographic observations of eclipsing variables is outside the scope of this section, and for fuller details the reader must be referred to other sources (Kopal, 1959, sec. VI. 9; or 1965a). Instead, our aim in this concluding part of section V-3 will be to present the available observational evidence of apsidal motions in close binary systems, in order to provide a basis for its confrontation with theoretical models of the internal structure of the stars.

This task should be preceded by a few preliminary remarks. First, a theory of apsidal motion as developed earlier in this section makes it evident that the secular advance of ω is a resultant of the action of *both* components, weighted in proportion of their respective contributions. Therefore, an empirical determination of the ratio P/U does not permit us to determine the values of the individual constants $(k_2)_i$ characterizing each component separately, but only of a certain weighted mean k_2 of them to defined below.

In order to do so, let us return to Equation (3.48) giving the theoretical values of the ratio P/U for rotational as well as second-harmonic tidal distortion of both stars, which can be rewritten as

$$\frac{P}{U} = c_i k_1 + c_2 k_2 \, , \tag{3.52}$$

where $k_i \equiv (k_2)_i$ and

$$c_i = \left\{ \left(\frac{\omega_i}{\omega_K} \right)^2 \left(1 + \frac{m_{3-i}}{m_i} \right) \frac{1}{(1-e^2)^2} + \right.$$

$$\left. + 15 \, \frac{m_{3-i}}{m_i} \, \frac{8 + 12e^2 + e^4}{8(1-e^2)^5} \right\} \left(\frac{R_i}{A} \right)^5 , \tag{3.53}$$

where ω_K stands for the Keplerian angular velocity (3.49). Consequently, the mean value $\bar{k}_2$ of the individual values of $k_{1,2}$ follows as

$$\bar{k}_2 = \frac{c_1 k_1 + c_2 k_2}{c_1 + c_2} = \frac{1}{c_1 + c_2} \, \frac{P}{U} , \tag{3.54}$$

in which the weights $c_{1,2}$ as given by (3.53) can be evaluated from the empirical evidence.

The ratios ω_i/ω_K occurging in (3.53) can sometimes be determined from direct spectroscopic evidence (such as line broadening; rotational effect within eclipses, etc.). Statistical investigations (e.g., Swings, 1936) have indicated long ago that, in eccentric binaries, the observed values of ω_i come close to the maximum (rather than mean) angular velocity ω_P of orbital revolution attained at the time of the periastron passage, given by

$$\omega_P^2 = \frac{1+e}{(1-e)^3} \, \omega_K^2 ; \tag{3.55}$$

and in what follows we shall adopt $\omega_{1,2} \equiv \omega_P$ as the best statistical approximation to reality.

The eccentric eclipsing systems in which apsidal advance has been established, or indicated, by the observations up to the present (1976) are listed in the accompanying Tables V-1 and V-2. The first table contains the systems for which apsidal advance has been disclosed by oscillations in the relative displacements of the minima; while in the second we have listed the relevant data for systems whose secondary minima are too shallow to disclose their relative positions in the cycle with sufficient accuracy, but which exhibit harmonic oscillations of the orbital period likely to be caused by apsidal motion (cf. sec. V-4).

The individual columns in these tables indicate, successively, (1) the designation of the system; (2) the spectral types of of its components (the data in square brackets indicating spectral types of invisible components, evaluated from the observed ratios of surface brightnesses of the two stars); (3) the orbital period (in days); (4) the ratios U/P of the period U of apsidal advance to that of the orbit; (5) the value of the orbital eccentricity e; (6) the mass-ratios m_2/m_1 (the values in square brackets indicating mass-ratios inferred from known ratios of the luminosities of the two components with the aid of the mass-luminosity relation; while those in parentheses were deduced from the fractional dimensions of their semi-detached components). Columns (7) and (8) list then the ratios $\omega_{1,2}/\omega_K$ (those in square brackets were approximated from Equation (3.55) in terms of the eccentricity e); columns (9) and (10) give the fractional dimensions $a_{1,2}$ of the respective components (for distorted stars these are defined as the radii of spheres having the same column as the actual component); columns (11) and (12) contain tabulations of the weight coefficients $c_{1,2}$ as defined by Equation (3.53); and, finally, the last column (13) lists the mean values $\bar{k}_2$ of the internal-structure constants $\bar{k}_{1,2}$ characteristic of each component, as defined by Equation (3.54).

The data on the 28 eclipsing systems listed in Tables V-1 and V-2 constitute the bulk of the present evidence on the internal structure of close binaries as disclosed by the observable rates of the advance of their apsidal lines. These data are of very unequal weights – ranging from well-established cases like Y Cyg or GL Car to systems like AR Cas, 380 Cyg or δ Ori, for which apsidal motion is merely indicated. On the other hand, unmistakeable evidence of apsidal advance is exhibited by quite a number of other eclipsing systems (such as XZ And, V 899 Aql, V 453 Cyg, UW Lac, AO Vel, DR Vul or DQ Vul), for which no geometrical (or other) elements are known so far to enable us to extract values of the apsidal-motion constants $\bar{k}_j$ from the observed ratios U/P.

In considering the quality of the data included in Tables V-1 and V-2, we may note that the reduced values of $\bar{k}_2$ are not too much affected by our incomplete knowledge of the ratios $\omega_{1,2}/\omega_K$ which we estimated from Equation (3.55); for the main contribution to the observed rate of apsidal advance stems from the tides. The principal source of the uncertainty of the $\bar{k}_j$'s goes back to that with which the fractional radii $R_{1,2}/A$ of the two components can be deduced from the light changes of the respective eclipsing systems. In this connection, it is important to re-emphasize that the unit of length in which the radii $R_{1,2}$ of the two stars are expressed is the semi-major axis A of the relative orbit of the respective system, and *not* the mean values of the radius-vector at the time of the eclipses, obtained from an analysis of their light curves.

In point of fact, the fictitious 'circular' fractional radii $r'_{1,2}$ of the two components, obtained by a conventional analysis of the light changes, are related with the true 'elliptical' fractional radii $a_{1,2}$ as defined by Equation (3.5) by an (approximate) equation of the form

$$r'_{1,2} = \frac{\sqrt{1-e^2}}{1+e\cos\omega}\, a_{1,2} \tag{3.56}$$

at the time of superior conjunction (when $v = 90° - \omega$); and

$$r''_{1,2} = \frac{\sqrt{1-e^2}}{1-e\cos\omega}\, a_{1,2} \tag{3.57}$$

for 'circular' radii $r''_{1,2}$ deduced from the minima at inferior conjunctions (when $v = 270° - \omega$).

The foregoing equations are sufficiently accurate for reductions of the observed 'circular' radii $r_{1,2}$ into the 'elliptical' $a_{1,2}$'s needed in the present section as long as the orbital eccentricity e is not too large (cf. e.g., Kopal, 1959; sec. VI. 10). For very eccentric orbits the relations between the $r_{1,2}$'s and $a_{1,2}$'s become, however, much more complicated – not only on account of the approximations inherent in the derivation of the foregoing Equations (3.56) and (3.57), but also because of the fact that the semi-major axis A of the relative orbit itself is subjected to periodic perturbations in the course of each orbital cycle, pointed out earlier in this section. Such perturbations will, in fact, render the ratio of the radii ('circular', as well as 'elliptical') of the two components to vary in the course of the eclipse; and this variation should be taken into account in solutions for the geometrical elements of eccentric eclipsing systems. The fact that this has so

TABLE V-1

Eclipsing systems exhibiting apsidal motions

Star	Spectra	P	U/P	e	m_2/m_1	ω_1/ω_K	ω_2/ω_K	a_1	a_2	c_1	c_2	$\bar{k}_1$
GL Car	B3 + B4	2.422	3800 ± 300	0.16 ± 0.01	1.0	1.4	1.4	0.217 ± 0.008	0.217 ± 0.008	0.0107	0.0107	0.0122
HH Car	B5 + B8	3.2315	75000	0.16	0.9	1	1	0.308	0.231	0.0495	0.0144	0.0021
AR Cas	B3 + A0	6.0665	25000 ± 2000	0.25 ± 0.01	0.25	2.2	1	0.204 ± 0.001	0.0663 ± 0.0004	0.00452	0.00012	0.0086
V346 Cen	B4 + B6	6.3227	11000 ± 1000	0.20 ± 0.01	1.0	1.5	1.5	0.23 ± 0.01	0.15 ± 0.01	0.0159	0.00187	0.0051
XX Cep	A8 + G6	2.3373	10000 ± 1200	0.14	0.22	1.3	1	0.213	0.24	0.0027	0.0656	0.0015
Y Cyg	O9.5 + O9.5	2.9963	5900 ± 500	0.14 ± 0.01	0.99 ± 0.01	1.7	1.7	0.208 ± 0.008	0.202 ± 0.008	0.00909	0.00793	0.0100
MR Cyg	A0 + F7	1.6770	12000 ± 2000	0.05	0.85	1.1	1.1	0.35 ± 0.01	0.25 ± 0.01	0.0786	0.0198	0.00085
V 380 Cyg	B1.5 + B3	12.4256	59000 ± 2400	0.219	0.57	1.6	1	0.325 ± 0.004	0.089 ± 0.001	0.0586	0.0002	0.00029
V 477 Cyg	A3 + F5	2.3470	54300	0.302 ± 0.002	0.68	1.0	1.0	0.147 ± 0.002	0.114 ± 0.002	0.00141	0.00083	0.0082
HS Her	B5 + A4	1.6374	3450	0.05	0.34	1.1	1.1	0.31 ± 0.01	0.17 ± 0.01	0.0192	0.0069	0.0111
CO Lac	B8.5 + A0	1.5422	10010 ± 120	0.028	0.82 ± 0.04	1.	1	0.251 ± 0.005	0.213 ± 0.007	0.0140	0.0089	0.0043
RU Mon	B9 + A0	3.5847	28900	0.376	0.96	2.4	2.4	0.130	0.126	0.00196	0.00180	0.0092
GN Nor		5.7034	31000	0.21	1	1.6	1.6	0.21 ± 0.01	0.21 ± 0.01	0.0104	0.0104	0.0016
FT Ori	A0 + A3	3.1504	60000 ± 12000	0.40	0.9	2.5	2.5	0.121 ± 0.006	0.124 ± 0.004	0.00148	0.00200	0.0048
δ Ori	B1 + B2	5.7325	9900 ± 2700	0.085 ± 0.004	0.38 ± 0.01	1.1	1.1	0.17 ± 0.01	0.17 ± 0.01	0.0012	0.0071	0.0121
AG Per	B5 + B7	2.0287	12900 ± 2000	0.067 ± 0.012	0.88 ± 0.03	1.1	1.1	0.211 ± 0.007	0.194 ± 0.007	0.00673	0.00559	0.0063
YY Sgr	A0 + A0	2.6285	46000 ± 3000	0.16 ± 0.01	0.9	1.4	1.4	0.168 ± 0.004	0.156 ± 0.004	0.0027	0.0022	0.0044
V 523 Sgr	A5 + A5	2.3238	33000 ± 4000	0.18 ± 0.02	1.0	1.5	1.5	0.22 ± 0.02	0.18 ± 0.01	0.0120	0.0044	0.0019
V 526 Sgr	A0 + A3	1.9195	27800 ± 2000	0.224	0.8	1.6	1.6	0.187	0.157	0.0050	0.0031	0.0044
V 2283 Sgr	A0 + A2	3.4714	59000 ± 8000	0.487	0.7	3.3	3.3	0.117	0.099	0.00191	0.00151	0.0050
α Vir	B2 + B3	4.0142	11200 ± 1000	0.15 ± 0.01	0.62 ± 0.08	2.3	2	0.20 ± 0.01	0.12 ± 0.01	0.0063	0.0010	0.0122
DR Vul	B7 + B8	2.2512	6140 ±	0.09 ± 0.01	0.95	1.2	1.2	0.223	0.208	0.0099	0.0077	0.0093

TABLE V-2

Eclipsing systems exhibiting period changes which may be due to apsidal motion

Star	Spectra	P	U/P	e	m_2/m_1	ω_1/ω_K	ω_2/ω_K	a_1	a_2	c_1	c_2	$\bar{k}_2$
Y Cam	A7 + gK1	3.3054	6600 ±	0.11	0.21	1.1	1	0.233 ± 0.007	0.228 ± 0.003	0.00337	0.0512	0.0027
RS CVn	F4 + gK0	4.7979	3300 ± 300	0.028	0.93 ± 0.02	1	1	0.092 ± 0.005	0.28 ± 0.01	0.00011	0.0202	0.0097
RZ Cas	A0 + gK1	1.1953	5600 ± 400	0.13	0.35	1.3	1	0.241 ± 0.002	0.284 ± 0.002	0.00059	0.0864	0.0019
W Del	A0 + gK0	4.8060	3870 ± 100	0.036 ± 0.002	0.21	1.4	1	0.153 ± 0.003	0.243 ± 0.002	0.00048	0.0627	0.0038
β Per	B8 + gK0	2.8673	4070 ± 20	0.010 ± 0.001	0.19	1.0	1	0.227 ± 0.002	0.236 ± 0.002	0.00244	0.0732	0.0036
TX UMa	B8 + gF2	3.0633	4100 ± 200	0.023 ± 0.001	0.30	1	1	0.158 ± 0.001	0.277 ± 0.001	0.00057	0.0785	0.0031

References to Tables V-1 and V-2

GL Car: Wijk, U. v., Rogerson, J. B., and Skumanich, A.: 1955, *Astron. J.* 60, 95 (photometric elements); Shapley, H. and Swope, H. H.: 1938, *Harvard Obs. Bull.*, No. 909; Sterne, T. E.: 1939, *Monthly Notices Roy. Astron. Soc.* **99**, 451, 662 (apsidal motion).

HH Car: O'Connell, D. J. K.: 1968, *Ricerche Astr. Specola Vaticana*, **7**, 399.

AR Cas: Huffer, C. M. and Collins, G. W.: 1962, *Astrophys. J. Suppl.* **7**, 351 (photometric elements); Batten, A. H.: 1961, *J. Roy. Astron. Soc. Canada*, **55**, 120 (spectroscopic elements and apsidal motion).

V 346 Cen: Dugan, R. S. and Wright, F. W.: 1939, *Princeton Contr.*, No. 19; and O'Connell, D. J. K.: 1939, *Riverview Coll. Publ.* **2**, 5; Kopal, Z.: 1940, *Harvard Obs. Circ.*, No. 443.

XX Cep: Iljasova, N.: 1946, *Engelhardt Obs. Bull.*, No. 24; Struve, O.: 1946, *Astrophys. J.* **103**, 76; Fresa, A.: 1953, *Mem. Soc. Astron. Ital.*, **24**, 341; Fresa, A.: 1956, *Mem. Soc. Astron. Ital.* **27**, 231; Lavrov, M. J.: *1957, Peremennye Zvjozdy* **12**, 21.

Y Cyg: Dugan, R. S.: 1931, *Princeton Contr.*, No. 12 (photometric elemments); Redman, R. O.: 1930, *Publ. Dominion Astrophys. Obs.*, **4**, 341; Struve, O., Sahade, J. and Zebergs, V.: 1969, *Astrophys. J.*, **129**, 59 (spectroscopic elements); Sterne, T. E.: 1939, *Monthly Notices Roy. Astron. Soc.*, **99**, 662; Luyten, W. J., Struve, O., and Morgan, W. W.: 1939, *Yerkes Obs. Publ.* **7**, 251 (apsidal motion).

MR Cyg: Kaminsky, B. I.: 1953, *Peremennye Zvjozdy*, **9**, 285; Hall, D. S. and Hardie, R. H.: 1969, *Publ. Astron. Soc. Pacific* **81**, 754.

V 380 Cyg: Kopal, Z.: 1940, *Harvard Obs. Circ.*, No. 441; Kopal, Z. and Shapley, M. B.: 1956, *Jodrell Brank Ann.* **1**, 141; Batten, A. H.: 1962, *Publ. Dominion Astrophys. Obs. Victoria*, **12**, 91; Semeniuk, I.: 1968, *Acta Astron.*, **18**, 1.

V 477 Cyg: Wallenquist, A.: 1949, *Uppsala Obs. Medd.*, No. 91; Rodono, M.: 1967, *Publ. Osserv. Astrofis. Catania*, No. 98; O'Connell, D. J. K.: 1970, in *Vistas in Astronomy* 12, 271; van den Heuvel, E. P. J: 1970 in *Stellar Rotation* (ed. A. Slettebak), pp. 178–86; Scarfe, C. D., Barlow, D. J. and Niehaus, R. I.: 1976 in *Astrophys. Space Sci.*; 3, 129.

HS Her: Hall, D. S. and Hubbard, G. S.: 1971, *Publ. Astron. Soc. Pacific* **83**, 459.

CO Lac: Semeniuk, I.: 1967, Acta Astron. 17, 223 (photometric elements); Smrak, J.: 1967, Acta Astron. 17,245 (spectroscopic elements).

RU Mon: Martynov, D. Ya.: 1965, Astron. Zhurn. **42**, 1209.

GN Nor: Kort, J. de: 1954, *Richerche Astr. Spec. Vaticana* 3, 119.

FT Ori: Cristaldi, S.: 1970, *Astron. Astrophys.* **5**, 228.

δ Ori: Luyten, W. J., Struve, O., and Morgan, W. W.: 1939, *Yerkes Obs. Publ.*, 7, 251; Pismis, P., Haro, G., and Struve, O.: 1950, *Astrophys. J.* **111**, 509; Worley, C. E.: 1955, *Publ. Astron. Soc. Pacific* **67**, 330.

AG Per: Kopal, Z. and Shapley, M. B.: 1956, *Jodrell Bank Ann.* **1**, 141; Semeniuk, I.: 1968, *Acta Astron.* **18**, 1.

YY Sgr: Keller, G. and Limber, D. N.: 1951, *Astrophys. J.* **113**, 637.

V 523 Sgr.: Jones, E. W.: 1938, *Harvard Obs. Bull.*, No. 909; Sterne, T. E.: 1939, *Monthly Notices Roy. Astron. Soc.* **99**, 662.

V 526 Sgr: O'Connell, D. J. K.: 1967, *Richerche Astr. Spec. Vaticana* **7**, 339.

V 2283 Sgr: Shapley, H. and Swope, H. H.: 1938, *Harvard Obs. Bull.*, No. 909; Sterne, T. E.: 1939, *Monthly Notices Roy. Astron. Soc.* **99**, 662; O'Connell, D. J. K. and Swope, H. H.: 1974, *Richerche Astron. Spec. Vaticana* **8**, 481 and 491.

α Vir: Luyten, W. J., Struve, O., and Morgan, W. W.: 1939, *Yerkes Obs. Publ.* **7**, 251; Struve, O: 1948, *Astrophys. J.* **108**, 154; Struve, O., Sahade, J., Huang, S. S., and Zebergs, V.: 1958, *Astrophys. J.* **128**, 310; Herbison-Evans, D., Hanbury Brown, R., Davis, J., and Allen, L. R.: 1971, *Monthly Notices Roy. Astron. Soc.* **151**, 161; Rucinski, S. M.: 1972, *Acta Astron.* **20**, 249; or Odell, A.P. 1974, *Astrophys. J.*, 192, 417.

DR Vul: O'Connell, D. I. K.: 1972, *Richerche Astron. Vatican*, **8**, 319.

Y Cam: Kopal, Z. and Shapley, M. B.: 1956, *Jodrell Bank Ann.* **1**, 143 (photometric elements); Struve, O., Horak, H. G., Cavanaggia, R., Kourganoff, V., and Colacevich, A.: 1960, *Astrophys. J.* **111**, 658 (spectroscopic elements); Plavec, M., Pěkný, Z., and Smetanová, M.: 1961, *Bull. Astron. Inst. Czech.* **12**, 117 (apsidal motion).

RS CVn: Keller, G. and Limber, D. B.: 1951, *Astrophys. J.* **113**, 637; Popper. D. M.: 1961, *Astrophys. J.* **133**, 148; Plavec, M.: 1960, *Bull. Astron. Inst. Czech.* **11**, 148; Plavec, M., Pěkný, Z., and Smetanová, M.: 1961, *Bull. Astron. Inst. Czech.* **12**, 117.

RZ Cas: Kopal, Z.: 1965, *in Advances in Astronomy and Astrophysics* **3**, 89.

W Del: Kopal, Z. and Shapley, M. B.: 1956, *Jodrell Bank Ann.* **1**, 143 photometric elements); Struve, O.: 1946, *Astrophys. J.* **104**, 253 (spectroscopic elements); Plavec, M.: 1959, *Bull. Astron. Inst. Czech.* **10**, 185; (1960), *Bull. Astron. Inst. Czech.* **11**, 148 (apsidal motion).

β Per: Kopal, Z. and Shapley, M. B.: 1956, *Jodrell Bank Ann.* **1**, 141 (photometric elements); Kopal, Z., Plavec, M., and Reilly, E. M.: 1960, *Jodrell Bank Ann.* **1**, 374; Plavec, M.: 1960, *Bull. Astron. Inst. Czech.* **11**, 148 (apsidal motion).

TX UMa: Kopal, Z. and Shapley, M. B.: 1956, *Jodrell Bank Ann.* **1**, 141 (photometric elements); Hiltner, W. A.: 1945, *Astrophys. J.* **101**, 108 (spectroscopic elements); Plavec, M.: 1960, *Bull. Astron. Inst. Czech.* **11**, 148 (apsidal motion).

far not been done constitutes an additional source of uncertainty in the values of $a_{1,2}$ listed in columns (9) and (10) of Tables V-1 and V-2; and it is only hoped that this uncertainty does not exceed the limits of the errors of $a_{1,2}$ listed in these tables.

The relation between the apsidal-motion constants k_2, deduced from the data listed in Tables V-1 and V-2, and the internal structure of the respective stars has already been investigated in section II-6 of this volume. In that section we found (cf. Equation (6.4) of Chapter II) that

$$k_j = \frac{j + 1 - \eta_i(a_1)}{2(j + \eta_j(a_1))} , \tag{3.58}$$

where $\eta_i(a_1)$ denotes the surface value of the solution of Equation (5.62) of Chapter II (obtainable by its integration throughout the interior); it is through $\eta_j(a)$ that the coefficients k_j depend on the internal structure of the stars. If both components of our binary were homogeneous, we have seen in section II-5 that $\eta_j = 0$, in which case

$$k_j = \frac{j+1}{2j} , \tag{3.59}$$

yielding $k_2 = 3/4$, $k_3 = 2/3$, $k_4 = 1/2$, etc. On the other hand, for a mass-point model $\eta_j(a_1) = j + 1$ leading to $k_j = 0$ for all j's.

A glance at the ultimate columns of Tables V-1 and V-2 discloses that the empirical values of k_2 for our systems lie between 0.0002 and 0.012 – indicating a much closer resemblance of the internal structure of the components of the corresponding systems to a mass-point model than to a homogeneous one. In this respect – we may note in passing – that while the observations of the close eclipsing system have consistently signposted the road to follow by the theoreticians, it was not without throwing in, now and then, fortuitous clues to mislead investigators prone to jump to conclusions. Thus at the time when a correct theory of apsidal motion due to mutual distortion of deformable components of close binary systems was established by Cowling (1938) and Sterne (1939), the majority of astronomers still clung to the belief stars were built up in accordance with the polytropic family of models. And – behold – the values of k_2 obtained from Equation (3.51) for GL Car and Y Cyg – the best established cases of apsidal motion known at that time – were both found to agree very closely with the theoretical value of k_2 for a polytropic gas sphere specified by the polytropic index $n = 3$ (Eddding-ton's 'standard model'); in fact, the observations for both GL Cas as well as Y Cyg turned out to be consistent, on this assumption, with the value of $n = 3.0 \pm 0.2$ (cf. Sterne, 1939).

This apparent vindication of the 'standard model' proved, however, to be short-lived. Already Sterne was careful to point out that his analysis . . . "does not, of course, have anything to say as to whether or not the components of Y Cygni are polytropes; it merely shows if the components are polytropes with the same index for each, then the observed rate of apsidal motion requires this index to be very nearly 3.02." (op. cit.). Subsequent discussions of all available data by Russell (1939), Luyten et al. (1939); or Kopal (1940) have disclosed that the 'effective polytropic indices' for other systems are scattered between 3 and 4.5 (corresponding to the ratios of central-to-mean densities of the respective components between 50 and 5000). These results demonstrated, conclusively and irrefutably, that considerable differences in internal structure of the stars must exist even for systems situated on the Main Sequence – and this was made clear by the evidence furnished by eclipsing variables at a time when a large majority of the theoreticians still held the view that all these stars should be homologous.

In more recent year, the bearing of the apsidal-motion evidence on our knowledge of the internal structure of the stars was re-discussed by Plavec (1960), Kopal (1965), Mathis (1967), Semeniuk and Paczynski (1968), Heasley (1971), Petty (1973), Mathis and Odell (1973), Odell (1974); Stothers (1974) and others; and its general outcome can be briefly summarized as follows:

(1) for evolved Main-Sequence stars of average or above-average masses (of spectral types O and B), the agreement between the observations and the current theories of the internal structure of such stars is generally good in most cases (except, perhaps, HH Car A or V380 Cyg A, which may currently find themselves in the 'Hertzsprung gap'); and can be made almost exact by (reasonable) assumptions on their chemical composition.

(2) for Main-Sequence stars of spectral types later than A0 (evolving more slowly) this agreement becomes distinctly worse – the actual stars appear to be two to three times more centrally-condensed than their models.

(3) For stars evolved away from the Main Sequence (subgiants!) one can scarcely speak so far of any agreement at all – though at least a part of this disagreement can be distributed between a possibly marginal quality of the observations and the deficiencies of adequate theoretical models. Insofar as one can draw any conclusions on the basis of so fragmentary an evidence, the empirical values of the apsidal-motion constants k_2 for subgiant components of semi-detached eclipsing systems appear to be much *smaller* than those expected on the basis of models currently available; though it may still be premature to put too much emphasis on more quantitative aspects of this fact.

V-4. Period Changes in Eclipsing Binary Systems

In section V-2 of this chapter we have investigated the first-order secular as well as periodic perturbations of the orbital elements Ω and i, arising from mutual distortion of rotating components of close binary systems; and in section V-3 we have done the same for e, and A. The aim of the present section will be to investigate, in a similar manner, the perturbations of the last remaining element: namely, of the *period* of the binary orbit. The importance of this task is underlined by the fact that – because of the repetitive nature of the light (or radial velocity) changes – the periods of close binary systems can be determined from the observations with a precision far surpassing that with which any other element can be ascertained from available observational evidence. The periods of many eclipsing binaries are indeed known to exhibit fluctuations of sometimes very complicated character. The object of the present section will not be to consider all causes which may produce such fluctuations – we shall return to them in several subsequent sections of this book – but rather *to investigate the fluctuations of orbital periods as caused by the perturbing function* $\mathfrak{R}_{12}$ defined by Equation (1.5) of this chapter – *both directly and indirectly* (i.e., through its effects on other elements whose variation may influence the duration of the apparent periods). We shall scrutinize also the methods by which the duration of the periods can actually be extracted from the observed data, and establish the exact form of the relationship between the instantaneous period and the observed epochs.

A. GENERALIZED LAW OF AREAS

Embarking on this task, our first step towards this objective should be to establish the relation between the time t and the true anomaly υ of the components in their relative orbit. For unperturbed elliptic motion such a relation is provided by Kepler's second law. In order to generalize this law to take account of the perturbations arising from distortion, let us differentiate the Equation (1.9) for the true anomaly υ with respect to the time, obtaining

$$\frac{d\upsilon}{dt} = \frac{\sin\upsilon}{1-e^2}\frac{de}{dt} + \frac{A\sqrt{1-e^2}}{r}\frac{dE}{dt}, \tag{4.1}$$

where a differentiation of Equation (1.7) for the eccentric anomaly E yields

$$\frac{dE}{dt} = \frac{A}{r}\left\{\frac{dM}{dt} + \sin E\,\frac{de}{dt}\right\}, \tag{4.2}$$

where

$$M = nt + \epsilon - \bar{\omega} \tag{4.3}$$

denotes the mean anomaly. In consequence,

$$\frac{d\upsilon}{dt} = \frac{A^2\sqrt{1-e^2}}{r^2}\left\{n + \frac{d\epsilon}{dt} - \frac{d\bar{\omega}}{dt}\right\} + \sin\upsilon\left\{\frac{1}{1-e^2} + \frac{A}{r}\right\}\frac{de}{dt}, \tag{4.4}$$

where the time-derivatives of A, e, ω and ϵ occurring on the right-hand side are, in turn, defined by Equations (1.12)–(1.15).

The foregoing Equation (4.4) represents the desired generalization of Kepler's 'law of the areas', to which it would indeed reduce if the orbital elements A, e, ω and ϵ were constant. In section V-1 of this chapter we have, however, seen that, in close binary systems, these elements are bound to vary with the time – in accordance with Equations (1.12)–(1.15) or (1.26)–(1.29) as a result of perturbations caused by our disturbing function $\mathfrak{R}_{12}$. Such perturbations will, in particular, affect the longitude ω of periastron from which the true anomaly υ is reckoned. In order to obviate complications arising from this source, let us replace υ by another true anomaly w measured from the inertial axis x, and related with the true anomalies u and υ (measured from the nodal and apsidal lines) by the equation

$$w = u + \Omega = \upsilon + \bar{\omega}. \tag{4.5}$$

If we replace ν on the left-hand side of (4.4) by $w - \bar{\omega}$, and insert for the time-derivatives of A, e, $\bar{\omega}$ and ϵ form (1.12)–(1.15), our generalized 'law of areas' will assume the more explicit form

$$\frac{1}{An}\frac{dw}{dt} = \frac{A\sqrt{1-e^2}}{r^2}\left\{1 - A\,\frac{1-e^2}{e}\frac{\partial\mathfrak{R}_{12}}{\partial e} - 2A^2\frac{\partial\mathfrak{R}_{12}}{\partial A}\right\} -$$

$$- \sin \nu \left\{ \frac{1}{1-e^2} + \frac{A}{r} \right\} \left\{ \frac{1-\sqrt{1-e^2}}{e} \frac{\partial \mathfrak{R}_{12}}{\partial \epsilon} + \frac{1}{\epsilon} \frac{\partial \mathfrak{R}_{12}}{\partial \omega} \right\} \sqrt{1-e^2} +$$

$$+ \frac{\tan \frac{1}{2} i}{\sqrt{1-e^2}} \frac{\partial \mathfrak{R}_{12}}{\partial i} + \frac{\sqrt{1-e^2}}{e} \frac{\partial \mathfrak{R}_{12}}{\partial e} \tag{4.6}$$

Now (to the first order in small quantities) the disturbing function $\mathfrak{R}_{12}$ as given by Equations (1.41) and (1.42) depends on the elements of the orbit only through the radius-vector r and be angle Θ_j between this radius-vector and the axis of the respective components, given (in accordance with Equation (2.80)) by

$$\cos \Theta_j \equiv \nu_1'' = \mathfrak{A}_j \sin u + \mathfrak{B}_j \cos u \, , \tag{4.7}$$

where the coefficients $\mathfrak{A}_j$ and $\mathfrak{B}_j$ are given by (2.82) and (2.83). Since, in accordance with Equations (1.17) and (1.19)–(1.20),

$$\frac{\partial r}{\partial A} = \frac{r}{A} \, , \quad \frac{\partial r}{\partial e} = -A \cos v \, , \tag{4.8}$$

$$\frac{\partial r}{\partial \bar{\omega}} = - \frac{A \, e \cos v}{\sqrt{1-e^2}} = - \frac{\partial r}{\partial \epsilon} \tag{4.9}$$

and

$$\frac{\partial r}{\partial i} = \frac{\partial r}{\partial \Omega} = 0 \, , \tag{4.10}$$

it follows that

$$\left\{ \frac{1-e^2}{e} \frac{\partial}{\partial e} + 2A \frac{\partial}{\partial A} \right\} r = 2r - \frac{1-e^2}{e} A \cos v, \tag{4.11}$$

$$\left\{ \frac{1-\sqrt{1-e^2}}{e} \frac{\partial}{\partial \epsilon} + \frac{1}{e} \frac{\partial}{\partial \omega} \right\} r = -A \sin v \tag{4.12}$$

and

$$\left\{ \frac{\tan \frac{1}{2} i}{\sqrt{1-e^2}} \frac{\partial}{\partial i} + \frac{\sqrt{1-e^2}}{e} \frac{\partial}{\partial e} \right\} r = - \frac{A \sqrt{1-e^2}}{e} \cos v \, , \tag{4.13}$$

it follows on insertion in Equation (4.6) that if the disturbing field of force were to possess a strict radial symmetry (i.e., $\mathfrak{R}_{12}$ to depend only on r), all terms on the right-hand side of (4.6) would add up to zero – leaving us with

$$\frac{\mathrm{d}w}{\mathrm{d}t} = \frac{nA^2}{r^2} \sqrt{1-e^2} \, , \tag{4.14}$$

which represents Kepler's second law of the two-body problem.

However, this will no longer be true if the angle Θ_j ceases to be constant and equal to $\frac{1}{2}\pi$, implying rotation about an axis constantly perpendicular to the orbital plane. Should this cease to be the case, we are faced with the need of evaluating the partial derivatives of

$$\frac{\partial \cos\Theta_j}{\partial x} = \frac{\partial \mathfrak{A}_j}{\partial x}\sin u + \mathfrak{A}_j \cos u \frac{\partial u}{\partial x} + + \frac{\partial \mathfrak{B}_j}{\partial x}\cos u - \mathfrak{B}_i \sin u \frac{\partial u}{\partial x} \tag{4.15}$$

with respect to all orbital elements x involved on the right-hand side of Equation (4.6).

In doing so, to a sufficient approximation (cf. Equations (2.143)) we can set $\phi_j - \Omega = \pm\pi$, which reduces (2.82) and (2.83) to

$$\mathfrak{A}_j = \sin(\theta_j + i) \qquad \text{and} \quad \mathfrak{B}_j = 0\,. \tag{4.16}$$

Secondly, for every element x concerned except ω we can write

$$\frac{\partial u}{\partial x} = \frac{\partial v}{\partial x}, \tag{4.17}$$

where, from (1.23),

$$\frac{\partial v}{\partial A} = 0\,, \quad \frac{\partial v}{\partial e} = \left\{\frac{1}{1-e^2} + \frac{A}{r}\right\}\sin v\,, \tag{4.18}$$

$$\frac{\partial v}{\partial \epsilon} = \frac{A^2\sqrt{1-e^2}}{r^2} = -\frac{\partial v}{\partial \omega}\;; \tag{4.19}$$

except for $x = \bar{\omega}$, when

$$\frac{\partial u}{\partial \bar{\omega}} = \frac{\partial v}{\partial \bar{\omega}} + 1 = 1 - \frac{A^2\sqrt{1-e^2}}{r^2}\,. \tag{4.20}$$

Lastly, of the partial derivatives $\partial\mathfrak{A}_j/\partial x$ the only non-zero one is, by (4.16),

$$\frac{\partial \mathfrak{A}_j}{\partial i} = \cos(\theta_j + i). \tag{4.21}$$

With the aid of all foregoing results we find that

$$\left\{\frac{1-e^2}{e}\frac{\partial}{\partial e} + 2A\frac{\partial}{\partial A}\right\}\cos\Theta_j = = \left\{\frac{2}{e} + \cos v\right\}\sin(\theta_j + i)\sin v\cos u\,, \tag{4.22}$$

$$\left\{\frac{1-\sqrt{1-e^2}}{e}\frac{\partial}{\partial \epsilon} + \frac{1}{e}\frac{\partial}{\partial \omega}\right\}\cos\Theta_j = = -\frac{\sin(\theta_j + i)}{1-e^2}\{2\cos v + e(1+\cos^2 v)\}\cos u\,, \tag{4.23}$$

$$\left\{\frac{\tan\frac{1}{2}i}{\sqrt{1-e^2}}\frac{\partial}{\partial i} + \frac{\sqrt{1-e^2}}{e}\frac{\partial}{\partial e}\right\}\cos\Theta_j = \tag{4.24}$$

$$= \frac{\cos(\theta_j + i) \tan \frac{1}{2} i}{\sqrt{1 - e^2}} \sin u + \left\{ \frac{\sin(\theta_j + i)}{e \sqrt{1 - e^2}} \right\} (2 + e \cos v) \sin v \cos u ,$$

which on insertion in the right-hand of Equation (4.6) yields Kepler's generalized 'law of areas' of the form

$$\frac{1}{An} \frac{dw}{dt} = \frac{A \sqrt{1 - e^2}}{r^2} - \frac{Q \sin^2 u}{r^3} , \tag{4.25}$$

where we have abbreviated

$$Q = \frac{1}{2} \tan \frac{1}{2} i \sum_{j=1}^{2} (k_2)_j \left(\frac{\omega_j^2 R_j^5}{G m_j} \right) \sin 2(\theta_j + i) . \tag{4.26}$$

It should be noted that the right-hand side of the foregoing Equation (4.25) is independent of tidal effects only as long as tides do not lag, and the axis of symmetry of all partial tides remains the radius-vector joining the centres of the two stars. Should tidal lag turn out to be appreciable, however, the situation would become quite complicated. Even if we were willing – like in the preceding section – to disregard a lag in latitude and to limit ourselves to that in longitude, the true anomaly on the right-hand side of Equation (4.6) would have to be augmented by the lag angle given by Equation (5.6); and the latter would introduce additional dependence on A, e, ω and ϵ which would have to be taken into account in carrying out differentiation with respect to them. Inasmuch as the results are complicated, and the lag angle very small (cf. section III-3) for close binaries at most evolutionary stages of their life, we shall not reproduce them in this place; but the reader should keep their effects in mind for systems in which one component (or both) is a white dwarf or a neutron star.

Subject to this proviso, an integration of the generalization (4.25) of Kepler's second law over 2π in w then yields the 'sidereal period' of a close binary system, one (or both) components of which rotate about an axis inclined to the orbital plane. There exists, however, no method by which the sidereal period of any such system could actually be measured. The observable period of a spectroscopic binary can be defined as a time-interval which elapses between two (superior or inferior) conjunctions. The timing of such conjunctions is, however, limited by the relatively low time-resolution of the observations of radial velocities (i.e., long exposures in photographic work); and even more so by the fact that, for spectroscopic binaries which happen to be eclipsing variables, the radial-velocity changes in the neighbourhood of conjunctions are complicated by the rotational effects (for their description cf., e.g., Kopal, 1959; sec. V. 2).

B. ORBITAL PERIOD: A DEFINITION

For eclipsing binary systems, on the other hand, *the orbital period can be defined as a time-interval between two successive* (primary, or secondary) *minima of light;* and since such minima can be timed by photometric observations with an accuracy far surpassing that of radial-velocity measurements, our present knowledge of period fluctuations in close binary systems is based mainly on such observations. In what follows we shall, therefore, limit ourselves to an analysis of the variation of 'synodic' periods consistent with the foregoing definition; and shall investigate the extent to which this period is bound to oscillate as a consequence of the perturbations of other elements established before in sections V-2 and V-3 of this chapter.

Before doing so we wish, however, to define first the exact significance of the 'moments of the minima'. *The light minima will,* by definition, *be expected to occur when the apparent separation of the centres of both components becomes a minimum.* The phase angles (i.e., true anomalies) of these minima can, in turn, be best expressed in terms of the geometry of our problem by a recourse to the fixed XYZ-system of space axes introduced in section IV-1. In that section the orientation of all axes was left wholly arbitrary; but in the subsequent section we were led to identify the Z-axis with the vector representing the total moment of momentum of our binary system to render XY its invariable plane. In order to remove the last arbitrariness from its definition, let us hereafter *assume the X-axis to be oriented so as to lie in a plane defined by the line of sight and the Z-axis* (which is perpendicular to the celestial sphere). Let, furthermore, the angle between the invariable (XY-)plane and a plane perpendicular to the line of sight of a distant observer be denoted by I. If so, the direction cosines of this line of sight, referred to fixed space-axes, obviously are ($\sin I$, 0, $\cos I$) and, consequently, the cosine l_0 of the angle between the radius-vector (of direction cosines a''_{i1} as defined by Equations (2.3)–(2.5) of Chapter IV), and the line of sight will obviously be given by

$$l_0 = a''_{11} \sin I + a''_{31} \cos I = L \sin u + M \cos u, \tag{4.27}$$

where we have abbreviated

$$\begin{aligned} L &= \sin i \cos I - \cos i \sin I \sin \Omega , \\ M &= \qquad\qquad + \qquad \sin I \cos \Omega . \end{aligned} \tag{4.28}$$

The apparent distance δ of the projected centres of the two components at any moment then follows from the equation

$$\begin{aligned} \delta^2 &= r^2 (1 - l_0^2) = \\ &= \left\{ \frac{A(1-e^2)}{1 + e\cos(u-\omega)} \right\}^2 \{1 - (L \sin u + M \cos u)^2\} ; \end{aligned} \tag{4.29}$$

and our task becomes to establish the value of u for which δ^2 becomes a minimum. Differentiating (4.29) we find this to be a root of the equation

$$\begin{aligned} &(L \sin u + M \cos u)(L \cos u - M \sin u)[1 + e \cos(u-\omega)] = \\ &= e \sin(u-\omega)\{1 - (L \sin u + M \cos u)^2\} , \end{aligned} \tag{4.30}$$

which for small values of e readily assumes either one of the two forms

$$\begin{aligned} u_p &= u_0 + 2n\pi - e \sin(u_0 - \omega) \cot^2 j + \cdots , \\ u_s &= u_0 + (2n+1)\pi + e \sin(u_0 - \omega) \cot^2 j + \cdots , \end{aligned} \tag{4.31}$$

appropriate for the primary or secondary minima (taking place in the neighbourhood of the superior or inferior conjunction). In these equations n stands for an arbitrary integer (including zero),

$$\sin u_0 = \frac{L}{\sqrt{L^2 + M^2}}, \qquad \cos u_0 = \frac{M}{\sqrt{L^2 + M^2}}; \tag{4.32}$$

and

$$\sin^2 j = L^2 + M^2 \tag{4.33}$$

defines the angle j between the orbital plane and the celestial sphere, such that

$$\cos j = \cos i \cos I + \sin i \sin I \sin \Omega \tag{4.34}$$

by (4.28). The reader should particularly note that, unlike the angle I between the celestial sphere and the invariable plane of our binary system which remains secularly constant,* the angle j between the actual position of the orbital plane and the celestial sphere must reflect all variations of Ω and i due to precession and nutation (cf. sec. IV-2) and, accordingly, is bound to oscillate between the limits $I \pm i$ in the period of nodal regression. Moreover, the angle u_0 as defined by Equations (4.32) assumes on expansion the more explicit form

$$u_0 = -\Omega + \sin i \cos \Omega \cot I + \frac{1}{4} \sin^2 i \sin 2\Omega + \cdots; \tag{4.35}$$

so that, consequently,

$$u_p + \Omega = 2n\pi + \sin i \cos \Omega \cot I + e \sin \bar{\omega} \cot^2 j + \cdots \tag{4.36}$$

and

$$u_s + \Omega = (2n + 1)\pi + \sin i \cos \Omega \cot I - e \sin \bar{\omega} \cot^2 j, \tag{4.37}$$

where $n = 0, 1, 2, \ldots$ etc.

Suppose that, in what follows, we shall define the orbital period as the time interval between two successive primary minima (the choice of the secondary minima would lead to quite analogous results). The duration of this interval can now be obtained by integrating the equation

$$n\,dt = \frac{(1-e^2)^{3/2}\,dw}{[1+e\cos(w-\bar{\omega})]^2} + \frac{Q}{A^2}\,\frac{\sin^2(w-\Omega)\,dw}{1+e\cos(w-\bar{\omega})}, \tag{4.38}$$

resulting from (4.25), between the limits

$$\begin{aligned} \omega_1 &= + \sin i_1 \cos \Omega_1 \cot I + e \sin \bar{\omega}_1 \cot^2 j_1 + \cdots \\ \omega_2 &= 2\pi + \sin i_2 \cos \Omega_2 \cot I + e \sin \bar{\omega}_2 \cot^2 j_2 + \cdots \end{aligned} \tag{4.39}$$

in considering these limits, it is of fundamental significance for us to realize that, unlike the elements A or e which exhibit (to the first order in small quantities) no secular perturbations, the angles Ω and ω are subject to such perturbations and, moreover, i oscillates in a period which is very long in conparison with that of the orbit. In consequence,

*Or its value may change but slowly in time, as the Sun and the respective eclipsing system alter their relative position in space, due to their peculiar motions in the Galaxy.

the values of $i_{1,2}$, $j_{1,2}$ or $\Omega_{1,2}$ will not be the same in w_1 and w_2, but will differ by the amount of their displacement in the course of a cycle.

Integrating Equation (4.38) between these limits and retaining, as before, only quantities of first order, we eventually establish that

$$\frac{P}{P_0} = \frac{1}{2\pi}\left\{w - 2e\sin(w - \bar{\omega}) + \cdots\right\}_{w_1}^{w_2} + \frac{Q}{2A^2}, \tag{4.40}$$

where P_0 denotes the period of the sidereal (undisturbed) orbit. Now by the trigonometric formula

$$\begin{aligned} &\sin(w_2 - \bar{\omega}_2) - \sin(w_1 - \bar{\omega}_1) = \\ &\quad = 2\sin\tfrac{1}{2}\{w_2 - w_1 + \bar{\omega}_1 - \bar{\omega}_2\}\cos\tfrac{1}{2}\{w_2 + w_1 - \bar{\omega}_1 - \bar{\omega}_2\}; \end{aligned} \tag{4.41}$$

and by (4.39), we find that

$$\begin{aligned} w_2 \pm w_1 &= 2\pi + e\,\{\sin\bar{\omega}_2\cot^2 j_2 \pm \sin\bar{\omega}_1\cot^2 j_1\} + \\ &\quad + \{\sin i_2\cos\Omega_2 \pm \sin i_1\cos\Omega_1\}\cot I. \end{aligned} \tag{4.42}$$

In section IV-3 we have, moreover, established that the longitude of the apsidal line in close binary systems advances at a uniform rate ν in such a way that if

$$\begin{aligned} \bar{\omega}_1 &= \bar{\omega}_0 + \nu t, \\ \bar{\omega}_2 &= \bar{\omega}_0 + \nu(t + P) = \bar{\omega}_1 + \nu P, \end{aligned} \tag{4.43}$$

then

$$\begin{aligned} \bar{\omega}_2 + \bar{\omega}_1 &= 2\bar{\omega}_1 + \nu P, \\ \bar{\omega}_2 - \bar{\omega}_1 &= \nu P, \end{aligned} \tag{4.44}$$

where

$$\nu \equiv \frac{2\pi}{U}; \tag{4.45}$$

U being the period of revolution of the apsidal line. We may also note that the difference $\sin(w_2 - \bar{\omega}_2) - \sin(w_1 - \bar{\omega}_1)$ on the right-hand side of (4.37) is multiplied by e which we have been regarding as a small quantity of first order. In consequence, the exact Equation (4.41) may be replaced, correctly to quantities of second order, by the approximation

$$\sin(w_2 - \bar{\omega}_2) - \sin(w_1 - \bar{\omega}_1) = -2\cos\bar{\omega}_0\sin\tfrac{1}{2}(w_2 - w_1 - \nu P) \tag{4.46}$$

revealing that, of the two combinations $w_2 \pm w_1$ represented by Equation (4.42), only the difference $w_2 - w_1$ will occur explicity in (4.40).

In order to evaluate it, let us (consistent with our present scheme of approximation) replace hereafter the cotangents of the angles I and j on the right-hand side of (4.42) by their cosines, and remember that

$$\sin\bar{\omega}_2 - \sin\bar{\omega}_1 = 2\sin\tfrac{1}{2}\nu P\cos(\omega_1 + \tfrac{1}{2}\nu P). \tag{4.47}$$

Moreover, $\cos j$ continues to be given by Equation (4.34), while Equations (2.163) and

(2.164) disclosed that, very approximately,

$$\begin{aligned} \sin i \sin \Omega \equiv r &= -\tilde{G} \sin y_1 t - \tilde{H} \sin y_2 t\,, \\ \sin i \cos \Omega \equiv s &= -\tilde{G} \cos y_1 t - \tilde{H} \cos y_2 t\,; \end{aligned} \tag{4.48}$$

and, therefore, in accordance with (2.109)

$$\sin^2 i = \tilde{G}^2 + 2\tilde{G}\tilde{H} \cos (y_1 - y_2)t + \tilde{H}^2, \tag{4.49}$$

where we have abbreviated

$$\tilde{G} \equiv a_1 \tilde{A}_1 + a_2 \tilde{A}_2 \qquad \text{and} \qquad \tilde{H} \equiv a_1 \tilde{B}_1 + a_2 \tilde{B}_2\,. \tag{4.50}$$

In these equations $y_{1,2}$ stand, as before, for the roots of the quadratic Equation (2.159), and the ratios $a_j \equiv (\gamma_j/\kappa_j)$ continue to be given by Equation (2.165). Lastly we may note that for small values of $\theta_{1,2} > i$ we may in the expression (4.26) for Q set

$$\tan \tfrac{1}{2} i \cong \tfrac{1}{2} \sin i = \tfrac{1}{2} \{\tilde{G} + \tilde{H} \cos (y_1 - y_2)\, t + \cdots\} \tag{4.51}$$

and (by (2.168) and (2.169))

$$\begin{aligned} \sin 2(\theta_j + i) &\cong 2\, (\sin \theta_j + \sin i) \\ &= 2\{\tilde{A}_j + \tilde{G} + (\tilde{B}_j + \tilde{H}) \cos (y_1 - y_2)\, t + \cdots\}\,; \end{aligned} \tag{4.52}$$

so that Q may be rewritten as

$$Q = Q^{(0)} + Q^{(1)} \cos (y_1 - y_2)\, t + \cdots, \tag{4.53}$$

where

$$Q^{(0)} = r^2\, \tilde{G} \sum_{j=1}^{2} (\gamma_j/n)\, (\tilde{A}_j + \tilde{G}) \tag{4.54}$$

and

$$Q^{(1)} = r^2 \sum_{j=1}^{2} (\gamma_j/n)\, (\tilde{A}_j \tilde{H} + \tilde{B}_j \tilde{G} + 2\tilde{G}\tilde{H}). \tag{4.55}$$

In consequence, Equation (4.40) can be written out in the more explicit form

$$\begin{aligned} 2\pi(P/P_0) &= 2\pi\, \{1 + Q^{(0)} + Q^{(1)} \cos (y_1 - y_2)\, t\} + \\ &\quad + \{\tilde{G} y_1 \sin y_1 t + \tilde{H} y_2 t\}\, P_0 \cos I + \\ &\quad + e\, \{\nu \cos^2 I \cos(\nu t + \bar{\omega}_{0,5}) + 2\nu \cos (\nu t + \bar{\omega}_0) \\ &\qquad - \tilde{G} y_1 \sin 2I \sin (\nu t + \bar{\omega}_0) \cos y_1 t \\ &\qquad + G^2 y_1 \sin^2 I \sin (\nu t + \bar{\omega}_0) \sin 2y_1 t \\ &\qquad - 2\tilde{G} y_1 \sin^2 I \sin (\nu t + \bar{\omega}_{0.5}) \sin y_1 t \end{aligned}$$

$$- 2\tilde{H} y_2 \cos I \cos (\nu t + \bar{\omega}_{0.5}) \sin y_2 t$$

$$+ \cdots \} P_0, \tag{4.56}$$

where the angles $\bar{\omega}_0$ and $\bar{\omega}_{0,5}$ are defined by the equations

$$\bar{\omega} - \Omega = \bar{\omega}_0 - \Omega_0 + (\nu + y_1) t \tag{4.57}$$

and

$$\bar{\omega}_{0.5} = \bar{\omega}_0 + \frac{1}{2} \nu P_0. \tag{4.58}$$

Within the scheme of approximation adopted in this section, Equation (4.56) represents the period variations of eclipsing binary systems caused – directly or indirectly – by mutual distortion of their components. With the exception of the terms factored by $Q^{(0,1)}$, all others represent *apparent* period changes, due to the particular way in which the times of the minima of eclipsing variables are observed, rather than to any intrinsic changes of the period of the sidereal orbit of the respective pair.

The period changes to be expected on the basis of Equation (4.56) are not subject to direct observational verification. They are, however, bound to influence the observed *times of the minima* $M(E)$, as observed at different epochs $E \equiv t/P_0$ for integral values of E. As is well known, the relation between $M(E)$ and $P(E)$ can be approximated by the differential equation

$$\frac{\mathrm{d}M}{\mathrm{d}E} = P(E), \tag{4.59}$$

which on integration yield

$$M(E) = \int P(E)\, \mathrm{d}E = \int (P/P_0)\, \mathrm{d}t. \tag{4.60}$$

If we evaluate this last integral for P/P_0 as given by Equation (4.56) and set $t = P_0 E$, the result can eventually be expressed in the form

$$\begin{aligned} M(E) - M_0 = {} & \{1 + Q^{(0)}\} P_0 E + \\ & + (P_0/2\pi) \{-a_1 \cos r_1 E - a_2 \cos r_2 E + \\ & + a_3 \sin (r_3 E + \bar{\omega}_{0.5}) \\ & + a_4 \cos (r_4 E + \bar{\omega}_{0.5}) \\ & - a_5 \cos (r_5 E + \bar{\omega}_{0.5}) \\ & + a_6 \cos (r_6 E + \bar{\omega}_{0.5}) \\ & - a_7 \cos (r_7 E + \bar{\omega}_{0.5}) \\ & - \tfrac{1}{2} a_8 \sin (r_8 E + \bar{\omega}_0) \\ & + \tfrac{1}{2} a_9 \sin (r_9 E + \bar{\omega}_0) \\ & + a_{10} \sin r_{10} E + \cdots \}, \end{aligned} \tag{4.61}$$

where the periods of the individual oscillations are given by

$$
\begin{aligned}
r_1 &= y_1 P_0 , \\
r_2 &= y_2 P_0 , \\
r_3 &= \nu P_0 , \\
r_4 &= (\nu + y_1) P_0 = r_3 + r_1 , \\
r_5 &= (\nu - y_1) P_0 = r_3 - r_1 , \\
r_6 &= (\nu + y_2) P_0 = r_3 + r_2 , \\
r_7 &= (\nu - y_2) P_0 = r_3 - r_2 , \\
r_8 &= (\nu + 2y_1) P_0 = r_4 + r_1 , \\
r_9 &= (\nu - 2y_1) P_0 = r_5 - r_1 , \\
r_{10} &= (y_1 - y_2) P_0 = r_1 - r_2 ,
\end{aligned}
\tag{4.62}
$$

and their amplitudes are

$$
\begin{aligned}
a_1 &= \tilde{G} \cos I, \\
a_2 &= \tilde{H} \cos I, \\
a_3 &= e\,(3 - \sin^2 I), \\
a_4 &= (r_1/r_4)\, e \tilde{G}\, (1 + \sin I - \text{cps}\, I, \\
a_5 &= (r_1/r_5)\, e \tilde{G}\, (1 - \sin I) \cos I, \\
a_6 &= (r_2/r_6)\, e \tilde{H} \cos I, \\
a_7 &= (r_2/r_7)\, e \tilde{H} \cos I, \\
a_8 &= (r_1/r_8)\, e \tilde{G}^2 \sin^2 I , \\
a_9 &= (r_1/r_9)\, e \tilde{G}^2 \sin^2 I , \\
a_{10} &= (2\pi Q^{(1)}/r_{10}) .
\end{aligned}
\tag{4.63}
$$

The foregoing Equation (4.61) with its coefficients as given by (4.62) and (4.63) constitutes the main outcome of the present section and reveals that, within the scheme of our approximation, the theoretical ephemeris $M(E)$ for the times of the light minima will contain a total of *ten* periodic terms arising from different causes. Of these, the *first* and *second* term (characterized by the amplitudes $a_{1,2}$ and periods $r_{1\ 2}$)are due solely to the *regression of the nodes* and its periodic inequality, while the *third* (of amplitude a_3 and period r_3) arises from the *apsidal advance.* The *tenth* term (a_{10}, r_{10}) is caused again by the *nutation* of the rotational axes of both components and the accompanying motion of the orbital plane. The *six* terms with periods from r_4 to r_9 may be regarded as *apse-node terms*, which arise from an interaction between apsidal advance and nodal regression or nutation of rotational axes.

An observational quest for empirical determination of the amplitudes a_j for the majority of the individual periodic terms on the right-hand side of Equation (4.61) has so far proved unavailing – largely due to the scarcity of observational data of the requisite precision. Period changes caused by a secular advance of the apsidal line – the rate of which can be independently ascertained from relative displacement (and other properties) of the alternative minima – have been amply verified for a number of eccentric systems (GL Car; Y Cyg; cf. Plavec *et al.*, 1960); and are strongly suggested in others (the 32-year oscillation of the orbital period of Algol; cf. Kopal *et al.*, 1960).

This is, however, not so far the case with terms caused by nodal regression. A careful discussion by Plavec (1960) failed to bring to light significant evidence for such terms, at least in the eclipsing systems considered by him. The reason of this failure cannot be insufficiently long time baseline of available observational evidence; for a period of nodal regression is generally shorter (cf. secs. V-2 and V-3 of this chapter) than that of apsidal advance. Therefore, the only reason why effects bound to be caused by such terms in periodic times of the minima are negligible can be the fact that *their amplitudes are too small for observational detection.*

This, if true, constitutes a fact of capital importance; and another instance of a situation when negative evidence may be more telling than a positive one. For in section V-2 of this chapter we found that secular dissipative processes operative in close binary systems tend to 'rectify' the axes of rotation of both components – from whatever positions they may have initially assumed – until the equatorial planes of both stars become co-planar with that of their orbit. A finite inclination between these planes could be permanently maintained, in the presence of tides, only if matter constituting the stars were inviscid. Frictionless fluid represents, however, only a mathematical abstraction which cannot be fulfilled exactly in any natural process. Now we know, from the results established in section V-2, that non-vanishing viscosity – no matter how small – will tend to make the Eulerian angles θ_i (describing the inclination of the rotational axis of each component to the invariable plane of the system) diminish exponentially in the course of time. If so, however, the coefficients $\tilde{G}$ and (still more) $\tilde{H}$ factoring all but one (that caused by the revolution of the apsides) terms on the right-hand side of Equation (4.62) will tend to zero with θ_1; and may be too small for observational detection at the present time.

The likeliest explanation of the negative outcome of Plavec's 1960 analysis is the fact that (at least in the systems considered by him) *the secular rectification of the axes of rotation by tidal friction has already progressed far enough to make their inclinations to the orbital plane insensible.* A contrary case would, in turn, imply that we are dealing with a *very young* system, in which secular action as dissipative processes did not have time yet to align the equatorial planes of both components with that of their orbit. In fact, *periodic fluctuations of the times of the minima due to nodal regression constitute the only way in which such a regression could be established from photometric observations;* and the most important periodic term to look for is the first – of amplitude $(P_o/2\pi)\,\tilde{G}\cos I$ and period $2\pi/y_1$. The amplitudes of the fourth and fifth – likewise factored by $\tilde{G}\cos I$ – go to zero with the orbital eccentricity e; and all remaining ones are proportional to $\tilde{G}^2$ or $\tilde{H}$. The amplitude of the first tends, however, to a finite limit even for circular orbits and small values of I; while its period should be several times shorter than that of apsidal advance – i.e., years rather than decades for most typical binaries – and yet long enough to distinguish it from that of orbital revolution without difficulty. Their presence in the ephemeris for the light minima should be the hallmark of a very young system; and its absence, an indication of approaching maturity.

C. RELATION BETWEEN ORBITAL PERIOD AND TIMES OF THE MINIMA

In conclusion of the present section, one retrospective remark of theoretical interest concerning the limits of validity of Equation (4.59) which we have used to relate the orbital period $P(E)$ with the times of the minima $M(E)$ may be added. If, as usual, we define the (instantaneous) period of an eclipsing variable as the difference between the moments of light minima taking place at two consecutive epochs E and $E + 1$, the functions $M(E)$ and $P(E)$ are bound to satisfy the linear *difference* equation

$$M(E+1) - M(E) = P(E) . \tag{4.64}$$

Both $M(E)$ and $P(E)$ so defined are, by their nature *discontinuous* functions of their argument and exist only for *integral values of E*. It is, unfortunately, impossible to utilize Equation (4.65) as it stands for a determination of the instantaneous periods in practice (as the consecutive minima can seldom – if ever – be observed with the requisite precision); instead, it is customary to rely on the *cumulative* effects of any period variation in as long a sequence of the minima as may be available for harmonic analysis. In order that we should be in a position to do so – or even to predict a single value of $M(E)$ at some future date from known values of $M(0)$ and $P(E)$ – a knowledge of the explicit solution of the foregoing difference Equation (4.64) is evidently prerequisite.

The exact solution of (4.64) can, to be sure, be written down at once in the form of a summation

$$M(E) = M(0) + \sum_{j=0}^{E-1} P(j) ; \tag{4.65}$$

but this is not only very unwieldy in use (calling as it does for an algebraic sum of E discrete terms – a very laborious task when E happens to be a large number), but becomes also of no avail if our professed aim is to deduce the form of the function $P(E)$ from the observed sequence of $M(E)$.

In order to investigate the mutual relationship of these two functions, let us extend the intuitive definitions of $M(E)$ and $P(E)$, and replace them by $\overline{M}(\overline{E})$ and $\overline{P}(\overline{E})$ regarded as functions of a *continuous* variable $\overline{E} \geqslant 0$, defined so that $\overline{M}(E) = M(E)$ and $\overline{P}(E) = P(E)$ for each integral value of E. For non-integral values of $\overline{E}$, $\overline{P}(\overline{E})$ may be any continuous function which coincides with $P(E)$ for integral values of the independent variable; and $\overline{M}(\overline{E})$ is then defined as a solution of the generalized difference equation

$$\overline{M}(\overline{E}+1) - \overline{M}(\overline{E}) = \overline{P}(\overline{E}) \tag{4.66}$$

for all $\overline{E} \geqslant 0$. Now let us assume that there exists an analytic solution of (4.66) with a radius of convergence greater than 1 for each $\overline{E} \geqslant 0$; if so, this solution should be expansible in a Taylor series of the form

$$\overline{M}(\overline{E}+1) = \overline{M}(\overline{E}) + \sum_{j=1}^{\infty} \frac{1}{j!} \frac{\mathrm{d}^j \overline{M}}{\mathrm{d}\overline{E}^j} , \tag{4.67}$$

and its insertion in (4.66) will convert the latter *difference* equation into the equivalent *differential* equation

$$\bar{P}(\bar{E}) = (e^{\frac{d}{d\bar{E}}} - 1)\,\bar{M} \tag{4.68}$$

of infinite order.

A particular solution of this differential equation with constant coefficients $(j!)^{-1}$ can be constructed by standard methods to assume the form

$$\bar{M}(\bar{E}) = \bar{M}(0) + \int_0^{\bar{E}} \bar{P}(\bar{E})\,d\bar{E} + \sum_{j=1}^{n} A_j\, e^{m_j \bar{E}} \int_0^{\bar{E}} e^{-m_j \bar{E}}\, \bar{P}(\bar{E})\,d\bar{E}, \tag{4.69}$$

where the constants A_j and m_j are defined as coefficients of the identity

$$\left\{\sum_{j=1}^{n} \frac{D^j}{j!}\right\}^{-1} = \sum_{j=1}^{n} \frac{A_j}{D - m_j}\,, \tag{4.70}$$

and $n \to \infty$.

The general solution of (4.68) can now be represented as the sum of the above particular solution and any (analytic) periodic function of period one. Only one initial condition is given to us, which specifies $M(0)$. In order that this condition be met, all additional periodic functions must vanish for $\bar{E} = 0$, but otherwise remain completely arbitrary. As we are interested only in what happens for integral values E of $\bar{E}$, no generality will clearly be lost if we choose all such functions to be identically zero. If so, however, all bars above E, M, or P may henceforward be dropped; and on evaluation of the A_j's and m_j's from (4.70) by repeated integration per partes Equation (4.69) may be reduced to

$$M(E) = M(0) + \int_0^E P(E)\,dE - \frac{1}{2}\,\{P(E) - P(0)\} + \tag{4.71}$$

$$+ \frac{1}{2!6}\,\{P'(E) - P'(0)\} - \frac{1}{4!30}\,\{P'''(E) - P'''(0)\} + \cdots,$$

which represents the explicit expression for $M(E)$ in terms of $P(E)$, just as the differential equation (4.68) represented the converse relation. A comparison of the foregoing Equation (4.71) with (4.60) reveals that the first two terms on the right-hand side of (4.71) represent the desired solution for $M(E)$ to the order of accuracy to which the difference equation (4.64) may be approximated by the differential equation (4.59) whose integral led to (4.70). The infinite series which follows the leading term on the right-hand side of (4.71) represents then the correction to (4.60) arising from the fact that the period P(E) is defined in terms of $M(E)$ by a difference, and not differential, equation.

We may add that the expansion on the right-hand side of Equation (4.71) can also be obtained, without recourse to (4.69), in the following alternative way. Let us assume that the solution of (4.64) is generally expressible in the form

$$M(E) = M(0) + \int_0^E P(E)\,\mathrm{d}E + \Delta(E)\,, \tag{4.72}$$

where $\Delta(E)$ stands for the error of approximating the solution of the difference equation (4.64) by that of the differential equation (4.59) of first order. By definition, therefore,

$$\Delta(E) = M(E) - M(0) - \int_0^E P(E)\,\mathrm{d}E\,, \tag{4.73}$$

where the difference $M(E) - M(0)$ follows exactly from (4.65) and the integral of $P(E)$ can be approximated with the aid of the well-known Euler-Maclaurin quadrature formula by

$$\int_0^E P(E)\,\mathrm{d}E = \frac{1}{2}P(0) + \sum_{j=1}^{E-1} P(j) + \frac{1}{2}P(E) - \sum_{j=1}^{n} \frac{B_{2j}}{(2j)!}\{P^{(2j-1)}(E) - P^{(2j-1)}(0)\}\,, \tag{4.74}$$

where the B_{2j}'s are the 'Bernoulli numbers', the first few of which are known to be

j	1	2	3	4	5 ...
B_{2j}	$\frac{1}{6}$	$-\frac{1}{30}$	$\frac{1}{42}$	$-\frac{1}{30}$	$\frac{5}{66}\cdots$

Therefore, by use of (4.65) and (4.74) Equation (4.73) assumes the more explicit form

$$\Delta(E) = \tfrac{1}{2}\{P(0) - P(E)\} + \sum_{j=1}^{n} \frac{B_{2j}}{(2j)!}\{P^{(2j-1)}(E) - P^{(2j-1)}(0)\}\,, \tag{4.75}$$

which, on insertion in (4.72) verifies (4.71).

What is the actual significance, in our present problem, of the correction ΔE *which should* be adjoined to the right-hand side of Equation (4.61)? As long as the variable part in $P(E)$ consists solely of periodic terms oscillating in the periods $2\pi P_0/r_j$, the coefficients of the individual terms in the integral (4.61) of $P(E)$ will be of the order of r_j^{-1}; while those of the successive derivatives $P^{(2j-1)}(E)$ in $\Delta(E)$ will be of the order of r_j^{2j-1}. In consequence, for small values of r_j (corresponding to long-periodic oscillations – such as are represented, in our present problem, by Equations (4.62)) the rate of convergence of the expansion on the right-hand side of (4.75) is likely to be so rapid that the retention of the first term alone – as we have done in Equation (4.60) and (4.61) – should guarantee sufficient accuracy. On the other hand, should our aim become to study the effects, on the orbital period, of short-periodic oscillations of the elements i, Ω, e, ω, or A (which were not discussed in the preceding sections) the correction term $\Delta(E)$ may become of a magnitude comparable with that of the integral of $P(E)$; and , if so, would have to be taken into account. The reason why we do not at present propose to study the short-periodic changes in the same detail as the long-periodic ones is merely the practical fact that these changes would be much more difficult to detect by such observations as are available for most eclipsing binary systems.

V-5. Effects of Variable Mass

In all preceding sections of this chapter we have been concerned with perturbations of the orbital elements in close binary systems, caused by mutual deformations of their components, on the assumption that their individual masses $m_{1,2}$ – and, therefore, the total mass $m_1 + m_2$ of the system – remain secularly constant. Such an assumption may indeed be fulfilled (at least, to a high degree of approximation) for some stars (of small masses) at all times; and for all stars at certain stages of their evolutionary course (for example, while they remain on the Main Sequence, or eventually reach their degenerate state). However, there is no room for doubt today that the masses of most stars in the sky cannot be regarded as remaining secularly constant on the nuclear time-scale; but that they undergo changes which may, at times, become relatively rapid – or even cataclysmic.

Such changes may, in principle, entail an increase of the star's mass as well as its loss. The *accretion* of mass by single stars from interstellar medium in which they may happen to be imbedded was in recent decades extensively discussed, in particular by the Cambridge school of cosmologists (Lyttleton, Hoyle, Bondi and Gold) then in the prime of their imagination. However, a general consensus of opinion on this subject tends to converge to a conclusion that – save under very special circumstances (such as very low differential velocity between the interstellar medium and accreting star), which are unlikely to be met except very rarely – accretion does not represent a process of general astrophysical significance.

With a *loss* of mass by the stars it is, however, a very different story. That all stars are bound to lose a small fraction of their mass by radiation has been clear since the early part of this century; and certain quantitative consequences of this fact drawn by members of the preceding generation of Cambridge cosmologists (Eddington, Jeans). More recently – since the pioneer work of Ambartsumian (1952) or Fessenkov (1952) it has gradually been realized that corpuscular radiation ('stellar wind') can remove much more mass from the reach of the star's gravitational field than nimble-footed photons alone could ever accomplish – especially during the earliest (Hayashi) stage of the star's life; or at later evolutionary stages, when stellar wind can temporarily develop the strength of a hurricane (capable of removing 10^{-6} or more of the Sun's mass from the star's reach per annum). Even more effective for mass removal are the temporary (though recurrent) losses of stability of the star's outer layers which can blow of off a part of their mass (10^{-4} to 10^{-3} $\odot$ per outburst) into space with velocities equal to those of the stellar winds (Novae) – not to speak of virtually complete dispersal of mass accompanying the suicides of stellar exhibitionists which end their lives as Supernovae.

All this is, of course, true of the stars in general – be these single or multiple. However, facts will be discussed in Chapter VIII indicating that particular types of stellar winds, or instabilities leading to mass loss, are stimulated by the gravitational dipole of a close binary – especially at advanced stages of their life, when one (or both) component(s) has evolved away from the Main Sequence. It is, in fact, virtually impossible for the components of close binary systems to retain their initial mass too long (on the nuclear time-scale) after the Main-Sequence stage has been passed; nor is it at all likely that their initial mass-ratios will remain unaltered in the course of the evolution. These facts should provide a sufficient motivation for us to explore the dynamical consequences of such phenomena on the structure and evolution of close binary systems. In the present section we shall confine our attention to the effects which a mass change in the masses m_1 and (or) m_2 of their components will exert of the elements A, e, ω of their plane orbits; while the evolutionary consequences of such changes will be considered later in Chapter VIII.

A. GENERALIZED EQUATIONS OF MOTION

In order to do so, let us specify first the exact form of the equations of motion appropriate for binary systems with variable mass. These are based on Newton's second law, making the rate of change of the momentum equal to the sum of all forces **F** (due to self-attraction, or any external field) acting on the system: i.e.,

$$\frac{\mathrm{d}}{\mathrm{d}t}(m\mathbf{V}) = \mathbf{F}\,, \tag{5.1}$$

where m stands for a sum of the masses of all particles constituting our body, and **V** is the (vector) velocity of its centre of mass. Equation (5.1) as it stands holds, however, good only for the assembly of *all* mass particles which were originally present. If, however, the mass of our body is variable, this implies that some of these particles will cease to belong to that body (or others may be accreted) in the course of time; while our aim is to study the motion of the centre of mass of the *remaining* particles, as specified by the vector velocity **V**. This however, cannot be described uniquely only by the derivative $\mathrm{d}m/\mathrm{d}t$ representing the rate of change of the mass of our body as a whole; because this rate is a scalar quantity and cannot, therefore, tell us anything about the momentum of the escaping (or incident) mass. In such a case, the equation safeguarding the conservation of the momentum appropriate for dynamical systems with variable mass assumes the form

$$m(t)\frac{\mathrm{d}V}{\mathrm{d}t} = \mathbf{F} + \mathbf{u}\frac{\mathrm{d}m}{\mathrm{d}t}\,, \tag{5.2}$$

where **u** represents the vector velocity of escaping (or incident) matter with respect to the centre of mass of our body.

Equation (5.2) was first obtained by Meščerskii (1893, 1902), in connection with his studies of motions of the rockets. Early astronomical applications were, however, concerned with the case of *isotropic* mass loss – i.e., ejection of matter in all directions which is lost to the system. If so, then

$$\sum \mathbf{u}\frac{\mathrm{d}m}{\mathrm{d}t} = 0 \tag{5.3}$$

for all particles of our body at all times; and Equation (5.3) reduces to

$$m(t)\,\frac{\mathrm{d}\mathbf{V}}{\mathrm{d}t} = \mathbf{F}\,, \tag{5.4}$$

a form considered somewhat earlier by Gyldén (1884).

If, on the other hand, the body in question were to accrete matter from a stationary cloud, then

$$\mathbf{u} = -\mathbf{V} \tag{5.5}$$

and Equation (5.2) would reduce to

$$\frac{\mathrm{d}}{\mathrm{d}t}\,\{m(t)\,\mathbf{V}\} = \mathbf{F}\,, \tag{5.6}$$

an equation connected in the literature with the name of Levi-Civita (1906).

It was Armellini (1953) who pointed out that the equations of Gyldén and Levi-Civita represent particular cases of Meščerskii's Equation (5.2); holding good for the case of isotropic mass decrease and increase, respectively. The two equations are, however, subject to a more general interpretation: namely, Levi-Civita's equation applies whenever the moments of incident (or escaping) mass is zero with respect to an inertial frame of reference; while Gyldén's equation holds good in any case when the momentum of such mass is equal to that which this mass would have were it attached to the moving body.

After these preliminary remarks, let us consider a dynamical system consisting of two bodies of masses $m_{1,2}$ which revolve around their common centre of gravity. If these masses were constant, their absolute orbits would be two similar conics, with foci at their common centre of mass. If, however, one (or both) lose mass isotropically, then (but only then) it is legitimate to set

$$\frac{\mathrm{d}\mathbf{V}}{\mathrm{d}t} \equiv \frac{\mathrm{d}^2\mathbf{r}}{\mathrm{d}t^2}\,, \tag{5.7}$$

where $\mathbf{r}$ stands for the radius-vector measured from the centre of mass of the respective body.

Since, moreover, the only external force $\mathbf{F}$ acting on each component of our binary system is the gravitational attraction of its mate, Gyldén's Equation (5.4) can be written out as

$$m_1\,\frac{\mathrm{d}^2\mathbf{r}_1}{\mathrm{d}t^2} = \frac{Gm_1m_2}{|\mathbf{r}_1 - \mathbf{r}_2|^3}\,(\mathbf{r}_2 - \mathbf{r}_1)\,, \tag{5.8}$$

and

$$m_2\,\frac{\mathrm{d}^2\mathbf{r}_2}{\mathrm{d}t^2} = -\,\frac{Gm_1m_2}{|\mathbf{r}_1 - \mathbf{r}_2|^3}\,(\mathbf{r}_2 - \mathbf{r}_1)\,, \tag{5.9}$$

where $\mathbf{r}_{1,2}$ are the radii-vectors of the masses $m_{1,2}$ from the centre of gravity of the system. Subtracting (5.8) from (5.9) we find that

$$\frac{\mathrm{d}^2\mathbf{r}}{\mathrm{d}t^2} = -\,\frac{Gm}{r^3}\,\mathbf{r}\,, \tag{5.10}$$

where $\mathbf{r} \equiv \mathbf{r}_2 - \mathbf{r}_1$ is the radius vector of the relative orbit of the two stars, and $m \equiv m_1 +$

m_2 stands for the total mass of the system which may (but need not) be a function of the time.

The foregoing Equation (5.10) can be rewritten as

$$\frac{\mathrm{d}}{\mathrm{d}t}\left(\mathbf{r}\times\frac{\mathrm{d}\mathbf{r}}{\mathrm{d}t}\right) = 0, \tag{5.11}$$

disclosing the orbits to be planar, and the areal velocity of motion constant. The orbits would, in fact, be Keplerian ellipses; with all three Kepler's laws continuing to hold good *without requiring the constancy of mass.* In point of fact, *the evolution of a binary system in the case of isotropic mass changes depends only on the total mass* m *of the system;* the effect of an isotropic decrease of mass is only to diminish the attractive force between its components.

Next, let us consider a case in which the components of a binary system *acquire* matter, which can be regarded as *stationary* with respect to our adopted system of coordinates. If so, Levi-Civita's Equation (5.6) can be shown to imply that, for the relative motion of our binary pair,

$$\frac{\mathrm{d}^2\mathbf{r}}{\mathrm{d}t^2} = -\frac{G(m_1+m_2)}{r^3}\mathbf{r} + \left\{\frac{\dot{m}_1}{m_1}\mathbf{V}_1 - \frac{\dot{m}_2}{m_2}\mathbf{V}_2\right\}, \tag{5.12}$$

where the dot signifies differentiation with respect to the time.

We may recall that, in the present case, the momentum of the mass falling on the two stars is equal to zero. Consequently, the total momentum of the system is conserved, and

$$m_1\mathbf{V}_1 + m_2\mathbf{V}_2 = \mathbf{C}, \tag{5.13}$$

where **C** is a constant vector. If we express now the second term on the right-hand side of Equation (5.12) in terms of the vectors $\mathbf{V}_2 - \mathbf{V}_1 \equiv \mathbf{V}$ and **C**, Equation (5.12) can be rewritten as

$$\frac{\mathrm{d}^2\mathbf{r}}{\mathrm{d}t^2} = -\frac{Gm}{r^3}\mathbf{r} - \frac{\mathbf{C}}{m}\left(\frac{\dot{m}_1}{m_1} - \frac{\dot{m}_2}{m_2}\right) - \frac{\mathbf{V}}{m}\left(\frac{m_2\dot{m}_1}{m_1} + \frac{m_1\dot{m}_2}{m_2}\right), \tag{5.14}$$

where, as before, $m \equiv m_1 + m_2$. The foregoing Equation (5.14) is identical with the one previously deduced by MacMillan (1919) or Markowitz (1933) in different ways. The structure of its right-hand side discloses the sensitivity of our problem to the way in which the mass is acquired by the system (i.e., the specific form of the time derivatives $m_{1,2}$).

Orbital trajectories described by the components of binary systems accreting (or losing) the mass isotropically can be constructed by an integration (analytical or numerical) of Equations (5.12) or (5.14) for adopted laws of dependence of the mass on the time. Barring exceptional cases in which the loss may become rapid (as, for example, if one component of our binary becomes a Nova – let alone a Supernova), the actual orbits are not expected to differ considerably from the Keplerian ellipses. If so, however, the possibility suggests itself again to consider the actual orbits – perturbed by the requisite mass loss – as osculating ellipses in the plane, the elements A, e and ω of which are functions of the time.

The equations governing their changes then continue to be of the form (1.10)–(1.15) or (1.24)–(1.29), in which the perturbing function $\mathfrak{R}_{12}$ must now be related to the requisite change in mass. It has been shown by Hadjidemetriou (1963) that the *perturbations arising from an isotropic variation of the mass of the system are derived from the function*

$$\mathfrak{R}_{12} = -\frac{1}{2}\frac{\mathrm{d}}{\mathrm{d}t}(\log m)\,\mathrm{V}\,, \tag{5.15}$$

equivalent to a force which propels the body in the direction of its orbit if m is negative. This force (per unit mass) is proportional to the velocity of motion, and to the logarithmic derivative of the combined mass $m \equiv m_1 + m_2$.

B. ISOTROPIC MASS LOSS

Before making use of (5.15) in (1.10)–(1.15) or (1.24) – (1.29) we should remember that it represents a force *tangential* to the orbit, which is related with the plane force components **R** and **S** introduced in section V-1 by the equations

$$\mathbf{R} = \frac{e \sin \upsilon}{\sqrt{1 + 2e\cos\upsilon + e^2}}\mathbf{T} - \frac{1 + e\cos\upsilon}{\sqrt{1 + 2e\cos\upsilon + e^2}}\mathbf{N} \tag{5.16}$$

and

$$\mathbf{S} = \frac{1 + e\cos\upsilon}{\sqrt{1 + 2e\cos\upsilon + e^2}}\mathbf{T} + \frac{e\sin\upsilon}{\sqrt{1 + 2e\cos\upsilon + e^2}}\mathbf{N}\,, \tag{5.17}$$

where the tangential component **T** is measured positively in the direction of motion; and the **N** component, normal to the direction of orbital motion, is taken positively inwards.

If the perturbing function (5.15) exerts its force in the direction of motion, the normal component $\mathbf{N} = 0$ in (5.16) and (5.17); and if so, Equations (1.15)–(1.24) for the perturbations of the elements of a plane orbit can be rewritten as

$$\frac{\mathrm{d}A}{\mathrm{d}t} = \frac{2}{n}\sqrt{\frac{1 + 2e\cos\upsilon + e^2}{1 - e^2}}\,\mathbf{T}, \tag{5.18}$$

$$\frac{\mathrm{d}e}{\mathrm{d}t} = \frac{2\sqrt{1 - e^2}\,(e + \cos\upsilon)}{nA\sqrt{1 + 2e\cos\nu + e^2}}\mathbf{T}\,, \tag{5.19}$$

$$\frac{\mathrm{d}\omega}{\mathrm{d}t} = \frac{2\sqrt{1 - e^2}\,\sin\upsilon}{nAe\sqrt{1 + 2e\cos\upsilon + e^2}}\mathbf{T}\,, \tag{5.20}$$

where $n^2 = Gm/A^3$, and

$$\mathbf{T} = -\frac{1}{2}V\frac{\dot{m}}{m}\,. \tag{5.21}$$

Since, moreover, for the elliptic motion

$$V^2 = Gm\left(\frac{2}{r} - \frac{1}{A}\right) = \frac{Gm}{A}\,\frac{1 + 2e\cos\upsilon + e^2}{1 - e^2}\,, \tag{5.22}$$

the foregoing Equations (5.18)–(5.22) can be re-stated in the form

$$\frac{\mathrm{d}A}{\mathrm{d}t} = -A^2\left\{\frac{2}{r} - \frac{1}{A}\right\}\frac{\dot{m}}{m}\,, \tag{5.23}$$

$$\frac{\mathrm{d}e}{\mathrm{d}t} = -(e + \cos\upsilon)\,\frac{\dot{m}}{m}\,, \tag{5.24}$$

$$\frac{\mathrm{d}\omega}{\mathrm{d}t} = -\frac{\sin\upsilon}{e}\,\frac{\dot{m}}{m}\,, \tag{5.25}$$

where the true anomaly υ can be obtained from the equation

$$\frac{\mathrm{d}\upsilon}{\mathrm{d}t} = \frac{nA^2\sqrt{1 - e^2}}{r^2} + \frac{\sin\upsilon}{e}\,\frac{\dot{m}}{m} \tag{5.26}$$

representing a generalization of Equation (4.14) for the case of variable mass.

The foregoing Equations (5.23)–(5.26) have first been derived by Hadjidemetriou (1963). They hold good, to be sure, only for isotropic loss of mass specified by Equation (5.15); but under these conditions certain of their integrals can be obtained at once. Thus Equations (5.23) and (5.24) can be combined to yield

$$m(1 - e^2)\,\frac{\mathrm{d}A}{\mathrm{d}t} - 2mAe\,\frac{\mathrm{d}e}{\mathrm{d}t} + A(1 - e^2)\,\frac{\mathrm{d}m}{\mathrm{d}t} = 0 \tag{5.27}$$

or

$$\frac{\mathrm{d}}{\mathrm{d}t}\,[mA(1 - e^2)] = 0\,, \tag{5.28}$$

which integrates into

$$mA(1 - e^2) = \text{constant}\,. \tag{5.29}$$

Other integrals of the system (5.23)–(5.25) must be sought either analytically in terms of appropriate expansions, or numerically. Extensive work in this field has been carried out by Hadjidemetriou (1963, 1966a, b) in several investigations to which the reader is referred for fuller details. For the adopted law of the loss of mass, of the three elements of the plane orbits, only the semi-major axis A exhibits secular increase with diminishing mass, in accordance with Equation (5.29). The orbital eccentricity e and longitude of periastron ω do not exhibit secular, but periodic perturbations of complicated nature.

What kind of physical processes can cause the stars to lose mass in an isotropic manner? The first and most obvious one is the loss of mass by *radiation.* For let the radiant energy output of a star be described by its luminosity L in erg per second. The mass lost in this way will evidently be given by the equation

$$\dot{m} = -\frac{L}{c^2}, \tag{5.30}$$

where c is the velocity of light; and if the latter is expressed in cm per second, the loss of mass $\dot{m}$ a given by Equation (5.30) will be in gram per second.

On the other hand, the luminosity L of a star is known to be a sensitive function of its mass m; and connected with it by a 'mass-luminosity' relation which can be empirically approximated (cf. sec. VIII-1) in the form

$$L = \alpha m^n, \tag{5.31}$$

where α and n are sensibly constant for stars not greatly different in structure (for the Main-Sequence stars, the empirical value of this exponent n lies between 3 and 4). A combination of (5.30) and (5.31) then yields for the logarithmic time-derivative of the mass loss the expression

$$\frac{\dot{m}}{m} = -\frac{\alpha}{c^2} m^{n-1}, \tag{5.32}$$

which can be used as such in Equations (5.23)–(5.26).

Equation (5.32) has been at the basis of the early cosmological work by Eddington (1926) or Jeans (1924); and was employed for investigations of the secular evolution of binary systems by Jeans (1925) or Smart (1925). The rate at which mass is being lost by stars through radiation is, however, so small as to add up to less than one per cent of the original mass over the entire nuclear time-scale. Corpuscular radiation can be very much more effective in removing matter from the gravitational reach of the star. Since, however, such matter escapes, in general, in the form of hot plasma, its emission would clearly cease to be isotropic in the presence of any magnetic dipole (let alone of a field of more complicated structure).

When, however, we come to consider the conditions obtaining in a close binary system, we face a still more complex situation. For even in the absence of any general magnetic field, the radiant flux emerging normally to the surface of each component distorted by rotation and tides will not be isotropic, but (for stars in radiative equilibrium) is bound to vary with direction in proportion to the local gravity (cf. sec. IV-4). The same would, moreover, be the case with corpuscular emission from any part of a distorted surface, whose 'ballistics' would be influenced by the combined field of the gravitational (or magnetic) dipole. And even if the velocity of ejection were much higher than that of escape from the gravitational field of the system (so that the trajectories of escape would deviate but little from straight lines), the secondary component would intercept a part of the corpuscular (or, for that matter, radiative) emission of its mate; and thus experience a 'radiation pressure'. The perturbation represented by it is given by the second term on the right-hand side of Equation (5.2), but the summation must be replaced by a double integral over that part of the surface of the star which is exposed to the radiation emitted by the other star.

In order to evaluate this integral, consider first the radiation contained between two co-axial cones, with angles φ and $\varphi + d\varphi$, with the vertex at the centre of the emitting star. If this radiation were isotropic, its contribution to the sum $\Sigma\, m\mathbf{u}$ would be a vector in the direction of $\mathbf{r}$, of the form $mu \cos \varphi$. The mass emitted by the star per unit solid angle will be $m/4\pi$ per second; so that, within the cone of semi-apex angle φ,

$$\Sigma\, \dot{m}\, \mathbf{u} = \{(mc/4) \cos \varphi \, \mathrm{d}\omega\} \, (\mathbf{r}/r) \,, \tag{5.33}$$

where the element of the solid angle

$$\mathrm{d}\omega = 2\pi \sin \varphi \, \mathrm{d}\varphi \,. \tag{5.34}$$

Inserting (5.34) in (5.33) and integrating the latter between the limits 0 and $\varphi_1 = \sin^{-1}(r_1/r)$, where r_1 denotes the radius of the emitting star, we find that the impulse I produced by the impact of radiation will be given by

$$I = \frac{\dot{m}c}{4} \left(\frac{r_1}{r} \right)^2 ; \tag{5.35}$$

and, accordingly, the equations of motion (5.8)–(5.9) of the two components will become

$$m_1 \frac{\mathrm{d}^2 \mathbf{r}_1}{\mathrm{d}t^2} = \frac{G m_1 m_2}{r^3} \mathbf{r} + \frac{\dot{m}_2 c r_1^2}{4 r^3} \mathbf{r} \,, \tag{5.36}$$

$$m_2 \frac{\mathrm{d}^2 \mathbf{r}_2}{\mathrm{d}t^2} = - \frac{G m_1 m_2}{r^3} \mathbf{r} - \frac{\dot{m}_1 c r_2^2}{4 r^3} \mathbf{r} \,, \tag{5.37}$$

from which the equation of the relative motion of two stars illuminating each other will again be of the form

$$\frac{\mathrm{d}^2 \mathbf{r}}{\mathrm{d}t^2} = - \frac{GM}{r^3} \mathbf{r} \,, \tag{5.38}$$

identical with (5.10), except that now

$$M = m_1 + m_2 + \frac{c}{4G} \left\{ \frac{\dot{m}_1 r_2^2}{m_2} + \frac{\dot{m}_2 r_1^2}{m_1} \right\} , \tag{5.39}$$

where $\dot{m}_{1,2}$ continues to be given by (5.32), and which for stars losing mass ($m_{1,2}$ negative) renders M less than the sum $m_1 + m_2$ – result which could alternatively be interpreted as a corresponding change in the constant G of gravitation.

This could indeed have been anticipated; for since the intensity of radiation is known to diminish with inverse square of the distance, the force exerted by the radiation pressure varies in exactly the same way as gravitational attraction – though with one important proviso: namely, that all radiation of the star of mass m_1 intercepted by its mate of mass m_2 and radius r_2 is contained within a cone of semi-apex angle $\sin^{-1} r_2$. This will, however, be the case only if (a) the illuminating component can be treated as a light point; and (b) the illuminated star can be regarded as a sphere.

Now neither of these assumptions can be rigorously fulfilled in close binary systems, whose components are distorted by rotation and tides, and whose fractional radii are too large to regard them as point sources of light. In point of fact, the illuminating radiation is not contained only within a cone with a vertex at the centre of the illuminating component which is tangent to the surface of the illuminated star, but *within a cone tangent to both stars.* In other words, radiation from star 1 will be incident (though in diminished amount) also on those parts of the surface of star 2 at which the apparent disc of star 1 is already partly set below the horizon.

The contribution of these 'penumbral' zones of star 2 to the impulse I can be ascertained from known results for the 'reflection effect' (cf., Kopal, 1959, sec. IV.6) at 'full' phase; and will contribute terms varying as inverse powers of the radius-vector higher than the second. These will, in turn, produce perturbations causing the relative orbits of the two stars to depart from the Keplerian ellipses; up to the present time these have not yet been investigated.

C. NON-ISOTROPIC MASS LOSS

So far in this section we have assumed the loss of mass suffered by either (or both) component in a close binary to be isotropic. In its simplest form – when the matter lost is ejected so rapidly as not to affect the orbital motion of the two stars after ejection – the problem was essentially solved by Jeans (1928). A loss of mass by radiation, considered in the preceding part of this section, constitutes the best example of such a process; and another may be violent ejection of mass in the course of Nova of Supernova outbursts (cf., e.g., Hadjidemetriou, 1966b); or its permanent outflow exhibited by the Wolf-Rayet stars – all with velocities of the order of 1000–5000 km s^{-1}. If, however, this velocity becomes of the order of 10–100 km s^{-1}, the ejected matter will not only affect the orbital motion of the binary star by its gravitational attraction, but may eventually be captured by either component to cause further complications.

Additional complications arise, moreover, if the process of ejection proves to be such as to render the corresponding mass loss *anisotropic.* This more general problem of non-isotropic ejection with a finite initial velocity from a star of finite dimensions has so far defied all attempts at its general solution; and consists of three separate aspects:

(1) the ejection of matter from the star;

(2) the orbital perturbations of the motion of the binary pair caused by moving matter which has separated itself from the two components; and

(3) the accretion (if any) of this matter by any one of these stars – be it a re-acquisition of mass by the component which previously lost it, or accretion by its mate.

The point (1) concerns essentially a study of the physical processes by which a star can eject mass from a given region of its surface with a given velocity – a process originating in the interior of the star in question, whose aspects may be complicated, in close binary systems, by the presence of a companion. Whathever this process may be, however, a loss (or acquisition) of a finite mass δm_1 by (say) the primary component must result in a shift in position of the centre of gravity of the mass m_1, as well as a corresponding change in the elements of its orbit.

Let us, in what follows, consider (with Piotrowski, 1964a, b) the ejection of matter to take place in the *plane* of the orbit, the period P of which is given by

$$P^2 = \frac{4\pi^2 A^3}{Gm}, \tag{5.40}$$

where, as before $m \equiv m_1 + m_2$ stands for the total mass of the system; and

$$p \equiv e \sin v, \qquad q \equiv e \cos v, \tag{5.41}$$

where e denotes, as before, the orbital eccentricity; and $v \equiv u - \omega$, the true anomaly measured from the longitude ω of periastron passage.

If so, it has been shown by Piotrowski (1964a,b) that the changes δP, δp and δq of the respective elements, caused by an accretion by m_1 of mass δm_1 will (correctly to terms of the first order in δm_1) be given by the equations

$$\frac{\delta P}{P} = \frac{\delta m_1}{m} - 3\frac{A}{r}\left\{1 - \left(\frac{h^2}{Gmr} + \frac{Gmrp^2}{h^2}\right)\frac{m_2}{m} + \right.$$
$$\left. + \left(1 + \frac{h^2}{Gmr}\right)\frac{x_1}{r} - \frac{py_1}{r} + \frac{pru_1}{h} + \frac{hv_1}{Gm}\right\}\frac{\delta m_1}{m_1}, \tag{5.42}$$

$$\delta p = -\left\{\left(1 + 2\frac{m_2}{m}\right) p + 2\frac{px_1}{r} - Gm\frac{p^2 y_1}{h^2} + \right.$$
$$\left. + \frac{hu_1}{Gm} + \frac{prv_1}{h}\right\}\frac{\delta m_1}{m_1}, \tag{5.43}$$

and

$$\delta q = -\left\{\left(1 + 2\frac{m_2}{m}\right)\frac{h^2}{Gmr} + 3\frac{h^2 x_1}{Gmr^2} - \right.$$
$$\left. - 2\frac{py_2}{r} + 2\frac{hu_1}{Gm}\right\}\frac{\delta m_1}{m_1}, \tag{5.44}$$

where r denotes the radius-vector of the relative orbit;

$$h \equiv \sqrt{GmA(1 - e^2)} \tag{5.45}$$

is the angular momentum of orbital motion per unit mass; and x_1, y_1 are the rectangular coordinates of the escaping mass δm_1 at the moment of ejection, while $u_1 \equiv \dot{x}_1$ and $v_1 \equiv \dot{y}_1$ represent the velocity components with which δm_1 has been ejected. The values of x_1, y_1 in the orbital plane are referred to the centre of mass of the binary system as the origin of coordinates; the x-axis being taken positively in the direction of the mass m_2.

The foregoing Equations (5.42)–(5.44) as they stand apply to the ejection ($\delta m_1 < 0$) of mass by the respective component. Should a mass be incident upon it $\delta m_1 > 0$. The same Equations (5.42)–(5.44) can also be used to ascertain the effects of a mass infall on the secondary component if the subscripts 1, 2 are interchanged, $\delta m_2 = -\delta m_1$, and the algebraic signs of x_2, y_2 and u_2, v_2 are altered from plus to minus or vice versa.

The changes δA, δe and $\delta\omega$ in the respective elements of plane orbit, caused by a loss of mass δm_1 occurring at a point of coordinates x_1, y_1 with the velocities u_1 and v_1, can be deduced from Equations (5.42)–(5.44) by a differentiation of the equations $A^3 = Gm(P/2\pi)^2$, $e^2 = p^2 + q^2$ and $\tan \upsilon = p/q$, given previously by Equations (5.40) and (5.41).

When we turn now to the task (2) concerning the influence of ejected matter on the motion of the components after ejection, the problem becomes one of the theory of perturbations, as discussed in the earlier parts of this section. If, however, we are interested only in the variations of A, e and P in a plane elliptic orbit, the problem can be reduced to the following form. Let

$$E \equiv \frac{Gm}{2A} \tag{5.46}$$

represent the mechanical energy per unit mass of the binary system; while h as given by (5.45) continues to stand for the angular momentum of its orbit. If so, it follows from the theory of Keplerian motion that the orbital elements A, e and P can be expressed in terms of E, h and m by the identities

$$A = \frac{Gm}{2E}, \tag{5.47}$$

$$e^2 = 1 + \frac{2\,Eh^2}{(Gm)^2} \tag{5.48}$$

and

$$P^2 = \frac{(\pi Gm)^2}{2E^3}. \tag{5.49}$$

A. logarithmic differentiation of the foregoing identities discloses that

$$\frac{\delta A}{A} = \frac{\delta m}{m} - \frac{\delta E}{E}, \tag{5.50}$$

$$\frac{\delta e^2}{1-e^2} = 2\,\frac{\delta m}{m} - \frac{\delta E}{E} - 2\,\frac{\delta h}{h}, \tag{5.51}$$

and

$$\frac{\delta P}{P} = \frac{\delta m}{m} - \frac{3}{2}\,\frac{\delta E}{E}. \tag{5.52}$$

During the time when the mass δm is in transit, the total mass m of the system does not change. It follows, moreover, from the principle of the conservation of energy and angular momentum that the changes δE and δh of the particle of mass δm_1 must appear in the expressions for the mechanical energy and momentum of the masses $m_{1,2}$ with opposite algebraic signs. The amounts of δE and δh for any particular case can be computed (cf. Kruszewski, 1966) from the known positions and velocities of the particles at the moments of their ejection and accretion from the equations

$$\frac{\delta E}{E} = -2A\left\{\frac{h^2x_1}{Gmr^3} - \frac{py_1}{r^2} + \frac{pu_1}{h} + \frac{hv_1}{Gmr} + \right. \tag{5.53}$$

$$\left. + \frac{m_1}{mr} + \frac{x_1}{r^2}\right\}\frac{\delta m}{m_1}$$

and

$$\frac{\delta h}{h} = \left\{2\frac{x_1}{r} + \frac{rv_1}{h} - \frac{Gmpy_1}{h^2}\right\}\frac{\delta m}{m_1} ; \tag{5.54}$$

and their knowledge permits us to specify the corresponding changes of $\delta A, \delta P$ and δe.

In particular, in the case of a circular orbit ($e = 0$), the addition of period variations caused by ejection, free flight and accretion leads (cf. Piotrowski, 1967) to the equations disclosing that

$$\frac{\delta P}{P} = \frac{3}{\mu(1-\mu)}\{(x_1+\mu)^2 + y_1^2 + (x_1+\mu)v_1 -$$

$$-y_1u_1 - (x_2+\mu-1)^2 - y_2^2 - (x_2+\mu-1)v_2 +$$

$$+ 2\mu - 1\}\frac{|\delta m_1|}{m} \tag{5.55}$$

when the mass δm_1 is transferred from m_1 to m_2. If, on the other hand, the ejected matter returns to m_1,

$$\frac{\delta P}{P} = \frac{3}{\mu(1-\mu)}\{(x_1+\mu)^2 + y_1^2 + (x_1+\mu)v_1 - y_1u_1 - (x_1'+\mu)^2 -$$

$$-y_1'^2 - (x_1'+\mu)v_1' + y_1'u_1'\}\frac{|\delta m_1|}{m}, \tag{5.56}$$

where (cf. sec. VI-1) the units have been chosen so that $A = Gm = 1$, and where we have abbreviated

$$\mu \equiv \frac{m_2}{m_1+m_2} . \tag{5.57}$$

Moreover, the primed coordinates x_1', y_1' and velocities u_1', v_1' in (5.56) refer to the position of the infall point, and velocity of impact, at which the particle of mass δm_1 returned to m_1.

Should we, lastly, be willing to regard the two components as mass-points, it would follow that

$$x_1 = -\mu, \qquad x_2 = 1-\mu ; \qquad y_1 = 0, \qquad y_2 = 0 ; \tag{5.58}$$

in which case Equation (5.55) would reduce further to

$$\frac{dP}{P} = \frac{3(2\mu-1)}{\mu(1-\mu)}\frac{|\delta m_1|}{m} = 3\left(1-\frac{m_1}{m_2}\right)\frac{|\delta m_1|}{m_1}, \tag{5.59}$$

an equation used by Kuiper (1941a) in his investigation of the system of β Lyrae, and subsequently re-derived in different forms by other investigators.

From Equation (5.59) it should follow that when $m_1 > m_2$ – i.e., when mass is transferred from the more massive component to the less massive one – the period of their orbit should decrease; while if the star losing mass is the less massive of the two, the converse should be true. Moreover, if in the course of this process one star merely acquires the mass lost by its mate, $\delta m = 0$. If so, however, Equation (5.50) makes it evident that what is true of the orbital period P should hold good also for their separation A : if $m_1 > m_2$, any mass transfer for which the sum $m_1 + m_2$ remains constant should *diminish* the distance between the components, and vice versa.

The probability of these predictions being fulfilled should, however, not be overestimated. First, their validity requires that $\delta m = 0$ – i.e., that in the process of transfer no mass is lost to the system. Secondly, the considerations which led to (5.55) were based on the assumed conservation of the angular momentum of orbital motion of the two componets, and took no account of the angular momenta stored in their axial rotation. This could be true if either the components could be legitimately treated as mass-points (in which case the angular momenta of axial rotation would be identically zero); or, for components of finite size, if the material constituting them were inviscid (which would preclude any momentum transfer between axial rotation and orbital motion).

In actual fact, inviscid fluid represents but a matematical abstraction; and in section III-4 we saw already that the viscosity of stellar material, far from negligible even for stars on the Main Sequence, can in more advanced evolutionary stages become very large. The implications of this fact for the tidal evolution of close binary systems will be taken up in fuller detail in section VIII-4 of the last chapter of this book. The problem becomes, however, even more involved when the mass of ejected particles constituting the mass $\delta m_{1,2}$ becomes such that collisions between them can no longer be ignored, and their ensemble begings to behave as a continuum. Our problem will then transcend the scope of rational mechanics, and becomes one of *hydrodynamics.* For, in reality, components of a close binary at the mass-loss stage of its evolution no longer revolve in empty space, but within a continuous fluid, generated partly (or wholly) by an escape of mass from the stars, which offers resistance to their motion and interacts with them in the framework of a single dynamical system. An outline of the problems confronting us in this connection will be given in section VIII-3 of this book; but their solution is still far in the future.

V-6. Perturbations by a Third Body

So far in this chapter we have been concerned with perturbations in close binary systems arising from the deformation of both components due to their axial rotation and mutual tidal action. Such perturbations are bound to be present in binaries consisting of deformable bodies, and their magnitude is governed by the degree of proximity of the components. Our treatment of the dynamics of close binary systems in this chapter would, however, be incomplete without a similar investigation of a wholly different class of perturbations which *may* be present if such a binary is attended by a *third body.*

Triple stellar systems in which a close pair is attended by a more distant companion are quite frequent in the sky – some ten per cent of all known binaries seem to posses such companions – and some are known to be even more complicated. Moreover, in a large majority of triple systems, the distance of the third component from the centre of gravity of the system is many times as large as the separation of the close pair; and its orbital period correspondingly longer.* Under such conditions, the third body will continue to move around the common centre of gravity in an elliptical orbit which is essentially undisturbed by the binary nature of its companion; and, besides, such perturbations as it may suffer from this cause are of little interest as they could not be observed. Furthermore, in typical systems of this nature, the angles of inclination between the orbital planes of the close and wide pairs are quite arbitrary, and may lie anywhere between 0 and 180°; and while the orbital eccentricity of the close pair will as a rule be small (generally less than 0.1), that of the wide orbit may be very much larger.

A. EQUATIONS OF THE PROBLEM

In order to set up the corresponding equations of motion, let $m_{1,2}$ denote the masses of the components of the close pair and m_3, that of the third body. Let, moreover, the plane of the wide orbit be adopted as one of reference; and the motion of the third body be referred to the centre of mass of the close pair as the origin of coordinates. In following the motion of the close pair, let that of the secondary component of mass m_2 be referred to the centre of mass m_1 (which will render the perturbations exerted by m_3 on the close pair a minimum).

Let, furthermore, the radii-vectors r, r' of the close and wide orbit be represented by

$$r = \frac{A(1-e^2)}{1+e\cos v} \tag{6.1}$$

$$r' = \frac{A'(1-e'^2)}{1+e'\cos v'}, \tag{6.2}$$

respectively, where A, A' denote the semi-major axes of the respective orbits of eccentricities e, e'; and v, v', the true anomalies of m_2 and m_3 measured from the longitudes ω,ω' of the pericenters (i.e., angles between the longitude Ω of the node and the position of the apsidal line of each orbit). The longitude of the node Ω is, as usual, defined as that of the line of intersection of the two orbital planes from a fixed direction in space. The angle of inclination between the two orbits will hereafter be denoted by i; while, lastly, $u = \omega + v$ or $u' = \omega' + v'$ will denote the longitudes (true anomalies) of m_2 and m_3 measured from Ω (rather than ω or ω') in the planes of their respective orbits.

* The most extreme well-known case where this is not so is represented by the eclipsing system λ Tauri, with the periods of the binary and triple orbit being 3.953 and 30.0 days, respectively – the latter being only less than 8 times as long as the former.

Consistent with our initial assumption that the orbit of m_2 around m_1 does not perturb the motion of m_3, its motion in our plane of reference will hereafter be regarded as a Keplerian ellipse characterized by *constant* values of the elements A', e' and ω' (though not ω'). The orbital elements A, e, ω, Ω, and i of the disturbed motion of the close pair may, however, be functions of the time, governed by a well-known set of six first-order variational equations; and our task will be to solve them.

In order to do so for unrestricted values of e or i, we find it expedient to change over from the elements of the orbit as dependent variables to a new set of areal velocities L, G, H, and the corresponding angles ℓ, g, h, introduced first by Delaunay (1867) in his celebrated theory of the motion of the Moon, which are related with the orbital elements by means of the equations*

$$\left.\begin{aligned} L &= nA^2, & \ell &= n(t-T),\\ G &= nA^2\sqrt{1-e^2}, & g &= \omega,\\ H &= nA^2\sqrt{1-e^2}\cos i; & h &= \Omega; \end{aligned}\right\} \tag{6.3}$$

where n denotes the mean daily motion of the close pair; and T, the time of the periastron passage of m_2 in its orbit; so that ℓ denotes, in effect, the mean anomaly of m_2 in its orbit.

As is well known, the advantage of the use of the Delaunay variables in place of the orbital elements themselves rests on the fact that their variational equations assume the *canonical form*

$$\left.\begin{aligned} \frac{dL}{dt} &= \frac{\partial S}{\partial \ell} & \frac{d\ell}{dt} &= -\frac{\partial S}{\partial L},\\ \frac{dG}{dt} &= \frac{\partial S}{\partial g}, & \frac{dg}{dt} &= -\frac{\partial S}{\partial G},\\ \frac{dH}{dt} &= \frac{\partial S}{\partial h}, & \frac{dh}{dt} &= -\frac{\partial S}{\partial H}, \end{aligned}\right\} \tag{6.4}$$

where t denotes the time; and S, the *disturbing function* which arises from the presence of a third body, and in the case that all three bodies can be regarded as mass-points, and can be shown (cf., e.g., Brown and Shook, 1933) to assume the form**

$$S = \tilde{G}m_3 \frac{r^2}{r'^3} \sum_{j=1}^{\omega} \frac{(m_1)^j - (-m_2)^j}{(m_1+m_2)^j} \left(\frac{r}{r'}\right)^{j-1} P_{j+1}(\sigma) + \frac{\tilde{G}(m_1+m_2)}{2A}, \tag{6.5}$$

where $\tilde{G}$ denotes the constant of gravitation***; r, r' stand, as before, for the radii-vectors of the close and wide orbits; and $P_j(\sigma)$ are the Legendre polynomials of σ, the cosine of the angle between them, as given by

* Brown (1936) employed a modified set of Delaunay variables, which we found to be less convenient for detailed work.

** The term on the right-hand side of (6.5) needs to be added to the disturbing function proper for the Delaunay variable ℓ to represent the mean anomaly (cf., e.g., Brouwer and Clemence, 1961; p. 291).

*** This symbol will be used in the remainder of this section to avoid confusion with the Delaunay variable G.

$$\left.\begin{aligned}\sigma &= \cos u \cos u' + \sin u \sin u' \cos i\\ &= \cos^2 \tfrac{1}{2} i \cos(u - u') + \sin^2 \tfrac{1}{2} i \cos(u + u')\end{aligned}\right\} \tag{6.6}$$

in terms of the true longitudes u, u' of m_2 and m_3 measured from Ω.

Once the system (6.4) of canonical equations has been solved, the actual elements of the orbit follow by an inversion of the definitions (6.3) of the Delaunay variables from

$$\left.\begin{aligned} a &= \sqrt{L/n}, & n(t-T) &= \ell\,,\\ e &= \sqrt{1-(G/L)^2}, & \omega &= g\,,\\ i &= \cos^{-1}(H/G); & \Omega &= h\,.\end{aligned}\right\} \tag{6.7}$$

Needless to say, the canonical equations (6.4) of our particular form of the three-body problem do not, in general, admit of any integrals in a closed form; and owing to their inherent non-linearity, any construction of their solutions can proceed only by successive approximations. In what follows we propose to divide this task into a separate investigation of:

(1) *short-range perturbations* of the close pair – secular or periodic (of periods comparable with that of its orbit) – for which the position of m_3 in space (and, therefore, r') can be regarded as *constant;* and

(2) *long-range perturbations* of the close pair, caused by the *motion* of m_3, whose orbital period will be regarded sufficiently long to justify replacement of all elements of the binary system by their *time-averages* over one orbital cycle.

In other words, the conditions stipulated under (1) will render our task to seek a solution of the corresponding *two-centre problem* – a problem which could be solved in a closed form (cf. Charlier, 1902; Samter, 1922) or, more recently, Langebartel (1965) or Anderle (1976) in terms of elliptic functions. This form is, however, so complicated that perturbation methods adopted in this section will lead to simpler results. On the other hand, a resort to time-averaging in the *three-body problem* listed under (2) will yield only *secular* (or long-periodic) perturbations of the elements of the respective orbit, in which we are primarily interested.

B. SHORT-RANGE PERTURBATIONS

In embarking on a discussion of short-range perturbations of the close orbit, we shall assume the mean value of the ratio r/r' to be sufficiently small to allow us to break off the expansion of the disturbing function S on the right-hand side of Equation (6.5) after the second term, retaining*

$$S = \frac{\tilde{G}m_3}{r'^2}\left\{r^2 P_2(\sigma) + \frac{m_1 - m_2}{m_1 + m_2}\,\frac{r^3}{r'}\,P_3(\sigma) + :::\right\} + \frac{\tilde{G}(m_1 + m_2)}{2A}, \tag{6.8}$$

* E. W. Brown in his work on long-periodic perturbations (Brown, 1936, 1937) considered only the first term of the above expression.

and with its aid let us evaluate the time-integral

$$I = \int S\,\mathrm{d}t\,, \tag{6.9}$$

the formulation of which will constitute the main part our task; for once we have obtained it, a solution for the individual Delaunay elements can evidently be obtained from (6.4) by a partial differentiation of I with respect to the requisite element.

Consistent with the assumed constancy of r' for short-range perturbations, the integral on the right-hand side of (6.9) can then be split up into

$$I = \frac{\tilde{G}m_3}{2r'^3}\left\{3J_2 - J_0 + \frac{m_1 - m_2}{m_1 + m_2}\,\frac{5J_3 - 3J_1}{r'} + \cdots\right\} + \frac{1}{2}\,n^2A^2t, \tag{6.10}$$

where we have abbreviated

$$\begin{aligned} J_0 &= \int r^2\mathrm{d}t, & J_1 &= \int r^3\sigma\,\mathrm{d}t\,, \\ J_2 &= \int r^2\sigma^2\mathrm{d}t, & J_3 &= \int r^3\sigma^3\mathrm{d}t. \end{aligned} \tag{6.11}$$

These integrals can be evaluated in a closed form if we replace the time t behind the integral sign by the eccentric anomaly E, related with t by Kepler's equation

$$E - e\sin E = n(t - T), \tag{6.12}$$

according to which

$$r = A(1 - e\cos E) \tag{6.13}$$

$$n\mathrm{d}t = (1 - e\cos E)\,\mathrm{d}E. \tag{6.14}$$

Moreover when we (remember that $u = \omega + \upsilon$ and that, consistent with our basic assumption, the longitude $\Omega + u'$ of the third body is to remain stationary in space) the quantity σ as defined by Equation (6.6) can evidently be rewritten as

$$\sigma = M\sin\upsilon + N\cos\upsilon \tag{6.15}$$

or

$$(1 - e\cos E)\,\sigma = M\sqrt{1 - e^2}\,\sin E + N(\cos E - e)\,, \tag{6.16}$$

where

$$M = -\sin\omega\cos u' + \cos\omega\sin u'\cos i \tag{6.17}$$

and

$$N = \cos\omega + \cos u' + \sin\omega\sin u'\cos i. \tag{6.18}$$

If, to a first approximation, we regard now the elements A, e, ω, Ω, i as well as u' behind the integral signs on the right-hand sides of Equations (6.11) to be constant, their literal evaluation yields

$$\begin{aligned}(n/A^2)\,J_0 = \left(1 + \frac{3}{2}\,e^2\right)E - 3e\left(1 + \frac{1}{4}\,e^2\right)\sin E \\ + \frac{3}{4}\,e^2\sin 2E - \frac{1}{12}\,e^3\sin 3E,\end{aligned} \tag{6.19}$$

$$(n/A^3)J_1 = M\sqrt{1-e^2}\left\{-\left(1+\frac{3}{4}e^2\right)\cos E + \frac{3}{4}e\left(1+\frac{1}{6}e^2\right)\cos 2E\right.$$

$$\left.-\frac{1}{4}e^2\cos 3E + \frac{1}{32}e^3\cos 4E\right\}$$

$$+N\left\{-\frac{5}{2}e\left(1+\frac{3}{4}e^2\right)E + \left(1+\frac{21}{4}e^2+\frac{3}{4}e^4\right)\sin E\right. \tag{6.20}$$

$$\left.-\left(\frac{3}{4}+e^2\right)e\sin 2E + \frac{1}{4}e^2\left(1+\frac{1}{3}e^2\right)\sin 3E - \frac{1}{32}e^3\sin 4E\right\},$$

$$(n/A^2)J_2 = \frac{1}{2}M^2(1-e^2)\left\{E - \frac{1}{2}e\sin E - \frac{1}{2}\sin 2E + \frac{1}{6}e\sin 3E\right\}$$

$$+\frac{1}{2}MN\sqrt{1-e^2}\left\{5e\cos E - (1+e^2)\cos 2E + \frac{1}{3}e\cos 3E\right\} \tag{6.21}$$

$$+\frac{1}{2}N^2\left\{(1+4e^2)E - \frac{1}{2}e(11+4e^2)\sin E\right.$$

$$\left.+\frac{1}{2}(1+2e^2)\sin 2E - \frac{1}{6}e\sin 3E\right\},$$

and

$$(n/A^3)J_3 = M^3(1-e^2)^{3/2}\left\{-\frac{3}{4}\cos E + \frac{1}{8}e\cos 2E + \frac{1}{12}\cos 3E\right.$$

$$\left.-\frac{1}{32}e\cos 4E\right\}$$

$$+3M^2N(1-e^2)\left\{-\frac{5}{8}eE + \frac{1}{4}(1+e^2)\sin E + \frac{1}{4}e\sin 2E\right.$$

$$\left.-\frac{1}{12}(1+e^2)\sin 3E + \frac{1}{32}e\sin 4E\right\}$$

$$+3MN^2\sqrt{1-e^2}\left\{-\frac{1}{4}(1+6e^2)\cos E + \frac{5}{8}e\left(1+\frac{2}{5}e^2\right)\cos 2E\right.$$

$$\left.-\frac{1}{12}(1+2e^2)\cos 3E + \frac{1}{32}e\cos 4E\right\}$$

$$+N^3\left\{-\frac{15}{8}e\left(1+\frac{4}{3}e^2\right)E + \frac{3}{4}\left(1+7e^2+\frac{4}{3}e^4\right)\sin E\right.$$

$$-e\left(1+\frac{3}{4}e^2\right)\sin 2E+\frac{1}{12}(1+3e^2)\sin 3E-\frac{1}{32}e\sin 4E\Bigg\}\,. \tag{6.22}$$

By insertion of the foregoing results in (6.10) the integral (6.9) can, therefore, be eventually reduced to the form

$$I=p_0E+\sum_{j=1}^{4}(p_j\sin jE+q_j\cos jE)+\frac{1}{2}(na)^2t\,, \tag{6.23}$$

where

$$\begin{aligned}p_0&=\frac{1}{2}k_1\,\{(1-e^2)P_2(M)+(1+4e^2)P_2(N)\}\\&\quad-\frac{5}{8}e\,k_2\,\{(1-e^2)NP_3(M)+(3+4e^2)P_3(N)\}\,.\end{aligned} \tag{6.24}$$

$$\begin{aligned}p_1&=-\frac{1}{4}e\,k_1\,\{(1-e^2)P_2(M)+(11+4e^2)P_2(N)\}\\&\quad+\frac{1}{4}k_2\,\{(1-e^4)NP_3(M)+(3+21e^2+4e^4)P_3(N)\}\,,\end{aligned} \tag{6.25}$$

$$\begin{aligned}p_2&=-\frac{1}{4}k_1\,\{(1-e^2)P_2(M)-(1+2e^2)P_2(N)\}\\&\quad+\frac{1}{4}e\,k_2\,\{(1-e^2)NP_3(M)-(4+3e^2)P_3(N)\}\,,\end{aligned} \tag{6.26}$$

$$\begin{aligned}p_3&=\frac{1}{12}e\,k_1\,\{(1-e^2)P_2(M)-P_2(N)\}\\&\quad-\frac{1}{12}k_2\,\{(1-e^4)NP_3'(M)-(1+3e^2)P_3(N)\}\,,\end{aligned} \tag{6.27}$$

$$p_4=\frac{1}{32}k_2\,\{(1-e^2)NP_3'(M)-P_3(N)\}\,; \tag{6.28}$$

and

$$\begin{aligned}(1-e^2)^{-1/2}q_1&=\frac{15}{4}e\,k_1MN-\frac{1}{4}k_2\,\{3(1-e^2)P_3(M)\\&\quad+(1+6e^2)MP_3'(N)\}\,,\end{aligned} \tag{6.29}$$

$$\begin{aligned}(1-e^2)^{-1/2}q_2&=-\frac{3}{4}k_1(1+e^2)MN+\frac{1}{8}e\,k_2\,\{(1-e^2)P_3(M)\\&\quad+(5+2e^2)MP_3(N)\}\,,\end{aligned} \tag{6.30}$$

$$\begin{aligned}(1-e^2)^{-1/2}q_3&=\frac{1}{4}e\,k_1MN-\frac{1}{12}k_2\,\{(1-e^2)P_3(M)\\&\quad-(1+2e^2)MP_3(N)\}\,,\end{aligned} \tag{6.31}$$

$$(1-e^2)^{-1/2}\ q_4 = -\frac{1}{32}\ e\,k_2\ \{(1-e^2)P_3(M) - MP_3'(N)\}\,, \tag{6.32}$$

where the P_j's denote the respective Legendre polynomials of M or N (with primes indicating first derivatives with respect to their arguments),and the constants $k_{1,2}$ are given by the equations

$$k_1 = \frac{\tilde{G}m_3A^2}{nr'^3} \tag{6.33}$$

and

$$k_2 = \left\{\frac{A}{r'}\ \frac{m_1-m_2}{m_1+m_2}\right\}k_1\,. \tag{6.34}$$

All that remains to be done now to specify the short-range perturbations of orbital elements of the close pair caused by the attraction a stationary third body of mass m_3 is to diffferentiate the expressions (6.7) with respect to the Delaunay variables concerned and insert from (6.4): in doing so we find that

$$\frac{\delta A}{A} = \frac{1}{2L}\ \frac{\partial I}{\partial \ell}\,, \tag{6.35}$$

$$\frac{\delta e}{e} = \frac{G^2}{L^2-G^2}\left\{\frac{1}{L}\ \frac{\partial}{\partial \ell} - \frac{1}{G}\ \frac{\partial}{\partial g}\right\}I\,, \tag{6.36}$$

$$(\sin i)\ \delta i = \frac{H}{G}\left\{\frac{1}{G}\ \frac{\partial}{\partial g} - \frac{1}{H}\ \frac{\partial}{\partial h}\right\}I, \tag{6.37}$$

and

$$\delta(nt - nT) = -\frac{\partial I}{\partial L}\,, \tag{6.38}$$

$$\delta\omega = -\frac{\partial I}{\partial G}\,, \tag{6.39}$$

$$\delta\Omega = -\frac{\partial I}{\partial H}\,, \tag{6.40}$$

where the variation δx in any one of the six orbital elements x corresponds to a time interval over which the integral (6.9) has been evaluated.

In order to establish the explicit forms of the foregoing Equations (6.35)–(6.40), let us differentiate the expression (6.23) for I partially with respect of the individual Delaunay variables X in accordance with the equation

$$\frac{\partial I}{\partial X} = \left(\frac{\partial p_0}{\partial X}\right)E + p_0\left(\frac{\partial E}{\partial X}\right) + \sum_{j=1}^{4}\left\{\left(\frac{\partial p_j}{\partial X} - jq_j\,\frac{\partial E}{\partial X}\right)\sin jE\right. \tag{6.41}$$

$$+\left(\frac{\partial q_j}{\partial X}+jp_j\frac{\partial E}{\partial X}\right)\cos jE\bigg\}.$$

The requisite derivatives of p_j and q_j with respect to the elements G, H and g, h follow by a straightforward partial differentiation of Equations (6.24)–(6.32), and assume the explicit forms

$$G\frac{\partial p_0}{\partial G}=k_1(1-e^2)\{P_2(M)-4P_2(N)\}$$
$$+\frac{5}{8}k_2\frac{1-e^2}{e}\{(1-3e^2)NP_3'(M)+3(1+4e^2)P_3(N)\} \tag{6.42}$$
$$-H\frac{\partial p_0}{\partial H},$$

$$H\frac{\partial p_0}{\partial H}=\frac{3}{2}k_1\{(1-e^2)M\mu_1+(1+4e^2)N\upsilon_1\}$$
$$-\frac{5}{8}k_2e\{[(1-e^2)P_3'(M)+(3+4e^2)P_3'(N)]\mu_1 \tag{6.43}$$
$$+15(1-e^2)MN\upsilon_1\},$$

$$\frac{\partial p_0}{\partial g}=\frac{15}{2}k_1e^2MN+\frac{5}{8}k_2e\{(1-e^2)[3-P_3'(M)]$$
$$-(1+6e^2)P_3'(N)\}M, \tag{6.44}$$

$$\frac{\partial p_0}{\partial h}=\frac{3}{2}(1-e^2)\left\{-k_1+\frac{25}{4}k_2eN\right\}M\mu$$
$$-\frac{1}{2}\{3k_1(1+4e^2)N-\frac{5}{4}ek_2[(1-e^2)P_3'(M)+$$
$$+(3+4e^2)P_3'(N)]\}\upsilon; \tag{6.45}$$

and, similarly,

$$G\frac{\partial p_1}{\partial G}=k_1\frac{1-e^2}{4e}\{(1-3e^2)P_2(M)+(11+12e^2)P_2(N)\}$$
$$+k_2\frac{1-e^2}{2}\{2e^2NP_3'(M)-(21+8e^2)P_3(N)\}-H\frac{\partial p_1}{\partial H}, \tag{6.46}$$

$$H\frac{\partial p_1}{\partial H}=-\frac{1}{4}k_1e\{(1-e^2)P_2'(M)\mu_1+(11+4e^2)P_2'(N)\upsilon_1\}$$
$$+\frac{1}{4}k_2\{[(1-e^4)P_3'(M)-(3-35e^2+4e^4)P_3'(N)]\upsilon_1 \tag{6.47}$$
$$+15(1-e^4)MN\mu_1\},$$

$$\frac{\partial p_1}{\partial g} = -\frac{15}{4} k_1 e(2 + e^2) MN + \frac{1}{4} k_2 N \{(1 - e^4)[15MN - \\ - P_3'(M)] - (3 - 35e^2 + 4e^4) P_3'(N)\}, \tag{6.48}$$

$$\frac{\partial p_1}{\partial h} = \frac{3}{4}(1 - e^2)\{k_1 e - 5k_2(1 + e^2)N\} M\mu \\ + \frac{1}{4}\{3k_1 e(11 + 4e^2)N - k_2[(1 - e^4)P_3'(M) - \\ - (3 - 35e^2 + 4e^4)P_3'(N)]\} v, \tag{6.49}$$

$$G\frac{\partial p_2}{\partial G} = -\frac{1}{2} k_1(1 - e^2)\{P_2(M) + 2P_2(N)\} \\ + k_2 \frac{1 - e^2}{4e}\{(4 + 3e^2)P_3(N) - (1 - e^2) NP_3'(M)\} - H\frac{\partial p_2}{\partial H}, \tag{6.50}$$

$$H\frac{\partial p_2}{\partial H} = -\frac{3}{4} k_1 \{(1 - e^2) M\mu_1 - (1 + 2e^2) Nv_1\} \\ + \frac{1}{4} k_2 e \{[(1 - e^2)P_3'(M) - (4 + 3e^2)P_3'(N)] v_1 \\ + 15(1 - e^2) MN\mu_1\}, \tag{6.51}$$

$$\frac{\partial p_2}{\partial g} = \frac{3}{4} k_1 e^2 MN + \frac{1}{4} eMk_2 \{(1 - e^2)P_3'(M) - \\ - (6 + e^2)P_3'(N) - 3(1 - e^2)\}, \tag{6.52}$$

$$\frac{\partial p_2}{\partial h} = \frac{3}{4}(1 - e^2)\{k_1 - 5k_2 eN\} M\mu \\ - \frac{1}{4}\{3k_2(1 + 2e^2) + k_1 e[(1 - e^2)P_3'(M) - (4 + 3e^2)P_3'(N)]\} v, \tag{6.53}$$

$$G\frac{\partial p_3}{\partial G} = k_1 \frac{1 - e^2}{12e}\{(3e^2 - 1)P_2(M) + P_2(N)\} \\ - \frac{1}{6} k_2(1 - e^2)\{2e^2 NP_3'(M) + 3P_3(N)\} - H\frac{\partial p_3}{\partial H}, \tag{6.54}$$

$$H\frac{\partial p_3}{\partial H} = \frac{1}{4} k_1 e\{(1 - e^2) M\mu_1 - Nv_1\}$$

$$- \frac{1}{12} k_2 \{[1 - e^4) P_3'(M) - (1 + 3e^2) P_3'(N)] v_1 \tag{6.55}$$

$$+ 15(1 - e^4) MN\mu_1 \} ,$$

$$\frac{\partial p_3}{\partial g} = - \frac{1}{4} k_1 e(2 - e^2) MN$$

$$- \frac{1}{12} k_2 M \{(1 - e^4) P_3'(M) - (3 + 3e^2 - 2e^4) P_3'(N) \tag{6.56}$$

$$- 3(1 - e^4)\} ,$$

$$\frac{\partial p_3}{\partial h} = \frac{1}{4} (1 - e^2) \{- k_1 e + 5k_2(1 + e^2) N\} M\mu$$
$$+ \frac{1}{4} \{k_1 eN + \frac{1}{3} k_2[(1 - e^4) P_3'(M) - (1 + 3e^2)P_3'(N)]\} v, \tag{6.57}$$

$$G \frac{\partial p_4}{\partial G} = k_2 \frac{1 - e^2}{32e} \{(3e^2 - 1) NP_3'(M) + P_3(N)\} - H \frac{\partial p_4}{\partial H} , \tag{6.58}$$

$$H \frac{\partial p_4}{\partial H} = \frac{1}{32} k_2 \{[(1 - e^2) P_3'(M) - P_3'(N)] v_1$$
$$+ 15(1 - e^2) MN\mu_1\} , \tag{6.59}$$

$$\frac{\partial p_4}{\partial g} = \frac{1}{32} k_2 Me \{(1 - e^2) P_3'(M) + (2e^2 - 3) P_3'(N) - 3(1 - e^2)\} , \tag{6.60}$$

$$\frac{\partial p_4}{\partial h} = - \frac{1}{32} k_2 e \{[(1 - e^2) P_3'(M) - P_3'(N)] v + 15(1 - e^2) MN\mu\}; \tag{6.61}$$

$$G \frac{\partial q_1}{\partial G} = \frac{15}{4e} k_1(2e^2 - 1) \sqrt{1 - e^2} MN$$

$$- \frac{1}{4} k_2 \sqrt{1 - e^2} \{9(1 - e^2) P_3(M) + (8e^2 - 11) MP_3'(N)\} \tag{6.62}$$

$$- H \frac{\partial q_1}{\partial H} ,$$

$$H \frac{\partial q_1}{\partial H} = \frac{15}{4} k_1 e \sqrt{1 - e^2} \{Mv_1 + N\mu_1\}$$

$$- \frac{1}{4} k_2 \sqrt{1 - e^2} \{[3(1 - e^2) P_3'(M) + (1 + 6e^2)P_3'(N)] \mu_1 \tag{6.63}$$

$$+ 15(1 + 6e^2) MNv_1\} ,$$

$$\frac{\partial q_1}{\partial g} = \frac{15}{4} k_1 e \sqrt{1-e^2}\,(M^2 - N^2) + \frac{1}{4} k_2 \sqrt{1-e^2}\, N\{(1-15e^2)P_3'(M) + (1+6e^2)[P_3'(N) - 3]\} \tag{6.64}$$

$$\frac{\partial q_1}{\partial h} = \frac{15}{4}\sqrt{1-e^2}\;\{k_1 e + k_2(1+6e^2)N\}\,\upsilon - \frac{1}{4}\sqrt{1-e^2}\;\{15k_1 eN + k_2[3(1-e^2)P_3'(M) + (1+6e^2)P_3'(N)]\}\,\mu, \tag{6.65}$$

$$G\frac{\partial q_2}{\partial G} = \frac{3}{4} k_1(1-3e^2)\sqrt{1-e^2}\,MN - k_2\frac{\sqrt{1-e^2}}{8e}\;\{(1-e^2)(1-2e^2)P_3(M) + (5-4e^2-8e^4)MP_3'(N)\} - H\frac{\partial q_2}{\partial H}, \tag{6.66}$$

$$H\frac{\partial q_2}{\partial H} = \frac{3}{4} k_1(1+e^2)\sqrt{1-e^2}\;\{M\upsilon_1 - N\mu_1\} + \frac{1}{8} k_2 e\sqrt{1-e^2}\;\{[(1-e^2)P_3'(M) + (5+2e^2)P_3'(N)]\,\mu_1 + 15(5+2e^2)MN\upsilon_1\}\,N, \tag{6.67}$$

$$\frac{\partial q_2}{\partial g} = \frac{3}{4} k_1(1+e^2)\sqrt{1-e^2}\;(N^2 - M^2) + \frac{1}{8}\, e\sqrt{1-e^2}\;\{(9+5e^2)P_3'(M) - (5+2e^2)P_3'(N)\}\,N, \tag{6.68}$$

$$\frac{\partial q_2}{\partial h} = \frac{3}{4}\sqrt{1-e^2}\,\{k_1(1+e^2) - \frac{5}{2} k_2 e(5+2e^2)M\}\,M\upsilon + \frac{1}{4}\sqrt{1-e^2}\;\{3k_1(1+e^2)N - \frac{1}{2} k_2 e(1-e^2)P_3'(M) + (5+2e^2)P_3'(N)]\}\,\mu, \tag{6.69}$$

$$G\frac{\partial q_3}{\partial G} = k_1(2e^2-1)\frac{\sqrt{1-e^2}}{4e}\;MN + \frac{1}{4} k_2\sqrt{1-e^2}\;\{(1-e^2)P_3(M) + (1-2e^2)MP_3'(N)\} - H\frac{\partial q_3}{\partial H}, \tag{6.70}$$

$$H\frac{\partial q_3}{\partial H} = \frac{1}{4}k_1 e\sqrt{1-e^2}\ (Mv_1 + N\mu_1)$$

$$+\frac{1}{2}k_2\sqrt{1-e^2}\ \{[(1-e^2)P_3'(M) - (1+2e^2)P_3'(N)]\,\mu_1 \tag{6.71}$$

$$-15(1+2e^2)MNv_1\},$$

$$\frac{\partial q_3}{\partial g} = \frac{1}{4}k_1 e\sqrt{1-e^2}\ (M^2 - N^2)$$

$$-\frac{1}{12}k_2\sqrt{1-e^2}\ \{(1+3e^2)P_3'(M) + \tag{6.72}$$

$$+(1+2e^2)\,[P_3'(N)+3]\}\,N,$$

$$\frac{\partial q_3}{\partial h} = \frac{1}{4}\sqrt{1-e^2}\ \{-k_1 e + 5k_2(1+2e^2)N\}\,Mv$$

$$-\frac{1}{12}\sqrt{1-e^2}\ \{3k_1 eN + k_2[(1-e^2)P_3'(M) - \tag{6.73}$$

$$-(1+2e^2)P_3'(N)]\}\,\mu,$$

$$G\frac{\partial q_4}{\partial G} = k_2(1-2e^2)\frac{\sqrt{1-e^2}}{32e}\{(1-e^2)P_3(M) - MP_3'(N)\} \tag{6.74}$$

$$-H\frac{\partial q_4}{\partial H},$$

$$H\frac{\partial q_4}{\partial H} = \frac{1}{32}k_2 e\sqrt{1-e^2}\ \{[(1-e^2)P_3'(M) - P_3'(N)]\,\mu_1 - 15MNv_1\}, \tag{6.75}$$

$$\frac{\partial q_4}{\partial g} = \frac{1}{32}k_2 e\sqrt{1-e^2}\ N\{(3-e^2)P_3'(M) - P_3'(N) + 3\}, \tag{6.76}$$

$$\frac{\partial q_4}{\partial h} = \frac{1}{32}k_2 e\sqrt{1-e^2}\ \{[(1-e^2)\mathrm{P}_3'(M) - P_3'(N)]\,\mu - 15MNv\}, \tag{6.77}$$

where the direction cosines M and N continue to be given by Equations (6.17)–(6.18); and where we have abbreviated

$$\left.\begin{aligned} \mu &= \sin\omega\ \sin u' + \cos\omega\cos u'\cos i = \frac{\partial M}{\partial \mu'}, \\ v &= -\cos\omega\sin u' + \sin\omega\cos u'\cos i = \frac{\partial N}{\partial \mu'}, \end{aligned}\right\} \tag{6.78}$$

and

$$\left.\begin{aligned} \mu_1 &= \cos\omega\sin u'\cos i, \\ v_1 &= \sin\omega\sin u'\cos i. \end{aligned}\right\} \tag{6.79}$$

Of the remaining expressions, all partial derivatives with respect to ℓ are identically zero (as neither p_j or q_j contain ℓ explicitly); but those with respect to L are non-vanishing and prove relatively most troublesome to evaluate, on account of the fact that p_j and q_j depend on L nor only through e, but also through k_1 and k_2. For, inasmuch as (by Kepler's second law) the mean daily motion n of the close pair is given by

$$n = \sqrt{\frac{\tilde{G}(m_1 + m_2)}{A^3}} = \frac{\tilde{G}(m_1 + m_2)}{L^3} \tag{6.80}$$

by virtue of the first one of Equations (6.3), the expressions (6.33)–(6.34) for the quantitites $k_{1,2}$ can be alternatively rewritten as

$$k_1 = \frac{m_3 L^7}{r'^3 \tilde{G}^3 (m_1 + m_2)^4} \tag{6.81}$$

and

$$k_2 = \frac{m_3(m_1 - m_2)L^9}{r'^4 \tilde{G}^4 (m_1 + m_2)^6}\,, \tag{6.82}$$

In order to ascertain the explicit forms of the partial derivatives of p_j and q_j with respect to L, let us rewrite the expressions (6.24)–(6.32) symbolically as

$$\left.\begin{aligned} p_j &= k_1 \frac{\partial p_j}{\partial k_1} + k_2 \frac{\partial p_j}{\partial k_2}\,, \qquad j \geqslant 0, \\ q_j &= k_1 \frac{\partial q_j}{\partial k_1} + k_2 \frac{\partial q_j}{\partial k_2}\,, \qquad j \geqslant 1. \end{aligned}\right\} \tag{6.83}$$

An inspection of (6.24)–(6.32) reveals that the partial derivatives of p_j and q_j (depending as they do on L, G, and H through e and $\cos i$ only) constitute homogeneous functions of L, G, and H of zero degrees; while k_1 and k_2 are of 7th and 9th degrees in L. Therefore, an application of Euler's theorem on homogeneous functions reveals that

$$\begin{aligned} L \frac{\partial}{\partial L}\left(k_i \frac{\partial p_j}{\partial k_i}\right) + G \frac{\partial}{\partial G}\left(k_i \frac{\partial p_j}{\partial k_i}\right) + H \frac{\partial}{\partial H}\left(k_i \frac{\partial p_j}{\partial k_i}\right) &= \\ = (2i+5) k_i \frac{\partial p_j}{\partial k_i}\,, \end{aligned} \tag{6.84}$$

for $i = 1,2; j = 0\ (1)\ 4$; and, similarly,

$$\begin{aligned} L \frac{\partial}{\partial L}\left(k_i \frac{\partial q_j}{\partial k_i}\right) + G \frac{\partial}{\partial G}\left(k_i \frac{\partial q_j}{\partial k_i}\right) + H \frac{\partial}{\partial H}\left(k_i \frac{\partial q_j}{\partial k_i}\right) &= \\ = (2i+5) k_i \frac{\partial q_j}{\partial k_i} \end{aligned} \tag{6.85}$$

for $i = 1,2$ and $j = 1\ (1)\ 4$. By adding the pairs of Equations (6.84) or (6.85) for $i = 1,2$

and taking note of (6.83), we find that, in accordance with Euler's theorem on homogeneous functions,

$$L \frac{\partial p_j}{\partial L} + G \frac{\partial p_j}{\partial G} + H \frac{\partial p_j}{\partial H} = 7p_j + 2k_2 \frac{\partial p_j}{\partial k_2} \tag{6.86}$$

for $j = 0\,(1)\,4$; and, similarly,

$$L \frac{\partial q_j}{\partial L} + G \frac{\partial q_j}{\partial G} + H \frac{\partial q_j}{\partial H} = 7q_j + 2k_2 \frac{\partial q_j}{\partial k_2} \tag{6.87}$$

for $j = 1(1)4$; from which the desired partial derivatives $\partial p_j/\partial L$ or $\partial q_j/\partial L$ can be evaluated algebraically if those with respect to G and H are known.

As the last step in our explicit evaluation of the expansion on the right-hand side of Equation (6.41), it remains for us to differentiate partially the eccentric anomaly E with respect to the appropriate Delaunay elements; and this can be easily accomplished with the aid of Kepler's equation (6.12). The latter, rewritten in terms of the Delaunay variables, readily assumes the form

$$LE - \sqrt{L^2 - G^2} \sin E = \ell L; \tag{6.88}$$

and its partial differentiation reveals that

$$\frac{\partial E}{\partial L} = \frac{e\ell + \sin E}{Aenr}, \tag{6.89}$$

$$\frac{\partial E}{\partial G} = -\frac{\sqrt{1-e^2}}{Aenr} \sin E = -\frac{\sin v}{A^2 en}, \tag{6.90}$$

and

$$\frac{\partial E}{\partial \ell} = \frac{A}{r}, \tag{6.91}$$

all other derivatives being identically zero.

By insertion of all the foregoing results represented by Equations (6.42)–(6.77) and (6.89)–(6.91) in (6.41), our task should be essentially complete. This insertion offers indeed no difficulty; but the explicit forms of the results (which should include all short-periodic perturbations) become too lengthy to be displayed here in full; and to obtain may be left as an exercise for the reader. Particular interest attaches, however, to the secular (short-range) perturbations $[\delta x]$ of the six Keplerian elements x, obtaining if the limits of integration on the right-hand side of Equation (6.9) have been extended over the entire cycle – i.e., over the interval $(0, 2\pi)$ in the eccentric anomaly E. If so, the only such perturbations will evidently arise from the terms $(nA)^2 t/2$ or p_0 on the right-hand side of Equation (6.23); and these will be worked out here in full.

In order to do so, let us note first that, inasmuch as neither na nor p_0 depends on the mean anomaly ℓ, it follows from Equation (6.35) that

$$[\delta a] = 0 \tag{6.92}$$

and, consequently, L becomes a constant. Furthermore, it follows from (6.36) that

$$\frac{[\delta e]}{e} = -\frac{G}{L^2 - G^2} \frac{\partial [I]}{\partial g} = -\frac{2\pi G}{L^2 - G^2} \frac{\partial p_0}{\partial g} ; \tag{6.93}$$

and an insertion for $\partial p_0/\partial g$ from (6.44) leads to

$$\begin{aligned} \frac{[\delta e]}{2\pi\sqrt{1-e^2}} = &-\frac{15}{2}\kappa_1 MNe + \frac{5}{8}\kappa_2[(1-e^2)P_3'(M) \\ &+ (1+6e^2)P_3'(N) - 3(1-e^2)]M, \end{aligned} \tag{6.94}$$

where

$$\kappa_1 = \frac{k_1}{na^2} = \frac{m_3}{m_1+m_2}\left(\frac{A}{r'}\right)^3 \tag{6.95}$$

and

$$\kappa_2 = \frac{k_2}{na^2} = \frac{m_3(m_1-m_2)}{(m_1+m_2)^2}\left(\frac{A}{r'}\right)^4 \tag{6.96}$$

are nondimensional constants.

Similarly, from (6.37), it follows that

$$\sin i\,[\delta i] = \frac{2\pi}{na^2\sqrt{1-e^2}}\left\{\cos i\,\frac{\partial p_0}{\partial g} - \frac{\partial p_0}{\partial h}\right\}, \tag{6.97}$$

which by virtue of (6.44) and (6.45) assumes the explicit form

$$\begin{aligned} \frac{\sqrt{1-e^2}\sin i}{2\pi}[\delta i] = &\frac{15}{2}\left\{\kappa_1 Ne + \frac{\kappa_2}{12}[-(1-e^2)P_3'(M)\right. \\ &\left. -(1+6e^2)P_3'(N) + 3(1-e^2)]\right\} Me\cos i \\ &+\frac{3}{2}(1-e^2)\left\{\kappa_1 - \frac{25}{4}\kappa_2 eN\right\} M\mu \\ &+\frac{1}{2}\nu\,\{3\kappa_1(1+4e^2)N - \frac{5}{4}e\kappa_2[(1-e^2)P_3'(M) + (3+4e^2)P_3'(N)]\}. \end{aligned} \tag{6.98}$$

Form (6.39)–(6.40) and (6.42)–(6.43) it follows, furthermore, that

$$\begin{aligned} \frac{[\delta\omega]}{2\pi} = -\frac{\partial p_0}{\partial G} = &-\kappa_1\sqrt{1-e^2}\,\{P_2(M) - 4P_2(N)\} \\ &- \frac{5\kappa_2\sqrt{1-e^2}}{8e}\{(1-3e^2)NP_3'(M) + 3(1+4e^2)P_3(N)\} \\ &- \frac{[\delta\Omega]}{2\pi}\cos i, \end{aligned} \tag{6.99}$$

$$\begin{aligned} \frac{[\delta\Omega]}{2\pi} = -\frac{\partial p_0}{\partial H} = &-\frac{3\kappa_1}{2\sqrt{1-e^2}}\{(1-e^2)M\cos\omega \\ &+ (1+4e^2)N\sin\omega\}\sin u' \end{aligned} \tag{6.100}$$

$$+\frac{5\kappa_2 e}{8\sqrt{1-e^2}}\{[(1-e^2)P_3'(M)+(3+4e^2)P_3'(N)]\cos\omega$$
$$+15(1-e^2)MN\sin\omega\}\sin u'$$

while, in accordance with (6.23) and (6.38),

$$\left.\begin{aligned}[\delta(nt-nT)] &= -\frac{\partial[I]}{\partial L}\\ &= -2\pi\frac{\partial p}{\partial L}-\frac{[t]}{2}\frac{\partial}{\partial L}(n^2A^2),\end{aligned}\right\}\qquad(6.101)$$

Since, however, $n^2=\tilde{G}(m_1+m_2)/A^3$ and $L=nA^2$, it follows that

$$nA=\frac{\tilde{G}(m_1+m_2)}{L}\qquad(6.102)$$

and, moreover,

$$\frac{1}{2}\frac{\partial}{\partial L}(n^2A^2)=-n.\qquad(6.103)$$

In consequence, Equation (6.101) reduces to

$$\begin{aligned}\frac{[\delta(nT)]}{2\pi} &= \frac{\partial p_0}{\partial L}\\ &= \frac{1}{na^2}\left\{7p_0+2k_2\frac{\partial p_0}{\partial k_2}-G\frac{\partial p_0}{\partial G}-H\frac{\partial p_0}{\partial H}\right\}\end{aligned}\qquad(6.104)$$

by (6.86), which on insertion form (6.24), (6.42) and (6.43) yields

$$\frac{[\delta(nT)]}{2\pi}=\frac{[\delta T]}{P}=\frac{5}{2}\kappa_1\{(1-e^2)P_2(M)+(3+4e^2)P_2(N)\}\qquad(6.105)$$
$$+\frac{5\kappa_2}{8e}\{(1-e^2)(1+6e^2)NP_3'(M)$$
$$+3(1+12e^2+8e^4)P_3(N)\}$$

for a ratio of the secular perturbation of the time T of the periastron passage to the orbital period $P=2\pi/n$.

It should be stressed that all the foregoing Equations (6.92)–(6.102) representing short-range secular perturbations of the elements of the close orbit are *exact* for any value of the eccentricities e and e' for the two orbits, or any angle i of inclination between them; their only limitation of their validity going back to our replacement of the infinite expansion (6.5) for the disturbing function S by its first two terms on the right-hand side of (6.8), and the neglect of the squares and higher powers of the coefficients κ_j. Indeed, the next approximation to obtain would be to allow for the variation of the orbital elements just found in integrating I with respect to the time on the right-hand side of Equation (6.9). In this way, terms which are quadratic and higher in the powers of $\kappa_{1,2}$ would eventually be obtained; but they would possess but little meaning unless higher-order terms in the expansion (6.5) of our disturbing function were simultaneously considered. To do so is possible indeed in principle (in point of fact, a consistent linear theory could still include terms of the order κ_3; but $\kappa_4 \approx \kappa_2^2$ would already be quadratic), but very cumbersome in practice; and the task of doing so is being gladly left over as a challenge for future investigations.

C. LONG-RANGE PERTURBATIONS

In order to insure whether the short-range secular perturbations of the orbital elements of the close pair, established at the end of the preceding section, are genuine secular perturbations, or whether they may prove to be long-periodic when the disturbing body of mass m_3 is allowed to move, we must proceed to investigate the *long-range perturbations* for which the longitude u' or v' of the third body becomes a function of the time.

The orbital period of the third body is usually very long in comparison with that of the close pair. Hence, it should be sufficient in the studies of long-range perturbations to omit all short-periodic terms of our disturbing function S, and set

$$S' = \frac{\tilde{G}(m_1 + m_2)}{2A} + \tilde{G}m_3 \frac{A^2}{r'^3} \left\{ \frac{3}{2} \bar{J}_2 - \frac{1}{2} \bar{J}_0 \right\} \tag{6.106}$$

$$+ \tilde{G}m_3 \frac{m_1 - m_2}{m_1 + m_2} \frac{A^3}{r'^4} \left\{ \frac{5}{2} \bar{J}_3 - \frac{3}{2} \bar{J}_1 \right\},$$

where

$$\bar{J}_0 = \left(1 + \frac{3}{2} e^2\right), \tag{6.107}$$

$$\bar{J}_1 = -\frac{5}{2} e \left(1 + \frac{3}{4} e^2\right) N, \tag{6.108}$$

$$\bar{J}_2 = \frac{1}{2} \{(1 - e^2) M^2 + (1 + 4e^2) N^2\}, \tag{6.109}$$

$$\bar{J}_3 = -\frac{15}{8} e \left\{(1 - e^2) M^2 + \left(1 + \frac{4}{3} e^2\right) N^2\right\} N, \tag{6.110}$$

are the *average* values of the J_j's as given by Equations (6.19)–(6.22) of the preceding section over a complete cycle of the close pair. It should be, however, observed that – unlike in the short-range case – the quantities M and N in the foregoing Equations (6.108)–(6.110) can no longer be considered constant (subject only to perturbations) in the course of one orbit of the third body, on account of the presence of $u' = \omega' + v'$ among their factors.

Similarly, by analogy with (6.10) we can now write

$$I' = \int S'\mathrm{d}t = \pi n A^2 + \tilde{G}m_3 \left\{ \left(\frac{3}{2} J_2' - \frac{1}{2} J_0' \right) A^2 + \frac{m_1 - m_2}{m_1 + m_2} \left(\frac{5}{2} J_3' - \frac{3}{2} J_1' \right) A^3 + \cdots \right\}, \tag{6.111}$$

where

$$\left.\begin{aligned} J_0' &= \int \frac{\bar{J}_0}{r'^3}\,\mathrm{d}t, & J_1' &= \int \frac{\bar{J}_1}{r'^4}\,\mathrm{d}t, \\ J_2' &= \int \frac{\bar{J}_2}{r'^3}\,\mathrm{d}t, & J_3' &= \int \frac{\bar{J}_3}{r'^4}\,\mathrm{d}t, \end{aligned}\right\} \tag{6.112}$$

As in the short-range case of the preceding section V-6A, we shall proceed to evaluate the above integrals to a linear approximation, in which the Delaunay variables L, G, H, etc. as well as the primed elliptical elements A', e', ω', etc. of the wide orbit can be considered as constants behind the integral signs in (6.112). If so, then a change of variables from the time t to the true anomaly v' of the third body measured from the apsidal line of its orbit, which can be effected by means of the relations

$$\frac{1}{r'} = \frac{1 + e'\cos v'}{A(1 - e'^2)}, \tag{6.113}$$

$$n'\mathrm{d}t = \frac{(1 - e'^2)^{3/2}\,\mathrm{d}v'}{(1 + e'\cos v')}; \tag{6.114}$$

n' denoting the mean daily motion of the third body. Moreover, the expressions M and N in (6.107) – (6.110) can obviously be rewritten as

$$\left.\begin{aligned} M &= m_1 \sin u' + m_2 \cos u', \\ N &= n_1 \sin u' + n_2 \cos u', \end{aligned}\right\} \tag{6.115}$$

where $u' = \omega' + v'$, and

$$\left.\begin{aligned} m_1 &= \cos\omega \cos i, & n_1 &= \sin\omega \cos i, \\ m_2 &= -\sin\omega; & n_2 &= \cos\omega; \end{aligned}\right\} \tag{6.116}$$

If so, the integrals on the right-hand side of Equations (6.112) can be expressed as

$$\begin{aligned} n'A'^3(1 - e'^2)^{3/2} J_0' &= \left(1 + \frac{3}{2} e^2\right) \int (1 + e'\cos v')\,\mathrm{d}v' \\ &= \left(1 + \frac{3}{2} e^2\right) S_0^1, \end{aligned} \tag{6.117}$$

$$n'A'^3(1-e'^2)^{3/2}\,J_2' = \frac{1}{2}(1-e^2)\int M^2(1+e'\cos v')\,dv'$$

$$+\frac{1}{2}(1+4e^2)\int N^2(1+e'\cos v')\,dv' \tag{6.118}$$

$$= \frac{1}{4}(1-e^2)\left\{(m_1^2+m_2^2)S_0^1 + 2m_1m_2C_2^1 + (m_2^2-m_1^2)S_2^1\right\}$$

$$+\frac{1}{4}(1+4e^2)\left\{(n_1^2+n_2^2)S_0^1 + 2n_1n_2C_2^1 + (n_2^2-n_1^2)S_2^1\right\},$$

$$n'A'^4(1-e'^2)^{5/2}\,J_1' = -\frac{5}{2}e\left(1+\frac{3}{4}e^2\right)\int N(1+e'\cos v')^2\,dv'$$

$$= -\frac{5}{2}e\left(1+\frac{3}{4}e^2\right)\{n_1C_1^2 + n_2S_1^2\} \tag{6.119}$$

and

$$n'A'^4(1-e'^2)^{5/2}\,J_3 = -\frac{15}{8}e(1-e^2)\int M^2N(1+e'\cos v')^2\,dv'$$

$$-\frac{15}{8}e\left(1+\frac{4}{3}e^2\right)\int N^3(1+e'\cos v')^2\,dv'$$

$$\begin{aligned} = -\frac{15}{32}e(1-e^2)\{&[3m_1^2n_1 + m_2(2m_1n_2+m_2n_1)]\,C_1^2\\ &+[3m_2^2n_2 + m_1(m_1n_2+2m_2n_1)]\,S_1^2\\ &-[m_1^2n_1 - m_2(2m_1n_2+m_2n_1)]\,C_3^2\\ &+[m_2^2n_2 - m_1(m_1n_2+2m_2n_1)]\,S_3^2\}\\ -\frac{15}{32}e\left(1+\frac{4}{3}e^2\right)\{&3(n_1^2+n_2^2)[n_1C_1^2+n_2S_1^2]\\ &-n_1(n_1^2-3n_2^2)\,C_3^2\\ &+n_2(n_2^2-3n_1^2)\,S_3^2\}, \end{aligned} \tag{6.120}$$

where we have abbreviated

$$C_j^i = \int (1+e'\cos v')^i \sin(ju')\,dv' \tag{6.121}$$

and

$$S_j^i = \int (1+e'\cos v')^i \cos(ju')\,dv'. \tag{6.122}$$

In particular,

$$S_0^1 = v' - e'\sin v', \tag{6.123}$$

$$S_2^1 = \frac{1}{2} e' \sin(v' + 2\omega') + \frac{1}{2} \sin(2v' + 2\omega') + \frac{1}{6} e' \sin(3v' + 2\omega'), \tag{6.124}$$

$$S_1^2 = e'\left\{v' + \frac{1}{2} e' \sin v'\right\} \cos \omega' + \left(1 + \frac{1}{2} e'^2\right) \sin(v' + \omega') + \frac{1}{2} e' \sin(2v' + \omega') + \frac{1}{12} e'^2 \sin(3v' + \omega'), \tag{6.125}$$

$$S_3^2 = \frac{1}{4} e'^2 \sin(v' + 3\omega') + \frac{1}{2} e' \sin(2v' + 3\omega') + \frac{1}{3}\left(1 + \frac{1}{2} e'^2\right) \sin(3v' + 3\omega') + \frac{1}{4} e' \sin(4v' + 3\omega') + \frac{1}{20} e'^2 \sin(5v' + 3\omega'); \tag{6.126}$$

$$C_2^1 = -\frac{1}{2} e' \cos(v' + 2\omega') - \frac{1}{2} \cos(2v' + 2\omega') - \frac{1}{6} e' \cos(3v' + 2\omega'), \tag{6.127}$$

$$C_1^2 = e'\left\{v' + \frac{1}{2} e' \sin v'\right\} \sin \omega' - \left(1 + \frac{1}{4} e'^2\right) \cos(v' + \omega') - \frac{1}{2} e' \cos(2v' + \omega') - \frac{1}{12} e'^2 \cos(3v' + \omega'), \tag{6.128}$$

$$C_3^2 = -\frac{1}{4} e'^2 \cos(v' + 3\omega') - \frac{1}{2} e' \cos(2v' + 3\omega') - \frac{1}{3}\left(1 + \frac{1}{2} e'^2\right) \cos(3v' + 3\omega') - \frac{1}{4} e' \cos(4v' + 3\omega') - \frac{1}{20} e'^2 \cos(5v' + 3\omega') \tag{6.129}$$

exactly; the constants of integration being consistently omitted.

If we insert the expressions J_j' as given now by Equations (6.117)–(6.120) in (6.111), the time integral of the disturbing function for long-range perturbations can be shown to assume the explicit form

$$\begin{aligned} I' = {} & k_1' \{(1-e^2)[(m_1^2+m_2^2)S_0^1 + 2m_1m_2C_2^1 + (m_2^2-m_1^2)S_2^1] \\ & + (1+4e^2)[(n_1^2+n_2^2)S_0^1 + 2n_1n_2C_2 + (n_2^2-n_1^2)S_2^1] \\ & - 2\left(\frac{2}{3}+e^2\right)S_0^1\} \\ & - k_2'e(1-e^2)\{[3m_1^2n_1 + m_2(2m_1n_2+m_2n_1)]C_1^2 \\ & + [3m_2^2n_2 + m_1(m_1n_2+2m_2n_1)]S_1^2 \\ & - [m_1^2n_1 - m_2(2m_1n_2+m_2n_1)]C_3^2 \\ & + [m_2^2n_2 - m_1(m_1n_2+2m_2n_1)]S_3^2\} \\ & - k_2'e\left(1+\frac{4}{3}e^2\right)\{3(n_1^2+n_2^2)(n_1C_1^2+n_2S_1^2) \\ & - n_1(n_1^2-3n_2^2)C_3^2 + n_2(n_2^2-3n_1^2)S_3^2\} \\ & + \frac{16}{5}k_2'e\left(1+\frac{3}{4}e^2\right)\{n_1C_1^2+n_2S_1^2\}, \end{aligned} \tag{6.130}$$

where we have abbreviated

$$k_1' = \frac{3}{8}\,\frac{\tilde{G}m_3a^2}{n'(1-e'^2)^{3/2}A'^3} \tag{6.131}$$

and

$$k_2' = \frac{25}{8}\,\frac{m_1-m_2}{m_1+m_2}\,\frac{Ak_1'}{A'(1-e'^2)}\;. \tag{6.132}$$

The variations δL, δG, δH, and δg, δh, $\delta \ell$ of the Delaunay elements for long-range perturbations of the close orbit can again be obtained by partial differentiation of I' with respect to the appropriate conjugate element. Since the mean longitude ℓ disappeared as a result of the averaging which led to (6.106), it follows once again from the first one of the Equation (6.4) of motion that

$$\delta L = 0\,; \tag{6.133}$$

and thus L proves to be constant in the long-range case as well. The long-range variations of other elements then follow from the equations

$$\left.\begin{aligned} \delta G &= \frac{\partial I'}{\partial g}, & \delta g &= -\frac{\partial I'}{\partial G}, \\ \delta H &= \frac{\partial I'}{\partial h}, & \delta h &= -\frac{\partial I'}{\partial H}; \end{aligned}\right\} \tag{6.134}$$

where I' has been given by Equation (6.130) above.

A partial differentiation of I' with respect to the individual Delaunay elements can be carried out in an analogous manner as in section V-1A in connection with the short-range perturbations, by remembering that

$$\left.\begin{aligned} m_1 &= (H/G)\cos g, \\ m_2 &= \quad -\sin g; \\ n_1 &= (H/G)\sin g, \\ n_2 &= \quad \cos g; \\ e &= \sqrt{1-(G/L)^2}. \end{aligned}\right\} \tag{6.135}$$

The primed elements of the third orbit continue to be regarded as constants – with the exception of its longitude ω' of the pericenter measured from the (varying) longitude of the node. In point of fact,

$$u' = \omega' + v' = \varpi' - \Omega + v', \tag{6.136}$$

where it is the longitude ϖ' of the pericenter of the orbit m_3 from a fixed direction in space which remains constant in time; and, hence, the angle

$$\omega' = \varpi' - \Omega = \varpi' - h \tag{6.137}$$

in Equations (6.124) – (6.129) depends linearly on h. The same is not true, however, of the true anomaly v' of m_3 in its orbit, which remains a known function of the time (given in terms of its mean anomaly ℓ' by the familiar expansions of elliptic motion) and independent of the Delaunay elements of the close orbit as long as ω' remains constant.

The actual evaluation of the variations (6.134) of the individual Delaunay elements by partial differentiation of (6.130) and subsequent evaluation of the close orbit by use of (6.35)–(6.40) is straightforward and offers no difficulty. The explicit forms of the results are, however, again too lengthy to be displayed here in full; and in what follows only the terms factored by k_1' will be given. These are:

$$\begin{aligned} (k_1')^{-1}\delta G = &- 5e^2(1-q^2)\sin 2\omega(v' + e'\sin v') \\ &- \frac{5}{4}e^2(1+q)^2 C(2\omega' - 2\omega) \\ &+ \frac{5}{4}e^2(1-q)^2 C(2\omega' + 2\omega), \end{aligned} \tag{6.138}$$

$$\begin{aligned} (k_1')^{-1}\delta H = &- \left(1 + \frac{3}{2}e^2\right)(1-q^2)C(2\omega') \\ &- \frac{5}{4}e^2(1-q)^2 C(2\omega' + 2\omega) \\ &- \frac{5}{4}e^2(1+q)^2 C(2\omega' - 2\omega), \end{aligned} \tag{6.139}$$

$$\begin{aligned} G(k_1')^{-1}\delta g = &[5q^2 - 1 + e^2 + 5(1 - e^2 - q^2)\cos 2\omega](v' + e'\sin v') \\ &+ \frac{3}{2}\left(1 - e^2 - \frac{5}{3}q^2\right) S(2\omega) \end{aligned} \tag{6.140}$$

$$+\frac{5}{4}(1-q)(1-q-e^2)S(2\omega'+2\omega)$$

$$+\frac{5}{4}(1+q)(1+q-e^2)S(2\omega'+2\omega),$$

$$\begin{aligned}H(k_1')^{-1}\,\delta h = &-[2+3e^2-5e^2\cos 2\omega]\,q(v'+e'\sin v')\\ &+\left(1+\frac{3}{2}e^2\right)q^2S(2\omega')-\frac{5}{4}e^2q(q-1)S(2\omega'+2\omega)\\ &-\frac{5}{4}e^2q(q+1)S(2\omega'-2\omega),\end{aligned} \tag{6.141}$$

where

$$\left.\begin{aligned}S(\alpha) &= e'\sin(v'+\alpha)+\sin(2v'+\alpha)+\frac{1}{3}e'\sin(3v'+\alpha),\\ C(\alpha) &= e'\cos(v'+\alpha)+\cos(2v'+\alpha)+\frac{1}{3}e'\cos(3v'+\alpha),\end{aligned}\right\} \tag{6.142}$$

and where we have abbreviated $q \equiv \cos i$.

The corresponding equations for the perturbations of the Keplerian elements of the close orbit follow from Equations (6.35)–(6.40) as

$$\delta A = 0, \tag{6.143}$$

$$\begin{aligned}(nA^2)(1-e^2)^{-1/2}\delta e = \frac{5}{4}k_1'e(v'+e'\sin v')\,\{&4(1-q^2)\sin 2\omega\\ &+(1+q)^2C(2\omega'-2\omega)-(1-q)^2C(2\omega'+2\omega)\},\end{aligned} \tag{6.144}$$

$$\begin{aligned}(nA^2)(1-e^2)^{1/2}(k_1')^{-1}\delta\omega = (v'+e'\sin v')\,[&e^2-1+5q^2\\ &+5(1-e^2-q^2)\cos 2\omega]\\ &+\frac{1}{2}(3+3e^2-5q^2)S(2\omega)\\ &+\frac{5}{4}(1-q)(1-q-e^2)S(2\omega'+2\omega)\\ &+\frac{5}{4}(1+q)(1+q-e^2)S(2\omega'-2\omega),\end{aligned} \tag{6.145}$$

$$\begin{aligned}nA^2(1-e^2)^{1/2}(k_1')^{-1}\delta\Omega = &-q(v'+e'\sin v')(2+3e^2-5e^2\cos 2\omega)\\ &+\left(1+\frac{3}{2}e^2\right)q^2S(2\omega')\end{aligned} \tag{6.146}$$

$$+\frac{5}{2}\,e^2q(1-q)\,S(2\omega'+2\omega)$$

$$-\frac{5}{2}\,e^2q(1+q)S(2\omega'-2\omega)\,,$$

$$\begin{aligned} nA^2(1-e^2)^{1/2}\,(k_1')^{-1}\sin i\,\delta i = &-5(v'+e'\sin v')\,e^2(1-q^2)\sin 2\omega \\ &-\frac{5}{4}\,e^2(1+q)^2C(2\omega'-2\omega) \\ &+\frac{5}{4}\,e^2(1-q)^2qC(2\omega'+2\omega) \\ &-\left(1+\frac{3}{2}\,e^2\right)\,(1-q^2)C(\omega'), \end{aligned} \tag{6.147}$$

and (by the same method as for short-range perturbations)

$$\begin{aligned} (nA^2)\delta(nT) = k_1'\,\Big\{&2(v'+e'\sin v')\,[(2+3e^2)\left(q^2-\frac{1}{3}\right) \\ &+5e^2(1-q^2)\cos 2\omega] \\ &+(2-3e^2)\,(1-q^2)S(2\omega') \\ &+\frac{5}{2}\,e^2(1-q)^2S(2\omega'+2\omega) \\ &+\frac{5}{2}\,e^2(1+q)^2S(2\omega'-2\omega)\,+G\delta g+H\delta h. \end{aligned} \tag{6.148}$$

A construction of the explicit form of the perturbations factored by k_2' with the aid of the preceding results offers no difficulty in principle; but the outcome proves too long to be printed here in full; and its derivation may be once more left as an exercise for the interested reader. As in section V-1A, however, particular interest attaches to the *secular* perturbations $[\delta x]$ of the six Keplerian elements x, obtained if the limits of integration on the right-hand side of Equation (6.111) have been extended over the entire orbit of the third body – i.e., over the interval $(0, 2\pi)$ in the true anomaly v' – and these will, in what follows, be evaluated explicitly in full.

In doing so we note from (6.123)–(6.129) that the only non-vanishing $[C_j^i]$'s and $[S_j^i]$,s are

$$\begin{aligned} [S_0^1] &= 2\pi, \\ [S_1^2] &= 2\pi e'\cos(\omega'-h)\,, \\ [C_1^2] &= 2\pi e'\sin(\omega'-h)\,; \end{aligned} \tag{6.149}$$

as a result of which

$$[I'] = 2\pi k_1' \left\{ \left(1 + \frac{3}{2}e^2\right)\left(q^2 - \frac{1}{3}\right) + \frac{5}{2}e^2(1 - q^2)\cos 2\omega \right\}$$

$$+ 2\pi e e' k_2' q \left\{ \left(1 + \frac{3}{4}e^2\right)\left(\frac{1}{5} + 3q^2\right)\sin\omega + \frac{7}{4}e^2(1 - q^2)\sin 3\omega' \right\} m_2'$$

$$- 2\pi e e' k_2' \left\{ \left(1 + \frac{3}{4}e^2\right)\left(\frac{11}{5} + q^2\right)\cos\omega + \frac{7}{4}e^2(1 - q^2)\cos 3\omega \right\} n_2' ; \quad (6.150)$$

where, by analogy with (6.116), we have abbreviated

$$\left.\begin{aligned} m_2' &= -\sin\omega' = -\sin(\bar{\omega}' - h), \\ n_2' &= \cos\omega' = \cos(\bar{\omega}' - h). \end{aligned}\right\} \quad (6.151)$$

On insertion for $m_{1,2}$ and $n_{1,2}$ from (6.116), the foregoing Equations (6.150) become readily differentiable with respect to the Delaunay elements to obtain their requisite variations from (6.134) and the latter can, in turn, be converted by (6.35)–(6.40) into secular perturbations of the respective Keplerian elements of the close orbit. On going through the necessary steps we eventually establish that

$$[\delta A] = 0, \quad (6.152)$$

$$\frac{[\delta e]}{(1 - e^2)} = 5\kappa_1 e(1 - q^2)\sin 2\omega -$$

$$- \kappa_2' e' \left\{ \left(1 + \frac{3}{4}e^2\right)\left[\left(q^2 + \frac{11}{5}\right)\sin\omega\cos\omega' - q\left(3q^2 + \frac{1}{5}\right)\cos\omega\sin\omega'\right]\right.$$

$$\left. + \frac{21}{4}(1 - q^2)\left[\sin 3\omega\cos\omega' - q\cos 3\omega\sin\omega'\right]\right\}, \quad (6.153)$$

$$\frac{[\delta i]}{e\sin i} = -5\kappa_1' eq\sin 2\omega$$

$$+ \kappa_2' e' \left\{ \left(1 + \frac{3}{4}e^2\right)\left[\left(3q^2 + \frac{11}{5}\right)\cos\omega\sin\omega' + 2q\sin\omega\cos\omega'\right]\right.$$

$$\left. + \frac{7}{4}e^2(1 - 3q^2)\cos 3\omega\sin\omega' + 2q\sin 3\omega\cos\omega' \right\}, \quad (6.154)$$

$$[\delta\omega] = \kappa_1' \left\{5q^2 - 1 + e^2 + 5(1 - e^2 - q^2)\cos 2\omega\right\}$$

$$- \kappa_2' e' \left\{ \frac{1 - e^2}{e}\left(1 + \frac{9}{4}e^2\right)\left[\left(\frac{11}{5} + q^2\right)\cos\omega\cos\omega' + q\left(\frac{1}{5} + 3q^2\right)\sin\omega\sin\omega'\right]\right.$$

$$+ e\left(1 + \frac{3}{4}e^2\right)q\left[2q\cos\omega\cos\omega' + 9\left(q^2 + \frac{1}{5}\right)\sin\omega\sin\omega'\right] + \quad (6.155)$$

$$+\frac{21}{4}e(1-e^2)(1-q^2)[\cos 3\omega \cos \omega' + q \sin 3\omega \sin \omega']$$

$$-\frac{7}{4}e^3 q[2q \cos 3\omega \cos \omega' + (3q^2-1)\sin 3\omega \sin \omega']\Big\},$$

$$[\delta\Omega] = -2\kappa_1' q\left(1+\frac{3}{2}e^2-\frac{5}{2}e^2\cos 2\omega\right)$$

$$+\kappa_2' e'\left\{e\left(1+\frac{3}{4}e^2\right)\left[2q\cos\omega\cos\omega' + 9\left(q^2+\frac{1}{5}\right)\sin\omega\sin\omega'\right]\right. \tag{6.156}$$

$$\left.-\frac{7}{4}e^3[2q\cos 3\omega\cos\omega' + (3q^2-1)\sin 3\omega \sin\omega']\right\},$$

and

$$n(1-e^2)^{-1/2}[\delta T] = \kappa_1'\left\{\left(\frac{7}{3}+e^2\right)(3q^2-1)+5(1+e^2)(1-q^2)\cos 2\omega\right\}$$

$$+\kappa_2' e'\frac{1-e^2}{e}\left\{\left(1+\frac{9}{4}e^2\right)\left[\left(\frac{1}{5}+3q^2\right)q\sin\omega\sin\omega' + \left(\frac{11}{5}+q^2\right)\cos\omega\cos\omega'\right]\right.$$

$$\left.+\frac{21}{4}e^2(1-q^2)[q\sin 3\omega\sin\omega' + \cos 3\omega\cos\omega']\right\} \tag{6.157}$$

$$+6\kappa_2' ee'\left\{\left(1+\frac{3}{4}e^2\right)\left[\left(\frac{1}{5}+3q^2\right)q\sin\omega\sin\omega' + \left(\frac{11}{5}+q^2\right)\cos\omega\cos\omega'\right]\right.$$

$$\left.+\frac{7}{4}e^2(1-q^2)[q\sin 3\omega\sin\omega' + \cos 3\omega\cos\omega']\right\},$$

where

$$\kappa_1' = \frac{3}{4}\,\frac{\pi\tilde{G}m_3}{nn'A'^3\sqrt{1-e^2)}(1-e'^2)^3}$$

$$= \frac{3}{4}\,\frac{m_3}{m_1+m_2+m_3}\left(\frac{n'}{n}\right)\frac{\pi}{\sqrt{1-e^2)}(1-e'^2)^3} \tag{6.158}$$

and

$$\kappa_2' = \frac{25}{8}\left(\frac{m_1-m_2}{m_1+m_2}\right)\frac{a\kappa_1'}{A'(1-e'^2)} \tag{6.159}$$

are nondimensional constants.

If the close and wide orbits happen to be coplanar (i.e., $q = 1$), the leading terms of the secular perturbations of the elliptical elements become

$$[\delta e] = -\tfrac{4}{5}\kappa_2' e'(1-e^2)(4+3e^2)\sin(\omega'-\omega), \tag{6.160}$$

$$[\delta i] = 0, \tag{6.161}$$

$$[\delta\omega] = 2\kappa_2'(2-2e^2+5e^2\sin^2\omega), \tag{6.162}$$

$$[\delta\Omega] = -\kappa_1'(2 + 3e^2 - 5e^2 \cos 2\omega), \tag{6.163}$$

and

$$n(1 - e^2)^{-1/2} [\delta T] = \tfrac{2}{3}\, \kappa_1'(7 + 3e^2)\,; \tag{6.164}$$

so that

$$[\delta\omega[= 2\kappa_1'(1 - e^2). \tag{6.165}$$

If, on the other hand, the close orbit happens to be circular (i.e., $e = 0$) but $q \neq 1$, the leading terms of the respective perturbations assume the form

$$[\delta e] = -\tfrac{1}{5}\, \kappa_2' e' \{(5q^2 + 11) \sin \omega \cos \omega' - q(15q^2 + 1) \cos \omega \sin \omega'\} \tag{6.166}$$

$$[\delta i] = 0, \tag{6.167}$$

but

$$[\delta\omega] = 2\kappa_1' (2 - 5 \sin^2 i \sin^2 \omega), \tag{6.168}$$

$$[\delta\Omega] = -2\kappa_1' \cos i, \tag{6.169}$$

$$n[\delta \mathrm{T}] = 2\kappa_1' \left(\tfrac{7}{3} - \sin^2 i - 5 \sin^2 i \sin^2 \omega \right), \tag{6.170}$$

and

$$[\delta\bar{\omega}] = 2\kappa_1'(2 - \cos i - 5 \sin^2 i \sin^2 \omega). \tag{6.171}$$

These equations represent the last resultsof our analysis which by now stands essentially complete. The results confirm that, to the order of accuracy we have been working, *the semi-major axis A of the close orbit is secularly constant, and subject only to periodic perturbations.* With regard to the *eccentricity* e and *inclination* i of the close orbit, Equations (6.153) and (6.154) disclose that, as e and i tend to zero, the secular motions of these elements vanish. It follows, therefore, that short-periodic 'secular' perturbations of these elements indicated by Equations (6.94) and (6.98) are in reality long-periodic (with periods equal to that of the third orbit and its submultiples), and only simulate secular trends as long as the third body is not allowed to move. However, our Equations (6.153) and (6.154) disclose that, *for $e > 0$ and $i > 0$, these elements undergo genuine secular perturbations;* whether these will result in an increase or decrease of the respective element depends on the algebraic sign of $\sin 2\omega$.

Of particular interest are the secular motions of the *apse* and the *node* – the former *advancing*, the latter *receding* at a rate which tends to the same limit as $e \to 0$ and $i \to 0$, in such a way that sum $\Omega + \omega$ of these angles (both measured from a fixed direction in space) tends to remain constant. In general, however, the periods U, V of revolution of the apsidal and nodal lines, expressed in terms of the orbital period P' of the disturbing body m_3, will be given by the ratios

$$\frac{P'}{U} = \|[\delta\omega]\| \quad \text{and} \quad \frac{P'}{V} = \|[\delta\Omega]\|\,, \tag{6.172}$$

with the values of $[\delta\omega]$ and $[\delta\Omega[$ as given by Equations (6.155) and (6.156).

As to the *periodic* perturbations of different elements, the existence of those principal families of such perturbations have been established: namely,

(1) short-periodic perturbations, of period equal to that of the the close orbit and its submultiples (argument jE), arising from the harmonic terms on the right-hand side of Equation (6.23);

(2) long-periodic perturbations, of period equal to that of the wide orbit and its submultiples (argument: jv'), going back to the harmonic terms of the functions C_j's and $S_j^!$'s as defined by Equations (6.123) to (6.129); and

(3) apse-node terms, arising from the secular motions of the angles ω and Ω, and oscillating with the periods U, V (or their sub-multiples and differences), as given by Equation (6.172).

D. EFFECTS OF THE LIGHT EQUATION

Not all orbital elements exhibit, to be sure, the perturbations of all these types. For instance, the semi-major axis a of the relative orbit of the close pair is (within the scheme of our approximations) found to be subject to neither secular nor long-periodic perturbations; and, as a result, the period P of the sidereal orbit of the two bodies of masses m_1 and m_2 will likewise remain constant. Should this close binary happen to be an eclipsing variable, its apparent period defined as a time interval between two successive light minima will, of course, be subject to complicated fluctuations arising from the secular motion of ω and Ω, and long-periodic oscillation of i, of the type already investigated in section V-2 of this chapter. Even if it were not, however, for such oscillations (whose periods would be of the order of U or V and, therefore, long in comparison with P'), the *apparent* (observed) period P of the light changes of an eclipsing binary which is accompanied by a third body would still not be constant on account of the *light equation* of our variable in its absolute orbit around the centre of gravity of the triple system; and the effects of such a light equation remain yet to be investigated.

In order to do so, consider a system of rectangular coordinates XYZ, with origin at the centre of mass of the triple system, oriented so that the Z-axis coincides with the line of sight and the XY-plane is then tangent to the celestial sphere. Let, moreover, z denote the distance of the centre of mass of the eclipsing system from the XY-plane. If this system is accompanied through space by a third body, the ensuing orbital motion will cause z to vary as

$$z = r \sin i' \sin(v' + \omega'), \tag{6.173}$$

where r stands for the radius-vector of the absolute orbit of the mass-centre of the close pair around that of the whole system; i', for the angle of inclination of this orbit to the celestial sphere; and (as before) v' denotes the true anomaly of the centre of mass of the system $m_{1,2}$, while ω' is the longitude of the periastron from the ascending node at which the orbit intersects the XY-plane. If, furthermore, the origin of coordinates itself is moving towards, or away from, the observer as a result of the space motion of the whole triple system with a (radial) velocity γ in the z-direction, the distance Z between this system and the observer at any time t becomes

$$Z = z_0 + \gamma(t - t_0) + r \sin i' \sin(v' + \omega'), \tag{6.174}$$

where z_0 denotes the initial value of Z at the time t_0 of a light minimum of the eclipsing system.

If, however, the distance Z between us and the eclipsing *system ceases to be constant, the time required for any light signal* (such as the occurrence of a minimum) *to reach us from our binary must obviously vary with* z, as a consequence of the 'light equation' in the absolute orbit of $m_{1,2}$ (i.e., of the fact that the light takes a finite time to traverse this orbit). In more specific terms, even if the sidereal period P_0 of the close pair were undisturbed by other sources, the ephemeris $M(E)$ of the light minima taking place at the epoch E (not to be confused with the eccentric anomaly) should contain a term $(Z - z_0)/c$ – where c, denotes the velocity of light – and assume the form

$$M(E) = M(0) + \left(1 + \frac{\gamma}{c}\right) P_0 E + \frac{r}{c} \sin i' \sin(v' + \omega'), \tag{6.175}$$

where we have set

$$t - t_0 = P_0 E. \tag{6.176}$$

Now, according to the known expansion of the elliptic motion,

$$r \sin v' = A_{12} \sum_{k=1}^{\infty} g_k(e') \sin n'k(t - T), \tag{6.177}$$

$$r \cos v' = A_{12} \sum_{k=1}^{\infty} h_k(e') \cos n'k(t - T) - \frac{3}{2} A_{12} e', \tag{6.178}$$

where A_{12} denotes the semi-major axis of the absolute orbit of the centre of mass of the eclipsing pair around that of the triple system; e', its eccentricity; n', the mean daily motion; T, the time of the periastron passage; and where we have abbreviated

$$g_k(e') \equiv 2\sqrt{1 - e'^2}\, \frac{J_k(ke')}{ke'} \quad \text{and} \quad h_k(e') \equiv \frac{2}{k^2} \frac{\mathrm{d} J_k(ke')}{\mathrm{d}e'} \;; \tag{6.179}$$

$J_k(ke')$ denoting the Bessel function of k-th order. Inserting (6.177) and (6.178) in (6.175) and making use of (6.176), we find that the ephemeris $M(E)$ for the times of minima of the eclipsing pair, as influenced by the 'light equation' due to its orbital mortion, will assume the more explicit form

$$\begin{aligned} M(E) &= M(0) + \left(1 + \frac{\gamma}{c}\right) P_0 E + K \left\{ \sum_{k=1}^{\infty} g_k(e') \cos \omega' \sin k\,(\nu E + \nu_0) \right. \\ &\quad \left. + h_k(e') \sin \omega' \cos k(\nu E + \nu_0) - \tfrac{3}{2}\, e' \sin \omega' \right\} \\ &= M(0) + \{1 + (\gamma/c)\}\, P_0 E + K \{\sin(\nu E + \nu_0 + \omega') \\ &\quad + \tfrac{1}{2}\, e' [2 \sin(2\nu E + 2\nu_0 + \omega') - 3 \sin \omega'] \end{aligned} \tag{6.180}$$

$$+ \tfrac{1}{8} e'^2 [18 \sin(3\nu E + 3\nu_0 + \omega')]$$

$$- \sin(4\nu E + \nu_0 + \omega') + \sin\omega' \cos(\nu E + \nu_0) + \cdots \},$$

where we have abbreviated

$$\begin{aligned} n'(t - t_0) &= 2\pi(P_0/P')E \equiv \nu E, \\ n'(t_0 - T) &= 2\pi(t_0 - T)/P' \equiv \nu_0; \end{aligned} \tag{6.181}$$

and

$$K = \frac{A_{12} \sin i'}{c} \quad ; \tag{6.182}$$

P' denoting (as before) the orbital period of the wide pair.

Suppose now that a sufficiently long sequence of the times of the minima has been observed, and analysed harmonically to yield a Fourier representation of the form

$$M(E) = M(0) + a_0 E + \sum_{k=1}^{\infty} \{a_k \sin(k\nu E) + b_k \cos(k\nu E)\}, \tag{6.183}$$

where the a_k's and b_k's as well as ν are empirical constants. If so, a comparison of the corresponding terms on the right-hand sides of Equations (6.180) and (6.182) reveals that

$$a_0 = \left(1 + \frac{\gamma}{c}\right) P_0 \tag{6.184}$$

and, for $k \geqslant 1$,

$$a_k = K \{g_k(e') \cos\omega' \cos k\nu_0 - h_k(e') \sin\omega' \sin k\nu_0\}, \tag{6.185}$$

$$b_k = K \{g_k(e') \cos\omega' \sin k\nu_0 + h_k(e') \sin\omega' \cos k\nu_0\}, \tag{6.186}$$

while

$$\nu = 2\pi(P_0/P'). \tag{6.187}$$

If we square now Equations (6.185)–(6.186) and add, we find that

$$a_k^2 + b_k^2 = K^2 \{g_k^2(e') \cos^2\omega' + h_k^2(e') \sin^2\omega'\}, \tag{6.188}$$

i.e., a relation form which ν_0 has been eliminated. Any pair of the ratios $(a_j^2 + b_j^2)\,(a_k^2 + b_k^2)$ for two different values of $j \neq k$ thus represent a system of two simultaneous equations for the determination of e' and ω'; whereupon ν_0 can be determined from any available ratio a_j/a_k or b_j/b_k; and, with e', ω', and ν_0 thus known, the value of $a_{12} \sin i' = cK$ can be obtained from any a_k or b_k. Therefore, the minimum number n of terms on the right-hand side of Equation (6.183) sufficient for the determination of a complete set of the elements $a_{12} \sin i$,, e', ω' and ν_0 (or T) of the orbit of the close pair around the mass centre of the triple system thus turns out to be *three*; and if more are available, the elements can be adjusted by least-squares or other suitable techniques.

A more detailed description of such techniques is outside the scope of this section. If, however, the orbital eccentricity e' happens to be small enough for its squares and higher powers to become ignorable, the procedure is somewhat simplied, and it can be shown that two terms of the summation on the right-hand side of (6.183) necessary for a determination of the orbital elements reduces them to *two.* These terms assume the forms

$$a_1 = K\cos(\nu_0 + \omega'), \qquad b_1 = K\sin(\nu_0 + \omega')\,; \tag{6.189}$$

$$a_2 = \tfrac{1}{2}\, e'K\cos(2\nu_0 + \omega'),\ b_2 = \tfrac{1}{2}\, e'K\sin(2\nu_0 + \omega'); \tag{6.190}$$

and their inversion yields

$$A_{12}\sin i' = c\sqrt{a_1^2 + b_1^2}, \tag{6.191}$$

$$e = 2\sqrt{\frac{a_2^2 + b_2^2}{a_1^2 + b_1^2}} \tag{6.192}$$

$$\omega' = \tan^{-1}\left\{\frac{(b_1^2 - a_1^2)\,b_2 + 2a_1a_2b_1}{(a_1^2 - b_1^2)\,a_2 + 2a_1b_1b_2}\right\}, \tag{6.193}$$

$$t_0 - T = \frac{P_0}{\nu}\tan^{-1}\left\{\frac{a_1b_2 - b_1a_2}{a_1a_2 + b_1b_2}\right\}, \tag{6.194}$$

$$P' = \frac{2\pi P_0}{\nu}, \tag{6.195}$$

as the solution of our problem.

If the orbital eccentricity e' becomes large, as revealed by an appreciable skewness of the function $M(E)$, the situation becomes more complicated. Fortunately, it is not necessary to go into it here, as it has been previously investigated exhaustively in connection with the determination of spectroscopic orbits of close binary systems from radial-velocity observations. Indeed, the radial velocity V of the centre of mass of the close pair revolving around that of the triple system should, by definition, be given by

$$V \equiv \frac{dZ}{dt} = c\left\{\frac{dM}{dt} - 1\right\} = c\left\{\frac{1}{P_0}\frac{dM}{dE} - 1\right\}. \tag{6.196}$$

As, moreover, to a sufficient approximation

$$\frac{dM}{dE} = P(E) \tag{6.197}$$

if the ratio P_0/P' is small, a combination of Equations (6.196) and (6.197) reveals that

$$\frac{V}{c} = \frac{P - P_0}{P_0}, \tag{6.198}$$

i.e., that *the systemic radial velocity V of our eclipsing binary is directly expressible in terms of the apparent period changes due to the revolution of its centre of gravity around* *that of the triple system,* and vice versa. *The observed variation of P(E) arising from this cause should, therefore, enable us to deduce from it all elements of such an orbit as are deducible from a single-spectrum radial velocity curve.* In point of fact, the variation of the orbital period due to the 'light equation' specifies the corresponding radial-velocity changes just as well as the spectroscope – in place of the frequency-shifts of vibrating atoms we observe the period changes of the eclipsing pair!

In order to obtain a more explicit form of (6.198) for the purpose of orbit determination, let us return to our previous Equation (6.174) for Z and differentiate it with respect to the time: we obtain

$$\frac{V}{c} \equiv \frac{P}{P_0} - 1 = \frac{\gamma}{c} + \frac{n'K}{\sqrt{1-e'^2}} \{\cos(v' + \omega') + e' \cos \omega'\}, \tag{6.199}$$

governing the systemic radial velocity (or period) changes of the eclipsing pair in terms of the elements of its orbit around the mass centre of the triple system. The methods – graphical or numerical – for extracting the elements of the spectroscopic orbit from such radial-velocity changes are too numerous and too well known to justify a review in this plane.* If the orbital eccentricity e' happens to be small, a direct solution of our problem as represented by Equations (6.191)–(6.195) should usually suffice. If, however, a plot of V against the time reveals appreciable skewness, recourse must be had to other methods (Lehmann-Filhés, Schwarzschild, Zurhellen, King, etc.) leading to elements which must, however, be regarded as preliminary in the first instance, and improved subsequently by least-squares corrections in the following manner.

Let a preliminary set of the elements

$$M_0 \equiv M(0), \quad \gamma, \quad K \equiv a_{12} \sin i'/c, e', \omega', \; n' \equiv 2\pi/P', \quad \sigma' \equiv n'T$$

obtained – graphically or otherwise – by preliminary methods be inserted in Equation (6.183) and lead to the residuals ΔM in the times of the minima, observed at the different epochs E. If so, a differentiation of the Equation (6.183) reveals that

$$\Delta M = \frac{\partial M}{\partial M_0} \Delta M_0 + \frac{\partial M}{\partial \gamma} \Delta\gamma + \frac{\partial M}{\partial K} \Delta K + \frac{\partial M}{\partial e'} \Delta e' + \frac{\partial M}{\partial \omega'} \Delta\omega' + \frac{\partial M}{\partial n'} \Delta n' + \frac{\partial M}{\partial \sigma'} \Delta\sigma', \tag{6.200}$$

where the Δ's denote the errors of the respective elements, multiplied by the coefficients

$$\frac{\partial M}{\partial M_0} = 1, \tag{6.201}$$

$$\frac{\partial M}{\partial \gamma} = t, \tag{6.202}$$

$$\frac{\partial M}{\partial K} = \frac{(1-e'^2)\sin(v' + \omega')}{1 + e' \cos v'} \tag{6.203}$$

* For their survey cf., e.g., R. G. Aitken, *Binary Stars*, New York 1935; Chapter VI.

$$\frac{1}{K}\frac{\partial M}{\partial e'} = \frac{\sin(v'\cos(v'+\omega')}{1+e'\cos v'} - \sin\omega', \tag{6.204}$$

$$\frac{1}{K}\frac{\partial M}{\partial \omega'} = \frac{(1-e'^2)\cos(v'+\omega')}{1+e'\cos v'}, \tag{6.205}$$

$$\frac{1}{K}\frac{\partial M}{\partial n'} = \frac{t}{K}\frac{\partial M}{\partial \sigma'} =$$

$$= \frac{t}{\sqrt{1-e'^2}}\{\cos(v' + \omega') + e'\cos\omega' + e'^2\sin v'\sin\omega'\} \tag{6.206}$$

As many equations of conditions of the form (6.200) are available as there are observed times of the minima; and their least-squares solution should furnish the most probable values of the corrections to be applied to the preliminary values of the respective elements in order to obtain their definitive set. After this has been done, the values of K, e' and n' (or P') then speficy, in particular, the value of the mass-function

$$\frac{m_3^3\sin^3 i',}{(m_1+m_2+m_3)^2} = \frac{n'^2(cK)^3}{G} \tag{6.207}$$

of the triple system, just as if its single-spectrum orbit were available.

Close eclipsing systems known to be accompanied by invisible third bodies (which disclose their presence by effects – both spectroscopic and photometric – exerted by their attraction on the motion of the eclipsing pair) are too many to be referred to here in any detail. A typical example is again Algol (β Persei), which consists of a semi-detached eclipsing system of period close to 2.867 days, attended by a third star (at the limit of spectroscopic visibility), the orbital period of which is 1.873 years. The presence of this body, and the period of its orbit, have been verified both spectroscopically by its effects on the systemic radial velocity of the close pair (McLaughlin, 1934) as well as photometrically – from the 'light equation' in the third orbit, causing the period of the eclipsing pair to fluctuate as the latter revolves around the centre of gravity of the triple system (Eggen, 1948; Kopal *et al.*, 1960).

More recently, the observed effects of this 'light equation' on the period of the X-ray flickers exhibeted by binaries like Cen X-3, Her X-1, or SMC X-1 (cf. sec. VIII-5C) have served as a basis for determinations (or, at least, estimates) of the masses and absolute dimensions of these binaries as listed in Table VIII-11. However, perhaps the most spectacular example of such effects disclosing the binary nature of a system was recently provided by the pulsar PSR 1913 + 16 (cf. Hulse and Taylor, 1975), for which the interval between successive pulses was found to oscillate in a period of 0.323 days (cf. again sec. VIII-5C) as a result of the motion of this pulsar around the centre of gravity of a binary system of which the pulsar constitutes one component.

BIBLIOGRAPHICAL NOTES

All developments outlined in section V-1 are classical; and for their proofs the reader can be referred, e.g., to F. Tisserand, *Traité de mécanique céleste*, vol. 1 (Gauthier-Villars, Paris 1889) and many other subsequent sources.

V-2: The perturbations of the elements which specify the position of the orbital plane in space, and which are inextricably coupled with the precession and nutation of the components in close binary systems constitute a subject much less well covered by the existing literature. The classical problem of the precession and nutation of rigid spheroids whose equatorial planes are inclined to that of their orbit has been treated by D. Brouwer (1946), and generalized to the case of deformable components distorted by rotation and tides by Z. Kopal (1972a) by a method largely followed in section V-2. For subsequent work see M. E. Alexander (1976).

V-3: A treatment of the perturbations in the elements in the orbital plane arising from the rotational as well as tidal distortion of the components, contained in section V-3, follows likewise closely an earlier treatment of the subject by Z. Kopal (1972b). The results for the rotational distortion section V-3 A) agree with those deduced previously by the writer (Z. Kopal 1959; sec. III-6) by a different method; while those arising from tidal distortion coincide with those by T.E. Sterne (1939).

A comparison of the predicted rates of apsidal advance with the data provided by the observations of a number of eclipsing systems characterized by eccentric orbits, as given in section V-3C, represents an up-dated version of a previous discussion of this problem by Z. Kopal (1965a).

V-4: A discussion of the period changes of eclipsing binary systems – both real (caused by tidal distortion) and apparent (caused by secular variations of othe elements) as given in sections V-4 A and V-4 B follow largely a treatment of the subject developed previously by Z. Kopal (1959; sec. II.7) and improved subsequently by M. Plavec (1960). A discussion of the exact relation between the orbital period and the times of the minima of eclipsing variables, as given in section V-4C, goes back to an earlier work by Z. Kopal and R. Kurth (1957).

V-5: For the dynamical effects of variable mass the fundamental work of recent date is that by J. Hadjidemetriou (1963, 1966, 1967), whose treatment of the subject has largely been followed in parts V-5A and 5B of this section. In particular, the long review paper by J. Hadjidemetriou (1967) contains a very complete list of references to the earlier work on this subject.

The principal contributors to the subjects discussed in part V-5C have been S. L. Piotrowski (1964a, b; 1967) and A.Kruszewski (1964a, b, c; 1966). Attention is again invited to Kruszewski's 1966 review memoir for a comprehensive summary of the state of the subject by that time, and an exhaustive list of references to earlier literature.

Of subsequent work, J. Hadjidemetriou (1969) extended his previous investigations to a non-isotropic loss of mass from a source coinciding in position with that of the inner Lagrangian point (L_1) of the restricted problem of three bodies (cf. sec. VI-1). The mass loss has again been restricted to occur only in the orbital plane of the binary system; and to follow in direction the tangent to the osculating cone which approximates the shape of the respective Roche equipotential at L_1 (cf. sec. VI-2C). Of more recent work cf. V. V. Radzievsky and L. P. Surkova (1973), or L. P. Surkova (1973).

One additional comment on the subject of the present section should be made in this place: namely, that concerning possible effects of mass loss (or transfer) on the observed apsidal motions in close binary systems discussed (for the case of constant mass) earlier in section V-3C. To what extent can the rate of apsidal advance deduced in section V-3B be affected by any change in mass of the constituent components? For *isotropic* loss of mass the motion of the apsidal motion was found to be governed by Equation (5.20) or (5.25) as a function of the proportional mass change given by the ratio of $\dot{m}/m$. However, for *non-isotropic* mass loss no general answer can be given; for the value of ω then depends not only on $\dot{m}$, but also on the rate in which a mass transfer influences the momentum of the system. In other words, it is not only the amount of mass lost (or gained), but also the mode of its transfer which must be known before the corresponding change in ω can be specified.

No proper investigation of this aspect has so far been made; and the attempt to do so by J. U. Cisneros-Parra (1970) was based on wholly hypothetical premises. However, at least the order of magnitude of apsidal advance should be that of the ratio $\dot{m}/m$. For systems consisting of components on the Main Sequence (cf. Table V-1) this should be utterly negligible in comparison with the effects produced by tides. For evolved systems (Table V-2) the ratios $\dot{m}/m$ can be expected (cf. sec. VIII-3) to be of the order of 10^{-6} to 10^{-5} per annum; but even so their effects on apsidal motion are likely to be no greater than the relativistic effects represented by Equation (3.51).

Of earlier papers dealing with the hydrodynamics of the accretion of matter by the stars from the surrounding gas cf. F. Hoyle and R. A. Lyttleton (1939); H. Bondi and F. Hoyle (1944); H. Bondi (1952); or R. A. Lyttleton (1972).

V-6: The perturbations of the orbital elements of a close binary caused by the presence of a distant third star have provided a rewarding ground for application of different types of lunar theories to this stellar problem of three bodies. The fundamental contributions to this subject, not restricted to small eccentricities or inclinations of the third orbit, are those of E.W. Brown (1936, 1937); and his method has been extended by Z. Kopal (1967) to quantities of second order to describe the perturbations of the close pair. The contents of section V-6 are based essentially on the work contained in this latter reference.

The variations of the orbital period of the close pair, caused by the 'light equation' in triple system and described in section V-6C of this chapter, have previously been studied by several investigators, including the present writer (Z. Kopal, 1959; sec. II.8). This latter source contains also references to all previous work on this subject, the results of which have been incorporated into the text of the present section.

CHAPTER VI

THE ROCHE MODEL

In Chapter II of this book we investigated in some detail the equilibrium forms of the components of close binary systems, of arbitrary internal structure, distorted by rotational and tidal forces, defined as surfaces over which the potential of all forces acting within the system remained constant. Such an approach to a study of the components ran, however, into difficulties increasing rapidly with increasing requirements of precision – so much so that within the scope of Chapter II we were unable to progress in our description of the surfaces beyond quantities second order in surficial distortion produced by equilibrium tides, and of third order in polar flattening caused by centrifugal force of axial rotation.

The principal reason why an extension of this work to terms of higher orders proved so difficult was due to our desire to make the results applicable to stars of any structure. The aim of the present chapter will, however, be to show that if this latter requirement is given up, and the density concentration of the stars constituting a binary system is allowed to approach infinity, their shape can be described in a closed *algebraic* form, which is *exact* for any double-star configuration consisting of centrally-condensed components *irrespective of their proximity or mass-ratio*. Such a model is generally known in the literature as the *Roche Model*; and the aim of the present chapter will be to summarize its relevant geometrical and other properties of interest for the students of close binary systems.

Although the Roche model of stellar configurations – named so in honour of Edouard Albert Roche (1820–1883), a distinguished French mathematician in whose writings it appeared more than a century ago (Roche, 1849, 1851, 1873) – has been with us for a long time, for more than the first eighty years it led a rather shadowy existence on the periphery of cosmology – as a limiting case of a configuration whose total mass is condensed into a point at its centre – occupying a position at the extreme opposite end of a family of models (Maclaurin, Jacobi) consisting of homogeneous incompressible fluids. As very little was known about the interval structure of the stars until at least the second decade of this century, the Roche model continued to stand as a sentinel at one end of the extreme range of configurations which could represent the stars through the time of Poincaré.

Since the second decade of this century we learned, however, largely through the work of Emden, Eddington and their contemporaries – that most part of the mass in the interiors of Main Sequence stars is mostly condensed near their centres, and overwhelmingly so for giants. Moreover, from numerical work carried out soon thereafter by Chandrasekhar (1933) we learned that the external form of such stars subject to distortion approaches that given by the Roche model to a truly astonishing degree of approximation (for the latest quantitative discussion of this aspect cf., e.g., Plavec, 1958; or Orlov,

1960) – a fact which renders quantitative properties of such a model of all the greater interest for the students of close binary systems.

This is particularly so if these happen to be eclipsing variables; for in such a case many of their geometrical as well as physical properties can be deduced from observations of their light changes on the basis of a theory of far-reaching exactitude. The aim of the present chapter will, therefore, be to summarize the salient points of this theory for the interested reader.

In section VI-1 which follows these introductory remarks, we shall be concerned with formal geometrical properties of the Roche model, and its relation with the zero-velocity surface of the restricted problem of two bodies. Section VI-2 will give a quantitative description of the geometry of the equipotential surfaces up to the Roche limit which represents the largest closed equipotential capable of containing the whole mass of the respective star. In section VI-3 we shall use this geometry as a basis for an analysis of the eclipse phenomena to be expected in binary systems which can be represented by the Roche model.

Finally, in the concluding section VI-4, we shall prepare the ground for studies of hydrodynamic gas streaming in close binary systems (to be discussed further in section VIII-3 by the introduction of 'Roche curvilinear coordinates', in which the Roche equipotentials ξ = constants play the role of the spheres r = const in spherical polars; while the angular variables η and ζ are orthogonal to the ξ-equipotentials. By virtue of the simplicity with which the distorted surface can be defined (or closely approximated by) in such coordinates (namely, ξ = constant) their use will facilitate greatly the formulation of the boundary conditions to be imposed on the gas flow, as well as many other tasks encountered in the theory of the light (or radial-velocity) changes of close eclipsing systems.

VI-1. Roche Equipotentials

In order to introduce the Roche equipotentials, let $m_{1,2}$ denote the masses of the two components of a close binary system; and R, the separation of their centres of gravity. Suppose, moreover, that the positions of these centres are referred to a rectangular system of Cartesian coordinates, with the origin at the centre of gravity of mass m_1 – the x-axis of which coincides with the line joining the centres of the two stars (i.e., the radius-vector of the relative orbit of the two masses); while the z-axis is perpendicular to the plane of the orbit. If so, the coordinates of the centre of gravity of the system are

$$\frac{m_2 R}{m_1 + m_2}\ ,\quad 0\ ,\quad 0\ ; \tag{1.1}$$

and the total potential Ψ of all forces acting at an arbitrary point $P(x, y, z)$ becomes expressible as

$$\Psi = G\frac{m_1}{r} + G\frac{m_2}{r'} + \frac{\omega^2}{2}\left\{\left(x - \frac{m_2 R}{m_1 + m_2}\right)^2 + y^2\right\}, \tag{1.2}$$

where

$$\begin{aligned} r^2 &= x^2 + y^2 + z^2, \\ r'^2 &= (R - x)^2 + y^2 + z^2, \end{aligned} \tag{1.3}$$

represent squares of the distance of P from the centres of gravity of the two components, and ω denotes the angular velocity of rotation of the system about an axis perpendicular to the orbital plane and passing through the centre of gravity of the system whose coordinates are given by (1.1). The first term on the right-hand side of Equation (1.2) represents the potential arising from the mass m_1; the second, the disturbing potential of its companion of mass m_2; and the third, the potential arising from the centrifugal force.

Let, furthermore, the angular velocity ω on the right-hand side of Equation (1.2) be identified with the Keplerian angular velocity

$$\omega_K^2 = \frac{G(m_1 + m_2)}{R^3} \tag{1.4}$$

of the system. If we insert (1.4) in (1.2) and, moreover, adopt m_1 as our unit of mass; R, as the unit of length while the unit of time is chosen so that $G = 1$, Equation (1.2) may be expressed in terms of spherical polar coordinates

$$\left.\begin{aligned} x &= r\cos\phi\sin\theta = r\lambda, \\ y &= r\sin\phi\sin\theta = r\mu, \\ z &= r\cos\theta \qquad\quad = r\nu, \end{aligned}\right\} \tag{1.5}$$

as

$$\xi = \frac{1}{r} + q\left\{\frac{1}{\sqrt{1 - 2\lambda r + r^2}} - \lambda r\right\} + \frac{q+1}{2} r^2(1 - \nu^2), \tag{1.6}$$

where

$$\xi \equiv \frac{R\Psi}{Gm_1} - \frac{m_2^2}{2m_1(m_1 + m_2)} \tag{1.7}$$

and

$$q \equiv \frac{m_2}{m_1} \tag{1.8}$$

are nondimensional parameters.

The surfaces generated by setting ξ = constant on the left-hand side of Equation (1.6) will hereafter be referred to as the *Roche Equipotentials*. The form of such equipotentials depends evidently on the value of ξ. If ξ is large, the corresponding equipotentials will consist of two separate ovals (see Figure 6-1) closed around each of the two mass-points; for the right-hand side of (1.6) can be large only if r (or $r' = \sqrt{1 - 2\lambda r + r^2}$) becomes small; and if the left-hand side of (1.6) is to be constant, so must be (very nearly) r or r'. Large values of ξ correspond, therefore, to equipotentials differing but little from spheres – the less so, the greater ξ becomes. With diminishing value of ξ the ovals defined by (1.6) become increasingly elongated in the direction of the centre of gravity of the system – until, for a certain critical value of ξ_1 characteristic of each mass-ratio, both ovals will unite in a single point on the x-axis to form a dumb-bell-like configuration (cf. again Figure 6-1) which we propose to call the *Roche Limit.** *For still smaller* values of ξ the con-

* Not to be confused with a concept used, under the same name, in older literature to signify the minimum distance at which a fluid satellite of infinitesimal mass can approach with impunity an oblate planet. This latter term, coined by G. H. Darwin in the latter half of the 19th century, has nothing to do with the 'Roche Limit' as defined in this section, as introduced under this name for the first time by Kopal (1955), and one which has subsequently become a household word among the workers in the subject.

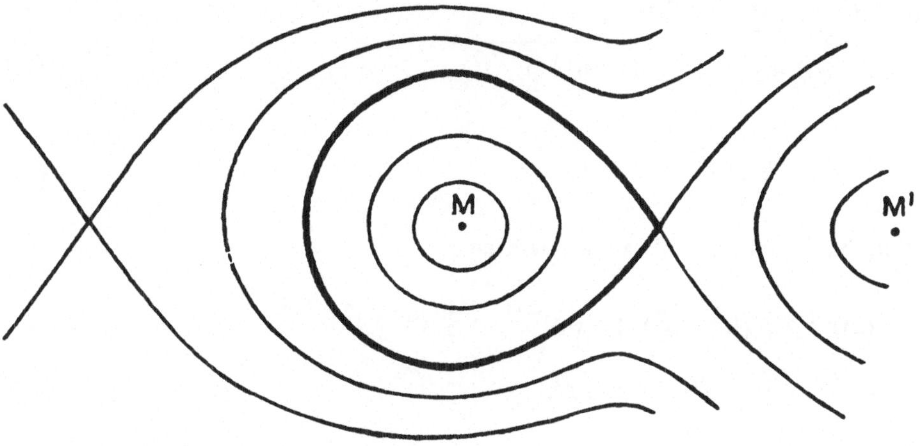

Fig. 6-1. Geometry of the Roche surfaces. (The Roche limit is marked by a heavy line.)

necting part of the dumb-bell opens up and the corresponding equipotential surfaces would envelop both bodies. This latter case is, however, of no direct interest to us in this connection; as for $\xi < \xi_1$ the two initially distinct bodies would coalesce in one and we should no longer have the right to speak of a binary system. In what follows we shall, therefore, limit ourselves to a study of the geometry of surfaces characterized by $\xi \geqslant \xi_1$.

Before we do so, however, it may be of interest to note the existence of certain transformations of coordinates – both real and complex – which render the form of the Roche potential (1.6) more symmetrical. Thus introducing a complex transformation

$$
\begin{aligned}
u^2 &= x + i(y^2 + z^2)^{1/2}, \\
s^2 &= x - i(y^2 + z^2)^{1/2}, \\
t^2 &= (z/u)^2;
\end{aligned} \tag{1.9}
$$

or its inverse

$$
\begin{aligned}
x^2 &= \frac{1}{4}(u^2 + s^2)^2, \\
y^2 &= -\frac{1}{4}(u^2 - s^2)^2 - u^2 t^2, \\
z^2 &= u^2 t^2,
\end{aligned} \tag{1.10}
$$

in place of (1.5) we find that

$$r = us, \qquad r' = \sqrt{(1-u^2)(1-s^2)}\ , \tag{1.11}$$

and

$$\lambda = \frac{u^2+s^2}{2us}, \qquad \nu = \frac{t}{s}, \tag{1.12}$$

we find that Equation (1.6) can be rewritten as

$$\xi = (us)^{-1} + q\ \{[(1-u^2)(1-s^2)]^{-1/2} - \tfrac{1}{2}\ (u^2+s^2)\}$$
$$+ \tfrac{1}{2}\ (q+1)u^2(s^2-t^2)\ . \tag{1.13}$$

Moreover, by setting

$$u^2 + s^2 = \quad 2\cosh E \cosh F \tag{1.14}$$

and

$$u^2 - s^2 = -2 \sinh E \sinh F, \tag{1.15}$$

we find that

$$r' = \cosh F - \cosh E, \tag{1.16}$$

but

$$r = \sqrt{\sinh^2 E + \cosh^2 F.} \tag{1.17}$$

On the other hand, a real transformation

$$1 - x = p^2 - q^2 - s^2, \qquad y = 2pq, \qquad z = 2ps \tag{1.18}$$

permits us to write

$$r'^2 = p^2 + q^2 + s^2; \tag{1.19}$$

but, unfortunately, this simplification results only at the price of more complicated expressions for r, λ, and ν.

After this digression prompted by analytical reasons, let us return to Equation (1.2) to inquire more closely into its meaning. The right-hand side of the expression (1.2) for Ψ can be regarded as a constant not depending on the time only if (a) the radius-vector R of the relative orbit of the two bodies m_1 and m_2 is constant (i.e., their orbit remains circular); and their angular velocity ω_1 of rotation mass m_1 is equal to the Keplerian angular velocity ω_K, given by (1.4), and the axis of rotation remains perpendicular to the orbital plane. These are the conditions which we established as sufficient in Chapter II for the tides distorting the components to be of the equilibrium type.

Should, however, the relative orbit of the two stars be eccentric – so that its radius-vector R becomes a function of the time t – the total potential Ψ must also depend on t. Moreover, the same is bound to be the case if the axis of rotation of either configuration becomes inclined to their orbital plane – regardless of whether or not their angular velocities $\omega_{1,2}$ of axial rotation are equal to ω_K.

Suppose that, in what follows, we focus our attention on the component of mass m_1, whose equipotential surfaces are distorted by tides raised by its companion regarded as a point mass m_2, and rotates about an arbitrarily oriented axis with an angular velocity ω_1. If so, Equation (1.2) for the potential Ψ should be replaced by

$$\Psi = G\frac{m_1}{r} + G\frac{m_2}{r'} - G\frac{m_2}{R^2}\,x' + \frac{Gm_2^2}{R(m_1+m_2)} + \\ + \frac{1}{2}\,\omega_1^2\,(x'^2 + y'^2)\,, \tag{1.20}$$

where the primed coordinates x', y', z' are defined so that $z' = 0$ represents the equatorial plane of the component rotating about the z'-axis with an angular velocity ω_1.

A transformation between the two sets of coordinates x, y, z and x', y', z' then continues to be given by the matrix equation (3.1) of Chapter IV, from which we find that

$$x'^2 + y'^2 = x^2 + y^2 - \{xz \sin\phi - yz\cos\phi\}\sin 2\theta - \\ - \{(x\sin\phi - y\cos\phi)^2 - z^2\}\sin^2\theta; \tag{1.21}$$

where θ and ϕ represent the Eulerian angles as shown on Figure 4-1. If $\theta = 0$ – i.e., if the axis of rotation of the configuration surrounding the mass m_1 is perpendicular to its orbital plane – then $x'^2 + y'^2 = x^2 + y^2$ and the potential $\Psi(x, y, z)$ as given by Equation (1.20) remains a function of the spatial coordinates, and does not depend explicity on the time, regardless of whether $\omega_1 \gtrless \omega_K$. This is easy to understand; for if, in such a case, $\omega_1 \neq \omega_K$, this fact will only alter its polar flattening by *constant* amount in the revolving set of coordinates x, y, z.

If, on the other hand, the axis of rotation of mass m_1 is inclined to its orbital plane (i.e., the Eulerian angle $\theta > 0$), Equation (1.21) makes it evident that the sum $x'^2 + y'^2$ on the right-hand side of (1.20) is bound to depend on the time t through the Eulerian angles

$$\phi = -\omega_K t \qquad \text{and} \qquad \psi = \omega_1 t, \tag{1.22}$$

in accordance with Equation (3.78) of Chapter III. In such a case, $\Psi \equiv \Psi(x, y, z; t)$ even in the case of synchronism between rotation and revolution. Should we, on the other hand, wish to rewrite the potential Ψ in terms of the primed coordinates x', y', z', then $\Psi = \Psi(x', y' z'; t)$ becomes a function of the time t if the axis of rotation of the mass m_1 is perpendicular to the orbital plane; and this dependence on time would vanish only if the rotation and revolution were synchronous.

What are the consequences which follow if the potential Ψ becomes a function of the time? Obviously the shape of the equipotential surfaces Ψ = constant will then change in the course of time; and, as a result, the particles constituting it will be made to move with finite velocity components, during each orbital cycle, in the revolving frame of coordinates x, y, z. If so, however, the terms containing the velocities in Equation (1.1) of Chapter II will no longer vanish identically; and, consequently, Equation (1.10) of that chapter will no longer furnish the free form of the two stars at any time with sufficient exactitude. This form will, in fact, become distorted by dynamical tides discussed in Chapter III. The radius-vector r' connecting the position of any arbitrary surface point with the origin will continue to be of the form (1.49) of Chapter II; but the amplitudes of its individual harmonics will become functions of the time.

The question can be asked; is it not possible to represent the free surface of the star by a series of instantaneous equipotentials $\Psi(t)$ = constant evaluated for discrete moments of the time? Theoretically such a procedure is not rigorously justifiable, because it would ignore the existence of the velocity components arising in a transition from the surface $\Psi(t_1)$ to $\Psi(t_2)$. In practice, however, the errors entailed in this process are likely to be negligible *if the star can adjust its form in any time to the instantaneous field of force:* and a necessary (though not sufficient) condition for this to happen is that *the period of free oscillation* (of the respective symmetry) *of such a configuration is short in comparison with that of the disturbing force* (i.e., the period of the orbits or its submultiples). This condition can be expected to be fulfilled for most types of the binary systems and their components. However, exceptions to this rule can be encountered in special cases; and the only way then to study the time-dependent distortion of the stars is by the methods developed in Chapter III.

One important consideration may, however, be emphasized which should greatly ease our problems. In section IV-3 we proved that, whatever may be the initial difference $\omega_1 - \omega_K$ in any particular close binary system, dissipative forces (due to the viscosity of stellar material) will tend to make the difference $|\omega_1 - \omega_K|$ tend asymptotically to zero. Moreover, in section V-2 we established the same result for the inclination θ of the equator to the orbital plane. Therefore, no matter from which initial state the system began to evolve, *after the lapse of a sufficiently long time* (whose length depends, in turn, on the absolute values of the viscosity of stellar matter – both plasma and radiative) *all systems should attain a 'steady state' of synchronous rotation about axes perpendicular to the plane of their orbit.* Observational indications that such a state has already been attained in the majority of known eclipsing systems have been discussed in section V-4; and if they are further confirmed by future evidence, a theory of the quilibrium figures of the stars – investigated in some detail in Chapter II – should indeed represent a safe guide for the interpretation of most phenomena observed in existing close binary systems.

A. SURFACES OF ZERO VELOCITY

Such being the case, let us return to the Roche equipotentials as introduced by Equation (1.2) at the beginning of this section, and compare them with another class of surfaces closely allied with them not only in form, but also physical meaning: namely, the Jacobian surfaces of zero velocity in the restricted problem of two bodies. In order to do so, let the masses $m_{1,2}$ separated by a constant distance R represent a gravitational dipole revolving around their common centre of mass with a constant angular velocity ω_K. If, moreover, we adopt R and $m_1 + m_2$ to serve as our unit of length and mass, respectively, while the unit of the time is chosen so that $G = 1$, the equations of motion of an arbitrary particle of negligible mass, in the system of coordinates x, y, z with origin at the centre of mass of the dipole $m_{1,2}$ and co-rotating with it so that $z = 0$ represents the plane of the binary orbit, are known to be of the form

$$\frac{d^2x}{dt^2} - 2\frac{dy}{dt} = \frac{\partial U}{\partial x}, \tag{1.23}$$

$$\frac{d^2y}{dt^2} + 2\frac{dx}{dt} = \frac{\partial U}{\partial y}, \tag{1.24}$$

$$\frac{d^2z}{dt^2} = \frac{\partial U}{\partial z}, \tag{1.25}$$

with

$$U = \frac{1-\mu}{r_1} + \frac{\mu}{r_2} + \frac{1}{2}(x^2 + y^2), \tag{1.26}$$

where $r_{1,2}$ denote the distances of the point from the centres of mass of $m_{1,2}$, and where we have abbreviated

$$\mu \equiv \frac{m_2}{m_1 + m_2} \tag{1.27}$$

Multiply now Equations (1.23)–(1.25) successively by $2(dx/dt)$, $2(dy/dt)$, $2(dz/dt)$ and add: the result can be readily integrated to disclose that

$$\left(\frac{dx}{dt}\right)^2 + \left(\frac{dy}{dt}\right)^2 + \left(\frac{dz}{dt}\right)^2 \equiv V^2 = 2U(x, y, z) - C, \tag{1.28}$$

where C denotes an integration constant. The surfaces over which the velocity V vanishes are then given by the equation

$$\frac{2(1-\mu)}{r_1} + \frac{2\mu}{r_2} + x^2 + y^2 = C, \tag{1.29}$$

which differs from Equation (1.2) defining the Roche equipotentials only by the units adopted and the position of the origin of coordinates.

Accordingly, *the Roche equipotentials prove to be identical with the surfaces of zero-velocity of the restricted problem of three bodies* in the circular case. Like the Roche equipotentials, for sufficiently large values of the constant C the zero-velocity surfaces as defined by Equation (1.29) constitute closed ovals surrounding each one of the two finite masses $m_{1,2}$. Any point of negligible mass can move inside these ovals in any manner whatever which is consistent with the Equations (1.23)–(1.25) of motions and the adopted initial conditions; but it cannot cross them; for its velocity vanishes over the surface defined by (1.29) and would become imaginary outside.

Moreover, the structure of Equations (1.23)–(1.27) discloses that a mass particle can actually approach at any point of the zero-velocity surface only in a direction normal to such a surface. For consider the equation

$$F(x, y, z) \equiv 2U - C = 0 \tag{1.30}$$

of the surfaces of zero velocity, as defined earlier by Equations (1.26)–(1.28). The derivatives

$$\frac{1}{2}\frac{\partial F}{\partial x}, \quad \frac{1}{2}\frac{\partial F}{\partial y}, \quad \frac{1}{2}\frac{\partial F}{\partial z}, \tag{1.31}$$

which are proportional to the direction cosines of a normal to the surfaces (1.30) of zero velocity, are obviously identical with the right-hand sides of the Equations (1.23)–(1.25) of motion. But the surfaces $F(x, y, z) = 0$ constitute also the loci at which the velocity components dx/dt, dy/dt and dz/dt vanish everywhere – so that Equations (1.23)–(1.25) reduce to

$$\frac{1}{2}\frac{\partial F}{\partial x} = \frac{d^2x}{dt^2}, \quad \frac{1}{2}\frac{\partial F}{\partial y} = \frac{d^2y}{dt^2}, \quad \frac{1}{2}\frac{\partial F}{\partial z} = \frac{d^2z}{dt^2}, \tag{1.32}$$

rendering the directions of acceleration orthogonal to the surfaces of zero velocity relative to the rotating system.

Over such surfaces, any particle which reaches them comes temporarily at rest. Since, however, the acceleration remains finite, the particle is bound to recoil back in a direction normal to the surface but opposite to that from which it came. This is true everywhere except if the velocities and acceleration of the particle vanish simultaneously; and this will obviously be the case at points the coordinates of which are roots of the equations

$$\frac{\partial F}{\partial x} = \frac{\partial F}{\partial y} = 0 \tag{1.33}$$

and

$$\frac{\partial F}{\partial z} = 0 \tag{1.34}$$

considered as a simultaneous system.

Equation (1.34) can obviously be satisfied only if $z = 0$; therefore, all 'double points' at which velocities *and* accelerations in the rotating frame of reference may vanish at the same time lie in the xy-plane. If a particle of infinitesimal mass is placed in positions specified by the roots of Equations (1.33), its coordinates will fullfil identically the differential equations (1.23)–(1.25) of motion and will, consequently, remain forever at rest in the revolving system of coordinates, unless disturbed by forces exterior to the system. These positions constitute the famous particular solutions of restricted problem of three bodies, the existence of which was discovered by J. L. Lagrange in the 18th century; and they are traditionally referred to as the *Lagrangian points* $L_{1,2...5}$.

The explicit form of Equations (1.33) and (1.34) discloses that all five Lagrangian points must be located in the plane $z = 0$ of the orbit of the gravitational dipole $m_{1,2}$. The first three (the 'collinear points' $L_{1,2,3}$) lie on the x-axis joining m_1 and m_2: the first one (L_1) occupying a position between m_1 and m_2 on the x-axis; while $L_{2,3}$ are situated ouside the radius-vector joining m_1 and m_2 on the side of the more massive, and less massive, component, respectively – in positions depending on the actual value of the mass-ratio. Lastly, the 'triangular points' $L_{4,5}$ form the vertices of equilateral triangles with m_1 and m_2 on either side of the x-axis, irrespective of the mass-ratio.

The actual positions of these five Lagrangian points in the xy-plane for different values of the mass-ratio will be investigated in subsequent parts of this chapter (cf. Tables VI-1 and 5); and the implications of their existence will be discussed in section VIII-3 of the concluding chapter of this book. For the present we wish to stress that any assembly of mass particles, moving in accordance with Equations (1.23)–(1.25) cannot escape a volume surrounding each one of the masses $m_{1,2}$ if the Jacobian constant C is large enough to render this volume closed. Moreover, each particle reaching its boundary must temporarily come to rest – and for this reason it constitutes a 'surface of equilibrium' coinciding with the Roche equipotentials Ψ = constant.

A physical meaning of this theorem should not, however, be overestimated; for its proof holds good only if particles moving inside closed volumes can do so without colliding with each other – in other words, if the pressure of the medium constituted by them is zero – which could be true for gas only at zero absolute temperature. A gas of finite temperature can approximate such a state only if the mean-free-path of its constituent particles were long in comparison with the dimensions of the respective configuration; and it is questionable whether or not a gas so rarefied could make the volume filled by it opaque.

At any rate, the questions arising in this connection cannot be answered by dynamical arguments alone; and we shall return to them later in this book. What we wish to point out here is, however, the fact that *the Jacobian integral* (1.28) of Equations (1.23) –(1.26) *exists only if the orbit of the two finite masses* $m_{1,2}$ *is circular,* and disappears as soon as its eccentricity e is greater than zero – no matter how small (cf. Kopal and Lyttleton, 1963). For $e > 0$, the Jacobian 'constant' C on the right-hand side of Equation (1.29) becomes, instead, the function of the time – just as the value of the Roche potential Ψ became under the same circumstances – and its value must be obtained by integration of Equations (1.23)–(1.25) for x, y and z as functions of the time.

In conclusion of the present section, one supplementary consideration should be pointed out. In setting up the expression (1.2) for the Roche potentials we assumed gravitational attraction and axial rotation to constitute the only forces shaping up the potential surfaces. If, however, in accordance with the underlying model the overwhelming part of the star's mass is to be condensed in close proximity of its centre, this mass should constitute an intense source of radiation; and the pressure produced by this radiation should exert an additional force which could influence the form of the Roche equipotentials.

In more specific terms, if we consider the star of mass (say) m_1 to constitute an intense source of radiation, the term Gm_1/r_1 representing the gravitational potential on the right-hand side of Equation (1.2) should be replaced by $G(1-\epsilon)m_1/r_1$, where

$$\epsilon = \frac{1}{4\pi c G m_1} \int_0^\infty k_\nu L_\nu \, d\nu; \qquad (1.35)$$

L_ν denoting the luminosity of the source as a function of the frequency ν; and k_ν, the absorption coefficient of the material in the envelope per unit mass. The sign of ϵ in the term $G(1-\epsilon)$ is negative because radiation pressure acts in a direction opposite to that of gravity.

A geometry of the Roche equipotentials in which the effective attraction of the mass m_1 of the radiating component has been reduced $(1-\epsilon)$-times by the effects of the radiation pressure was recently investigated in some detail by Schuerman (1972) or Kondo *et al.* (1975). His work indicates that, with increasing value of the nondimensional parameter ϵ, the shape of the corresponding equipotentials may soon cease to possess much similarity with those obtaining in the absence of radiation; and, in particular, closed equipotentials soon disappear from the real domain. Whether or not actual binaries exist in which radiation pressure may become competitive with gravity in shaping up the equipotential surfaces in the proximity of their mass centres is an open question; but the possibility should be kept in mind (particularly for systems with x-ray components; cf. sec. VIII-5C).

VI-2. Geometry of Roche Surfaces

Equation (1.6) of the Roche equipotentials represents an implicit function defining, for given values of ξ and q, r as a function of λ and ν. When it has been rationalized and cleared of fractions, the result is an algebraic equation of *eight* degree in r, whose analytical solution presents unsurmountable difficulties. In the case of pure rotational distortion (obtaining if $q = 0$), Equation (1.6) can be reduced to a cubic solvable in terms of circular functions. In the case of a pure tidal distortion ($\omega = 0$) Equation (1.6) becomes a quartic, which could also be solved for r in a closed form (though its solution would be very much more involved). In the general case of rotational *and* tidal distortion interacting, however, any attempt at an *exact* solution of (1.6) for r becomes virtually hopeless, and approximate solutions must inevitably be sought.

A. RADIUS AND VOLUME

In order to obtain them, let us begin by expanding the radical $(1 - 2\lambda r + r^2)^{-1/2}$ on the right-hand side of (1.6) in terms of the Legendre polynomials $P_j(\lambda)$. Doing so and removing fractions we find it possible to replace (1.5) by

$$(\xi - q)r = 1 + q \sum_{j=2}^{\infty} r^{j+1} P_j(\lambda) + nr^3(1 - r^2), \tag{2.1}$$

where we have abbreviated

$$n = \frac{q+1}{2} . \tag{2.2}$$

If r is small in comparison with unity (i.e., if the linear dimensions of the equipotential surfaces are small in comparison with our unit of length R), the second and third terms on the right-hand side of (2.1) may be neglected in comparison with unity – in which case, to a first approximation,

$$r_0 = \frac{1}{\xi - q} . \tag{2.3}$$

This result asserts that if ξ is large, the corresponding Roche equipotential will differ but little from a sphere of radius r_0.

Suppose now that

$$r_1 = r_0 + \Delta' r = r_0 \left(1 + \frac{\Delta' r}{r_0}\right) \tag{2.4}$$

should represent our next approximation to r. Inserting it in (2.1) we find that

$$1+\frac{\Delta' r}{r_0} = 1+q\sum_{j=1}^{\infty} r_0^{j+1}P_j(\lambda)+nr_0^3(1-\nu^2) \tag{2.5}$$

where, in small terms on the right-hand side, r was legitimately replaced by r_0. The foregoing equation then yields

$$\frac{\Delta' r}{r_0} = q\sum_{j=2}^{4} r_0^{j+1}P_j(\lambda)+nr_0^3(1-\nu^2) \tag{2.6}$$

correctly to quantities of the order of r_0^5 (i.e., as far as squares and higher terms of first-order distortion remain negligible).

In order to improve upon this approximation let us set, successively,

$$r_2 = r_1+\Delta'' r = r_0\left\{1+\frac{\Delta' r}{r_0}+\frac{\Delta'' r}{r_0}\right\}, \tag{2.7}$$

$$r_3 = r_2+\Delta''' r = r_0\left\{1+\frac{\Delta' r}{r_0}+\frac{\Delta'' r}{r_0}+\frac{\Delta''' r}{r_0}\right\}, \tag{2.8}$$

$$\vdots \qquad \vdots$$

$$r_{j+1} = r_j+\Delta^{(j+1)}r = r_0\left\{1+\sum_{i=0}^{j}\frac{\Delta^{(i+1)}r}{r_0}\right\}, \tag{2.9}$$

where

$$\frac{\Delta^{(j+1)}r}{r_0} = q\sum_{k=3}^{3(N-1)}(r_i^k-r_{i-1}^k)P_{k-1}(\lambda)+n(r_i^3-r_{i-1}^3)(1-\nu^2), \tag{2.10}$$

$3(N-1)$ denoting the highest power of r_0 to which Equation (2.9) represents a correct solution for r. We may note that, in general, the leading terms of the expression (2.10) for $(\Delta^{(i+1)}r)/r_0$ will be of $3(i+1)$st degree in r_0; and, similarly, the difference $r_i^k-r_{i-1}^k$ in higher terms on the right-hand side of (2.10) will be of the order of r_0^{3i+k}.

Suppose that, in what follows, we wish to construct the explicit form of an approximate solution of Equation (2.1), in the form of (2.8), correctly to (say) quantities of the order of $(\Delta''' r)/r_0$ – which should, therefore, differ from the exact solution of (2.1) at most in quantities of the order of r_0^{12}. By use of the expression already established for $(\Delta' r)/r_0$ the explicit forms of $(\Delta'' r)/r_0$ and $(\Delta''' r)/r_0$ can sucessively be found*; and their insertion in (2.8) leads to the equation

$$\begin{aligned}\frac{r-r_0}{r_0} &= r_0^3\{qP_2+n(1-\nu^2)\}\\ &+r_0^4\{qP_3\}\\ &+r_0^5\{qP_4\}\\ &+r_0^6\{qP_5+3[qP_2+n(1-\nu^2)]^2\}\end{aligned} \tag{2.11}$$

* For fuller details of this process, cf. Kopal (1954b, 1959).

$$+ r_0^7 \{qP_6 + 7q[qP_2 + n(1-\nu^2)]P_3\}$$
$$+ r_0^8 \{qP_7 + 8q[qP_2 + n(1-\nu^2)]P_4 + 4q^2P_3^2\}$$
$$+ r_0^9 \{qP_8 + 9q[qP_2 + n(1-\nu^2)]P_5 + 9q^2P_3P_4$$
$$+ 6[qP_2 + n(1-\nu^2)]^3 + 6[q^3P_2^3 + n^3(1-\nu^2)^3]\}$$
$$+ r_0^{10} \{qP_0 + 10q[qP_2 + n(1-\nu^2)]P_6 + 5q^2[P_4^2 + 2P_3P_5]\}$$
$$+ 45q[qP_2 + n(1-\nu^2)]^2P_3\}$$
$$+ r_0^{11} \{qP_{10} + 11q[qP_2 + n(1-\nu^2)]P_7 + 11q^2[P_3P_6 + P_4P_5]$$
$$+ 55q[qP_2 + n(1-\nu^2)]^2P_4$$
$$+ 55q^2[qP_2 + n(1-\nu^2)]P_3^2\} + \cdots,$$

where we have abbreviated $P_j \equiv P_j(\lambda)$, and which represents the desired approximate solution of Equation (2.1) for r as a function of λ and ν in the form of an expansion in ascending powers of r_0 (as defined by Equation (2.10)).

The volume V of a configurations whose radius-vector r is given by the foregoing Equation (2.11) will be specified by

$$V = \frac{2}{3} \int_{-1}^{1} \int_{-\sqrt{1-\lambda^2}}^{\sqrt{1-\lambda^2}} \frac{r^3 \, d\lambda \, d\nu}{\mu} . \tag{2.12}$$

where $\mu^2 = 1 - \lambda^2 - \nu^2$. By virtue of the algebraic identity

$$r^3 \equiv r_0^3 \left\{1 + \frac{r - r_0}{r_0}\right\}^3 \tag{2.13}$$

we find it convenient to express the integrand in (2.19) in terms of (2.18) as a function of λ and ν. This integrand will, in general, consist of a series of terms of the form $\lambda^m \nu^n/\mu$, factored by constant coefficients; therefore, the entire volume V will be given by an appropriate sum of partial expressions V_n^m of the form

$$V_n^m = \int_{-1}^{1} \int_{-\sqrt{1-\lambda^2}}^{\sqrt{1-\lambda^2}} \frac{\lambda^m \nu^n}{\mu} \, d\lambda \, d\nu. \tag{2.14}$$

These expresssions vanish (on grounds of symmetry) if either m or n is an odd integer. If, however, both happen to be even and such that $m = 2a$ and $n = 2b$, an evaluation of the foregoing integrals readily reveals that

$$V_{2b}^{2a} = \frac{\sqrt{\pi}\,\Gamma(a + \frac{1}{2})\,\Gamma(b + \frac{1}{2})}{\Gamma(a + b + \frac{3}{2})} , \tag{2.15}$$

where Γ denotes the ordinary gamma function. As, accordingly,

$$\int_{-1}^{1} \int_{-\sqrt{1-\lambda^2}}^{\sqrt{1-\lambda^2}} \frac{P_j(\lambda) d\lambda \, d\nu}{\sqrt{1 - \lambda^2 - \nu^2}} = \left\{ \begin{matrix} 2\pi & \text{if} & j = 0, \\ 0 & \text{if} & j > 0, \end{matrix} \right\} \tag{2.16}$$

and

$$\int_{-1}^{1}\int_{-\sqrt{1-\lambda^2}}^{\sqrt{1-\lambda^2}} \frac{\nu^{2j}\,d\lambda\,d\nu}{\sqrt{1-\lambda^2-\nu^2}} = \frac{2\pi}{j+1}, \tag{2.17}$$

we eventually find that the volume of a configuration whose surface is a Roche equipotential will be given by

$$\begin{aligned} V = \tfrac{4}{3}\pi r_0^3\{1 &+ \tfrac{12}{5}q^2r_0^6 + \tfrac{15}{7}q^2r_0^8 + \tfrac{18}{9}q^2r_0^{10} + \cdots \\ &+ \tfrac{22}{7}q^3r_0^9 + \tfrac{157}{7}q^3r_0^{11} + \cdots \\ &+ 2nr_0^3 + \tfrac{32}{5}n^2r_0^6 + \tfrac{176}{7}n^3r_0^9 + \cdots \\ &+ \tfrac{8}{5}nqr_0^6 + \tfrac{296}{35}nq(2q+n)r_0^9 \\ &+ \tfrac{26}{35}nq(q+3n)r_0^{11} + \cdots\}, \end{aligned} \tag{2.18}$$

correctly to quantities of the order up to and including r_0^{11}. With n and r_0 as given by Equations (2.2) and (2.3) the volume V becomes an explicit function of ξ and q alone and can be tabulated in terms of these parameters.

B. ROCHE LIMIT

It was pointed out already in section VI-1 that a diminution of the value of the constant ξ on the left-hand side of Equation (1.6) will cause the respective Roche equipotentials to expand from nearly spherical configurations to ovals of increased elongation in the direction of the attracting centre until, for a certain critical value of ξ characteristic of each mass-ratio, these ovals unite in a single point on the line joining their centres. Such configurations represent the largest *closed* equipotentials capable of containing the whole mass of the respective components, and will hereafter be referred to as their *Roche limits.* Any star filling its Roche limit will therefore be termed a *contact component;* and a binary system consisting of a pair of such components, a *contact system.* The fact that close binaries in which one, or both, components have attained their Roche limits actually exist in considerable numbers* adds importance to a study of the geometry of Roche limits in binary systems of different mass-ratios.

In order to do so, our first task should be to specify the values of ξ for which the two loops of the critical equipotential (cf. Figure 6-2) develop a common point of contact at P_1; but its determination presupposes a knowledge of the position of P_1 on the x-axis. The location of this point is characterized by the vanishing of the gravity due to all forces – which means that, at that point,

$$\xi_x = \xi_y = 0. \tag{2.19}$$

* Cf. Kopal (1955, 1956, 1959).

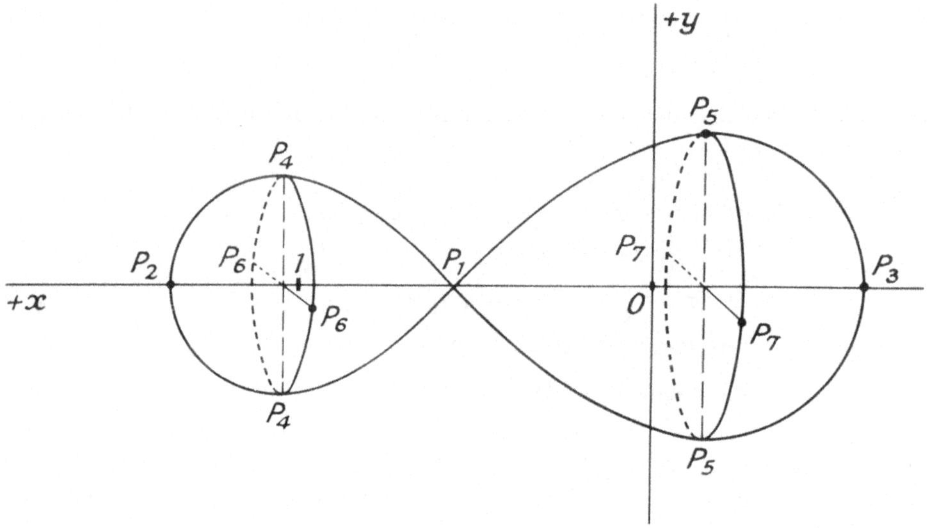

Fig. 6.-2. Schematic view of a contact binary at the Roche limit. In order to exhibit the geometry of this model, the diagram has not been drawn to scale for any specific mass-ratio; and certain features of it (such as the distance of the P_5P_7-plane from the origin) have been exaggerated.

Now a differentiation of (1.6), rewritten in terms of rectangular coordinates, with respect to x and y yields

$$\xi_x = -xr^{-3} + q\{(1-x)(r')^{-3} - 1\} + 2nx, \tag{2.20}$$

$$\xi_y = -y\{r^{-3} + q(r')^{-3} - 2n\}, \tag{2.21}$$

$$\xi_z = -zr^{-3} - qz(r')^{-3}, \tag{2.22}$$

where $r^2 = x^2 + y^2 + z^2$ and $r'^2 = (1-x)^2 + y^2 + z^2$ continue to be given by Equations (1.3) and $2n = q + 1$ in accordance with (2.2).

The partial derivative ξ_y vanishes evidently everywhere along the x-axis; but the vanishing of ξ_x renders the x-coordinate of P_1 to be a root of the equation

$$x^{-2} - x = q\{(1-x)^{-2} - (1-x)\}. \tag{2.23}$$

which, after removal of the fractions, assumes the form

$$(1+q)x^5 - (2+3q)x^4 + (1+3q)x^3 - x^2 + 2x - 1 = 0. \tag{2.24}$$

For $q = 0$ the foregoing equation would evidently reduce to

$$(1-x)^3(x^2 + x + 1) = 0, \tag{2.25}$$

the value $x = 1$ becoming a triple root. Therefore, for small values of q, the root x_1 of Equation (2.23) which is interior to the interval $0 < x < 1$ should be approximated by the expansion

$$x_1 = 1 - w + \frac{1}{3} w^2 + \frac{1}{9} w^3 + \cdots \tag{2.26}$$

in terms of the auxiliary parameter

$$w^3 = \frac{q}{3(1+q)} \; ; \tag{2.27}$$

and more accurate values of of x_1 can further be obtained by the method of differential corrections.

Once a sufficiently accurate value of x_1 has thus been obtained, the actual value of ξ corresponding to our critical equipotential follows as

$$\xi_1 \equiv \xi(x_1, 0, 0). \tag{2.28}$$

Moreover, the points $P_{4,5}$ in the xy-plane (see again Figure 6-2) are evidently characterized by the vanishing of the derivative dy/dx at the Roche limit. Their coordinates $x_{4,5}$ and $y_{4,5}$ can, therefore, be evaluated by solving the simultaneous system

$$\left.\begin{aligned} &\xi(x, y, 0) = \xi_1 , \\ &\xi_x(x, y, 0) = 0 \; ; \end{aligned}\right\} \tag{2.29}$$

and once the values of $x_{4,5}$ have thus been found, the z-coordinates of points $P_{6,7}$ in the xz-plane (cf. Figure 6-2) follow as roots of a single equation

$$\xi(x_{4,5}, 0, z) = \xi_1 . \tag{2.30}$$

The accompanying Table VI-1 lists five-digit values of ξ_1, x_1; $x_{4,5}$, $y_{4,5}$; and $z_{6,7}$ for Roche limits appropriate for 15 discrete values of the mass-ratio.

It may further be noticed that if, in place of ξ_1, we introduce a new constant C_1 as defined by the equation

$$C_1 = \frac{2\xi_1}{1+q} + \left(\frac{q}{1+q}\right)^2 = 2(1-\mu)\xi_1 + \mu^2, \tag{2.31}$$

where we have abbreviated

$$\mu = \frac{q}{1+q} = \frac{m_2}{m_1 + m_2} , \tag{2.32}$$

the values of C_1 remain largely invariant with respect to the mass-ratio, and sensibly equal to 4 provided that q does not depart greatly from unity. This is demonstrated by an inspection of the tabulation of C_1 as given in column (2) of the following Table VI-2. In consequence, the corresponding simple expression

$$\xi_1 = 2(1+q) - \frac{q^2}{2(1+q)} = \frac{4-\mu^2}{2(1-\mu)} \tag{2.33}$$

is found to approximate the exact values of ξ_1 within 1% if $1 \geqslant q \geqslant 0.5$, or within 10% for the wider range $1 \geqslant q > 0.1$.

TABLE VI-1[a]

q	x_1	ξ_1	x_4	$\pm y_4$	x_5	$\pm y_5$	$\pm z_6$	$\pm z_7$
1.0	0.50000	3.75000	1.01134	0.37420	- -0.1134	0.37420	0.35621	0.35621
0.8	0.52295	3.41697	1.01092	0.35388	−0.01168	0.39501	0.33770	0.37491
0.6	0.55234	3.06344	1.01029	0.32853	−0.01198	0.42244	0.31431	0.39909
0.4	0.59295	2.67810	1.00926	0.29465	−0.01213	0.46189	0.28260	0.43278
0.3	0.62087	2.46622	1.00847	0.27204	−0.01204	0.49015	0.26123	0.45599
0.2	0.65856	2.23273	1.00735	0.24233	−0.01163	0.52989	0.23294	0.48714
0.15	0.68392	2.10309	1.00656	0.22280	−0.01117	0.55774	0.21425	0.50781
0.1	0.71751	1.95910	1.00552	0.19746	−0.01034	0.59609	0.18991	0.53451
0.05	0.76875	1.78886	1.00397	0.15979	−0.00859	0.65804	0.15366	0.57291
0.02	0.82456	1.65702	1.00245	0.11992	−0.00618	0.73070	0.11522	0.61434
0.01	0.85853	1.59911	1.00165	0.09613	−0.00457	0.77779	0.09231	0.62867
0.005	0.88635	1.56256	1.00110	0.07689	−0.00327	0.81807	0.07379	0.64170
0.001	0.93231	1.52148	1.00041	0.04550	−0.00137	0.88816	0.04361	0.65762
0.0002	0.96001	1.50737	1.00015	0.02678	−0.00052	0.93264	0.02566	0.66348
0	1.00000	1.50000	1.00000	0.00000	0.00000	1.00000	0.00000	0.66667

a The data collected in this table are taken from Kopal (1959). More extensive tabulations of the same parameters for q = 1.00 (−0.02) 0.10 to 5*D* have been prepared by Plavec and Kratochvíl (1964). Cf. also ten Bruggencate (1934).

TABLE VI-2[a]

q	C_1	$(r_0)_1$	$(r_0)_2$	V_1	V_2	$(r^*)_1$	$(r^*)_2$	v_1	v_2
1	4.00000	0.36363	0.36363	0.22704	0.22704	0.37845	0.37845	0.072267	0.072267
0.8	3.99417	0.38212	0.34528	0.26459	0.19374	0.39825	0.35896	0.075799	0.069377
0.6	3.96993	0.40594	0.32199	0.31974	0.15665	0.42420	0.33441	0.081422	0.066485
0.4	3.90749	0.43896	0.29025	0.40923	0.11444	0.46057	0.30115	0.091184	0.063726
0.3	3.84744	0.46163	0.26876	0.48148	0.09089	0.48622	0.27892	0.099619	0.062683
0.2	3.74900	0.49195	0.24018	0.59399	0.06492	0.52147	0.24933	0.113443	0.061996
0.15	3.67456	0.51201	0.22121	0.68002	0.05079	0.54552	0.22973	0.124462	0.061967
0.1	3.57027	0.53789	0.19612	0.80715	0.03564	0.57760	0.20414	0.141308	0.062385
0.05	3.40962	0.57509	0.15931	1.0289	0.01910	0.62626	0.16584	0.17193	0.06384
0.02	3.24945	0.61087	0.11974	1.2700	0.007961	0.67179	0.12387	0.2062	0.06462
0.01	3.16665	0.62928	0.09606	1.4656	0.004042	0.70465	0.09882	0.2356	0.06497
0.005	3.10959	0.64203	0.07686	1.5950	0.002038	0.7248	0.07865	0.2551	0.06520
0.001	3.03992	0.65769	0.04549	1.868	0.0004114	0.764	0.04614	0.298	0.06554
0.0002	3.01414	0.66350	0.02697	2.067	0.0000826	0.790	0.02702	0.329	0.06575
0	3.00000	0.66667	0.00000	2.26663	0.0000000	0.81488	0.00000	0.36075	0.065843

a After Kopal (1959).

The mean radii $(r_0)_{1,2}$ of the two components of contact systems become (consistent with Equations (2.3) and (2.31)) equal to

$$(r_0)_{1,2} = \frac{2(1-\mu)}{C_1 - (1+\mu)^2 + 1}, \tag{2.34}$$

where, for the primary component, $0 \leqslant \mu \leqslant 0.5$; while, for the secondary, $0.5 \leqslant \mu \leqslant 1$. Alternatively, we may fall back on Equation (2.3) and, by inserting for ξ_1 from (2.28) write

$$(r_0)_1 = \frac{2x_1}{2 + 2qx_1^3(1-x_1)^{-1} + (q+1)x_1^3}; \tag{2.35}$$

while $(r_0)_2$ is obtainable from the same expression if we replace x_1 by $1 - x_1$ and q by its reciprocal. The values of $(r_0)_{1,2}$ so determined are listed as functions of the mass-ratio in columns (3) and (4) of Table VI-2. Having evaluated them, we are in a position to invoke Equation (2.18) for expressing the volumes $V_{1,2}$ of contact components – the reader will find them tabulated in columns (5) and (6) of Table VI-2 – while columns (7) and (8) list the equivalent radii $(r^*)_{1,2}$ of spheres having the same volume as the respective contact component. The penultimate and ultimate columns of Table VI-2 then contain the quantities

$$v_{1,2} = \frac{\omega^2}{2\pi G\rho_{1,2}} = \frac{2}{3}\left\{1 + \frac{m_{2,1}}{m_{1,2}}\right\}(r^*)_{1,2}^3, \tag{2.36}$$

where ω denotes the (Keplerian) angular velocity of axial rotation of each component and $\rho_{1,2}$, their respective mean densities.

The series on the right-hand side of the volume Equation (2.18) – which are at the basis of our numerical data as given in columns (5) – (10) – converge with satisfactory rapidity if the masses of the two components are not too unequal, but fail to do so if the mass of one component becomes very much larger than the other. In order to attain adequate representation of the radii and volumes in such cases, asymptotic solutions of Equation (2.1) must be sought as $\mu \to 0$ or 1.

In order to do so, we find it advantageous to rewrite (2.1) in the alternative form

$$(1 - 2\lambda r + r^2)\{(1-\nu^2)r^3 - 2\lambda\mu r^2 + (\mu^2 - C_1)r + 2(1-\mu)\}^2 = 4\mu^2 r^2, \tag{2.37}$$

where C_1 as well as μ are defined by Equations (2.31) and (2.32); and consider first the case of negligible disturbing mass, when $\mu = 0$. As long as quantities of the order of μ^2 remain ignorable, Equation (2.37) will admit of a real solution only if

$$(1-\nu^2)r^3 - 2\lambda\mu r^2 - C_1 r + 2(1-\mu) = 0. \tag{2.38}$$

For small values of μ, the solution of this latter equation can be sought in the form

$$r = S_{10} + S_{12}\mu + \cdots, \tag{2.39}$$

where $S_{10}, S_{11}, \ldots$ are defined by the equations

$$(1-\nu^2)S_{10}^3 - C_1 S_{10} + 2 = 0, \tag{2.40}$$

$$3(1-\nu^2)S_{10}^2 S_{11} - C_1 S_{11} - 2 = 2\lambda S_{10}^2, \tag{2.41}$$

etc., whose solutions become

$$S_{10} = 2\left\{\frac{C_1}{3(1-\nu^2)}\right\}^{1/2} \sin\tfrac{1}{3}\sin^{-1}\frac{3}{C_1}\sqrt{\frac{3(1-\nu^2)}{C_1}} \tag{2.42}$$

and

$$S_{11} = \frac{2(1+\lambda S_{10}^2)}{3(1-\nu^2)S_{10}^2 - C_1}, \tag{2.43}$$

respectively.

Equation (2.39) with its coefficients as given by (2.42) and (2.43) will closely approximate the form of the primary component of a contact system which is very much more massive than the secondary. Its first term S_{10} defines obviously the form of a Roche equipotential distorted by centrifugal force alone. If $\mu \to 0$, $\xi_1 \to 1.5$ and $C_1 \to 3$, in which case the parametric equation of the corresponding critical equipotential assumes the neat form

$$r = \frac{2}{\sqrt{1-\nu^2}}\{\sin\frac{1}{3}\cos^{-1}\nu\}, \tag{2.44}$$

and its volume V_1, in accordance with Equation (2.12), becomes

$$\begin{aligned} V_1 &= \tfrac{32}{3}\pi\int_0^1 (1-\nu^2)^{-3/2}\sin^3(\tfrac{1}{3}\cos^{-1}\nu)\,d\nu \\ &= \tfrac{4}{3}\pi\left\{3\sqrt{3} - 4 + 3\log\frac{3(\sqrt{3}-1)}{\sqrt{3}+1}\right\} = 2.26663\ldots \end{aligned} \tag{2.45}$$

It is this foregoing value, rather than the one which would follow from a straightforward application of (2.18), which has been used to complete the last entry in column (5) of Table VI-2.

If the primary component accounts thus for most part of the total mass of our contact binary system, the volume of the secondary must clearly tend to zero. The form of its surface will, in turn, be given by an asymptotic solution of Equation (2.37) as $\mu \to 1$. Let us, therefore, expand this solution in a series of the form

$$r = S_{20}(1-\mu) + S_{21}(1-\mu)^2 + \cdots, \tag{2.46}$$

inserting it in (2.37) we find the vanishing of the coefficients of equal powers of $(1-\mu)$ to require that

$$S_{20} = \frac{2}{C_1 - 3}, \qquad S_{21} = -\left\{2 + \frac{\lambda}{C_1 - 3}\right\}S_{20}^2, \tag{2.47}$$

etc. An application of Equation (2.12) reveals, moreover, that the volume V_2 of the respective configuration should be approximated by

$$V_2 = \tfrac{4}{3}\pi\,\{(1-\mu)^3 S_{20}^3 - 6(1-\mu)^4 S_{20}^4 - \ldots\}, \tag{2.48}$$

and the radius r_2^* of a sphere of equal volume becomes

$$r_2^* = (1-\mu)S_{20} - 2(1-\mu)^2 S_{20}^2 + \cdots \tag{2.49}$$

A glance at the second column of Table VI-2 reveals that, as $q \to 0$, $C_1 \to 3$ and, as a result, the product $(1 - \mu)S_{20}$ tends to become indeterminate for $\mu = 1$. In order to ascertain its limiting value, let us depart from the Equation (2.31) which, on insertion of ξ_1 from (2.28) assumes the form

$$C_1 = \frac{2(1-\mu)}{1-x_1} + \frac{2\mu}{x_1} + (1+\mu+x_1)^2, \tag{2.50}$$

with the root x_1 approximable by means of (2.26) where, by (2.32),

$$3w^2 = 1 - \mu. \tag{2.51}$$

Inserting (2.26) in (2.50) we find that, within the scheme of our approximation,

$$C_1 = 3(1 + 3w^2 - 4w^3 - \cdots), \tag{2.52}$$

so that

$$(1-\mu)S_{20} = \frac{2w}{3-4w} + \cdots, \tag{2.53}$$

and, therefore,

$$r_2^* = \frac{2}{3}w - \frac{32}{27}w^3 + \cdots . \tag{2.54}$$

In consequence, it follows from (2.36) that, for a secondary component of vanishing mass

$$\nu_2 = \frac{1}{3}\left(\frac{2}{3}\right)^4 \left\{1 - \frac{16}{3}w^2 + \cdots\right\}. \tag{2.55}$$

An inspection of the last two columns of Table VI-2 reveals that, for the primary (more massive) component, the value of ν_1 increases monotonously with diminishing mass-ratio m_2/m_1 from 0.07227 for the case of equality of masses to 0.36075 for $m_2 = 0$, at which point the primary component becomes rotationally unstable and matter begins to be shed off along the equator if axial rotation were any faster. On the other hand, for the secondary (less massive) component the values of ν_2 diminish with decreasing mass-ratio from 0.07227 until, as $m_2 \to 0$, the value of $2^4/3^5$ has been attained.

C. GEOMETRY OF THE ECLIPSES

The data assembled in the foregoing section on the geometry of contact configuration lead to a number of specific conclusions regarding the eclipse phenomena to be exhibited by such systems. For suppose that a contact binary whose both components are at their Roche limits is viewed by a distant observer, whose line of sight does not deviate greatly from the x-axis of our model as shown on Figure 6-2. If so, then in the neighbourhood of either conjunction one component is going to eclipse the other, and the system will exhibit a characteristic variation in brightness. If, in turn, the observed light variation is analysed for the geometrical elements, the fractional 'radii' $r_{1,2}$ of the two components should (very apuroximately) be identical with the quantities $y_{4,5}$ as listed in columns (5) and (7) of Table VI-1. In Table VI-3, we have, accordingly, listed four-digit values of the sums $r_1 + r_2$ as well as the ratios r_2/r_1 of the 'radii' of such contact components as functions of their mass-ratio.

An inspection of this tabulation reveals that, within the scheme of our approximation, *the sum $r_1 + r_2$ of fractional radii of both components in contact binary systems is very nearly constant* and equal to 0.75 ± 0.01 for a very wide range of the mass-ratios q; whereas the *ratio r_2/r_1 decreases monotonically with diminishing value of q*. Therefore,

TABLE VI-3

q	$r_1 + r_2$	r_2/r_1
1	0.7484	1.0000
0.9	0.7486	0.9495
0.8	0.7489	0.8959
0.7	0.7496	0.8389
0.6	0.7510	0.7777
0.5	0.7529	0.7112
0.4	0.7565	0.6379
0.3	0.7622	0.5550
0.2	0.7722	0.4573
0.15	0.7805	0.3995
0.1	0.7935	0.3312

a photometric determination of the sum $r_1 + r_2$ – which, unfortunately, represents nearly all that can be deduced with any accuracy from an analysis of light curves due to shallow partial eclipses – cannot be expected to tell us anything new about contact systems; or, in particular, about their mass-ratios. It is the ratio of the radii r_2/r_1 whose determination would provide a sensitive photometric clue to the mass-ratio of a contact system. This underlines the importance of photometric determination of the ratios of the radii of contact binary systems; but owing to purely geometrical difficulties this important task of light curve analysis is, unfortunately, not yet well in hand.

Suppose next that a contact binary system, consisting of two components at their Roche limits, is viewed by a distant observer from an arbitrary direction. What will be the range of such directions from which this observer will see both bodies mutually eclipse each other during their revolution? In order to answer this question, let us replace the actual form of the corresponding Roche limit by an *osculating cone* which is tangent to it at the point of contact P_1. The equation of this cone may readily be obtained if we expand the function $\xi(x, y, z)$ of Roche equipotentials in a Taylor series, in three variables, about P_2.

The first partial derivatives ξ_x, ξ_y and ξ_z have already been given by Equations (2.20) –(2.22) in the preceding part of this section. Differentiating these equations further we find that

$$\xi_{xx} = (3x^2 - r^2)r^{-5} + q\{(3(1-x)^2 - r'^2\}(r')^{-5} + q + 1. \tag{2.56}$$

$$\xi_{yy} = (3y^2 - r^2)r^{-5} + q\{3y^2 - r'^2\}(r')^{-5} + q + 1, \tag{2.57}$$

$$\xi_{zz} = (3z^2 - r^2)r^{-5} + q\{3z^2 - r'^2\}(r')^{-5}, \tag{2.58}$$

$$\xi_{xy} = 3xyr^{-5} - 3q(1-x)y(r')^{-5}, \tag{2.59}$$

$$\xi_{xz} = 3xzr^{-5} - 3q(1-x)z(r')^{-5}, \tag{2.60}$$

$$\xi_{yz} = 3yzr^{-5} + 3qyz(r')^{-5}. \tag{2.61}$$

We note that all first (as well as mixed second) derivatives of ξ vanish at P_1. Hence, a requirement that the sum of nonvanishing second-order terms should add up to zero provides us with the desired equation of the osculating cone in the form

$$(x - x_2)^2\,(\xi_{xx})_1 + y^2(\xi_{yy})_1 + z^2(\xi_{zz})_1 = 0, \tag{2.62}$$

where

$$\left.\begin{aligned} (\xi_{xx})_1 &= 2p + q + 1, \\ (\xi_{yy})_1 &= -p + q + 1, \\ (\xi_{zz})_1 &= -p, \end{aligned}\right\} \tag{2.63}$$

in which we have abbreviated

$$p \equiv x_1^{-3} + q(1 - x_1)^{-3}. \tag{2.64}$$

The direction cosines l, m, n, of a line normal to the surface of this cone clearly are

$$l, m, n = \{f_\xi, f_y, f_z\} \div \{f_\xi^2 + f_y^2 + f_z^2\}^{1/2}, \tag{2.65}$$

where $f(\zeta, y, z)$ stands for the left-hand side of Equation (2.62) and $\zeta \equiv x - x_1$. Moreover, the direction cosines of the axis of this cone in the same coordinate system are $(1, 0, 0)$. Consequently, the angle ϵ between any arbitrary line on the surface of the osculating cone and its axis will be defined by the equation

$$\cos\left(\frac{1}{2}\pi - \epsilon\right) = l \tag{2.66}$$

or, more explicitly,

$$\tan^2\epsilon = -\left\{\frac{(\xi_{yy})_1\, y^2 + (\xi_{zz})_1\, z^2}{(\xi_{yy})_1^2\, y^2 + (\xi_{zz})_1^2\, z^2}\right\}(\xi_{xx})_1, \tag{2.67}$$

where the values of ξ_{xx}, ξ_{yy}, and ξ_{zz} at P_1 are given by Equations (2.63).

Suppose now that the orbital xy-plane of the two components is inclined at an angle i to a plane perpendicular to the line of sight (i.e., one tangent to the celestial sphere at the origin of the coordinates), and that ψ_1 denotes the angle of the first contact of the eclipse (as measured from the moment of superior conjunction). If so, then obviously

$$\left.\begin{aligned} \cos\epsilon = x &= \cos\psi_2 \sin i, \\ y &= \sin\psi_1 \sin i, \\ z &= \cos i, \end{aligned}\right\} \tag{2.68}$$

in Equation (2.67), and the latter can be simplified to disclose that

$$\delta^2 = \frac{az^2}{a-1}\left\{\frac{a+2-4\delta^2}{a+2-3\delta^2}\right\}, \tag{2.69}$$

where

$$\delta^2 = y^2 + z^2 = \sin^2\psi_1\,\sin^2 i + \cos^2 i \tag{2.70}$$

denotes the apparent projected distance between the centres of both components at the moment of first contact of the eclipse, and

$$a = \frac{q+1}{p} = \frac{q+1}{x_1^{-3} + q(1-x_1)^{-3}}\,. \tag{2.71}$$

All the foregoing results of this section have been based on the approximation of the Roche lobes in contact by an osculating cone at P_1. While this should always constitute a legitimate basis for computations of the limits of eclipses by the less massive component of a contact pair (in the sense that its surface is always interior to the common osculating cone), Chanan *et al.* (1976) called attention recently to the fact that the surface of the more massive component can actually 'overflow' this cone by amounts increasing with the disparity in masses of the two stars.

In order to demonstrate this, let us expand – with Chanan *et al.* (1976) – the Roche equipotentials $\Psi(x,y,z)$ of a contact loop in het proximity of the point P_1 of coordinates $x_1, 0, 0$ in a Taylor series of the form

$$\begin{aligned}\Psi(x,y,z) = \Psi(x_1,0,0) - \frac{1}{2}(2p+q+1)\zeta^2 + \\ + \frac{1}{2}(p-q-1)y^2 + \frac{1}{2}pz^2 + s\zeta^3 \\ - \frac{3}{2}s\zeta y^2 - \frac{3}{2}s\zeta z^2 + \cdots,\end{aligned} \tag{2.72}$$

correctly to terms of third order in x, y, z, where – as before, $\zeta \equiv x - x_1$ and where we have abbreviated

$$s = x_1^{-4} - q(1-x_1)^{-4}. \tag{2.73}$$

Over an equipotential surface Ψ = constant and, therefore, $\Psi(x, y, z) = \Psi(x_1, 0, 0)$. If, moreover, we confine our attention to an intersection of these equipotentials with the plane $z = 0$, Equation (2.72) can be solved for y in terms of ζ in the form

$$\begin{aligned}y^2 &= \frac{(2p+q+1)\zeta^2 - 2s\zeta^3}{p-q-1-3s\zeta} \\ &= \frac{2p+q+1}{p-q-1}\zeta^2 + \frac{4p+5q+5}{(p-q-1)^2}s\zeta^3 + \cdots.\end{aligned} \tag{2.74}$$

If we truncate the expansion on the right-hand side of the foregoing equation to its first term, we obtain the osculating cone identical with Equation (2.69) above. The next term, factored by an odd power of ζ, will change sign as $x \gtrless x_1$: for $x < x_1$ (i.e., in the direction of the less massive star) it will be negative and, hence, the actual value of

y^2 will be less than that appropriate for the osculating cone. For $x > x_1$ the converse will, however, be the case; and the actual Roche surface will overflow the osculating cone.

The extent to which this is the case cannot, in general, be established analytically, and recourse must be had to numerical computation. This has recently been done by Chanan *et al.* (1976), from whose paper the data given in the accompanying Table VI-4 have been excerpted. In the conical approximation, *the maximum duration of eclipse* (i.e., the maximum value of ψ_1) will obtain if the line of sight lies in the orbital plane and, therefore, $z = \cos^2 i = 0$: if so, Equation (2.69) discloses that

$$\cos^2 \psi_{max} = \tfrac{1}{3}\,(1-a). \tag{2.75}$$

TABLE VI-4
Eclipse and osculating cone angles in degrees as a function of mass ratio for a binary in which both components fill their Roche lobes

q	ψ_{max}	ψ_{cone}	i_{min}	i_{cone}
1.00	57.31	57.31	34.45	34.45
0.80	57.35	57.32	34.43	34.45
0.60	57.49	57.35	34.33	34.44
0.40	57.88	57.43	34.07	34.42
0.20	59.00	57.64	33.34	34.35
0.10	60.56	57.92	32.39	34.27

This is always bound to be true if a contact component is the less massive one; while for the more massive component the values appropriate for each mass ratio are as given in column (2) of the same table.

A glance at these data discloses that *the values of* ψ_{max} *are remarkably insensitive to the mass-ratio* – a fact of considerable significance for the student of close binary systems. For the variation of light exhibited by systems whose components fill in the largest Roche lobes capable of containing their mass (in particular, eclipsing systems of the W UMa-type) in the course of an orbital cycle is so smooth and continuous that it is next to impossible to detect by a mere inspection of their respective light curves just where eclipses may set in. Our present analysis has now supplied a theoretical answer, namely, *the light changes of an eclipsing system will be unaffected by eclipses for all phase angles in excess of ± 60° even if both components are in actual contact* – at least as long as their mass-ratios do not become less than 1 : 10 (for greater disparity in masses this limit will continue to increase; cf. Chanan *et al.*, 1976). Therefore, the light changes exhibited at phases ± 30° (or more) around each quadrature should be due solely to the proximity effects associated with both stars, and may be analysed as such without fear of interference from eclipse phenomena.

The closed form of Equation (2.69) invites another question which can be asked in this connection: namely, what is the *minimum inclination* of the orbital plane to the celestial sphere below which no eclipses may occur even for contact binary systems? This minimum will be attained if the eclipse becomes incipient at the time of the conjunctions (i.e, if $\psi_1 = 0$). Under these circumstances, Equation (2.69) implies that

$$\cos^2 i_{\min} = \frac{2+a}{3-a} \tag{2.76}$$

in the conical approximation; though more elaborate computations are necessary to establish this limit for $q \ll 1$.

The corresponding results are tabulated in the last two columns of Table VI-4. An inspection of these data reveals that – again almost regardless of the mass ratio – *no binary system can exhibit eclipses if its orbit is inclined to the celestial sphere by less than 33°–34°*. For values of i greater than this limit eclipses may occur (and must occur for contact binary systems) of durations ψ_1 connected with i by Equation (2.69) in the conical approximation. For mass ratios $q \ll 1$ the relation between the two becomes again more involved; and for its tabulation the reader is referred to Chanan *et al.* (1976).

D. EXTERNAL ENVELOPES

In the foregoing parts of this section we have been concerned with various properties of Roche equipotentials when $\xi > \xi_1$, and later we investigated the geometry of limiting double-star configurations for which $\xi = \xi_1$. The aim of the present section will be to complete our analysis of the geometrical properties of the Roche model by considering what happens when $\xi < \xi_1$. In introductory part VI-1 of this chapter we inferred on general grounds that, if $\xi < \xi_1$, the dumb-bell figure which originally surrounded the two components will open up at P_1 (cf. again Fig. 6-1), and the corresponding equipotentials will enclose *both* bodies.

When will these latter equipotentials containing the total mass of our binary system cease to form a *closed* surface? A quest for the answer will take us back to Equation (2.20) defining the partial derivative ξ_x. We may note that the right-hand side of this equation is positive when $x \to \infty$, but becomes negative when $x = 1 + \epsilon$, where ϵ denotes a small positive quantity. It becomes positive again as $x \to 0$, and changes sign once more for $x \to -\infty$. Since ξ_x is finite and continuous everywhere except at $|x| = \infty$ and for $r = 0$ or $r' = 0$, it follows that it changes sign *three* times by passing through zero at points x_1, x_2, x_3, whose values are such that

$$\left.\begin{array}{ll} \text{(a)} & 0 < x_1 < 1, \\ \text{(b)} & \quad x_2 > 1, \\ \text{(c)} & \quad x_3 < 0; \end{array}\right\} \tag{2.77}$$

and of these, only the first one has been evaluated so far in this section, and its numerical values listed in column (2) of Table VI-1.

An evaluation of the remaining roots $x_{2,3}$ offers, however, no greater difficulty. In embarking upon it we should merely keep in mind that, regardless of the sign of x, the distances r and r' as defined by Equations (1.3) are *positive* quantities. Thus, unlike in case (a) – when, by setting $r = x$ and $r' = 1 - x$, we were led to define x_1 as a root of Equation (2.24) – in case (b), when $x_2 > 1$, we must set $r = x$ but $r' = x - 1$; and in case (c), when $x_3 < 0$, $r = -x$ and $r' = 1 - x$. After doing so and clearing the fractions we may verify that the equation $\xi_x = 0$ in the case of (b) and (c) assumes the explicit form

$$(1+q)x^5 - (2+3q)x^4 + (1+3q)x^3 - (1+2q)x^2 + 2x - 1 = 0 \tag{2.78}$$

and

$$(1+q)x^5 - (2+3q)x^4 + (1+3q)x^3 + x^2 - 2x + 1 = 0, \tag{2.79}$$

respectively.

For $q = 0$, the former Equation (2.78) becomes identical with (2.24) and reduces to (2.25) admitting of $x = 1$ as a triple root. Hence, for small values of q, the root $x_2 > 1$ of the complete Equation (2.78) should be expansible as

$$x_2 = 1 + \left(\frac{\mu}{3}\right)^{1/3} + \frac{1}{3}\left(\frac{\mu}{3}\right)^{2/3} + \frac{1}{9}\left(\frac{\mu}{3}\right) + \cdots \tag{2.80}$$

in terms of fractional powers of $\mu \equiv q/(q+1)$. Similarly, Equation (2.79) reduces for $q = 0$ to

$$(x-1)^2\,(x^3+1) = 0, \tag{2.81}$$

admitting of only one negative root (namely, -1). In consequence, the negative root x_3 of (2.79) should, for small values of μ, be approximable in terms of integral powers of μ by an expansion of the form

$$x_3 = -1 + \frac{7}{12}\,\mu - \frac{1127}{20736}\,\mu^3 + \cdots . \tag{2.82}$$

The approximate values of x_2 and x_3 as obtained from (2.80)–(2.82) may, moreover, be subsequently refined to any degree of accuracy by differential corrections or any other standard method.

Once sufficiently accurate values of $x_{2,3}$ have thus been established, the values of ξ corresponding to equipotentials which pass through these points can be ascertained from the equation

$$\xi_{2,3} = \xi(x_{2,3}, 0, 0); \tag{2.83}$$

while the corresponding values of $C_{2,3}$ then can be found from

$$C_{2,3} = 2(1-\mu)\xi_{2,3} + \mu^2 . \tag{2.84}$$

A tabulation of five-digit values of $x_{2,3}$ and $C_{2,3}$ is given in columns (2) – (5) of the accompanying Table VI–5. It may also be noticed that, to a high degree of approximation

$$\xi_3 \doteq \frac{3}{2} + 2q - \frac{q^2}{2(1+q)} \tag{2.85}$$

or

$$C_3 \doteq 3 + \mu ; \tag{2.86}$$

while, somewhat less accurately,

$$(x_2 - 1)^2 = 1 - x_3^2 . \tag{2.87}$$

A comparison of the values of $C_{2,3}$ as given in Table VI-5 with those of C_1 from Table VI-2 reveals that, for all values of $q > 0$,

$$C_1 > C_2 \geqslant C_3. \tag{2.88}$$

For any value of C within the limits of the inequality $C_1 > C > C_2$ the corresponding equipotential will surround the whole mass of the system by a common *external envelope*, which may enclose the common atmosphere of the two stars. For $C = C_2$, this envelope will develop a conical point P_2 (at which $\xi_x = \xi_y = \xi_z = 0$) at $x = x_2$ – i.e., behind the centre of gravity of the less massive component (see Figures 6-1 or 2); and if $C < C_2$, the respective equipotentials will open up at P_2. For $C = C_3$, a third conical point P_3 develops behind the centre of gravity of the more massive component at $x = x_3$; and if $C < C_3$, the equipotentials will become open at both ends. Their intersection with the xy-plane will then no longer represent a single closed curve, but will split up in two separate sections (symmetrical with respect to the x-axis), closing gradually around two points which make equilateral triangles with the centres of mass of the two components. The coordinates of such points are specified by the requirements that $r = r' = 1$; consequently, $x = 0.5$ and $y = \pm \sqrt{3}/2$. These triangular points represent also the loci at which our equipotentials vanish eventually from the real plane – if (consistent with Equations (1.6) and (2.31)) their constants C reduce to

$$C_{4,5} = 3 - \mu + \mu^2. \tag{2.89}$$

The values of $C_{4,5}$'s as given by this equation are listed in column (6) of Table VI-5 for $1 > q > 0$, and represent the lower limits attainable by these constants; for if $C < C_{4,5}$, the equipotential curves ξ = constant in the xy-plane become imaginary, and thus devoid of any further physical significance.

TABLE VI-5[a]

q	x_2	C_2	$-x_3$	C_3	$C_{4,5}$
1.0	1.69841	3.45680	0.69841	3.45680	2.75000
0.8	1.66148	3.49368	0.73412	3.41509	2.75309
0.6	1.61304	3.53108	0.77751	3.35791	2.76563
0.4	1.54538	3.55894	0.83180	3.27822	2.79592
0.3	1.49917	3.55965	0.86461	3.22675	2.82249
0.2	1.43808	3.53634	0.90250	3.16506	2.86111
0.15	1.39813	3.50618	0.92372	3.12959	2.88658
0.1	1.34700	3.45153	0.94693	3.09058	2.91735
0.05	1.27320	3.34671	0.97222	3.04755	2.95465
0.02	1.19869	3.22339	0.98854	3.01961	2.98077
0.01	1.15614	3.15344	0.99422	3.00990	2.99020
0.005	1.12294	3.10301	0.99710	3.00498	2.99504
0.001	1.07089	3.03838	0.99942	3.00099	2.99900
0.0002	1.04108	3.01387	0.99988	3.00020	2.99980
0	1.00000	3.00000	1.00000	3.00000	3.00000

a The data collected in this table are taken from Kopal (1959).
For other tabulations of these quantities – in particular, for very small values of the parameter $\mu = q/(q+1)$ – cf., Rosenthal (1931), Kuiper and Johnson (1956), Szebehely (1967), or Kitamura (1970).

VI-3. The Roche Coordinates

In the second part of this chapter we outlined the principal geometrical features characteristic of the Roche equipotentials and – in particular – of the largest closed equipotentials capable of containing their total mass, which we call the 'Roche limit'. In section VI-2 we expressed the external form of such equipotentials, in spherical solar coordinates, by Equation (2.11) requiring 4, 15 and 45 separate harmonic terms to describe the shape of a Roche equipotential correctly to terms of the first, second and third order in r_0, respectively. For, (say) $q = 1$, $r_0 = 0.378$ at the Roche limit (cf. Table VI-2); which means that even a retention of 45 terms on the right-hand side of Equation (2.11) will not permit us to describe the form of the respective equipotential more accurately than within errors of the order of one unit of the fourth decimal place.

On the other hand, an implicit equation of the form

$$\xi(r, \lambda, \nu) \equiv r^{-1} + r'^{-1} - \lambda r + r^2(1 - \nu^2) = 3.75,$$

obtained by setting $q = 1$ in Equation (1.6), will define the same surface exactly in much simpler algebraic form. Inasmuch as such equipotentials can represent to a high degree of approximation the surface of the components in close binary systems, and the form of such a surface figures prominently in many tasks encountered in double-star astronomy (such as a determination of the light- or velocity-curves of close eclipsing systems, hydrodynamic studies of the gas streams in such systems, etc.), the idea suggests itself to introduce a *curvilinear system* of coordinates – hereafter referred to as the *Roche coordinates* – in which one (the ξ-) coordinate is defined by the Roche equipotential surfaces of the form (1.6) given to us in closed algebraic form (and playing thus the generalized role of the radial coordinate in spherical polars); while the angular Roche coordinates η, ζ are defined by the requirement that they be *orthogonal* to ξ as well as with respect to each other.

The differential equation generating the surfaces $\xi(x, y, z) = $ constant is known to be of the form

$$\xi_x \mathrm{d}x + \xi_y \mathrm{d}y + \xi_z \mathrm{d}z = 0; \tag{3.1}$$

while the equations generating surfaces which are orthogonal to those produced by (3.1) are given by

$$\frac{\mathrm{d}x}{\xi_x} = \frac{\mathrm{d}y}{\xi_y} = \frac{\mathrm{d}z}{\xi_z}. \tag{3.2}$$

These latter relations constitute two independent first-order differential equations, which together with (3.1) constitute a complete set of orthogonality conditions whose integration constants represent the Roche coordinates whose net we propose to construct.

In more specific terms, the relations in question constitute a set of simultaneous partial differential equations of the form

$$\left.\begin{aligned} \xi_x\eta_x + \xi_y\eta_y + \xi_z\eta_z &= 0,\\ \xi_x\zeta_x + \xi_y\zeta_y + \xi_z\zeta_z &= 0,\\ \eta_x\zeta_x + \eta_y\zeta_y + \eta_z\zeta_z &= 0, \end{aligned}\right\} \tag{3.3}$$

where the subscripts x, y, z signify partial differentiation with respect to the particular variable. If so, a transformation of the metric element

$$(\mathrm{d}x)^2 + (\mathrm{d}y)^2 + (\mathrm{d}z)^2 = h_1^2(\mathrm{d}\xi)^2 + h_2^2(\mathrm{d}\eta)^2 + h_3^2(\mathrm{d}\zeta)^2, \tag{3.4}$$

will be specified by the coefficients

$$h_1^{-2} = \xi_x^2 + \xi_y^2 + \xi_z^2\,, \tag{3.5}$$

$$h_2^{-2} = \eta_x^2 + \eta_y^2 + \eta_z^2, \tag{3.6}$$

and

$$h_3^{-2} = \zeta_x^2 + \zeta_y^2 + \zeta_z^2\ . \tag{3.7}$$

Moreover, the direction cosines of a line normal to the surfaces ξ = constant, η = constant and ζ = constant are given by the ratios

$$l_1 = h_1\xi_x, \quad m_1 = h_1\xi_y, \quad n_1 = h_1\xi_z; \tag{3.8}$$

$$l_2 = h_2\eta_x, \quad m_2 = h_2\eta_y, \quad n_2 = h_2\eta_z; \tag{3.9}$$

$$l_3 = h_3\zeta_x, \quad m_3 = h_3\zeta_y, \quad n_3 = h_3\zeta_z; \tag{3.10}$$

respectively.

The foregoing Equations (3.1) to (3.10) hold good for any triply-orthogonal set of curvilinear coordinates ξ, η, ζ (cf. Darboux, 1910; or Forsyth, 1912). In what follows, we propose to restrict their choice by an identification of its ξ-coordinate with the Roche potential (1.6) – so that surfaces ξ = constant become identical with those of equal potential of centrally-condensed stars – and our task will be to specify the form of the corresponding surfaces of constant η and ζ.

A. ROTATIONAL PROBLEM

In order to approach the general problem raised in the preceding section in steps, consider first the case when $q = 0$ and when the sole reason of the departure of the Roche equipotentials from spherical form is axial rotation with a (normalized) angular velocity, the square of which we shall (consistent with Equation (2.2)) denote by $2n$. If so, the Roche potential (1.6) reduces to

$$\xi(r, \nu) = \frac{1}{r} + nr^2(1 - \nu^2), \tag{3.11}$$

where $\nu \equiv \cos\theta$; and the partial first derivatives of ξ with respect to x, y, z – already known from Equations (2.20)–(2.22) – now reduce to

$$\left.\begin{aligned} \xi_x &= -x(r^{-3} - 2n), \\ \xi_y &= -y(r^{-3} - 2n), \\ \xi_z &= -z(r^{-3})\,. \end{aligned}\right\} \tag{3.12}$$

The lines that are orthogonal to the equipotentials defined by Equation (3.11) are known to be given by the differential equations

$$\frac{dx}{\xi_x} = \frac{dy}{\xi_y} = \frac{dz}{\xi_z} . \tag{3.13}$$

Since, however, by (3.12)

$$\frac{\xi_x}{\xi_y} = \frac{x}{y} \tag{3.14}$$

regardless of the extent of the distortion, the first part of Equations (3.13) readily yields

$$\frac{dx}{dy} = \frac{x}{y}, \tag{3.15}$$

and integrates to

$$\frac{x}{y} = \text{constant} = \cot \eta \text{ (say)}. \tag{3.16}$$

Accordingly, the surfaces η = constant turn out to be identical with the meridional planes ϕ = constant of the rotationally distorted Roche model; and the three-dimensional Roche coordinates in this case prove to constitute a 'Lamé family' (cf. Darboux, 1910) of the respective triply-orthogonal set.

The properties of the remaining surfaces ζ = constant which are orthogonal to ξ and η can then be obtained by an integration of the orthogonality conditions (3.1)–(3.3). Since, however, by (3.16)

$$\left.\begin{aligned} (r \sin \eta)\, \eta_x &= \mu^{-1}, \\ (r \sin \eta)\, \eta_y &= -\lambda\mu^{-1}, \\ (r \sin \eta)\, \eta_z &= 0 , \end{aligned}\right\} \tag{3.17}$$

and, in addition, $\zeta_\lambda = 0$ on account of the rotational symmetry of our configuration, Equation (3.3) happens to be automatically satisfied without imposing any restriction on ζ. Therefore, in order to determine the latter Equation (3.2) alone needs to be integrated; and in doing so (cf. Kopal, 1971b) we find it to be satisfied by an expression of the form

$$\cos \zeta = \nu \sum_{j=0}^{\infty} (2n)^j r^{3j} X_j(\nu), \tag{3.18}$$

where

$$X_0(\nu) = 1, \qquad X_1(\nu) = -\frac{1}{3}(1 - \nu^2); \tag{3.19}$$

while, for $j > 1$, all subsequent $X_j(\nu)$'s can be generated with the aid of the recursion formula

$$3j\, X_j + (1 - \nu^2)\, \{(\nu X_{j-1})' - 3(j-1) X_{j-1}\} = 0, \tag{3.20}$$

where primes denote differentiation with respect to ν.

Accordingly, to quantities of second order in n, Equation (3.18) can be written out as

$$\cos\zeta = \nu\{1 - \frac{2}{3}nr^3(1-\nu^2) - \frac{4}{9}n^2r^6(1-\nu^2) + \cdots\}; \tag{3.21}$$

and, by (3.16)

$$\eta = \phi. \tag{3.22}$$

Equations (3.11) and (3.22) together with (3.18) constitute the exact form of triply-orthogonal set of Roche coordinates ξ, η, ζ for the rotational problem expressed in terms of the spherical-polar coordinates r, θ, ϕ (or r, λ, ν); with (3.20) representing a second approximation to (3.18). Moreover, by Equations (3.5) and (3.6) we find that the metric coefficients

$$h_1(r,\nu) = -\frac{r^2}{\sqrt{1-4nr^3(1-\nu^2)+n^2r^6(1-\nu^2)}}, \tag{3.23}$$

$$h_2(r,\nu) = r\sqrt{1-\nu^2}, \tag{3.24}$$

exactly; while, to a second-order approximation, equation (3.7) leads to

$$h_3(r,\nu) = r\{1 + \frac{2}{3}nr^3(1-2\nu^2) + \frac{2}{9}n^2r^6(4-24\nu^2+21\nu^4) + \cdots\}. \tag{3.25}$$

Moreover, the direction cosines

$$l_1 = \lambda\{1 - 2nr^2\nu^2 - 6n^2r^6\nu^2(1-\nu^2) + \cdots\}, \tag{3.26}$$

$$m_1 = \mu\{1 - 2nr^3\nu^2 - 6n^2r^6\nu^2(1-\nu^2) + \cdots\}, \tag{3.27}$$

$$n_1 = \nu\{1 + 2nr^3(1-\nu^2) + 2n^2r^6(1-\nu^2)(2-3\nu^2) + \cdots\}; \tag{3.28}$$

$$l_2 = -\frac{\mu}{\sqrt{1-\nu^2}} \tag{3.29}$$

$$m_2 = \frac{\lambda}{\sqrt{1-\nu^2}}, \tag{3.30}$$

$$n_2 = 0; \tag{3.31}$$

and

$$l_3 = \frac{\lambda\nu}{\sqrt{1-\nu^2}}\{1 + 2nr^3(1-\nu^2) + 2n^2r^6(1-\nu^2)(2-3\nu^2) + \cdots\}, \tag{3.32}$$

$$m_3 = \frac{\mu\nu}{\sqrt{1-\nu^2}}\{1 + 2nr^3(1-\nu^2) + 2n^2r^6(1-\nu^2)(2-3\nu^2) + \cdots\}, \tag{3.33}$$

$$n_3 = -\sqrt{1-\nu^2}\{1 - 2nr^3\nu^2 - 6n^2r^6\nu^2(1-\nu^2) + \cdots\}. \tag{3.34}$$

The foregoing Equations (3.11), (3.18) and (3.22) express the Roche curvilinear coordinates ξ, η, ζ and the associated metric coefficients or direction cosines in terms of the polar coordinates r, θ, and ϕ. On the other hand, from Equation (2.11) it follows that, in our case

$$r = r_0 \{1 + nr_0^3(1-\nu^2) + 3n^2 r_0^6 (1-\nu^2)^2 + \cdots\}, \tag{3.35}$$

where

$$r_0 = \frac{1}{\xi}; \tag{3.36}$$

and by an inversion of (3.21) it follows that

$$1-\nu^2 = \sin^2\zeta \left\{1 - \frac{4}{3} nr_0^3 \cos^2\zeta - \frac{8}{9} n^2 r_0^6 \cos^2\zeta\,(7 - 8\cos^2\zeta) + \cdots \right\}, \tag{3.37}$$

it can be shown (cf. again Kopal, 1970, 1971) that, correctly to quantities of second order in n,

$$h_1(\xi, \zeta) = -r_0^2 \{1 + 4nr_0^3 \sin^2\zeta - \frac{4}{3} n^2 r_0^6 \sin^2\zeta\,(22 - 85\sin^2\zeta) + \cdots, \tag{3.38}$$

$$h_2(\xi, \zeta) = r_0 \sin\zeta \{1 - \frac{1}{3} nr_0^3 (2 - 5\sin^2\zeta) + \frac{1}{9} n^2 r_0^6 (2 - 50\sin^2\zeta + 75\sin^4\zeta) + \cdots\} \tag{3.39}$$

,
and

$$h_3(\xi, \zeta) = r_0 \{1 - \frac{1}{3} nr_0^3 (2 - 7\sin^2\zeta) + \frac{1}{9} n^2 r_0^6 (2 - 88\sin^2\zeta + 145\sin^4\zeta) + \cdots\}; \tag{3.40}$$

while the direction cosines l_j, m_j, n_j given by (3.26)–(3.34) in terms of spherical polar coordinates can be rewritten (with the aid of (3.35) to (3.37)) in terms of the Roche coordinates ξ, η, ζ if required with equal ease.

B. TIDAL PROBLEM

If a configuration built up in accordance with our Roche model is non-rotating (i.e., $n = 0$) but distorted by tides ($q > 0$), the expression (1.6) for its potential will reduce to

$$\xi = \frac{1}{r} + q \left\{ \frac{1}{\sqrt{1 - 2\lambda r + r^2}} - \lambda r \right\}; \tag{3.41}$$

and its derivatives, to

$$\left.\begin{aligned}\xi_x &= -xr^{-3} + q\{(1-x)r'^{-3} - 1\},\\ \xi_y &= -y\{r^{-3} + qr'^{-3}\},\\ \xi_z &= -z\{r^{-3} + qr'^{-3}\};\end{aligned}\right\} \tag{3.42}$$

in accordance with (2.20)–(2.22). Since now

$$\frac{\xi_y}{\xi_z} = \frac{y}{z} \tag{3.43}$$

in place of (3.14) of the rotational problem, the second part of Equation (3.13) now leads to

$$\frac{\mathrm{d}y}{\mathrm{d}z} = \frac{y}{z}, \tag{3.44}$$

and integrates to

$$\frac{y}{z} = \text{constant}; \tag{3.45}$$

so that only one integral of the system (3.1)–(3.8) needs to be obtained to specify the triply-orthogonal system of Roche coordinates associated with tidal distortion, of which (by virtue of (3.45)) one surface will again constitute a family of planes.

On performing the requisite integration (for its details, cf. Kopal, 1970) we find that, correctly to terms of *first* order in surficial distortion,

$$\eta = \cos^{-1}\lambda - \frac{q}{\sqrt{1-\lambda^2}} \sum_{j=2}^{4} \frac{r^{j+1}}{j+1} P_j'(\lambda), \tag{3.46}$$

while, constant with (3.46)

$$\zeta = \cos^{-1} \frac{\nu}{\sqrt{1-\lambda^2}} \quad . \tag{3.47}$$

The geometrical meaning of the surfaces defined by constant values of the curvilinear coordinates η and ζ are simple: the equation η = constant represents a family of distorted cones passing through the origin and symmetrical with respect to the X-axis (i.e., one of symmetry of the tidal distortion); while the surfaces ζ = constant are planes passing through the X-axis and perpendicular to the plane $X = 0$.

The metric coefficients $h_{1,2,3}$ associated with the transformation (3.4) in the case of the tidal distortion continue to be given by Equation (3.5)–(3.7), in which we insert the requisite partial derivatives of (3.41), (3.46) and (3.47). The results disclose that

$$h_1(r, \lambda, \nu) = r^2\left\{1 + q \sum_{j=2}^{4} j r^{j+1} P_j(\lambda)\right\}, \tag{3.48}$$

$$h_2(r, \lambda, \nu) = r\left\{1 + q \sum_{j=2}^{4} \frac{r^{j+1}}{j+1} [j(j+1)P_j(\lambda) - \lambda P_j'(\lambda)]\right\}, \tag{3.49}$$

and

$$h_3(r, \lambda, \nu) = r\sqrt{1-\lambda^2}; \tag{3.50}$$

the last of which is exact, but the first two represent first-order approximations.

Moreover, the direction cosines of the normals to the surfaces ξ = constant, η = constant and ζ = constant follow again from Equations (3.8)–(3.10) as

$$l_1 = \lambda - q(1-\lambda^2)\sum_{j=2}^{4} r^{j+1}P_j'(\lambda), \tag{3.51}$$

$$m_1 = \mu\left\{1 + q\lambda\sum_{j=2}^{4} r^{j+1}P_j'(\lambda)\right\}, \tag{3.52}$$

$$n_1 = \nu\left\{1 + q\lambda\sum_{j=2}^{4} r^{j+1}P_j'(\lambda)\right\}; \tag{3.53}$$

$$l_2 = \sqrt{1-\lambda^2}\left\{1 + q\lambda\sum_{j=2}^{4} r^{j+1}P_j'(\lambda)\right\}, \tag{3.54}$$

$$m_2 = -\frac{\mu}{\sqrt{1-\lambda^2}}\left\{\lambda - q\sum_{j=2}^{4} r^{j+1}(1-\lambda^2)P_j'(\lambda)\right\}, \tag{3.55}$$

$$n_2 = -\frac{\nu}{\sqrt{1-\lambda^2}}\left\{\lambda - q\sum_{j=2}^{4} r^{j+1}(1-\lambda^2)P_j'(\lambda)\right\}; \tag{3.56}$$

and

$$l_3 = 0, \tag{3.57}$$

$$m_3 = \frac{\nu}{\sqrt{1-\lambda^2}}, \tag{3.58}$$

$$n_3 = -\frac{\mu}{\sqrt{1-\lambda^2}} \tag{3.59}$$

the last three of which are again exact.

On the other hand, if we wish to express these quantities in terms of Roche curvilinear rather than spherical-polar coordinates, this task can be accomplished by noting from (2.11) that, to the first order in small quantities,

$$r(\xi, \eta) = r_0\left\{1 + q\sum_{j=2}^{4} r_0^{j+1}P_j(\lambda_0)\right\} \tag{3.60}$$

and

$$1-\lambda^2 = (1-\lambda_0^2)\left\{1 + 2q\sum_{j=2}^{4}\frac{r_0^{j+1}}{j+1}\lambda_0 P_j'(\lambda_0)\right\}, \tag{3.61}$$

where

$$r_0 \equiv \frac{1}{\xi - q} \quad \text{and} \quad \lambda_0 \equiv \cos \eta. \tag{3.62}$$

Inserting (3.60) and (3.61) in (3.48)–(3.50) we easily establish that, to the same order of accuracy,

$$h_1(\xi, \eta) = r_0^2 \left\{ 1 + q \sum_{j=2}^{4} (j+2) r_0^{j+1} P_j(\lambda_0) \right\}, \tag{3.63}$$

$$h_2(\zeta, \eta) = r_0 \left\{ 1 + q \sum_{j=2}^{4} \frac{r_0^{j+1}}{j+1} [(j+1)(j+2) P_j(\lambda_0) - P'_{j+1}(\lambda_0)] \right\}, \tag{3.64}$$

$$h_3(\xi, \eta) = r_0 \sqrt{1 - \lambda_0^2} \left\{ 1 + q \sum_{j=2}^{4} \frac{r_0^{j+1}}{j+1} P'_{j+1}(\lambda_0) \right\}; \tag{3.65}$$

and the direction cosines (3.51)–(3.59) can be rewritten in terms of ξ and η with the aid of (3.60) and (3.61) if needed with equal ease.

C. DOUBLE-STAR PROBLEM

In the preceding two sections of this part of our chapter we established the existence and approximate form of Roche curvilinear coordinates associated with the separate effects of rotation and tides. In actual double stars the effects of rotational and tidal distortion are, however, of the same order of magnitude, and their first-order effects are additive. Therefore, the Roche coordinates associated with the components of close binary systems revolving around the common centre of gravity should be obtained by an integration of the complete system (3.1)–(3.3) of orthogonality conditions for the Roche potential ξ as given by Equation (1.6).

In sections 3A and B we integrated this system for the separate cases of $q = 0, n > 0$ and $n = 0, q > 0$ – and, in each case, one Roche surface associated with the angular coordinates turned out to be a plane – a fact which reduced the order of the differential system (3.1)–(3.3) by one; and reduced the corresponding set of curvilinear coordinates to constitute a Lamé family. In the general double-star case this will, however, no longer be true: the surfaces over which all three orthogonal coordinates ξ, η, ζ = constant will be curved; and the question arises whether or not, under these conditions, the differential systems (3.1)–(3.3) continue to be integrable for the general form of $\xi(r, \theta, \phi)$ as given by Equation (1.6).

In order to answer the question, let us return to Equations (3.1)–(3.3) and eliminate between them the derivatives of either η or ζ (on acccount of the symmetry, the choice is immaterial). Eliminating η we obtain the following determinantal equation.

$$\begin{vmatrix} \xi_x & \zeta_x & \zeta_x\xi_{xx} + \zeta_y\xi_{xy} + \zeta_z\xi_{xz} \\ \xi_y & \zeta_y & \zeta_x\xi_{xy} + \zeta_y\xi_{yy} + \zeta_z\xi_{yz} \\ \xi_z & \zeta_z & \zeta_x\xi_{xz} + \zeta_y\xi_{yz} + \zeta_z\xi_{zz} \end{vmatrix} = 0 \tag{3.66}$$

between first and second partial derivatives of ξ and ζ, to which we should adjoin the relation (3.2) between first derivatives of the same variables.

The foregoing Equation (3.66) is homogeneous and quadratic in $\zeta_{x,y,z}$. We can, therefore, solve (3.66) together with (3.2) for the *ratios* of $\zeta_{x,y,z}$ in the form

$$\zeta_x : \zeta_y : \zeta_z = L : M : N, \tag{3.67}$$

where L, M, N are functions, not only of the first derivatives of ξ as given by Equations (2.20)–(2.22), but also of its second derivatives (2.56)–(2.61).

In order that the function $\zeta(x, y, z)$ be integrable from (3.1)–(3.3), it is necessary (cf. Darboux, 1910; or Forsyth, 1912) that the total differential equation

$$L\,\mathrm{d}x + M\,\mathrm{d}y + N\,\mathrm{d}z = 0 \tag{3.68}$$

be integrable; and its condition of integrability is known to be of the form

$$L(M_z - N_y) + M(N_x - L_z) + N(L_y - M_x) = 0, \tag{3.69}$$

where the subscripts x, y, z denote again partial differentiation with respect to the respective variable.

If a family of surfaces represented by the equation $\xi(x, y, z)$ = constant be capable of generating an integrable triply-orthogonal system, it is both necessary and sufficient that the generating function ξ should satisfy the above partial differential Equation (3.69) of fourth degree and third order. An evaluation of the left-hand side of (3.69) for ξ as given by (1.6) was recently undertaken by Kopal and Ali (1971). The result disclosed that this equation is indeed satisfied if either n or q is zero (i.e., in the case of pure rotational or tidal distortion), but *not* if both n and q are nonvanishing – as they indeed are in the double-star problem. Therefore, the general Roche potential of the form (1.6) – which constitutes a scalar sum of the effects arising from rotation and tides – renders the orthogonality conditions (3.1)–(3.3) nonintegrable analytically because it fails to satisfy the integrability condition (3.69).

It may be noted that the situation we meet is analogous to that encountered in the restricted problem of three bodies, of which ξ represents indeed the potential. As is well known, for $q = 0$ and $n > 0$, the three-body problem reduces to that of two bodies which is analytically integrable in terms of elementary functions; while if $n = 0$ and $q > 0$, the three-body problem reduces to a 'two-centre' case which can likewise be integrated in terms of elliptic functions. However, since the days of Poincaré (1892) we know that if both n and q are different from zero, the restricted problem of three bodies does *not* admit of any integral (algebraic or transcendental) other than that of the energy – which our Roche potential (1.6) represents a constant velocity surface.

It appears, therefore, that when we set out to construct a triply-orthogonal system of curvilinear coordinates of which the complete Roche potential (1.6) of the double-star problem is a member, the fact that this form of ξ fails to fulfil Equation (3.69) precludes the existence of any analytic integrals of our orthogonality conditions (3.1)–(3.3) other than ξ in the same way as Poincaré's theorem has ruled out the existence of any non-trival integrals in the restricted problem of three bodies other than the energy integral; and numerical integrations represent the only avenue of approach towards their construction.

Extensive and accurate numerical integrations of the system (3.1)–(3.3) have indeed been recently carried out by Kitamura (1970) in such cases when this system can be reduced to ordinary differential equations. Such a possibility arises indeed in the planes $X = 0$, $Y = 0$ or $Z = 0$ where the families of the surfaces defined by ξ = constant, η = constant, or ζ = constant intersect the Cartesian planes in the form of two-dimensional curves, which can be generated by numerical solutions of ordinary differential equations in two variables.

Before we do so, however, let us consider in more detail the geometrical characteristics of the curves obtaining by an intersection of the surfaces $\xi(x, y, z)$ with the xy-plane. Figure 6-1 has given us a schematic sketch of such curves; and a more detailed presentation of the profiles of the functions $\xi(x, y, 0)$ in the plane $Z = 0$ can be found on the accompanying Figures 6-3 to 6-6 drawn to scale for four values of the mass-ratio $q = 1.0$ (-0.2) 0.4; while Figure 6-7 contains a sketch of a three-dimensional representation of the surfaces $\xi(x, y, z) = \xi_{1,2,\ldots 5}$ passing through the five Lagrangian points discussed in section VI-1.

Moreover, in the same plane $(Z = 0) = 0$; and if so, the orthogonality conditions (3.1)–(3.3) reduce to the single requirement that

$$\xi_x \eta_x + \xi_y \eta_y = 0, \tag{3.70}$$

disclosing that the curves $\eta(x, y)$ = constant in the xy-plane should be generated by one-parameter solutions of the ordinary differential equation

$$\frac{\mathrm{d}x}{\mathrm{d}y} = \frac{\xi_x}{\xi_y} = \frac{x(1 - r_1^3)r_2^3 - q(1 - x)r_1^3(1 - r_2^3)}{y[(1 - r_1^3)r_2^3 + qr_1^3(1 - r_2^3)]}, \tag{3.71}$$

where

$$r_1^2 = x^2 + y^2 \qquad \text{and} \qquad r_2^2 = (1 - x)^2 + y^2. \tag{3.72}$$

In the immediate proximity of the origin, approximate analytical solutions of the linerarized Equation (3.70) were constructed by Kopal (1969); but it was not till Kitamura (1970) embarked on extensive numerical integration of the exact (nonlinear) Equation (3.71) that the topological properties of the family of curves defined by the equation $(x, y, 0)$ = constant have emerged from his work; and their principal features can be summarized as follows.

For large values of ξ – in the close proximity of the two finite centres of mass 1 and q – the curves η = constant depart but slightly from straight lines passing through the two centres; and so they are at great distances from our gravitational dipole, where they are tangent to the quasi-circular potential 'curtains' surrounding the system from outside. In the intermediate regions of the plane $Z = 0$, however, both the inner and outer tangent lines curve along a common envelope, and pass through the Lagrangian equilateral-triangle points $L_{4,5}$. However, the positions of the collinear Lagrangian points $L_{1,2,3}$ can be reached with the limiting η = constant lines which coincide with the X- or Y-axis. This remains true for every mass-ratio $0 \lesssim q \lesssim 1$.

These results make it evident that, in the immediate proximity of the mass-centres of the two components – or at sufficiently great distance from them, the $\xi\eta$-Roche coordinates tend asymptotically to plane-polar coordinates in which ξ plays the role of the radial; and η, of the angular, coordinate. However, in the intermediate region the situation becomes very much more complicated – as shown on Figures 6–8 to 6–14 representing, in graphical form, the results of Kitamura's numerical integrations of Equations (3.71) for seven different values of the mass-ratio q ranging from 1 to 0.01.

The mapping of the principal features in the XY-plane on to the $\xi\eta$-plane is illustrated on the accompanying Figure 6-15. The mass-points m_1, m_2 as well as the potential 'curtain' surrounding the whole system will be represented on the $\xi\eta$-plane by the abscissae $\xi = \infty$. Closed ovals surrounding m_1, m_2 will be represented by the abscissae $\xi > \xi_1$, where ξ_1 stands for the (normalized) value of the potential at which the closed loops surrounding m_1 and m_2 coalesce at the inner Lagrangian point L_1, the position of which is specified on the upper part of Figure 6–15 by $\xi = \xi_1$ and $\eta = 0°$. For $\xi < \xi_1$ (lower part of the diagram) the abscissae ξ = constant will represent dumbbell figures seen in the XY-plane on Figures 6-3 to 6-6, which eventually open up at the Lagrangian points L_2 $(\xi_2, 0°)$ and $L_3(\xi_3, \pi)$; while the abscissa ξ_0 represents the location on our diagram of the equilateral points $L_{4,5}$.

If we turn now to the plane $Y = 0$, the curves $(x, 0, z)$ in the XZ-plane must satisfy the orthogonality condition

$$\xi_x \zeta_x + \xi_z \zeta_z = 0; \tag{3.73}$$

and as such will be generated by the solutions of the first-order nonlinear differential equation

$$\frac{dx}{dz} = \frac{\xi_x}{\xi_z} = \frac{x(1 - r_1^3)r_2^3 - q(1 - x)r_1^3(1 - r_2^3)}{z(1 + qr_1^3)}, \tag{3.74}$$

with

$$r_1^2 = x^2 + z^2 \quad \text{and} \quad r_2^2 = (1 - x)^2 + z^2. \tag{3.75}$$

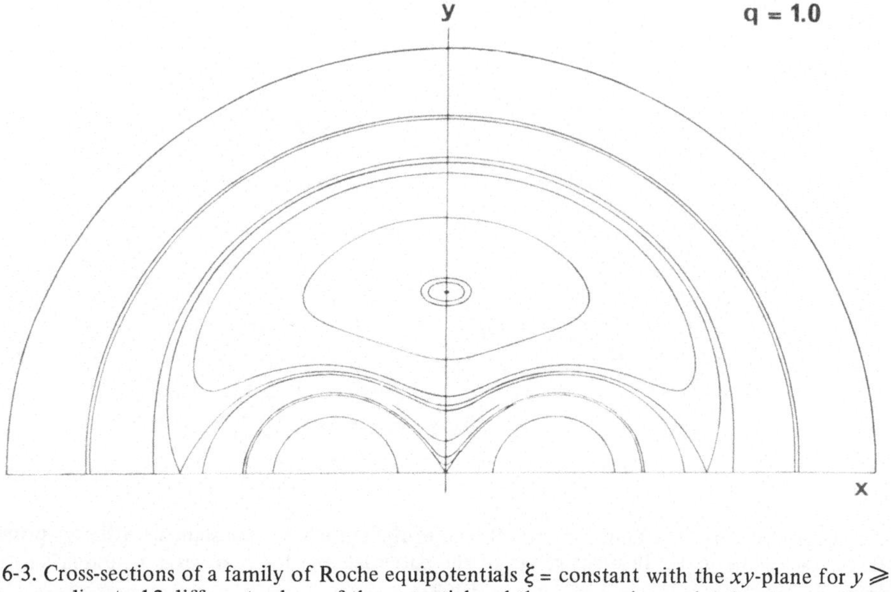

Fig. 6-3. Cross-sections of a family of Roche equipotentials ξ = constant with the xy-plane for $y \geqslant 0$, corresponding to 12 different values of the potential and the mass-ratio $q = 1$ (after Kopal, 1959).

This equation has likewise been integrated extensively by Kitamura (1970) numerically, and the profiles of the corresponding curves $\zeta(x, 0, z)$ in the XZ-plane are shown on the accompanying Figures 6–16 to 6–22; while Figure 6-23 gives a representation of the surfaces $\zeta(x, y, z)$ in three-dimensional perspective based on numerical integrations.

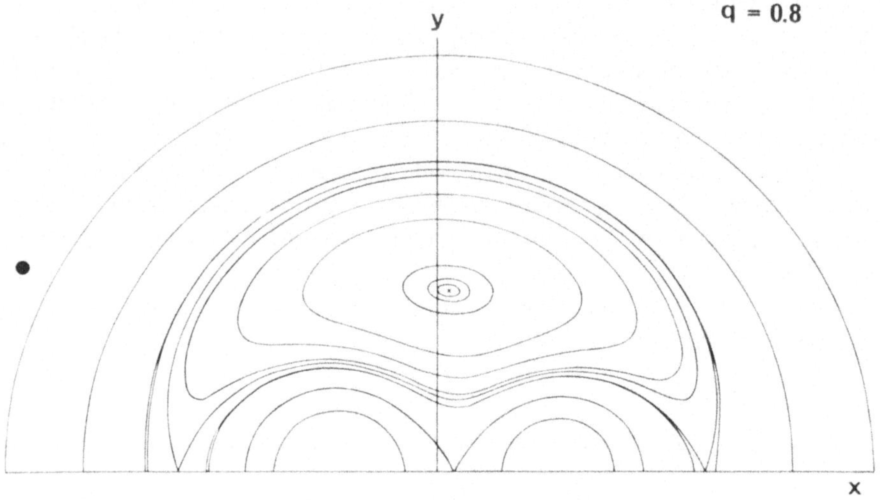

Fig. 6-4. Cross-sections of a family of Roche equipotentials ξ = constant with the xy-plane for $y \geqslant 0$, corresponding to 11 different values of the potential, and the mass-ratio $q = 0.8$ (after Kopal, 1959).

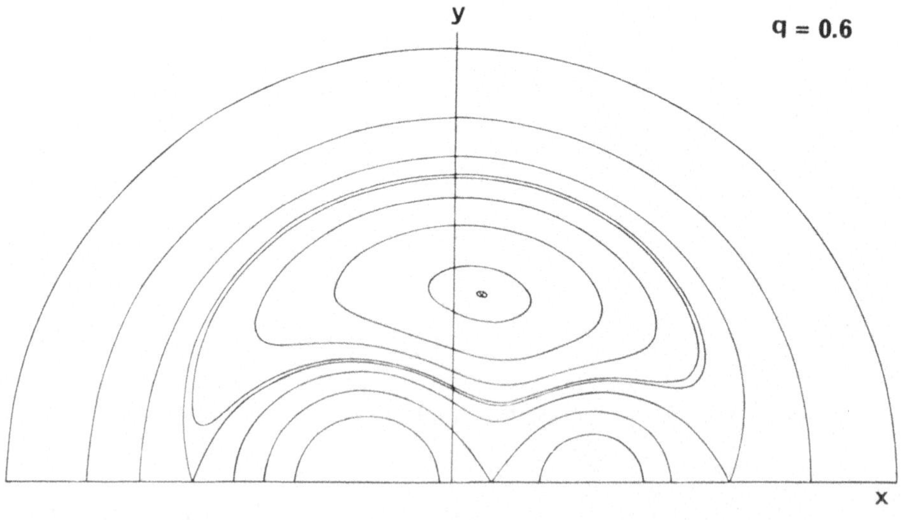

Fig. 6-5. Cross-sections of a family of the Roche equipotentials ξ = constant with the xy-plane for $y \geqslant 0$, corresponding to 10 different values of the potential, and the mass-ratio $q = 0.6$ (after Kopal, 1959).

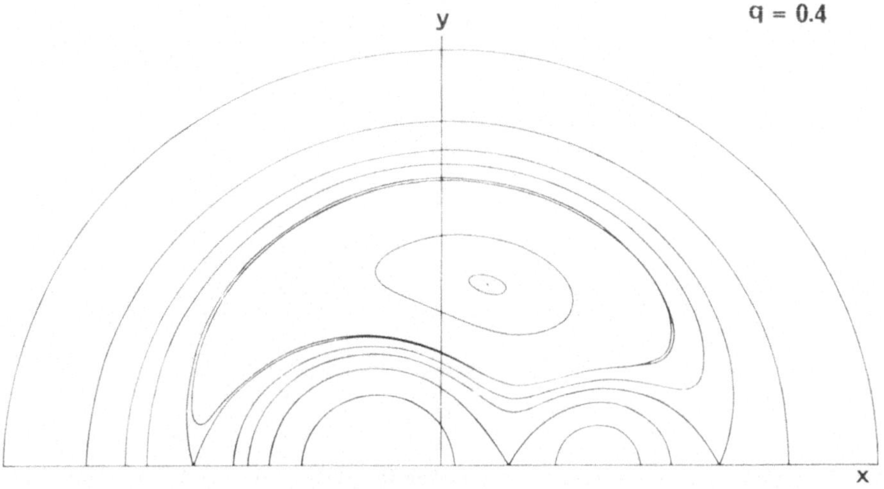

Fig. 6-6. Cross-sections of a family of the Roche equipotentials ξ = constant with the xy-plane for $y \geqslant 0$, corresponding to 8 different values of the potential, and the mass-ratio $q = 0.4$ (after Kopal, 1959).

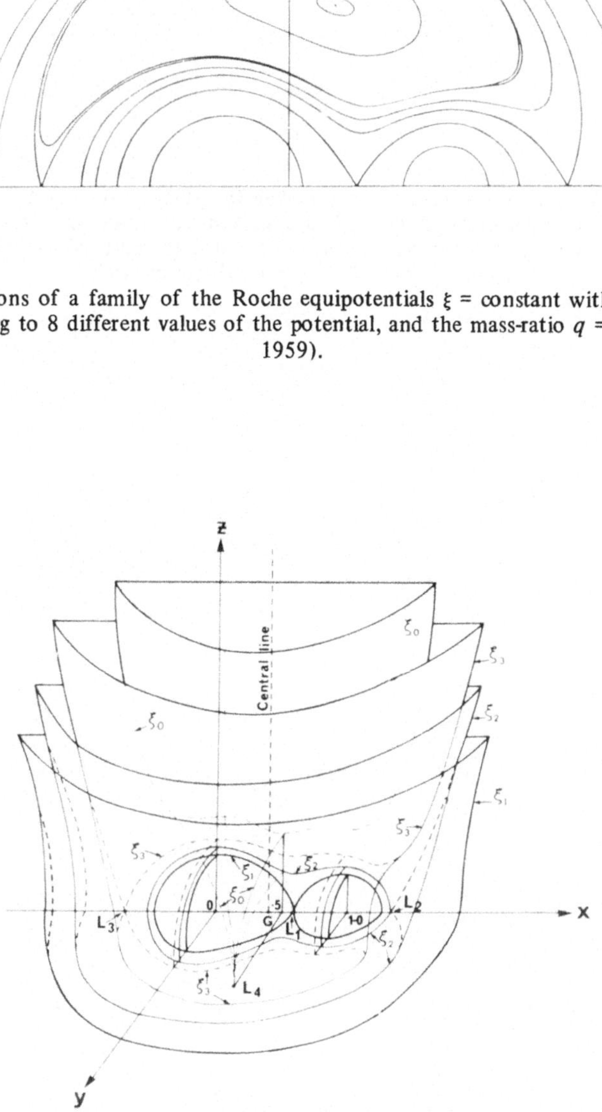

Fig. 6.7. A schematic view of the Roche equipotentials ξ = constant in three-dimensional perspective, corresponding to the values which the potential assumes at different Lagrangian points of the configuration (after Kitamura, 1970).

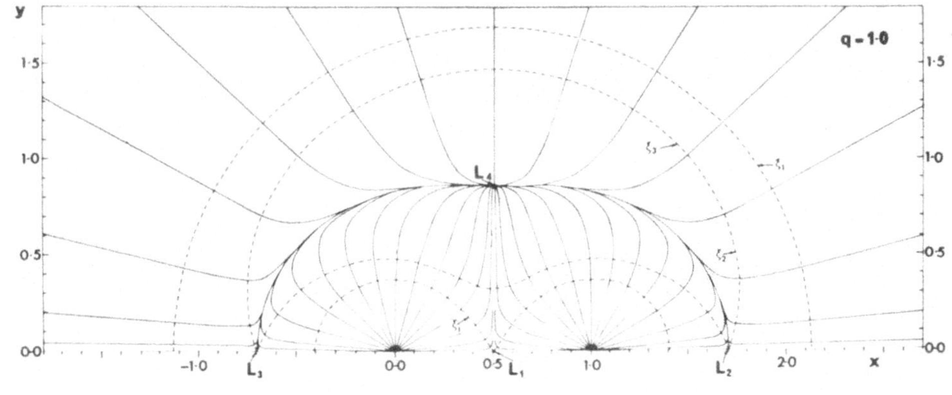

Fig. 6-8. The geometry of the curves η = constant in the xy-plane for $y \geqslant 0$, corresponding to the mass-ratio $q = 1$. The trajectories of $\eta(x, y, 0)$ are marked with full curves; while the dotted curves represent the equipotentials ξ = constant for the values which the potential assumes at the Lagrangian collinear points $L_{1,2,3}$ (after Kitamura, 1970).

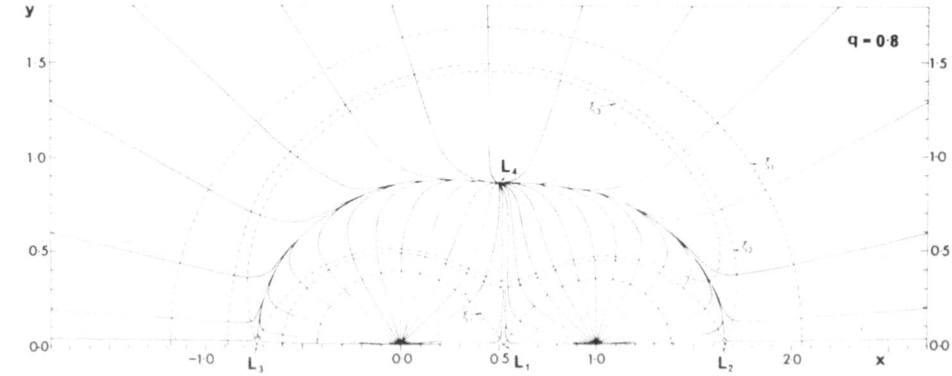

Fig.6-9. The geometry of the curves η = constant in the xy-plane for $y \geqslant 0$, corresponding to the mass-ratio $q = 0.8$. The trajectories of $\eta(x, y, 0)$ are marked with full curves; while the dotted curves represent the equipotentials ξ = constant for the values which the potential assumes at the Lagrangian collinear points $L_{1,2,3}$ (after Kitamura, 1970).

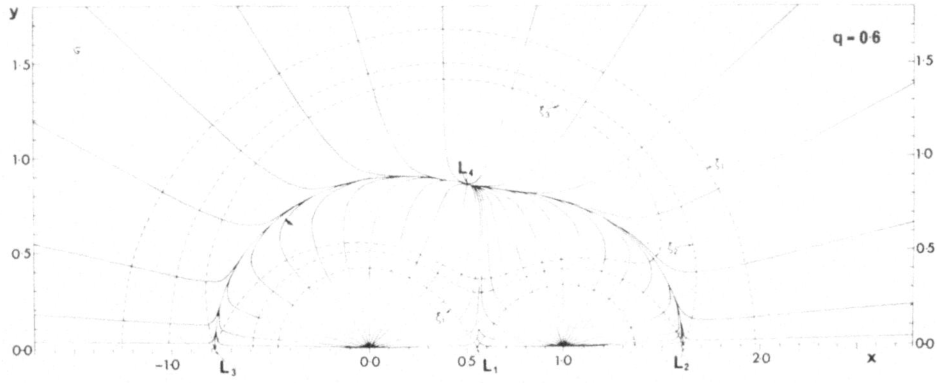

Fig. 6-10. The geometry of the curves η = constant in the xy-plane for $y \geqslant 0$, corresponding to the mass-ratio $q = 0.6$. The trajectories of $\eta(x, y, 0)$ are marked with full curves; while the dotted curves represent the equipotentials ξ = constant for the values which the potential assumes at the Lagrangian collinear points $L_{1,2,3}$ (after Kitamura, 1970).

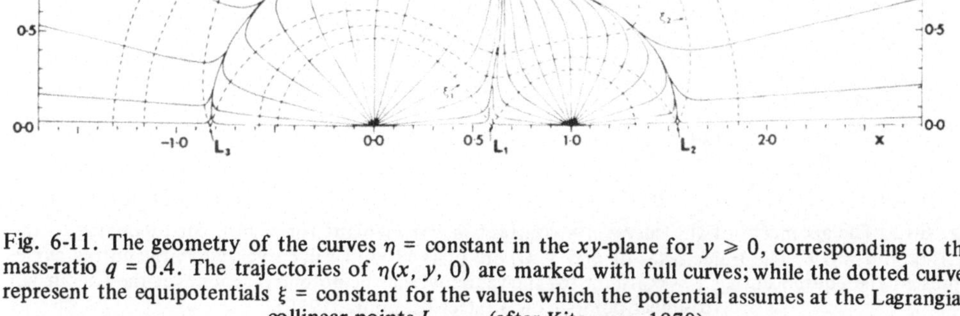

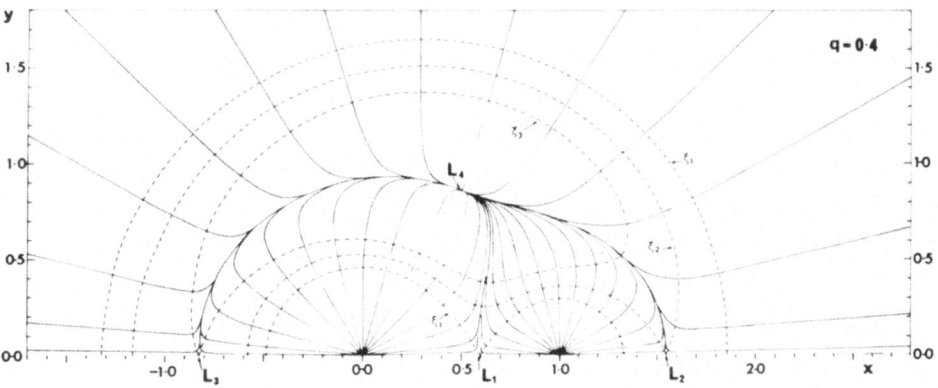

Fig. 6-11. The geometry of the curves η = constant in the xy-plane for $y \geqslant 0$, corresponding to the mass-ratio $q = 0.4$. The trajectories of $\eta(x, y, 0)$ are marked with full curves; while the dotted curves represent the equipotentials ξ = constant for the values which the potential assumes at the Lagrangian collinear points $L_{1,2,3}$ (after Kitamura, 1970).

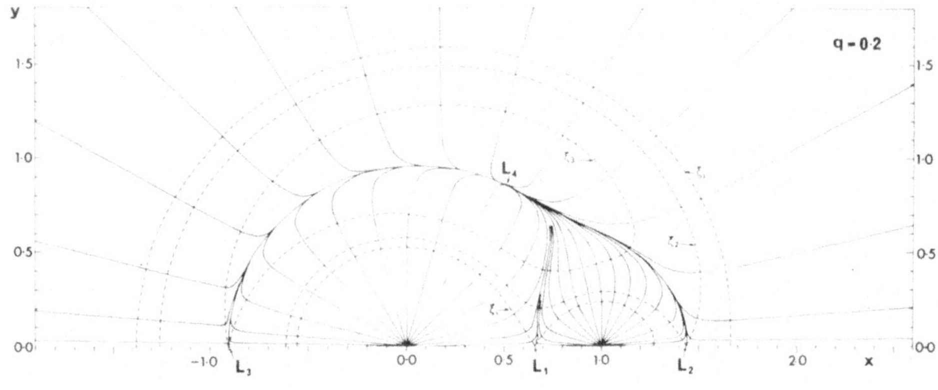

Fig. 6-12. The geometry of the curves η = constant in the xy-plane for $y \geqslant 0$, corresponding to the mass-ratio $q = 0.2$. The trajectories of $\eta(x, y, 0)$ are marked with full curves; while the dotted curves represent the equipotentials ξ = constant for the values which the potential assumes at the Lagrangian collinear points $L_{1,2,3}$ (after Kitamura, 1970).

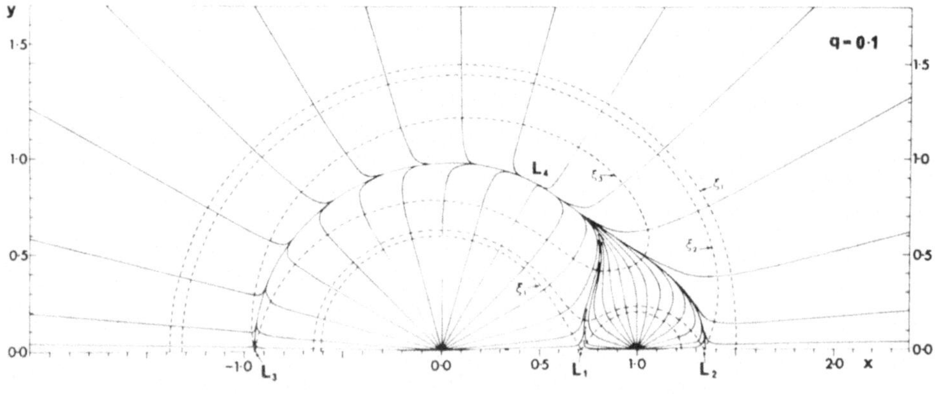

Fig. 6-13. The geometry of the curves η = constant in the xy-plane for $y \geqslant 0$, corresponding to the mass-ratio $q = 0.1$. The trajectories of $\eta(x, y, 0)$ are marked with full curves; while the dotted curves represent the equipotentials ξ = constant for the values which the potential assumes at the Lagrangian collinear points $L_{1,2,3}$ (after Kitamura, 1970).

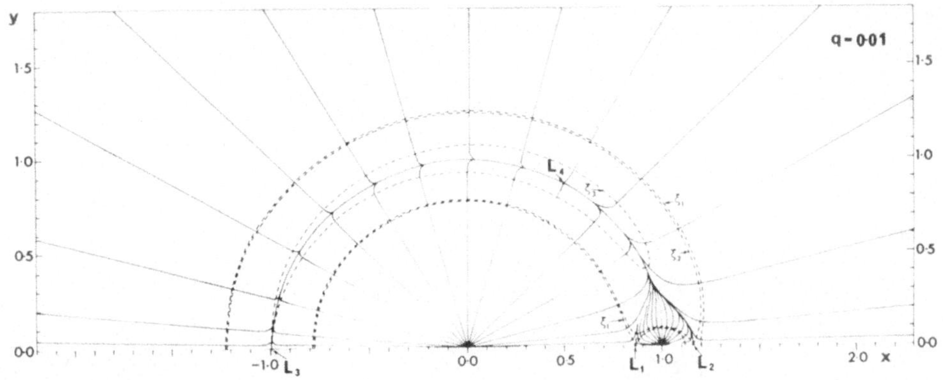

Fig. 6-14. The geometry of the curves η = constant in the xy-plane for $y \geqslant 0$, corresponding to the mass-ratio $q = 0.01$. The trajectories of $\eta(x, y, 0)$ are marked with full curves; while the dotted curves represent the equipotentials ξ = constant for the values which the potential assumes at the Lagrangian collinear points $L_{1,2,3}$ (after Kitamura, 1970).

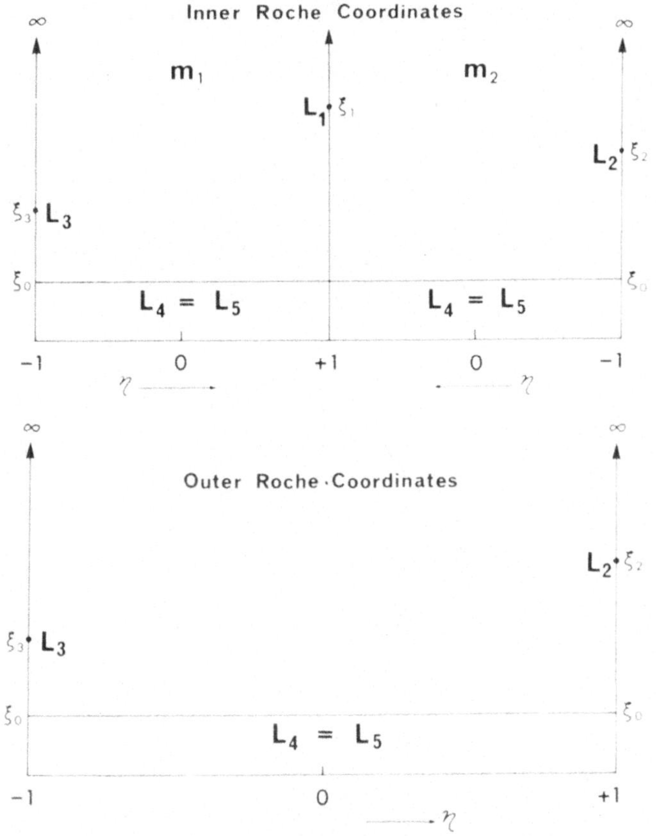

Fig. 6-15. Mapping of the positions of the Lagrangian points $L_{1,..,5}$ of the XY-plane on to the $\xi\eta$-plane in the Roche curvilinear coordinates (a) outside, and (b) within, the envelope of the η = constant curves (after Kopal, 1969; and Kitamura, 1970).

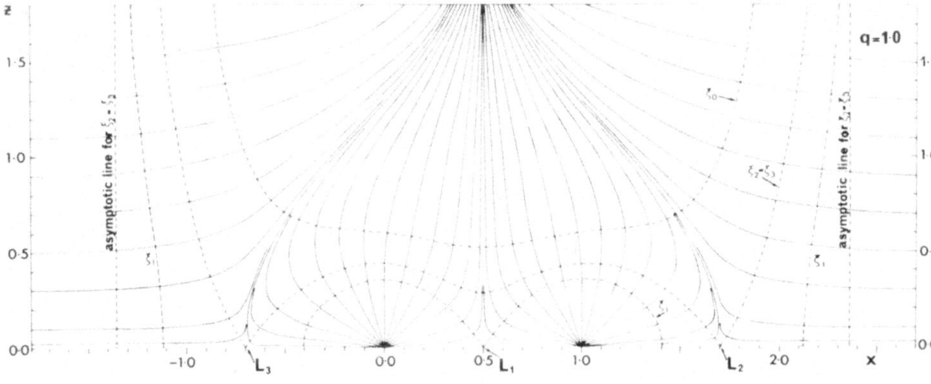

Fig. 6-16. The geometry of the curves ζ = constant in the xz-plane for $z \geqslant 0$, corresponding to the mass-ratio $q = 1$. The trajectories of $\zeta(x, 0, z)$ are marked with full curves; while dotted curves represent the equipotentials ξ = constant for the values which the potential assumes at the Lagrangian collinear points $L_{1,2,3}$ (after Kitamura, 1970).

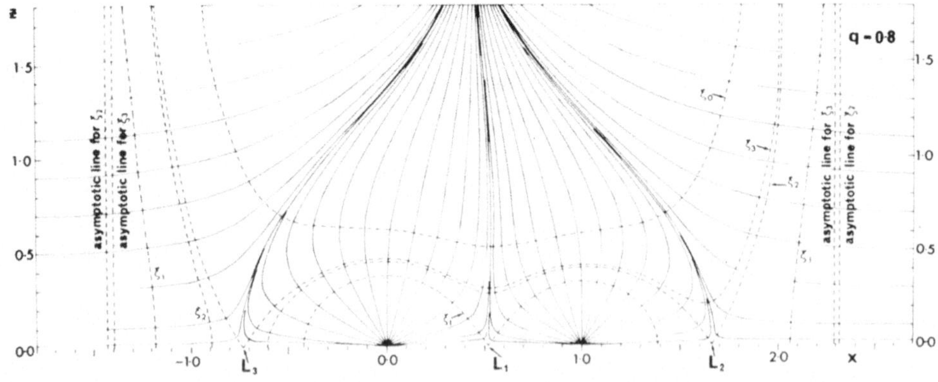

Fig. 6-17. The geometry of the curves ζ = constant in the xz-plane for $z \geqslant 0$, corresponding to the mass-ratio $q = 0.8$. The trajectories of $\zeta(x, 0, z)$ are marked with full curves; while dotted curves represent the equipotentials ξ = constant for the values which the potential assumes at the Lagrangian collinear points $L_{1,2,3}$ (after Kitamura, 1970).

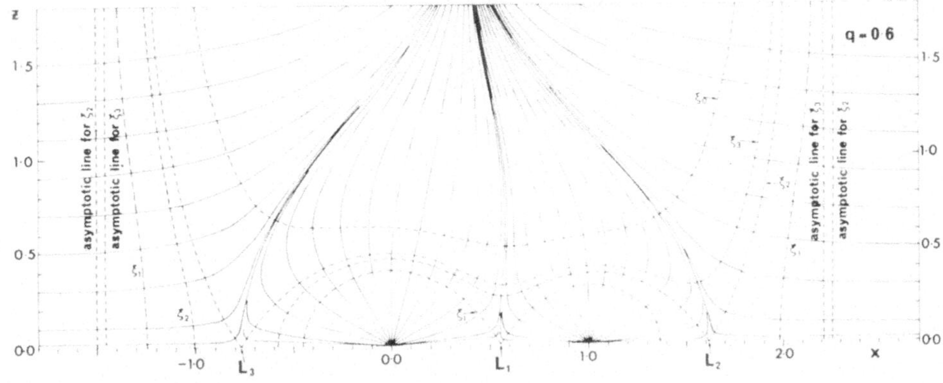

Fig. 6-18. The geometry of the curves ζ = constant in the xz-plane for $z \geqslant 0$, corresponding to the mass-ratio $q = 0.6$. The trajectories of $\zeta(x, 0, z)$ are marked with full curves; while dotted curves represent the equipotentials ξ = constant for the values which the potential assumes at the Lagrangian collinear points $L_{1,2,3}$ (after Kitamura, 1970).

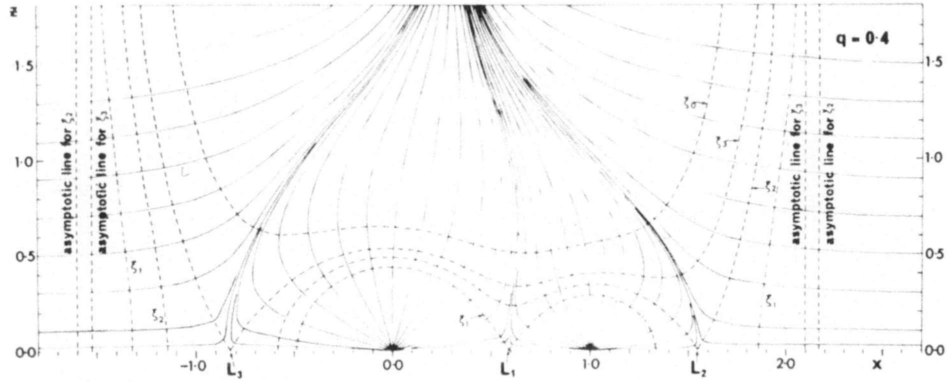

Fig. 6-19. The geometry of the curves ζ = constant in the xz-plane for $z \geqslant 0$, corresponding to the mass-ratio $q = 0.4$. The trajectories of $\zeta(x, 0, z)$ are marked with full curves; while dotted curves represent the equipotentials ξ = constant for the values which the potential assumes at the Lagrangian collinear points $L_{1,2,3}$ (after Kitamura, 1970).

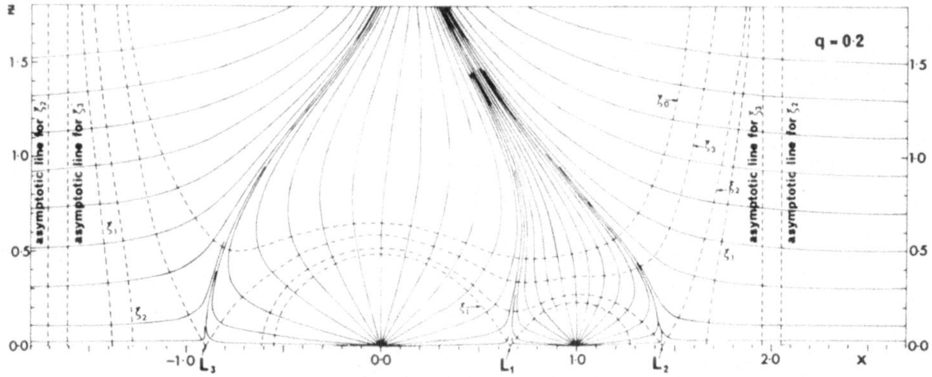

Fig. 6-20. The geometry of the curves ζ = constant in the xz-plane for $z \geqslant 0$, corresponding to the mass-ratio $q = 0.2$. The trajectories of $\zeta(x, 0, z)$ are marked with full curves; while dotted curves represent the equipotentials ξ = constant for the values which the potential assumes at the Lagrangian collinear points $L_{1,2,3}$ (after Kitamura, 1970).

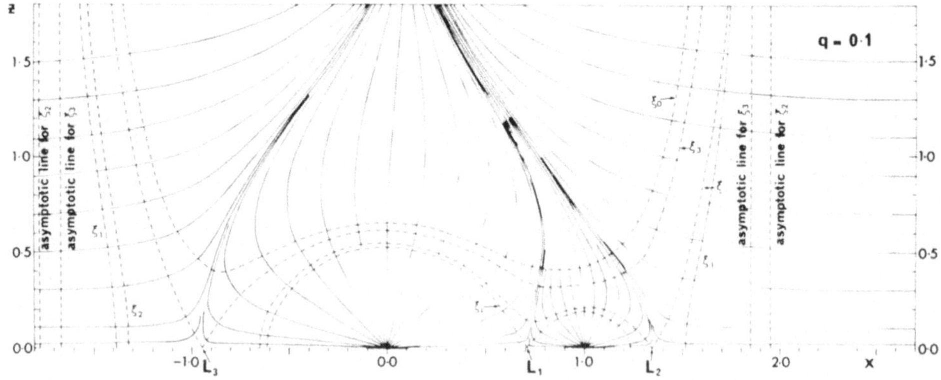

Fig. 6-21. The geometry of the curves ζ = constant in the xz-plane for $z \geqslant 0$, corresponding to the mass-ratio $q = 0.1$. The trajectories of $\zeta(x, 0, z)$ are marked with full curves; while dotted curves represent the equipotential ξ = constant for the values which the potential assumes at the Lagrangian collinear points $L_{1,2,3}$ (after Kitamura, 1970).

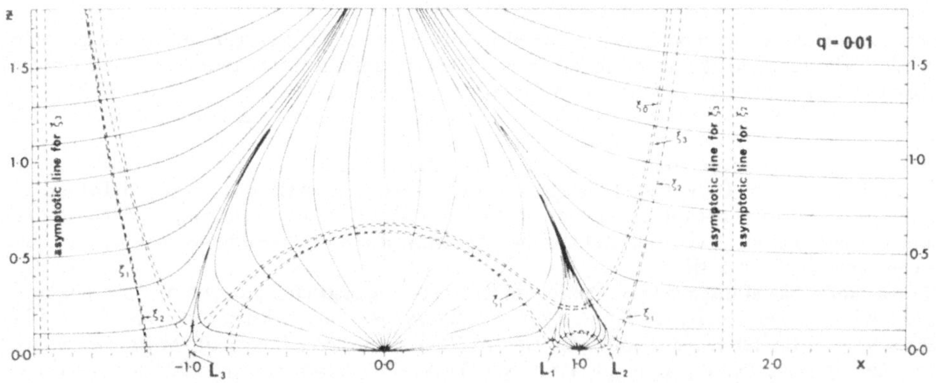

Fig. 6-22. The geometry of the curves ζ = constant in the xz-plane for $z \geqslant 0$, corresponding to the mass-ratio $q = 0.01$. The trajectories of $\zeta(x, 0, z)$ are marked with full curves; while dotted curves represent the equipotentials ξ = constant for the values which the potential assumes at the Lagrangian collinear points $L_{1,2,3}$ (after Kitamura, 1970).

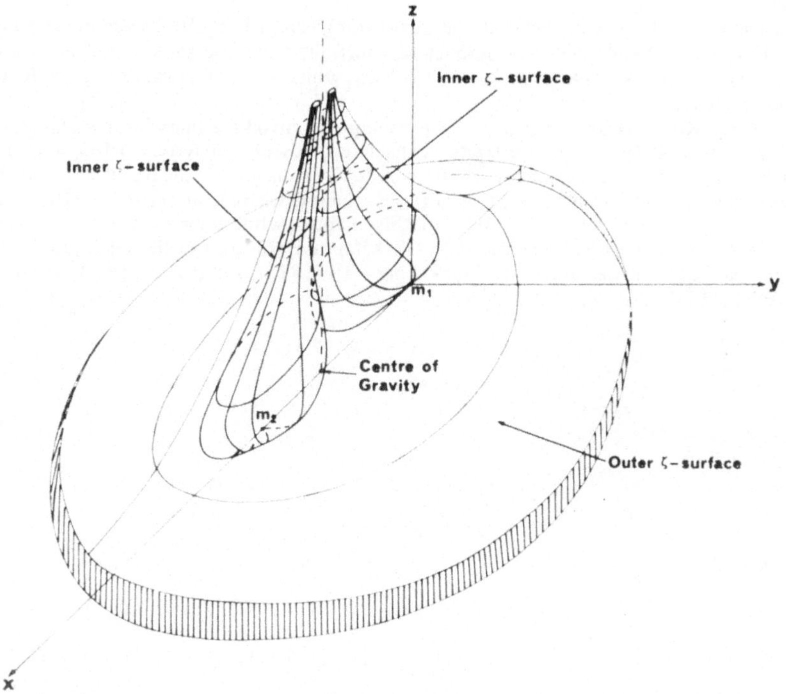

Fig. 6-23. Three-dimensional representation of the surfaces $\zeta(x, y, z)$ = constant (after Kitamura, 1970).

BIBLIOGRAPHICAL NOTES

The concept of the Roche model, as introduced in this chapter, owes its origin to an investigation by Edouard Roche, entitled 'La figure d'une masse fluide, soumise a l'attraction d'un point éloigné', published in *Mémoires de l'Academie des Sciences de Montpellier*, vol. **1**, pp. 243ff and 333ff; **2**, pp. 21ff; 1849–1851; cf. also Roche's memoir in the *Annales* of the same Academy, vol. **8**, pp. 235ff, 1873.

The complex transformation (1.9) of the cross-sections of Roche equipotentials with the plane $z = 0$ was introduced by G. W. Hill (1878); and that represented (likewise for $z = 0$) by (1.18) is due to T. Levi Civita (1906) – both in connection with their work on the Jacobian integral of the restricted problem of three bodies. A generalization of their transformations to three-dimensional space was later given by Kopal (1959, sec. III.1).

For a more general discussion of the relevant parts of the restricted problem of three bodies cf., e.g., A. Wintner, *The Analytical Foundations of Celestial Mechanics,* Princeton Univ. Press, 1941, Chapter VI (and the references quoted therein); or V. G. Szebehely, *Theory of Orbits: The Restricted Problem of Three Bodies,* Academic Press, New York, 1967; Chapters 4 and 10. The non-existence of the Jacobian energy integral of the eccentric restricted problem (with the two finite masses describing elliptic orbit) was proved by Z. Kopal and R. A. Lyttleton (1963).

Most part of the discussion and results given in section 2 of this chapter goes back to earlier studies of the geometrical properties of the Roche model by the present writer (Kopal, 1954), and subsequently elaborated in other sources (Kopal, 1959, Chapter III; or Kopal, 1972). The term 'Roche limit' as used in section 2-C was also introduced by Kopal (1954, 1955, 1956); and is not to be confused with the same term previously used for a completely different concept (i.e., the limit of stability of an infinitesimal satellite revolving around an oblate spheroid).

The form of the Roche equipotentials whose components do not rotate with the Keplerian angular velocity of orbital revolution has first been investigated by Z. Kopal (1956) to the degree of accuracy to which the squares and higher powers of the inclinations of the equators of either components to their orbital plane can be neglected. Kopal's results were subsequently generalized for arbitrary inclination of the equators to the orbital plane by M. Plavec (1958) and D. N. Limber (1963); see also A. Kruszewski (1966; pp. 240–241).

More extensive tabulations of some of the quantities given in Table VI-1 were prepared by M. Plavec and P. Kratochvíl (1964); while an important contribution to the subject matter of section 2-D was recently made by G. A. Chanan *et al.* (1976) from which the numerical data given in Table VI-4 have been excerpted.

As regards the 'Roche coordinates', the first attempt to introduce them in practical use goes back to H. K. Prendergast (1960). This investigator considered, however, a system orthogonal only in the *xy*-plane, without determining the shape of the curves ζ = constant. A theory of such coordinates – in two as well as three dimensions – had to await subsequent work by Kopal (1969, 1970) which constitutes the basis of the contents of section 3 of this chapter; while their topological properties were clarified by extensive numerical integrations by M. Kitamura (1970). For the separability of variables in the general case cf. Z. Kopal and A. K. M. Sekender Ali (1971); and (for the purely rotational problem) also Kopal (1971b).

CHAPTER VII

STABILITY OF THE COMPONENTS OF CLOSE BINARY SYSTEMS

In Chapter II of this volume we outlined the methods for determination of the equilibrium form of the components of close binary systems in any external field of force; and in sections II-1 and 2 of Chapter III we set up equations which govern the time-response of such configurations to an arbitrary type of disturbance. The nature of such disturbances may be connected with fluctuations in external gravitational field – such as those causing forced oscillations (dynamical tides) discussed in sections III-2 and 3 – or may arise from evolutionary changes in structure of the respective stars. These may be slow or abrupt (such as the onset of a nuclear reaction in the interior, or of the ionization of hydrogen or helium in the outer layers); and the configuration must respond to them on an appropriate time-scale.

The nature of this response bears vitally on the question of *stability* of the respective configuration, and its discussion will constitute the principal objective of this chapter. In more colloquial terms, the term 'stability' will hereafter be used to imply 'immunity from disturbance' – in the sense that, if stable, the configuration disturbed by an arbitrary cause or impulse will return to its original state after a lapse of time; while 'instability' will mean a transition from one stable state to another (or, in extreme case, the dispersal or collapse of the entire configuration).

In this sense, no star – binary or single – can be considered as stable on the nuclear time-scale; for the end product of its evolution will be completely different from its initial state. However, in many phases of the lives of most stars – their evolution is a very slow process, which can be represented by a succession of quasi-equilibrium states characterized by ignorable values of all time-derivatives of the variables of stars. In doing so, we are neglecting the consequences of a small imbalance between the amount of energy generated in the interior and radiated away through its surface which constitutes the motive power of stellar evolution. However, as long as the time-scale of a transient response to a given disturbance is short in comparison with that of gravitational contraction (let alone of nuclear time-scale), such a procedure should be entirely legitimate.

In such a case, the problem at issue consists of ascertaining whether or not an equilibrium configuration of given form or structure finds itself in a state which is stable with respect to a given type of disturbance. If it were not, a star would advertise this fact by exhibiting phenomena (such as the variability of its light or size) which could be detected by the observations. Their absence discloses, therefore, the existence of stability on the time-scale of the observations; but this fact itself may impose certain constraints on the structure of the star exposed to a given field of force – or, for binary stars, to certain limits in the fractional size of the constituent components, or their mass-ratio. The aim of the present chapter will be to point out the specific form of these constraints, and to relate them with the physical properties of the respective system, as well as with the internal structure of its components.

VII-1. Criteria of Stability

In order to endow the concept of stability introduced in the opening part of this chapter with a more precise meaning, let us consider the most general form of the equations safeguarding the conservation of momentum of the stars or stellar systems, of which a double star represents the simplest example. Since the days of Lagrange we have known that if T denotes the kinetic energy of a self-graviting configuration (relative to its centre of mass); and W, the potential energy arising form a given shape and distribution of mass in the interior (representing the amount of work which would be performed by our body on its contraction from infinite distension to its present state), in the case of any *conservative* dynamical system the two must at all times satisfy the relation

$$\frac{\mathrm{d}}{\mathrm{d}t}\left(\frac{\partial T}{\partial \dot{q}_i}\right) - \frac{\partial T}{\partial q_i} + \frac{\partial W}{\partial q_i} = 0, \qquad i = 1, 2, 3, \ldots; \tag{1.1}$$

where the q's represent the 'generalized coordinates' characterizing the state of our configuration; and t, the time (with $\dot{q}$ denoting the time derivatives of the respective coordinate). Dissipative forces – if any – will exert their influence on the right-hand side of (1.1), rendering this equation non-homogeneous.

Moreover, the kinetic and potential energy of a fluid configuration of mass m_1 can, in general, be expressed as

$$T = \frac{1}{2}\int_0^{m_1} \mathbf{u}^2\, \mathrm{d}m \tag{1.2}$$

and

$$W = -\int_0^{m_1} \Omega\, \mathrm{d}m, \tag{1.3}$$

i.e., as mass integrals of the velocity $\mathbf{u}$ of individual particles in the interior, and of the potential function Ω arising from the mass, as introduced already in section II-1. As the $\mathbf{u}$'s are vectors, the three-dimensional Equation (1.1) is equivalent to a system of three scalar equations, each (for $q_I \neq 0$) of second order in t. The Eulerian equations (1.4)–(1.6) of Chapter III or Equations (1.108)–(1.110) of Chapter IV constitute merely particular examples of the Lagrangian equations (1.1) in scalar form.

As regards the choice of the generalized coordinates q appropriate for our present problems, in Chapter II we established that the radius-vector r of an equipotential surface of the component of a close binary system, distorted by rotation and tides, can generally be expressed in the form

$$\frac{r}{a} = \sum_{j=0}^{\infty}\sum_{k=0}^{j} f_j^{(k)}(a, t)\, Y_j^k(\theta, \phi), \tag{1.4}$$

where the amplitudes $f_j^{(k)}(a, t)$ of the individual harmonics

$$Y_j^k(\theta, \phi) = \left.\begin{matrix}\sin k\phi\\ \cos k\phi\end{matrix}\right\} P_j^k(\cos\theta) \tag{1.5}$$

are defined as solutions of Clairauts differential equations in terms of the mean radius a of the respective equipotential of a configuration of given structure.

Let, in what follows, *the amplitudes $f(a, t)$ be identified with the generalized coordinates q of the Lagrangian equation* (1.1). The amplitudes so chosen will depend on the relative dimensions of the respective component, its internal structure, as well as the mass-ratio of the binary systems in the manner already investigated in Chapter II. The most general type of distortion associated with spherical harmonics of second order should, accordingly, be described by five such 'degrees of freedom'; those associated with the third harmonic, by seven degrees; and, in general, the harmonics Y_j^k of j-th degree, with $2j + 1$ permissible values of q_i. The total number of the degrees of freedom implied in the expansion on the right-hand side of Equation (1.4) would, of course, be infinite; though can be made finite if we break off the series with harmonics Y_n^k of certain maximum order n (or select only a certain set of these for subsequent consideration).

After these preliminaries we wish to return to Equation (1.1) and consider two distinct classes of solutions which are consistent with it. First, let us assume that the kinetic energy T of our configuration is so small in comparison with its potential energy W that the first two terms involving T on the left-hand side of Equation (1.1) can be ignored in comparison with the third, which depends on W only. In such a case, Equation (1.1) reduces to a system of equations of the form

$$\frac{\partial W}{\partial q_i} = 0, \qquad i = 1, 2, 3, \ldots \tag{1.6}$$

constituting a set of necessary conditions for W to be *stationary* with respect to any change of the q_i's.

If, in addition, such a stationary state can be identified with a *minimum*, the respective configuration will be termed 'secularly stable' – a term aiming to underline the fact that a configuration in quest of such a stability will tend towards it by slow gravitational adjustment on the Kelvin time-scale; and the velocity components of the corresponding adjustments should indeed be small enough to justify the neglect of the kinetic energy T in comparison with the potential energy W underlying a transition from (1.1) to (1.6). Incidentally, the same fact renders also Equation (1.1) immune to any action of dissipative forces that would arise from the viscosity of stellar material; for no matter how large this viscosity may be, its effects (being linear and homogeneous in the velocity components) will go to zero with $\mathbf{u}$.

The absolute magnitude of these components can be estimated from a known duration of the time τ_c in which a self-gravitating configuration of mass m, radius R, and losing energy at a rate L per unit time, needed to contract to this form from a state of infinite distension, and which is equal to

$$\tau_c = \frac{Gm^2}{RL}, \tag{1.7}$$

where G denotes the gravitational constant. For a star of the mass of the Sun ($m = 1.99 \times 10^{33}$ g) and of solar dimensions ($R = 6.96 \times 10^{10}$ cm) losing energy at a solar rate of 3.93×10^{33} erg s^{-1}, the Kelvin contraction time-scale τ_c turns out to amount to 9.66×10^{14} s or 30.6 m.y. Accordingly, a change of (say) ± 10% of the present radius of the Sun would call, on this time-scale, for a (radial) velocity component of the order of 10^{-5} cm

s^{-1} – making the smallness of the velocity components generated by a purely gravitational adjustment in shape or structure of the respective configuration manifest.

In a more general case – when the terms involving both T and W in Equation (1.1) are of comparable importance – the stability of a configuration requires that motions representing its response to an arbitrary disturbance be bounded. Configurations satisfying this condition are deemed to be *dynamically stable* – on a time-scale τ_ν equal to the fundamental period in which the respective configuration is capable of responding to any such disturbance; and which can be shown (cf. Ledoux and Pekeris, 1941) to be of the order of

$$\tau_\nu = \left(\frac{I}{W}\right)^{1/2} \cong \left(\frac{R^3}{Gm}\right)^{1/2}, \tag{1.8}$$

where I represents the moment of inertia of the respective star about its centre of mass. For solar values of R and m, τ_ν turns out to be a quantity of the order of 1.6×10^3 s or 26.5 min; and similarly so far stars different from our Sun – be they larger or smaller.

A great disparity in duration of the 'contractional' and 'vibrational' times-scales renders the 'secular' and 'dynamical' stability' quite distinct concepts – mathematically as well as physically. A presence of dynamical instability would imply that a configuration so afflicted must signal its distress by transient phenomena on a time-scale comparable with (1.8)–lasting (typically) from hours to weeks – which could thus be easily ascertained from the observations. Configurations which do not exhibit such phenomena are, therefore, bound to be dynamically stable; though whether or not they are also secularly stable cannot be borne out by observations lasting years or even decades; and can be ascertained on the basis of mathematical analysis of their models by methods expounded in Section VII-2.

Thus unlike the dynamical stability of the stars (be they single or double) – existence (or non-existence) of which can often be detected by the observer – that of the secular stability can be ascertained only by the mathematician. The observer may note that stars (or systems) which are so unstable are not encountered in the sky (or only rarely so); but to understand their absence in terms of the secular instability of their respective structures constitutes a task to which we propose to address ourselves in the next section.

VII-2. Secular Stability

Consistent with our plans outlined in the preceding section of this chapter, a self-gravitating fluid configuration in hydrostatic equilibrium will be regarded secularly stable if its form, as well as structure, is such that its potential energy

$$W(q_1, q_2, \ldots\ldots q_n) = \text{minimum}\ ; \tag{2.1}$$

so that the change of its generalized coordinates q_i by an arbitrary amount δq_i would result in an increase of W for any i. The *necessary* conditions for this to be the case are represented by Equations (2.1), requiring that partial derivatives of W with respect to $i = 1, 2, 3, \ldots, n$ be all equal to zero. The task confronting us in this section will be to establish the explicit form of the conditions which are *sufficient* for this purpose; and, conversely, whose violation would render such a configuration unstable.

In order to do so, let us assume that, in the state of equilibrium, the generalized coordinates q_i describing the potential energy assume (say) the values

$$(q_i)_0 = a_i , \tag{2.2}$$

and that, in the neightbourhood of equilibrium,

$$q_i = a_i + \delta q_i . \tag{2.3}$$

If, moreover, the potential energy $W(q_i)$ represents a function which is *analytic* for $q_i = a_i$, it should allow, at that point, a Taylor expansion of the form

$$W(q_i) = W(a_i) + \left(\frac{\partial W}{\partial q_i}\right)_{a_i} \delta q_i + \frac{1}{2}\left(\frac{\partial^2 W}{\partial q_r \partial q_s}\right)_{a_r a_s} \delta q_r \delta q_s + \cdots , \tag{2.4}$$

in which, for sufficiently small δq_i's, the remainder of the series on the right-hand side of Equation (2.4) can be ignored. Moreover, in the state of equilibrium,

$$\left(\frac{\partial W}{\partial q_i}\right)_{a_i} = 0 \tag{2.5}$$

by Equation (1.6); so that, in the neighbourhood of equilibrium,

$$W(q_i) = W(a_i) + \tfrac{1}{2}\, W_{rs} \delta q_r \delta q_s + \cdots , \tag{2.6}$$

where W_{rs} will hereafter denote the partial derivatives of W with respect to q_r and q_s at $q_i = a_i$.

Now, in accordance with a theorem first enunciated by Lagrange and rigorously proved by Dirichlet (1861), *the necessary and sufficient condition for the equilibrium configuration to be secularly stable is that $W(a_i)$ represents an absolute minimum of $W(q_i)$.* This will be the case provided that the *quadratic form*

$$\delta W = \tfrac{1}{2}\, W_{rs} \delta q_r \delta q_s \tag{2.7}$$

be *positive definite* for all values of the variables δq_r and δq_s.*

To establish the conditions for which the homogeneous quadratic function (2.7) remains positive, let us introduce a linear transformation of variables

* Should, per chance, all W_{rs} be identically zero, the existence of a minimum would require the positivity of the first group of the nonvanishing partial derivatives of W of lowest *even* order on the right-hand side of Equation (2.4).

$$\left.\begin{aligned}\delta q_1 &= c_{11}\theta_1 + c_{12}\theta_2 + \cdots + c_{1n}\theta_n\,,\\ \delta q_2 &= c_{21}\theta_1 + c_{22}\theta_2 + \cdots + c_{2n}\theta_n\,,\\ &\vdots \\ \delta q_n &= c_{n1}\theta_1 + c_{n2}\theta_2 + \cdots + c_{nn}\theta_n\,,\end{aligned}\right\} \tag{2.8}$$

such that the determinant

$$\|c_{ij}\| \equiv \lambda > 0; \tag{2.9}$$

and determine the c_{ij}'s in such a way that the right-hand side of the quadratic function (2.7) will reduce to a sum of the squares of the form

$$\delta W = \tfrac{1}{2}\,(b_1\theta_1^2 + b_2\theta_2^2 + \cdots + b_n\theta_n^2), \tag{2.10}$$

where the 'coefficients of stability' b_i are given by the determinantal equation of the form

$$\begin{vmatrix} b_1 & 0 & \ldots & 0 \\ 0 & b_2 & \ldots & 0 \\ \vdots & & & \\ 0 & 0 & \ldots & b_j \end{vmatrix} = \lambda^2 \begin{vmatrix} W_{11} & W_{12} & \ldots & W_{ij} \\ W_{21} & W_{22} & \ldots & W_{2j} \\ \vdots & & & \vdots \\ W_{j1} & W_{j2} & \ldots & W_{jj} \end{vmatrix} \tag{2.11}$$

equivalent to

$$b_1 b_2 \ldots b_j = \lambda^2 \Delta j \qquad \text{for } j = 1, 2, 3, \ldots, n, \tag{2.12}$$

in which $W_{ij} \equiv W_{ji}$, rendering the Hessian determinant Δn of W on the right-hand side of Equation (2.11) diagonally symmetrical.

The coefficients b_j are by no means defined by Equation (2.11) uniquely, for in Equation (2.8) we introduced n^2 arbitrary constants c_{ij} to remove $n(n-1)/2$ cross-products $\delta q_r \delta q_s$ from the quadratic form (2.7) and since $n^2 > n(n-1)/2$, this can be accomplished in an infinity of ways. However, whichever particular transformation we employ, the number of the coefficients b_i which are positive or negative are in each case the same, so that it is always possible to choose the c_{ij}'s in such a way that $\|c_{ij}\| \equiv \lambda = 1$ – a convention which we shall hereafter adopt.

Let us, however, return to Equation (2.10). Since the squares θ_i^2 of all new coordinates on its right-hand side are necessarily positive, *the quadratic form δW as defined by Equation (2.10) will be positive definite if all the coefficients b_i of stability are also positive* – i.e., if all minors Δj of the Hessian on the right-hand side of Equation (2.11) are positive for $j = 1, 2, 3, \ldots, n$. This will be true if

$$b_1 = W_{11} > 0, \tag{2.13}$$

$$b_1 b_2 = \begin{vmatrix} W_{11} & W_{12} \\ W_{21} & W_{22} \end{vmatrix} > 0, \tag{2.14}$$

$$b_1 b_2 b_3 = \begin{vmatrix} W_{11} & W_{12} & W_{13} \\ W_{21} & W_{22} & W_{23} \\ W_{31} & W_{32} & W_{33} \end{vmatrix} > 0, \tag{2.15}$$

and, in general

$$b_1 b_2 b_3 \ldots b_n = \begin{vmatrix} W_{11} & W_{12} & \ldots & W_{1n} \\ W_{21} & W_{22} & \ldots & W_{2n} \\ \vdots & \vdots & & \vdots \\ W_{n1} & W_{n2} & \ldots & W_{nn} \end{vmatrix} > 0. \tag{2.16}$$

Therefore, a positivity of all b_j's constitutes a condition which is *sufficient* to ensure secular stability of the respective configuration, but *not* necessary for the purpose. For even should one (or more) of the b_j's become negative, it may still be possible to maintain the positivity of δW (which constitutes *the* necessary and sufficient condition for stability) by imposing suitable restrictions on the range of the θ_i^2's which factor the negative b_i's. Without such restrictions, a change in sign of one or more b_i's implies, in general, secular instability; though stability may be recovered by a removal of the degree of freedom giving rise to it. A restriction in the number of the degrees of freedom may indeed restore stability – if such a move can be justified on physical grounds; and conformance to such a restriction may represent the price of survival of the respective quasi-equilibrium stage as a link in secular evolution of the respective configuration (or system). On the other hand, an increase in the number of the degrees of freedom will never remove an instability which develops when $\Delta_n = 0$.

The stage at which this happens marks the 'bifurcation point' in secular evolution of the respective configuration, at which stability will be lost (or gained). If so, it may pass over to another family of configurations, characterized by a different Hessian Δ_n, which becomes identical with the former one at the bifurcation point. However, the properties of this new family are *not* defined by those of the former; and must, in general, be anticipated on other physical grounds.

Thus far we have considered the criteria for stability of *a stationary* configuration in hydrostatic equilibrium and did not take account of the fact that such a configuration may be endowed by *axial rotation.* To generalize our stability criteria for such a case, in what follows, let H and T denote the angular momentum and kinetic energy relative to a fixed (inertial) frame of reference, while H_R and T_R represent the same quantity relative to a frame rotating with an angular velocity ω about one (say, the z-)axis. Moreover, let C be the moment of inertia about this axis, and

$$\begin{aligned} u &= \dot{x} - y\omega, \\ v &= \dot{y} + x\omega, \\ w &= \dot{z}, \end{aligned} \tag{2.17}$$

be the rectangular velocity components relative to the inertial frame of reference.

If so, then the total kinetic energy T can be expressed in the form of the following three mass integrals:

$$
\begin{aligned}
T &= \frac{1}{2}\int (u^2 + v^2 + w^2)\,\mathrm{d}m \\
&= \frac{1}{2}\int (\dot{x}^2 + \dot{y}^2 + \dot{z}^2)\,\mathrm{d}m + \omega \int (x\dot{y} - y\dot{x})\mathrm{d}m + \frac{1}{2}\,\omega^2 \int (x^2 + y^2)\mathrm{d}m \\
&= T_R + \omega H_R + \frac{1}{2}\,\omega^2 C;
\end{aligned}
\tag{2.18}
$$

and, similarly, the angular momentum

$$H = H_R + \omega C. \tag{2.19}$$

If we temporarily disregard possible internal motions in our body, and set the *body* velocity components with respect to the *space* axes

$$\dot{x} = \dot{y} = \dot{z} = 0 \tag{2.20}$$

on the right-hand side of Equations (2.18), the expressions for T and H as given by Equations (2.18) and (2.19) will reduce to

$$T = \frac{1}{2}\,\omega^2 C \qquad \text{and} \qquad H = \omega C, \tag{2.21}$$

respectively.

On the other hand, the Lagrangian equations of motion (1.1) are known to admit of the integral

$$T + W = \text{constant}, \tag{2.22}$$

which, on insertion for T from Equations (2.21) yields

$$W + \frac{1}{2}\,\omega^2 C = \text{constant}. \tag{2.23}$$

In the state of equilibrium,

$$\delta\left(W + \frac{1}{2}\,\omega^2 C\right) = 0, \tag{2.24}$$

which yields

$$\delta W = -\frac{1}{2}\,\delta(\omega^2 C). \tag{2.25}$$

Therefore, the potential energy to be minimized for the rotating configurations becomes (for constant ω) equal to

$$W + \delta W = W - \frac{1}{2}\,\omega^2 C, \tag{2.26}$$

and this expression should replace W in the preceding Equations (2.1)–(2.16) of this section.

The minimization of Equation (2.26) should result in a linear series of equilibrium configurations in which the angular velocity ω is allowed to vary sufficiently slowly for the velocity components of such motion to remain negligible, and the configuration retain its equilibrium state. If, however, we restrict such variations to those in which the angular momentum H remains secularly constant, then on elimination of ω between Equations (2.21) the kinetic energy

$$T = \frac{H^2}{2C}, \tag{2.27}$$

which on insertion into the energy integral of Equation (2.22) requires that, in this case, the expression to be minimized will assume the form

$$W + \frac{H^2}{2C} = \text{constant}. \tag{2.28}$$

The corresponding series of configurations for which this is true will be stable provided that this expression in minimized for constant values of H.

Since, under these conditions,

$$\begin{aligned} \delta\left(W + \frac{H^2}{2C}\right) &= \delta W - (H^2/2C^2)\delta C = \delta W - \frac{1}{2}\,\omega^2 \delta C \\ &= \delta\left(W - \frac{1}{2}\,\omega^2 C\right) - \omega^2 \delta C, \end{aligned} \tag{2.29}$$

by $H = \omega C$ and $\delta\omega = -\,\omega(\delta C/C)$ it follows that the two stability criteria differ only in amounts of the order of $\omega^2 \delta C$.

A. ROTATIONAL PROBLEM

Having specified the function (2.28) which must be a minimum if a configuration rotating freely in space is to be secularly stable, our next task is to establish the appropriate expressions for the potential energy W and the moment of inertia C about the axis of rotation in terms of the generalized coordinates of our problem. If we identify the latter with the amplitudes $f_j(a)$ of the expansion on the right-hand side of Equation (2.1) of Chapter II, these can be obtained (numerically or otherwise) as particular solutions of the differential equations (2.33)–(2.35) of that chapter, subject to the boundary conditions (2.39)–(2.41) correctly to quantities of third order in surficial distortion; and their generalization by Kopal and Mahanta (1974) can be used to push the accuracy to quantities of fourth order.

Having done so, we are in a position to proceed with the evaluation of the potential energy W of our configuration from the equation

$$W = -\frac{1}{2}\int_0^{m_1} \Omega \mathrm{d}m, \tag{2.30}$$

where Ω denotes the potential arising from the mass and where, in accordance in with Equation (3.44),

$$\Omega = 4\pi G \sum_{j=0}^{\infty} \left\{\frac{r'^j E_j + r'^{-j-1} F_j}{2j+1}\right\} P_j(\cos\theta) = \tag{2.31}$$

$$= 4\pi G \sum_{j=0}^{\infty} \beta_j(a) P_j(\cos\theta),$$

whereas the mass element is

$$dm = \frac{\rho}{3} \frac{\partial r'^3}{\partial a} da \sin\theta \, d\theta \, d\phi, \tag{2.32}$$

with $r'(a, \theta, \phi)$ as given by the expansion (2.1) of Chapter II.

The coefficients $\beta_j(a)$ in the second expansion on the right-hand side of Equation (2.31) are already known to us from section II-2, for they become identical with the left-hand side of Equations (2.21)–(2.23) for $j = 2$, 4, and 6. Therefore, only that for $j = 0$ remains yet to be evaluated, and this can be done in the same manner by which those for $j > 0$ have already been established. Doing so we find that, correctly to terms of the *fourth* order in surficial distortion.

$$\begin{aligned}\beta_0(a) = E_0 &+ \frac{2}{25} a^2 E_2 \left\{ f_2 + \frac{1}{7} f_2^2 - \frac{1}{5} f_2^3 + \frac{2}{7} f_2 f_4 \right\} \\ &+ \frac{4}{81} a^4 E_4 \left\{ f_4 + \frac{27}{35} f_2^2 \right\} \\ &+ \frac{F_0}{a} \left\{ 1 + \frac{2}{5} f_2^2 - \frac{4}{105} f_2^3 + \frac{43}{175} f_2^4 + \frac{2}{9} f_4^2 - \frac{4}{35} f_2^2 f_4 \right\} \\ &+ \frac{3F_2}{25a^3} \left\{ -f_2 + \frac{4}{7} f_2^2 - \frac{78}{35} f_2^3 + \frac{8}{7} f_2 f_4 \right\} \\ &+ \frac{5F_4}{82a^5} \left\{ -f_4 + \frac{54}{35} f_2^2 \right\},\end{aligned} \tag{2.33}$$

where, correctly to quantities of fourth order,

$$E_0 = \int_a^{a_1} \rho \frac{\partial}{\partial a} \left\{ a^2 \left[\frac{1}{2} + f_0 + \frac{1}{2} f_0^2 + \frac{1}{10} f_2^2 + \frac{1}{18} f_2^4 \right] \right\} da. \tag{2.34}$$

Since, within the scheme of this approximation,

$$f_0 = -\frac{1}{5} f_2^2 - \frac{2}{105} f_2^3 - \frac{1}{9} f_4^2 - \frac{2}{35} f_2^2 f_4 + \cdots \tag{2.35}$$

to safeguard the constancy of mass, it follows on its insertion in Equation (5.5) that

$$E_0 = \int_a^{a_1} \rho \frac{\partial}{\partial a} \left\{ a^2 \left[\frac{1}{2} - \frac{1}{10} f_2^2 - \frac{2}{105} f_2^3 + \frac{1}{50} f_2^4 - \frac{1}{18} f_4^2 - \frac{2}{35} f_2^2 f_4 \right] \right\} da \tag{2.36}$$

as a generalization of (2.13). The function F_0 continues to be given by Equation (2.17) exactly, whereas for $j > 0$ the E_j's and F_j's are given by (2.14)–(2.16) and (2.18)–(2.20) of section II-2 as they already represent an approximation sufficient for the present purpose.

Moreover, to the requisite approximation,

$$r'^3 = a^3\left\{1 + 3\left[f_2 + \frac{2}{7}f_2^2 - \frac{9}{35}f_2^3 + \frac{4}{7}f_2f_4\right]P_2 + 3\left[f_4 + \frac{18}{35}f_2^2\right]P_4 + \cdots\right\}, \quad (2.37)$$

so that if we decompose (5.1) into

$$W = W_0 + W_2 + W_4, \quad (2.38)$$

it follows from (5.3) and (5.8) that

$$W_0 = -8\pi^2 G\int_0^{a_1} \beta_0(a)\,\rho a^2\,\mathrm{d}a, \quad (2.39)$$

$$W_2 = -\frac{8}{5}\pi^2 G\int_0^{a_1} \beta_2(a)\left\{(3+\eta_2)f_2 + \frac{2}{7}(3+2\eta_2)f_2^2 - \right. \quad (2.40)$$

$$\left. - \frac{9}{35}(3+3\eta_2)f_2^3 + \frac{4}{7}(3+\eta_2+\eta_4)f_2f_4\right\}\rho a^2\,\mathrm{d}a\,,$$

and

$$W_4 = -\frac{8}{9}\pi^2 G\int_0^{a_1} \beta_4(a)\left\{(3+\eta_4)f_4 + \frac{18}{35}(3+2\eta_2)f_2^2\right\}\rho a^2\,\mathrm{d}a, \quad (2.41)$$

consistently to quantities of fourth order in rotational distortion.

We note that, within this scheme of approximation, only the amplitudes f_2 and f_4 occur explicity in the integrands on the right-hand side of the preceding Equations (2.39)–(2.41) together with their logarithmic derivatives η_j but not f_6. Therefore, to evaluate the W_j's – by quadratures or otherwise – only Equations (2.33) and (2.34) of section II-2 need to be solved for the boundary conditions (2.39)–(2.41) in terms of a suitably chosen set of generalized coordinates q_i. In order to do so, *let us identify the generalized coordinates q_i with the surface values $f_j(a_1)$ of the respective amplitudes of distortion.* In the preceding parts of this section we found it possible to express W correctly to quantities of *fourth* order in terms of *two* such coordinates, i.e.,

$$q_2 \equiv f_2(a_1) \quad \text{and} \quad q_4 \equiv f_4(a_1) \quad (2.42)$$

only. It is clear from the structure of Equations (2.38)–(2.41) that the potential energy W does not depend on the first powers of q_2 or q_4, but only on their squares and higher powers (or cross-products). In Section II-2 we determined q_2 and q_4 correctly to quantities of third order. Therefore, q_2^2 or q_4^2 will commence to be affected by errors inherent in our scheme of approximation which are of *fifth* order in surficial distortion. Accordingly, within the scheme of fourth-order approximation, the potential energy W of our rotating configuration can be expressed as

$$W = A_0 + A_2 q_2^2 + A_3 q_2^3 + A_3 q_2^3 + \cdots +$$
$$+ B_2 q_4^2 \qquad + B_4 q_2^2 q_4 + \cdots, \tag{2.43}$$

where

$$A_0 = -8\pi^2 G \int_0^{a_1} \rho \left\{ \int_a^{a_1} \rho a \, da + \frac{1}{a} \int_0^a \rho a^2 \, da \right\} a^2 \, da \tag{2.44}$$

represents the potential energy of our configuration in its nonrotating state. For $j > 0$, the coefficients A_j and B_j can be ascertained by factorization of the individual W_j's as given by Equations (2.39)–(2.41).

The actual process of this factorization should indeed offer no difficulty. By introducing a nondimensional variable

$$\varphi_j(a) = \frac{f_j(a)}{f_j(a_1)} \tag{2.45}$$

constrained so that

$$0 < \varphi_j(a) \leqslant 1, \tag{2.46}$$

we can immediately rewrite Equation (2.36)

$$E_0 = \int_0^{a_1} \rho a da - \frac{q_2^2}{10} \int_a^{a_1} \rho \frac{\partial}{\partial a} (a^2 \varphi_2^2) da - \frac{2q_2^3}{105} \int_a^{a_1} \rho \frac{\partial}{\partial a} (a^2 \varphi_2^3) \, da$$
$$+ \frac{q_2^4}{50} \int_a^{a_1} \rho \frac{\partial}{\partial a} (a^2 \varphi_2^4) da - \frac{q_4^2}{18} \int_a^{a_1} \rho \frac{\partial}{\partial a} (a^2 \varphi_4^2) \, da \tag{2.47}$$
$$- \frac{2q_2^2 q_4}{35} \int_a^{a_1} \rho \frac{\partial}{\partial a} (a^2 \varphi_2 \varphi_4^2) \, da,$$

and similarly for other E_j's and F_j's. The process itself is quite straightforward, but the final results for the A_j's and B_j's are too long to be quoted here in full; and their formulation may be left as an exercise for the interested reader.

The potential energy W represents, to be sure, only one part of the expressions (2.26) or (2.28) whose minimization should ensure the stability of our configuration. The second part involves the moment of inertia C about the axis of rotation. The explicit form of the latter can, however, be established in terms of our f_j's with equal ease: namely, in accordance with Equation (6.63) of Chapter II we have

$$C = \frac{1}{5} \int_0^{a_1} \rho \frac{\partial}{\partial a} \left\{ \int_0^{\pi} \int_0^{\pi} (r')^5 \sin^3 \theta \, d\theta \, d\phi \right\} da, \tag{2.48}$$

where

$$\sin^2 \theta = \frac{2}{3} \left\{ 1 - P_2(\cos \theta) \right\} \tag{2.49}$$

and, within the scheme of our approximation,

$$r'^5 = a^5 \Big\{1 + f_2^2 + \frac{10}{12} f_2^3 - \frac{13}{35} f_2^4 + \frac{5}{9} f_4^2 + \frac{10}{7} f_2^2 f_4 + 5\left[f_2 + \frac{4}{7} f_2^2 + \frac{2}{35} f_2^3 - \frac{184}{1155} f_2^4 + \frac{200}{693} f_4^2 + \frac{8}{7} f_2 f_4 + \frac{72}{77} f_2^2 f_4\right] P_2 + \text{harmonics of higher orders.}\Big\} \tag{2.50}$$

On carrying out the requisite integrations with respect to the angular variables, we find that

$$\begin{aligned} C &= \frac{8\pi}{15}\int_0^{a_1} \rho \frac{\partial}{\partial a}\Big\{a^5\Big[1 - f_2 + \frac{3}{7} f_2^2 + \frac{44}{105} f_2^3 - \frac{7}{33} f_2^4 - \frac{8}{7} f_2 f_4 + \frac{185}{693} f_4^2 + \frac{38}{77} f_2^2 f_4\Big]\Big\} da \\ &= \frac{8\pi}{15}\int_0^{a_1} \rho a^4 \Big\{5 - (5+\eta_2) f_2 + \frac{3}{7}(5 + 2\eta_2) f_2^2 + \frac{44}{105}(5 + 3\eta_2) f_2^3 \\ &\quad - \frac{7}{33}(5 + 4\eta_2) f_2^4 + \frac{185}{693}(5 + 2\eta_4) f_4^2 - \frac{8}{7}(5 + \eta_2 + \eta_4) f_2 f_4 \\ &\quad + \frac{38}{77}(5 + 2\eta_2 + \eta_4) f_2^2 f_4 \Big\} da \end{aligned} \tag{2.51}$$

$$C = C_0 + C_1 q_2 + C_2 q_2^2 + C_3 q_2^3 + C_4 q_2^4 + D_2 q_4^2 + D_3 q_2 q_4 + D_4 q_2^2 q_4, \tag{2.52}$$

$$C_0 = -\frac{8}{3}\pi \int_0^{a_1} \rho a^4 da, \tag{2.53}$$

$$C_1 = -\frac{8}{15}\pi \int_0^{a_1} \rho \varphi_2 (5 + \eta_2) a^4 da, \tag{2.54}$$

$$C_2 = \frac{24}{105}\pi \int_0^{a1} \rho \varphi_2^2 (5 + 2\eta_2) a^4 da, \tag{2.55}$$

$$C_3 = \frac{352}{1575}\pi \int_0^{a_1} \rho \varphi_2^3 (5 + 3\eta_2) a^4 da, \tag{2.56}$$

$$C_4 = -\frac{56}{495}\pi \int_0^{a_1} \rho \varphi_2^4 (5 + 4\eta_2) a^4 da, \tag{2.57}$$

and

$$D_2 = \frac{296}{2079}\pi\int_0^{a_1} \rho\varphi_4^2(5 + 2\eta_4)a^4 da \tag{2.58}$$

$$D_3 = -\frac{64}{105}\pi\int_0^{a_1} \rho\varphi_2\varphi_4(5 + \eta_2 + \eta_4)a^4 da, \tag{2.59}$$

$$D_4 = \frac{304}{1155}\pi\int_0^{a_1} \rho\varphi_2^2\varphi_4(5 + 2\eta_2 + \eta_4)a^4 da. \tag{2.60}$$

The reader may note that, unlike the potential energy W, the moment of inertia C depends *linearly* on the generalized coordinate q_2 (though not on q_4).

B. DOUBLE-STAR PROBLEM

So far we have considered the potential energy of a single configuration distorted solely by centrifugal force arising from axial rotation. But if the star represented by it were to constitute a component of a close binary system, the two stars will be distorted not only by axial rotation, but also by the tides; and, moreover, their potential energy will depend not only on their shape and internal structure, but also on their position with respect to the centre of mass of the system.

Let W denote, accordingly, the total potential energy of a binary system; and $\Omega_{1,2}$ the potential functions of the constituent components. If, moreover, $dm_{1,2}$ denote the mass elements of the respective stars, then

$$\begin{aligned} -W &= \tfrac{1}{2}\int(\Omega_1 + \Omega_2)(dm_1 + dm_2) \\ &= \tfrac{1}{2}\int \Omega_1\, dm_1 + \tfrac{1}{2}\int \Omega_1\, dm_2 + \tfrac{1}{2}\int \Omega_2\, dm_1 + \tfrac{1}{2}\int \Omega_2\, dm_2, \end{aligned} \tag{2.61}$$

where

$$W_i \equiv -\tfrac{1}{2}\int \Omega_i\, dm_i, \qquad i = 1, 2, \tag{2.62}$$

denotes the potential energy of each component; and

$$W_{12} \equiv -\tfrac{1}{2}\left\{\int \Omega_1\, dm_2 + \int \Omega_2\, dm_1\right\}, \tag{2.63}$$

the potential energy arising from interaction of the two stars. It can be shown (cf., e.g., MacMillan, 1958) that if the potentials $\Omega_{1,2}$ are continuous everywhere (though their derivatives may be discontinuous over the boundary surface of each star), satisfy the Laplace equation $\nabla^2\Omega_{1,2} = 0$ on the exterior, and vanish at infinity, the two integrals on the right-hand side of Equation (2.63) are equal; and their sum represents the potential energy W_{12} arising from the fact that the two components are situated at a finite distance from each other.

In order to obtain more explicit expressions for the respective mass-integrals of Ω_i, advantage can be taken of the fact that, in the interior of each configuration, Ω_i is bound to satisfy Poisson's equation

$$\nabla^2 \Omega_i + 4\pi G \rho_i = 0, \tag{2.64}$$

where ρ_i denotes the internal density of the respective configuration; and G, the gravitation constant. If we multiply now the foregoing equation by Ω_i and integrate with respect to the volume element $d\tau_i$ related with the mass element dm_i by

$$dm_i = \rho_i \, d\tau_i , \tag{2.65}$$

we easily find that

$$\int \Omega_i \, dm_i = -\frac{1}{4\pi G} \int \Omega_i \nabla^2 \Omega_i d\tau_i , \tag{2.66}$$

the limits of which can temporarily be regarded as arbitrary.

On the other hand, it follows from Green's theorem that

$$\int \Omega_i \nabla^2 \Omega_j \, d\tau_i + \int \operatorname{grad} \Omega_i \operatorname{grad} \Omega_j \, d\tau_i = \int \Omega_i \operatorname{grad} \Omega_j \, dS_i , \tag{2.67}$$

where dS_i stands for the surface element of the respective configuration – an expression which for $i = j$ reduces to

$$\int \Omega_i \nabla^2 \Omega_i \, d\tau_i = \int \Omega_i \operatorname{grad} \Omega_i \, dS_i - \int (\operatorname{grad} \Omega_i)^2 \, d\tau_i. \tag{2.68}$$

The limits of integration in the foregoing equation are so far arbitrary. Since, however, Ω_i has been defined so as to vanish at infinity, the foregoing Equation (2.68) reduces to

$$\int^{\infty} \Omega_i \nabla^2 \Omega_i \, d\tau_i = -\int^{\infty} (\operatorname{grad} \Omega_i)^2 \, d\tau_i \tag{2.69}$$

if integration is extended over the entire space.

On returning to (2.66) we note that – inasmuch as $\rho_i \equiv 0$ outside each respective configuration – the limits of integration can likewise be extended over the entire space rather than over the actual volume without altering the value of the mass integral on the left-hand side. Therefore, by a combination of (2.66) and (2.69) it follows from (2.62) that

$$W_i = -\frac{1}{8\pi G} \int^{\infty} (\operatorname{grad} \Omega_i)^2 \, d\tau_i \tag{2.70}$$

where, as is well known,

$$(\operatorname{grad} \Omega_i)^2 = \left(\frac{\partial \Omega_i}{\partial x}\right)^2 + \left(\frac{\partial \Omega_i}{\partial y}\right)^2 + \left(\frac{\partial \Omega_i}{\partial z}\right)^2 ;$$

x, y, z being the Cartesian coordinates with respect to which $\Omega_i(x, y, z)$ is referred, and

$$d\tau = dx \, dy \, dz. \tag{2.71}$$

Furthermore, by (2.63) and (2.67) for $i = 1, j = 2$, it follows likewise that

$$W_{12} = \frac{1}{8\pi G} \int \Omega_1 \nabla^2 \Omega_2 \, d\tau_1 + \frac{1}{8\pi G} \int \Omega_2 \nabla^2 \Omega_1 \, d\tau_2 =$$

$$= -\frac{1}{4\pi G}\int^{\infty}\left\{\frac{\partial\Omega_1}{\partial x}\frac{\partial\Omega_2}{\partial x} + \frac{\partial\Omega_1}{\partial y}\frac{\partial\Omega_2}{\partial y} + \frac{\partial\Omega_1}{\partial z}\frac{\partial\Omega_2}{\partial z}\right\} \mathrm{d}x\,\mathrm{d}y\,\mathrm{d}z, \quad (2.72)$$

since both $\mathrm{d}\tau_1 = \mathrm{d}\tau_2 = \mathrm{d}x\,\mathrm{d}y\,\mathrm{d}z$ for integration over the entire space.

On insertion of the foregoing results in (2.61) we find the total potential energy of our binary configuration to be expressible in the form

$$W = -\frac{1}{8\pi G}\int^{\infty}\left\{\left(\frac{\partial\Omega}{\partial x}\right)^2 + \left(\frac{\partial\Omega}{\partial y}\right)^2 + \left(\frac{\partial\Omega}{\partial z}\right)^2\right\} \mathrm{d}x\,\mathrm{d}y\,\mathrm{d}z, \quad (2.73)$$

in which we have abbreviated

$$\Omega = \Omega_1 + \Omega_2, \quad (2.74)$$

and where

$$\left(\frac{\partial}{\partial x}\right)^2 + \left(\frac{\partial}{\partial y}\right)^2 + \left(\frac{\partial}{\partial z}\right)^2 \equiv \left(\frac{\partial}{\partial r}\right)^2 + \left(\frac{1}{r}\frac{\partial}{\partial\theta}\right)^2 + \left(\frac{1}{r\sin\theta}\frac{\partial}{\partial\phi}\right)^2 \quad (2.75)$$

in spherical polar coordinates r, θ, ϕ.

In order to evaluate this total potential energy, we have to fall back on the contents of sections II-3 and II-4 in which we established the explicit forms of the gravitational potentials $\Omega_{1,2}$ of both components. The potential Ω arising from the mass of tidally-distorted components can – in accordance with Equations (3.16) and (3.31)–(3.32) of Chapter II – be expressed as

$$\Omega = 4\pi G\sum_{j=0}^{\infty}\left\{\frac{r'^j E_j + r'^{-j-1}F_j}{2j+1}\right\}P_j(\lambda) =$$

$$= 4\pi G\sum_{j=0}^{\infty}(\mathfrak{E}_j + \mathfrak{F}_j)\,r'^j P_j(\lambda). \quad (2.76)$$

For $j = 0$, within the scheme of our approximation,

$$\mathfrak{E}_0 = E_0 = \int_a^{a_1}\rho\,\frac{\partial}{\partial a}\,\{a^2[\tfrac{1}{2} - \tfrac{1}{10}f_2^2 - \tfrac{1}{14}f_3^2]\}\,\mathrm{d}a \quad (2.77)$$

by (3.29), and

$$\mathfrak{F}_0 = \{1 + \tfrac{4}{5}f_2^2 + \tfrac{5}{7}f_3^2\}\frac{F_0}{a} - \frac{f_2}{5}\frac{F_2}{a^3} - \frac{f_3}{7}\frac{F_3}{a^4} =$$

$$= \{1 + \tfrac{1}{5}(\eta_2 + 1)f_2^2 + \tfrac{1}{7}(\eta_3 + 1)f_3^2\}\frac{F_0}{a} \quad (2.78)$$

by (3.48) of Chapter II while, for $j > 0$, the constancy of the total potential $\Psi(r', \lambda)$ over the distorted surface requires that

$$\mathfrak{E}_j + \mathfrak{F}_j + \frac{C_j}{4\pi G} = 0, \quad (2.79)$$

of which Equations (3.38)–(3.44) represent explicit forms for $j = 1(1)7$.

In accordance with Equation (2.2), the potential energy W_i of a tidally-distorted configuration should then be given by

$$W_i = -\frac{1}{2}\int^{m_i} \Omega_i \, dm_i =$$

$$= -\frac{1}{6}\int_0^{a_i}\int_0^{\pi}\int_0^{2\pi} \rho\Omega_i \frac{\partial r'^3}{\partial a} da \sin\theta \, d\theta \, d\phi =$$

$$= -\frac{2\pi G}{j+3}\int_0^{a_i}\int_0^{\pi}\int_0^{2\pi} \rho \sum_{j=0}^{\infty} \{\mathfrak{E}_j + \mathfrak{F}_j\} P_j(\lambda) \frac{\partial r'^{j+3}}{\partial a} da \sin\theta \, d\theta \, d\phi. \quad (2.80)$$

On expanding the expression (3.16) for r' in which we inserted for f_0 from (3.33), and setting

$$W_i = \sum_{j=0}^{\infty} W_i^{(j)} \quad (2.81)$$

we find that

$$W_i^{(0)} = -8\pi^2 G\int_0^{a_i} \beta_0(a)\,\rho a^2 da, \quad (2.82)$$

where, correctly to quantities of second order,

$$\beta_0(a) = \int_a^{a_i} \rho \frac{\partial}{\partial a}\{a^2[\tfrac{1}{2} - \tfrac{1}{10}f_2^2 - \tfrac{1}{14}f_3^2]\}\, da +$$

$$+ \frac{1}{a}\{1 + \tfrac{1}{5}(\eta_2 + 1)f_2^2 + \tfrac{1}{7}(\eta_3 + 1)f_3^2\}\int_0^a \rho a^2 da\,; \quad (2.83)$$

while, for $j > 0$, on insertion from (2.79) we find that

$$W_i^{(j)} = \frac{2\pi(c_j)_i}{2j+1}\int_0^{a_1} (\eta_j + j + 3)\,\rho a^{j+2} f_j \, da \quad (2.84)$$

for $j = 2(1)4$.

In addition to the effects arising from pure tidal distortion, an interaction between rotation and tides will contribute to the gravitational potential of the respective configuration second-order terms of the form

$$\Omega = 4\pi G \sum_{n=0}^{\infty}\left\{\frac{a^n U_n + a^{-n-1} V_n}{2n+1}\right\}, \quad (2.85)$$

where the U_n's and V_n's continue to be given by Equations (4.43) and (4.44) of Chapter II. The only parts of U_n and V_n which will not be annihilated by integration over the whole sphere with respect to the angular variables are those factored by harmonics of zero order (i.e., for $j = k = 0$). By use of Equations (4.45)–(4.60) it can be shown that the 'mixed' terms arising by interaction between rotation and tides will contribute to the total gravitational energy of our configuration an amount equal to

$$W_i^{(0)} = 16\pi^2 G \int_0^{a_i} \rho \frac{\partial}{\partial a}\left[a^2\left(g_0 - \frac{h_2}{10}\right)\right] da +$$

$$+ \frac{a^4}{9}\int_a^{a_i} \rho \frac{\partial}{\partial a}\left[\frac{1}{a^2}\left(\frac{h_2}{10} - \frac{h_4}{11}\right)\right] da + \frac{a^6}{13}\int_a^{a_i} \rho \frac{\partial}{\partial a}\left[\frac{1}{a^4}\left(\frac{h_4}{11}\right)\right] da +$$

$$+ \frac{1}{a}\int_0^{a} \rho \frac{\partial}{\partial a}\left[a^3\left(g_0 - \frac{h_2}{10}\right)\right] da + \frac{1}{9a^4}\int_0^{a} \rho \frac{\partial}{\partial a}\left[a^7\left(\frac{h_2}{10} - \frac{h_4}{11}\right)\right] da +$$

$$+ \frac{1}{13a^6}\int_0^{a} \rho \frac{\partial}{\partial a}\left[a^9 \frac{h_4}{11}\right] da \Big\} a^2 da, \qquad (2.86)$$

where the amplitudes $g_0(a)$ and $h_j(a)$ can be obtained from Equations (4.94), (4.101) and (4.103) by a solution of the differential Equations (4.82) and (4.83) for the constituent functions G_0 and $H_{2,4}$, with appropriate boundary conditions.

A sum of the terms on the right-hand sides of Equations (2.81) and (2.89) particularized for each component of mean radii $a_{1,2}$ does not represent yet the total potential energy W of the respective binary system. In order to obtain the latter, we must augment the sum $W_1 + W_2$ of potential energies arising from the mass of each configuration distorted by rotation and tides by the interaction terms W_{12} as given by Equation (2.63). The evaluation of the latter offers, however, but little additional difficulty. In order to demonstrate this, let us turn our attention to the evaluation of an integral of (say) Ω_2 with respect to the mass element dm_1. As the secondary component of mass m_2 lies (in a detached double-star system) outside the mass m_1, the interior potential U vanishes when $r_2 > a_2$ (i.e., outside the volume occupied by mass m_1); so that, accordingly, Ω_2 will reduce to the 'tide-generating' potential $V'(r)$ given already by Equation (3.11) of Section II-3.

Such being the case, within the scheme of our approximation

$$\int_0^{m_1} \Omega_2\, dm_1 = \frac{1}{3}\int_0^{a_1}\int_0^{\pi}\int_0^{2\pi} \rho V_t^{(2)}(r') \frac{\partial r'^3}{\partial a}\, da \sin\theta\, d\theta\, d\phi =$$

$$= \sum_{j=2}^{7} \frac{c_j}{j+3}\int_0^{a_1}\int_0^{\pi}\int_0^{2\pi} \rho P_j(\lambda) \frac{\partial r'^{j+3}}{\partial a}\, da \sin\theta\, d\theta\, d\phi =$$

$$= 4\pi c_0 \int_0^{a_1} \rho a^2 da + \sum_{j=2}^{4} \frac{4\pi c_j}{2j+1}\int_0^{a_1} (\eta_j + j + 3)\rho a^{j+2} f_j da =$$

$$= G\frac{m_1 m_2}{R}\left\{1 + \frac{m_1}{m_2}\sum_{j=2}^{4} (k_j)_2 \left(\frac{a_2}{R}\right)^{2j+1}\right\} +$$

$$+\sum_{j=2}^{4} \frac{4\pi G m_2}{(2j+1)\,R^{j+1}} \int_0^{a_1} (\eta_j + j + 3)\,\rho a^{j+2}\, f_j\, da \tag{2.87}$$

on insertion for the c_j's from (3.12)–(3.15). As the leading terms of the amplitudes f_j of tidal distortion are of the order of magnitude of R^{-j-1}, the foregoing expression is exact up to terms of the order of R^{-8}. The first group of terms on the right-hand side of the last part of Equation (2.87) would alone represent the net result if the primary component could be regarded as a mass-point (i.e., if $a_1 = 0$).

Moreover, a similar integral of Ω_1 with respect to the mass element dm_2 can be obtained from Equation (2.87) by a mere interchange of indices – an operation which yields

$$\int_0^{m_2} \Omega_1\, dm_2 = \frac{G m_1 m_2}{R}\left\{1 + \frac{m_2}{m_1}\sum_{j=2}^{4} (k_j)_2 \left(\frac{a_1}{R}\right)^{2j+1}\right\} +$$

$$+\sum_{j=2}^{4} \frac{4\pi G m_1}{(2j+1)\,R^{j+1}} \int_0^{a_2} (\eta_j + j + 3)\,\rho a^{j+2}\, f_j\, da, \tag{2.88}$$

where f_j now stands for the amplitudes of first-order tidal distortion (and η_j, for its logarithmic derivative) of the secondary component; while ρ represents the latter's internal density distribution. If, lastly, we insert the outcome of Equations (2.87) and (2.88) in (2.63), we find that the potential energy W_{12} of mutual interaction assumes the neat form

$$W_{12} = -\frac{G m_1 m_2}{R}\left\{1 + \frac{1}{2}\sum_{i=1}^{2}\sum_{j=2}^{4} (k_j)_i \frac{m_{3-i}}{m_i}\left(\frac{a_i}{R}\right)^{2j+1}\right\} -$$

$$-\sum_{i=1}^{2}\sum_{j=2}^{4} \frac{2\pi (c_j)_i}{2j+1} \int_0^{a_i} (\eta_j + j + 3)\,\rho a^{j+2} f_j\, da. \tag{2.89}$$

The first part of this result is classical, and represents the potential energy of tidal interaction of a finite configuration and a mass-point; while the second part cancels with the expressions for $W_i^{(j)}$ as given by Equation (2.84).

The total potential energy W of our tidally-distorted system reduces, therefore, to the sum

$$W = W_1^{(0)} + W_1^{(0)} - \frac{G m_1 m_2}{R}\left\{1 + \frac{1}{2}\sum_{i=1}^{2}\sum_{j=2}^{4} (k_j)_i \frac{m_{3-i}}{m_i}\left(\frac{a_i}{R}\right)^{2j+1}\right\}, \tag{2.90}$$

where the $W_i^{(0)}$'s continue to be given by (2.82).

If, in addition, our configuration is distorted also by centrifugal force arising from axial rotation, the $W_i^{(0)}$'s as given by Equation (2.82) should be augmented by the addition of the 'mixed' terms (2.80) due to an interaction between rotation and tides, as well as by purely rotational terms established already in section VII-2B.

In order to investigate the *secular stability* of so distorted a binary system by means of the criteria set up in section VII-1 all we need to do is to evaluate again all second partial derivatives (both straight and mixed) of the complete expresssion for the potential energy W of the system with respect to the generalized coordinates q_j of our problem.

VII-3. Dynamical Stability

The stability criteria used in the preceding section assumed any adjustment of the form (or internal structure) of the components of binary systems to take place so slowly that the kinetic energy of motion representing such an adjustment can be regarded as negligible in comparison with their potential energy. This is tantamount to saying that, in the course of such an adjustment, our star will pass through a sequence of equilibrium configurations specified by a certain set of 'stability coefficients' characteristic of each stage; and it is only so long as they all remain positive that we know the state of minimum potential energy has been attained.

Suppose, however, that our star has been disturbed from its (permanent or temporary) state of equilibrium by an event producing response on the 'vibrational' time-scale (1.7 (1.7), such that the kinetic energy of motion can no longer be ignored in comparison with the potential energy on the left-hand side of (1.1). If so, the latter can be shown to assume the more explicit form deduced already in section III-1; and linearized (for small oscillations) in section III-2. The solutions of *homogeneous* forms of such equations of motion – in the absence of any external field of force – are subject to boundary conditions requiring the solutions to remain bounded at the centre as well as on the surface. If, moreover, we assume that such solutions should be *harmonic* in the time, the equations of motion in their linearized form can generally be made consistent with the requisite boundary conditions only for a certain discrete set of characteristic time-exponents which, in the presence of dissipative forces (viscosity), will generally be complex numbers with negative real parts* – a condition sufficient to ensure dynamical stability of the respective system (cf. Liapounov, 1908).

The linearized equations of motion governing small oscillations of the stars of given structure have already been derived in section III-2, and the reader is referred to it for fuller details. The principal limitation of applicability of this theory for our present purposes is, however, the fact that throughout section III-2 we were concerned solely with oscillations (radial or nonradial) *about a sphere* – i.e., we assumed that the equilibrium form of the oscillating configuration was spherical. This is admissible – indeed, inevitable – only as long as we consider a single star to be alone in space; and the external gravitational field to be zero. However, in *close binary systems the equilibrium form of the components cannot be spherical;* but is bound to be distorted by axial rotation and mutual tidal attraction. Consequently, in order to be able to make use of the equations of motion deduced in section III-1 for investigations of dynamical stability of the components of close binary systems, we must first reduce them to a form governing free oscillations of

* Negative because of the dissipation of kinetic energy into heat. If dissipation were absent and, consequently, the characteristic exponents were purely imaginary, the stability of harmonically oscillating systems would be neutral; and could not be clarified on the basis of linearized theory alone.

the respective configurations which are, not spherical, but distorted in the prevalent field force.

In contrast with a truly enormous volume of work carried out in the past half century to study small oscillations – radial or nonradial – of stellar structures about *spherical* state of equilibrium (for a partial survey of it, cf., e.g., Ledoux, 1958), investigations of similar motions about *distorted* states of equilibrium are conspicuous by their almost complete absence. And yet the equilibrium forms of distorted stars are well-known to us from work summarized already in Chapter II. Moreover, in Chapter VI we pointed out that the equilibrium forms of centrally-condensed configurations (approximating real stars to a high degree of approximation) come very close to those of the 'Roche model' discussed extensively in that chapter. In order to facilitate that discussion, in section VI-3 we introduced a set of curvilinear 'Roche coordinates' ξ, η, ζ, defined so that ξ represents the (normalized) potential generated by the mass of the respective star; and η, ζ are angular coordinates defining surfaces which are orthogonal to ξ. In what follows we wish to return to these coordinates to study in their terms harmonic oscillations of the stars about equilibrium surfaces defined by ξ = constant (rather than r = constant).

In order to do so , let u_ξ, u_η, u_ζ denote the curvilinear components of the velocity vector $\mathbf{u}$ in the direction of increasing ξ, η, ζ defined by the equations

$$u_\xi = h_1\dot{\xi} = l_1\dot{x} + m_1\dot{y} + n_1\dot{z}\,, \tag{3.1}$$

$$u_\eta = h_2\dot{\eta} = l_2\dot{x} + m_2\dot{y} + n_2\dot{z}\,, \tag{3.2}$$

$$u_\zeta = h_3\dot{\zeta} = l_3\dot{x} + m_3\dot{y} + n_3\dot{z}\,, \tag{3.3}$$

where dots stand for the time-derivatives of the respective variables, and the direction cosines l_j, m_j, n_j (j = 1, 2, 3) continue to be defined by Equations (3.8)–(3.10) of Chapter VI.

If so, the fundamental Eulerian equations of inviscid flow can be rewritten in the scalar form as*

$$h_1\frac{\mathrm{D}u_\xi}{\mathrm{D}t} + \frac{1}{\rho}\frac{\partial P}{\partial \xi} = \frac{\partial \Omega}{\partial \xi}\,, \tag{3.4}$$

$$h_2\frac{\mathrm{D}u_\eta}{\mathrm{D}t} + \frac{1}{\rho}\frac{\partial P}{\partial \eta} = 0\,, \tag{3.5}$$

$$h_3\frac{\mathrm{D}u_\zeta}{\mathrm{D}t} + \frac{1}{\rho}\frac{\partial P}{\partial \zeta} = 0\,, \tag{3.6}$$

where t denotes the time; P, the pressure inside the fluid; ρ, the density; and Ω, the total potential of forces acting upon any element of the fluid; and where (cf., e.g., Lamb, 1932) the Lagrangian components of the accelerations on the left-hand sides of Equations (3.4)–(3.6) in the Roche curvilinear coordinates are expressible as

* These equations are, strictly speaking, true as they stand only in a stationary (non-rotating) systems of coordinates. Should – as will be the case of close binary systems – the X-axis of our coordinate system be allowed to rotate with the radius-vector joining the centres of the two components, the left-hand sides of Equations (3.4)–(3.6) should be augmented to the respective components of the vector product $2\boldsymbol{\omega} \times \mathbf{u}$ arising from the Coriolis force.

$$\begin{aligned}\frac{\mathrm{D}u_\xi}{\mathrm{D}t} &= \frac{\partial u_\xi}{\partial t} + \frac{u_\xi}{h_1}\frac{\partial u_\xi}{\partial \xi} + \frac{u_\eta}{h_2}\frac{\partial u_\xi}{\partial \eta} + \frac{u_\zeta}{h_3}\frac{\partial u_\xi}{\partial \zeta} + \\ &+ \frac{u_\xi}{h_1}\left\{\frac{u_\xi}{h_1}\frac{\partial h_1}{\partial \xi} + \frac{u_\eta}{h_2}\frac{\partial h_1}{\partial \eta} + \frac{u_\zeta}{h_3}\frac{\partial h_1}{\partial \zeta}\right\} - \\ &- \frac{1}{h_1}\left\{\frac{u_\xi^2}{h_1}\frac{\partial h_1}{\partial \xi} + \frac{u_\eta^2}{h_2}\frac{\partial h_2}{\partial \xi} + \frac{u_\zeta^2}{h_3}\frac{\partial h_3}{\partial \xi}\right\},\end{aligned} \tag{3.7}$$

$$\begin{aligned}\frac{\mathrm{D}u_\eta}{\mathrm{D}t} &= \frac{\partial u_\eta}{\partial t} + \frac{u_\xi}{h_1}\frac{\partial u_\eta}{\partial \xi} + \frac{u_\eta}{h_2}\frac{\partial u_\eta}{\partial \eta} + \frac{u_\zeta}{h_3}\frac{\partial u_\eta}{\partial \zeta} + \\ &+ \frac{u_\eta}{h_2}\left\{\frac{u_\xi}{h_1}\frac{\partial h_2}{\partial \xi} + \frac{u_\eta}{h_2}\frac{\partial h_2}{\partial \eta} + \frac{u_\zeta}{h_3}\frac{\partial h_2}{\partial \zeta}\right\} - \\ &- \frac{1}{h_2}\frac{u_\xi^2}{h_1}\frac{\partial h_1}{\partial \eta} + \frac{u_\eta^2}{h_2}\frac{\partial h_2}{\partial \eta} + \frac{u_\zeta^2}{h_3}\frac{\partial h_3}{\partial \zeta},\end{aligned} \tag{3.8}$$

$$\begin{aligned}\frac{\mathrm{D}u_\zeta}{\mathrm{D}t} &= \frac{\partial u_\zeta}{\partial t} + \frac{u_\xi}{h_1}\frac{\partial u_\zeta}{\partial \xi} + \frac{u_\eta}{h_2}\frac{\partial u_\zeta}{\partial \eta} + \frac{u_\zeta}{h_3}\frac{\partial u_\zeta}{\partial \zeta} + \\ &+ \frac{u_\zeta}{h_3}\left\{\frac{u_\xi}{h_1}\frac{\partial h_3}{\partial \xi} + \frac{u_\eta}{h_2}\frac{\partial h_3}{\partial \eta} + \frac{u_\zeta}{h_3}\frac{\partial h_3}{\partial \zeta}\right\} - \\ &- \frac{1}{h_3}\left\{\frac{u_\xi^2}{h_1}\frac{\partial h_1}{\partial \zeta} + \frac{u_\eta^2}{h_2}\frac{\partial h_2}{\partial \zeta} + \frac{u_\zeta^2}{h_3}\frac{\partial h_3}{\partial \zeta}\right\}.\end{aligned} \tag{3.9}$$

As is well known, the Eulerian equations (3.4)–(3.6) of motion safeguard the conservation of *momentum* of the respective dynamical system. To safeguard the conservation of *mass*, we must invoke the use of the equation of continuity

$$\frac{\mathrm{D}\rho}{\mathrm{D}t} + \rho\Delta = 0\,, \tag{3.10}$$

where

$$\Delta = \frac{1}{h_1 h_2 h_3}\left\{\frac{\partial}{\partial \xi}(h_2 h_3 u_\xi) + \frac{\partial}{\partial \eta}(h_3 h_1 u_\eta) + \frac{\partial}{\partial \zeta}(h_1 h_2 u_\zeta)\right\}; \tag{3.11}$$

stands for the divergence of the velocity vector, and which represents a single scalar equation between the variables already involved in (3.4)–(3.6).

Equations (3.4)–(3.6) and (3.10) constitute, in fact, a simultaneous system of four differential equations for five dependent variables – $u_{\xi,\eta,\zeta}$, P, ρ – which is as yet not sufficient for a complete specification of any one of them. In order to render our problem determinate, we must adjoin to it one additional equation relating the same variables; and this can be done by expressing the conservation of *energy* in a form appropriate for the problem. If, in particular, this energy can be conserved with the aid of an 'equation of state' requiring the changes of P and ρ to be adiabatic, the fifth equation needed to complete our system is known to assume the form

$$\frac{\mathrm{D}P}{\mathrm{D}t} = a^2 \frac{\mathrm{D}\rho}{\mathrm{D}t}, \tag{3.12}$$

where

$$a^2 = \gamma \frac{P}{\rho} \tag{3.13}$$

denotes the square of the local velocity of sound (γ being the ratio of specific heats at constant pressure and volume of the respective fluid). Moreover, inasmuch as the pressure P and density ρ are scalar quantities, their Lagrangian time-derivatives can be expressed by

$$\frac{\mathrm{D}}{\mathrm{D}t}(P, \rho) = \left\{ \frac{\partial}{\partial t} + \frac{u_\xi}{h_1}\frac{\partial}{\partial \xi} + \frac{u_\eta}{h_2}\frac{\partial}{\partial \eta} + \frac{u_\zeta}{h_3}\frac{\partial}{\partial \zeta} \right\}(P, \rho); \tag{3.14}$$

and on insertion of (3.14) into (3.10) and (3.12) a specification of the equations of our problem in terms of the Roche coordinates ξ, η, ζ is now complete.

A. VIBRATIONS OF THE ROCHE MODEL

The general equations of hydrodynamics in Roche coordinates offer considerable advantages over their Cartesian form for treatment of many problems – in particular, of those in which the boundary conditions can be simplified in their terms. For example, any specific property (such as the constancy of pressure or density, or the vanishing of the gradients of any one of the dependent variables) to be enforced over the distorted surface of a star can be localized with the aid of only one of the three Roche spatial coordinates (e.g., ξ). In particular, to describe the free surface of a distorted component in a close binary system in terms of rectangular or polar coordinates may be a matter of some complexity (cf. Equation (2.11) of Chapter VI); but in the Roche coordinates it is sufficient to do so by setting ξ = constant (say, ξ_0). Or, in problems connected with fluid flow of gas in such systems, the boundary conditions which require the vanishing of motion normal to the surface of a star reduce to $u_\xi(\xi_0) = 0$.

The aim of the present section will, however, be to demonstrate such advantages on one particular example: namely, on the problem of *vibrational stability* of self-gravitating gaseous configurations, the surface of which can be represented by Roche equipotentials (1-6) of Chapter VI. Suppose that a configuration of this shape becomes disturbed –by whatever process – from its state of equilibrium; and its consequent change in size or form will render its dependent variables functions of the time. Under which conditions can such motions remain periodic?

In order to embark on an effort to answer this question, assume first that all effects caused by the initial disturbance are small enough for quantities of the order of their squares and higher powers to be ignorable. Secondly, let us assume that the pressure P and density ρ can be expressed as

$$P(\xi, \eta, \zeta; t) = P_0(\xi) + P'(\xi, \eta, \zeta; t), \tag{3.15}$$

$$\rho(\xi, \eta, \zeta; t) = \rho_0(\xi) + \rho'(\xi, \eta, \zeta; t), \tag{3.16}$$

where P_0 and ρ_0 describe the respective properties of our configuration in its equilibrium state (depending on ξ only); while P', ρ' represent their changes brought about by motion with the velocity components $u_{\xi,\eta,\zeta}$. If, moreover, P', ρ' as well as $u_{\xi,\eta,\zeta}$ can be regarded as small quantities of first order, our fundamental system (3.4)–(3.6) of Eulerian equations can be *linearized* to yield

$$h_1 \frac{\partial u_\xi}{\partial t} + \frac{1}{\rho_0}\frac{\partial P_0}{\partial \xi} - \frac{\partial \Omega}{\partial \xi} + \frac{1}{\rho_0}\left\{\frac{\partial P'}{\partial \xi} - \frac{\rho'}{\rho_0}\frac{\partial P_0}{\partial \xi}\right\} = 0\,, \tag{3.17}$$

$$h_2 \frac{\partial u_\eta}{\partial t} + \frac{1}{\rho_0}\frac{\partial P_0}{\partial \eta} + \frac{1}{\rho_0}\left\{\frac{\partial P'}{\partial \eta} - \frac{\rho'}{\rho_0}\frac{\partial P_0}{\partial \eta}\right\} = 0, \tag{3.18}$$

$$h_3 \frac{\partial u_\zeta}{\partial t} + \frac{1}{\rho_0}\frac{\partial P_0}{\partial \zeta} + \frac{1}{\rho_0}\left\{\frac{\partial P'}{\partial \zeta} - \frac{\rho'}{\rho_0}\frac{\partial P_0}{\partial \zeta}\right\} = 0. \tag{3.19}$$

Since, moreover, consistent with (3.15) and (3.16)

$$\frac{\partial P_0}{\partial \eta} = \frac{\partial P_0}{\partial \zeta} = 0 \tag{3.20}$$

and, in the equilibrium state,

$$\frac{\partial P_0}{\partial \xi} = \rho_0 \frac{\partial \Omega}{\partial \xi} = k\rho_0, \tag{3.21}$$

where the constant k can be normalized to one, the foregoing system (3.17)–(3.19) of linearized Eulerian equations in Roche coordinates readily reduces to

$$\rho_0 h_1 \frac{\partial u_\xi}{\partial t} = \rho' - \frac{\partial P'}{\partial \xi}, \tag{3.22}$$

$$\rho_0 h_2 \frac{\partial u_\eta}{\partial t} = -\frac{\partial P'}{\partial \eta}, \tag{3.23}$$

$$\rho_0 h_3 \frac{\partial u_\zeta}{\partial t} = -\frac{\partial P'}{\partial \zeta}, \tag{3.24}$$

while the linearized Equations (3.24) of continuity and (3.26) of adiabatic changes of state similarly reduce to

$$\frac{\partial \rho'}{\partial t} + \frac{u_\xi}{h_1}\frac{\partial \rho_0}{\partial \xi} + \rho_0 \Delta = 0 \tag{3.25}$$

and

$$\frac{\partial P'}{\partial t} + \frac{u_\xi}{h_1}\rho_0 + a^2\rho_0 + a^2\rho_0\Delta = 0, \tag{3.26}$$

respectively.

In order to eliminate P' and ρ' from Equations (3.22)–(3.24), differentiate the latter with respect to t, and insert for $\partial P'/\partial t$ and $\partial\rho'/\partial t$ from (3.25)–(3.26); the result can be expressed as

$$\frac{\partial^2}{\partial t^2}(h_1 u_\xi) = \frac{\partial}{\partial \xi}\left\{a^2\Delta + \frac{u_\xi}{h_1}\right\} + a^2 A\,\Delta, \tag{3.27}$$

$$\frac{\partial^2}{\partial t^2}(h_2 u_\eta) = \frac{\partial}{\partial \eta}\left\{a^2\Delta + \frac{u_\xi}{h_1}\right\}, \tag{3.28}$$

$$\frac{\partial^2}{\partial t^2}(h_3 u_\zeta) = \frac{\partial}{\partial \zeta}\left\{a^2\Delta + \frac{u_\xi}{h_1}\right\}, \tag{3.29}$$

where (consistent with (3.27)) $a^2 = \gamma P_0/\rho_0$, and where we have abbreviated

$$A = \frac{1}{\rho_0}\frac{\partial \rho_0}{\partial \xi} - \frac{1}{\gamma P_0}\frac{\partial P_0}{\partial \xi} = \frac{\partial}{\partial \xi}\log \rho_0 P_0^{-1/\gamma}; \tag{3.30}$$

in a linearized problem both a^2 and A are functions of ξ alone.

If, moreover, the motion we anticipate is to be *harmonic* – such that

$$\frac{\partial^2}{\partial t^2} \equiv -\tilde{\nu}^2, \tag{3.31}$$

where $\tilde{\nu}^2$ denotes a positive frequency of oscillatory motion, our system (3.27)–(3.29) further simplifies to

$$\psi_\xi + \tilde{\nu}^2 h_1 u_\xi + a^2 A\,\Delta = 0, \tag{3.32}$$

$$\psi_\eta + \tilde{\nu}^2 h_2 u_\eta = 0, \tag{3.33}$$

$$\psi_\zeta + \tilde{\nu}^2 h_3 u_\zeta = 0, \tag{3.34}$$

where we have abbreviated

$$\psi \equiv a^2\Delta + h_1^{-1} u_\xi, \tag{3.35}$$

and the divergence Δ continues to be given by Equation (3.25).

Accordingly, the foregoing system (3.32)–(3.34) consists of three simultaneous partial differential equations in ξ, η, ζ of second order; so that the problem represented by them is one of *sixth* order. However, if the distortion of the Roche equipotentials ξ is caused by either rotation or tides alone, it is easy to show that this order becomes diminished to *four*. Thus in the case of distortion caused by axial rotation with constant angular velocity $\omega^2 \equiv \eta$, by (3.16)

$$u_\eta = \omega h_2 \tag{3.36}$$

and, therefore,

$$\frac{\partial u_\eta}{\partial t} = 0\,. \tag{3.37}$$

Since, moreover, P' does not depend on η in the case of rotational-spheroid symmetry, Equation (3.33) is satisfied by its both sides vanishing identically – in which case the system (3.32)–(3.34) reduces to

$$\psi_\xi + \tilde{\nu}^2 h_1 u_\xi + a^2 A \Delta = 0, \tag{3.38}$$

$$\psi_\zeta + \tilde{\nu}^2 h_3 u_\zeta = 0, \tag{3.39}$$

where Ψ has already been defined by (3.35),

$$\begin{aligned}\Delta \equiv & \frac{\partial}{\partial\xi}\left(\frac{u_\xi}{h_1}\right) + \left(\frac{\partial}{\partial\xi}\right)\log|h_1h_2h_3| \frac{u_\xi}{h_1} \\ & + \frac{\partial}{\partial\zeta}\left(\frac{u_\zeta}{h_3}\right) + \left(\frac{\partial}{\partial\zeta}\right)\log|h_1h_2h_3| \frac{u_\zeta}{h_3},\end{aligned} \tag{3.40}$$

and where the metric coefficients $h_{1,2,3}$ (ξ, ζ) in the neighbourhood of the origin continue to be given by the expansions on the right-hand sides of Equations (3.38)–(3.40) of Chapter VI.

On the other hand, in the case of purely tidal distortion of a non-rotating configuration, it is the third Equation (3.34) which vanishes identically, leaving us with the fourth-order system

$$\psi_\xi + \tilde{\nu}^2 h_1 u_\xi + a^2 A \Delta = 0, \tag{3.41}$$

$$\psi_\eta + \tilde{\nu}^2 h_2 u_\eta = 0, \tag{3.42}$$

where

$$\begin{aligned}\Delta \equiv & \frac{\partial}{\partial\xi}\left(\frac{u_\xi}{h_1}\right) + \left(\frac{\partial}{\partial\xi}\log|h_1h_2h_3|\right)\frac{u_\xi}{h_1} \\ & + \frac{\partial}{\partial\eta}\left(\frac{u_\eta}{h_2}\right) + \left(\frac{\partial}{\partial\eta}\log|h_1h_2h_3|\right)\frac{u_\eta}{h_2}.\end{aligned} \tag{3.43}$$

In order to convert the partial differential equations (3.32)–(3.34) into ordinary ones, let us seek – as we did in section III-2A – to separate the variables on the assumption that harmonic oscillations of our distorted configurations possess spheroidal symmetry in the Roche coordinates, i.e., that it is possible to express their velocity components $u_{\xi,\eta,\zeta}$ in the form

$$u_\xi = u(\xi)\mathfrak{P}_j(\eta, \zeta), \tag{3.44}$$

$$u_\eta = v(\xi)\frac{\partial\mathfrak{P}_j}{\partial\eta}, \tag{3.45}$$

$$u_\zeta = \frac{v(\xi)}{\sin\eta}\frac{\partial\mathfrak{P}_j}{\partial\zeta}, \tag{3.46}$$

where $u(\xi)$ and $v(\xi)$ are suitable functions of the potential ξ, and the $\mathfrak{P}_j(\eta, \zeta)$'s represent the 'Roche harmonics' – i.e., the angular parts of separable solutions of the Laplace equation

$$\nabla^2 \mathfrak{L} \equiv \frac{1}{h_1 h_2 h_3}\left\{\frac{\partial}{\partial \xi}\left(\frac{h_2 h_3}{h_1}\frac{\partial}{\partial \xi}\right)+\frac{\partial}{\partial \eta}\left(\frac{h_1 h_3}{h_2}\frac{\partial}{\partial \eta}\right)+\frac{\partial}{\partial \zeta}\left(\frac{h_1 h_2}{h_3}\frac{\partial}{\partial \zeta}\right)\right\}\mathfrak{L} = 0 \tag{3.47}$$

in the Roche coordinates, where

$$\mathfrak{L}(\xi, \eta, \zeta) \equiv D_j(\xi)\,\mathfrak{P}_j(\eta, \zeta)\,. \tag{3.48}$$

Such harmonics should be associated with the equipotential surfaces ξ = constant in a similar manner as the spherical harmonics $Y_j^i(\theta, \phi)$ of section III-2 are with the spheres r = constant. In point of fact, the ordinary spherical harmonics should be regarded as limiting cases of more general family of Roche harmonics; and, as $\xi \to \infty$, the latter obviously reduce to the former in the immediate neighbourhood of the origin.

A more complete mathematical theory of the formal properties of the Roche harmonics (such as their orthogonality, etc.) has not yet been developed beyond first steps (cf. Kopal, 1972; Roach, 1975); and much remains to be done before this entire subject can be placed on more solid analytical foundations. However, it is obvious from the above definition that if the amplitudes $u(\xi)$ and $v(\xi)$ of velocity components $u_{\xi,\eta,\zeta}$ on the right-hand sides of Equations (3.44)–(3.46) can be regarded as small quantities of first order, it is legitimate to replace the Roche harmonics $\mathfrak{P}_j(\eta, \zeta)$ in (3.48) by ordinary surface harmonics $P_j(\eta, \zeta)$ of the same variables to obtain expressions for $u_{\xi,\eta,\zeta}$ describing first-order distortion of the oscillating surfaces. More detailed expressions for the Roche harmonics will be required only when our theory will have been extended to terms of second and higher orders in surficial distortion; but a systematic development of such a theory constitutes a task for the future.

VII-4. Concluding Remarks

In conclusion of this chapter devoted to a discussion of the stability of the components of close binary systems – a number of retrospective observations on the present state of the subject should be pointed out. First, as the reader no doubt already gathered, this chapter constitutes much less complete an account of its subject than is true of the preceding parts of this book. An incomplete state of our knowledge on this subject is, in turn, due to a greater (though unavoidable) complexity of the underlying mathematics, which has no doubt acted as a deterrent for many investigators to enter this field and advance it to a significant extent. As a result, more than before we had to restrict ourselves in this chapter to a mathematical formulation of problems still awaiting solution; and this state of affairs may persist for some time.

Perhaps the most important question in need of adequate clarification concerns the relation betweeen the dynamical and secular stabilities. Their mathematical definitions are quite different; but how are they physically related? A quest for an answer left behind it a somewhat chequered history throughout the first half of the present century. Thus, for example, Jeans (1928) expressed an opinion that "... as the physical conditions of a system gradually change, secular instability necessarily sets in before ordinary" (i.e., dynamical) "stability. Thus for problems of cosmogony it is secular instability alone which is of interest. A system never attains to a configuration in which ordinary instability comes into operation, since secular instability must always have previously intervened." (op. cit., p. 199).

Jeans probably based this conclusion on the facts known to him concerning the stability of Maclaurin spheroids of homogeneous incompressible liquid – the only model adequately investigated up to that time – which are known to become secularly unstable when the eccentricity e of their meridian cross-sections attain the value of 0.8127; but retain ordinary stability until $e = 0.9529$ (Bryan, 1888) – thus losing indeed secular stability before ordinary (dynamical) instability sets in. But the Jacobian ellipsoids (on which secular stability descended from the Maclaurin spheroids at the point of bifurcation) were found to lose ordinary stability at the same time as secular stability (Cartan, 1922); and there is not doubt that the linear descendent of these ellipsoids – the pear-shaped figure – whose secular instability was proved by Poincaré (1885), Liapounov (1904, 1908) as well as Jeans (1916), is also ordinarily unstable.

For configurations consisting of heterogeneous compressible fluids, general theorems of this nature are so far unavailing. But when the number of their degrees of freedom is restricted to remain *finite*, it can be shown (cf., e.g., Lyttleton, 1953; pp. 21–33) that if an *equilibrium configuration is secularly stable, it is also dynamically stable;* while *if such a configuration is secularly unstable, it may or may not be dynamically unstable* – the outcome depending on the number of negative coefficients b_i which brought about secular instability: if their number is even, the configuration can remain dynamically stable; but will lose also this kind of stability if their number is odd. Thus although dynamical stability cannot vanish before secular stability does so, and may persist even after secular stability has been lost, both types of stability may also vanish at the same time. An example of the first situation is met at the bifurcation point of the linear series of Maclaurin spheroids; while the bifurcation point of the Jacobian ellipsoids provides an example of the other situation.

Also from another point of view secular stability would appear to be the more important of the two. If (as has invariably been the case so far) dynamical stability is investigated on the basis of linearized equations of oscillatory motion, the appearance of incipient instability cannot in general furnish any indication of the new steady state to which the system will eventually move. Linearized equations cannot describe more than the initial stages of this motion. As such, they are incapable of dealing, e.g., with 'self-excited oscillations', which would arise in close binary systems in the case of a *resonance* of any one of the characteristic frequencies of free oscillations of either component with that of their orbital motion; while the eccentricity (no matter how small) of such an orbit provides an ever-present source of excitation for forced oscillations of tidal origin.

On the other hand, the condition of secular stability, requiring the quadratic form (2.7) for δW to be positive definite for positive values of the coefficients b_i of stability, is independent of the magnitudes θ_i of disturbances of the state of equilibrium; and its validity is *not* restricted to small values of θ_i. Therefore, an applicability of this test does not require the new position of a stable equilibrium to be in the proximity of the unstable one. In other words, while an effort to ascertain the existence of ordinary (dynamical) stability focuses attention on the stability of small oscillations of our system in the proximity of its state of equilibrium by methods outlined in this section, a stability of this figure of equilibrium around which the system may (but need not) oscillate is of the secular kind; and can be investigated by the methods of section VII-2. Both methods are likely in most cases to furnish the same verdict; and differ mainly in the ways in which the respective questions are mathematically formulated; but whether or not this is always bound to be true constitutes a question to which no final answer can be given at this time.

BIBLIOGRAPHICAL NOTES

The problems connected with the stability of self-gravitating configurations of arbitrary structure belong among mathematically most difficult aspects of the subject discussed in this volume; and although these problems attracted due attention of mathematicians and astronomers for more than a century, the progress has been far from rapid. For recent accounts of such work (concerned almost exclusively with the stability of single stars) cf., e.g., P. Ledoux (1958, 1965).

On the mathematical side, the entire problem (in particular, that of secular stability of the stars) has been profoundly influenced by the work of two men – Henri Jules Poincairé (1854–1912), and Alexander Mikhailovich Liapounov (1857–1918) – who rigorously established the principal stability criteria expounded in this chapter. Although Liapounov's work was largely contemporary with Poincaré's (commencing in 1884), it did not become more widely known till after its translation into French twenty years later (cf., Liapounov, 1904 and 1908). Both often attacked the same problems; though differed in their methods: while Poincaré made wide use of geometrical and topological concepts, Liapounov was a pure analyst; though his analysis was no less penetrating than that of his great French contemporary.

The ideas developed by Poincaré (1885) in his great memoir on secular stability were subsequently summarized by him in his *Leçons sur les figures d'equilibre d'une masse fluide* (1903); and by P. Appell in vol. 4 of his *Traité de mécanique rationelle* (1921). This latter source contains also a full list of references to earlier works on this subject.

Simultaneously with Poincaré, the subject attracted a lasting attention of G. H. Darwin (1845–1912); and a considerable number of his papers (Darwin, 1901, 1902, 1908) dealing with the stability of celestial bodies can be found in vol. 3 of his *Scientific Papers* (London, 1910). In contrast with Poincaré or Liapounov – who were primarily pure mathematicians – Darwin's work was marked by a much more pragmatic approach to his problems: he was content to study the stability phenomena by mathematical methods most convenient for the purpose regardless of their novelty or elegance; and did not hesitate to embark on onerous numerical work – as marks his investigations of the stability of the pear-shaped figures of equilibrium. That such an approach can be sometimes too blunt for the purpose was illustrated by his investigations (Darwin, 1901, 1902, 1908) in which he claimed the pear-shaped figure to be secularly stable. Shortly after the publication of this result, Liapounov (1903, 1912) announced his proof of the instability of this figure; and several years later Darwin's pupil J. H. Jeans (1916) demonstrated that Liapounov's result was indeed correct. Jeans's own work in this subject was likewise summarized by him in two volumes; *Problems of Cosmogony and Stellar Dynamics* (1919), and *Astronomy and Cosmogony* (1928); though some of his conclusions were later found to be ill-founded by R. A. Lyttleton in his book on *The Stability of Rotating Liquid Masses* (1953). Cf. also, in this connection, W. S. Jardetzky, *Theories of Figures of Celestial Bodies* (1958).

The criteria for secular stability given in section VII-2 go back to Dirichlet (1861) and Poincaré (1885); the condition (2.29) appears to have been used first by K. Schwarzschild (1897). The application of Clairaut's theory of Chapter II to an investigation of secular stability of heterogeneous configurations in rapid rotation, as given in section II-A, follows Kopal (1973); while subsequent treatment of the binary systems consisting of tidally-distorted configurations (section II-B) follows Kopal (1974).

An investigation of the dynamical stability of distorted configurations, whose equilibrium form is approximated by the Roche model (consisting of central mass-point surrounded by an envelope tenuous enough for its self-attraction to be ignorable), in terms of the Roche curvilinear coordinates introduced in section 3 of Chapter VI – as given in section VII-3 – has been initiated by Kopal (1970, 1972). For a discussion of the 'Roche harmonics' cf. section IV-B of Z. Kopal (1972); or G. F. Roach (1975).

CHAPTER VIII

ORIGIN AND EVOLUTION OF BINARY SYSTEMS

"The study of Algol variables should bring us to the very threshold of the question of stellar evolution, and to the heart of not a few greatest cosmical problems."

A. W. Roberts (1902)

As is well known, the evolution of the stars constitutes a process which (barring exceptional stages) proceeds at so slow a pace that no changes arising from it become perceptible on the human time scale. However, atomic and nuclear processes which motivate such an evolution are now sufficiently well understood to enable us to predict the past and the future of self-gravitating configurations of stellar mass and size with a fair probability throughout most (though not yet all) parts of their lives. In particular, it is generally recognized now that the evolution of the stars constitutes – in effect – an initial-value problem, the course of which is governed by the star's initial mass, chemical composition, and (to a lesser extent) angular momentum. Of these the mass and momentum are known to remain unaltered for long evolutionary epochs; while the mean composition changes continuously and irreversibly in the course of time as a result of gradual depletion of hydrogen and other kinds of nuclear fuel in the interiors of the stars.

No theories of phenomena of this degree of complexity can, of course, remain for long on the right track without adequate observational tests at different stages of stellar evolution. Unfortunately, single stars are not particularly useful for this purpose because of the impossibility of ascertaining their vital statistics needed to confront theory with the observations. A random sample of stars selected anywhere in our neighbourhood would contain objects of very different age as well as of initial mass and composition – and, therefore, constitute a tri-variate manifold very difficult to interpret.

As we stressed already in the Preface, a veritable *royal road* for this purpose is provided, however, by close binary systems, which represent associations of stars of (practically) the same age and initial composition, and whose properties at any stage reflect the effects of *differential evolution* going back to an initial difference in mass. Moreover, the fact that such associations last to perpetuity, and cannot be dissolved within the age of a galaxy by any known process (except, in extreme cases of massive pairs, by the explosion of a supernova), makes close binaries almost ideal tools for tracing the evolution of their components virtually from cradle to grave.

It will be the aim of the present chapter to examine, from this point of view, known physical properties of close binary systems, in order to establish the extent to which they are consistent with theoretical expectations based on what we know of the evolution of their components as individuals. This constitutes indeed a fascinating story – not more than 20 years old and still full of gaps – but one which is bound to play a central role in all efforts at a fuller understanding of the entire subject of stellar evolution. While in some parts of it theory and observations appear to be in fair agreement, there are others where this is manifestly not the case. It is the latter than are likely to enrich our knowledge with unexpected developments which have not previously been anticipated; and on which shall focus attention in different parts of this chapter for this reason.

The subject matter of this chapter will be divided into five parts. In the first (section VIII-1) we shall give a broad outline of the present state of our knowledge of the nuclear evolution of single stars, with no regard yet for their binary nature. Section VIII-2 will then contain a survey of the observed facts on close binary systems and their classification, which will provide a basis for their comparison with theoretical consequences if the components could be regarded as single. It is not till in Sections VIII-3 and 4 that we get really to grips with the actual close binary system and consider the effects of their tidal interaction in addition to their nuclear evolution; for it is precisely this interaction which renders two discrete stars a system. Lastly, in section VIII-5 we shall consider possible modes of origin of such systems, as well as their ultimate ends – ends which for very massive (albeit very rare) systems will confront us with some of the most energetic as well as weirdest manifestations of matter in our entire Universe.

VIII-1. Evolution of the Stars

A great diversity of stars have been established by observation to exist in the sky; and a variety of those encountered in close binary systems has already been described in Chapter I. The main task confronting us in this chapter will be to demonstrate that this variety is more apparent than real, and to reduce it by an identification of common elements and principles which are responsible for it.

Some are suggested by the observations directly – such as a remarkable uniformity in stellar masses: for more than 99% are comprised within the limits 0.1–10 $\odot$ – of which the bulk lies between 0.5–5 $\odot$ – thus differing no more than a factor 10; and a factor 100 would account for almost all stellar individuals known to us at the present time. Furthermore, the chemical composition of most stars (or, more accurately, of their atmospheres which are accessible to quantitative spectroscopic analysis) appears likewise to be remarkably uniform; with the lighter elements constituting most part of their mass.

On the other hand, the dimensions of the stars differ among each other by not less than factor 10^4; and their luminosities – the principal tell-tale characteristics of the stars – by 10^8 or even more. We know now that this variety is due almost exclusively to the fact that each star with a mass greater than (say) 0.1 $\odot$ undergoes profound changes in its evolution; and it is the evolutionary changes, bound to occur in the course of time, that are primarily responsible for the observed diversity of the stars.

The ages of the stars in the sky are known to differ greatly: while those of the oldest generation (Population II) originated some 10^{10} yrs ago (being approximately twice as old as our Sun), others are being born at the present time from collapsing clouds of gas. Moreover, we know that (for reasons illustrated below) the rate of the evolution of any star – once born – depends vitally on its mass. Therefore, each star encountered in the sky may differ from any other in its evolutionary stage (and the manifestations which accompany this evolution) not only because it may have been born at a different time, but also because it may have been evolving (on account of different mass) since at a very different rate.

The only way to unravel this variety is, therefore, to reconstruct a sequence of successive evolutionary stages of a given star theoretically, by a solution of the equations safeguarding the conservation of momentum, mass, and energy, the formulation of which was outlined already in Section III-1 – for different types of initial conditions.

The primary cause motivating stellar evolution is the continuous *loss of energy* by radiation, whose rate defines the observed luminosity. The principle of the conservation of energy requires, in turn, that this radiant energy must be defrayed from some other energy source possessed initially by the star, and on which it must be able to draw in case of need. This energy can be gravitational, thermal, and nuclear; or again kinetic and associated with the angular momentum which a new-born star inherited from the turbulent state of a gas cloud from which it originated. The total amount of energy stored in these sources is large, but not inexhaustible; and the rate at which it can be deployed controls the star's luminosity. When, in particular, any of these sources becomes (temporarily, or permanently) exhausted, another one must take its place; and their interplay governs the whole course of the evolution.

Most of the energy of the stars – in particular, the nuclear (and, to a lesser extent, thermal) energy sources – reside in the deep interior where, most part of the star's mass is concentrated. It is in this interior – far from the reach of observations – that the stately drama of stellar evolution runs its appointed course. Most part of the light which makes the star shine in the sky is born there; though none of it can reach the surface in its original from; being gradually transformed in its energy spectrum by a passage through layers overlying the star's core; and of the corpuscular radiation, only neutrinos may reach us form stellar interiors without appreciable loss of their original identity. A more detailed penetration of the 'engine room' of the stars is, therefore, possible only with the aid of a suitable mathematical theory describing the physical processes likely to take place under prevalent conditions.

These processes can, in turn, be described qualitatively in the following terms. At the commencement of the star's life, the new-born configuration has not yet attained high enough internal temperature to ignite any but ephemeral exothermic nuclear reaction involving Li, Be and B; and its thermal energy store is no match for contraction dominating this phase of the star's life. The energy source at this stage is, therefore, gravitational, and due to a conversion of potential energy into heat. Moreover, the outward flow of heat generated by contraction occurs through material (gaseous) transfer due to convection. In point of fact, at this stage the equilibrium of the stars may be convective throughout their interiors – the only stage of their evolution where this is likely to occur. The duration of this stage is of the order of the Helmholtz-Kelvin (HK) time-scale, the order of magnitude of which has already been indicated by Equation (1.7) of Chapter VII.

As soon as this gravitational contraction has reached the stage at which the central temperature exceeded 10^7 deg, other nuclear reactions begin to take place, in the course of which hydrogen – by far the most abundant constituent of the stellar material at that time – becomes converted into helium by different cycles (the nature of which depends on the total mass of the star). A conversion of hydrogen into helium constitutes the most efficient reaction for a generation of radiant energy from the mass-defect of the α-particles. As, moreover, hydrogen represents also the most abundant nuclear fuel of stellar endowment, the time during which the luminosity of the stars may be maintained in this manner represents by far the longest period in any star's life – a time in the course of which the production and expenditure of energy (i.e., the star's luminosity) are more neatly balanced than at any subsequent stage of its celestial career.

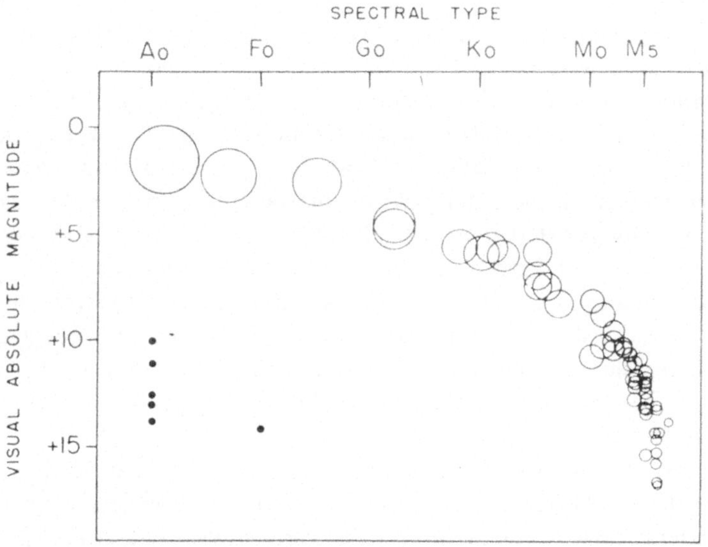

Fig. 8-1. H-R diagram for stars nearer than five parsec. Approximate relative diameters are indicated by open circles except for the white dwarfs (dots). (After van de Kamp, 1969.)

If we plot the principal vital statistics of the star – namely, their energy output (or absolute brightness) against the temperature (or spectral type) to which the surface is heated by the outward flux of radiation passing through it per unit time – we obtain the well-known Hertzsprung-Russell (HR) diagram, as shown (for nearby stars) on the accompanying Figure 8-1. Most stars on this diagram are located on a fairly narrow diagonal strip running downwards from left to right – called the Main Sequence (MS). This sequence represents, in fact, a locus of points occupied by hydrogen-burning stars; and their concentration along the MS attests to the length of time which the stars can spend at this stage. While the star thus defrays its radiation losses from this source, its position in the HR-diagram moves only a little upward towards the right; and the smallness of this shift accounts for the narrow width of the Main Sequence.

How long can such a stage last? Computations show (and observations confirm) that the position of a star on the MS is determined essentially by its mass; the more massive the star, the higher its luminosity. In point of fact, this luminosity is known to vary (approximately) with a k-th power of the mass m, where k is a number between 3 and 4. Since the total energy production is proportional to the first power fo the mass, it follows that the lifetime of a star on the MS should be proportional to m^{1-k}. In more specific terms, while stars of the masses not much exceeding that of the Sun can linger on the Main Sequence for some milliards (10^9) of years, and several milliards for stars less massive than the Sun; while stars with masses of 10 $\odot$ or more on the upper part of the

MS – the high-luminosity stars visible even at great distances – may spend in this stage only a few million (10^6) years. Most of these ornaments of our night sky represent, therefore, severely transient phenomena, shining on the background of a substrate of their less massive and spendthrift sisters.

How long can this process last? Since the pioneer work of Schoenberg and Chandrasekhar (1942) we know that, when the hydrogen gets exhausted in the central parts of the star which consist predominantly of helium, nuclear energy production comes to a temporary standstill; and the flux of radiation in the respective region can no longer counteract the gravitational effects of self-attraction. As a result, the core of the star begins to contract: – as a result, its potential energy begins to diminish; but a conservation of potential energy in the configuration as a whole requires that this decrease is compensated by an increase in the outer parts of the star (the 'mirror effect') causing their expansion. As seen by an external observer, the star begins to grow in size – essentially in concert with the core shrinkage on the Helmholtz-Kelvin time-scale – but as its radius grows, the total luminosity changes but little. As this means that the total flux of energy derived from gravitational contraction of the core now leaves the star through a surface of increasing size, the flux per unit area diminishes. This, in turn, implies that the surface temperatures of the star will diminish; and in the HR-diagram the star will move away from the MS to the right.

Such a process cannot, however, proceed to arbitrary lengths. In 1960, the Japanese astrophysicists Hayashi (1961) and Kiyokawa discovered independently by computation that an upper limit exists – in the form of a nearly straight line in the HR-diagram as shown (schematically) on Figure 8-2 – which no stellar configuration can cross if it is to remain dynamically stable. As, moreover, this line is inclined at a fairly sharp angle to the general direction of the Main Sequence, the stars of low masses find less room between the two for their evolution in the HR-diagram than the more massive configuration populating the upper part of the MS; and, as a result, their evolution may take a different course.

For stars of appreciable mass (say, $m \geqslant 3$ ⊙) the conversion of hydrogen into helium on the MS takes place predominantly through the carbon-nitrogen cycle, the efficiency of which is more sensitive to the temperature than that of the proton-proton reaction dominant in stars of more moderate masses; accordingly, the slope of the Main Sequence in the HR-diagram changes somewhat in the transition region (cf. Figure 8-1). After the hydrogen abundance in the central parts drops below approximately 12%, the role of nuclear reactions as principal sources of energy radiated away by the star comes temporarily to a halt.

Each time when this happens, the star is bound to revert to Helmholtz-Kelvin contraction, providing luminosity at the expense of the potential energy of its core. In the course of such a process the star crosses rapidly the gap between the MS and the Hayashi Limit in the HR-diagram on the HK-time scale. In the course of its contraction, the star's core (still consisting of ideal gas) becomes hotter – so much so, in fact, that the hydrogen still left in the outer parts of the star may begin to burn in a shell surrounding the core, and thus provide a limited amount of nuclear energy – but not sufficient to arrest continuing shrinkage of the core and consequent expansion of its envelope towards the Hayashi Limit.

At this limit, the temperature in the core should attain approximately 100×10^6 deg.

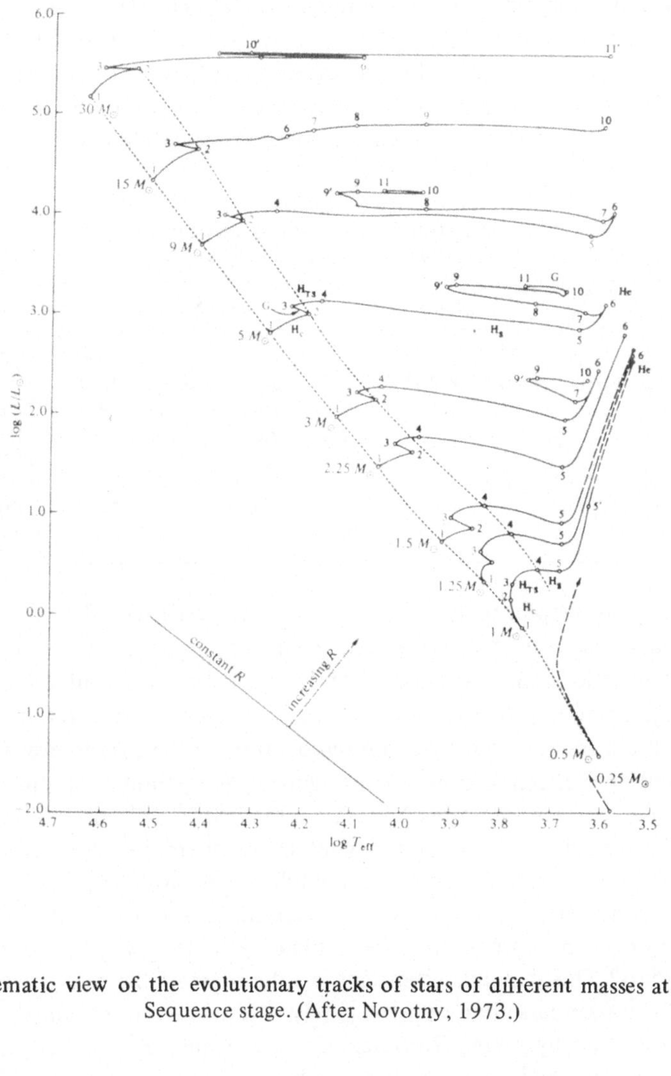

Fig. 8-2. A schematic view of the evolutionary tracks of stars of different masses at the post-Main-Sequence stage. (After Novotny, 1973.)

When this happens, the commencement of the next chapter of stellar nucleogenesis is reached, in the course of which helium – now the most abundant element contained in the core – gets converted into carbon by another set of exothermic nuclear reactions. On account of the smaller packing-fraction of the respective nuclei these are, however, no longer as efficient energy producers as hydrogen was on the Main Sequence. A conversion of helium into carbon delivers indeed much less radiant energy than that which sustained the stars in the prime of their life; but while it lasts, it will enable the stars to enjoy a brief period of 'Indian summer' before the eventual approach of decrepitude.

A more quantitative picture of these processes, which has emerged from systematic efforts to solve the equations of stellar structure for stars of different mass, is summarized in the accompanying Table VIII-1, based on the computations by Iben (1965, 1966, 1967). Its individual columns contain the times t (in units of 10^7 yr) and radii R (in solar units of 6.96×10^5 km) which stars of the masses 3, 5 and 9 ⊙ will attain in the course of evolution outlined in the preceding paragraphs. Although these data may no longer represent the latest state of the subject, they remain the most comprehensive ones we possess up to this time; and should still remain closely representative of the actual situation.

TABLE VIII-1
Theoretical Evolution of Massive Stars
(after Iben, 1965 and 1966)

Evolutionary phase	$m = 3$ ⊙		$m = 5$ ⊙		$m = 9$ ⊙	
	t	R	t	R	t	R
Contraction to Main Sequence	0.25×10^7 yr	1.8 ⊙	0.058×10^7 yr	2.40 ⊙	0.015×10^7 yr	3.4 ⊙
H-burning (core)	22.12	2.9	6.55	4.3	2.11	6.7
Gravit. contraction	1.04	2.8	0.22	4.2	0.061	6.5
H-burning (envelope)	1.03	15	0.14	6.3	0.009	13
Envelope expansion	0.45		0.075		0.015	
Red giant stage	0.42	33	0.049	74	0.007	242
He-burning (core)	7.32	28	1.71	61	0.41	90

What is likely to happen to the stars after the last stage recorded in Table VIII-1? At the conversion of carbon (and subsequent elements) into heavier nuclei by successive absorption of α-particles capable of penetrating their respective Coulomb barriers, periods of short-lived gravitational contraction are bound to recur; and – as long as compression caused by such contraction can raise the temperature in the core to ever increasing levels – will heat up its central regions to the ignition temperature of the next perishable nuclear fuel. Thus at temperatures between $600–700 \times 10^6$ deg, carbon itself should largely be transformed into neon and magnesium by thermonuclear processes (involving α-captures) in thermodynamic equilibrium; while temperatures in excess of 10^9 deg are necessary for a synthesis of the elements between silicon and iron.

Can, however, such temperatures be actually attained by gravitational contraction in stellar interiors? This depends essentially on the mass of the respective configuration. The stars with masses less than approximately 0.08 ⊙ can never ignite by gravitational contraction even their hydrogen. In order to ignite helium at 100×10^6 deg, a mass of at least 1.5 ⊙ is needed; and elements heavier than neon can be produced by equilibrium thermonuclear processes only in very massive stars ($m > 10$ ⊙).

For stars of moderate masses (1–2 ⊙), situated (after the initial contraction) on the lower half of the Main Sequence, the evolution proceeds at first along similar lines as for more massive objects – only on a very different time-scale. If the initial HK-contraction to the MS takes only 10^5 yr for a star of mass of 10 ⊙ (and even less for more massive stars); for one of the mass of our Sun (1 ⊙) it takes already some 3×10^7 yr; and becomes comparable with (or longer than) the age of the Galaxy (10^{10} yr) if $m = 0.1$ ⊙ or less.

An even more important difference between the evolution of more massive and less massive stars becomes apparent when we come to consider the time-scale of their nuclear evolution. If the hydrogen burning can maintain a star on the Main Sequence for some 20×10^6 yr if its mass is equal to 10 ⊙ (and less than a million years for stars with masses between 25–30 ⊙), for a star of one solar mass its H-nuclear time-scale becomes already of the order of 10^{10} yr (our Sun, of age 4.6×10^9 yr, has so far exhausted but less than one-quarter of its initial hydrogen supply). Such stars in our Galaxy – including those of the oldest Population II – will not have had yet a chance to evolve away from the Main Sequence (as is borne out indeed by the HR-diagrams of the globular clusters); and those with masses less than 0.1 ⊙ are unlikely ever to reach it.

As we shall see in subsequent sections of this chapter, these facts bear vitally on an interpretation of phenomena observed in close binary systems in terms of their evolution. Before, however, we approach this subject, we should point out the existence of further essential differences between the evolution of massive and less massive stars. First, a glance at Figure 8-2 makes it obvious that a star evolving from the lower part of the Main Sequence does not have much elbow-room to move in the HR-diagram to the right off the MS before it reaches the Hayashi line; and, therefore, a continuing evolution can send it only *upwards* along this line. Such a course will increase both its radius and luminosity, and decrease its surface temperature; thus endowing the star with external characteristics commonly associated with *red giants.* And while this goes on, the density of the contracting core in the interior becomes so high as to render the electron component of stellar plasma *degenerate.*

This incipient degeneracy entails, in turn, two important consequences. First, it will prevent the contraction to bring about a rapid rise in temperature (as it would do in ideal gas),and thus delays the onset of eventual helium burning. Secondly, the pressure of degenerate gas will tend to become independent of the temperature; and, moreover, it will increase to such an extent as to arrest gravitational contraction and stabilize the entire stellar structure without the intervention of a new nuclear source of energy. The energy which keeps the star shining is then produced in a shell surrounding the core, where the temperature is just high enough to maintain the hydrogen burning; and as hydrogen becomes gradually exhausted, the radius of the burning shell gradually increases and encloses an essentially helium core of increasing dimensions.

When the mass of this core has approached about a half of the mass of the Sun, its helium does not commence to smoulder or burn, but detonates almost instantaneously (in a few seconds) to convert a large part of its material into carbon. The physical cause of this 'helium flash' is a very high thermal conductivity of the degenerate electron gas. While in an ideal gas constituting cores of more massive stars, a less efficient energy transport will maintain a finite gradient of temperature throughout the configuration, a core consisting of degenerate gas is virtually isothermal; and when conditions ripe for the combustion of helium have been obtained, its conversion occurs almost instantaneously throughout the interior – liberating in this process an amount of radiation comparable with that sent out by an entire galaxy in the same number of seconds.

Intriguing is the fact that this large amount of radiation does not penetrate readily outwards; so that the effects of such a flash are not directly observable: its onset will merely have increased the internal energy store inside the star which seeps out slowly through layers surrounding the helium core. Computations show that, immediately after the helium flash, the evolutionary track of the star in the HR-diagram actually doubles back from the helium-flash locus as from a cusp; and the observed luminosity decreases for a time before eventual recovery.

The physical reason of this phenomenon is the fact that the onset of the helium flash has lessened the degeneracy of the electron gas in the core – a fact which makes it possible for the core to expand and resume a conversion of the remaining helium into carbon, as in more massive stars. In the course of this process the evolutionary track of the star in the HR-diagram may oscillate horizontally for some time, before it eventually returns to the Hayashi line.

After the completion of the helium burning, the bulk of the mass of the core consists of carbon (and oxygen) – surrounded by a shell in which the helium outside the core still continues to burn; and closer to the surface a hydrogen-burning shell may still be operative. Both shells slowly expand, and move the zones of nuclear fire increasingly closer to the surface. Inside the core the nuclear burning is, however, over; which means that the mass of the core must resume gravitational contraction; and this will cause an increasing degeneracy of the bulk of its mass.

When the mass of such a degenerate core attains 0.6–0.9 ⊙, its properties differ but little from those of a *white dwarf* – except that its structure, thus 'prefabricated' in the interior, is still hidden from outside view by a thin (but opaque) layer consisting of hydrogen and helium. The configuration as a whole continues to climb upwards along the Hayashi line in the HR-diagram; and to an external observer would appear as a large cool star of advanced spectral type and relatively high luminosity ('red giant'). These properties imply an increase of radiation pressure and diminution of gravity in outer layers; and a growing inability of gravitation to keep hold on the star's mass will be further impaired by a new source of energy now appearing in the outer layers: namely, a recombination of plasma back to neutral gas (mainly hydrogen).

When this energy becomes greater than the gravitational binding force, the respective layers of the star become unstable and will be ejected into space. We know several classes of stellar objects in the sky which may correspond to this evolutionary stage, and which differ in their external manifestations mainly in the time-scale on which these phenomena occur. If the velocities of ejection are relatively low (of the order of 10 km s^{-1}), the respective configurations may be known to us as planetary nebulae (whose central stars resemble indeed strikingly white dwarfs). For ejection velocities of the order of $10^2 - 10^3$ km^{-1}, they may appear as variables of the U Gem- or SS Cyg-type, whose more extreme types are represented by the recurrent Novae. Moreover, if the processes of mass loss simulate periodic rather than relaxation oscillations, the unstable layers of the star may oscillate semi-periodically around a state of quasi-neutral equilibrium – such as may be identified with red variable stars of the Mira Ceti- or RV Tau-type.

And this completes, in brief, a schematic outline of the evolution of single stars, in so far as can be reconstructed within the framework of our present knowledge of physical processes governing the structure of self-gravitating gas spheres of astronomical dimensions. Several facts should, however, be kept in mind in contemplating this picture. First, in all underlying computations no attention was paid to the third independent variable characteristic of stellar structure: namely, its angular momentum corresponding to axial rotation. Although fast rotation can exert effects which observationally are very significant, it has so far been taken into account – if at all – only as a perturbation of the spherical form; and even that only for Main Sequence stars. A more complete theory of stellar structure will obviously have to include a study of the (non-rigid) rotation of the star in the corse of its evolution; but such a theory is still largely lacking.

Secondly, throughout the developments outlined in this section we have ignored a possibility of the *mixing* of the material in stellar interiors. In internal regions which are in convective equilibrium, this material should generally be regarded as well-mixed by convective currents and its chemical composition should, therefore, be homogeneous. This should, in particular, be true of newly formed stars contracting towards the Main Sequence (which may be convective throughout their interiors); or of the convective cores of the stars of moderate to large masses. But how much material mixing can there be across the interface between convective cores and radiative envelopes? If such a mixing were to occur, its effects could profoundly influence the evolutionary picture outlined earlier in this section. But this constitutes again a subject on which we can but speculate so far, because of a very incomplete state of its investigation.

Third, throughout all previous discussion it has been assumed that the entire history of stellar evolution can be reconstructed from a series of consecutive *equilibrium models* of the stars, constructed by a solution of equations in which the velocity components of mass motions have been ignored. Such a procedure should, in general, be legitimate during those periods of stellar life where energy expenditures are being defrayed from nuclear sources (in particular, at the Main Sequence stage). It becomes, however, more questionable at the time of Helmholtz-Kelvin contraction (especially during its more rapid stages); and quite untenable in certain 'instability strips' of the HR-diagram – between the Main Sequence and the Hayashi Limit – where at least the outer layers of the star cannot maintain hydrostatic equilibrium continuously if disturbed from it by arbitrary perturbation.

Such perturbations are, in turn, known to set off periodic light variations invoked by oscillations in the temperature and absorption coefficient in the outer layers of such stars, which modulate the outflow of radiant energy from the interior and thus act as a 'light rheostat' gating the energy outflow – even though they entail motions of very little mass. Such phenomena are at the basis of the light fluctuations of Cepheid variables which are strictly periodic; or of long-period variables of the Mira Ceti-type of longer periods and larger light changes.

While the light fluctuations of Cepheid variables are highly regular in periods ranging from a few hours to a few weeks (for very massive stars), and the amplitudes of their light changes are generally moderate (corresponding to a factor 2–3 in total intensity), the variables of Mira Ceti-type are only semi-periodic (with average periods from several months to a few years); and the amplitudes of their visible light changes may attain 5–10 mag (corresponding to factors of 100–10 000 in intensity). These large fluctuations in visible light are, to be sure, due mainly ot the fact that the spectral distribution of the light of stars as cool as the long-periodic variables (of types *M, R* or *S*) is apt to drift with varying surface temperature in and out of the region to which our eyes are sensitive (or our terrestrial atmosphere transparent). The total luminosity of such stars varies very much less than the extent of its visible range would indicate – no more, in general, than that of the Cepheid variables – and their variability constitutes (like for the Cepheids) a largely superficial phenomenon.

All types of variability considered so far in this section apply to, and can be performed by, stellar configurations of spherical form. But in any attempt to use a theory based on the assumption of spherical symmetry to predict the evolution of the components of close binary systems we must keep in mind that – by definition – the latter cannot remain spherical, but must depart from spherical form to an extent depending on their fractional dimensions and mass-ratio. The ways in which they will do so have already been investigated in Chapter II; and the aim of this chapter will be to consider the extent to which a mutual symbiosis of two stars in the form of a close binary system can make the evolution of its components *differ* from that which each star would undergo on its own (without regard to its mate), as outlined in this section.

The reasons which can invoke such differences are indeed several. A departure of the components from spherical form, caused by the proximity of their companion, may be treated as a perturbation of spherical structure if this proximity is not too close; but *not* if the dimensions of the components are of the same order of magnitude as their separation. In such cases it would be necessary to solve the equations of stellar structure in the actually prevalent force field; and this has so far never been done.

Secondly, the occurrence of dynamical tides in the system (cf. Chapter III) will provide both components with an additional source of internal energy through viscous friction, causing dissipation of kinetic energy into heat (cf. sections III-4 and IV-3). Such a source is likely to be negligible in deep interiors of the component stars (where tides are low), but may be of importance in their outer layers, which are more directly accessible to observation. Such tides may also influence (or alter) the type of equilibrium of these layers – from radiative to convective or vice versa. It appears unlikely (cf. Seguin, 1976) that dynamical tides will either produce, or inhibit, convection to a linear approximation as long as the tidal distortion can be treated as a small perturbation whose squares can be ignored). However, for close binaries whose tidal distortion grows large the situation remains unexplored.

Lastly, we stressed already earlier in this section that, in advanced (post-Main Sequence) stages of stellar evolution, most stars are likely to *lose mass;* and in close binary systems this loss may be influenced by the gravitational field of the companion to such an extent that a part of the mass, which would be lost if the star were single, may be captured by its mate and transferred on to its surface – or even exchanged at a later evolutionry stage. If and when such a process becomes operative, it can affect profoundly subsequent evolution of the system, in a manner quite distinct from the evolution of single stars; and a major part of subsequent sections of this chapter will be devoted to a critical discussion of such possibilities, and to their confrontation with the observed data.

VIII-2. Classification of Close Binary Systems

In the opening section of this chapter we have described the reasons for a great diversity in physical properties of the stars encountered in the sky; and in the first chapter of this book we described the diversity of cases in which such stars team up to form binary systems. We mentioned that, up to the present, more than four thousand close binaries have been identified as such by virute of their characteristic variations of light and (or) radial velocity. The material available on these systems – constituting no doubt only a minute fraction of the total galactic supply – has served as a basis for numerous statistical studies of close binary systems and, in particular, for their appropriate *classification.*

As immediate objective of such a classification may be merely to group together systems with certain common apparent characteristics. In order for such a classification to possess a physical meaning, however, the characteristics on which it is based should not be ambiguous or concerned with mere superficial qualities, but with properties sufficiently fundamental to render each group a definite physical association. In point of fact, while the immediate aim of a classification may be a purely pragmatic, and one of convenience for practical purposes, its ultimate aim should be to throw light on the origin and generic relation of different groups encountered among close binary systems, and to identify their proper place within the general framework of stellar evolution.

Let us begin to examine, from this point of view, the systems of classification proposed so far for close binary systems. In doing so, we shall be led almost from the outset to confine our attention to eclipsing variables. The reason is not only the fact that binaries which reveal their nature by characteristic variation of light greatly outnumber those for which radial velocity alone is known to vary; but also because an inspection (or a more appropriate analysis) fo their observed light curves can reveal many more facts concerning each particular system than could be deciphered from the radial velocity curves. However, it goes without saying that both eclipsing and mere spectroscopic binaries constitute the same physical group, and differ in their observable manifestations only by accidental orientation of their orbits in space.

It has been customary, in the past, to divide eclipsing variables into three principal groups – with Algol, β Lyrae, and W Ursae Maioris as their respective prototypes. It is, however, easy to demontrate that such a classification – still current in some catalogues today (e.g., Kukarkin *et al.*, 1969) – lacks any physical basis and its use can be explained only on historical grounds. Its principal distinguishing characteristics – namely, the presence or absence of photometric ellipticity effect between minima – is the natural resultant of a combination of fractional dimensions, masses and relative luminosities of the two components; and can be predicted from them in terms of a theory expounded in Chapter II. Statistical studies disclose, moreover, no kind of break in the frequency-distribution of fractional radii, or mass – and luminosity-ratios among eclipsing variables for which these properties are known. As a result, systems of the β Lyrae-type merge gradually with increasing separation of the components into the Algol-type systems; and any division between them depends merely on the precision with which the amount of photometric ellipticity can be ascertained from the observations.

This fact is, moreover, bound to render any such classification not only inexact, but ambiguous as well; for the dependence of the photometric ellipticity of eclipsing variables (causing curvature of the light curve between minima) on the relative luminosities of their components – which are, generally, of different spectral types – makes this ellipticity dependent also on the effective wavelength of observation. In order to illustrate this point consider, for instance, the well-known eclipsing systems of Algol and *u* Herculis. The former – a prototype of its group – exhibits a light curve which is virtually free from any ellipticity in visible light. On the other hand, the light curve of *u* Her is conspicuously convex between minima – thus rendering the system one of β Lyrae-type according to conventional classification.

Geometrically, however, both these systems are very similar: in both the primary (more luminous) component of earlier spectral type is several times more massive than its mate, but slightly smaller; its form departs, therefore, relatively little from a sphere. On the other hand, their secondary components – being larger and much less massive – are conspicuously distorted; so much, in fact, as effectively to fill the largest closed equipotentials capable of containing their mass. The principal difference between the two systems reduces, indeed, to a difference in the surface brightnesses of their primary and secondary components: whereas, for Algol, the ratio of surface brightnesses J_1/J_2 of its components (B8 + K0) is close to 18 in blue light, for *u* Her (B3 + B7) it proves to be only 2.5; and since the ratios of the radii k in both systems are very nearly equal (k = 0.95 for Algol, and 1.05 for *u* Her), a disparity in the values of J_1/J_2 reflects, therefore, almost wholly a corresponding disparity in luminosities of the two components. The secondaries in both systems are highly distorted; but on account of its small fractional luminosity, the secondary (K0) component of Algol can influence the visible light of this system very much less than does the secondary (B7) component of *u* Her; and hence a conspicuous difference in the shape of their light curves.

A disparity in surface brightness of two stars of dissimilar temperature and spectral type is, however, known to diminish with increasing wavelength of light in which observations are made. A recourse to Planck's law discloses that the value of $J_1/J_2 = 18$ in the blue (λ 4500 Å) should decrease to 2.5 at about $\lambda = 2.7\mu$ – a wavelength still accessible to modern infrared detectors. Therefore, an infrared light curve of Algol observed through the atmospheric 'window' between 2.0 and 2.5 μ should markedly resemble that of *u* Her, and thus earn to Algol a claim for admission to the β Lyrae group. Moroever, a generalization of this point should make it clear that, whereas eclipsing systems consisting of components of similar spectral types should exhibit similar changes when observed in light of any wavelength, those consisting of stars of dissimilar spectra may exhibit very different light curves at different effective wavelengths if their components are also dissimilar in form. These facts show conclusively that a simple division of eclipsing systems into Algol- and β Lyrae-type stars is not only inaccurate for lack of a suitable physical division between the two types, but also ambiguous; for the same system may pose as a member of one group or another if observed in different spectral domains.

A. CHARACTERISTIC PARAMETERS

In order to avoid the ambiguities inherent in the use of a mere form of the light curves for the purpose of classification, let us begin by asking ourselves the following question: *how many parameters are necessary and sufficient for a complete specification of the geometrical form of both components in close binary systems?* This form should, in principle, be specified by the nature of the forces (centrifugal, or tidal) acting on the surface; and provided that the free periods of non-radial oscillations (cf. sec. III-2) of both components are short in comparison with that of their orbit, the distortion of the components should be governed by the equilibrium theory of tides (Chapter II). The level surfaces of constant density then coincide with those of constant potential; and the boundary of zero density becomes a particular case of surfaces over which the potential arising from all forces remains constant.

An exact theory of the form of such surfaces for stars of arbitrary internal structure has not so far been developed. If, however, the density concentration in the components of binary systems is sufficiently high for their attraction to be approximable by that of a pair of central mass-points, the total potential of forces acting on any other point has already been investigated in Chapter VI under the name of *Roche equipotentials,* as defined by Equation (2.2) of that chapter. If, moreover, in that equation we identify the angular velocity ω of rotation about an axis perpendicular to the orbital plane with the Keplerian angular velocity as given by Equation (1.4) of Chapter VI, the outcome assumes the more explicit form

$$(1+q)\,C = 2r^{-1} + 2q(r'^{-1} - \lambda r) + \\ + (1+q)r^2(1-\nu^2) + q^2(1+q)^{-1}, \tag{2.1}$$

where the radii-vectors r, r' as well as the direction cosines λ, ν continue to be given by Equations (1.3) and (1.5) of Chapter VI,

$$q = \frac{m_{2,1}}{m_{1,2}} \tag{2.2}$$

denotes the mass-ratio of the system; and

$$C_{1,2} = \frac{2A\,\Psi_{1,2}}{G(m_1 + m_2)} \tag{2.3}$$

the normalized value of the surface potential (the 'Roche constant'). Provided only that the density concentrations in the interiors of the components of our binary system are sufficiently high (as there is every physical reason to expect, except for white dwarfs or neutron stars), the Roche equipotentials (2.1) should approximate the actual shape of such bodies to a very high degree of accuracy *regardless of their proximity.*

The form of the Roche equipotentials (2.1), generated by setting C = constant, will evidently depend on the adopted values of C and q. We have already seen in section VI-2 that if C is large, the corresponding equipotentials consist of *two* separate ovals enclosing each one of the two mass-centres, and differ but little from spheres – the less so, the greater the value of C. With diminishing C, these ovals increase in size and become elongated in the direction of the centre of gravity of the binary system – until, for a certain critical value of C_0 characteristic of each mass-ratio, both ovals come into contact at the Lagrangian point L_1 on the axis joining the centres of the two components, to form a dumb-bell like configuration (see Figure 6.1), which we called the *Roche limit*.

For still smaller values of $C < C_0$, the connecting part of the dumb-bell would open up, and a single equipotential surface enclose both bodies – thus depriving us of the possibility of regarding Equation (2.1) as the representation of a binary system. However, for each value of $C \geqslant C_0$ and q, Equation (2.1) defines two detached equipotentials which approximate the forms of centrally-condensed components of a close binary system to a remarkable degree. It is, moreover, clear that, given the mass-ratio, the fractional size and form of each component will be uniquely specified by a *single* value of C. Therefore, the question raised at the outset turns out to admit of the following answer: the minimum number of parameters sufficient for a complete geometrical description of both components of a close binary system is *three*: namely, the values of C_1, C_2 and q. Each one of the values of $C_{1,2}$ introduced in (2.1) defines, to be sure, a pair of equipotentials for any given q, of which only the one enveloping the centre of mass of the respective component is relevant. A trio of the values of $C_{1,2} \geqslant C_0$ and q

can, however, be made to specify the complete geometry of a close binary very much more economically as well as accurately than could be accomplished – more artificially – by any number of particular semi-axes; moreover, the quantities C_1, C_2 and q possess the additional advantage of a direct and simple physical meaning.

A determination of the mass-ratio q can be accomplished spectroscopically from the measured amplitudes of the radial-velocity curves of both components – or, in the absence of spectroscopic observations, by a number of indirect methods. On the other hand, the values of $C_{1,2}$ are not directly observable, nor can they be readily deduced from an analysis of the light curves of eclipsing binary systems. A knowledge of the coordinates of any point $P(x, y, z)$ on the surface of the respective Roche equipotential would, however, be sufficient for a unique specification of C. In actual practice, it is expedient to utilize a pair of points at the intersection of the $\pm y$-axis with the respective equipotential; for the y-coordinate of such points is (essentially) identical with the fractional radii $r_{1,2}$ of the two components, obtainable from an analysis of the light curves of the respective eclipsing systems by well-known methods (cf., e.g., Kopal, 1959; Chapter VI).

Since, then, along the y-axis, $y = r_{1,2}$ while $x = 0$ or 1 (again within an insensible error) and $z = 0$, it follows from (2.1) that

$$C_i = \frac{2(1 + \mu_i)}{r_i} + \frac{2\mu_i}{\sqrt{1 + r_i^2}} + r_i^2 + \mu_i^2, \tag{2.4}$$

where

$$\mu_i = \frac{m_{3-i}}{m_1 + m_2} \tag{2.5}$$

and $i = 1,2$,. These equations can, therefore, be used to deterine the values of $C_{1,2}$ for all eclipsing systems whose geometrical elements $r_{1,2}$ and mass-ratios are known.

Suppose that we do so, and compare the values of $C_{1,2}$ so obtained with the Roche limit values C_0 appropriate for the respective mass-ratio. The outcome disclosed (cf. Kopal, 1955) that an overwhelming majority of known eclipsing systems was found to belong to one of the following three groups.

I. *Detached Systems*

The volumes of both components in systems of this group are significantly *smaller* than their Roche limits (i.e., $C_1 > C_0$ and $C_2 > C_0$). Prototypes: β Aurigae or U Ophiuchi.

II. *Semi-Detached Systems*

The primary (more massive) components are distinctly smaller than their Roche limits, but their secondaries appear to fill exactly the largest closed equipotentials capable of containing their whole mass (i.e., $C_1 > C_0$ but $C_2 = C_0$ within the limits of observational errors). Prototype: Algol.

III. *Contact Systems*

Both components appear to fill completely the respective lobes of their Roche limits, and may actually overflow them (i.e., $C_1 = C_2 \leqslant C_0$). Prototype: W Ursae Maioris.

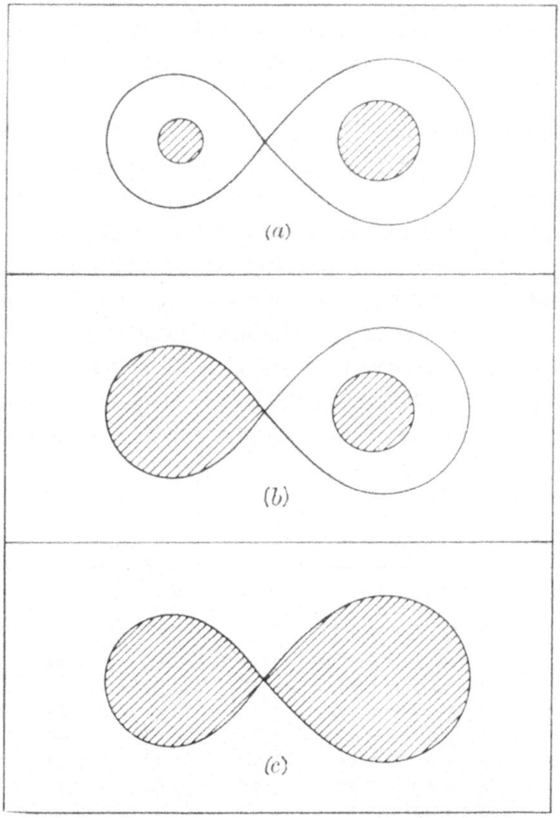

Fig. 8-3. A schematic view of the three principal types of close binary systems: (a) detached systems; (b) semi-detached systems; (c) contact binaries.

The geometry of these three types is illustrated on the accompanying Figure 8-3, drawn to scale for a mass-ratio of $m_2/m_1 = 0.6$.

The physical significance of this classification, and probable reasons why most known close binary systems conform to these types, will be discussed in the next section of this chapter.

VIII-3. Nuclear Evolution of Close Binary Systems

In the first section of this chapter we have outlined our present views on the evolution – both gravitational and nuclear – of single stars in the sky; and in section VIII-2 we introduced the principal types of binary stars identified among known eclipsing variables. The aim of the present section will be to subject this observational evidence broadly outlined in the preceding section to a critical discussion in terms of the evolutionary story of section VIII-1. Our aim will be, in particular, to identify the evolutionary stage of different types of close binary systems in the Hertzsprung-Russell diagram – stages not uniquely specified by their present position in the HR-diagram alone; for we know that their evolutionary tracks in later stages of stellar evolution can go repeatedly into reverse or cross; and some parts of these tracks remain still untrodden even by the theoreticians.

In doing so we shall commence with the simplest case presented to us by Nature – that obtaining when the fractional dimensions of both components are well inside their respective maximum Roche lobes ('detached systems'). We shall show that such a geometry is characteristic primarily of stellar pairs in their young age or early maturity – on the Main Sequence or before. Semi-detached systems with one 'contact' component we shall already regard as at least partly evolved past that stage and an account of them in section VIII-3B will constitute one of the most exciting (and perplexing) parts of our story. On the other hand, the evolutionary stage of 'contact' systems of the W UMa-type (discussed in section VIII-3C) remain still largely enigmatic – a fact embarrassing all the more because of the abundance of these stars in space – at least in the neighbourhood of our Sun.

TABLE VIII-2
Absolute properties of selected detached eclipsing systems

Star	Period	Spectra	m_1	m_2	R_1	R_2	M_1	M_2
V 805 Aql	$2^{d}.408$	A2 + A6	1.85 ⊙	1.50 ⊙	2.16 ⊙	1.84 ⊙	$0^{m}.7$	$1^{m}.6$
σ Aql	1.950	B8 + B9	6.8	5.4	4.2	3.3	−1.9	−0.9
TT Aur	1.333	B3 + B7	6.7	5.3	3.3	3.2	−3.1	−1.8
WW Aur	2.525	A7 + F0	1.92	1.90	1.92	1.90	1.7	2.0
AR Aur	4.135	B9 + A0	2.55	2.30	1.82	1.82	0.3	0.6
β Aur	3.960	A2 + A2	2.33	2.25	2.48	2.27	−0.1	0.2
SZ Cam	2.698	O9.5 + B2	21	6	10.1	4.5	−6.1	−3.7
AH Cep	1.775	B0 + B0.5	16.5	14.2	6.06	5.50	−5.6	−5.2
Y Cyg	2.996	O9.5 + O9.5	17.4	17.2	5.9	5.9	−6.0	−6.0
V 477 Cyg	2.347	A3 + F5	2.4	1.6	1.5	1.2	1.7	3.7
YY Gem	0.814	M1 + M1	0.64	0.64	0.62	0.62	7.7	7.7
RX Her	1.779	A0 + A1	2.1	1.9	2.1	1.8	0.3	0.7
TX Her	2.060	A5 + F1	2.1	1.8	1.65	1.6	1.7	2.5
UV Leo	0.600	G0 + G1	1.36	1.25	1.21	1.20	4.0	4.1
U Oph	1.677	B5 + B6	5.30	4.65	3.4	3.1	−2.4	−1.9
V 451 Oph	2.197	A0 + A1.5	2.3	1.9	2.4	1.9	0.0	0.8
AG Per	2.029	B5 + B7	5.1	4.5	2.98	2.74	−2.3	−1.7

In contrast with the problems presented by a system possessing contact components, those consisting of detached stars are almost classical in their simplicity; although they too do not lack many points of interest which we shall now proceed to discuss.

A. DETACHED SYSTEMS

As the name suggests, a close binary system will be regarded as belonging to this group if the free surfaces of *both* components are *detached* from (i.e., smaller than) their respective Roche limits. A list of typical examples of systems which spectroscopically-determined mass-ratios has been compiled in the following Table VIII-2. The headings of most of its individual columns are self-explanatory (the symbols $M_{1,2}$ stand for the absolute bolometric magnitude of the respective components; while their masses $m_{1,2}$ and (mean) radii $R_{1,2}$ have been expressed in solar units.

The material collected in this table is representative of the different properties of this class of stars, though by no means exhaustive. All data have been taken from the *Catalogue of the Elements of Eclipsing Binary Systems* by Kopal and Shapley (1956), to which the reader is referred for the sources of the underlying observational data. Although this source is now 20 years old (and a large amount of new or improved data have become available since that time), we shall adhere to the use of this reference for two reasons: first, because no homogeneous catalogue of similar scope has appeared since that time; and, secondly, because an incorporation of new data would not only disturb its homogeneity, but also expose us to a possible bias in selection. As it is, Table VIII-2 contains the data on the absolute properties of the individual components of 17 typical systems of this class, to which several more could be added at this time.

A discussion of these data discloses that the components of binary systems of fractional dimensions well inside their Roche limits cluster closely around the Main Sequence (see Figure 8-4), and conform closely to an empirical mass-luminosity and mass-radius relations (Figures 8-5 and 8-6) appropriate for single stars.

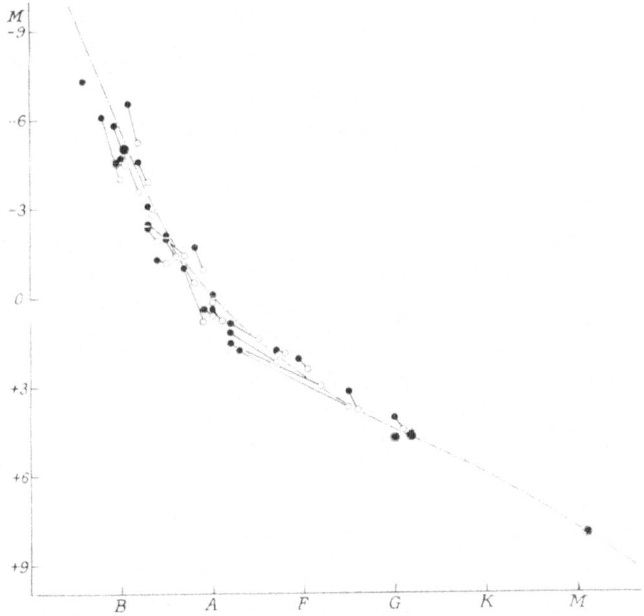

Fig. 8-4. Hertzsprung-Russell diagram of detached binary systems. Full circles denote the primary (more massive) components; open circles, the secondary components. Broken line represents the standard Main Sequence for single stars.

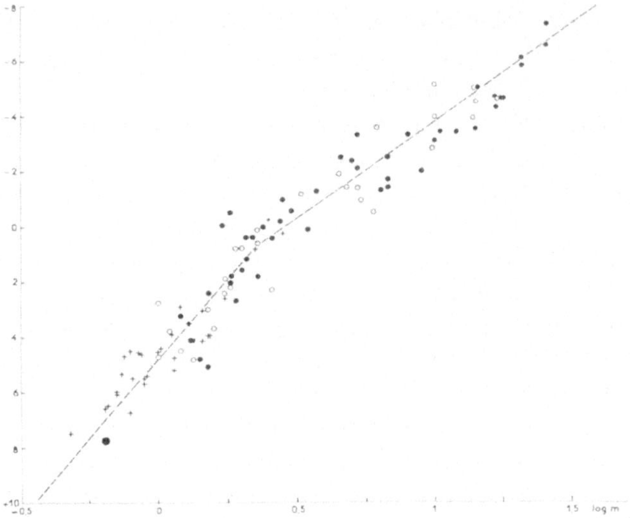

Fig. 8-5. Mass-luminosity relations for detached components of binary systems. Full circles denote the detached components (primary or secondary) of close binary systems; open circles, the primary components of semi-detached systems. The crosses indicate the components of visual binaries of known masses and luminosities. Broken lines represent statistical relations as given by Equations (3.1) and (3.2).

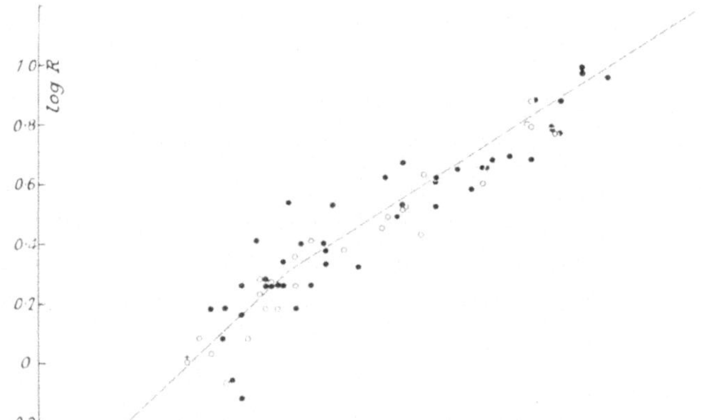

Fig. 8-6. Mass-radius relation for detached components of close binary systems. Full circles denote the positions of the primary (more massive) components; open circles, those of the secondary (less massive) components. Broken lines represent the statistical relations given by Equations (3.1) and (3.2).

Both these relations seem piecewise linear within the limits of observational errors of the underlying data; but their slopes indicate a distinct systematic change in statistical behaviour of more massive and less massive stars in the neighbourhood of $m \sim 1.5-2\odot$ – a change connected no doubt with a transition from the proton-proton reaction to the carbon-nitrogen cycle as the dominant source of nuclear energy in these stars. For stars with masses $m > 2\odot$ we have, to a sufficient approximation, an empirical relation

$$\log m = 0.45 - 0.143\,M = 1.57 \log R - 0.15; \tag{3.1}$$

while for less massive stars ($m \leqslant 2\odot$)

$$\log m = 0.42 - 0.086\,M = 1.02 \log R, \tag{3.2}$$

irrespective of whether the stars in question happen to be the primary or secondary components.

As a result of these properties, the more massive components in binary systems of this type are bound to be the *larger* of the two, and of *earlier* spectral type. If, therefore, such systems happen to be eclipsing variables, their deeper minima will be due to eclipses of the *transit* type. The orbits of well-separated binaries of this type are frequently *eccentric* (as we have seen in section V-3C, the orbital eccentricity e of Y Cyg in Table VII-2 is equal to 0.14; and for V 477 Cyg, $e = 0.30$). Their *orbital periods* are generally *constant* or oscillate periodically as a result of dynamical situation obtaining in the system.

As we already know, such oscillations can be caused by apsidal motion in eccentric binary systems, or by the precession of rotating components whose equators are inclined to the orbital plane of the system (cf. sec. V-4); and others can again be produced if a binary system revolves in spacce around the common centre of gravity (due to the 'light equation' in the wide orbit; cf. sec. V-6D). But in no case known so far do we meet, among detached systems, with *irregular* period changes which are so common in systems of semi-detached type: in point of fact, their absence is a distinguishing a mark of such systems as is the detached nature of their components.

In detached systems characterized by circular orbits, *the angular velocity of axial rotation of their components is as a rule identical* (within the limits of observational errors) *with the Keplerian angular velocity of orbital revolution* (cf., e.g., Olson, 1968; or Nairai, 1972); and takes place always in the direction of orbital motion. *No case of retrograde rotation* (which could be detected spectroscopically from the sign of the 'rotational effect' before or after the moments of conjunction) *has so far been found in any system.* In *eccentric* systems, however, the components are found as a rule to rotate *faster* than the average velocity of their orbital revolution (cf., e.g., Swings, 1936). The extreme cases of such a situation are provided by AR Cas ($e = 0.25$) or α CrB ($e = 0.33$), whose primary component has been found (spectroscopically) to rotate approximately 10 times faster than the mean (Keplerian) velocity of orbital revolution. However, the relation between the velocity of axial rotation and orbital eccentricity still remains only statistical; though the impulse received through tides (which are highest at the time of periastron passage) may offer at least a part of the reason behind it.

While the velocity of axial rotation of the stars (be these single or double) can be estimated from the Doppler broadening of the observed lines in their spectra (and determined more precisely if the angle of equatorial inclination is known from a photometric evidence); or from the magnitude of the spectroscopic 'rotational effect' deforming the radial-velocity curves before and after conjunctions (cf., e.g., Kopal, 1969; sec. V. 2); the actual position of the rotational axis in space can be indicated by the asymmetry of the rotational effect, but not determined uniquely from it. However, we pointed out already (cf. sec. V-2) that if the equatorial plane of either component is not coplanar with that of the orbit, its axis of rotation cannot remain fixed in space, but must precess – under the tidal influence of its mate – in a period depending on the mass-ratio, fractional size, and internal structure of the respective star.

Moreover, we have also shown (sec. V-4) that this precessional motion may become observable through a difference to which it gives rise betwen the 'sidereal' and 'synodic' period of revolution – a difference which can be resolved into periodic terms affecting the observed times of the minima of eclipsing variables. The *frequencies* of such terms depend on the rate of precession and nutation of the rotational axes of the two stars, while their *amplitudes* are proportional to the inclination of the equatorial planes of the components to that of their orbit. The rate of the precessional motion (i.e., of the regression of the nodes) should, moreover, be *higher* than that with which the apsidal line advances – its period being (for moderately close systems) of the order of years rather than decades. It should, therefore, be feasible to separate the two effects in the observed data, and identify the terms arising from precession. However, as we already mentioned in section V-4B, the observations have so far failed to bear out any effects of this nature in the available data. From this we can only conclude that the amplitudes of these terms are too small for observational detection; and this can be true only if, in such systems, *the equatorial planes of both components are aligned with that of their orbit within a degree or less.*

In section V-2 we have established that such an alignment should indeed be the result of s secular evolution of close binary systems due to tidal friction. However, in section III-4 we also found that, as long as the stellar material consists of hydrogen-helium plasma, the viscosity of such a plasma is insufficient to bring about so close an alignment in the time vouchsafed on the Main Sequence to stars of moderate or large masses by their diminishing hydrogen contents. And if so, we are inadvertently led to a conclusion that *the observed parallelism between equatorial and orbital planes must have already been established in the pre-Main Sequence past of the respective systems;* when the contracting configurations were convective throughout their interiors. Convective motions on such a scale would be almost necessarily turbulent; and turbulent viscosity can be so much larger than that of quiescent gas or plasma (being equal roughly to the product of gas (plasma) viscosity times the Reynolds number of the respective turbulent flow; and this number may be anywhere between $10^4 – 10^6$ or larger) that dissipative phenomena produced by it would have been very much more effective to attain their aims even in contraction times 100 times shorter than those which the star can subsequently spend on the Main Sequence (see Table VIII-1).

As an example of eclipsing systems which may be at present in this stage, we could quote BM Ori (cf. Hall and Garrison, 1969; Huang, 1975, or Popper and Plavec, 1976) and others. It is among such systems that we primarily should look for indications – photometric or spectroscopic – of precessional motion; since they are still too young to have given the dissipative processes a chance to accomplish their ends; and if they did not rotate about axes perpendicular to the plane of their orbit at the time of their origin, they may persist in doing so for some (but not too long a) time.

This is not the place to open up the question of the origin of close binary systems – a subject whose discussion we postpone till section VIII-5 of this chapter. Be it what it may, however, it is very probable that the components of ordinary close pairs evolved side by side in the course of their terminal contraction towards the Main Sequence, and joined the MS as a pair of stars of essentially the same age and homogeneous chemical composition (well-mixed because of widespread convection). If so, however, their subsequent evolution could embark on different evolutionary tracks only an account of a possible initial difference in mass. While on the Main Sequence, the mass of a star – be it single or double – should remain constant within better than 1% of its initial value (as the losses of mass by radiation – both electromagnetic and corpuscular – will scarcely attain that value).

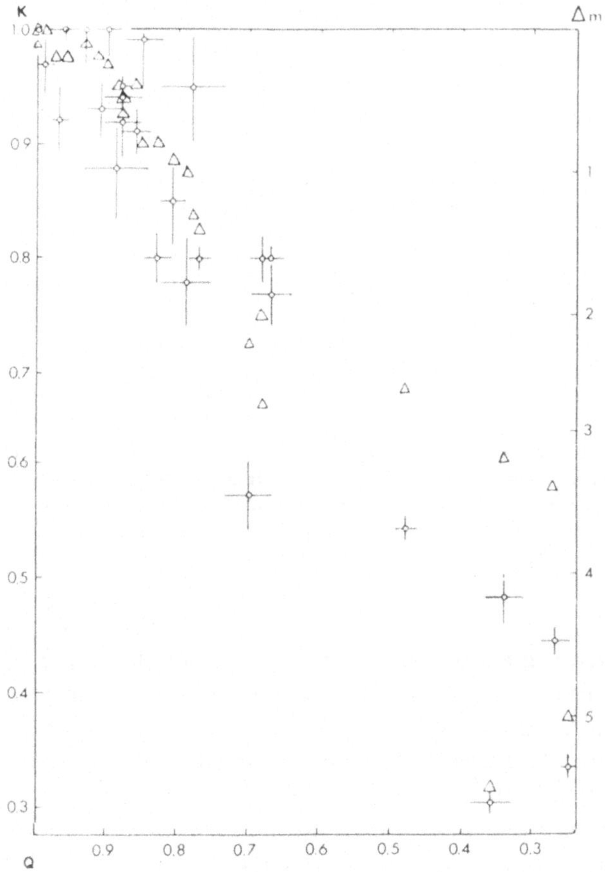

Fig. 8-7. A plot of the ratio k of radii (left) and the difference Δm in magnitudes (right) of the components of close binary systems against their mass-ratios. The ratios k of radii have been marked with open circles (with crosses indicating the uncertainty in both coordinates), while triangles denote the respective bolometric magnitude differences for the same stars. Only systems with both components on the Main Sequence have been included in this diagram.

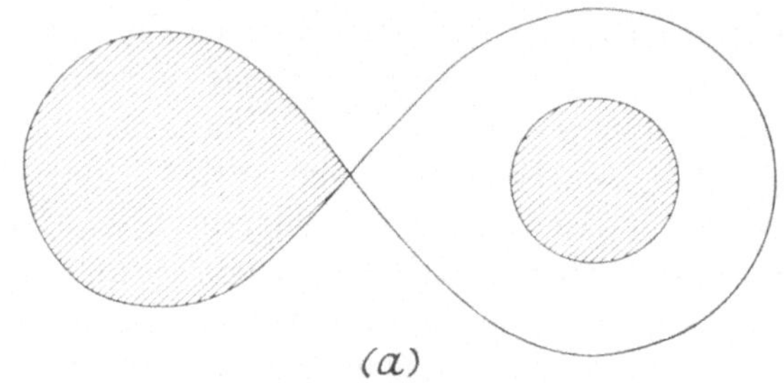

(a)

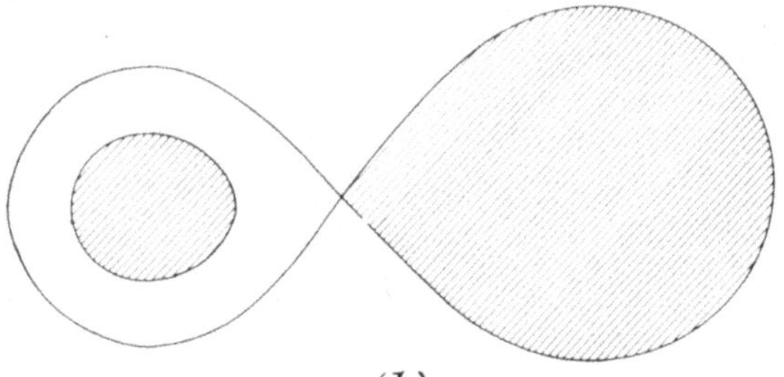

(b)

Fig. 8-8. A schematic view of two possible types of semi-detached systems. The diagrams (drawn to scale for a mass-ratio $m_2/m_1 = 0.4$) represent semi-detached systems in which (a) the secondary (less massive), or (b) the primary (more massive) component is in contact with its Roche limit.

Should, however, the masses of the components of a close binary be initially different, they should remain so at least for the rest of their Main-Sequence age. This age depends, of course, on the magnitude of the star's mass: the more massive it is, the more rapidly it should consume its hydrogen in the core (on account of a higher temperature produced by self-compression in the interior) and outdistance its less massive mate in its evolutionary course. For pairs of stars which are equal in mass the evolutionary tracks in the HR-diagram should, however, be identical; and stars of initially the same mass, radius and luminosity should retain their similarity also throughout the rest of their evolution.

In order to verify the extent to which this is actually the case at least on the Main Sequence, attention is invited to the accompanying Figure 8-7, on which the ratios of the ratios of the radii $R_2/R_1 \equiv k$ and of the luminosities (or, rather, differences Δm of the bolometric magnitudes) are plotted against the mass-ratios $m_2/m_1 \equiv q$ for well-observed

Main-Sequence systems taken from Table II of the Kopal-Shapley 1956 *Catalogue.* An inspection of this diagram discloses that the values of both k and Δm are strongly correlated with q. If $q < 1$, a dispersion in both parameters becomes considerable; and reflects, no doubt, the effects of differential evolution in pairs of different age. But as q tends to unity, both k and the luminosity ratio tend separately to 1 (especially the latter, as indicated by Δm going to zero) – leaving no room for doubt that *the components of the same mass remain alike also in all other characteristics* – i.e., that their evolution follows indeed an identical course.

This outcome may perhaps not sound surprising, and is indeed in accord with the expectations raised in section VIII-1. However, when we turn next to the pairs of stars of which at least one has left the Main Sequence, it will be a very different story.

B. SEMI-DETACHED SYSTEMS

Let us attempt now to follow the course of further evolution of the components of close binary systems, which unrolls when an incipient shortage of hydrogen in their cores should cause them to evolve to the right of the Main Sequence in the HR-diagram, as outlined already in section VIII-1. If both the mass and chemical composition of the components were initially the same, their whole subsequent evolution should follow an identical course not only on the Main Sequence, but also beyond it; and the two stars should remain alike in their external characteristics (radius, spectrum) as long as their masses are equal. If, however, these masses had been originally unequal, the more massive and luminous component is bound to reach the stage of incipient hydrogen shortage *ahead* of its less luminous mate; and to succumb to the symptoms of advancing age the more rapidly, the greater the disparity in masses.

These symptoms have already been described in section VIII-1, and may be recalled in a few words. When the remaining hydrogen supply in the core has diminished below (approximately) 12% by mass, the core should begin to shrink; and the potential energy released by such a shrinkage should make the outer parts of the star to expand (the 'mirror effect'). If the star were single, such a process could continue unchecked as long as the core keeps providing the necessary potential energy; and the extent to which this should be theoretically the case is illustrated by the data presented in Table VIII-1.

However, in close binaries the proximity of the companion will surround the expanding star with an invisible barrier – in the form of the Roche limit – defining the largest closed volume capable of containing the star's mass. If and when a given star has reached this limit, its further growth in size is likely to get *arrested;* and a continuing tendency to expand may bring about an actual *loss of mass.*

Which types of binary systems are likely to fall into such a predicament? The answer is provided by the data compiled (after Cester, 1967) in Table VIII-3, giving the orbital periods (in days) for systems whose primary components of mass m_1 (in $\odot$) can, in the course of their post-Main Sequence evolution, just attain the size of the respective Roche limit for different values of the mass-ratio m_2/m_1.

A construction of this table is simple. In accordance with Kepler's third law, $P_l^2 = 4\pi^2 A^3/G(m_1 + m_2)$, where the sum $m_1 + m_2$ is specified, for each entry, by the arguments of tabulation used in Table VIII-3. On the other hand, such data as are contained in Table VIII-1 will indicate the maximum absolute radii R_j which a star of mass m_j is likely to attain in the course of its post-Main Sequence evolution. If such a maximum radius is to become identical with the Roche limit of the respective component in a system characterized by the mass-ratio q, the ratio R_j/A should become (in effect) identical with the values of $y_{1,2}$ as listed in columns. (4) and (6) of Table VI-1. The absolute value of $A = R_j/y_j$, based on this identity, then specifies together with the sum $m_1 + m_2$ the value of P_l.

TABLE VIII-3

Orbital periods P_l (in days) of binary systems in which the primary component of mass m_1 (in solar units) just fills its respective Roche limit.

$\frac{m_2}{m_1}$	m_1 (in ⊙)				
	1	3	5	9	15
1.0	20.78	39.15	101.83	448.9	1231.7
0.9	20.51	38.64	100.51	443.1	1215.8
0.8	20.20	38.05	98.97	436.2	1197.1
0.7	19.82	37.34	97.13	428.2	1173.9
0.6	19.37	36.49	94.91	418.4	1148.1
0.5	18.81	35.44	92.19	406.4	1115.2
0.4	18.11	34.12	88.75	391.2	1073.6
0.3	17.19	32.39	84.25	371.4	1019.1
0.2	15.92	29.99	78.01	343.6	943.7

TABLE VIII-4

Absolute properties of semi-detached eclipsing systems

Star	Period	Spectra	m_1	m_2	R_1	R_2	M_1	M_2
	d							
U Cep	2.493	B8 + G8	2.9 ⊙	1.4 ⊙	2.4 ⊙	3.9 ⊙	$-0.^m6$	$2.^m3$
U CrB	3.452	B5 + (A5)	6.5	2.5	3.5	5.5	−2.4	−0.9
u Her	2.051	B3 + B7.5	7.9	2.8	4.5	4.3	−3.8	−2.1
β Per	2.867	B8 + K0	5.2	1.0	3.6	3.8	−1.0	2.7
V Pup	1.454	B1 + B3.5	16.6	9.8	6.0	5.3	−5.1	−3.9
U Sge	3.381	B9 + G2	6.7	2.0	4.1	5.4	−1.4	1.2
V 356 Sgr	8.896	B3 + A2	12	4.7	5.0	13	−3.9	−3.2
μ^1 Sco	1.446	B3 + B7	14.0	9.2	4.8	5.3	−4.2	−3.3
TX UMa	3.063	B8 + F2	2.8	0.85	2.16	3.79	−0.4	2.1
Z Vul	2.455	B3 + A2.5	5.4	2.3	4.7	4.7	−3.5	−1.0
RS Vul	4.478	B5 + (F9)	4.6	1.4	3.9	5.3	−2.6	0.9

Should, for given values of m_1 and m_2, the period P actually observed be longer than the limiting period P_l listed in Table VIII-3, neither component can ever reach the size of the Roche limit in the course of its evolution; and system is bound to remain of the 'detached' type for all times. This is generally true of all binaries which we call 'wide'. On the other hand, if $P < P_l$, the system is bound to become semi-detached at some stage of its evolution, even through at other stages it may be detached.

To give some specific example, consider the long-period close binary ζ Aurigae, whose primary component of mass $m_1 = 21$ ⊙ is a supergiant of spectral type cK5, attended by a

B6-star of mass $m_2 = 9.3$ ⊙. The latter remains still a Main-Sequence star; for its smaller mass makes it trail in evolution behind its more massive mate, evolving already to the right of the Main Sequence in the HR-diagram towards the Hayashi Limit. The orbital period $P = 972.2$ days is, however, shorter than the value of $P_l \cong 1500$ days, extrapolated from Table VIII-3 for the respective of m_1 and m_2. Therefore, although by its relatively small present fractional radius $r_1 = 0.225$ (cf. Kopal, 1946b) the cK5-component of ζ Aurigae fits in very well inside its Roche lobe for the mass-ratio $q = 0.45$ in spite of its large absolute size ($R_1 \sim 200$ ⊙) – which makes ζ Aur a detached system – the inequality $P_l > P$ discloses that, in the course of time, the system is bound to become semi-detached – a state in which present cK5 component will attain the size of Antares or Betelgeuse. Moreover, in view of the large masses involved, this is likely to happen in the relatively near future. A similair fate is also in store for all 'detached' systems listed in Table VIII-2. Their orbital periods P, as listed in column (2) of that table, are very much shorter than the limiting periods P_l from Table VIII-3 for masses $m_{1,2}$ as given in columns (3) and (4) of Table VIII-2; which means that while the components of all these systems are at present well inside their respective Roche limits, all are bound to become semi-detached in due course. Such a prediction is based on only one assumption: namely, that the distance A between the two components remains secularly constant (for limits within which this can be true cf. the forthcoming section VIII-4).

Semi-detached systems with one component at the Roche limit are encountered in the sky in considerable numbers – Algol (β Per) representing a typical example – and the relevant astronomical data on this and other well-known systems of this type are listed in the accompanying Table VIII-4, taken from the same source as those of Table VIII-2. An examination of these data discloses, however, that – unfortunately for our expectations – *the contact component happens to be the wrong star!*

Let us explain this somewhat embarrassing situation in more specific terms. A theory of nuclear evolution of (unmixed) stars insists – and we have no reason to doubt it – that, in the course of such an evolution of close binary systems, it is the more massive component which should be in the lead; with the secondary (less massive) star trailing behind it in its evolutionary stage. Moreover, we likewise have no reason to doubt that when a star begins to evolve away from the Main Sequence, it does so (as detailed in section VIII-1) to the right of it, along an evolutionary track which will take it to the domain of giants or subgiants.

The existence of such components in close binary systems of the Algol type has been known to us for a long time. However, it was not till twenty years ago that the present writer (Kopal, 1954, 1955) and, independently, Crawford (1955) pointed out that (in most of such systems) the fractional dimensions of subgiant components coincide with those of their Roche limits within the limits of observational errors – thus constituting configurations for which Kopal (1955) coined the term of 'semi-detached systems'. But the most striking feature of such systems has turned out to be the fact that – in every single case known to us so far – *it is the less massive component which appears to fill its Roche limit, while its more massive mate remains well interior to it.* This 'evolutionary paradox' has been with us for more than 20 years, and continues to remain one; for in spite of a large amount of work towards its clarification, its causes are still not satisfactorily understood; and the aim of the present section will be to explain why we hold this to be the case.

Theoretical reasons which lead us to believe that stars in post-Main-Sequence stage are likely to divest themselves of a large amount of their mass were outlined already in section VIII-1; and observations of binary systems – not necessarily close – bear this out in a very convincing manner. For consider the system of Sirius at our veritable doorstep in space (only 2.67 pc away) – a wide (optical) binary in which a normal Main-Sequence star of spectral class A2V and mass 2.24 ⊙ is attended by a white dwarf of mass 1.07 ⊙. We have every reason to believe that white dwarfs constitute a terminal evolutionary stage of stars of moderate or small mass (<1.4 ⊙), which can be attained after hydrogen has been completely exhausted in the interior, and the electron gas of its material has become degenerate. The Population I (or disc) stars with initial masses smaller than the Chandrasekhar limit could not have attained the white-dwarf stage in the past $5-7 \times 10^9$ yr because of the slow rhythm of their evolution; so that the present population of white dwarfs must represent the remnants of stars with initial masses greater than 2, which lost their surplus and thus were able to reach the degenerate state some time after they left the Main Sequence.

That this is what must have happened is clearly demonstrated by the existence of binary systems such as that of Sirius. It is well-nigh certain that both components of the Sirius system originated at virtually the same time, and from material of very much the same chemical composition. Therefore, if the present Sirius B has already attained the white-dwarf stage while its (presently) brighter mate still lingers on the Main Sequence, it must have progressed through its evolutionary course at a more rapid pace. The present mass of Sirius B is, however, smaller than that of its mate (1.1 ⊙ as against 2.3 ⊙); and if (as we have every reason to believe) the speed on the evolutionary race-track increases rapidly with the star's mass, the present properties of the Sirus system can be understood only if we assume that – initially and throughout most part of its past – *the mass of the present Sirius B must have been larger than that of Sirius A*; and that their original mass-ratio must have been *reversed* some time after the present white dwarf evolved away from the Main Sequence and suffered a large loss of mass.

Sirius represents, moreover, not the only system in our neighbourhood in space delivering this message; another is Procyon – another nearby visual binary only slightly more distant (3.47 pc) consisting of a Main-Sequence star of spectral type F5 and mass 1.74 ⊙, attended by a white dwarf of mass only 0.63 ⊙. Again, the less massive component corresponds to a more advanced evolutionary stage – a situation suggesting a large previous loss of mass by the present white dwarf.

Moreover – and this is essential – both Sirius as well as Procyon are systems so wide that their initially more massive components could have evolved as single stars – oblivious of the presence of their companions – even if their initial mass was of the order (say) 10 ⊙. The orbital periods of Sirius and Procyon are known to be 49.9 and 40.7 years (i.e., between 15 000–18 000 days) respectively. These place both systems away outside the range of the data asssembled in Table VIII-3 – a fact demonstrating that for no reasonable initial mass (or mass-ratio) could the initially more massive star expand at any stage of its evolution even to approach (let alone attain) its Roche limit. In other words, a *loss of mass* on a massive scale, which must have preceded the formation of the present white-dwarf components of Sirius or Procyon, *had nothing to do with the Roche limits* of these systems.

It is true that the limiting values of periods P_l listed in Table VIII-3 were calculated on the assumption that the semi-major axis A of the system remains constant in the course of evolution. In section V-5 we have seen that a transfer of an (infinitesimal) amount of mass from a more massive component to a less massive one should decrease the orbital period as well as the distance A between them; but as soon as the mass-ratio has been reversed the converse should be true. However, the present semi-major axes of the relative orbits of these binaries are 20.2 AU for Sirius, and 15.8 AU for Procyon – i.e., close to that of the heliocentric orbit of the planet Uranus for Sirius, and only somewhat less for Procyon. For separations as large as this any exchange of mass between the components is highly unlikely (the angular diameter of Sirius A as seen at present from Sirius B is only 2.4 minutes of arc!); and it is only too probable that the mass of which Sirius B once divested itself escaped into interstellar space.

In *close* binary systems the situation can, of course, be very different; and in order to reconcile the observed features of semi-detached systems with a theory that the primary cause of post-Main Sequence expansion is the incipient hydrogen shortage, Crawford (1955) put forward a hypothesis – subsequently popularized by Hoyle (1955) under the term 'dog-eats-dog' – of *mass transfer* between the two components; *assuming that the part of the material of the more massive component which*, in the course of its post-Main Sequence expansion, *overflows its Roche limit will be captured by the secondary component*, and thus augment its mass.

According to this hypothesis, it is indeed the more massive component which begins to expand first to reach the Roche limit (cf. Figure 8-8, top); and by continuing expansion transfers the overflow on to the former secondary, which then becomes the more massive (and luminous) of the two – while the erstwhile primary (and now secondary) component hovers around its (shrinking) Roche limit to appear as a subgiant (cf. Figure 8-8, bottom). Indeed, as was argued at that time by the present writer (Kopal, 1954), "The observed clustering of the secondaries around their Roche limits can only mean that these stars are *secularly expanding*; for any star can approach this limit from below, but cannot exceed it. Once the maximum distension permissible" (by the Roche limit) "has been attained, a continuous tendency to expand would be bound to bring about a *secular loss of mass* through the conical end of the critical equipotential" (op. cit., p. 685).

At the time when these views were first put forward, they lacked any theoretical foundations; and, in particular, the Crawford-Hoyle hypothesis of mass transfer was invoked as a *deus ex machina* to 'save the phenomena' – and not a necessary one at that for the purpose. In addition, the more quantitative aspects of such a proposition have proved to be truly formidable. In order to demonstrate this consider, for example, the system of Algol consisting of two components: the present primary, of spectral type B8, possesses a mass of 5.2 $\odot$; while the mass of the secondary (a subgiant of spectral type K0) is only 1.0 $\odot$; making up a total mass of $5.2 + 1.0 = 6.2$ $\odot$. An equalization of the mass-ratio would have entailed a loss to the former primary component of 2.1 $\odot$ (or 68% of the total); while a reversal of the mass-ratio would have necessitated a transfer of 81% of total mass in the course of the process; and similarly for other well-known systems.

Is such a mass loss possible in reality? Can the rest of the star survive the cataclysm and behave as the observed subgiants? And why do we not observe some systems at the stage of this mass transfer? In fact, why do we not observe at least some systems in which the primary (more massive) component is at (or in the neighbourhood of) its Roche limit (cf. Figure 8-8, top)? The probability of discovery of such configurations as eclipsing systems would be greater than in any other stage. These and other questions have rendered the hypothesis of mass-transfer in close binary systems dubious from the outset; and while some of the more hypothetical features have been rendered more plausible by subsequent theoretical work, others still remain unanswered; and in what follows our task will be to point these out in some detail.

The question on why we may not observe systems in transient phase represented on the top half of Figure 8-8 may have been partly answered by Morton (1960) and Smak (1962), who pointed out that the branch of stellar evolution which could take the system through that phase unrolls on the Kelvin time-scale – and with sufficient rapidity for few if any stars to be 'caught in the act' in such a sample of data as we possess today; the respective absolute time-scales for stars of different masses have already been quoted in Table VIII-1 based on subsequent computations by Iben (1965, 1966). This view has, in turn, provided the basis for subsequent theoretical work of the Czechoslovak (Plavec *et al.*), German (Kippenhahn *et al.*) and Polish (Paczynski *et al.*) astronomers, aiming to elaborate more quantitative aspects of such a scheme. The main results of this work have already been summarized recently by Plavec (1968, 1970) and Paczynski (1971); so that it is unnecessary to repeat them in this place. In what follows we propose, instead, to confine our attention to other intriguing aspects of the problem which have *not* been discussed in the summaries just quoted (or, for that matter, anywhere else); and to point out what remains yet to be done before a satisfactory reconstruction of the evolutionary trends in the post-Main Sequence stars can be placed on a more secure basis.

The first question arises when we turn our attention to close binaries at the post-Main Sequence stage with a mass-ratio close to one – or, rather, a lack of them. In section 3A of this chapter we stressed (cf. Figure 8-7) that detached systems with components on the Main Sequence bear out the theoretical expectations that components with equal masses should follow also identical evolutionary tracks, and continue to remain alike in their external characteristics – such as the size or spectral type. These expectations should not, however, be limited to the time which such systems spend on the Main Sequence. Their validity should extend beyond the Main-Sequence stage; and there to give rise to phenomena which we wish briefly to discuss in this place.

In evolutionary views based on the hypothesis of mass-transfer it has been customarily assumed that the initially more massive component exhausts its hydrogen first, expands rapidly to its Roche limit, and transfers overflow of mass on to the secondary component, which still has room to spare inside its own Roche lobe to receive it. What happens, however, when two components of the same mass, age and composition arrive (as they should) at the stage of incipient hydrogen shortage simultaneously? Their subsequent expansion should take them to their respective Roche limits of the same size, at the same time. But if any one of them feels impelled to transfer mass (by whichever process) on to its mate, it will find itself the recipient of a similar contribution from the opposite direction. As a result, no differential change should occur in the external characteristics of both components: their masses could intermingle, but the mass-ratio should remain unchanged.

The existence of numerous pairs of Main-Sequence stars with mass-ratios indistinguishable from unity leads us to look with anticipation for their predecessors which should already have left the Main Sequence – but we shall do so in vain! Well-known binaries WW Aur, AR Aur or β Aur can be regarded as prototypes of systems we have in mind in the intermediate part of the Main Sequence; but they extend from the O-stars (Y Cyg) down to the M-dwarfs (YY Gem); all these constitute two-spectra binaries with spectroscopically-established mass-ratios. And once they (and others like them) begin to expand at their post-Main Sequence stage, they should become observationally even more conspicuous than they were before; for with increasing fractional size non-eclipsing binaries should begin to eclipse, and partial eclipses should deepen in the course of time.

A pair of dwarfs like YY Gem ($m_1 + m_2 = 1.2 \odot$ would still need a very long time to attain this stage; but binaries like β Aur ($m_1 + m_2 = 4.6 \odot$) or AR Aur ($m_1 + m_2 = 4.9 \odot$) – let alone Y Cyg ($m_1 + m_2 = 35 \odot$) should do so in the foreseeable future. Others must have done it before; and as the probability of their discovery should increase – not diminish – as a result of post-Main Sequence expansion, we should expect them to be well represented in our catalogues. In reality, not a single one has been discovered so far. Why don't we find, off the Main Sequence, any obvious descendants of systems plotted in the upper left corner of the diagram reproduced on Figure 8-7?

This constitutes an embarrassing question, and the answer still eludes us. One could argue that in none of the above-mentioned systems the mass-ratio need be exactly one; and may differ from unity by amounts within the limits of observational errors (say, 1%). Let, for the sake of argument, the marginally more massive component arrive at the critical stage first. Since its subsequent expansion to the Roche limit occurs on the Kelvin time-scale (say, a hundred times faster than the nuclear time-scale), the marginally more massive component may already find itself on the Roche limit (or even beyond) before the less massive one will begin to follow suit. Moreover, if layers which exceeded the Roche limit are captured by the expanding secondary, the mass-ratio may begin to differ more significantly from unity. This could account indeed for a sudden extinction of unit mass-ratios at the post-Main Sequence stage; but whether or not it actually does so is unsure. For the mass-influx on a marginally less massive component already on the brink of subsequent expansion would, in turn, accelerate its evolution and make it to return quickly borrowed mass to its mate with interest – so that after a short spell of time the situation would still be the same.

That, when the system begins to evolve from the Main Sequence, is is the more massive component which takes the lead is attested by the well-known detached binary ζ Aur (cK5 + B6) which we described already in the earlier part of this section. However, as soon as the expanding component attains (or comes close to) the Roche limit, observational guidance is suddenly lost; and we come face to face with our 'evolutionary paradox', which many investigators of the past 20 years attempted to explain away in different ways – mostly in terms of a mass transfer between the components.

We shall have more to say on such a possibility in subsequent parts of this section. Here we wish to stress that, strictly speaking, such a hypothesis is unnecessary; for even if the mass of the expanding component which overflows the Roche limit it lost to the system rather than being captured by its mate, the final outcome would still be a semi-detached system of the type (b) on Figure 8-8; only the time in which this would occur would be longer. Besides, let us have a look again at Figures 8-5 and 8-6, presenting empirical data on the mass-luminosity and mass-radius relations for detached components of binary systems of known absolute properties – both close and wide (visual). Open circles (O) on both diagrams mark the respective properties of primary (detached) components of semi-detached eclipsing systems which, according to the mass-transfer hypothesis, should be former secondaries rejuvenated by a massive influx of matter from its mate –

a transfusion which should have increased its mass 2-3 times or more. So drastic a transfusion should have affected profoundly also the internal structure of the rejuvenated star; but Figures 8-5 and 8-6 disclose no appreciable difference in external characteristics of the stars which have not yet received any mass from their companions, and of those which have already done so. Could the stars of comparable mass, radius, and luminosity but separated in their evolutionary stage by so profound a cataclysm, remain indistinguishable from each other (i.e., be built on the same model)? As Eliza Doolittle may have put it (cf. Shaw, 1912) "not bloody likely".

Under these circumstances, the question can be asked: is there any possibility of producing the predominant type of semi-detached binaries, as shown on Figure 8-8(b) without the intermediary of the model represented by Figure 8-8(a) – i.e., without the need of any mass transfer between the two components? The answer is – maybe. The intermediary stage shown on Figure 8–8(a) becomes unavoidable only if the material inside the stars is not mixed (by convection or otherwise); so that hydrogen consumed in the core (constituting only a few per cent of the star's mass) is not being replenished from the cooler layers surrounding the core.

Suppose, however, that the decreasing amount of hydrogen in the core of the more massive component can be freshened up from the outside supplies untouched by nuclear fire; while, for some reason, this does not occur inside the secondary (or, if it does, not to the same extent). If so, it should indeed be possible for the primary (more massive) component to outlast on the Main Sequence its less massive mate, the interior of which is less well mixed. Such a possibility has been previously suggested by Zahn (1967); and deserves further attention. The problem will be mainly one of the time-scale: for example, would the total hydrogen supply in the mass 5.2 $\odot$ of Algol A outlast – at its current energy output ($L = 229\ \odot$) – the hydrogen supply inside the core of its unmixed mate of mass 1.0 $\odot$? If we assume, for the sake of argument, that the entire mass of Algol A consists of hydrogen available for combustion, while only the core (of mass 0.12 $\odot$) is available for this purpose to its mate, and that the (present) luminosities of both components are in the ratio 30 to 1, then the ratio $(5.2/0.12 \times 30) = 1.4$ is indeed greater than one.

The longer life of the more massive star on the Main Sequence would, therefore, be possible on energetic grounds; but the argument is oversimplified; and whether or not this actually happens remains to be clarified by future investigations.

Such being the case, let us return to the alternative possibility of explaining the paradoxical nature of observed semi-detached systems in terms of a mass-transfer hypothesis. First, let us stress the magnitude of the problem confronting us in this connection, and use Algol again as an illustrative example. If the present Algol B of spectral class K0 was initially the more massive of the two, but was subsequently demoted to its present secondary role by a mass loss acquired by its mate, Plavec (1968, p. 209) showed that such a transfer would have amounted to not less than 2.1–2.5 $\odot$ over a time-interval of the order of 10^4 to 10^5 yr – corresponding to mass loss at a rate of $10^{-4}\ \odot$ per annum. Compare this with a typical rate of mass loss of $10^{-6}\ \odot \mathrm{yr}^{-1}$ for the Wolf-Rayet stars, or of 10^{-5} $\odot\ \mathrm{yr}^{-1}$ for novae (averaged over their explosive and quiescent stages) – let alone with other spectroscopically observable phenomena indicative of much smaller rates of mass loss (cf., e.g., Deutsch, 1969).

The rates are now to be confronted with the hard fact that no binary systems are known to be exhibiting a mass loss at anything like the required rate; or to be passing through the 'transient phase' on Figure 8-8 at the present time. We know well over one hundred semi-detached eclipsing binaries of the type represented on the lower part of Figure 8-8 ('stable state'). However, not a single case has yet been discovered of the 'transient phase' – a fact suggesting that the duration of this latter phase is less than 1% of the former (much less still if we stop to consider that observational selection would favour discovery of close pairs in the 'transient' over 'stable' state; and also that, in the former case, the system should be intrinsically – and, therefore, also apparently – much brighter and more conspicuous).

If we wish to account for this fact – and to conceal the existence of such systems in the sky during their 'transient' phase – we have to assume this phase to be so severely transient that only one star in several hundreds may be seen in the act at any particular time. Theoretical considerations (Morton, 1960; and others), disclosing that the process can be accomplished on the gravitational (Kelvin) time-scale, would not rule out this to be within the bounds of possibilities. However, the shorter the time which this active phase is to last, the greater amount of mass which would have to be lost per unit time to accomplish the same task in the end; and the greater this mass, the more conspicuous other phenomena could become which would attract the attention of the observer; thus making it increasingly difficult for the respective binary to pass through the transient stage unobtrusively enough not to be caught in the act.

In order to appreciate the importance of such phenomena, let us consider the physical state of matter which may be expelled from the star. Since it is originating from the outer layers (presumably close to the Roche limit), it should consist predominantly of hydrogen (with some proportion of helium); and – as soon as the pressure drops outside the star – be ionized almost completely by the radiation field of the primary (hotter) component. For a mass loss at a rate of 10^{-4} ⊙ yr^{-1} and the exciting star to be of A- or B-type, the ratio of HI/HII should (from the theory of HII-regions) be of the order of magnitude of $10^{-3} - 10^{-4}$ or even smaller. In other words, *in close binary systems with components of early to intermediate spectra we find ourselves deep inside a highly ionized HII region;* and the gas within it would radiate like an HII-region emitting bright lines. Moreover, for the rate of mass loss under consideration, the emission lines produced by these regions should be observable; and their presence in the spectra should identify the binaries which happen to find themselves in this predicament.

The situation which we should meet in binary systems at a rapid mass-loss stage is not unlike that existing in stars of Wolf-Rayet type, which were also discovered spectroscopically three-quarters of a century before the first one of them was found to eclipse. *The principal means of detection of binaries at a rapid mass-loss stage should, therefore, be objective-prism surveys of the sky rather than photometric measures;* and the well-nigh total absence on them of stellar objects which could be assigned to that stage should give rise to increasing concern.

That free gas is present in close binary systems with contact components has been attested by spectroscopic observations for a long time. The lines in the Algol spectrum, discovered by Miss Barney (1923), whose Doppler shifts did not share those of the two components; or a conspicuous asymmetry of the radial-velocity changes of U Cep (a system whose components revolve in circular orbits) discovered by Carpenter (1930), heralded the advent of new era in double-star research, to which astronomers of the 1930's preferred to adopt an ostrich-like attitude – until such a posture was made untenable by the pioneer work of Otto Struve and his school between 1940–1960. We know now that similar phenomena occur predominantly in systems in which at least one component has evolved appreciably away from the Main Sequence; and it is tempting to connect such phenomena with this fact.

This being said, however, it should be made equally clear that the gas streams so observed are utterly inadequate by themselves to bring about a loss (or transfer) of mass necessary to exchange the role of the primary and secondary components with sufficient rapidity to escape detection in the act. In order to demonstrate this, let us depart from the fact that the curves of growth of the line spectra of such streams correspond to 10^{10}–10^{13} neutral hydrogen atoms per cm^3 (cf., e.g., Batten, 1970; or Biermann, 1971; and references quoted therein). Moreover, the average cross-section of such streams are of the order of 10^{20}–10^{21} cm^2; and their velocities (relative to the revolving frame of reference) are of the order of 10^6 to 10^7 cm s^{-1}.

Therefore, a mass transport by such streams should be given by $(10^{10}–10^{13}) \times 1.67 \times 10^{-24}$ g cm^{-3} times the cross-section of the flow of 10^{20}–10^{21} cm^2 times its velocity of 10^6–10^7 cm s^{-1} – i.e., 10^{13} to 10^{16} g s^{-1} or 10^{-12} to 10^{-9} $\odot$ yr^{-1}. These figures refer, to be sure, only to hydrogen which happens to be in neutral state. The total amount of hydrogen involved will be obtained if we multiply the rates of 10^{-12} to 10^{-9} $\odot$ yr^{-1} by the ionization ratios of HII/HI, which can be very large (depending on the spectral type of the exciting star); but whether these are large enough to bring the foregoing figures anywhere near 10^{-4} $\odot$ yr^{-1} must be considered separately in each particular case.

Besides, there are other problems. While the neutral component of circumstellar hydrogen will absorb (or emit) in discrete frequencies, the overwhelming fraction of ionized hydrogen will produce *scattering* as well as *polarization* of light incident upon it from different directions in the course of an orbital cycle. The magnitude of these effects would, of course, depend on the density of the circumstellar hydrogen; and this (for a given flux) on the velocity of gas motion. The only stars among close binaries which are known to eject mass in anything like the requisite amount – namely, the Wolf-Rayet stars – do so with velocities between 1000–2500 km s^{-1} (cf., e.g., Underhill, 1969); and these numbers, coupled with the dimensions of their envelopes, lead to their spatial densities of the order of 10^{-11} g cm^{-3}.

Ionized gas of this density endows the envelopes of Wolf-Rayet stars with a significant opacity even at a considerable distance from the surface – as evidenced by anomalous widths of the minima observed when the Wolf-Rayet component of a binary system eclipses its mate (for V 444 Cygni, cf. Kron and Gordon, 1943; Kopal, 1944). If, now, the velocity with which matter is lost by a Wolf-Rayet star were diminished by a factor of (say) 10, the density 10^{-11} g cm^{-3} of the flux would have to be increased by the same factor to keep the efficiency of transfer at the same level. An ionized gas so dense along a path comparable with the dimensions of the system would, however, then become so opaque as to give rise to optical phenomena which should render such a system all the more conspicuous; thus greatly aggravating our already considerable difficulties over the problem of why no such systems (other than Wolf-Rayet stars) are known.

But let us, for the moment, abandon such a question at this unsatisfactory state, and turn to consider the *dynamics* of the process by which the evolved components of close binary systems are likely to divest themselves of superfluous mass. We pointed out already earlier that neutral gas streams are observed only (or, at least, preferentially) in close binary systems possessing contact components; and the presumption is, therefore, strong that their source is related with the evolving component at (or near) its Roche limit.

The traditional reason in favour of such a view has been a diminished acceleration to be found there. As the star expands, the value of the total potential prevalent over each surface diminishes; and so does the gravitational acceleration. But while the surface potential should remain constant all over the star at any stage of its expansion, its gradient (i.e., the gravitational acceleration) does not. When the star has ultimately attained its Roche limit, the surface potential attains a minimum value it can possess for any closed configuration; while the gravitational acceleration, varying over the surface, and diminishing in the direction of its mate, actually vanishes at the inner Lagrangian point L_1 (cf. sec. VI-1). This fact by itself should not, to be sure, cause any matter to escape from the vicinity of such a point; for the Roche limit and any point on it (including L_1) represents a strictly *static* property of the model. However, a smallness of the gravitational acceleration in the neighbourhood of L_1 (zero at L_1) makes it certainly easier for any small perturbation to remove a mass element from there than from any other part of the star's surface.

This circumstance has prompted many investigators (Kopal, 1956, 1959; Gould, 1957, 1959, Plavec and Křiž, 1965; and others) to consider the hypothetical outflow of mass from a contact configuration as a problem of particle mechanics within the framework of the restricted problem of three bodies (cf. sec. VI-1A). In order to do so, however, it is necessary first to specify the *boundary conditions* of the problem – i.e.,

(1) to localize the region on the star's surface from which mass is being lost;

(2) to determine the velocity with which the atoms cross the effective boundary surface; and

(3) to estimate the density (or flux) of the moving material.

As is well known, the system of differential equations of the restricted problem of three bodies (of which the two components play the role of the finite masses; and the escaping particle, that of a mass point) is one of sixth order. Consequently, six boundary conditions – i.e., the position and velocity components of the escaping particle – must be specified to render any solution determinate; and of these at best only three (i.e., the radial-velocity component, and the position of the region of escape in the plane of the celestial sphere if the system happens to be an eclipsing variable) can be inferred from the observations; the remaining ones are to be estimated as best as one can from a model one adopts for the case.

In order to do so, all investigators in the past considered any ejection to take place from the Lagrangian point L_1 (whose position is uniquely specified by the mass-ratio of the system; cf. Table VI-1) in the plane of the orbit; with an arbitrary velocity and in an arbitrary direction. The subsequent history of the particle so ejected must then be followed by numerical integration of the respective equations of motion; and the possible outcome may be (a) escape of the particle from the system; (b) its re-capture (after one or more revolutions) by the body which emitted it; (c) a capture by its mate; or (d) its retention in circum-stellar space in a (simply or multiply) periodic orbit – direct or retrograde.

Most investigators in the past confined their attention to the establishment of initial conditions leading to the case (c) – i.e., to a mass transfer from one component to another.

In particular, following an earlier suggestion by Kopal (1956), Piotrowski (1964b) and Kruszewski (1963, 1964b,c) investigated the circumstances of ejection caused by non-synchronism between axial rotation and orbital revolution – in which case a particle of finite momentum may be ejected from L_1 in the direction of a tangent to the respective branch of the Roche limit (depending on whether the rotation is faster or slower than the revolution).

In retrospect, however, this mechanical approach proved to be (at best) a detour, rather than progress, towards the solution of the underlying physical problem. The reason is the fact that the use of the equations of the restricted problem of three bodies could be physically justified *only if the mean free path of the particles ejected by the expanding star were long in comparison with the dimensions of the flow* (which are of the same order of magnitude as those of the system as a whole); so that mutual *collisions* of moving particles can be disregarded. The mean free path is, in turn, related (by the kinetic theory of gases) with the *density* of particles in the space between the stars; and this density will determine whether or not the respective assembly of particles could produce any observable effects.

If the escaping material were hydrogen, helium (or indeed any light element), it is easy to show that *the mean free path of particles in a gaseous medium dense enough for photometric or spectroscopic detection must be by many orders of magnitude smaller than the dimensions of binary systems.* In other words, gas envelopes capable of impressing observable features in the composite spectra of such systems are no mere 'exospheres' in which individual particles move along ballistic trajectories, but constitute in a very real sense *extended atmospheres;* and, therefore, *hydrodynamics* rather than particle mechanics must be used as a basis of any investigation of their motion.

The first investigator to have realized this was Prendergast (1960), followed by subsequent investigators such as Biermann (1971), Prendergast and Taam (1974), Sorensen *et al.* (1975), Lubow and Shu (1975), and others. If the progress of this effort was slow, and individual contributions to it far between, this is due to intrinsic difficulties of the problem; and formidable difficulties are yet to be overcome before meaningful comparisons between theory and observations can be made.

Let us mention what follows at least some of them – if alone only as a goal which we must aim to accomplish. First, spectroscopic observations indicate that (at least the neutral) gas streams in close binary systems move with velocities of the order of 100 km s^{-1}; while the velocity of sound in (hydrogen) gas at a temperature of 10 000 deg (a typical ionization temperature indicated by the spectra) should be close to 11.8 km s^{-1}. If so, however, it follows that motions of gas in between the stars and surrounding the system should be *hypersonic* and correspond to Mach numbers of the order of 10 or more. This fact, in turn , *rules out* any possibility of *linearization* of the equations of hydrodynamic motion; for in doing so we would rule out possible occurrence of *shock waves* whose effects may play an important role in any comparison between theory and observations.

Secondly, inasmuch as an overwhelming fraction of the circumstellar gas in close binary systems with at least one component of early or intermediate spectral type must be *ionized*, its viscosity μ is bound to be very high (cf. sec. III-4); and (in view of its low density ρ) *the kinematic viscosity* μ/ρ *may be enormous.* Therefore, it would be plainly unrealistic to treat the motions as inviscid; the terms factored by the kinematic viscosity may, in fact, be the dominant ones in the equations of motion.

Third, the Reynolds numbers of the respective motions are so large that such motions are probably *turbulent* as well, and characterized by high turbulent viscosity into the bargain. Therefore, fluid motions in gaseous envelopes of close binary systems may represent the case of a *hypersonic flow in viscous turbulent media* – about the worst accumulation of attributes one can ascribe to any flow – to be treated in three-dimensional space, with the time constituting the fourth independent variable (dispensable only if the flow were steady). It is only the solutions of partial differential equations (subject to appropriate boundary conditions) governing such flows which can really tell us something definite about the proportion of matter ejected by the components of close binary systems that can be captured by their mates, remain in circulation within the system, or be lost to the system altogether.

It should, furthermore, be stressed that the *forces causing ejection are internal to each star;* and before the gas will disengage itself from the gravitational field of the parent star, its companion merely stands by as a largely passive onlooker. If the velocity of ejection is of the order of 10^3 km s^{-1} (as it is in the Wolf-Rayet stars of Novae), there is no doubt that matter so ejected will escape from the system altogether; and the same should be true of any particles carried away by 'stellar winds' (cf., e.g., Siscoe and Heinemann, 1974). The role of the latter in subgiant component of semi-detached eclipsing systems, posessing external convective zones of considerable depth, should be particularly emphasized (cf. Lauterborn and Weigert, 1972). *Hydrodynamical motions which could transfer mass from one star to another should, in general, be characterized by Mach numbers less than one;* at least that seems indicated by the outcome of extensive numerical integrations of individual particle trajectories (whose periodic orbits can represent, to be sure, only limiting cases of steady-state hydrodynamic flow, obtaining when the flow density is allowed to approach zero); cf. also Plavec *et al.* (1973) or Bielicki *et al.* (1974).

In the face of such a situation – in which emphatic nonlinearities of the problem should make us wary of simplifications which could cripple its context (remember how many consequences of linearized or inviscid hydrodynamics have in the past earned for themselves the epithets of 'paradoxa'!) – it is apparent that the students of the dynamics of close binary systems (and their computing machines) are unlikely to run out of work in the foreseeable future. Also, in retrospect, it is easy to see how uncertain have been the foundations of most part of the work of the last 10–15 years on the evolution of close binary systems based on the familiar hypothesis of 'mass-exchange' between the components, and reviewed in publications already referred to (cf. again Plavec, 1968, 1970; or Paczynski, 1971). Even the terminology of that case was inexact; for what most of these investigators had in mind was a transfer of mass from one component to another; not a mutual exchange between them*.

But what can we say about the possibility of transfer itself, or its efficiency? The answer is contained in the systems of equations which have not yet been solved; and (in view of a large number of the 'degrees of freedom' represented by their boundary con-

* At least, according to Hoyle (1955; p. 200) "There is the possibility that the predatory star will be forced to make amends for its former behaviour by returning material to the (at present) fainter star, which it robbed of mass so unfeelingly in the past. In the interest of cosmic justice it is to be hoped that this happens; but whether it does or not is unsure."

ditions) it is hopeless merely to guess at the outcome; for the probability of a guess being right is very small. Most previous investigators have assumed – by way of such a guess – that all material of an expanding star (evolving in blissful ignorance of the presence of its mate) which overflows the Roche limit (existing only in the presence of the secondary) will be transferred *in toto* to the mate; and all alternative processes of its disposal disregarded.

Why, one may ask, was such a series of particular assumptions compounded before a more realistic theory of the phenomena could give them their blessing? The answer is, of course, simple: these *ad hoc* assumptions were made because their protagonists wanted to compute something definite in advance of a more general solution of the problem; and this they did with the same zeal with which their predecessors integrated trajectories of the restricted problem of three bodies in the same (misguided) effort to elucidate the problem of a mass-transfer between components of close binary systems on an inadequate physical basis.

That this could happen may, in retrospect, seem odd; but not altogether without historical precedent. For who would forget the delightful episode from the times of Catherine the Great – when that Empress, concerned about the atheistic propaganda spread by Denis Diderot at her Court, commissioned Leonard Euler to silence the turbulent philosopher? In a public debate staged for the purpose, Euler challenged Diderot with the famous statement: "Monsieur, $e^{i\pi} = -1$; donc Dieu existe; respondez!" Diderot could not; and so lost his case by default. Had he been a better mathematician, he could, of course, have turned the tables and asked; what do these statements have to do with each other?

But if we were to replace de Moivre's theorem with the 'Roche limit', and the existence of God with a 'mass-exchange' as the conjuring words of the day, should we not encounter a somewhat similar situation? These terms do have, to be sure, *something* to do with each other; for otherwise why should the fractional dimensions of subgiant components in semi-detached systems cluster around their Roche limits? But the mechanism connecting them is still but incompletely understood. Computations based on the assumption that mass transfer from one component to the other constitutes the sole mechanism which governs the evolution of close binaries at post-Main Sequence stage are of value only in so far as they throw light on certain particular aspects of the problem – just as computations of particle trajectories may have helped to elucidate others. But, considered by themselves and out of context of the underlying hydrodynamical problem, they do *not* provide any secure basis on which more detailed comparisons with the observations could meaningfully be attempted.

Or – to underline another motive and to give word to a more recent historian of our science (Fernie, 1969) in this connection – "The definite study of the herd instinct of astronomers is yet to be written, but there are times when we resemble nothing so much as a herd of antelope, heads down in tight parallel formation, thundering with firm determination in a particular direction across the plain. At a given signal from the leader we whirl about and, with equally firm determination, thunder off in quite a different direction, still in tight parallel formation." A not very complimentary picture of ourselves; but containing more than a grain of truth. Do we indeed do too much running with heads down, and not thinking enough en route as we thunder across the plains?

But, with this in mind, let us turn now our backs to further theoretical problems in this section, and consider instead any help or guidance which we can derive from the existing observations. In earlier part of this section we stressed the fact that the first evolutionary steps into the post-Main Sequence future are expected to occur so rapidly (on the Kelvin time-scale) that no system is likely to be caught in this stage at any particular time. Within the framework of the Hertzsprung-Russell diagram this should imply that no systems so evolving should be caught within the Hertzsprung gap of this diagram, as stars destined to pass through it will do so very rapidly. But are any eclipsing systems known which would be caught in this gap – en route to the semi-detached model of the type (b) on Figure 8-8? While we are not in a position to answer this question with any final validity, as a certain group of systems exists – called RS CVn-type stars after their best known representative – where we suspect that this may be the case.

The prototype itself – RS CVn – shares with a small group of similar stars (RT Lac, AR Lac, UX Ari and a few other – less certain – members) several intriguing properties which should be pointed out in this place. By their geometry alone, these eclipsing systems should be classified as detached: for the fractional dimensions of their components are well inside the Roche limits appropriate for their masses; and, moreover, their mass-ratios – unlike those for typical semi-detached systems – are very close to one (those of RS CVn itself, for instance, or of AR Lac, does not deviate from unity to any significant extent).

However, in other respects again the stars of this type exhibit characteristics common to semi-detached systems. Their orbital periods – far from remaining constant – fluctuate conspicuously and erratically as do those of systems containing contact components. But it is not only the period of such systems which exhibits irregular fluctuations; the shape of their light curves continues to change significantly between minima as well as within eclipses (cf., e.g., Theokas, 1977) – a fact making any determination of the geometric elements from an analysis of such light curves a very difficult task.

And – the most telling fact of all – although the masses of the components of RS CVn and other stars akin to it are closely the same, their dimensions and spectra are quite different. Thus the old charm that equally masssive stars are also equal in size and luminosity – so manifest on the Main Sequence – seems to lose its power for stars of this group. Their exact positions in the Hertzsprung–Russell diagram are difficult to ascertain with sufficient accuracy, because of the notorious difficulties in deducing the geometrical elements of as bad photometric actors as these stars, which never remember their lines too long. Nevertheless, it is at least possible (if not probable) that some (if not all) of them are situated within the Hertzsprung gap; and are in in the state of rapid evolution.

Are we witnessing, in these stars, systems evolving from the Main Sequence on the Kelvin time-scale, at the stage of rapid mass loss or transfer? The photometric irregularities of their light curves could indeed be explained by absorption effects of dense streams of turbulent gas whirling inside these systems; but their spectra show no evidence (for instance, in the form of emission lines associated with neither component) that such is the case. On the other hand, the fact that several of the systems of this group exhibit variable radio emission at centimetre wavelengths (cf. Hjellming and Blankenship, 1973; Gibson *et al.*, 1975; Owen *et al.*, 1976) suggests a presence of plasma cloud surrounding the components (or the system) and emitting synchrotron radiation; but whether the observed radio emission is a general property of RS CVn-type binaries or a special characteristic of particular systems exhibiting it is not yet clear. Moreover, if the RS CVn-type stars were binaries in the stage of rapid mass loss (or transfer), the number of already known cases of this type would be much too large for the anticipated duration of this phase of their evolution; so that their evolutionary status remains still in doubt.

Another group of close binaries with properties which overlap the lines of classification outlined in section VIII-2 are systems whose secondary components are 'undersize' subgiants – not far from their Roche limits, but distinctly interior to them (cf. Kopal, 1959; sec. VII.4). As prototypes of this group we may regard the systems of S Cnc, TV Cas, or Z Her. Binaries of this type resemble in most respects 'normal' semi-detached systems discussed earlier in this section; in particular, they share with them a large disparity in masses of the two components and irregular fluctuations of their orbital periods. But their secondary (less massive) components – exhibiting otherwise all spectroscopic and other properties of the subgiants – are characterized by fractional dimensions *smaller* than those of their respective Roche limits; and the question arises as to the significance of this defect.

Three possibilities appear open to account for this fact: namely, that 'undersize' subgiants are (a) contracting to the Main Sequence; (b) expanding in post-Main Sequence stage towards the Roche limit; or (c) contracting from this limit at a later stage of their evolution. It should be noted that both processes (a) and (b) should proceed on the gravitational (Kelvin) time-scale; making it equally unlikely to catch any system in this stage at a particular time. The time-scale of process (c) is less certain, but probably also short. On the other hand, the existence of such systems – especially if established in larger numbers – may signify that the post-Main Sequence evolution towards the Roche limit does not unroll as rapidly as simplified theory outlined in section VIII-1 may lead us to expect (particularly in view of the complications arising from tidal evolution, which we shall discuss in the next section).

In order to avoid the embarrasment from the existence of such systems in numbers greater than those theoretically expected, recent discussions (cf., e.g., Hall, 1972, 1974) tended to minimize their number. Such a process requires, however, a considerable caution if it is not to lead us astray. The question as to whether any subgiant component of a close binary system actually fills the largest closed volume capable of containing its whole mass, or is interior to it, can be answered only by observational evidence of the requisite precision. The fact that the precision of this evidence can only be finite makes it, however, difficult to distinguish between the propositions that (a) all secondary components in semi-detached systems are in actual contact with their Roche limits; or (b) a certain fraction of them approaches this limit within a small margin. The observational difference between them may be very small – yet very important for the understanding of the evolutionary stage in which these systems find themselves at the present time.

A third anomalous group of close binaries of the semi-detached type is formed by systems of anomalously low masses, of which the eclipsing variable R CMa can be regarded as a prototype; and T LMi, XZ Sgr or S Vel as representative members (to which several others could be added at the present time). All these are single-spectrum, partially eclipsing systems whose distinguishing characteristics are abnormally small values of their spectroscopic mass-functions. In point of fact, for all systems mentioned above the value of the mass-function $f(m) \equiv m_2^3 \sin^3 i/(m_1 + m_2)^2$ turns out to be between 0.001 and 0.01 $\odot$ – indicating that the masses of their components are very small – too small for their observed luminosities.

Consider, for example, the system of R CMa, of period 1.136 days, for which the latest spectroscopic determination of the mass function by Galeotti (1970) led to the value of $f(m) = 0.0020\odot$. According to the most recent photometric orbit by Sato (1971), the angle i of orbital inclination should be close to 80°; and, accordingly, the mass of the secondary component

$$m_2 = f(m)\,(1+q)^2 \csc^3 i = 0.0021\,(1+q)^2\,\odot, \tag{3.3}$$

where $q \equiv m_1/m_2$ denotes the mass-ratio of the system.

The fractional radius r_2 of the secondary component of R CMa (deduced from an analysis of its light changes during partial eclipses) is, according to Sato (1971), equal to 0.219; and if the latter value represents (on contact hypothesis) also the fractional dimensions of the Roche limit enclosing the mass m_2, it would correspond (cf. Table VI-2) to a mass-ratio $q = 7.7$; and the masses of the two components would then follow from the foregoing Equation (3.3) as

$$m_1 = 1.2\,\odot \quad \text{and} \quad m_2 = 0.16\,\odot\,. \tag{3.4}$$

respectively. It should, moreover, be stressed that the hypothesized contact nature of the secondary component leads to *maximum* possible masses for both components; for if the secondary were detached (and, therefore, the actual dimensions of the Roche limit enclosing the mass m_2 larger than Sato's value of r_2 quoted above), the mass-ratio q would thereby be diminished; and so would both masses (especially m_1). A similar disparity in masses of the components of XZ Sgr was established by Smak (1965).

The absolute bolometric magnitudes of the components of R CMa (of apparent visual magnitude of 5.9 at a distance of 30 parsec from us; see Table I-1) are +3.6 and +6.2, respectively; and their radii $R_1 = 1.7\,\odot$ and $R_2 = 1.2\,\odot$. The primary component of R CMA appears, therefore, to be overluminous for its mass by approximately one magnitude; but the overluminosity of the secondary component of mass $0.16\,\odot$ is enormous – being some six magnitudes brighter than the Main-Sequence secondary component of the wide (visual) binary Krüger 60 of comparable mass – and so is its size! The position of R CMa B is, therefore, to be sought high above the Main Sequence along the Hayashi line on Figure 8-2; and a star of so small a mass could not have got there within the lifetime of the Galaxy* unless it was much more massive in the past, and *lost* much of it by escape into space rather than by any transfer on to its mate (which itself has too small a mass for its luminosity). Was this mass lost in the course of a repeated mass transfer between the components (a real exchange this time) – a process proposed many years ago by Kopal (1959; p. 538) or Smak (1961, 1962), during which the identity of the primary and secondary components changes more than once – or was it ejected with velocities in excess of that of escape from the gravitational field of the system?

No definitive answer to this question can so far be given. However, the view which keeps coming to the forefront of our mind with increasing insistence is the probability that *the most important process influencing* – if not actually dominating – *the evolution*

* With its space velocity of 55 km s^{-1} relative to the Sun, R CMa may possibly be a Population II star; but all facts of the case are not yet clear.

of close binary systems in post-Main-Sequence stage is not a mere mass-transfer from one component to another, but a loss of mass removed (hydrodynamically or otherwise) *from the gravitational confines of the system.* It is understandable that a massive system like μ^1 Sco ($m_1 + m_2 = 23 \odot$) could have reached its present semi-detached state in some 70×10^6 yr since it came into being as a member of the Scorpio-Centaurus association (cf. Blaauw, 1952); according to the model calculations by Iben (1965, 1966) it could have done so in less than 25×10^6 yr. The same conclusion continues to be true of (say) u Her ($m_1 + m_2 = 10 \odot$), and probably also of Algol ($m_1 + m_2 = 6 \odot$), without encountering any serious difficulty with the nuclear time-scale of their evolution.

It is, however, difficult to expect that 10^9 yr is long enough a time for systems like Z Dra, RX Hya, TY Peg or RT Per – all of which possess combined masses smaller than $2 \odot$ (cf. Kopal and Shapley, 1956) – to have become semi-detached (which they are) as a result of thermonuclear hydrogen depletion in the interior, unless the mass of the whole system was very much larger in the past than it is today. And it is quite impossible, in a Galaxy no more than 10^{10} yr old, to expect semi-detached systems like R CMa, T LMi or S Vel, of combined mass 1–1.5 $\odot$, to have attained that stage by hydrogen depletion unless they once had been very much more massive. The total mass defects of these systems are large – indicative that, once upon a time, the masses of their components should have been 3–5 times as large (or even more for stars like R CMa B and others). But if so, are we not meeting really the same situation encountered previously with (say) Algol B, which (as we already suspected before) divested itself of about 80% of its original mass some time in the past? In the case of Algol, we had an option between loss and transfer to account for the missing part; but in the case of (say) R CMa B we have no option other than an outright loss. Could, however, the same be then true of Algol (and other stars similar to it) as well?

In conclusion, while climbing up along the Hayashi line with R CMa B, we should be ready to answer another question which may arise in the reader's mind in this connection: namely, can components of close binaries at this stage of their evolution wander into the 'instability strip' of the HR-diagram to become Cepheids or long-period variables? Certainly no systems are known where this would be the case; and the reason is the fact that – thanks to the Roche limit – *the expansion of the evolving component is not allowed to continue to the stage at which pulsational instability would set in* (cf. Weigert, 1967). In wide binaries which remain detached even at the maximum evolutionary expansion of their component – so that their surfaces will never get in contact with their Roche limits – this can happen. Wide binaries (with orbital periods much in excess of those listed in Table VIII-3) containing Cepheid variables as one of their components are indeed known (cf., e.g., Abt, 1959; or Evans, 1968); and at least one visual binary (ADS 11524) contains a long-period variable (X Oph; cf. Fernie, 1959). None of these is, however, ever likely to eclipse; and, therefore, they do not belong properly to the scope of this book.

C. CONTACT SYSTEMS

The length of the account given in the preceding part of this section of the evolutionary problems of close binary systems in post-Main-Sequence stages might almost suggest that we already have enough problems on hand, in this connection, to need to extend their range even further – by bringing to the scene other groups of binaries which have not been discussed so far. Yet our survey of the evolutionary processes in such systems would be seriously incomplete if it were to terminate at the semi-detached stage – without at least an outline of the properties of another group of close binaries, the position of which in the general framework of stellar evolution continues to pose a challenge: namely, *eclipsing variables of the W Ursae Maioris type* – so called after their best known (albeit neither brightest, nor nearest) representative.

Binaries of this type represent a remarkably compact group of similar properties, possessing a number of well-defined characteristics which can be briefly summarized as follows:

(1) The orbital periods of W UMa-type systems are shorter than one day; and for a large majority are comprised between 6–12 h; therefore, with one or two exceptions (RT And or UV Leo) shorter than those of any detached or semi-detached system now known. Moreover, the periods of most systems of the W UMa-type are not constant, but fluctuate (like they do for semi-detached systems) appreciably in an erratic manner. The eccentricities of their orbits are as a rule negligible.

(2) The amplitudes of light variations of the stars of W UMa-type range from a few tenths to just over a full magnitude (largely regardless of the effective wavelength of observations); and are, moreover, by no means limited to the times of the eclipses: they extend continuously over the entire cycle – a fact indicative of an extreme proximity of their components.

(3) The light curves of many W UMa-type systems are distinctly *asymmetric* with respect to the conjunctions; and the sense of asymmetry may fluctuate in the course of time – thus making their analysis for geometrical elements difficult. However, in so far as this can be done by methods now available, the outcome strongly suggests that the masses of *both* components of such systems fill completely their respective Roche limits – a characteristic which has earned these stars a designation of *contact systems* – and may actually exceed them ('oversize' components). In consequence, the *primary* (more massive) component is bound to be the *larger* of the two.

(4) The spectra of the components in systems of the W UMa-type correspond to those *dwarf* stars (luminosity class V) and exhibit all characteristics of the relatively advanced classes F and G – predominantly, but not exclusively so. Thus V 701 Sco – a typical variable of this group in most other respects (with a period $P = 0.762$ days) – exhibits a (mean) spectrum B5; and EM Cep ($P = 0.806$ d) of even earlier spectra (B0.5 + B1) can likewise be provisionally classified in this group (cf. Johnston, 1970).

The observed period-colour relation for W UMa-type variables exhibits a near-discontinuity in the proximity of periods close to half a day, and a well-established discontinuity at $P = 0.65$; while for $P > 0.65$d the period-colour relation ceases to be unique. Moreover, the colour of the system as a whole corresponds closely to that of a Main-Sequence star of mass and luminosity equal to those of the two components combined (Eggen, 1961, 1967).

(5) For a majority of known systems, the times of the primary (deeper) minima lie on the ascending branch of the radial-velocity curves of the luminous components – a fact which signifies that more massive and luminous stars possess lesser surface brightness (i.e., a later spectrum). If so, the *primary* (deeper) minima are, therefore, due to *occultation*

eclipses; and *secondary* minima are *transits.*

(6) The mass-ratios of the components in W UMa-type systems are (as in semi-detached systems) frequently quite different from unity – 0.08 for AW UMa (Mochnacki and Doughty, 1972a; Mauder, 1972; or Lucy, 1973), or 0.09 for ϵ CrA (Mauder, 1972) – reminiscent of mass-ratios previously encountered in semi-detached systems of R CMa-type. Unlike in the latter, however, the primary components in W UMa-type systems are (if anything) *underluminous* for their masses; and appear to lie systematically *below* the statistical mass-luminosity relation valid for Main-Sequence stars as their secondaries are above it.

(7) The masses of the components of W UMa-type systems are small to moderate (0.5 to 1.5 $\odot$); and their absolute magnitudes range mostly between +4 and +6 mag. Such stars cluster loosely around the Main Sequence; but the dispersion of their positions is considerable; and the slope of the mass-luminosity relation for the individual components is *not* the same as that for single Main Sequence stars (cf. Osaki, 1965).

(8) A surprising number of W UMa-type systems happen to be components of wide binaries – such as *i* Boo (ADS 9494B), AK Her (ADS 10498A), AM Leo (ADS 8024A) – or form common-proper motion pairs (such as VW Cep with HD 199476, or W UMa itself with BD +55° 1351). Many occur also in galactic clusters – such as TX Cnc in Praesepe, IR Car in NGC 3532, LW or BH Cen in IC 2994, EM Cep in NGC 7160, V 701 Sco in NGC 6383, AH Vir in Wolf 630, or EP to ES Cep in NGC 188.

A tabulation of the absolute properties of a representative sample of eclipsing systems of this type is reproduced in the accompanying Table VIII-4, which contains a list of the geometrical as well as physical characteristics of their components. With one exception (TX Cnc, for which we adopted the data given by Whelan et al, 1973) the tabulated values have been taken from Table IV of Kopal and Shapley's 1956 *Catalogue*; and the headings of the individual columns are self-explanatory.

These characteristics (in particular, the relatively low luminosities $M_{1,2}$ listed in the last two columns) do not render the W UMa-systems very conspicuous in the sky. Although two (*i* Boo and VW Cep) belong among our nearest binary neighbours (see Table I-1) – *i* Boo being the nearest eclipsing variable to us in space – only the latter is (just) visible to the naked eye; and not more than 30 are brighter than the 10th apparent magnitude. Yet when due account is taken of the effects of observational selection, it transpires that *eclipsing variables of W UMa-type outnumber all other forms of close binaries many times per unit volume of galactic space.*

In the proximity of the Sun, Shapley (1948) found the W UMa-type stars to be 10–30 times as numerous as all other types of eclipsing variables taken together. Subsequently, Kraft (1965) or Eggen (1967) found a somewhat smaller preponderance; but still, according to Eggen's estimate, one out of 1000–2000 stars of the same spectral range is an eclipsing variable of W UMa-type – a frequency corresponding to about two such binaries (predominantly with periods of 6–10 h) per million cubic parsec. But whichever estimate we adopt, there seems no escape from the conclusion that *the W UMa-type stars represent the commonest type of binary systems in the space around us;* and their abundance alone renders the problem of their origin and evolutionary stage all the more important challenging.

The fact that the W UMa-stars cluster (albeit loosely) around the Main Sequence suggests that they are hydrogen-burning stars; and since their dimensions as well as

TABLE VIII-5
Geometrical properties and absolute dimensions of the contact eclipsing, system of W UMa-type

Star	Period	Spectra	m_2/m_1	C_0	r_1	r_2	m_1	m_2	R_1	R_2	ρ_1	ρ_2	M_1	M_2
	d												m	m
AB And	0. 332	G5 + G4	0.62 ± 0.04	3.97 ± 0.01	0.42 ± 0.01	0.33 ± 0.01	1.65 $\odot$	1.03 $\odot$	1.20	0.95 $\odot$	1.0 $\odot$	1.2 $\odot$	4. 5	4. 8
i Boo	0.268	G2 + F9	0.50 ± 0.01	3.944 ± 0.003	0.441 ± 0.002	0.312 ± 0.002	1.35	0.68	0.98	0.70	1.4	2.0	4.6	5.1
TX Cnc	0.383	G0 + G2	0.60 ± 0.05	3.88 ± 0.01	0.44 ± 0.01	0.35 ± 0.01	1.0	0.62	1.1	0.88	0.76	0.91	3.8	3.9
VW Cep	0.278	K1 + G6	0.32 ± 0.04	3.86 ± 0.03	0.48 ± 0.02	0.28 ± 0.01	1.44	0.47	1.1	0.62	1.2	1.9	5.1	5.8
TW Cet	0.317	G5 + G4	0.53 ± 0.04	3.95 ± 0.01	0.43 ± 0.01	0.32 ± 0.01	1.3	0.69	1.05	0.77	1.1	1.4	5.0	5.4
RZ Com	0.339	K0 + G9	0.48 ± 0.05	3.94 ± 0.02	0.44 ± 0.01	0.31 ± 0.01	1.7	0.82	1.2	0.85	0.91	1.2	4.7	5.3
YY Eri	0.321	G5 + G4	0.65 ± 0.06	3.98 ± 0.01	0.42 ± 0.01	0.33 ± 0.01	0.76	0.50	0.88	0.71	1.1	1.3	5.1	5.4
SW Lac	0.321	G3 + G2	0.85 ± 0.09	4.00 ± 0.00	0.39 ± 0.01	0.36 ± 0.01	0.97	0.83	0.93	0.86	1.1	1.3	4.8	4.9
V 502 Oph	0.453	G2 + F9	0.40 ± 0.03	3.91 ± 0.01	0.46 ± 0.01	0.30 ± 0.01	1.85	0.74	1.6	1.01	0.48	0.69	3.6	4.3
ER Ori	0.423	G1 + G2	0.61 ± 0.11	3.97 ± 0.02	0.42 ± 0.02	0.33 ± 0.02	0.46	0.28	0.90	0.71	0.62	0.76	4.7	5.3
U Peg	0.375	F3 + F3	0.80 ± 0.07	3.99 ± 0.00	0.40 ± 0.01	0.35 ± 0.01	1.35	1.1	1.2	1.1	0.84	0.91	3.4	3.6
RZ Tau	0.416	F0 + F0	0.54 ± 0.05	3.96 ± 0.01	0.43 ± 0.02	0.32 ± 0.01	1.8	0.97	1.4	1.1	0.63	0.80	2.7	3.2
W UMa	0.334	F8 + F7	0.50 ± 0.02	3.94 ± 0.01	0.441 ± 0.004	0.312 ± 0.03	1.30	0.65	1.11	0.79	0.94	1.24	4.1	4.7
AH Vir	0.408	K0 + G6	0.42 ± 0.05	3.92 ± 0.01	0.46 ± 0.01	0.30 ± 0.01	2.0	0.84	1.5	0.98	0.60	0.86	4.2	4.8

luminosities likewise tend to equal in systems with equally massive components (cf. Binnendijk, 1965) underlines their affinity to detached systems on the Main Sequence (cf. sec. VIII-3A). Geometrically, however, W UMa-type systems are far from being detached: in point of fact, an interpretation of heir light curves is impossible except on the basis of a working hypothesis that both components fill completely the largest closed Roche lobes capable of containing their mass – and thus constitute, not detached, but *contact* (or, possibly, 'oversize') *systems.*

Let us inquire more closely into the meaning of the most important photometric message which these stars send us in the form of their light changes – particularly the continuous variations of light caused by the distortion of figure; these are photometrically more important not only because (unlike eclipses) they never cease, but also because their amplitudes are often larger than those of additional light changes that may arise from the eclipses. These amplitudes are, indeed, so large that they can be reconciled with a contact model of the respective systems only if we assume (cf. Kopal, 1968b) that *photometric effects of geometrical distortion are augmented by a very large amount of gravity-darkening* – on the average, in excess of that appropriate for the case of radiative transfer of total flux of light (see sec. IV-4A) by a factor close to two.

This fact poses, however, a real problem; for stars of physical properties of the components of W UMa-type systems are expected to possess convective sub-photospheric layers of some depth; and it has been shown by Lucy (1967) that, in such a case, the coefficient of gravity darkening should be only approximately one-third of that appropriate for the case of radiative transfer.* And this, in turn, confronts us with the following dilemma: either the sub-photospheric layers of W UMa-type stars are not in convective equilibrium, and the radiative gravity-darkening is augmented in excess to von Zeipel's value by some process as yet unknown, or *the components in such systems possess fractional dimensions in excess of their Roche limits* – in other words, they represent 'oversize' stars which are not merely in contact with each other at the inner Lagrangian point L_1, but are surrounded by a common envelope, overflowing the Roche limit, which is optically thick and endows the whole configuration with the form of a dumb-bell. In such a case, we should more properly regard the W UMa-type stars not as close binaries, but as highly elongated single configurations.

Unfortunately, the photometry of the light curves does not permit us to discriminate between these two possibilities by an analysis of the light changes within eclipses; since for dumb-bell configurations no reliable theory is so far on hand which should enable us to predict light variations caused by axial rotation of highly elongated body and to compare them with the observations. The problem must, therefore, be left temporarily in abeyance at this tantalizing stage; and, in what follows, we shall turn our attention on another facet of the problem concerning the evolutionary stage of these enigmatic configurations. Are binaries of the W UMa-type young or evolved; and what is their past or the future?

That stars of this group must represent very close systems was obvious from their light curves ever since W UMa was discovered at the commencement of this century (cf. Müller and Kempf, 1903). However, the first investigator who identified them with a pair of

* The conventional coefficient τ of gravity darkening is connected with Lucy's parameter β by the relation $\tau = 4\beta$.

contact configurations in the sense used in this book appears to have been Kuiper (1941); and he also happened to advance first the argument which proved, much more recently, to offer the point of departure to investigations which we wish now briefly to describe: namely, why such systems with components of unequal masses cannot be identified with zero-age Main Sequence stars in dynamical equilibrium.

The argument is based on a comparison of the mass-radius relation for single, chemically homogeneous stars with the conditions to be fulfilled if the surface of the binary as a whole is a common equipotential. Let $m_{1,2}$ and $R_{1,2}$ denote the masses and (mean) radii of the two components, related by the identity

$$\frac{R_2}{R_1} = \left(\frac{m_1}{m_1}\right)^{\alpha}, \tag{3.5}$$

where

$$\alpha \equiv \frac{\log (R_2/R_1)}{\log (m_2/m_1)} \equiv \frac{\log k}{\log q} . \tag{3.6}$$

For systems consisting of a pair of stars in contact with their Roche limits the values of k and q should, however, be related by the geometry of the common equipotential as given in columns (1) and (3) of Table VI-3 from which we find that

q	α
0.9	0.4918
0.8	0.4926
0.7	0.4925
0.6	0.4922
0.5	0.4917
0.4	0.4906
0.3	0.4890
0.2	0.4861
0.15	0.4836
0.1	0.4799

i.e., that the value of α remains remarkably constant and equal to 0.49* over a wide range of q. On the other hand, for the Main Sequence stars with masses less than $2\odot$ Equation (3.6) yields $\alpha = (1.02)^{-1} = 0.98$. It seems, therefore, that such stars cannot satisfy the contact condition (3.5) unless the masses of the two components are equal. From this fact Kuiper (1941a) concluded that contact binaries cannot exist as zero-age Main Sequence stars with mass-ratios differing from unity. We pointed out, however, already that the majority of known systems of W UMa-type possess components the masses of which are very unequal; and if these do form contact systems, two alternative explanations can be advanced to account for this fact: either they have evolved to their present state from zero-age detached binaries on the Main Sequence, or they do not constitute equilibrium configurations.

Kuiper himself favoured the latter alternative. However, more recently Lucy (1968) pointed out that contact binaries with components of unequal masses can exist in equi-

* Kuiper (1941a) adopted a value of $\alpha = 0.46$ on the basis of less accurate computations.

librium at zero age if the system is surrounded by a convective envelope with a common adiabatic constant. If, moreover, this envelope exterior to the Roche limit is optically thick – as is strongly suggested by the amplitude of the light changes due to proximity effects – one can retain the equilibrium model as a basis for an interpretation of the observations.

Lucy's ideas were subsequently developed further by several investigators (Hazlehurst, 1970; Moss and Whelan, 1970; Moss, 1971, 1972; and many others referred to in Bibliographical Notes at the end of this chapter); but a satisfactory agreement between theory and all aspects of observational evidence has not yet been attained – leaving the idea that the W UMa-type stars are young (i.e., unevolved) systems surrounded by common convective envelopes still in the realm of hypotheses. The only reliable way of advancing the solution of the problem is obviously to appeal to the observations; and Nature has indeed provided us with some clues which we shall attempt to follow.

The most important clue so provided concerns the membership of several systems of W UMa-type in galactic clusters of known ages. The cluster IC 2994 which provides a celestial home for BH and LW Cen is extremely young (cf. Thackeray, 1964) – with stars later than B3 still contracting towards the Main Sequence. The two variables just quoted appear to be among them (Eggen, 1967); and should, therefore, be not more than 5×10^6 yr old. On the other hand, EM Cep in NGC 7160 or V 701 Sco in NGC 6383 are among the absolutely brightest stars of these clusters; and their positions in the HR-diagram (cf. Figure 15 of Eggen, 1967) indicate that they already evolved away from the Main Sequence. Lastly, TX Cnc in the Praesepe cluster cannot be younger than $6–9 \times 10^8$ yr which is the estimated age of that cluster (cf. van den Heuvel, 1969; or Mauder, 1972). As the mass of the primary component of TX Cnc should be very close to that of the Sun (1.0 ⊙ according to Whelan *et al.*, 1973), its contraction to the Main Sequence should be accomplished in no more than 3×10^7 yr. This star must, therefore, have been on the Main Sequence of the Praesepe population already for several hundred million years – though it can still be regarded as 'unevolved' (since its nuclear evolution in less than 10^9 yr would not at least have progressed very far). But this cannot manifestly be the case with W UMa-type variables discovered in really old galactic clusters like M67 or NGC 188 (cf. Efremov *et al.*, 1964; or Kurochkin, 1965) whose ages are not less than 5 to 10×10^9 yr (Eggen and Sandage, 1969).

In point of fact – with possible exceptions of B-type W UMa-stars like EM Cep or V 701 Sco – a large majority of known systems of this type are likely to be of at least moderate age. Their kinematic properties (cf. Schatzman and Rigal, 1954; Rigal, 1955; Artiukhina, 1964) and a moderate concentration towards the galactic plane (Popov, 1964) indicate that these stars do not belong to the extreme Population I, but rather are members of the more intermediate disc-type population. On the other hand, their ages are certainly not sufficient for attaining a contact stage by a post-Main Sequence evolution of formerly detached systems – their masses are too small to enable them to accomplish such a feat in less than 10^{10} yr. Besides, an adequate backlog of systems from which the present W UMa-type could have evolved is conspicuous by its absence. It has already been pointed out by Kraft (1967) that the numbers of known W UMa-type stars exceeds those of detached systems on the Main Sequence by so wide a margin that the former cannot be regarded as the offspring of the latter – the hypothetical parents are just too few!

Besides, the age of the W UMa-type systems poses also the following fundamental problem: if these age are of the order of 10^9 yr, their frequency in the sky could be so high because so many were formed in the early stage of the disc of our Galaxy by a process as yet unspecified. But it they are only 10^8 (let alone 10^7) yr old, what has happened to the W UMa-type stars of preceding generations? In which guise are they known to us today? Kraft (1962, 1967) saw them in the present subdwarfs with explosive components (cf. sec. VIII-5B); while Hazlehurst and Meyer-Hofmeister (1973) suggested that W UMa-type binaries may eventually coalesce to form single rapidly-rotating A-stars. Neither view can, however, claim the status of anything more than that of a tentative hypothesis, awaiting the verdict of future investigations.

As regards the process by which such systems could have come into being, the only reasonable explanation would be an assumption that *their initial contraction towards the Main Sequence had been arrested at a stage in which we see them today* – i.e., as condensations around preexisting gravitational *dipoles*, extending to (or beyond) their present Roche limits. Their current luminosity is no doubt being defrayed by hydrogen burning; but the cores of both components may be so close to each other that their envelopes overlap. For semi-detached systems such a possibility was ruled out, because no conceivable reason could have arrested a contraction of their secondary (less massive) components at the Roche limit – at which they cluster so conspicuously. In contrast, for W UMa-type systems we cannot derive as yet any assurance from the observed light curves whether their components are in contact with their respective Roche limits, or actually 'oversize'. The latter alternative would, at present, seem to be more probable of the two; and as long as this is the case, the contraction option for an explanation of their present state cannot be ruled out.

VIII-4. Tidal Evolution of Close Binary Systems

In the preceding section of this chapter we have endeavoured to understand the principal features of the phenomena exhibited by close binary systems, and described schematically in section VIII-3, in terms of the evolutionary theories for single stars as outlined in the first section of this chapter. A limited success of this effort – which we did not try to conceal from the reader or camouflage by plausible but unproven assertions – is bound to give rise to second throughts about a possible significance of such an effort. May it not be indicative of the fact that some essential features, necessary for a fuller understanding of the problem, have as yet been left out of the picture?

In order to appreciate the reasons for this concern, let us recall that investigations of the subject carried out by virtually all previous investigators (cf., e.g., Plavec, 1970; p. 969) have been based on the following three assumptions.

(1) Both components evolve as single stars and independently of each other until the more massive one reaches the Roche limit.

(2) In the course of subsequent evolution, the total mass as well as angular momentum of the system remain preserved – i.e., material may be exchanged between the components, but not lost to the system.

(3) The components are assumed to remain spherical, revolve in circular orbits and rotate with the Keplerian angular velocity of orbital revolution.

These assumptions have, of course, been made for the sake of convenience of treatment rather than their intrinsic reasonableness; but what can they have in common with the actual behaviour of close binary systems? Very little; for they altogether ignore the fact that *the components of close binaries are partners in the same dynamical system;* and as such *are bound to influence each other in the course of their evolution.* This influence may not be very important in well-separated detached systems on the Main Sequence; and this is why their observed properties have been found in such a good agreement with the theory as detailed in section 3A of this chapter. However, for evolved systems discussed in sections 3B and C the interaction phenomena between the components may become not only important, but dominant; and cannot be disposed of by arguments of hand-waving nature – a gymnastics employed not too infrequently to chase away uncomfortable thoughts about problems which we are not yet in a position to face.

Consider, for instance, the internal contradictions inherent in the above-mentioned premises. How can a star rotate about an axis and retain spherical shape? It can be argued that polar flattening due to slow rotation may be small enough to be treated as a perturbation; so that the postulated spherical symmetry may constitute only a harmless false assumption. This may indeed be the case for detached systems, where the distortion of the components is often small enough for its consequences on the internal strucure to be ignorable (or treated as a small perturbation). It will, however, manifestly not be the case when the star commences to approach the Roche limit – which (as we have seen in Chapter VI) deviates from a sphere by quantities of zero order. Incidentally, how does a spherical configuration attain a Roche limit which is manifestly non-spherical? It cannot obviously do it everywhere at the same time; and if so, what are the consequences of this fact? Besides, for dynamical systems isolated in space it is not sufficient to conserve only their orbital momenta. What happens with the rotational momenta of the components in the course of their expansion to the Roche limit? If these are also conserved (to keep the total momentum of the respective system constant), the components cannot obviously continue to rotate with the same angular velocity – Keplerian or any other.

There is, moreover, yet another aspect of the problem which must be kept in mind in this connection: namely, the limits to the validity of the 'mirror effect' in the evolution of stars at the post-Main Sequence stage, according to which potential energy liberated by contraction of hydrogen-poor core is used up to expand the outer layers of the star. It is this effect which is responsible for the expansion of the evolving stars to the subgiant or giant stage (as illustrated by the data given in Table VIII-1); and its validity is inevitable in single stars which cannot communicate dynamically with their environment. However, in close binary systems the *tidal interaction* between the components can redistribute potential energy (as well as momentum) between them and their orbit; and thus undermine the literal validity of the 'mirror effect' on which the data given in Table VIII-1 have been based.

As long as the relative orbit of the two stars is circular, and rotation about axes perpendicular to the orbital plane has been synchronized with the revolution, the tides remain of the equilibrium type (cf. Chapter II of this book) and entail no motions in the frame of reference co-rotating with the star. If, however, any one of the aforementioned conditions is violated, dynamical tides arise (cf. Chapter III) which cause differential motions in the star's mass: an inclination of the equator to the orbital plane will produce motions along the meridians; while nonsynchronism between rotation and revolution will cause the tides to sweep around the equator in lingitude, following (with appropriate lag) the motion of the attracting mass. Lastly, a finite eccentricity of the orbit will cause the

j-th harmonic tides to ebb and fall with the inverse $(j + 1)$st power of the instantaneous radius-vector.

As long as the material constituting the stars could be regarded as inviscid, the crest of each tide would always remain oriented in the direction of the attracting body. However, as soon as the viscosity of stellar material (gas, plasma, turbulence) makes itself felt, this will cease to be true (cf. secs. II-3 and 4); and tides begin to lag in proportion of the viscosity. The gravitational attraction on a tidal bulge which (because of tidal lag) becomes asymmetric with respect to the radius-vector will give rise to *torques*, whose dynamical effects have been discussed at some length in Chapter V.

Such torques transfer also angular momentum; and, moreover, the kinetic energy of axial rotation can be partly dissipated by tides into heat, and partly converted into kinetic as well as potential energy of the orbital motion. Nonradial oscillations of tidal origin, caused by orbital eccentricity, are likewise made to lag in phase behind orbital motion by viscous dissipation – in the sense that tides do not attain maximum height when the two components are closest to each other, but only some time after the periastron passage. Since, however, such oscillations are always symmetrical with respect to the radius vector, no angular momentum is transferred by their action. On the other hand, they likewise tend to dissipate kinetic energy into heat, and thus lose mechanical energy to the system.

In spite of this somewhat complicated situation, certain quantities exist which should be *preserved* by close binary systems over long intervals of time. The most important of these is their *total momentum* of the system – representing a sum of the rotational momenta of the components and of the orbital momentum of relative motion. As we stressed repeatedly in earlier parts of this book, close binaries constitute isolated dynamical systems in space – essentially immune to gravitational interaction with other stars by an extreme scarcity of close encounters between them. Therefore, their total initial momenta should remain preserved in the course of time to a far-reaching exactitude as long as their mass remains the same. Barring a change in mass, the only 'leakage' of momentum is possible by its removal by radiation. A fractional loss of momentum by photos of escaping light is, however, altogether negligible for most stars part of their life, and will hereafter be ignored.

Another quantity which should likewise be preserved in the course of (albeit shorter) time is the total *energy* of a binary system – representing a sum of potential, kinetic and thermal energy of the constituent components as well as the kinetic energy of their orbit. Radiation, to be sure, constitutes a more serious leakage of energy than was the case with the momentum. The possibility, furthermore, exists that, in addition to the energy loss through electromagnetic radiation, some energy is being carried away (at least in very close systems with ultra-short periods) also by gravitational waves, the existence of which is inherent in any field theory of gravitation. The magnitude of this loss, and its effects on the orbital periods of dwarf binaries has already been the subject of several investigations (cf., e.g., Kraft et al., 1962; Peters and Mathews, 1963; Braginski, 1965; or Paczynski, 1967), on the basis of the linearized equations of general relativity. It has, however, been shown more recently (cf. Ehlers *et al.*, 1976) that a linearized theory does not provide an adequate basis for a quantitative determination of the magnitude of the energy loss through gravitational waves; and that no formula derived so far for this purpose is really consistent with the general theory of relativity. As long as this appears to be the case, a more specific consideration of the effects of gravitational radiation by close binary systems would be clearly premature, and will be left out of this book.

Notwithstanding these circumstances, a continuous exchange between different types of energy mentioned above should, however, be operative as a result of evolutionary changes in the interiors of the constituent components of close binary systems, as well as between them as a result of dissipative processes connected with dynamical tides; for the latter must tend gradually to degrade kinetic energy into heat through the medium of viscosity. As a result, the kinetic (and potential) energy of close binary systems whose components interact through viscous dynamical tides should secularly diminish, while the total momentum remains constant. Therefore, *a ratio of energy-to-momentum in any given system can, in principle, serve as an indicator of its evolutionary stage.*

The first investigator who voiced this idea almost a century ago was Darwin (1879). He considered, to be sure, only a very simplified case of a binary system in which only one component was of finite size (and angular momentum), but was constrained to remain spherical in form; while its companion was regarded as a mass-point of finite mass but zero angular momentum. The orbit of two such bodies was, moreover, constrained to remain circular.

Within the scheme of such an approximation Darwin showed that, for systems of constant momentum, the states of maximum and minimum energy coincided with those of synchronism between rotation and revolution; moreover, he proved that these were the only cases in which synchronism between rotation and revolution could be encountered. Therefore, such tidal evolution as may occur in the course of time must be accomodated between these two extremes – in which the bulk of the kinetic energy of the system is stored in the axial rotation of the components and their orbital motion, respectively – at least as long as the mass of the system remains conserved.

So simplified a model of a binary system which Darwin used would, of course, no longer satisfy a contemporary student of the subject. In order to generalize it, in the present section we shall allow both components to possess finite amounts of angular momenta; and their orbit to be characterized by an arbitrary eccentricity. In addition, a process will be outlined by which the effects of mutual tidal distortion of both components can be taken into account in the formation of the expression for the total energy as well as momentum. Moreover, profiting from our earlier work in section IV-3 on the effects of dynamical tides on the axial rotation of the components, we shall be able to treat the angular velocities $\omega_{1,2}$ of axial rotation of the two components as well as the semi-major axis A or the eccentricity e of their relative orbit as dependent variables of the problem (cf. sec. V-3). The same should, in principle, be true also of the absolute dimensions (or, more specifically of the radii of gyration) of the constituent stars as they change in the course of their evolution. As, however, these are governed by systems of equations different from those treated in this book, that aspect of the problem will remain outside the scope of the present volume; though it offers a challenging opportunity for further work which will have to be undertaken before our understanding of the problem of the evolution of close binary systems can be put on a more satisfactory basis.

A. ENERGY AND MOMENTUM

In order to investigate the range through which a binary system of constant total momentum can evolve in the course of a continuing decrease of its kinetic and potential energy, consider first the total amount of momentum which must be conserved in the course of time. Let, as before, $m_{1,2}$ denote the masses of the two components (in general, we shall assume $m_1 \geqslant m_2$); and A, the semi-major axis of their relative orbit of eccentricity e. If so, the angular momentum H_0 associated with the orbital motion of the two components about their common centre of gravity is given by

$$H_0 = \frac{m_1 m_2}{m_1 + m_2} \left(r^2 \frac{dv}{dt} \right), \tag{4.1}$$

where r denotes the instantaneous radius-vector and v, the true anomaly measured from the periastron passage. Kepler's second law of (unperturbed) elliptic motion, of the form

$$r^2 \frac{dv}{dt} = \sqrt{G(m_1 + m_2)A(1 - e^2)} \quad , \tag{4.2}$$

where G denotes the constant of gravitation, enables us to reduce (4.1) to

$$H_0^2 = G \left\{ \frac{m_1^2 m_2^2}{m_1 + m_2} \right\} A(1 - e^2); \tag{4.3}$$

which by virtue of Kepler's third law

$$\Omega^2 = \frac{G(m_1 + m_2)}{A^3} \tag{4.4}$$

can be rewritten as

$$H_0 = G^{2/3} m_1 m_2 (m_1 + m_2)^{-1/3} \Omega^{-1/3} (1 - e^2)^{1/2}. \tag{4.5}$$

The angular moments $H_{1,2}$ of axial rotation of the two components are obviously given by

$$H_i = C_i \omega_i + C_i \Omega, \qquad i = 1,2 \;, \tag{4.6}$$

where $\omega_{1,2}$ stand for the angular velocities of axial rotation of the respective configurations; and $C_{1,2}$, for their moments of inertia about the axis of rotation; while $C_{1,2}$ represent the contributions to rotational momenta arising from the tides which the two components mutually raise on each other.

In what follows we shall assume that the axes of rotation of both components are perpendicular to the plane of their orbit. If so, neither of the two Keplerian laws as represented by Equations (4.2) and (4.4) is subject to any perturbations of first order caused by mutual distortion of the components in close binary systems; while the values of $C_{1,2}$ as well as $C_{1,2}$ are, to this order of accuracy, given by

$$C_i = m_i \mathfrak{H}_i^2 + \frac{4(k_2)_i \omega_i^2 R_i^5}{9G} + \ldots \tag{4.7}$$

and

$$C_i = \frac{2}{3} m_{3-i} (k_2)_i R_i^2 \left(\frac{R_i}{r}\right)^3 + \ldots, \tag{4.8}$$

$i = 1,2$; where $\mathfrak{H}_i$ denotes the radius of gyration of the respective component of external radius R_i, defined by the equation

$$m_i \mathfrak{H}_i^2 = \frac{8}{3} \pi \int_0^{R_i} \rho a^4 \mathrm{d}a; \tag{4.9}$$

while the 'apsidal-motion' constants $(k_j)_i$, associated with j-th spherical harmonic distortion, are given by Equation (6.4) – or, for sufficiently high degrees of central condensation (cf. Equation (6.9) of Chapter II),

$$(k_j)_i = \frac{4(j+2)\pi}{(2j+1) m_i R_i^{2j+1}} \int_0^{R_i} \rho a^{2j+3} \mathrm{d}a\,. \tag{4.10}$$

The total angular momentum H of the system, which should be secularly conserved to a very high degree of exactitude, will then be given by the equation

$$H = H_0 + H_1 + H_2, \tag{4.11}$$

where H_0 and $H_{1,2}$ can be inserted from (4.5) and (4.6). In order to reduce this expression to a nondimensional form, let the combined mass $m_1 + m_2$ of the system be hereafter adopted as our unit of mass, and L as the unit of length. If, moreover, τ_0 denotes the corresponding unit of the time, Equation (4.11) can be rewritten more explicitly as

$$H = \frac{(m_1 + m_2)L^2}{\tau_0} \left\{ \frac{G^{2/3} m_1 m_2}{(m_1 + m_2)^{4/3}} \frac{\sqrt{1-e^2}}{L^2} \tau_0^{4/3} (\tau_0 \Omega)^{-1/3} + \right. \tag{4.12}$$

$$\left. + \frac{C_1 \tau_0 \omega_1}{(m_1 + m_2)L^2} + \frac{C_2 \tau_0 \omega_2}{(m_1 + m_2)L^2} + \frac{(\bar{C}_1 + \bar{C}_2)(\tau_0 \Omega)}{(m_1 + m_2)L^2} \right\}.$$

In order to simplify this expression, let us set

$$\frac{\tau_0^{4/3} G^{2/3} m_1 m_2}{(m_1 + m_2)^{4/3} L^2} = 1, \tag{4.13}$$

yielding

$$\tau_0 = \frac{(m_1 + m_2) L^{3/2}}{G^{1/2} (m_1 m_2)^{3/4}} \tag{4.14}$$

as our unit of time, and

$$\frac{(m_1 + m_2)L^2}{\tau_0} = \{G(m_1 m_2)^{3/2} L\}^{1/2} \tag{4.15}$$

as the unit of momentum.

If, lastly, we set

$$(\tau_0 \Omega)^{-1/3} = x, \tag{4.16}$$

$$\tau_0 \omega_1 = y, \tag{4.17}$$

$$\tau_0 \omega_2 = z; \tag{4.18}$$

and

$$\frac{C_{1,2}}{(m_1 + m_2)L^2} = \kappa_{1,2} \tag{4.19}$$

while

$$\frac{\overline{C_1} + \overline{C_2}}{(m_1 + m_2)L^2} \tag{4.20}$$

Equation (4.12) can be rewritten as

$$h = x\sqrt{1 - e^2} + \lambda x^{-3} + \kappa_1 y + \kappa_2 z, \tag{4.21}$$

where h denotes now the normalized value of the total momentum H, and x, y, z as well as $\kappa_{1,2}$, λ are nondimensional variables and parameters.

Having formulated the complete expression for the total angular momentum (rotational plus orbital) of a close binary system, let us proceed to formulate similarly a complete expression for the total energy of the system – kinetic, potential, and thermal. Let, in what follows, T_0 denote the kinetic energy of orbital motion, and $T_{1,2}$ that of axial rotation. As is well known, the kinetic energy of the orbital motion can be generally represented by an expression of the form

$$T_0 = \frac{1}{2} m_1 (\dot{x}_1^2 + \dot{y}_1^2 + \dot{z}_1^2) + \frac{1}{2} m_2 (\dot{x}_2^2 + \dot{y}_2^2 + \dot{z}_2^2), \tag{4.22}$$

where $x_{1,2}$, $y_{1,2}$, $z_{1,2}$ denote the rectangular coordinates of the centres of gravity of the two components of mass $m_{1,2}$ in a system whith origin at the centre of gravity of the binary; and dots denote their time-derivatives.

In order to evaluate the kinetic energy of relative orbital motion, we shift the origin of coordinates from the centre of gravity of the system to that of one (say, the primary) component by setting

$$x_2 - x_1 = x, \quad y_2 - y_1 = y, \quad z_2 - z_1 = z, \tag{4.23}$$

while, by the integrals of the mass centre,

$$m_1 x_1 + m_2 x_2 = m_1 y_1 + m_2 y_2 = m_1 z_1 + m_2 z_2 = 0. \tag{4.24}$$

Eliminating $x_{1,2}, y_{1,2}, z_{1,2}$ from (4.22) with the aid of (4.23) and (4.24) we find that

$$T_0 = \frac{1}{2}\,\frac{m_1 m_2}{m_1 + m_2}(\dot{x}^2 + \dot{y}^2 + \dot{z}^2), \tag{4.25}$$

where

$$\dot{x}^2 + \dot{y}^2 + \dot{z}^2 \equiv V^2 \tag{4.26}$$

represents the square of the velocity V of motion of the secondary in its relative orbit around the primary star. On the other hand, by a well-known integral of the problem of two bodies,

$$V^2 = G(m_1 + m_2)\left\{\frac{2}{r} - \frac{1}{A}\right\}; \tag{4.27}$$

and as the mean value of r^{-1} by the laws of elliptic motion is given by

$$\frac{1}{P}\int_0^P \frac{dt}{r} = \frac{1}{A} \tag{4.28}$$

over the orbital cycle of period P, the kinetic energy of orbital motion

$$T_0 = \frac{Gm_1m_2}{2A} = \frac{1}{2}\,G^{2/3}\,m_1m_2(m_1+m_2)^{-1/3}\Omega^{2/3} \tag{4.29}$$

by (4.4); while the kinetic energy of axial rotation

$$T_i = \tfrac{1}{2}\,C_i\,\omega_i^2 + \tfrac{1}{2}\,\overline{C}_i\,\Omega^2. \tag{4.30}$$

The total potential energy W of our binary system (equal to the negative value of gravitational energy liberated when the configurations contract from infinite distension to their present size) can be expressed as

$$-W = W_0 + W_1 + W_2, \tag{4.31}$$

where W_0 represents the relative potential energy of the two bodies, while $W_{1,2}$ stands for the separate potential energy of each component. As is well known, the relative potential energy of the two bodies can be expressed as

$$W_0 = \frac{Gm_1m_2}{r}\{1 + S_{12}\}, \tag{4.32}$$

where, correctly to quantities of first order in surficial distortion (cf. Equations (1.5) and (1.41)–(1.42) of Chapter V),

$$S_{12} \equiv \frac{r\,\mathfrak{R}_{12}}{G(m_1+m_2)} \sum_{i=1}^{2}\sum_{j=2}^{4}(k_j)_i\,\frac{m_{3-i}}{m_i}\left(\frac{R_i}{r}\right)^{2j+1} - \sum_{i=1}^{2}\frac{\omega_i^2}{2\pi G\rho_i}(k_2)_i\left(\frac{R_i}{r}\right)^2, \tag{4.33}$$

in which ρ_i stands for the mean density of the respective configurations; and $(k_j)_i$, for their internal-structure constants as defined by Equation (6.4) of Chapter II.

In addition, the potential energies of the individual components are given by the mass-integrals

$$\left.\begin{aligned} W_i &= G\int_0^{m_i} \frac{m(a)}{a}\,\mathrm{d}m\,, \\ \mathrm{d}m &= 4\pi\rho a^2\,\mathrm{d}a\,\sin\theta\,\mathrm{d}\theta\,\mathrm{d}\phi\,, \end{aligned}\right\} \tag{4.34}$$

which depend only on the internal distribution of density ρ, and are unaffected by distortion to the first order in small quantities.

The internal thermal energy E_T of the individual components is given by

$$E_T = \int_0^{m_i} \frac{P}{\rho}\frac{\mathrm{d}m}{1-\gamma}\,, \tag{4.35}$$

where P denotes the pressure inside the respective configuration; and γ, the ratio of specific heats of stellar material. This thermal energy (being fed mainly by sources of nuclear origin) is – like the potential energy W_i – obviously unaffected by orbital parameters (except at the time of Helmholtz-Kelvin contraction). The only exception is the heat source arising from viscous friction of tidal origin; and the latter – being quadratic in the velocity components of tidal displacements – can be regarded as a small quantity of second order. The same is, moreover, true of the kinetic energy of nonradial oscillations produced by tides in systems with eccentric orbits, which will likewise be hereafter disregarded.

Within the scheme of this approximation – and if we omit parts like W_i or $(E_T)_i$ which are independent of the orbital elements – the total energy E of the system should be expressible by

$$\begin{aligned} E = T + W = \frac{1}{2}\{C_1\omega_1^2 + C_2\omega_2^2 + (\overline{C}_1 + \overline{C}_2)\,\Omega^2\} \;+ \\ + \frac{1}{2}\frac{m_1 m_2}{m_1 + m_2}A^2\Omega^2 - \frac{Gm_1m_2}{A}\{1 + \overline{S}_{12}\}\,, \end{aligned} \tag{4.36}$$

in which we replaced r^{-1} by its time-average A^{-1} over a cycle, and where $\overline{S}_{12}$ represents the average value of (4.33), in which

$$\frac{1}{P}\int_0^P \frac{\mathrm{d}t}{r^6} = \frac{1 + 3e^2 + \frac{3}{8}e^4}{A^6(1-e^2)^{9/2}}\,, \tag{4.37}$$

$$\frac{1}{P}\int_0^P \frac{dt}{r^8} = \frac{1+\frac{15}{2}e^2+\frac{45}{8}e^4+\frac{5}{16}e^6}{A^8(1-e^2)^{13/2}}, \tag{4.38}$$

$$\frac{1}{P}\int_0^P \frac{dt}{r^{10}} = \frac{1+14e^2+\frac{105}{4}e^4+\frac{35}{4}e^6+\frac{35}{128}e^8}{A^{10}(1-e^2)^{17/2}}, \tag{4.39}$$

are rapidly-increasing functions of orbital eccentricity, exact for any value of $0 \leqslant e < 1$, and identical with A^{-2j} times the Hansen's coefficients $X_0^{-2j,0}(e), j = 3, 4, 5$, introduced already by Equation (3.3) of Chapter V.

To begin with, let us disregard the small term involving S_{12} on the right-hand side of (4.36), and on elimination of A rewrite this latter equation as

$$2E = C_1\omega_1^2 + C_2\omega_2^2 + (\bar{C}_1 + \bar{C}_2)\Omega^2 - \\ - m_1 m_2 \left\{\frac{G^2\Omega^2}{m_1+m_2}\right\}^{1/3} \tag{4.40}$$

In order to normalize this equation in an appropriate manner, let us (consistent with (4.14)) adopt

$$\frac{(m_1+m_2)L^2}{\tau_0^2} = \frac{G(m_1m_2)^{3/2}}{(m_1+m_2)L} \tag{4.41}$$

as our unit of energy, while all others remain the same as before. If so, Equation (4.40) can obviously be rewritten in the nondimensional form

$$2E = \lambda x^{-6} - x^{-2} + \kappa_1 y^2 + \kappa_2 z^2, \tag{4.42}$$

where the variables x, y, z as well as the parameters $\kappa_{1,2}$ and λ continue to be given by Equations (4.16)–(4.20).

B. EVOLUTION WITH CONSTANT MOMENTUM

In order to ascertain the range through which a sum of the kinetic and potential energy of a binary system can vary while preserving the constancy of the momentum, let us depart from the normalized Equations (4.21) and (4.42) for the momentum and the energy; and regard the system to be wide enough not only for the distortion S_{12} occurring on the right-hand side of Equation (4.36), but also the parameter λ as negligible to a first approximation. In other words, we shall regard both components – though of finite size – to remain spherical. If, accordingly, the rotational momenta of both components, as given by Equation (4.7) reduce to their leading terms $m_1\mathfrak{H}_1^2$, the remaining parameters $\kappa_{1,2}$ of our problem can be regarded as constant; and Equations (.4.21) with (4.42) simplified (for circular orbits) to

$$h = x + \kappa_1 y + \kappa_2 z, \tag{4.43}$$

$$2E = -x^{-2} + \kappa_1 y^2 + \kappa_2 z^2, \tag{4.44}$$

which will constitute the basis for our present work in this section.

If we eliminate x from the equation for the energy with the aid of that for the momentum,

$$2E = \kappa_1 y^2 + \kappa_2 z^2 - (h - \kappa_1 y - \kappa_2 z)^{-2}, \tag{4.45}$$

and the conditions for the extrema

$$\frac{\partial E}{\partial y} = \frac{\partial E}{\partial z} = 0 \tag{4.46}$$

will be fulfilled if

$$y = z = (h - \kappa_1 y - \kappa_2 z)^{-3}. \tag{4.47}$$

Setting $z = y$ in the foregoing equation we find that

$$y = [h - (\kappa_1 + \kappa_2)y]^{-3}, \tag{4.48}$$

and similarly (if we wish) for z. If, moreover, we abbreviate

$$h - (\kappa_1 + \kappa_2)y = y, \tag{4.49}$$

so that

$$y = \frac{h - y}{\kappa_1 + \kappa_2}, \tag{4.50}$$

Equation (4.48) assumes the form

$$y^4 - h\,Y^3 + (\kappa_1 + \kappa_2) = 0\,. \tag{4.51}$$

If, lastly, we set

$$Y = (\kappa_1 + \kappa_2)^{1/4}\,Y' \tag{4.52}$$

and

$$h = (\kappa_1 + \kappa_2)^{1/4}\,h', \tag{4.53}$$

Equation (4.49) can be reduced to a biquadratic

$$Y'^4 - h'\,Y'^3 + 1 = 0, \tag{4.54}$$

containing only one constant arbitrary parameter. This equation will be fundamental to all arguments which we shall develop in this section.

In order to solve the foregoing biquadratic, let us introduce an auxiliary quantity ρ defined as a solution of

$$\rho^3 - 4\rho = h'^2. \tag{4.55}$$

If so, then

$$Y'^4 - h'Y'^3 + 1 = \left\{ Y'^2 + 2Y'\left(\frac{\rho^{3/2} - h'}{4}\right) + \frac{\rho^{3/2} - h'}{2\rho^{1/2}} \right\} \times$$
$$\times \left\{ Y'^2 - 2Y'\left(\frac{\rho^{3/2} + h'}{4}\right) + \frac{\rho^{3/2} + h'}{2\rho^{1/2}} \right\} ; \tag{4.56}$$

which by virtue of the identity

$$(\rho^{3/2} + h')(\rho^{3/2} - h') = 4\rho \tag{4.57}$$

allows us to rewrite our biquadratic in the form of the product

$$\left\{\left[Y' + \frac{1}{4}(\rho^{3/2} - h')\right]^2 + \left[\frac{1}{4}(\rho^{3/2} - h')\sqrt{1 + 2h'\rho^{-3/2}}\right]^2\right\} \times$$
$$\times \left\{\left[Y' - \frac{1}{4}(\rho^{3/2} + h')\right]^2 + \frac{1}{4}\left[(\rho^{3/2} + h')\sqrt{1 - 2h'\rho^{-3/2}}\right]^2\right\} = 0. \tag{4.58}$$

This result makes it obvious that the nature of the roots of this biquadratic depends on the value of h'. If

$$h' > \frac{4}{3^{3/4}}, \tag{4.59}$$

our biquadratic admits of two real and distinct roots of the form

$$Y'_{1,2} = \frac{1}{4}(\rho^{3/2} + h')\ \{1 \pm \sqrt{2h'\rho^{-3/2} - 1}\}, \tag{4.60}$$

and two complex roots (which are of no physical interest); while if

$$h' < 4/3^{3/4}, \tag{4.61}$$

all four roots of (3.12) are purely imaginary. Accordingly, by (4.50), (4.52) and (4.60),

$$y_{1,2} = \frac{h}{\kappa_1 + \kappa_2} - \frac{\rho^{3/2} + h'}{4(\kappa_1 + \kappa_2)^{3/4}}\ \{1 \pm \sqrt{2h'\rho^{-3/2} - 1}\}; \tag{4.62}$$

and, by (4.47),

$$z_{1,2} = y_{1,2}\,. \tag{4.63}$$

Consistent with the extremal conditions (4.46) which led to our biquadratic, the roots (4.62) and (4.63) obviously correspond to the states of maximum and minimum energy consistent with a given momentum h. It is, furthermore, easy to show that each one of the real roots of our biquadratic corresponds to the state of synchronism between rotation and revolution of the two components. For, in accordance with Equations (4.16)–(4.18) of the preceding section,

$$\frac{\omega_1}{\Omega} = x^3 y \quad \text{and} \quad \frac{\omega_2}{\Omega} = x^3 z\,; \tag{4.64}$$

and if $\omega_1 = \omega_2 = \Omega$, we should have

$$x^3 y = 1 \quad \text{and} \quad x^3 z = 1 \,. \tag{4.65}$$

If, however, we insert in these equations for x from (4.43) and remember that – in accordance with (4.47) – $y = z$ at the extrema, the foregoing Equations (3.23) will reduce to (4.47) obtaining as a consequence of (4.46). In other words, *the conditions (4.46) requiring that the energy E be maximum or minimum for constant momentum h are consistent with Equations (4.65).* Both extrema correspond, therefore, to synchronism between rotation and revolution; the maximum corresponding to the case in which the bulk of the total momentum H is represented by the angular momenta $H_{1,2}$ of axial rotation; and the minimum, to a case in which most part of H has been transferred to the orbital momentum H_0. In the former case, the components are close to each other and rotate fast; in the latter, an increase in the size of the orbit has rendered the orbital momentum the dominant part of the whole; but the axial rotation of both components has been slowed down to the mean daily motion in a wide orbit. Moreover, in the limiting case $h'_{\min} = 4/3^{3/4}$, these two states coalesce into one and leave no room for evolution from one into the other; while if $h' < 4/3^{3/4}$, synchronism between rotation and revolution can never be attained.

Before we proceed to confront these consequences of the foregoing simple theory with the observations, let us consider still a simpler case to which our preceding analysis will reduce if one of the two coefficients $\kappa_{1,2}$ (say, κ_2) reduces to zero or a quantity very small in comparison with κ_1 – as it may happen even in systems whose components are comparable in mass, if the secondary's internal density concentration becomes so high that its radius of gyration $\mathfrak{H}_2 \to 0$. If so, all of our foregoing analysis not only continues to hold good for $\kappa_2 = 0$, but can be readily generalized to cover the case of arbitrary eccentricity e of the relative orbit of the two stars.

For if $e > 0$, Equation (4.43) follows then from (4.21) as

$$h = x\sqrt{1 - e^2} + \kappa_1 y \,; \tag{4.66}$$

and putting this through the mill of our procedure, we recover the biquadratic Equation (4.54) with its roots (4.60) provided that the foregoing definitions (4.52) and (4.53) for Y' and h' are replaced by

$$Y = \kappa_1^{1/4} (1 - e^2)^{-1/4}\, Y' \tag{4.67}$$

and

$$h = [\kappa_1 (1 - e^2)]^{1/4}\, h' \,. \tag{4.68}$$

After this digression, let us return to establish the maximum and minimum values of the energy E attainable for a given value of h (or h'). Since these extrema correspond to the two roots (4.62) for $y_{1,2}$ while $z_{1,2} = y_{1,2}$ and, by (4.65), $x_{1,2} = y_{1,2}^{-1/3}$ on insertion in (4.44) it follows that

$$2E_{1,2} = -y_{1,2}^{2/3} + (\kappa_1 + \kappa_2) y_{1,2}^2 \,, \tag{4.69}$$

where $y_{1,2}$ is given by (4.62). Moreover, since for $h' > 4/3^{3/4}$ the cubic equation (4.55) admits of only one real root of the form

$$\rho = \frac{4}{\sqrt{3}} \cosh \frac{1}{3} \phi , \tag{4.70}$$

where

$$\cosh \phi = \frac{3\sqrt{3}}{16} h'^2 = \left(\frac{h'}{h'_{\min}}\right)^2 , \tag{4.71}$$

it follows that the factor $\rho^{3/2}$ which occurs repeatedly in Equation (4.62) will be given by

$$\rho^{3/2} = (8/3^{3/4}) \cosh^{3/2} \frac{1}{3} \phi = h' \left\{ 1 + 3\, (h'_{\min}/h')^2 \cosh \frac{1}{3} \phi \right\}^{1/2} \tag{4.72}$$

The two values of $E_{1,2}$ as given by Equation (4.69) are distinct (and such that $E_1 \leqslant E_2$) provided that $h' > 4/3^{3/4}$ in accordance with (4.59); and we wish now briefly to consider what happens as $E_2 \to E_1$. For $E = E_2$, the axial rotation of both components starts in synchronism with their revolution in close proximity of each other. As $E < E_2$, the revolution begins to slow down as a result of increasing separation; and the products

$$x^3 y = \kappa_1^{-1} x^3 (h - x - \kappa_2 z) = (h - \kappa_1 y - \kappa_2 z)^3 y, \tag{4.73}$$

$$x^3 z = \kappa_2^{-1} x^3 (h - x - \kappa_1 y) = (h - \kappa_1 y - \kappa_2 z)^3 z , \tag{4.74}$$

obtained by an insertion of (4.43) in (4.65), will be maximized when

$$x = \frac{3}{4} (h - \kappa_1 y) \text{ or } \frac{3}{4} (h - \kappa_2 z) \tag{4.75}$$

and

$$x = 3\kappa_1 y \text{ or } 3\kappa_2 z \tag{4.76}$$

Moreover, the sum $\kappa_1(x^3 y) + \kappa_2(x^3 z)$ will be maximized for

$$x = \frac{3}{4} h, \tag{4.77}$$

implying that

$$[\kappa_1(x^3 y) + \kappa_2(x^3 z)]_{\max} = \frac{1}{3} \left(\frac{3}{4} h\right)^4 . \tag{4.78}$$

If, for example, κ_2(or κ_1) were equal to zero – such as would be the case if one component could be regarded as a mass-point – the maximum number of axial rotations per orbital revolution (i.e., the maximum number of 'days' in a 'month' would be given by the ratios

$$(\omega_1/\Omega)_{\max} \equiv (x^3 y)_{\max} = \frac{1}{3\kappa_1}\left(\frac{3}{4}h\right)^4 \tag{4.79}$$

or

$$(\omega_2/\Omega)_{\max} \equiv (x^3 z)_{\max} = \frac{1}{3\kappa_2}\left(\frac{3}{4}h\right)^4, \tag{4.80}$$

respectively.

C. COMPARISON WITH OBSERVATIONS

Returning to the main theme of our analysis, we note that Equations (3.17) and (3.27) in the preceding part of this section impose certain restraints on the physical properties of close binary systems which we wish to confront now with values furnished by the actual observations. In order to do so, let us first specify completely our system of units – of which a choice of the unit L of length has so far been left open. For investigations of any processes during which the total momentum H as given by Equations (4.11) or (4.12) is conserved, the possibility suggests itself to normalize our variables in such a way as to render its normalized value

$$h = 1. \tag{4.82}$$

If so, however, then it follows from Equation (4.15) that

$$L = \frac{H^2}{G(m_1 m_2)^{3/2}}, \tag{4.83}$$

which inserted in (4.14) yields

$$\tau_0 = \frac{(m_1 + m_2)H^3}{G^2(m_1 m_2)^3} \tag{4.84}$$

as our unit of time; while from (4.16)

$$x = \left\{\frac{G m_1^2 m_2^2 A}{(m_1 + m_2)H^2}\right\}^{1/2}, \tag{4.85}$$

rendering x^2 proportional to the semi-major axis A of the relative orbit. Lastly, by (4.19), the constants

$$\kappa_{1,2} = \left\{\frac{G^2(m_1 m_2)^3}{(m_1 + m_2)H^4}\right\} C_{1,2}. \tag{4.86}$$

Notice that the time-invariance of these units requires a constancy, not only of the sum $m_1 + m_2$ which has been adopted as the unit of mass, but of m_1 and m_2 separately.

With these facts in mind let us return now to the inequality (4.59), disclosing – in the light of (4.53) and (4.82) that

$$(\kappa_1 + \kappa_2)^{-1/4} \geqslant 4/3^{3/4} \tag{4.87}$$

if synchronism between rotation and revolution is to be possible at all. If, in particular, $\kappa_1 = \kappa_2$, it would follow that

$$\kappa_{1,2} \leqslant \frac{27}{512}, \tag{4.88}$$

or, by (4.86) that the moments of inertia $C_{1,2}$ of the two components

$$C_{1,2} \leqslant \frac{27}{512} \left[\frac{G^2(m_1 m_2)^3}{(m_1 + m_2)H^4} \right]^{-1} \tag{4.89}$$

If the density concentrations of the two components are pronounced and their radii of gyration small, the foregoing relations (4.88) and (4.89) should represent strong inequalities; but with decreasing central condensations (or increasing size) of the components the limit given by the equality sign may be approached.

Next, let us consider the limiting values $E_{1,2}$ of the total energy expressed in terms of

$$\frac{m_1 + m_2}{\tau_0^2} L^2 = \frac{G^2(m_1 m_2)^3}{(m_1 + m_2)H^2} \tag{4.90}$$

This fractional energy has already been defined by Equation (4.69) which for $h = 1$ can be rewritten as

$$2E_{1,2} = (y_{1,2}/h'^2)^2 - y_{1,2}^{2/3}, \tag{4.91}$$

where (by (4.62))

$$y_{1,2} = h'^4 \left\{ 1 - \frac{1}{4} (\alpha + 1) \left[1 \sqrt{2\alpha^{-1} - 1} \right] \right\}, \tag{4.92}$$

in which we have abbreviated

$$\alpha = \frac{\rho^{3/2}}{h} = 2\beta^{-1} \cosh^{3/2} \frac{1}{3} (\cosh^{-1} \beta^2) \tag{4.93}$$

and

$$\beta = h'/h'_{\min}. \tag{4.94}$$

For the minimum value $h'_{\min} = 4/3^{3/4}$ in accordance with (4.59), $\alpha = 2$. On the other hand, as $h' \to \infty$ (corresponding to both components being regarded as mass-points), Equation (4.55) makes it evident that $\rho^{3/2}/h' \equiv \alpha \to 1$. Therefore, as

$$h'_{\min} \leqslant h' \leqslant \infty, \tag{4.95}$$

the ratio α will be constrained to obey the inequality

$$2 \geqslant \alpha \geqslant 1. \tag{4.96}$$

Accordingly, the values of $y_{1,2}$ will vary from the double-root $(4/3)^3$ to $y_1 = 1, y_2 = \infty$ at $\alpha = 1$. Moreover, since

$$\cosh^{-1}\beta^2 = \log\{\beta^2 + \sqrt{\beta^4 - 1}\} \tag{4.97}$$

and, therefore,

$$2\cosh\tfrac{1}{3}\cosh^{-1}\beta^2 = [\beta^2 + \sqrt{\beta^4 - 1}]^{1/3} + [\beta^2 + \sqrt{\beta^4 - 1}]^{-1/3}; \tag{4.98}$$

it follows from (4.93) that

$$\begin{aligned} \alpha &= \left\{\frac{\beta^2 + \sqrt{\beta^4 - 1}}{2\beta^2}\right\}^{1/2} \left\{1 + [\beta^2 + \sqrt{\beta^4 - 1}\,]^{-2/3}\right\}^{3/2} \\ &= 1 + 6\,(4\beta)^{-4/3} + 6\,(4\beta)^{-8/3} - 36\,(4\beta)^{-12/3} + 70\,(4\beta)^{-16/3} + \cdots; \end{aligned} \tag{4.99}$$

or, inversely,

$$(4\beta)^{-4/3} = \frac{1}{6}(\alpha - 1) - \frac{1}{36}(\alpha - 1)^2 + \frac{1}{27}(\alpha - 1)^3 - \frac{35}{972}(\alpha - 1)^4 + \dots. \tag{4.100}$$

Moreover, on insertion of these results in (4.92) it follows that

$$\left.\begin{aligned} y_1 &= 1 - 3\,(4\beta)^{-8/3} + 116\,(4\beta)^{-12/3} + \dots \\ &= 1 - \frac{1}{12}(\alpha - 1)^2 + \frac{14}{27}(\alpha - 1)^3 + \dots \end{aligned}\right\} \tag{4.101}$$

and

$$\begin{aligned} y_2 &= 3^{-3}\,(4\beta)^4\,\{1 - 3(4\beta)^{-4/3} + 3\,(4\beta)^{-8/3} - 9\,(4\beta)^{-12/3} + \dots\} \\ &= 3^{-3}\,(4\beta)^4 \left\{1 - \frac{1}{2}(\alpha - 1) + \frac{1}{6}(\alpha - 1)^2 - \frac{1}{8}(\alpha - 1)^3 + \dots\right\}, \end{aligned} \tag{4.102}$$

correct for powers of β up to, and including, the -4th.

A four-digit tabulation of α as a function of β, constructed on the basis of Equation (4.98), is contained in the accompanying Table VIII-6; while Table VIII-7, contains the values of β, $y_{1,2} \equiv z_{1,2}$ (cf. Equation (4.92)); $x_{1,2}$ (from Equation (4.65)) and $E_{1,2}$ (Equation (4.91)) corresponding to $\alpha = 1(0.1)2$. The product $h'x$ satifies the same bi-quadratic equation (4.54) as y' – i.e.,

$$x_{1,2}^4 - x_{1,2}^3 + h'^{-4} \equiv 0 \tag{4.103}$$

which, combined with (4.65), furnishes a linear relation

$$y_{1,2} = h'^4(1 - x_{1,2}) \tag{4.104}$$

between the roots of our biquadratic listed in Table VIII-7.

TABLE VIII-6

β	α	β	α
1	2.00000	5	1.11234
1.25	1.74105	5.5	1.09877
1.5	1.57869	6	1.08782
1.75	1.46904	7	1.07134
2	1.39087	8	1.05961
2.25	1.33278	9	1.05088
3	1.22052	10	1.04416
3.5	1.18221	20	1.01746
3.5	1.8	50	1.00513
4	1.5199	100	1.00204
4.5	1.12956	∞	1.00000

TABLE VIII-7

α	β^{-1}	y_1	y_2	x_1	x_2	E_1	E_2
1	0	1	∞	1	0	−0.5	$+\infty$
1.1	0.183503	1.00052	7942.96	0.999880	0.050119	−0.500114	3753.38
1.2	0.305967	1.00252	972.682	0.999073	0.100927	−0.500374	388.171
1.3	0.411945	1.00914	278.682	0.996935	0.153075	−0.501495	96.7427
1.4	0.508526	1.02200	112.404	0.992792	0.207211	−0.503622	32.9111
1.5	0.598840	1.04371	54.2524	0.985844	0.264156	−0.507080	12.7951
1.6	0.684632	1.07892	29.1307	0.975000	0.325000	−0.512484	5.09788
1.7	0.767001	1.13539	16.6722	0.958557	0.391443	−0.520642	1.80988
1.8	0.846689	1.22996	9.83963	0.933333	0.466667	−0.532981	−0.327986
1.9	0.924225	1.41217	5.73489	0.891327	0.558673	−0.552622	−0.336486
2.	1.000000	2.37037	2.37037	0.750000	0.750000	−0.592592	−0.592592

With these results in our possession let us turn now to real binary systems such as we know to exist in the sky, and inquire about the stage which they may have already reached in the course of their evolution. In order to do so we must, of course, be in possession of all requisite basic data, definining not only the orbital momentum of the system (expressible in terms of the masses $m_{1,2}$ of the two components and their separation A), but also the rotational momenta of the two stars (i.e., their moments of inertia $C_{1,2}$ and angular velocities $\omega_{1,2}$ of axial rotation); and it is the latter two quantities which are more difficult to come by.

The angular velocities $\omega_{1,2}$ can, in principle, be measured from the rotational broadening of the spectral lines of the respective components (or, for eclipsing system, from the extent of the 'rotational effect' within minima). Such observations as are available (for their recent survey see, e.g., Slettebak, 1970) indicate that the components in most known close binaries describing orbits which deviate but little from circles rotate essentially in synchronism with their revolution. Noticeable deviations from synchronism, with components rotating *faster* than they revolve, are encountered only in systems with markedly eccentric orbits (such as α CrB, $e = 0.33$; AR Cas, $e = 0.25$; or Y

Cyg, $e = 0.14$; the latter being a very young system) and are not fully covered by the discussion contained in this section.

Therefore, in what follows we shall restrict the subject of our inquiry to the following question: are the data available to us on close binary systems consistent with an assumption that these systems have already attained their state of maximum distension – in which $\omega_{1,2} = \Omega$ and the total energy of the system is minimum for given total momentum?

In an attempt to answer this question, let us return to Equation (4.4) from which, by hypothesis,

$$x_1^2 = \frac{G(m_1 m_2)^2 A}{(m_1 + m_2)H^2}, \tag{4.105}$$

where x_1 is the root whose numerical values are listed in column (5) of Table VIII-7. For rotation of both components which is synchronized with their revolution it follows from Equations (4.3), (4.4), (4.6) and (4.11) that

$$H = H_0 + \Omega\,(C_1 + C_2) = (1 + \nu)H_0\,, \tag{4.106}$$

where H_0 continues to be given by (4.3), $C_{1,2}$ by (4.7), and ν represents a nondimensional parameter

$$\nu = \frac{\Omega(C_1 + C_2)}{H_0} = \left(1 + \frac{m_1}{m_2}\right) h_1^2 + \left(1 + \frac{m_2}{m_1}\right) h_2^2\,, \tag{4.107}$$

where $h_{1,2}$ denote the fractional radii of gyration of the two components expressed in terms of the semi-axis A of the relative orbit.

The radii of gyration of our configurations could be evaluated by quadrature from (4.9) if we knew the internal structure of both stars; or determined empirically for systems exhibiting evidence of apsidal advance, from the rate of which the constants k_2 can be deduced (cf. sec. V-3C). That these constants should be related with the fractional radii of gyration $\mathfrak{H}/R$ of the respective stars is evident from a comparison of Equation (4.9) with Equation (6.9) of Chapter II, which shows that (at least for configurations of pronounced central condensation) k_2 represents merely a moment of inertia of higher order. Numerical work by Motz (1952) disclosed that, very approximately, the relationship between $\mathfrak{H}/R$ and k_2 assumes the form

$$\log(\mathfrak{H}/R) = 0.24 \log k_2 - 0.13\,, \tag{4.108}$$

which, rewritten in terms of the ratios

$$\frac{\mathfrak{H}_i}{A} = h_i \quad \text{and} \quad \frac{R_i}{A} = a_i, \quad i = 1,2, \tag{4.109}$$

transforms into a relation disclosing that

$$\log h_i = \log a_i + 0.24 \log k_2 - 0.13\,, \tag{4.110}$$

where both the fractional radii $a_{1,2}$ and apsidal-motion constants k_2 can be deduced from an appropriate kind of photometric evidence (cf. sec. V-3C). And once this has been done, the values of the ratio ν of the sum of the rotational momenta of both components to the orbital momentum of the system can be evaluated from (4.106).

If we combine now (4.104) with (4.105) and insert for H_0 from (4.3), we arrive at a simple relation

$$x = \frac{1}{1+\nu}\,, \tag{4.111}$$

which in combination with the numerical data compiled in Table VIII-2 discloses that, in the limiting case of greatest distension (i.e., $x = x_1$) corresponding to *minimum* energy per given total momentum, the maximum value of $x_1 = 1$ corresponds (by Equation (4.111)) to $\nu = 0$ and, by (4.107), to zero gyration radii (i.e., infinite density concentration) of both components; while the minimum value of $x_1 = \frac{3}{4}$ corresponds to $\nu = \frac{1}{3}$. Therefore, in order that our binary should have reached the synchronous orbit of maximum distension, it is necessary that

$$0 \leqslant \nu \leqslant \frac{1}{3}\,. \tag{4.112}$$

On the other hand, an excess of the actual value of ν over one-third would – in the light of our analysis – be sufficient to justify a conclusion that the respective system is still in the process of dynamical evolution, and has not yet reached its ultimate stage of maximum distension.

As regards the initial stage of *maximum* energy for given momentum – corresponding to the case when $x = x_2$ – the numerical values of this latter parameter are constrained by the inequality

$$\frac{3}{4} > x_2 > 0 \tag{4.113}$$

and, therefore,

$$\frac{1}{3} < \nu < \infty\,. \tag{4.114}$$

The limit $\nu = \infty$ corresponds then to the case of infinite disparity between the masses of the two components, one of which (say, m_2) becomes zero. In such a case, the secondary's rotational momentum as well as its orbital momentum (for the centre of mass of the primary component of mass m_1 coincides with that of the orbit) also become zero; the total momentum (4.105) of the system consists only of $C_1\omega_1$.

Close binary systems for which sufficiently reliable observational data are on hand for computation of the ratios ν of the rotational to the orbital momenta are listed in the accompanying Table VIII-8. It includes six eclipsing systems of the 'detached' type, and four 'semi-detached' systems – all characterized by an orbital eccentricity small enough for its squares and higher powers to be negligible. The individual columns of Table VIII-8 indicate successively: (1) the system; (2) the spectra of its components; (3) orbital eccentricity; (4) the mean apsidal-motion constant $\bar{k}_2$ of the system; (5) and (6), the mean fractional radii $a_{1,2}$ of the respective components; (7) and (8), their fractional radii $h_{1,2}$ of gyration evaluated with the aid of Equation (4.109); (9), the mass-ratio of the system; and (10), the parameter ν as given by (4.106).

The basic data for $a_{1,2}$ and m_2/m_1 have been taken from Kopal and Shapley's

Catalogue (1956), augmented by subsequent spectroscopic data by Batten (1967); while the mean apsidal-motion constants k_2 have been taken from Tables V-1 and V-2. The ultimate column of Table VIII-8 then contains the values of ν following from Equation (4.106) from the data listed in the preceding columns – with the exception of that for Y Cyg, which was multiplied by the factor 1.75 giving the excess of its spectroscopically determined angular velocity of axial rotation (cf. Luyten *et al.*, 1939) over the Keplerian angular velocity of orbital revolution.

TABLE VIII-8

Comparison with Observations

Star	Spectra	e	$\bar{k}_2$	a_1	a_2	h_1	h_2	m_2/m_1	ν
				Detached Systems					
RS CVn	F4 + K0 IV	0.033	0.0097	0.092	0.28	0.022	0.068	0.93	0.0094
		± 7	± 17	± 5	± 1			± 2	
GL Car	B3 + B4	0.16	0.0122	0.217	0.217	0.056	0.056	1.0	0.0125
		± 1	± 19	± 8	± 8				
Y Cyg	O9.5 + O9.5	0.14	0.0100	0.208	0.202	0.051	0.049	0.99	0.0100
		± 1	± 13	± 8	± 8			± 1	
CO Lac	B8.5 + A0	0.028	0.0043	0.252	0.213	0.050	0.043	0.82	0.0089
		3	± 15	± 25	± 21			± 7	
V 451Oph	A0 + [A2]	0.025	0.0082	0.212	0.170	0.049	0.045	0.83	0.0090
		± 10	± 8	± 21	± 17			± 1	
AG Per	B5 + B7	0.067	0.0063	0.211	0.194	0.051	0.048	0.88	0.0090
		± 1	± 20	± 7	± 7			± 3	
				Semi-detached Systems					
RZ Cas	A0 + gK1	0.013	0.0043	0.241	0.284	0.048	0.057	0.35	0.0132
		± 6	± 20	± 2	± 2			± 1	
W Del	A0 + gK0	0.036	0.0038	0.153	0.243	0.030	0.048	0.21	0.0080
		± 2	± 19	± 3	± 2			± 1	
β Per	B8 + gK0	0.010	0.0029	0.227	0.236	0.041	0.043	0.19	0.0127
		± 1	± 10	± 2	± 2			± 1	
TX UMa	B8 + gF2	0.023	0.0031	0.158	0.277	0.029	0.051	0.30	0.0070
		± 1	± 1	± 1	± 1			± 2	

A glance at the data listed in the last column of Table VIII-8 discloses at once that, in all systems considered by us, *the sum of the angular momenta of axial rotation turns out to be less than half a percent of the angular momentum of orbital motion;* and, therefore, a synchronism between rotation and revolution is possible – though not necessary – for each system listed in our table. Some – like Y Cygni, for instance – may not have yet attained it; but none are probably very far from it at the present time. In other words, these (and probably most other known) close binary systems have already come pretty close to the 'end of their tether'; and tidal evolution alone can no longer increase the present separation of their components to any appreciable extent.

But, on the other hand, close binary systems could scarcely have originated in the form of such limiting configurations to begin with, and must at one time have been much closer together. That this must have been so if their origin is to be sought in any kind of fission of an originally single configuration is self-evident. But the same would also be true if they originated by capture; for a dissipation of the requisite amount of kinetic energy could have been accomplished only at a very close range – close to actual contact. From whichever direction we attempt to approach our problem, therefore, we find that the initial state of the system should have been close to our 'state 2' (i.e., synchronous rotation at a close range); the initial degree of proximity being limited mainly by the physical dimensions of the constituent components; and 'state 1' could have been approached only by some kind of evolutionary process.

A manifest lack of intermediate cases (for which the ratio ν would be, say, greater than 0.1) suggests that the process of dynamical evolution through viscous tides, which tends to increase separation, may be very effective and operate on a (cosmically) short time-scale. Such a possibility should focus attention on the quantitative aspects of the time-scale of tidal evolution in close binary systems – an aspect to which we shall next turn our attention.

D. TIDAL EVOLUTION

In the preceding part of this section we investigated the range within which the total energy of a binary system can vary for the constant value of its total momentum. This analysis has, however, left the actual process by which the energy E can vary within the interval $E_2 \geqslant E \geqslant E_1$ completely open; and its is not till this has been specified that we can assign this progression any definite time-scale. The aim of this last part of the present section will be to investigate the processes which can cause the total energy of the system to vary – monotonically or otherwise – within the permitted interval between $E_{1,2}$; and to attempt to trace the evolutionary course of the individual components for the case in which the total momentum of the respective binary system remains preserved in the course of time.

In order to investigate the specific nature of constraints imposed by the postulated constancy of momentum on the evolution of close binaries, let us depart from the expression for H as given by Equation (4.11) and assume – for the time being – that the masses $m_{1,2}$ of both components as well as their internal structure (i.e., the moments of inertia $C_{1,2}$) remain constant in time. Let us, moreover, assume that the relative orbit of the two components is nearly circular – so that quantities of the order to the squares (or time-derivatives) of its eccentricity e can hereafter be ignored. If so, a time-derivative of (4.11) can be expressed as

$$\dot{H} = C_1\dot{\omega}_1 + C_2\dot{\omega}_2 + \left\{\frac{Gm_1^2m_2^2}{(m_1+m_2)A}\right\}^{1/2}\dot{A} = 0\,, \qquad (4.115)$$

and represents an algebraic relation which the time-derivatives $\dot{\omega}_{1,2}$ and $\dot{A}$ must satisfy in the course of time.

An obvious way in which Equation (4.115) can be satisfied is to set $\dot{\omega}_{1,2} = 0$ and $\dot{A} = 0$ – corresponding to a secularly stationary case – in which all terms in (4.115) vanish identically. Such a solution would, indeed, be admissible if the fluid (and, therefore, deformable) components of a close binary were to consist of inviscid gas, in which case there would be no dissipation of energy through dynamical tides; and the tidal bulge produced by the attraction of the companion would follow exactly the direction of the attracting force.

Inviscid fluids represent, however, only a mathematical abstraction which is not fulfilled exactly by any material in nature; and so are conservative dynamical systems in which the conditions $\dot{\omega}_{1,2} = 0$ and $\dot{A} = 0$ could be fulfilled for any length of time. All materials constituting celestial bodies possess a certain viscosity (which, in turbulent media or for degenerate configurations, can become very large), capable of degrading kinetic energy into heat; and dissipative forces arising from viscosity are bound to render both time-derivatives $\dot{A}$ as well as $m_{1,2}$ different from zero in the course of time – the latter, because of the dissipation of kinetic energy of axial rotation through viscous tides; and the former, because viscosity will render the crest of the tidal wave to deviate from the direction of the attracting force, and thus give rise to the phenomenon of 'tidal lag'.

Let, in what follows, ϵ denote the angle by which the axis of symmetry of a tidally-distorted configuration deviates from the radius-vector joining the centres of the two stars on account of viscosity. If so, and if we restrict ourselves (to begin with) to tides characterized by second-harmonic symmetry, then we know from the equations for the perturbation of the Keplerian elements in the plane of the orbit (cf. Equations (3.25) and (3.45) of Chapter V) that

$$\frac{dA}{dt} = 6\,\frac{G(m_1+m_2)}{nA^2}\left\{\frac{m_2}{m_1}\left(\frac{R_1}{A}\right)^5 (k_2)_1\,\epsilon_1 + \frac{m_1}{m_2}\left(\frac{R_2}{A}\right)^5 (k_2)_2\,\epsilon_2 + \ldots\right\}, \tag{4.116}$$

where $(k_2)_{1,2}$ are 'apsidal-motion' constants associated with the second-harmonic tidal distortion, as given by Equation (6.4) of Chapter II, and depending on the internal structure of the respective component; $\epsilon_{1,2}$, the angles of tidal lag of each star; and n, their mean daily motion, given by

$$\Omega^2 = n^2 = \frac{G(m_1+m_2)}{A^3}. \tag{4.117}$$

If we insert (4.116) in (4.115), we find the former to be satisfied if

$$C_i\,\dot{\omega}_i + 6\left(\frac{Gm_{3-i}^2 R_i^5}{A^6}\right)(k_2)_i\,\epsilon_i = 0, \quad i = 1, 2. \tag{4.118}$$

These conditions are sufficient, not necessary.

On the other hand, in our earlier discussion of axial rotation of viscous configurations (cf. sec. IV-3) we established that, within the scheme of our appoximation, for configurations in radiative equilibrium

$$C_i\,\dot{\omega}_i + \frac{152}{35}\,\pi(\omega_i - n)\,\frac{\omega_i^2 m_{3-i}}{Gm_i^2 A^3}\int_0^{R_i} \mu r^8\,dr = 0, \tag{4.119}$$

$i = 1, 2$, where $\mu(r)$ denotes the distribution of viscosity in the interior; and

$$C_i \dot{\omega}_i + \frac{4}{5} \pi \mu_{ci} \left(\frac{m_{3-i}}{m_i} \right) \frac{R_{ci}^6}{A^3} (\omega_i - n) = 0 \tag{4.120}$$

if the configuration possesses a convective core of radius R_{ci}, with a viscosity $\mu_{ci} \equiv \mu(R_{ci})$ at the interface.

Eliminating the terms $C_i\omega_i$ between Equations (4.118) and (4.119)–(4.120) we find that, in the case of radiative equilibrium, the angle ϵ_i of tidal lag should be given by

$$\begin{aligned} \epsilon_i &= \frac{76}{105} \left(\frac{\pi \omega_i^2 A^3}{G^2 m_i^2 m_{3-i} R_i^5} \right) \frac{\omega_i - n}{(k_2)_i} \int_0^{R_i} \mu r^8 \, dr \\ &= \frac{19}{84} \left(\frac{\omega_i^2 (\omega_i - n) A^3}{G^2 m_1 m_2} \right) \frac{\int_0^{R_i} \mu r^8 \, dr}{\int_0^{R_i} \rho r^7 \, dr} \; ; \end{aligned} \tag{4.121}$$

while in the convective case,

$$\begin{aligned} \epsilon_i &= \frac{2 \pi A^3}{15\, G m_1 m_2} \left(\frac{R_{ci}^6}{R_i^5} \right) (\omega_i - n) \frac{\mu_{ci}}{(k_2)_i} \\ &= \frac{R_{ci}^6 A^3 (\omega_i - n) \mu_{ci}}{24\, G m_i \int_0^{R_i} \rho r^7 \, dr} \; ; \end{aligned} \tag{4.122}$$

implying that $\epsilon_i > 0$ if $\omega_i > n$ and vice versa.

Let us now introduce the foregoing expressions (4.121) and (4.122) for tidal lag in Equation (4.116) for A in which we make use of (4.117); the result will be

$$\begin{aligned} \frac{dA^{7/2}}{dt} = \frac{76\pi}{5} \sqrt{\frac{m_1 + m_2}{G^3}} \Bigg\{ & \frac{\omega_1^2 (\omega_1 - n)}{m_1^3} \int_0^{R_1} \mu r^8 \, dr + \\ & + \frac{\omega_2^2 (\omega_2 - n)}{m_2^3} \int_0^{R_2} \mu r^8 \, dr \Bigg\} \end{aligned} \tag{4.123}$$

for radiative equilibrium, and

$$\begin{aligned} \frac{dA^{7/2}}{dt} = \frac{14\pi}{5} \sqrt{\frac{m_1 + m_2}{G}} \Bigg\{ & \frac{\mu_{c1} R_{c1}^6}{m_1^2} (\omega_1 - n) + \\ & + \frac{\mu_{c2} R_{c2}^2}{m_2^2} (\omega_2 - n) \Bigg\} \end{aligned} \tag{4.124}$$

in the presence of a convective core. We may note that A is positive of $\omega_i' > n$ (i.e., if the 'day' is shorter than the 'month'), and negative if the opposite is the case.

If we normalize all variables in the foregoing equations in accordance with the scheme adopted in section 4C we find that Equation (4.123) appropriate for radiative equilibrium can be rewritten as

$$\frac{dx^7}{d\tau} = f_1 y_1^2 (y_1 - x^{-3}) + f_2 y_2^2 (y_2 - x^{-3}) , \tag{4.125}$$

and Equation (4.124) applicable to convective equilibrium as

$$\frac{dx^7}{d\tau} = f_1' (y_1 - x^{-3}) + f_2' (y_2 - x^{-3}) , \tag{4.126}$$

where

$$f_i = \frac{76\,\pi}{5HL_0^6} \left(\frac{m_1 + m_2}{m_i}\right) \frac{m_1 m_2}{m_i^2} \int_0^{R_i} \mu r^8 \, dr , \tag{4.127}$$

and

$$f_i' = \frac{14\,\pi}{5HL_0^3} \left(\frac{m_1 m_2}{m_i^2}\right) \mu_{ci} R_{ci}^6 \tag{4.128}$$

are nondimensional parameters, in which we have abbreviated

$$L_0 = \frac{(m_1 + m_2) H^2}{G(m_1 m_2)^2} ; \tag{4.129}$$

and τ denotes the normalized time.

Moreover, Equations (4.119) and (4.120) governing axial rotation can be similarly normalized to

$$x^6 \frac{dy_i}{dt} + g_i y_i^2 (y_i - x^{-3}) = 0 \tag{4.130}$$

for radiative equilibrium, and

$$x^6 \frac{dy_i}{dt} + g_i' (y_i - x^{-3}) = 0 \tag{4.131}$$

for convective core, where

$$g_i = \frac{76\,\pi}{35HL_0^6} \left(\frac{m_1 + m_2}{m_i}\right) \frac{m_{3-i}}{m_i} \left(\frac{1}{\kappa_i} \int_0^{R_i} \mu r^8 \, dr\right) \tag{4.132}$$

and

$$g_i' = \frac{2\pi}{5HL_0^3}\left(\frac{m_{3-i}}{m_i}\right)\frac{\mu_{ci}R_{ci}^6}{\kappa_i}\quad, \tag{4.133}$$

represent another set of nondimensional parameters, in which the κ_i's continue to be given by Equation (4.86).

Equations (4.125) and (4.130) or (4.126) and (4.131) constitute sets of simultaneous differential equations of second order, which govern the tidal evolution of close binary systems motivated by viscous friction; and these provide us with the means for bridging the gap between the two extremes of synchronous rotation investigated in section 4C. In order to utilize them to this end, all that is necessary is to integrate (numerically or otherwise) Equations (4.125)–(4.126) and (4.130)–(4.131) from the initial conditions

$$x = x_2 \qquad \text{and} \qquad y_i = x_2^{-3} \tag{4.134}$$

for values of x_2 tabulated in column (6) of Table VIII-7 as a function of α corresponding to the state of maximum positive energy, in the direction of diminishing energy – until a state been reached when these equations are fulfilled again for

$$x = x_1 \qquad \text{and} \qquad y_i = x_1^{-3}; \tag{4.135}$$

for which the axial rotation of both components becomes once more synchronized with that of revolution in an orbit of maximum distension.

The energy dissipation through tidal friction is bound to be operative in all close binaries, at the rate required by the viscosity of matter constituting their components. This rate is likely to be very high for stars contracting towards the Main Sequence, which find themselves in the state of convective equilibrium throughout their interiors. This convection is almost certain to be turbulent; and the viscosity associated with its turbulence can be very large. It is, we repeat, at this early stage of evolution that synchronization of rotation and revolution of the components, and circularization of their orbits, may have largely been accomplished.

As long as both components of detached binary systems remain on the Main Sequence, their tidal evolution should slow down considerably. The reason is the fact that, with the advent on the Main Sequence, the internal convection in the stars should die down everywhere except, for massive stars ($m > 2$ ⊙), in the cores surrounding their centres. However, the fractional dimensions of such cores are generally small; and dynamical tides raised in them by the companion are unlikely to be of much importance for tidal evolution of the system – in spite of the fact that the turbulent viscosity may remain high there. For most parts of the star now find themselves in the state of radiative equilibrium, and the main source of viscosity remains that of the hydrogen-helium plasma (cf. sec. III-4) constituting their outer layers (where tides are high), augmented (in very massive stars) by radiative viscosity of the photon gas.

However, soon after the ageing component of a close binary system has left the Main Sequence and embarks on its evolution towards its Roche limit, a convective layer begins to form on the surface (where the tides attain their maximum height) and its depth will increase rapidly towards the interior with increasing size of the star. Moreover, the flow caused by this convection will again be turbulent; and the effective viscosity of the turbulent outer layers may become as high as in the pre-Main Sequence past. As a result, the heat generated by viscous friction may once more become of importance for the evolutionary course of the star; and its effects will make themselves felt on the same Helmholtz-Kelvin time-scale. It is there effects which may cause an evolving component of a close binary in its semi-detached state to behave differently from what it would have done had it been single; and the difference in structure due to this cause may further alter its radius of gyration to an appreciable extent.

Apart from this influence on the radii of gyration as defined by Equation (4.9), a more serious limitation of the formulation of the tidal evolution of close binary systems, as outlined in this section, is inherent in the assumed constancy of mass of both components. While this assumption is indeed fulfilled to a high degree of approximation (within an error of less than 1%) while these stars remain on the Main Sequence, compelling reasons were listed in section VIII-3 for subsequent loss of mass at post-Main Sequence stage. At that stage (or in certain parts of it) the mass of the contact component could, indeed, become a sensitive function of the time; and related with the amount of momentum which the system will lose as a result of the transfer of mass between the components, or its escape from the system.

Needless to say, such a transfer or loss of mass will affect also all other elements of the orbit – including the distance A between the two components and, through it, the orbital period P which is best defined by the observations. Within the domain of rational mechanics, the corresponding variations of the elements (including P) could be investigated by procedures outlined in section V-5. However, in the hydrodynamical case, for reasons explained in section VIII-3B we are still far from this goal; and as long as this continues to be so, our knowledge of the tidal evolution of close binary systems as outlined in this section may still be seriously incomplete – leaving ample scope for further research.

It is for the third time that we have met, in this volume, with the problems of period variation in close binary systems – previously encountered in sections V-4 to 6 in other connections. In section VIII-3C we stressed the fact that conspicuous period fluctuations become the rule, rather than the exception, in systems containing contact components; and this alone suggests that these period changes may be connected with a transfer or loss of mass expected at this evolutionary stage. But, on the other hand, a secular change of mass would be much more likely to produce *monotonic* variations (i.e., increase or decrease) of orbital period over time intervals of at least 10^3-10^4 yr; while the observations disclose the existence of *irregular* (or, at any rate, very complicated) variations over much shorter time-intervals of the order of 10–100 yr.

This apparent contradiction would be impossible to account for on the basis of rational mechanics, where the changes of P should be connected with those of other elements by algebraic expressions, and instantaneously respond to their changes. Not so in hydrodynamics, where the variations in P need not be related instantaneously with those of other elements; and a lag can develop between them which, through its nonlinear terms, may indeed simulate the manifestations of 'noise' over limited intervals of time. The present writer ventures to express an opinion – no more than a working hypothesis so far – that *the fluctuations in orbital periods of semi-detached systems*, observed over time intervals of the order of 10–100 yr, *bear out the effects of nonlinear noise in the hydrodynamical machinery of tidal evolution* – its 'backlash' so to speak – rather than the direct first-order effects of a secular change in mass.

For as long as a mass flow does not actually reverse its direction, its first-order effects should make the orbital period either increase or decrease monotonously (though not necessarily at a uniform rate) over much longer intervals of time than those covered by the observations. The fluctuations actually observed can be hydrodynamically accounted for only by terms of higher orders; and the fact that this 'noise' has so far been detected only in the period P, and not other elements of the orbit, is due to the superior precision with which P can be established from the observational data. But all this remains so far only a tentative hypothesis; and much work in the nonlinear domain of our problem may be necessary to put it on a more creditable basis.

VIII-5. Beginnings and Ends

Our long narrative on different physical and dynamical properties of close binary systems, to which this book has been largely dedicated, has brought us at last to face some questions whose consideration was appropriately deferred to the end of this book: namely, how do binary stars originate in the sky, and how do they end?

A. ORIGIN OF CLOSE BINARIES

To commence our story at the beginning, the origin of a large majority of known close binary systems is no doubt to be sought in terms of the same general processes which result in the formation of the stars – be they single or multiple. In recent decades it has been generally recognized that the stars originate by a gravitational collapse of cosmic gas clouds with enough mass to give birth to, not only one or a few, but to hundreds or thousands of individual stars, as a part of the same creative act occurring within (astronomically) very short time. If the total energy of such a group is positive (i.e., if the sum of the kinetic energies of its individual members exceeds the potential energy of the group as a whole), the outcome of the process will be a fleeting 'association' of stars, likely to dissolve in a relatively short time; while if the total energy of the new-born group is negative, we have a 'cluster' with a life expectancy of a different order of magnitude.

On account of their short life-times, the mere 'associations' of stars are relatively few in number. Several are, however, known (cf. Table I-2) to contain close binaries; and some of these have proved to be eclipsing variables. Thus AG Persei – a 'detached' system listed in Table VIII-2 – is a member of ζ-Persei association (cf. Blaauw, 1952; Delhaye and Blaauw, 1953). The equally well-known eclipsing binary BM Orionis – together with ν^1 Ori A, recently discovered as an eclipsing variable by Lohsen (1975) and also a member of the Trapezium – belong to the Orion complex; while the semi-detached system of μ^1 Sco, listed in Table VIII-4, belongs according to Blaauw (1946) to the Scorpio-Centaurus stream.

The ζ-Persei association is quite young (of an estimated age of 1.3×10^6 yr; and the fact that its components of masses close to 5 $\odot$ are still on the Main Sequence and detached from their Roche limits is in accordance with the evolutionary data compiled in Table VIII-1. The Orion association is even younger; and it is probable that BM Ori is still in its pre-Main Sequence stage (cf. Hall and Garrison, 1969). On the other hand, the

Scorpio-Centaurus stream represents a much older group of stars – Blaauw (1946) estimated its age to 70×10^6 yr – and, in the light of the evolutionary date given in Table VIII-1, a semi-detached state of a binary with a total mass of 23 ⊙, would be long overdue.

As far as the Orion complex is concerned, the presence of close (eclipsing) binaries among its members is undoubted. However, we cannot fail to note that their percentage among the absolutely brightest members of this complex appears to be much lower (possibly by as much as a factor 10) than the percentage of close binaries among stars of the same spectra in the galactic substrate. Penston's recent (1973) catalogue of stars belonging to this complex failed to list any other eclipsing variable; and although it is difficult to estimate the extent to which this may be due to observational selection, one cannot get rid of an impresssion that this disparity may be genuine, and due to one distinguishing mark of the Orion complex which is its very low age. If real, does it mean that the binary component of its population originated otherwise than its single stars?

Whether the process which led to their formation was simultaneous condensation of matter around a pair of mass centres so close that the two stars entered the HR-diagram already as components of a binary system; or whether such binaries originated as the result of a capture of two stars at a time when the density of stars per unit volume in a newly-formed cluster or association was sufficiently high (for a capture requires the interaction of a minimum of three bodies) remains unclear – yet one indication exists by which Nature seems to be lending us a helpful hand: namely, the fact that *in all known binary systems* in which the facts can be checked, *the components rotate in the same direction as they revolve;* and never in the retrograde sense. That this would have to be so is far from obvious; and in what follows we shall attempt to explain why.

As is well known, the axial rotation of the stars can be ascertained (at leaast statistically) from the Doppler broadening of their spectral lines produced by rotational motion; but its actual velocity as well as direction can be established only in close binary systems which, by virtue of a chance inclination of their orbital plane with respect to the line of sight, happen to become eclipsing variables. In such cases, the necessary data can be extracted from the light curves to enable us to determine the absolute value of the velocity of rotation on the equator, as well as its sign, from the 'rotational effect' exerted by axial rotation on the measured radial velocity of the respective component. Many eclipsing systems (among them Algol; cf. McLaughlin, 1924) have been found to exhibit this rotational effect to a conspicuous degree; and in every single case the sense of rotation proved to be *direct* – never retrograde (as it is, in the solar system, with the planets Venus or Uranus).

Now should close binary systems constitute chance associations of two stars that originated by capture, the initial directions of their axes of rotation in space could be expected to exhibit random distribution on the celestial sphere. We know, of course, from section V-2 that this could not remain so long; for the dissipative forces continuously operative in close binaries secularly tend to 'rectify' these axes so as to make the equators and orbit co-planar. But – and this is essential – for initial inclinations θ anywhere between 90° and 180° tidal friction would have brought about co-planarity with rotation in the *retrograde* direction; while for inclinations between 0° and 90° the opposite should be the case. Since not a single instance of retrograde rotation has been detected so far, we are driven to the conclusion that *if close binaries originated by capture, the directions of their poles were not initially distributed at random.*

Conversely, should the collapse of the primordial gas cloud from which stars originated have occurred in such a way as to result in an *arbitrary* orientation of the stellar axes of rotation, then *close binaries could not have originated by capture*, but by some other process by which the initial axes of rotation were more nearly aligned (or, at least, their north poles confined to one hemisphere). Moreover, should it turn out that new-born associations contain a smaller percentage of close binaries than the substrate in which they will eventually be dissolved, it may become necessary to consider alternative hypotheses which may lead to their formation; and the most obvious one would be fission.

Theories attributing the origin of close binary systems to the fission of a single star in rapid rotation posess a very distinguished pedigree going back to Kant and Laplace; and in a more recent past were elaborated, in particular, by Darwin (1910) and Jeans (1919). Although the quantitative aspects of these theories have subsequently become the targets of criticisms (cf., e.g., Lyttleton, 1953) many of which are indeed well-founded, they do not dispose of the fission theory as such, but only of the form in which this theory was elaborated by Darwin and Jeans.

In historical retrospect, it is clear to us now that why stars cannot be regarded as homogeneous and incompressible (in fact, liquid) at any stage of their evolution – assumptions to the contrary having been made by Darwin or Jeans for sake of mathematical tractability of the underlying problem. We know today that real stars depart very widely from such a simplistic model; though very little is known about the effects of non-uniform rotation on the internal structure if the rotation becomes fast enough for centrifugal force to become comparable with self-attraction. So extreme a case of stellar hydrodynamics as a fission, under stress, of an originally single star in two components of comparable mass has never yet been properly formulated – let alone solved; and as long this remains so, it is impossible to rule out the fission theory as an alternative explanation of the origin of close binary systems by any final verdict.

In the absence of such a verdict on the part of a theory, can some light be thrown on the problem by the observations? In order to do so, we wish to invite the reader's attention to the data collected in Table VIII-9 containing the geometrical elements of detached eclipsing systems, of absolute properties given in Table VIII-2. The contents of the individual columns of Table VIII-9 is self-explanatory; $r_{1,2}$ stand for the fractional radii of the two components (i.e., the absolute radii $R_{1,2}$ expressed in terms of their separation A; and $C_{1,2}$ are the values of the normalized Roche constants as defined by Equation (1.7) of Chapter VI. Perhaps the most significant testimony borne out by these data is the fact that, in detached systems consisting of Main-Sequence stars, the Roche constants $C_{1,2}$ characteristic of the surface potentials of both components appear to be sensibly equal. Although the absolute values of such potentials can vary by more than a factor 10 from one end of the Main Sequence to another, the observed ratios C_1/C_2 do not seem to deviate from unity by more than ± 10%, and their mean turns out to be equal to 1.007 ± 0.013; the standard deviation of the individual values of C_1/C_2 from this mean being ± 0.055. In more general terms, *the masses and absolute dimensions of the individual components in close binary systems of the detached type appear to be approportioned in such a way that the potentials over their free surfaces are approximately the same* – respective of their masses of absolute dimensions.

Is this near-equality of surface potentials of the Main-Sequence components of detached binary systems a mere consequence of their evolutionary stage, or does it constitute a lingering reminiscence of their origin – for if such systems originated by fission, the surface potentials of both components at the time of their separation would naturally be the same? In an attempt to answer this question, suppose – for the sake of argument – that any vestige of initial conditions in the present characteristics of the individual components has been obliterated by the passage of time, and that their present properties are those of any pair of typical Main Sequence stars of the same masses. If so, the ratio of their Roche constants should (if we ignore the distortion) be given by an approximate equation of the form

$$\frac{C_1}{C_2} = \frac{m_1}{m_2}\frac{R_2}{R_1}, \tag{5.1}$$

which by use of the statistical Equations (3.1) and (3.2) for the empirical mass-radius relations of the Main Sequence stars discloses that

$$\begin{aligned} \log\frac{C_1}{C_2} &= 0.57\log\frac{R_1}{R_2}, \quad m > 2\odot; \\ &= 0.02\log\frac{R_1}{R_2}, \quad m \ll 2\odot. \end{aligned} \tag{5.2}$$

From these equations it transpires that, for stars of small masses ($\ll 2\odot$), the ratios C_1/C_2 should indeed come very close to unity for a pair of stars of arbitrary radii $R_{1,2}$ – regardless of their origin or age. However, for more massive stars ($m > 2\odot$) this should no longer be statistically true; and a more specific test of the underlying assumption becomes possible.

In order to carry it out, we have evaluated the theoretical ratios of C_1/C_2 from the foregoing Equation (5.2), based on the assumed random association of average Main-Sequence stars, and listed them in column (8) of Table VIII-9 after the observed values of this ratio as given in col. (7); the ultimate column (9) of this table then contains the corresponding *O-C* – differences. With the exception of a still somewhat anomalous case of SZ Cam, the mean value of the theoretical C_1/C_2's turns out to be 1.044 ± 0.014, as compared with their 'observed' mean of 1.007 ± 0.013. The respective O-C – differences of column (9) are, accordingly, of a systematic character (mostly negative). The observational material presently at our disposal is, unfortunately, not large enough to enable us to assess the significance of this discrepancy; but future investigations along these lines, based on more extensive material, may furnish valuable information on the past history of close binaries of the detached type.

B. TERMINAL STAGES: SUBDWARF BINARIES

In the first section of this chapter we outlined the probable way in which stars of different masses are likely to evolve in post-Main Sequence stage; and underlined the fact that stars of moderate masses are likely to 'pre-fabricate' in their interiors degnerate cores of properties approaching those of white dwarfs. It is still next to impossible to follow by computations the exact sequence of models in which the ageing configuration will gradually (or cataclysmically) divest itself of its superstructure of diminishing mass, eventually to lay bare the degenerate core in its interior. Are there, however, any cases known to us from observations which could empirically document this particular part of the terminal evolution of the stars? The answer is indeed in the affirmative; and the aim of the remaining parts of this section will be to summarize the relevant evidence.

TABLE VIII-9
Geometrical properties of detached eclipsing systems

Star	r_1	r_2	r_2/r_1	C_1	C_2	$(C_1/C_2)_c$	$(C_1/C_2)_c$	O-C
V 805 Aql	0.191 ± 0.006	0.163 ± 0.006	0.85 ± 0.04	6.9 ± 0.3	6.9 ± 0.3	1.00 ± 0.04	1.00 ± 0.01	0.00
σ Aql	0.278 ± 0.011	0.218 ± 0.009	0.78 ± 0.04	5.1 ± 0.2	5.5 ± 0.3	0.93 ± 0.03	1.15 ± 0.01	−0.22
TT Aur	0.283 ± 0.014	0.275 ± 0.028	0.97 ± 0.13	5.1 ± 0.3	4.7 ± 0.5	1.08 ± 0.07	1.02 ± 0.03	−0.06
WW Aur	0.161 ± 0.001	0.160 ± 0.001	1.00 ± 0.01	7.6 ± 0.2	7.4 ± 0.2	1.03 ± 0.02	1.00 ± 0.00	+0.03
AR Aur	0.098 ± 0.001	0.098 ± 0.001	1.00 ± 0.02	11.9 ± 0.5	11.0 ± 0.4	1.08 ± 0.05	1.00 ± 0.00	+0.08
β Aur	0.142 ± 0.005	0.130 ± 0.004	0.92 ± 0.03	8.4 ± 0.3	8.9 ± 0.3	0.94 ± 0.04	1.05 ± 0.01	−0.11
SZ Cam	0.425 ± 0.003	0.188 ± 0.001	0.44 ± 0.01	4.3 ± 0.3	4.6 ± 0.3	0.93 ± 0.07	1.60 ± 0.06	−0.67
AH Cep	0.324 ± 0.005	0.294 ± 0.008	0.91 ± 0.02	4.5 ± 0.1	4.6 ± 0.1	0.98 ± 0.02	1.06 ± 0.00	−0.08
Y Cyg	0.208 ± 0.008	0.202 ± 0.008	0.97 ± 0.03	6.7 ± 0.5	6.8 ± 0.5	0.99 ± 0.06	1.02 ± 0.01	−0.03
V 477 Cyg	0.124 ± 0.002	0.099 ± 0.002	0.80 ± 0.01	10.6 ± 0.2	9.7 ± 0.2	1.09 ± 0.04	1.14 ± 0.01	−0.05
YY Gem	0.156 ± 0.001	0.156 ± 0.001	1.00 ± 0.02	7.7 ± 0.2	7.7 ± 0.2	1.00 ± 0.03	1.00 ± 0.00	0.00
RX Her	0.217 ± 0.004	0.190 ± 0.004	0.88 ± 0.05	6.1 ± 0.3	6.3 ± 0.3	0.97 ± 0.04	1.07 ± 0.02	−0.10
TX Her	0.156 ± 0.003	0.154 ± 0.003	0.99 ± 0.04	8.1 ± 0.4	7.3 ± 0.5	1.11 ± 0.05	1.01 ± 0.01	−0.10
UV Leo	0.295 ± 0.006	0.293 ± 0.005	1.00 ± 0.03	4.8 ± 0.6	4.6 ± 0.4	1.04 ± 0.13	1.00 ± 0.00	−0.04
U Oph	0.263 ± 0.002	0.247 ± 0.002	0.94 ± 0.01	5.2 ± 0.2	5.2 ± 0.2	1.00 ± 0.04	1.04 ± 0.00	−0.04
V 451 Oph	0.209 ± 0.004	0.167 ± 0.003	0.80 ± 0.02	6.4 ± 0.3	6.8 ± 0.3	0.94 ± 0.05	1.14 ± 0.00	−0.20
AG Per	0.211 ± 0.007	0.194 ± 0.007	0.92 ± 0.03	6.2 ± 0.2	6.2 ± 0.2	1.00 ± 0.03	1.01 ± 0.01	−0.01

The beginning of this story, commencing with the discovery of UX UMa in 1931, has already been told in Chapter I; and, as is usual with major advances in any branch of science, its sequal was rather slow to unroll. The discovery, by Sanford (1949), of the binary nature of T CrB (a recurrent Nova of 1866 and 1946) came also still somewhat too early for its implications to be fully appreciated at that time. It was not till in the mid-1950's – on the heels of the discovery by Merle Walker (1956) that another Nova (N Her 1934) is a close eclipsing system with a period of 4h 39min, followed by discoveries of the binary nature of the cataclysmic variables SS Cyg (Joy, 1956) as well as of U Gem (Kraft, 1962) which proved also to be an eclipsing variable (Krzeminski, 1965) – that the subject has at last come into its own. The aim of the present part of this section will, therefore, be to give an account of the contemporary state of this subject in the framework of our general views on the course of the advanced stages of stellar evolution.

A list of typical systems belonging to this class has been compiled in Table VIII-10; and the headings of its individual columns are self-explanatory. The division of its contents into three groups is, to some extent, arbitrary. In the group of UX UMa-stars we listed systems in which the amplitudes of light variations due to mutual eclipses of their components exceed those of non-eclipsing nature; although Z Cha, EM Cyg or RW Tri could also be regarded as U Gem-type stars.

The role of observational selection is difficult to assess for stars of UX UMa-type; on account of their generally low absolute brightness their sample we know is heavily selected with regard to the distance. This, however, is much less the case for systems containing Novae which, by virtue of their relatively high absolute brightness at the time of their outburst, attract appropriate attention at much greater distance. Although the number of such systems observable (photometrically or spectroscopically) for short-period variations of light or radial velocity during their quiescent stage is still limited, it is already apparent that if as many of these eclipse as the sample on hand now indicates (i.e., 4–5 out of 7), then – unless their orbital planes are preferentially orientated with respect to the line of sight (which is most unlikely) – the masses as well as fractional dimensions of both components should, in general, be very unequal – an expectation which will indeed be borne out by a closer inspection of the relevant facts.

TABLE VIII-10
Close Binaries with Subdwarf Components

Star	Period (binary)	Nature of binary	Average Interval between outbursts	Range of apparent light changes during outbursts
	d			m m
Z Cha	0.0745	ecl	66^d	12.0 – 13.5
EM Cyg	0.2909	ecl	–	11.9 – 14.4
VV Pup	0.0697	ecl	–	14.6 – 17.1
RW Tri	0.3219	ecl	–	13.5 – 16.0
UX UMa	0.1967	ecl	–	12.7 – 13.8
		2. Systems containing cataclysmic variables (SS Cyg or U Gem-type)		
RX And	0.2117	–	14?	10.3 – 13.6
AE Aqr	0.7007	–	–	10.4 – 12.0
SS Aur	0.1458	–	560	10.5 – 14.8
Z Cam	0.3898	ecl	22	10.2 – 14.5
SS Cyg	0.2762	–	52	8.2 – 12.1
U Gem	0.1769	ecl	102	8.8 – 14.2
EX Hya	0.0682	ecl?	–	11.1 – 14.1
VW Hyi	0.0743	–		
RU Peg	0.3708	–	66	9.0 – 13.1
		3. Systems containing Novae		
V 603 Aql (N 1918)	0.1385	–	–	–1.1 – 10.8
T Aur (N 1891)	0.2042	ecl	–	4.1 – 15.8
T CrB (N 1866, 1946)	227.5	ecl	2900?	2.0 – 10.8
DQ Her (N 1934)	0.1938	ecl	–	1.3 – 15.4
GK Per (N 1901)	0.685[a]	ecl?	–	0.2 – 14.0
RR Pic (N 1925)	0.1451	–	–	1.2 – 12.8
WZ Sge (N 1913, 1946)	0.0567	ecl	12000?	7.0 – 15.5

[a] After Paczynski (1965)

The outstanding characteristic of the binary systems listed in Table VIII-10 are, however, their very short orbital periods which are – with one exception (T CrB, on which more will be said later) – shorter than one day; for a large majority less than half a day; and, for the recurrent Nova WZ Sge, only 1 h and 21.5 min! This, together with their observed spectra and relatively low luminosities, renders their components to be *subdwarfs* – well below the Main Sequence of hydrogen-burning stars; and of more advanced evolutionary stage.

Secondly, all these systems (with a possible exception of some of those in the first group listed in Table VIII-10) contain components prove to exhibit *spasmodic outbursts of light*, at semi-regular intervals ranging from months for U Gem-type stars to centuries for recurrent Novae, which within the time of days to months can increase the normal brightness of the respective star by a factor ranging from $10–10^2$ for stars of the U Gem-type to $10^3 - 10^4$ for Novae.

Spectroscopic phenomena disclose, moreover, that these light outbursts are caused by *eruptions,* in the course of which the star so afflicted divests itself of a certain fraction of its mass – with velocities which range from several hundred to a few thousand kms per second – of which obviously only very little can be captured by its more sedate companion; and the rest will escape into space.

The total amount of mass involved in such individual escapades is difficult to estimate from the observed spectroscopic phenomena with any accuracy for systems of the SS Cyg or U Gem-type; but is no doubt small in comparison with 10^{-5} to 10^{-4} $\odot$ per outburst estimated for ordinary Novae. Therefore, a great many successive outbursts would be necessary to diminish the mass of the respective star by (say) 1% – let alone more.

The amplitude of the outbursts of light accompanying this mass loss increases with increasing interval of the time elapsed between successive outbursts; for systems of SS Cyg or U Gem the range of light changes is typically 1–3 mag separated by intervals of a few months; increasing to 4–6 mag for recurrent Novae (like N Sge 1913 and 1946, or N CrB 1866 and 1946) whose outbursts recur after decades; and possibly centuries for ordinary Novae. All systems listed not only in part 2, but also in part 3 of Table VIII-10 exhibit outbursts which are no doubt recurrent; and the fact that (with the exception of T CrB and WZ Sge) no other Nova listed in Table VIII-10 (3) exhibited more than one outburst so far is due to the fact that, for conspicuous Novae of large amplitudes (like N Per 1901 or N Aql 1918), the time interval between successive eruptions may be too long – from centuries to millenia (cf., e.g., Payne-Gaposchkin, 1957).

The principal gift of the binary nature of these systems to us is an offer of the possibility of establishing the values of the masses and dimensions of their components in absolute units. As is well known, the radial velocity observations of at least one component can (together with a knowledge of the orbital period) furnish the absolute value of the 'mass-function' $f(m) \equiv m_2^3 \sin^3 i/(m_1 + m_2)^2$; and if the lines of both components are measurable, we can establish the individual values of $m_{1,2} \sin^3 i$ in (say) solar units. If, moreover, the inclination i of the orbital plane to the celestial sphere can be deduced from the eclipses, the observations can furnish the absolute values of the masses $m_{1,2}$ as well as radii $R_{1,2}$ of the two components if their fractional values $r_{1,2}$ (expressed in terms of the separation of the two stars) could be deduced from an analysis of the light changes arising from eclipses.

In the form of the data now available Nature has dangled several clues before us which are as important as they are elusive. Of the U Gem-type stars, the two-spectra binaries AE Aqr, SS Cyg, RU Peg (or EY Cyg) do not eclipse; and for Z Cam eclipses are only indicated. The system of U Gem eclipses all right, but constitutes a single-spectrum binary. In every case listed in Table VIII-10, the systems in question consist of a late-type star of spectral type G5–K0 – usually (though not always) more massive of the two – attended by a hot companion, akin to a white dwarf, the spectrum of which can be described (in most cases) as Be; and which in every single case is the very much smaller of the two (with absolute dimensions less than 0.1 ⊙).

Let, in what follows, the red (larger) component be referred to as the primary, and the small hot dwarf as the secondary one. If so, it follows that, for SS Cyg (two-spectra, non-eclipsing binary) $m_1 \sin^3 i = 0.20$ ⊙ and $m_2 \sin^3 i = 0.18$ ⊙; and for RU Peg, $m_1 \sin^3 i = 0.32$ ⊙ and $m_2 \sin^3 i = 0.27$ ⊙. For he eclipsing systems U Gem (Krzeminski, 1965) $m_1 = 1.3$ ⊙, $m_2 = 1.2$ ⊙; and for Z Cam (Kraft *et al.*, 1969), $m_1 \sin^3 i = 0.60$ ⊙, $m_2 \sin^3 i = 0.5$ ⊙.

As far as the systems listed in Table VIII-10 (c) are concerned, the masses of the components of DQ Her (cf. Kraft, 1964) appear to be close to $m_1 = 0.20$ ⊙; $m_2 = 0.12$ ⊙; and those of WZ Sge (Krzeminski and Smak, 1971) $m_1 = 0.025$ ⊙, $m_2 = 0.35$ ⊙; while for T CrB (Kraft, 1958), $m_1 \sin^3 i = 2.6$ ⊙, $m_2 \sin^3 i = 1.9$ ⊙; and, for GK Per (Paczynski, 1965) $m_1 \sin^3 i = 0.06$ ⊙, $m_2 \sin^3 i = 0.69$ ⊙ and $R_2 = 0.026$ ⊙ (Krzeminski, 1965; Smak, 1971); while for WZ Sge, $R_1 = 0.08$ ⊙ and $R_2 = 0.17$ ⊙, respectively (Krzeminski and Smak, 1971).

With the exception of the recurrent Nova T CrB (whose binary period is also some 10^3 times as long as those of all other known objects of this class), the masses of all post-Novae turned out be *smaller* than that of the Sun; and very much smaller in size. Moreover, among the systems of U Gem-type, the primary (red) components appear to possess fractional dimensions comparable with those of their Roche limits; while those of the early-type secondaries are well inside those limits. In terms of the terminology introduced in section VIII-2 the sub-dwarf binaries should, therefore, be classified as *semi-detached*; with the late-type stars representing contact components.

Which one of these two components happens to be the seat of the observed eruptions? In the case U Gem-type stars – in which eclipses have been observed repeatedly during the outbursts – there is no doubt that it is the primary (red) component, whose spasmodic eruptions give rise to the type of variability which first attracted attention to these systems. It is probable, that these eruptions are connected with the unstable star occupying the largest closed volume capable of containing its mass in the state of equilibrium; and the binary nature may facilitate an escape of mass by lowering the gravity over the hemisphere facing the early-type companion. The frequent occurrence of binaries among cataclysmic variables of SS Cyg- or U Gem-type is certainly not accidental; and – on grounds of probability alone – the binary nature appears to be a necessary prerequisite for stimulating eruptions of the components in a certain stage of their evolution.

Many models of such eruptive systems have been proposed in recent years by several investigators. We shall not discuss them in any detail in this place; for, in the face of many complexities presented to us by the observations, we regard much of them to be distinctly premature. Instead, we propose to limit ourselves to an enumeration of several *caveats* addressed to more courageous workers labouring in this part of the vineyard – both at present and in the future:

(1) In constructing models of given systems, remember the wisdom of 'Occam's razor' not to compound hypothetical features or mechanisms of a scenario beyond necessity ("entia non sunt multiplicanda praeter necessitatem"). An explanation of n observed phenomena in terms of m special assumptions represents surely nothing else but an outright guess without any real information contents *if* $m = n$; and the probability that 'there is something to it' increases only as $m \ll n$.

(2) Beware of a temptation to conjure up arbitrary 'scenarios' of your own making, without sufficient physical reasons behind it. Conjectures are certainly free to wander far and wide over their spacious domain; but they are very unlikely to bring home conviction – let alone truth. In problems characterized by very many degrees of freedom (largely under discussion in this chapter) it is virtually hopeless to try to arrive at their solution by any kind of guess work based on mere intuition and unsupported by adequate physical theory. Yet it is also true that the minds of only too many theoreticians seem to work well only in one or two 'normal' modes; and are capable of a considerable amount of self-deception in their efforts to convince themselves (and others) that observations of unexpected phenomena can be made to fit into one or the other of their preconceived categories.

(3) Never forget that the only proper framework for rational interpretation of phenomena exhibited by explosive binary systems is *hydrodynamics* (or hydromagnetics) of radiating media, and *not* particle mechanics. If the motions of atoms or ions exhibited in the course of such events could be legitimately described by collisionless particle trajectories, the density of gas consisting of such particles would be too low by many orders of magnitude for any kind of observational detection.

(4) Remember that the concept of Roche limit as defined in Chapter VI is of significance only for *equilibrium* configurations, and becomes physically irrelevant in the presence of *motion* (or, again, of a strong field of radiation). Pay respect to the Roche limit whenever it is deserved, but do not abuse its concept by allowing it to degenerate into a 'folklore' in the literature on close binaries, as would happen if we were to stretch its validity to cases where it no longer applies – such as those involving gas motions, where it is the full-dress Jacobi integral (1.28) of Chapter VI which should take its place (or the Bernoulli integral in the case of fluid flow).

(5) Remember that a mass loss (or transfer) in close binary systems is related with any period changes by equations which are (cf. sec. V-5) of *vector* (not scalar) form. In other words, a knowledge of the amount of mass lost (or transferred) does not, by itself, specify uniquely as yet any period change. In order to do so, we need to know not only the magnitude of the mass involved, but also the velocity and direction of escape, as well as the locus of the point from which matter may have escaped.

These considerations do not, of course, apply only to sub-dwarf binaries; but in no other branch of double-star astronomy had they been sinned against more. If you wish to consider the effects of 'shells', 'rings', 'jets', 'mass exchange', etc. ask yourself first: how could such formations or processes have come into being; and if they did, what should prevent their dissolution? Is a formation of given type necessary, or only optional, for producing the desired effects; and is it used only because one cannot prove that it does not exist? Remember that the lack of a proof of non-existence should never be confused with a proof of the existence! To do otherwise in our considerations might easily earn our subject an epithet of 'new astrology' in the minds of our more critical professional confrères.

After this digression let us return to subdwarf binary systems whose component may be a Nova. In this case, the observations on hand cannot decide alone which of the two components (of early or late spectral type) is the one suffering from Nova-itis. For U Gem-type stars this can be settled by observations of the eclipses at the time of the outbursts. As, however, the last binary Novae flared up in 1946 – i.e., before the advent of modern high-speed photoelectric techniques – we have as yet no idea what happens with the light changes due to eclipses during outbursts; and a long time may yet elapse before the necessary evidence is safely in our hands.

In the meantime, the weight of the current opinion (cf., e.g., Crawford and Kraft, 1956; or Paczynski, 1965) inclines to the view that, in this case, it is the secondary (early-type) component in subdwarf systems which is primarily responsible for the celestial fireworks which we observe as Novae – some 20 or 30 of them in the Galaxy per annum (cf. Arp, 1956). The cause of this process seems connected with the gradual formation of a degenerate helium core in the interior of the hot star – a process its companion. As we already mentioned in section 3B of this chapter, the velocity of overflow must be moderate or low (of the order of 10 km s^{-1}) if matter is to be captured by the hot companion which, in turn, responds to it by ejection of shells with velocities 10^2 times higher – so that very little of the borrowed mass will ever be paid back by this process.

A fuller discussion of hydrodynamical aspects of such a process is outside the scope of this section. We may, however, note that the masses of both components of T CrB (one of which must be the recurrent Nova) are well above the limit of 1.3 ⊙ for an electron-degenerate configuration (cf., e.g., Schatzman, 1958); so that – at least in this case – the post-Nova cannot be regarded as degenerate. On the other hand, the components of WZ Sge – an equally recurrent Nova – possess masses well below this limit (the same being true of the blue component of DQ Her) and could have already attained the degnerate state. If, however, Nova-itis represents a symptom of this process, why do stars like DQ Her or WZ Sge need to go on exploding at all?

Moreover, as has first been pointed out by Walker (1956) and confirmed since by others (Nelson, 1976) that the system of DQ Her (Nova Her 1934) – in addition to the light changes in a period 4h and 39s of the binary orbit – exhibits also a regular flicker with a period of only 71.0654 s; and since this flicker disappears during eclipses of the hot star, there is no doubt that it originates in the latter. Its 71-sec period is of the right order of magnitude for a free period of radial oscillations of a degenerate configuration of this star's size and mass; and may be indicative of latent stress which from time to time is released by Nova outbursts. But whether or not this is actually the case only the future can tell; and for the next outburst of DQ Her we may have to wait until well into the 21st century.

What is the evolutionary position of these stars within the general framework of stellar evolution? The decisive word rests again with the mass. From the data already quoted it transpires that (barring exceptions like T CrB) the total mass of subdwarf systems is generally so small that (unless they are much older than the disc of our Galaxy) *the could not have reached their present stage unless their mass in the past has been very much higher than it is at present.*

In other words, we find ourselves again face to face with the requirement of a large loss of mass in the past to account for the existence of systems we observe at present.

The mechanism capable of bringing about so far-reaching a mass loss continues to remain obscure. It could not, in particular, be the ejection of shells from Novae or recurrent Nova-like variables if the cause of instability is to be sought in the onset of degeneracy; for the main part of the mass loss should already have been accomplished before the star diminished its mass below 1.3 ⊙ to commence its career as a Nova. In the course of such a career the star could only have completed what commenced already at the pre-degenerate stage by methods less spectacular, but more effective.

The same argument largely weakens also the more recent attempts by Kraft (1962, 1967) to identify the W UMa-type stars discussed in section VIII-3C as progenitors of the subdwarf systems discussed in this section. It is true that, as regards orbital period, mass, and angular momentum, a resemblance between the two groups of close binaries is indeed considerable. But, on the other hand, this very fact renders a transformation of other characteristics (radii or spectra) of one of these groups into another more difficult on the nuclear time-scale available for the purpose. It is much more likely that the real progenitors of the subdwarf binaries were stars considerably more massive; and if so, these could not have been the present W UMa-type systems, which are again evolving too slowly for the purpose on account of their low masses.

What, then, are the remaining alternatives? Where should we seek the hypothetical progenitors of the present subdwarf binaries which, we repeat, could not have reached that state within the age of our Galaxy had they been born with the masses they now possess? The answer may be: among the stars which were still *single* on the Main Sequence, and whose cores split up by fission at the post-Main Sequence stage. The reasons underlying the formation and development of small massive cores in the interiors of evolving stars have already been described in sec. VIII-1. If, in the course of such a process, the angular momentum of the core was at least approximately preserved, its contraction would have entailed an accelerated rotation – until the limit could have been reached at which the core would split up in two parts, revolving now in the interior around the common centre of gravity.

This 'seeding' of an incipient binary in the interior of a single star need not, at first, have produced phenomena conspicuous on the surface, since the common envelope surrounding both components could have long remained optically thick. However, once the core has split up in two parts, the visible surface of such a configuration need no longer have remained spherical; nor the distribution of brightness over it radially-symmetrical. Whether or not any phenomena arising in this connection could become photometrically (or spectroscopically) observable – or are, indeed, related to actual stellar variability encountered in the respective parts of the HR-diagram – constitutes a question a closer discussion of which is outside the scope of this section; but we may end up by reiterating (cf. Kopal, 1965b; Gorbatsky, 1975) that *there may exist, not one, but two distinct epochs at which binary stars are formed:* one is at the pre-Main Sequence stage, when stars originate by condensation from pre-existing gas and dust; and the other at the post-Main Sequence stage, when a contracting stellar core may split up by fission. A large majority of binaries – both close and wide – with components on (or above) the Main Sequence no doubt can trace origin to their pre-stellar past. Subdwarf binaries below the Main Sequence constitute the principal exception: they could not have been double on the Main Sequence, and presumably originated subsequently by a process whose details remain yet to be reconstructed.

C. X-RAY BINARIES AND BLACK HOLES

One of the most dramatic events in double-star astronomy of recent years as surely been the discovery of close binaries with components emitting powerfully in the domain of X-rays. The X-ray astronomy of discrete sources in the sky is barely more than 15 years old; and the accidental discovery of the first such source (Sco X-1) in 1962 was already described on p. 7 of Chapter I. An identification of this source with a 12th mag. star with Nova-like spectrum (cf. Sandage *et al.*, 1966) added further interest to the case. But as long as the principal tools of exploration of the sky in the domain of X-rays were rockets and balloons, which could provide glimpses of the sky in this strange light lasting only minutes at a time, the progress – although full of promise – was bound to be slow.

The launch, in December 1970, of the Uhuru satellite – the first spacecraft devoted entirely to X-ray astronomy – marked the beginning of a new era of the subject; to which the latter-day Solar Orbiting Observatories and other spacecraft made further important contributions. But the path of X-ray and double-star astronomy became intimately allied only with the discovery, by Schreier *et al.* (1972), that one of the X-ray sources discovered with the Uhuru satellite coincided in position with a previously known eclipsing binary HZ Herculis – a close system with a period of 1.70017 days; and subsequent discoveries of X-ray sources in other binaries (eclipsing or not) have gradually led to a realization that *a large part of known discrete X-ray sources in the sky are components of close binary systems,* of particular properties which we shall proceed to describe.

Table VIII-11
X-ray Binary Systems
1. Optical components

Designation	Spectrum	App. vis. mag.	m_o (in ☉)	R_o (in ☉)
Cen X-3 = Krzeminski's stars	O6.5I	13	20	10
Cyg X-1 = V 1357 Cyg	O9.7I	8.8 – 8.9	21	19
Her X-1 = HZ Her	F3	12.8 – 15.1	2.2	3.8
Sco X-1 = V 818 Sco	Nova-like	12 – 13	–	–
SMC X-1 = Sanduleak 160	B0.5I	13	12?	12
3U 1700-37 = HD 153919	O6.5	6.7	60?	24
Vel X-1 = HD 77581	B0.5I	6.9	23	28

2. X-ray components

Designation	Nature of binary	Period (binary)	Period (flicker)	m_x (in ☉)	L_x (in ☉)
Cen X-3 = 3U 1118-60	ecl	$2^d.873$	$4^s.842$	2.4	16000
Cyg X-1 = 3U 1956+35	–	5.601	–	12	3700
Her X-1 = 3U 1653+35	ecl	1.7002	1.2378	1.3	1800
Sco X-1 = 3U 1617-15	–	0.7873[a]	–	2	2000
SMC X-1 = 3U 0115-37	ecl	3.8922	0.7149	2.5	37000
HD 153919 = 3U 1700-37	ecl	3.4126	–	2.4	16000
Vel X-1 = 3U 9000-40	ecl	8.9721	284	1.7	850

[a] The period is after Gottlieb *et al.* (1975). Luytyi (1975) deduced for the period a value of 3.39309 days, which is almost exactly equal to five times that given by Gottlieb *et al.*

The preceding Table VIII-11 contains a summary of the observational data on systems whose one component is an X-ray star, known up to the end of 1976. The table is divided into two parts: the first summarizing the relevant data on the optical companions in such pairs; and the second, on their X-ray mates. Only reliably known cases have been included in these tabulations; those still regarded as marginal – such as Cep X-4 = HD 206267 (P = 3.7 days), Circinus X-1 (P = 16.5 days, possibly eclipsing), Cygnus X-2 = V 1341 Cyg (P = 0.68 d) or the radio-source Cyg X-3 = 3U 2030 + 40 (a possibly eclipsing system with a period P = 0.2 d) – have been left out of our tables because of an incomplete state of our knowledge of some aspects of the observational data. The meanings of most columns in Table VIII-11 are self-explanatory; the 3U-numbers of the X-ray sources refer to a catalogue by Giaconni *et al.* (1974).

Table VIII-11 contains only the bare essentials of interest to double-star astronomy. A fuller discussion of all other properties of these systems is outside the scope of this section; and the reader desirous of getting acquainted with them can be reffered to several recent summaries of the relevant data (for the latest such summary cf. Gursky and Schreier, 1975; or Boldt and Kondo, 1976). Perhaps the most significant feature of the data listed in Table VIII-11 is the enormous luminosities of the X-ray components (of the order of 10^3–10^4 $\odot$) emitted mainly in energies between 1–10 keV, and exceeding by orders of magnitude the energy output of these objects in the optical domain. If these X-ray luminosities were to be interpreted as due to thermal radiation of stars of estimated dimensions, surface temperatures of the order of 10^7 deg would be required to account for the observed flux; otherwise their origin is to be sought in non-thermal processes.

As far as the masses and absolute dimensions of these systems are concerned, the following facts have been gathered so far. Reliable single-spectrum orbits are available only for the (apparently) brightest objects of this class – like Cyg X-1 (cf., e.g., Bolton, 1974); and these provide merely a relation between the masses of the two components and the inclinations i of their orbital plane with respect to the line of sight. For no X-ray component are any spectroscopic observations available so far – nor can any be expected in the foreseeable future. However, the periodic flicker which some of them exhibit (we do not call it pulses; as the mechanism of their emission may be quite different form that operative in radio pulsars) provides an independent way of establishing – through their periodic fluctuations observed in the course of each cycle (the 'light equation'; cf. sec. V-6D) – a single-spectrum orbit of the X-ray component around the centre of mass of the respective binary system.

If such orbits were available independently for *both* components, the mass-ratio of the system could be established from the amplitudes of the respective radial-velocity changes. Unfortunately, for no known systems of this type this has been possible so far; as either one, or the other part of the relevant evidence is missing. However, a combination of such data as are currently on hand (including limits on orbital inclination i imposed by eclipse phenomena) has led to the estimates of mases and absolute dimensions of the individual components as are listed in the respective columns of Table VIII-11. The reader should only keep in mind that, on account of various approximations involved in the formations of our estimates, the uncertainty of these data can amount the several units of their last places.

In so far as the double-star aspects of X-ray binaaries are concerned, the data available so far lend themselves for the following conclusions:

(1) The optical components of most known X-ray binaries exhibit a *continuous* variation of light throughout the entire orbital cycle, which have (at optical frequencies) nothing to do with eclipses. In some – for instance, HZ Herculis (cf. Petro and Hiltner, 1973) – the observed variation of light can be described almost completely by the first harmonic in true anomaly – a fact disclosing unmistakeably that it is caused by the reflection effect (i.e., irradiation of the optical component by the X-ray flux from its companion, which is absorbed and re-radiated at optical frequencies). Others – like Cen X-3 (Petro, 1975) or Vela X-1 (Petro and Hiltner, 1974) – exhibit distinct evidence of a double-wave in their light changes, indicative of the action of second-harmonic *tides;* and if so, the masses of their X-ray companions should not be very much smaller than those of the optical stars.

(2) In each case when the orbital inclination i in such systems is sufficient to give rise to eclipses, those of the X-ray components prove to be *total*. Moreover, within the time-resolution of the X-ray telescopes employed for such observations, the disappearance of the X-ray star behind the limb of its optical mate is practically *instantaneous* – a fact disclosing that the *size* of the source of the X-rays must be *very small* in comparison with the dimensions of the system.

(3) The unusually long duration of total eclipses in X-ray domain (in 3U 1700–37, 1.1 days out of the orbital period of 3.4 days) rules out the possibility that the X-ray source might be confined to a small spot on the surface of a more normal star. It discloses that *the X-ray star as a whole must be very small*, and revolve in close proximity of the surface of its mate.

(4) The very compact nature of the X-ray sources seems also attested (albeit indirectly) by a very high frequency of their *flicker*. A discussion of the physical causes of this flicker is wholly outside the scope of this section (cf., e.g., Rappaport and Joss, 1977; or other sources). We merely wish to note that whereas in the system of DQ Herculis its recurrent Nova component (akin in its physical properties to white dwarfs) exhibits a flicker of 71-sec period, those of the X-ray stars (with the exception of Vel X-1) are of the order of 100 times shorter; and whatever their cause may be (free oscillations or axial rotation), they indicate smaller dimensions (or greater compactness) of the respective objects.

Indeed, as a result of this and other arguments, as hypothesis has repeatedly been put forward in recent years that *the X-ray components in close binaries are neutron stars produced by explosions of Supernovae*. Such a hypothesis appears, on the first sight, to offer many attractive features – the most important being a provision of ample source for the requisite energy of X-ray emission by acquisition of hydrogen from its early type companion.

It may be instructive, in this connection, to evaluate the amount of energy which can be released by one hydrogen atom (or proton) incident on the surface of a white dwarf or a neutron star. In the course of such an infall in the gravitational field of a mass M, a particle of mass m will acquire the velocity

$$V^2 = \frac{2GM}{r}, \tag{5.3}$$

where r signifies a distance of the mass-point m from the centre of gravity of the attracting mass M, and $G = 6.672 \times 10^{-8}$ cm^3 g s^2 denotes the constant of gravitation. Accordingly, the kinetic energy E of the particle m will become

$$E = \frac{1}{2} mV^2 = \frac{GmM}{r} \tag{5.4}$$

Assume now that the attracting star possesses a mass equal to that of our Sun, or 1.985×10^{33} g; and that the impinging particle is a proton of mass $m = 1.673 \times 10^{-24}$ g. Its energy E should then become equal to

$$E = \frac{221.6}{r} \text{ erg} = \frac{1.393 \times 10^{14}}{r} \quad eV, \tag{5.5}$$

corresponding to the frequency ν or the wavelength λ of the respective quantum of light given by

$$E = h\nu = \frac{hc}{\lambda}, \tag{5.6}$$

where h stands for the Planck constant of 6.626×10^{-27} erg s and c, for the velocity of light of 2.998×10^{10} cm s^{-1}. On insertion of these values in Equations (5.5) and (5.6) we find that, for $r = 10^4$km $= 10^9$ cm (a typical value for the radius of a white dwarf), the wavelength of light follows from (5.6) as $\lambda = 8.965 \times 10^{-10}$ cm or 0.089 Å (corresponding to moderately hard X-rays of quantum energy 139 keV); while for $r = 100$ km $= 10^7$ cm (a typical radius of a neutron star), the kinetic energy of the impinging particle would correspond to $\lambda = 0.00089$ Å of quantum energy 13.9 MeV (or moderately hard γ-rays).

The bulk of the X-rays emitted by their sources in close binary systems is known to be confined to the domain between 2–10 keV's (the γ-rays presumably associated with Cyg X-1, to the vicinity of 1 MeV) and could, therefore, be produced by impacts one-tenth as energetic as those considered above. On the other hand, the total amount of energy emitted by known X-ray stars range between 10^{35}–10^{38} erg s – many thousand times as much as that emitted in optical frequencies – and attaining 10^{40} erg s^{-1} in momentary γ-ray bursts if we accept (see p. 482), a tentative identification of Cyg X-1 with the mysterious Vela-sources. Therefore, an influx of 10^{17}–10^{20} g s^{-1} of hydrogen on to a neutron star should, in principle, be sufficient to account for the luminosity of the X-ray stars (and 10^{22} g s^{-1} for the hypothetical γ-bursts from Cyg X-1) if the kinetic energy released by impact per particle could be converted without lossess into X- or γ-rays sent out by these stars.

In reality, of course, we are not concerned with impacts of individual particles, but with deceleration of gas streams producing heat through the medium of shock waves. A mass transfer envisaged in such operations (10^{-9} to 10^{-6} $\odot$ yr^{-1}) is certainly in the range of expected possibilities. Moreover, the number of X-ray binaries in the sky is also compatible with the observed frequency of Supernova outbursts within the Galaxy. By a comparison of the luminosity 10^{35}–10^{38} erg s^{-1} of individual X-ray binaries with the total luminosity of the Galaxy in the 1–10 keV range, Gursky and Schreier (1975) arrived at an estimate of the total number of galactic X-ray binaries to be no greater than 100.

On the other hand, a glance at the data compiled in the upper half of Table VIII-11 discloses that (with the exception of HZ Her, to which we could probably add Sco X-1: or Cyg X-2 and X-3) the optical components of X-ray binaries are OB-supergiants of extreme Population I and luminosity class I – with masses between 10–20 ⊙. Moreover, if their X-ray companions of equal age have already evolved into neutron stars, their initial masses must have been at least 30–40 ⊙, whose lifetime from cradle to grave (i.e., from a protostar to a neutron stage) should be of the order of $10^5 - 10^6$ yr. If, therefore, all known X-ray components of close binary systems – representing the offspring of fomerly supermassive stars – are no older, their estimated number of 100 in the Galaxy could be accounted for by Supernova explosions in binary systems at a rate of one in 1000–10000 yr – which is significantly smaller than current estimates of the total Supernova production per Galaxy; and, to this extent, our hypothesis seems consistent with the observed facts.

However, when we come to consider the dynamical aspects of a Supernova explosion in close binary systems, the situation becomes very much less clear. If, as has been suggested (cf. Arnett, 1969), the cause of a Supernova explosion is a detonation of the degenerate carbon core of a massive star (or any similar process), a large part of the mass of the exploding star will be ejected from it instantaneously and isotropically with velocities far in excess of that of escape from the gravitational field of the system. Under these conditions it can be shown (cf. Hadjidemetriou, 1966b; or, more recently, Khabarin, 1975) that the eccentricity e of the orbit of the system will *increase* – the more so, the larger the mass lost, and the more violent the explosion. The latter would certainly not last more than a small fraction of one orbital cycle; and if so, less than a half of the mass of the Supernova could be lost if the system is to retain closed orbit; if more than a half were lost, the system would get dissolved and its components would henceforth travel through space alone.

Conjectures that Supernova outbursts in close binary systems may be responsible for the 'runaway stars' of high space velocities (the higher, the closer and (or) more massive the parent binary had been) were previously advanced by Zwicky (1957) and Blaauw (1961) or Boersma (1961); or, more recently, by Beckenstein (1976). Should, on the other hand, the circumstances of explosion have been such that the orbital eccentricity after the mass loss will be less than unity, the orbit of the system should remain closed (albeit eccentric, and of much longer period); and the (formerly) less massive components will drag along the Supernova remnant with it through space.

Having received the full blast of Supernova ejecta in the course of the outburst, the outer layers of the companion should be enriched with an injection of heavier elements produced in the course of Supernova nucleogenesis; and these may leave an imprint in its spectrum. But the principal hallmark left behind by such an event should be a marked *increase in orbital eccentricity* produced by sudden mass loss. Even if the original eccentricity e of the respective binary before explosion was insignificant, in the post-explosion state it could be anything between $0 < e < 1$ provided that the system remains double; or greater than 1 if the explosion dissolved the system. But *no known X-ray binary shows evidence of significant orbital eccentricity* – certainly not in excess of 0.01 – although it would be relatively easy to establish for systems whose X-ray components exhibit periodic flicker.

This fact is indeed difficult to reconcile with an assumption that the compact components in X-ray binary systems were produced by Supernova explosions entailing an appreciable loss of mass; and may force us to consider the possibilities that either the X-ray components are not neutron stars, or that neutron stars may be produced also by processes of non-explosive nature. Attempts to escape from this dilemma by an appeal to secular effects of tidal friction (cf. Lea and Margon, 1973; or Lecar *et al.*, 1976) have so far been unconvincing. It is true (cf. sec. V-3B) that, in general, tidal friction can bring about a secular decrease of e through the medium of 'tidal lag'. This process operates, however, far too slowly in systems containing post-Supernovae unless turbulent viscosity of their mates is enormous; and in early-type supergiants in radiative equilibrium this is most unlikely to be the case.

In the event of a relatively slow mass loss – which we considered in section VIII-3B in connection with the evolution of semi-detached systems at post-Main Sequence stage – it is impossible to say (at least without a prior analysis of the case) whether as secular increase of e caused by a mass loss or transfer (cf. sec. V-5) will prove to be more important than a secular diminution of e by tidal friction (sec. V-3B). However, in supergiant systems which suffered an explosive loss of mass the outcome is not in doubt: only its early-type star can be regarded as a sufficient source of tidal friction (the minute dimensions of tis companion make it virtually immune to tides); and its plasma viscosity is far too low to make its effects felt over the time-scale of $10^6 - 10^7$ yr.

These are, to be sure, not the only problems facing the students of X-ray binary systems. A perpetual source of frustration has been the fact that most of these variables are poor actors who dislike repetitive performance. Thus Her X-1 – the first X-ray source identified with an eclipsing variable (HZ Her) – sometimes fails to emit any X-rays at all (as do SMC X-1 or Cir X-1); and worse: for even the optical variability of HZ Her disappeared for years at a time. Now and then that system stopped eclipsing altogether, for reasons only to be guessed at (such as a change in orbital inclination due to precession and nutation discussed in section V-4; or varying transparency of the outer layers of the eclipsing star?). In addition, HZ Her exhibits optical evidence of a 35-day period possibly associated with orbital precession (Deeter and Boynton, 1976; Chevalier, 1976); while Thomas (1974) believes to have detected apsidal motion in Cen X-3 from fluctuations of the period of this system (cf. again sec. V-4); though the orbital eccentricity indicated by the observations appears to be only 0.03 ± 0.01 (p.e.) – i.e., hardly a significant one.

Summarizing this part of our discussion we may say that observations of unquestionable validity – such as virtually instantaneous disappearance of the X-ray source behind the limb of the optical companion, as well as the unusually long duration of the eclipses – have demonstrated that the disparity in size of the two stars must be very large; and that, therefore, the X-ray source must constitute very compact objects. At the same time, however, these observations cannot impose anything more than an upper limit on the size of the X-ray source; and such limit could be perfectly met by a star of the size of an ordinary white dwarf.

A temptation to identify this source with a body smaller still – such as a neutron star – is motivated by theoretical, rather than observational, reasons: namely, by the need to provide for the observed X-ray emission by a conversion of the gravitational energy into light. Even if it were the case – and this remains still a hypothesis – we have shown above that a gravitational field of a dense white dwarf is entirely adequate for the purpose: to invoke to this end a neutron star is like using a steamhammer to crush eggs. The principal reason why the majority of the binary X-ray sources are *not* likely to be neutron stars is (as was already mentioned) the circularity of their orbits. For if the production of a neutron star entails a sudden loss of a large amount of mass, this should (in most cases) result in a dissolution of the former binary, and in the creation of very eccentric orbits among the survivors. Now if the present X-ray sources (no doubt compact; and far ahead of tis optical mate in its evolutionary stage) underwent already the ordeal of a Supernova explosion, they should have lost more than 90–95% of its former mass in this process: how – we must ask again – could they have done so and still retained essentially circular orbits?

Perhaps the most intriguing X-ray binary listed in Table VIII-11 is the system of V 1357 Cyg (Cyg X-1), for several reasons. Its optical component (HDE 226 868 = V 1357 Cyg) is an early-type supergiant of spectral class O9.7 I – obviously a very young star – the mass of which can be (conservatively) estimated to not less than 20 ⊙. Its binary period of $P = 5\overset{d}{.}59982 \pm 0\overset{d}{.}0.00004$ is not unusual for stars of this type; and the system does not eclipse. Its light changes in the optical domain (cf. Lyutyi, *et al.*, 1973, 1974; Hilditch and Hill, 1974) exhibit periodic oscillations with an amplitude of less than 0.1 magn, and are no doubt caused by the ellipticity of the early-type component. This fact indicates again that the system must be a close one; and if so, the absence of eclipses necessitates that $i \ll 90°$. In accordance with the limits on eclipses for contact systems established in section VI-2C, we shall hereafter adopt the round value of $i = 30°$ (the real i may be still smaller).

The X-ray component of Cyg X-1 shows no periodic flicker to offer a clue to the mass-function of the system; but a single-spectrum orbit of V 1357 Cyg by Bolton (1975) disclosed that

$$\frac{m_x^3 \sin^3 i}{(m_o + m_x)^2} = 0.22 \pm 0.01 \odot . \tag{5.7}$$

If, in this equation, we set $m_o = 20 \odot$ and $i = 30°$, it would follow that $m_x = 12 \odot$ – rendering the X-ray component of V 1357 Cyg by far the most massive star listed in Table VIII-11 (2); and the conjecture has repeatedly been voiced that, in this case, the X-ray source is no longer a neutron star, but a *black hole.*

As is well known, the concept of a black hole goes back to a consequence of the general theory of relativity which, for sufficiently compact configurations (contracted to a configuration of dimensions inferior to the Schwarzschild radius) permits space to close in on so dense a body to prevent any communication between its interior and the outer space. As such a configuration cannot send out either electromagnetic or gravitational waves (cf. Zeldovich and Novikov, 1967), it can become observable only through its gravitational attraction on neighbouring masses – such as a nearby component of a close binary system – or, possibly, the ambient gas (cf. Shakura and Sunyaev, 1973). Therefore, in practice, 'black holes' can be detected if – and only if – they happen to be components of close binary systems. A search for them among known single-spectrum binaries (cf., e.g, Trimble and Thorne, 1969; Gibbons and Hawking, 1971; Batten and Olowin, 1971; Gott, 1971) has proved so far fruitless; and claims of alleged identifications put forward in 1971–72 on behalf of ϵ Aur (Cameron, 1971; Wilson, 1971a), β Lyr Devinney, 1971; Wilson, 1971b; Kondo *et al.*, 1972) or BM Ori (Wilson, 1972) have also proved unconvincing – as, in each of these cases, the observed phenomena admit also of alternative explanations by less exotic means.

Of the remaining candidates for this role, the strongest contender is the X-ray component of V 1357 Cyg. It differs from all other known X-ray stars not only by an absence of periodic flicker (though its X-ray emission continues to oscillate in an irregular manner) but also by a very much larger mass (probably no less than 10 ⊙), which stimulates greater self-compression. Whether or not it has actually attained the dimensions comparable with the Schwarzschild radius of a black hole cannot be settled by available observations. The known facts are, at best, not inconsistent with such an interpretation; though a statement to this effect is, of course, still far from a positive proof.

One additional fact has, however, emerged recently which – if confirmed – would greatly strengthen the claim of Cyg X-1 to the role of a black hole: namely, its tentative identification (Strong and Klebesadel, 1976) with the transient source of γ-rays (with energies between 0.15 to 1.5 MeV) discovered by the Vela satellites. A story of the discovery of this source in 1967 represents one of the most exciting episodes of high-energy astrophysics of the last decade; but any fuller account of it is wholly outside the scope of this book. However, what brought it in intimate touch with double-star astronomy has been a realization that not only the position of this γ-ray source coincides – within the limits of its observational errors (which are still considerable) – with that of Cyg X-1, but also that the observed fluctuations of γ-ray emission appear to be correlated with similar changes of the X-ray output from V 1357 Cyg.

If these coincidences will prove to be genuine, it is tempting to associate the observed γ-ray emission with an interaction between two plasma streams incident on a black hole. While infall on to a neutron star may generate the X-rays characteristic of most binaries listed in Table VIII-11, higher accelerations necessary to produce γ-rays becomes possible in the proximity of a black hole. And if so, γ-ray emission should be the principal guide to a discovery of such objects; though only one so far appears to be consistent in position with that of a well-known X-ray source.

The consequences of an association of the γ-ray bursts and Cygnus X-1 – if confirmed – would be truly fascinating. Since the distance of Cyg X-1 is known (from spectroscopic parallax of its optical companion) to be close to 2600 parsec, the γ-ray bursts recorded by the Vela and other satellites (including Apollo 16 on its return form the Moon in April 1972) would correspond to a energy output of the source close to 10^{40} erg s^{-1} – i.e., 2 million times the total energy output of our Sun; and about 100 times the X-ray output of Cyg X-1. The γ-ray flux during outbursts would, therefore, be well in excess of the object's Eddington limit (i.e., the amount of radiation which, in steady state, would blow away the surface of a star by radiation pressure).

While – we repeat – V 1357 Cyg is so far the only known X-ray source which may emit also in the domain of γ-rays (a circumstance suggestive of the presence of a black hole), another object came recently to light where the existence of a black hole is at least under suspicion: namely, the binary pulsar PSR 1913 + 16 discovered by Hulse and Taylor in 1974 (cf. Hulse and Taylor, 1975). Unlike the X-ray pulsars in Cen X-3, Her X-Xz1 or SMC X-1 characterized by periods of the order of one second, this radio-pulsar exhibited a mean period of only 0.059030 ± 0.000001 s which, moreover, did not remain constant (or secularly increased as for many other objects of this type) but turned out to oscillate between $0^s.058967$ and $0^s.059045$ in a period of 0.3230 of a day.

Such an oscillation immediately suggested a periodic motion of the pulsar in the line of sight; and on interpretation in terms of the 'light equation' described in section V-6D it led to the single-spectrum elements of the binary system given by

$$\left.\begin{aligned} P &= 0\overset{d}{.}32300 \\ a_1 \sin i &= 1.00 \pm 0.02 \odot , \\ e &= 0.62 \pm 0.01 , \end{aligned}\right\} \tag{5.8}$$

and

$$\frac{m_2^3 \sin^3 i}{(m_1 + m_2)^2} = 0.13 \pm 0.01 \odot ; \tag{5.9}$$

where the index 1 refers to the pulsar; and 2, to its companion.

A search for the optical component of the pair (Bernacca *et al.*, 1975) has so far proved unavailing, and no eclipses were noted; so that both the angle i of orbital inclination as well as the mass-ratio of the system remain arbitrary. The only additional constraint which has emerged since from accumulating observations was the establishment of the rate of apsidal advance at $4\overset{\circ}{.}24 \mp 0\overset{\circ}{.}04$ per annum – corresponding to a ratio $U/P = 95900 \pm 800$ of the period U of the revolution of the apsides to that of the orbit.

The absolute dimensions of the pulsar component are (by analogy with those of other known pulsars) likely to be of the order of 10^2 km, while the separation of the components will be no less than 10^6 km. If, therefore, the invisible component were a neutron star or a black hole, the entire amount of observed apsidal advance would have to be relativistic, and if so, Equation (3.51) of Chapter V would furnish an additional constraint requiring that

$$\frac{m_1 + m_2}{A(1 - e^2)} = 1.64, \tag{5.10}$$

where $m_{1,2}$ as well as $A \equiv a_1 + a_2$ are expressed in solar units.

Since, moreover,

$$A = (a_1 \sin i)\left(1 + \frac{m_1}{m_2}\right) \csc i = 1.00 \left(1 + \frac{m_1}{m_2}\right) \csc i \odot , \tag{5.11}$$

and $e = 0.62$, it follows from (5.10) that

$$m_2 = (1.01 \pm 0.02) \csc i \odot . \tag{5.12}$$

Therefore, the mass of the invisible companion cannot be smaller than 1 $\odot$. If, moreover, we insert (5.12) in the mass-function (5.9), we find that the total mass of the system

$$m_1 + m_2 = (2.81 \pm 0.04) \odot ; \tag{5.13}$$

so that the mass of the pulsar

$$m_1 = (2.81 - 1.01 \csc i) \odot . \tag{5.14}$$

As long as the angle i of inclination remains unknown, the values of m_1 and m_2 cannot be separated from it. However, if $\sin i$ is not much less than 1, it would follow that $m_1 =$ 1.8 $\odot$ and $m_2 = 1.0 \odot$, the pulsar being the more massive of the two. Should, however,

$i \ll 90°$, this proportion could of course be reversed; though the positivity of m_1 restricts i to be greater than 21°.

All these computations have been based on the assumption that the total amount (4°.24) of the yearly apsidal advance is of relativistic origin. This would be the case if the secondary component were likewise a compact object (white dwarf, neutron star, or a black hole). On account of the requisite mass, the black hole would be an admissible candidate only if csc $i \gg 1$; other objects would impose smaller restrictions on the actual value of i. However, if the secondary were not a compact object (for instance, if it were a helium Main Sequence star), its oblateness caused by axial rotation, together with tides raised upon it by its mate, could account for at least a part of the observed apsidal advance. Whether or not a hypothetical mass loss (or transfer) could do the same (cf. sec. V-5) would depend not only on the total flux of mass involved, but also on the circumstances of such a loss; and whether this would result in a an advance or regression of the apsides would depend on the particular way in which a transfer would take place.

How does the binary PSR 1913 + 16 with its pulsar component compare with the X-ray binaries discussed earlier in this section? First, we cannot fail to note that the total mass of the pulsar binary, as given by Equation (5.13), is very much smaller than those of all X-ray binaries. The pulsar emits, of course, no X-rays; and its pulse period is very much shorter than that of the X-ray flicker of stars like Her X-1 or SMC X-1 – indicating a smaller size and more compact nature of the respective object. Secondly, the pulsars are know to be much more closely connected with the explosions of Supernovae; and PSR 1913 + 16 may well be the remnant of one that exploded in the past. The fact that its present pulse period of 0.05903 s is longer only than that of the Crab pulsar which originated in a relatively recent past suggests that PSR 1913 + 16 too may not be a very old object. Moreover, the facts that (unlike for X-ray binaries) its orbit is characterized by a conspicuous eccentricity ($e = 0.62$), and its total mass ($m_1 + m_2 = 2.8$ ⊙) is so small, lend support to a conjecture that while the X-ray stars are still awaiting their explosion, the binary pulsar PSR 1913 + 16 already had it.

But the real enigma of the PSR 1913 + 16 system is its elusive secondary component, of which there is so far no trace. Was it originally the more massive star, which for this reason collapsed in its present state ahead of its mate? If so, the binary system today is a very wide one; and whatever may have been the case in the past, its components no longer interact with each other otherwise than by their revolution around the common centre of gravity. Is this why the secondary is now so quiet – resting in its gravitational coffin as a black hole?

But do black holes exist? A perusal of literature on the subject of the past ten years leaves one with an impression that everyone believes in their existence – the observers, because they consider them a theoretical necessity; and the theoreticians, because they regard them as observed facts. In no case discussed in this section has empirical evidence for black holes been more than circumstantial – sufficient, perhaps, to indict, but not to convict – and the confident prediction by Novikov and Thorne (1973) that . . . "the clear discovery of black holes, we expect, is only a few months or year away" is still awaiting its fulfilment.

It is certainly true that no one would feel impelled to postulate black holes on the strength of empirical evidence alone if the general theory of relativity would not predict that such formations may exist. But the theory does not insist that they must exist; it allows only for such a possibility; and whether or not this particular prediction is actually borne out in reality can be learned only from the wide-open book of the heavens.

If the present writer may venture a personal comment on this situation, it is to record his basic distrust in the validity of *any* physical theory *in the proximity of its singularities.* In this connection, the concept of a 'black hole' did not enter the arena of human thought only with the advent of the general theory of relativity, but almost two centuries before – when Laplace stopped to consider the significance of the singularity of Newton's law at $r = 0$; and wondered what would happen to the motion of two mass-points as their distance tends to zero, and the force of mutual attraction becomes infinite. At a certain finite distance $r = \epsilon$ between them the velocity of free fall should become equal to that of light (which Laplace knew from previous measurements by Roemer); and Laplace noted that no light could reach us from points moving so fast.

Today – after a lapse of two centuries – we may smile indulgently at Laplace's argument, knowing that his 'mass-points' constituted but a mathematical abstraction; and that before the critical distance ϵ is approached, other forces (of which Laplace had no inkling) take over which are incomparably more powerful than gravity; and which can exchange attraction for repulsion.

In the fullness of time, Newton's law of gravitation was replaced by the field equations of general relativity, which describe the metric properties of space in a different (and less heuristic) way. However, the observational verification of their superior description of nature (relativistic advances of planetary perihelia!) concerned domains far from the singularities of the underlying non-linear equations; and because of their non-linearity, an extrapolation of their validity to the immediate neighbourhood of singularities becomes – in effect – an act of faith.

Tertullian's famous dictum "Credo quia absurdum" may apply not only to ancient theology, but to its more modern successors as well – in whichever garb they may parade before us. In particular, at least until it proves possible to reconcile relativity with quantum mechanics in the form of a unitary theory which would embrace phenomena both small and large within the same logical fold, it is difficult to concede that the scientific advances of the past century represent more than very small steps in the direction of a fuller understanding of the fundamental questions of cosmology.

There are, moreover, other indications in the sky auguring that not all is well with too dogmatic an application of the currently recognized 'laws of Nature' to the Universe around us. For example, an orthodox interpretation of red shifts observed in the spectra of distant galaxies as 'cosmological' leads to an 'age of the Universe' too short to accommodate the lifetimes of old Population II stars (or stellar systems consisting of them) – a situation reminiscent of one concerning the age of the Earth in the latter part of the 19th century. Then too a physicist (Lord Kelvin), in a paper entitled 'The Doctrine of Uniformity in Geology briefly Refuted' (Thompson, 1866) and others peremptorily dismissed the views of the more empirically-minded geologists that the Earth must be much older than seemed admissible on the (then popular) contraction theory of the origin of the solar system; and he did so by a mathematical argument which was logically correct, but (as it eventually turned out) inapplicable to reality.

Kelvin's dilemma (or, for that matter, ours today face to face with the expansion of the Universe as deduced from the red shifts) was, to be sure, nothing in comparison with that presented to his contemporaries of the 17th century by Archbishop James Ussher of

Armagh (1581–1675), who in 1658 identified (from biblical sources) the year of the creation as 4004 B.C. – a view accepted without demur by Isaac Newton (1728). This confronted the public with a cosmic time-scale on which only 1656 years would have elapsed from the time of Adam to that of the Deluge (and of which the life of Methusalem alone would have occupied 969 years).

But this was not enough; for even the time of the year of the Creation was likewise fixed with the aid of the Bible. Thus the Venerable Bede, writing in the 8th century (and supported in the 13th century by Vincent of Beauvais) was of the opinion that the Creation must have been accomplished (for agricultural reasons) in the spring; while many references to water during the week of Genesis led again others to conjecture that an equinoctial season (September?) would have been more appropriate for the occasion.

One scholar who subscribed fully to this view was Dr. John Lightfoot (1602–1675), Master of St. Catherine's College and Vice-Chancellor of the University of Cambridge. In endorsing Ussher's year of 4004 B.C. for the date of the Creation, he concluded that Man himself was created that year on Sunday, October 23rd, at nine o'clock in the morning (GMT). This latter precision owed, to be sure, nothing to biblical revelation. It says, however, much for Lightfoot's high opinion of his professional calling that he attributed the dawn of mankind to the date and time of the commencement of the academic year at Cambridge! It is true that not all scholars of the 17th or 18th century accepted the views of these high priests of the church or universities; and although days had gone by that time when unorthodox interpretations of the Bible were being purged in the fires of the stake, it was the view of the leaders of the Establishment well into the 19th century that exponents of such heterodoxy would be accorded a similar treatment on the day of the Last Judgment.

Archbishop Ussher and his fellow-Christians arrived at their quaint chronology by too literal an interpretation of their sacred scripts as representing the only source of infallible knowledge. But have we really become much wiser since that time? It is true that our present sacred cosmological texts are no longer written in Greek or Hebrew, but in terms of the tensor calculus – another type of esoteric scripts which gains additional respect by being intelligible only to the initiated. Is, however, the contents of these texts likely to put us in a more intimate contact with the ultimate reality than Ussher and his likewise-minded contemporaries were in the 17th century?

No doubt yes; for science advanced greatly since that time; but not necessarily so in the proximity of the singularities of its description of Nature in space or time. Surely what current cosmology claims to have happened in the first 10^{-30}th of a second after the moment of 'Big Bang' – as pronounced by the latter-day saints of contemporary scientific theology – will no doubt one day be regarded by posterity with the same indulgent smile as the one which we bestow today on the cosmology of Archbishop Ussher (or, not to pick up only on the theologians, of Johannes Kepler).

And what is true of our approach to the time of creation applies likewise to one to the states of extreme density of contracting matter – leading to the portals of 'black holes' from which there is no return for any particle unfortunate enough to have once passed through them. The behaviour of matter under densities encountered in white dwarfs ($10^5 - 10^8$ g cm^{-3}) is now familiar enough to astronomers and physicists alike. Properties of matter compressed to the state of a neutron fluid (cca 10^{13} g cm^{-3}) may be indirectly studied from certain phenomena exhibited by the pulsars; but there is no such approach to the properties of matter inside black holes. Suffice it to say that, on order to have a black hole, the escape velocity from its gravitational field would have to be equal to that of light; and if so, a configuration of the mass of the Earth would have to be compressed to a sphere less than 1 cm in size!

In hearing contemporary cosmologists discourse on the history of the Universe in the first 10^{-30} of a second, or on the ultimate state of matter at density of 10^{27} g cm^{-3}, some more whimsical onlookers may perhaps wonder about the real progress of our knowledge since the days when our erudite ancestors at medieval universities used to hold disputations on such topics as the number of the angels who may simultaneously dance on the tip of a needle.

Needless to say, in situations so exceptional (or states of matter so extreme) neither theory nor observations can provide much guidance (for no one so far has seen an angel any more than the interior of a black hole). It is true that the behaviour of matter in the proximity of a black hole can indicate to observations a possible presence of such a singularity (cf., e.g., Shakura and Sunyaev, 1973); but great care must be exercised in the interpretation of such observations – if we are not to be led astray by scenarios of our own making.

Therefore, to those prepared to take a more detached view of the present state of our knowledge, our parting advice on such problems (as well as many others discussed in earlier parts of this book) is to remember that the one distinguishing hallmark of a true student of science is the ability to keep an open mind; for only by a never-ending re-examination of the validity of all we hold true now can a further progress of science be forever assured.

BIBLIOGRAPHICAL NOTES

VIII-1: The number of investigations – both theoretical and numerical – devoted in recent decades to the subject of stellar evolution is truly enormous – too many for individual quotation in this place. This section contains only the merest outline of the underlying theory, in so far as it may be of help to understanding the developments in subsequent sections. A good and more detailed account of the subject matter can be found, e.g., in E. Novotny's *Introduction to Stellar Atmospheres and Interiors* (Oxford Univ. Press, 1973), part three.

VIII-2: The classification of close binary systems expounded in this section is due to Z. Kopal (1955); and has since become generally adopted by all investigators. It was in that paper that the terms 'detached' or 'semi-detached' binaries, 'undersize subgiants', or the 'Roche limit' in the sense used in this book*, and other now familiar terms, made their first appearance in the astronomical literature. This classification was subsequently refined (along similar lines) by M. Plavec (1964).

VIII-3: The ideas on the evolution of close binaries expounded in this section have their beginning in an investigation by the present writer (Kopal, 1956), subsequently developed by a great number of investigators. For investigations preceding that time cf. the Bibliographical Notes accompanying Chapter VII of Kopal (1959). Of subsequent papers of summarizing nature cf. Z. Kopal (1957, 1958), M. Plavec (1967a, b), B. Cester (1967), Z. Kopal (1971a) and many others; the most comprehensive reviews of the subject being those of M. Plavec (1968, 1970) and B. Paczynski (1971).

* Not to be confused with the same term introduced by Darwin, and subsequently used by Jeans, to denote the minimum distance at which a fluid satellite of infinitesimal mass may approach with impunity a rotating oblate spheroid.

An interpretation of the observed clustering of the secondary components of semi-detached eclipsing systems at their Roche limits as evidence of their expansion was proposed by Z. Kopal (1954) and independently by J. A. Crawford (1955). Quantitative investigations of this phenomenon commenced with D. C. Morton (1960), and have been continued by J. Smak (1961, 1962), B. Paczynski (1966, 1967a, b, c), B. Paczynski and J. Ziolkowski (1967), R. Kippenhahn, H. C. Thomas and A. Weigert (1965, 1966), R. Kippenhahn and A. Weigert (1967), R. Kippenhahn, K. Kohl and A. Weigert (1967), P. Giannone, K. Kohl and A. Weigert (1968), R. Kippenhahn *et al.* (1968), M. Plavec (1967a, b, c, 1968a, b), S. Křiž (1968, 1969), S. Refsdahl and A. Weigert (1969), J. Ziolkowski (1969), M. Plavec, *et al.* (1969), P. Harmanec (1970), J. Horn (1970), J. Horn *et al.* (1969, 1970); and others.

For the evolution of detached systems cf. D. S. Hall (1968, 1973, 1974, 1975); and on 'undersize' subgiants, J. V. Field (1969) or R. C. Barnes (1974). For R CMa-type systems cf. Z. Kopal (1956 or 1959, sec. VII-4) or J. Smak (1962); and subsequent individual references as quoted in the text.

For investigations of mass transfer or loss in terms of particle mechanics cf. Z. Kopal (1956; 1959 sec. VII. 5), N. L. Gould, (1957, 1959), M. Plavec and S. Kříž (1965), A. Kruszewski (1964b, 1967); or, more recently, L. Angeletti (1976). For hydrodynamical treatment of the respective phenomena cf. Z. Kopal (1958), K. H. Prendergast (1960), P. Biermann (1971), G. T. Bath (1972), B. Warner and W. L. Peters (1972), K. H. Prendergast and R. E. Taam (1974), W. Y. Chau *et al.* (1973), S. A. Sorensen *et al.* (1974, 1975); G. T. Bath (1975), B. P. Flannery (1975) or S. H. Lubow and F. H. Shu (1975, 1976).

The term 'contact binary' as employed in section 3C appears to have been introduced in astronomical literature by G. P. Kuiper (1941a), through in a sense somewhat different from that used subsequently. Whereas we now regard a contact component (or binary) as a star(s) whose surface(s) coincide with their respective Roche limits for the given mass-ratio, Kuiper's definition . . . "does not mean that mere contact exists, but a common envelope as well" (op. cit., p. 137). In other words, Kuiper's 'contact binaries' were systems in which the Roche limit was exceeded by a common envelope surrounding both stars – or, in our present terminology, systems with 'oversize' components.

For a review of the literature on this subject before 1958 cf. Bibliographical Notes to ec. VIII.6 of Kopal (1959). For a discussion of the kinematic properties of W UMa-type stars cf. E. Schatzman and J. L. Rigal (1954); J. L. Rigal (1955); N. M. Artiukhina (1964) or M. V. Popov (1964). For membership of such systems in galactic clusters cf. O. J. Eggen 91967), or O. J. Eggen and A. Sandage (1969).

A development of the theory of zero-age contact binaries with common convective envelopes was initiated by L. Lucy (1968a, b; 1973) and developed further by J. Hazlehurst (1970, 1974), D. L. Moss and J. A. J. Whelan (1970, 1973), D. L. Moss (1971, 1973), S. W. Mochnacki and N. A. Doughty (1972a, b), P. Biermann and H. C. Thomas (1972), J. A. J. Whelan (1972), J. Hazlehurst and E. Meyer-Hofmeister (1973), S. P. Worden and J. A. J. Whelan (1973), J. Whelan, S. W. Mochnacki and S. P. Worden (1973, 1974), R. E. Wilson and P. Biermann (1976) and others. In perusing all this literature, the reader should keep in mind the quotation from Fernie (1969) reproduced on p. 426.

It should, furthermore, be remembered that these models were developed to account for contact systems of zero-age; while, in reality, only some (rare) W UMa-type stars can be regarded as such; others (like TX Cnc in Praesepe) cannot be much less than 10^9 yr old; and similar systems in NGC 188 or M 67 must be several times older still. Although the hydrogen depletion in stars of moderate to small masses proceeds byt slowly on the nuclear scale, their common (convective) envelopes of W UMa-type systems should be far from zero-age; and the consequences of such an assumption should, therefore, be taken with several grains of salt.

For the origin, strucutre, and absolute properties of W UMa-type systems cf. Y. Osaki (1965), I. Okamoto and K. Sato (1970), H. Mauder (1972); and for their light changes, E. L. Robinson (1972), S. M. Rucinski (1973, 1974), R. A. Breinhorst and M. Reinhardt (1974), or A. Yamasaki (1975).

VIII-4: A treatment of the problem of tidal evolution of close binary systems as given in this section follows the lines initiated by G. H. Darwin (1879a), and subsequently elaborated by Z. Kopal (1972c). Of more recent papers on this subject cf. M. E. Alexander (1973), A. Z. Dolginov and D. G. Yakovlev (1975), or L. G. Green and E. K. Kolchin (1975).

VIII-5: For recent studies (both analytical and numerical) of the possibility of formation of binary systems by capture of cf. T. A. Agekian *et al.* (1969); P. Mansbach (1970); R. S. Harrington (1970); S. J. Aarseth (1970, 1971); or V. G. Szebehely (1974).

For a theory of the 'rotational effect' in eclipsing binary systems cf. Chapter V of Z. Kopal (1959) and references quoted therein. For a recent comprehensive survey of the observational data cf. *Stellar Rotation* (ed. by A. Slettebak), D. Reidel Publ. Co., 1970.

A view that close binary systems may originate by fission of a rapidly rotating configuration was raised to the rank of a mathematical theory by the work of G. H. Darwin (1901, 1902, 1908; see also vol. 3 of his *Scientific Papers*, Cambridge Univ. Press, 1910) and J. H. Jeans (1916; see also his Adams prize essay on *The Problems of Cosmology and Stellar Dynamics*. Cambridge Univ. Press, 1919). For cogent criticisms of some views held by these investigators cf. Chapter X of R. A. Lyttleton's *The Stability of Rotating Fluid Masses* (Cambridge Univ. Press, 1953).

For more recent expositions of the fission theory cf. I. W. Roxburgh (1966), J. Ostriker (1970) and N. R. Lebovitz (1972). The near-equality of potentials prevalent over the surfaces of detached components of close binary systems was first pointed out by Z. Kopal (1959; sec. VII.3). Cf. also E. M. Drobyshevski (1974).

For comprehensive discussions of the physical properties of subdwarf binaries cf. (in addition to references quoted in the rext) the summarizing articles by R. P. Kraft (1963), J. Smak (1971b) or V. G. Gorbatsky (1975).

For investigations of the system U Gem cf. W. Krzeminski (1965), J. Smak (1969, 1970, 1971a, 1976). The possibility of formation of subdwarf binaries in post-Main Sequence stage was first put forward by Z. Kopal (1965b), and discussed further by V. G. Gorbatsky (1975). Cf. also P. Biermann and H. C. Thomas (1973).

Although the subject of X-ray binaries is less than 15 years old, its literature since 1962 has grown truly ernormous; and only a merest outline of it could have been quoted in the text. The reader who wishes to follow its sources in greater detail is, however, well served by existing literature. The following volumes deserve special mention:

X-and Gamma-Ray-Astronomy
(ed. by H. Bradt and R. Giacconi)
IAU Symp. 55; D. Reidel Publ. Co., 1973 (cf. in particular, Part I.2, R. P. Kraft pp. 36–50; and Part II.13, N. I. Shakura and R. A. Sunyaev, pp. 155–163).

X-Ray Astronomy
(ed. by R. Giacconi and H. Gursky), Astrophysics and Space Science Library No. 43; D. Reidel Publ. Co., 1974.

Neutron Stars, Black Holes, and Binary X-Ray Sources
(ed. by H. Gursky and R. Ruffini), Astrophysics and Space Science Library No. 48; D. Reidel Publ. Co., 1975. Cf., in particular, pp. 175–213 (H. Gursky and E. Schreier), pp. 221–234 (P. E. Boynton); and po. 235–255 (R. P. Kraft);

Variable Stars and Stellar Evolution
(ed. by V. E. Sherwood and L. Plaut), *IAU Symp.* 67; D. Reidel Publ. Co., 1975 (cf., in particular, Parts 7 and 9).

The most comprehensive summary of both astronomical as well as instrumental aspects of the subject, complete up to (approximately) the end of 1975 can be found in the *Proceedings of a Symposium on X-Ray Binaries*, held at NASA's Goddard Space Flight Center between October 20–22; 1975; and published in 1976 as NASA Special Publ. No. 389 under the editorship of Y. Kondo and E. Boldt. This publication contains 108 original papers and several hundred references to collateral texts.

For the most recent review of the subject cf. Apparao and Chitre (1976).

The most comprehensive summary concerning the physics of black holes can be found in a volume entitled *Black Holes*, containing a collection of lectures delivered at the Summer School of Theoretical Physics at Les Houches in 1972, and published (under this title) by Gordon and Breach Science Publishers in London and New York (1973). Cf., in particular, the article by I. D. Novikov and K. S. Thorne on po. 343–450, containing a great number of collateral references.

Of other sources cf. also the volume on *Neutron Stars, Black Holes and Binary X-Ray Sources* (ed. by H. Gursky and R. Ruffini) already referred to.

Of more recent literature on X-ray binaries and related subjects (not referenced in the above-quoted volumes) cf. M. Milgrom and E. E. Salpeter (1975), J. A. de Freitas Pacheco (1975), A. F. Illarionov and R. A. Sunyaev (1975), R. E. Taam and J. Faulkner (1975), P. R. Amnuel and O. H. Guseinov (1976), B. Cester and M. Mezzetti (1976), and others.

Ever since the discovery, in 1974, of the fact that the pulsar PSR 1913 + 16 is in a binary orbit by R. A. Hulse and J. H. Taylor (1975), this system has become the subject of a considerable literature; of which we wish to mention the papers by A. R. Masters and D. H. Roberts (1975); K. Brecher

(1975); L. W. Esposito and E. R. Harrison (1975); C. M. Will (1975); H. D. Hari Dass and V. Radhakrishnan (1975); V. A. Brumberg *et al.* (1975); B. P. Flannery and E. P. J. van den Heuvel (1975); H. M. van Horn, S. Sofia, M. P. Savedoff, J. G. Duthie and R. A. Berg (1975); L. L. Smarr and R. Blandford (1976); T. Daishido *et al.* (1976); and others.

REFERENCES

Aarseth, S. J.: 1970, *Astron. Astrophys.* **9**, 64.
Aarseth, S. J.: 1971, *Astrophys. Space Sci.* **13**, 324.
Abt, H. A.: 1959, *Astrophys. J.* **130**, 769.
Abt, H. A.: 1961, *Astrophys. J. Suppl.* **6**, 37.
Abt, H. A. and Hunter, J. H.: 1972, *Astrophys. J.* **136**, 381.
Abt, H. A., Barnes, R. C., Biggs, E. S., and Osmer, P. S.: 1965, *Astrophys. J.* **142**, 1604.
Abt, H. A., Levy, A., Baylor, S. G., Hayward, R., Jewsbury, C. P., and Snell, C. M.: 1970, *Astrophys. J.* **159**, 919.
Agekian, T. A., Anosova, J. P., and Bezgubova, B. N.: 1969, *Astrofizika* **5**, 637.
Aikman, G. C.: 1971, J. Roy. Astr. Soc. Canada **65**, 173.
Aitken, R. G.: 1932, *New General Catalogue of Double Stars* (Carnegie Inst. of Washington Publ. No. 417).
Aitken, R. G.: 1935, *The Binary Stars*, McGraw Hill, New York.
Albada, T. S. van: 1968, *Bull. Astron. Inst. Neth.* **20**, 47.
Alexander, M. E.: 1973, *Astrophys. Space Sci.* **23**, 459.
Alexander, M. E.: 1976, *Astrophys. Space Sci.* **45**, 105.
Ambartsumian, V. A.: 1937, *Astron. Zh.* **14**, 207.
Ambartsumian, V. A.: 1952, *Trans. IAU*, **8**, 665.
Amnuel, P. R. and Guseinov, D. H.: 1976, *Astrophys. Space Sci.* **45**, 283.
Anderle, P.: 1976, *Bull. Astron. Inst. Czech.* **27**, 118.
Anderson, C. M., Stoeckly, R., and Kraft, R. P.: 1966, *Astrophys. J.* **142**, 681.
Angeletti, L.: 1976, *Astrophys. Space Sci.* **44**, 23.
Apparao, K. and Chitre, S. M.: 1976, *Space Sci. Rev.*, **19**, 281.
Appell, P.: 1921, *Traité de mécanique rationelle* (Gauthier-Villars, Paris), vol. IV.
Arend, S.: 1950, *Comm. Obs. Roy. Uccle*, No. 20.
Armellini, G.: 1953, *Atti Accad. Naz. Lincei*, (8) **14**, 727.
Arnett, W. D.: 1969, *Astrophys. Space Sci.* **5**, 180.
Arp, H. C.: 1956, *Astron. J.*, **61**, 15.
Artiukhina, N. M.: 1964, *Per. Zvjozdy* **15**, 127.
Baade, W. and Swope, H. H.: 1965, *Astron. J.* **70**, 212.
Bachmann, P. J. and Hershey, J. L.: 1975, *Astron. J.* **80**, 836.
Banks, J.: 1783, as quoted in C. A. Lubbock, *The Herschel Chronicle* (Cambridge Univ. Press, 1933), p. 184.
Barnes, R. C.: 1974, *Publ. Astron. Soc. Pacific* **86**, 195.
Barney, I.: 1923, *Astron. J.* **35**, 95.
Bath, G. T.: 1972, *Astrophys. J.* **173**, 121.
Bath, G. T.: 1975, *Monthly Notices Roy. Astron. Soc.* **171**, 311.
Batten, A. H.: 1967, *Publ. Dominion Astrophys. Obs.* **13**, No. 8.
Batten, A. H.: 1970, *Publ. Astron. Soc. Pacific* **82**, 574.
Batten, A. H.: 1973, *Binary and Multiple Star Systems*, Pergamon Press, London.
Batten, A. H. and Olowin, R. P.: 1971, *Nature (Phys. Sci.)* **234**, 341.
Beckenstein, J. D.: 1976, *Astrophys. J.* **210**, 544.
Beliavsky, S.: 1933, *Per. Zvjozdy* **4**, 196.
Berglund, F.: 1938, *Lund Obs. Medd.*, **II, No. 99.**
Bernacca, P. L., Ciatti, F., Guzzi, F., Sedmak, G., Campisi, I. E., and Treves, A.: 1975, *Astron. Astrophys.* **40**, 327.
Bertrand, J.: 1873, *Compt. Rend. Acad. Sci. Paris* **77**, 849.
Bielicki, M., Piotrowski, S., and Ziolkovski, K.: 1974, *Astrophys. Space Sci.* **26**, 173.
Biermann, P.: 1971, *Astron. Astrophys.* **10**, 205.
Biermann, P. and Thomas, H. C.: 1972, *Astron. Astrophys.* **16**, 60.

Biermann, P. and Thomas, H. C.: 1973, *Astron. Astrophys.* **23**, 55.
Binnendijk, L.: 1965, *Kl. Veröffentl. Remeis Sternw. Bamberg* **4**, 36.
Blaauw, A.: 1946, *Groningen Publ.*, No. 52.
Blaauw, A.: 1952, *Bull. Astron. Inst. Neth.* **11**, 405.
Blaauw, A.: 1961, *Bull. Astrom. Inst. Neth.* 15, 265.
Boersma, J.: 1961, *Bull. Astron. Inst. Neth.* **15**, 291.
Böhm-Vitense, E.: 1954, *Astrophys. J.* **120**, 271.
Boldt, E. and Kondo, Y.: 1976, *X-Ray Binaries* (NASA SP-389), Washington, D.C., pp. 1–757.
Bolton, C. T.: 1975, *Astrophys. J.* **200**, 269.
Bondi, H.: 1952, *Monthly Notices Roy. Astron. Soc.* **112**, 195.
Bondi, H. and Hoyle, F.: 1944, *Monthly Notices Roy. Astron. Soc.* **104**, 273.
Bos, H. W. van den: 1928, *Mem. Copenhagen Acad.* **12**, No. 2.
Braginski, V. B.: 1965, *Uspekhi Fiz. Nauk* **86**, 433.
Brecher, K.: 1975, *Astrophys. J.* **195**, L 113.
Breinhorst, R. A. and Reinhardt, M.: 1974, *Acta Astron.* **24**, 377.
Brooker, R. A. and Olle, T. W.: 1955, *Monthly Notices Roy. Astron. Soc.* **115**, 101.
Brosche, P.: 1964, *Astron. Nachr.* **288**, 33.
Brouwer, D.: 1946, *Astron. J.* **52**, 57.
Brouwer, D. and Clemence, G. M.: 1961, *Celestial Mechanics*, Academic Press, New York.
Brown, E. W.: 1936, *Monthly Notices Roy. Astron. Soc.* **97**, 116, 388.
Brown, E. W. and Shook, C. A.: 1933, *Planetary Theory*, Cambridge Univ. Prss.
Bruggencate, P. ten: 1934, *Z. Astrophys.* **8**, 344.
Brumberg, V. A., Zeldovich, Ya. B., Novikov, I. D., and Shakura, N. I.: 1975, *Astron. Zh. Letters*, **1**, 5.
Bryan, G. H.: 1888, *Phil. Trans. Roy. Soc. London (A)* **190**, 187.
Budding, E.: 1973, *Astrophys. Space Sci.* **22**, 87.
Bullard, E. C.: 1948, *Monthly Notices Roy. Astron. Soc. (Geophys. Suppl.)* **5**, 186.
Callendreau, O.: 1889, *Ann. Observ. Paris* **19** E.
Cameron, A. G. W.: 1971, *Nature (Phys. Sci.)* **229**, 178.
Carpenter, F.: 1930, *Astrophys. J.* **72**, 205.
Cartan, E.: 1922, *Bull. Sci. Math.* **46**, 332.
Cester, B.: 1967, *Nuovo Cimento Suppl.* **5**, 1089.
Cester, B. and Mezzetti, M.: 1976, *Astrophys. Space Sci.*, **45**, 337.
Chanan, G. A., Middleditch, J., and Nelson, J. E.: 1976, *Astrophys. J.* **208**, 512.
Chandrasekhar, S.: 1933, *Monthly Notices Roy. Astron. Soc.* **93**, 539.
Chandrasekhar, S.: 1944, *Astrophys. J.* **99**, 54.
Chandrasekhar, S.: 1961, *Hydrodynamic and Hydromagnetic Stability*, Oxford Univ. Press.
Chandrasekhar, S. and Krogdahl, W.: 1942, *Astrophys. J.* **96**, 151.
Chapman, S.: 1954, *Astrophys. J.* **120**, 151.
Charlier, C. V. L.: 1902, *Die Mechanik des Himmels*, Veit and Co., Leipzig.
Chau, W. Y., Chia, T. T., and Henriksen, R. N.: 1973, *Astrophys. Letters* **14**, 221.
Chevalier, R. A.: 1976, *Astrophys. Letters* **18**, 35.
Cisneros-Parra, J. U.: 1970, *Astron. Astrophys.* **8**, 141.
Clairaut, A. C.: 1743, *Théorie de la figure de la terre, tirée des principles de l'hydrostatique*, Paris.
Coulson, J., Ledoux, P., and Simon, R.: 1956, *Bull. Soc. Roy. Sci. Liège* **25**, 144.
Cowling, T. G.: 1938, *Monthly Notices Roy. Astron. Soc.* **98**, 734.
Cowling, T. G.: 1941, *Monthly Notices Roy. Astron. Soc.* **101**, 367.
Crawford, J. A.: 1955, *Astrophys. J.* **121**, 71.
Crawford, J. A and Kraft, R. P.: 1956, *Astrophys. J.* **123**, 44.
Daishido, T., Nomoto, K., and Sugimoto, D.: 1976, *Progr. Theor. Phys. Japan* **55**, 314.
Darboux, G.: 1910, *Leçons sur les systèmes orthogonaux et les coordonnées curvilignes* (Gauthier-Villars, Paris), Chapitre II.
Darwin, G. H.: 1879, *Phil. Trans. Roy. Soc. London* **170**, 1–35, 447–530.
Darwin, G. H.: 1879a, *Proc. Roy. Soc. London* **29**, 168.
Darwin, G. H.: 1880, *Phil. Trans. Roy. Soc. London* **171**, 713–891.
Darwin, G. H.: 1900, *Monthly Notices Roy. Astron. Soc.* **60**, 82.
Darwin, G. H.: 1901, *Phil. Trans. Roy. Soc. London, (A)* **198**, 301–334.
Darwin, G. H.: 1902, *Phil. Trans. Roy. Soc. London, (A)* **200**, 251–314.
Darwin, G. H.: 1908, *Phil. Trans. Roy. Soc. London (A)* **208**, 1–19.
Darwin, G. H.: 1910, *Scientific Papers* (Cambridge Univ. Press), vol. 3.
Deeter, J. E. and Boynton, P. E.: 1976, *Astrophys. J.* **210**, L 133.

Delaunay, C. E.: 1867, *Mem. Acad. Sci. Paris* **29**.
Delhaye, J. and Blaauw, A.: 1953, *Bull. Astron. Inst. Neth.* **12**, 72.
De Marcus, W. C.: 1958, *Astron. J.* **63**, 2.
De Sitter, W.: 1924, *Bull. Astron. Inst. Neth.* **2**, 97.
Deutsch, A. J.: 1969, in *Mass Loss from the Stars* (ed. by M. Hack), (D. Reidel Publ. Co.), pp. 1 ff.
Devinney, E. J.: 1971, *Nature (Phys. Sci.)* **233**, 110.
Dickens, R. J., Kraft, R. P., and Krzeminski, W.: 1968, *Astron. J.* 73, 6.
Dirichlet, L.: 1861, *J. Crelle* 58, 209.
Dolginov, A. Z. and Yakovlev, D. G.: 1975, *Astrophys. Space Sci.* **36**, 31.
Dommanget, J.: 1970, *Comm. Obs. Roy. Belg. Uccle (B)*, No. 56.
Drobyshevski, E. M.: 1974, *Astron. Astrophys.*, **36**, 409.
Ebbighausen, E. G.: 1970, *Publ. Astron. Soc. Pacific* **82**, 349.
Eddington, A. S.: 1925, *Observatory* **48**, 73.
Eddington, A. S.: 1926, *Internal Constitution of the Stars* (Cambridge Univ. Press), secs. 198–200.
Efremov, Y. N., Kholopov, P. N., Kukarkin, B. V., and Sharov, A. S.: 1964, *Inf. Bull. Var. Stars*, No. 76.
Eggen, O. J.: 1948, *Astrophys. J.* **108**, 1.
Eggen, O. J.: 1961, *Roy. Obs. Bull.*, No. 31.
Eggen, O. J.: 1967, *Mem. Roy. Astron. Soc.* **70**, 111.
Eggen, O. J. and Sandage, A.: 1969, *Astrophys. J.* **158**, 669.
Ehlers, J., Rosenblum, A., Goldberg, J. N., and Havas, P.: 1976, *Astrophys. J.* **208**, L77.
El-Shaarawy, M. B.: 1974, Ph. D. Thesis (Univ. of Manchester), unpublished.
Esposito, L. W. and Harrison, E. R.: 1975, *Astrophys. J.* **196**, L1.
Evans, D. S.: 1968, *Quart. J. Roy. Astron. Soc.* **9**, 388.
Evrard, L.: 1951, *Ann. Astrophys.* **14**, 17.
Fernie, J. D.: 1959, *Astrophys. J.* **130**, 611.
Fernie, J. D.: 1969, *Publ. Astron. Soc. Pacific* **81**, 707.
Fessenkov, V. G.: 1952, *IAU Trans.* **8**, 702.
Field, J. V.: 1969, *Monthly Notices Roy. Astron. Soc.* **144**, 419.
Flannery, B. P.: 1975, *Monthly Notices Roy. Astron. Soc.* **170**, 325.
Fletcher, E. S.: 1964, *Astron. J.* **69**, 357.
Forsyth, A. R.: 1912, *Lectures on Differential Geometry of Curves and Surfaces*, Cambridge Univ. Press.
Freitas Pacheco, J. A. de: 1975, *Astrophys. Space Sci.* **32**, 205.
Galeotti, P.: 1970, *Astrophys. Space Sci.* 7, 87.
Gaposchkin, S. I.: 1940, *Publ. Amer. Astron. Soc.* **10**, 52.
Geary, J. C. and Abt, H. A.: 1970, *Astron. J.* **75**, 718.
Geyer, E. H.: 1967, *Z. Astrophys.*, **66**, 16.
Giacconi, R., Murray, S., Gursky, H., Kellogg, E., Schreier, E., Matilsky, T., Koch, D., and Tananbaum, H.: 1974, *Astrophys. J. Suppl.* 27, 37 (No. 237).
Giannone, P., Kohl, K., and Weigert, A.: 1968, *Z. Astrophys.* **68**, 107.
Gibbons, G. W. and Hawking, S. W.: 1971, *Nature (Phys. Sci.)* **232**, 465.
Gibson, D. M., Hjellming, R. M., and Owen, F. N.: 1975, *Astrophys. J.* **200**, L99.
Gliese, W.: 1969, *Veröffentl. Astron. Rechen-Inst. Heidelberg*, **Nr. 22**.
Goldreich, P.: 1966, *Rev. Geophys.* **4**, 441.
Goodricke, J.: 1783: *Phil. Trans. Roy. Soc. London* 73, 474.
Gorbatsky, V. G.: 1975, *Astrophys. Space Sci.* **33**, 325.
Gott, J. R.: 1971, *Nature (Phys. Sci.)* **234**, 342.
Gottlieb, E. W., Wright, E. L., and Liller, W.: 1975, *Astrophys. J.* **195**, L 33.
Gould, N. L.: 1957, *Publ. Astron. Soc. Pacific* **69**, 541.
Gould, N. L.: 1959, *Astron. J.* **64**, 136.
Grant, G.: 1955, *Astrophys. J.* **122**, 566.
Green, L. C. and Kolchin, E. K.: 1975, *Astrophys. J. Suppl.* **28**, 449 (No. 271).
Gursky, H. and Schreier, E.: 1975, in *Variable Stars and Stellar Evolution* (ed. by V. E. Sherwood and L. Plaut) *IAU Symp.* 67, 414–463.
Gyldén, H.: 1873, *Nova Acta Reg. Soc. Sci. Uppsala* **8**, No. 3.
Gyldén, H.: 1884, *Astron. Nachr.* **109**, 1.
Hadjidemetriou, J.: 1963, *Icarus* **2**, 440.
Hadjidemetriou, J.: 1966a, *Icarus* **5**, 34.
Hadjidemetriou, J.: 1966b *Z. Astrophys.* **63**, 116.

Hadjidemetriou, J.: 1967, in *Adv. Astron. Astrophys.* (ed. by Z. Kopal) Academic Press, New York. **5**, 131.
Hadjidemetriou, J.: 1969, *Astrophys. Space Sci.* **3**, 31 and 330.
Hagihara, Y.: 1972, *Celestial Mechanics* (MIT Press, Cambridge, Mass), vol. 2.
Hall, D. S.: 1968, *Publ. Astron. Soc. Pacific*, **80**, 477.
Hall, D. S.: 1973, *Publ. Astron. Soc. Pacific*, **85**, 478.
Hall, D. S.: 1974, *Acta Astron.* **24**, 215.
Hall, D. S.: 1975, *Acta Astron.* **25**, 95.
Hall, D. S. and Garrison, L. M.: 1969, *Publ. Astron. Soc. Pacific*, **81**, 771.
Hari Dass, N. D. and Radhakrishnan, V.: 1975, *Astrophys. Letters* **16**, 135.
Harmanec, P.: 1970, *Astrophys. Space Sci.* **6**, 497.
Harper, W. E.: 1937, *Publ. Dominion Astrophys. Obs.* **7**, 1.
Harrington, R. S.: 1970, *Astron. J.* **75**, 1140.
Hartle, J. B.: 1967, *Astrophys. J.* **150**, 1005.
Hayashi, C.: 1961, *Publ. Astron. Soc. Japan* **13**, 450.
Hazlehurst, J.: 1970, *Monthly Notices Roy. Astron. Soc.* **149**, 129.
Hazlehurst, J.: 1974, *Astron. Astrophys.* **36**, 49.
Hazlehurst, J.. and Meyer-Hofmeister, E.: 1973, *Astron. Astrophys.*, **24**, 379.
Hazlehurst, J. and Sargent, W. L. W.: 1959, *Astrophys. J.* **130**, 276.
Heasley, J. N.: 1971, *Astrophys. J.* **163**, 345.
Heintz, W. D.: 1967, *Comm. Roy. Obs. Belg. Uccle (B)*, No. 17, p. 49.
Heintz, W. D.: 1969, *J. Roy. Astron. Soc. Canada* **63**, 275.
Herschel, W.: 1782, *Phil. Trans. Roy. Soc. London* **72**, 82.
Herschel, W.: 1783, in an informal report entitled 'Observations upon Algol' but not printed till it appeared in *The Scientific Papers of Sir William Herschel*, London 1912, vol. 1, p. cvii.
Herschel, W.: 1802, *Phil. Trans. Roy. Soc. London* for 1802; pp. 477–528.
Herschel, W.: 1803, *Phil. Trans. Roy. Soc. London* for 1803; pp. 339–382.
Heuvel, E. P. J. van den: *Publ. Astron. Soc. Pacific* **81**, 815.
Higgins, T. P. and Kopal, Z.: 1968, *Astrophys. Space Sci.* **2**, 352.
Hilditch, R. and Hill, G.: 1974, *Monthly Notices Roy. Astron. Soc.* **168**, 543.
Hill, G. W.: 1878, *Amer. J. Math.* **1**, 129 and 245.
Hjellming, R. M.: 1972, *Nature (Phys. Sci.)* **238**, 52.
Hjellming, R. M. and Blankenship, L.: 1973, *Nature (Phys. Sci.)* **243**, 81.
Horn, J.: 1970, *Astrophys. Space Sci.* **6**, 492.
Horn, J., Kříž, S., and Plavec, M.: 1969, *Bull. Astron. Inst. Czech.* **20**, 193.
Horn, J., Kříž, S., and Plavec, M.: 1970, *Bull. Astron. Inst. Czech.* **21**, 45.
Hoyle, F.: 1940, *Proc. Cambr. Phil. Soc.* **36**, 325 and 424.
Hoyle, F.: 1941, *Monthly Notices Roy. Astron. Soc.* **101**, 227.
Hoyle, F.: 1955, *Frontiers of Astronomy*, Heinemann, London.
Hoyle, F. and Lyttleton, R. A.: 1939, *Proc. Cambr. Phil. Soc.*. **35**, 592 and 405.
Huang, S. S.: 1975, *Astrophys. J.* **195**, 127.
Huffer, C. M. and Collins, G.W.: 1962, *Astrophys. J. Suppl.* **7**, 351.
Hughes, V. A. and Woodsworth, A.: 1972, *Nature (Phys. Sci.)* **236**, 42.
Hulse, R. A. and Taylor, J. H.: 1975, *Astrophys.* **J. 195**, L51.
Iben, I.: 1965, *Astrophys. J.* **141**, 993; 142 and 1447.
Iben, I.: 1966, *Astrophys. J.* **143**, 483, 505 and 515.
Iben, I.: 1967, *Astrophys. J.* **147**, 624 and 650.
Illarionov, A. F. and Sunyaev, R. A.: 1975, *Astr. Zh. Letters* **1**, 11.
James, R. and Kopal, Z.: 1963, *Icarus* **1**, 442.
Jardetzky, W. S.: 1958, *Theories of Figures of Celestial Bodies*, Interscience Publ., New York.
Jaschek, C. and Gomez, A. E.: 1970, *Publ. Astron. Soc. Pacific* **82**, 809.
Jaschek, C.and Jaschek, M.: 1957, *Publ. Astron. Soc. Pacific*, **69**, 546.
Jeans, J.H.: 1916, *Phil. Trans. Roy. Soc. London* **217** A, 7.
Jeans, J. H.: 1919, *Monthly Notices Roy. Astron. Soc.* **79**, 330.
Jeans, J. H.: 1924, *Monthly Notices Roy. Astron. Soc.* 84, 2, 912, 1912.
Jeans, J. H.: 1925, *Monthly Notices Roy. Astron. Soc.* **85**, 917.
Jeans, J. H.: 1926, *Monthly Notices Roy. Astron. Soc.* **86**, 328 and 444.
Jeans, J. H.: 1928, *Astronomy and Cosmogony*, Cambridge Univ. Press.
Jeffers, H. M., Bos, W. H. v. d., and Greeby, F. M.: 1963, *Lick Obs. Annals* vol. **21**.
Jeffreys, H.: 1924, *The Earth*, Cambridge Univ. Press.
Jeffreys, H.: 1953, *Monthly Notices Roy. Astron. Soc.* **113**, 97.

Johnston, K.: 1970, *Publ. Astron. Soc. Pacific* **82**, 1093.
Joy, A. H.: 1956, *Astrophys. J.* **124**, 317.
Jurkevich, I.: 1970, in *Vistas in Astronomy* (ed. by A. Beer) Pergamon Press, **12**, 63.
Kamp, P. van de: 1969, *Publ. Astron. Soc. Pacific* **81**, 5.
Kamp, P. van de, Smith, S. M., and Thomas, A.: 1951, *Astron. J.* **55**, 251.
Kaula, W. M.: 1963, *J. Geophys. Res.* **68**, 4959 and 4967.
Kaula, W. M.: 1964, *Rev. Geophys.* **2**, 661.
Khabarin, Yu. G.: 1975, *Astron. Zh.* **52**, 57.
Kholopov, P. N.: 1971, *Astron. Circ. USSR Acad. Sci.*, No. 601.
Kippenhahn, R.: 1954, *Z. Astrophys.* **35**, 165.
Kippenhahn, R., 1955, *Z. Astrophys.* **38**, 166.
Kippenhahn, R. and Weigert, A.: 1967, *Z. Astrophys.* **65**, 251.
Kippenhahn, R., Thomas, H. C., and Weigert, A.: 1965, *Z. Astrophys.* **61**, 241.
Kippenhahn, R., Thomas, H. C., and Weigert, A.: 1966, *Z. Astrophys.* **64**, 373.
Kippenhahn, R., Kohl, K., and Weigert, A.: 1967, *Z. Astrophys.* **66**, 58.
Kippenhahn, R., Thomas, H. C., and Weigert, A.: 1968, *Z. Astrophys.* **69**, 265.
Kirillova, T. and Pavlovskaya, E. D.: 1963, *Astron. Zh.* **40**, 131.
Kitamura, M.: 1970, *Astrophys. Space Sci.* **7**, 272.
Kondo, Y, McCluskey, G. E., and Houck, T. E.: 1972, *Nature (Phys. Sci.)* **240**, 119.
Kondo, Y., McCluskey, G. E., and Gulden, S. L.: 1975, in *X-Ray Binaries* (NASA SP-389), pp. 499–511.
Kopal, Z.: 1938, *Monthly Notices Roy. Astron. Soc.* **99**, 266.
Kopal, Z.: 1941a, *Astrophys. J.* **94**, 145.
Kopal, Z.: 1941c, *Ann. New York Acad. Sci.* **41**, 13.
Kopal, Z.: 1942, *Proc. Amer. Phil. Soc.* **85**, 399.
Kopal, Z.: 1944, *Astrophys. J.* **100**, 204.
Kopal, Z.: 1946a, *An Introduction to the Study of Eclipsing Variables* (Harvard Obs. Mono., No. 6), Harvard Univ. Press, Cambridge, Mass.
Kopal, Z.: 1946b, *Astrophys. J.* **103**, 310.
Kopal, Z.: 1948a, *Astrophys. J.* **108**, 46.
Kopal, Z.: 1948b, *Proc. U.S. Nat. Acad. Sci.* **34**, 377.
Kopal, Z.: 1950a, *Computation of the Elements of Eclipsing Binary Systems* (Harvard Obs. Mono. No. 8) Cambridge, Mass.
Kopal, Z.: 1950b, *Proc. U.S. Nat. Acad. Sci.* **36**, 72.
Kopal, Z.: 1953, *Monthly Notices Astron. Soc.* **113**, 769.
Kopal, Z.: 1964, *Jodrell Bank Ann.* **1**, 37.
Kopal, Z.: 1955, *Ann. Astrophys.* **18**, 379.
Kopal, Z.: 1956, *Ann. Astrophys.* **19**, 298.
Kopal, Z.: 1957a, *Trans. IAU* **9**, 611.
Kopal, Z.: 1957b, in *Non-Stable Stars* (by G. H. Herbig), *IAU Symp.* **3** (Cambridge Univ. Press), p. 123 ff.
Kopal, Z.: 1958a, *Astron. Nachr.* **284**, 169.
Kopal, Z.: 1958b, *Asiago Obs. Contr.*, No. 95.
Kopal, Z.: 1959, *Close Binary Systems*, Chapman-Hall and John Wiley, London and New York.
Kopal, Z.: 1960, *Figures of Equilibrium of Celestial Bodies*, Univ. of Wisconsin Press, Madison, Wisc.
Kopal, Z.: 1965a, in *Adv. Astron. Astrophys.* (ed. by Z. Kopal), Academic Press, New York **3**, 89.
Kopal, Z.: 1965b, *Kl. Veröffentl. Remeis Sternw. Bamberg* **4**, 52.
Kopal, Z.: 1967, *Icarus* **6**, 298.
Kopal, Z.: 1968a, *Astrophys. Space Sci.* **1**, 179.
Kopal, Z.: 1968b, *Astrophys. Space Sci.* **1**, 284.
Kopal, Z.: 1968c, *Astrophys. Space Sci.* **1**, 411.
Kopal, Z.: 1968d, *Astrophys. Space Sci.* **2**, 23.
Kopal, Z.: 1968e, *Astrophys. Space Sci.* **2**, 48.
Kopal, Z.: 1968f, *Astrophys. Space Sci.* **2**, 166.
Kopal, Z.: 1969, *Astrophys. Space Sci.* **5**, 360.
Kopal, Z.: 1970, *Astrophys. Space Sci.* **8**, 149.
Kopal, Z.: 1971a, *Publ. Astron. Soc. Pacific*, **83**, 521.
Kopal, Z.: 1971b, *Astrophys. Space Sci.* **10**, 328.
Kopal, Z.: 1972a, *Astrophys. Space Sci.* **16**, 3.
Kopal, Z.: 1972b, *Astrophys. Space Sci.* **16**, 347.
Kopal, Z.: 1972c, *Astrophys. Space Sci.* **17**, 161.

Kopal, Z.: 1972d, *Astrophys. Space Sci.* **18**, 287.
Kopal, Z.: 1973, *Astrophys. Space Sci.* **24**, 145.
Kopal, Z.: 1974, *Astrophys. Space Sci.* **27**, 389.
Kopal, Z. and Kurth, R.: 1957, *Z. Astrophys.* **42**, 90.
Kopal, Z. and Lanzano, P.: 1973, *Astrophys. Space Sci.* **23**, 425.
Kopal, Z. and Lyttleton, R. A.: 1963, *Icarus* **1**, 455.
Kopal, Z. and Kamala Mahanta, M.: 1974, *Astrophys. Space Sci.* **30**, 347.
Kopal, Z., Plavec, M. and Reilly, E.: 1960, *Jodrell Bank Ann,* **1**, 374.
Kopal, Z. and Sekender Ali, A. K. M.: 1971, *Astrophys. Space Sci.* **11**, 423.
Kopal, Z. and Shapley, H. B.: 1956, *Jodrell Bank Ann.* **1**, 141.
Kraft, R. P.: 1958, *Astrophys. J.* **127**, 625.
Kraft, R. P.: 1962, *Astrophys. J.* **135**, 408.
Kraft, R. P.: 1963, in *Adv. Astron. Astrophys.* (ed. by Z. Kopal) (Acad. Press, New York), **2**, 43.
Kraft, R. P.: 1974, *Astrophys. J.* **139**, 457.
Kraft, R. P.: 1965, *Astrophys. J.* **142**, 681 and 1588.
Kraft, R. P.: 1967, *Publ. Astron. Soc. Pacific* 79, 395.
Kraft, R. P., Mathews, J., and Greenstein, J. L.: 1962, *Astrophys. J.* **136**, 312.
Kraft, R. P., Krzeminski, W., and Mumford, G. S.: 1969, *Astrophys. J.* **158**, 589.
Kreiner, J. M.: 1974, *Acta Cosmolog.* **4**, 99.
Kříž, S.: 1968, *Bull. Astron. Inst. Czech.* **19**, 248.
Kříž, S.: 1969, *Bull. Astron. Inst. Czech.* **20**, 127.
Krogdahl, W.: 1942, *Astrophys. J.* **96**, 124.
Kron, G. E.: 1941, *Astrophys. J.* **93**, 133.
Kron, G. E. and Gordon, K. C.: 1943, *Astrophys. J.* **97**, 311.
Kruszewski, A.: 1963, *Acta Astron.* **13**, 106.
Kruszewski, A.: 1964a, *Bull. Acad. Polonaise des Sci. (math.-astr.-phys.)* **12**, 317.
Kruszewski, A.: 1964b, *Acta Astron.* **14**, 231.
Kruszewski, A.: 1964c, *Acta Astron.* **14**, 241.
Kruszewski, A.: 1966, in *Adv. Astron. Astrophys.* (ed. by Z. Kopal), (Academic Press, New York), **3**, 89.
Kruszewski, A.: 1967, *Acta Astron.* 17, 297.
Krzeminski, W.: 1965, *Astrophys. J.* **142**, 1051.
Krzeminski, W.: 1973, *Inf. Bull. Var. Stars,* No. 2612.
Krzeminski, W. and Smak, J.: 1971, *Acta Astron.* **21**, 133.
Kuiper, G. P.: 1941a, *Astrophys. J.* 93, 133.
Kuiper, G. P.: 1941b, *Publ. Amer. Astron. Soc.* **10**, 206.
Kuiper, G. P.: 1942, *Astrophys. J.* **95**, 201.
Kuiper, G. P. and Johnson, J. R.: 1956, *Astrophys. J.* **123**, 90.
Kukarkin, B. V., Kurochkin, N. E., Medvedeva, G. I., Perova, G. I., Kholopov, P. N., Efremov, Yu. N., Kukarkina, N. P., Fedorovich, V. P., and Frolov, M. S.: 1969, *General Catalogue of Variable Stars,* Moscow, USSR.
Kurochkin, N. E.: 1965, *Inf. Bull. Var. Stars,* No. 79.
Kurochkin, N. E. and Kukarkin, B. V.: 1966, *Astron. Zh.* **43**, 83.
Labeyrie, A., Bonneau, D., Stachnik, R. V., and Gezari, D. Y.: 1974, *Astrophys. J.* **194**, L 147.
Lamb, H.: 1932, *Hydrodynamics* (6th ed.), Cambr. Univ. Press, Chapter V (appendix)
Landau, L. D. and Lifschitz, E. M.: 1959, *Fluid Mechanics* (transl. by J. B. Sykes), Pergamon Press, Oxford.
Langebartel, R. G.: 1965, NASA Techn. Rept. TN-D-2939.
Lanzano, P.: 1962, *Icarus* **1**, 121.
Lanzano, P.: 1968, *Astrophys. Space Sci.* **1**, 92.
Lanzano, P.: 1973, *Astrophys. Space Sci.* **20**, 71.
Lanzano, P.: 1974, *Astrophys. Space Sci.* **29**, 161.
Lanzano, P.: 1975, *Astrophys. Space Sci.* **37**, 173.
Laplace, P. S.: 1825, *Mécanique Céleste,* Paris vol. 5.
Latyshev, I. N.: 1974, *Astron. Zh.* **51**, 786.
Lauterborn, D. and Weigert, A.: 1972, *Astron. Astrophys.* **18**, 294.
Lea, S. M. and Margon, B.: 1973, *Astrophys. Letters* **13**, 33.
Lebovitz, N. R.: 1965, *Astrophys. J.* **142**, 229 and 1257.
Lebovitz, N.R.: 1970, *Astrophys. Space Sci.* **9**, 398.
Lebovitz, N. R.: 1972, *Astrophys. J.* **175**, 171.
Lecar, M., Wheeler, J. C., and McKee, Ch. F.: 1976, *Astrophys. J.* **205**, 556.

Ledoux, P.: 1958, in *Handbuch der Physik* (ed. by S. Flügge), Julius Springer Verlag, **51**, 605–688.
Ledoux, P.: 1965, in *Stars and Stellar Systems* (ed. by L. H. Aller and D. B. McLaughlin) Univ. of Chicago Press, pp. 499–574.
Ledoux, P. and Pekeris, C. L.: 1941, *Astrophys. J.* **94**, 124.
Legendre, A. M.: 1793, *Mémoires de mathématique pour 1789*, Paris.
Levi-Civita, T.: 1906, *Acta Math.* **30**, 305.
Levi-Civita, T.: 1937, *Amer. J. Math.* **59**, 225.
Levy, M.: 1888, *Comptes Rendus Acad. Sci. Paris* **106**, 1270, 1314, 1375.
Liapounov, A. M.: 1903, *Mém. de l'acad. impérial de St. Petersbourg* (8) **14**, Pt. 7.
Liapounov, A. M.: 1904, *Ann. de la faculté de sci. Toulouse* ser II, vol. **6**.
Liapounov, A. M.: 1908, *Mém. de l'acad. impérial de St. Petersbourg* (8) **22**, Pt. 5.
Liapounov, A. M.: 1912, *Mém. de l'acad. impérial de St. Petersbourg (8)* **29**, Pt. 3.
Lichtenstein, L.: 1933, *Gleichgewichtsfiguren rotierender Flüssigkeiten*, Julius Springer Verlag, Berlin.
Limber, D. N.: 1963, *Astrophys. J.* **138**, 1112.
Linnell, A. P.: 1950, *Harvard Obs. Circ.*, No. 455.
Linnell, A. P. and Proctor, D. D.: 1970, *Astrophys. J.* **161**, 1045; **162**, 683.
Linnell, A. P. and Proctor, D. D.: 1971, *Astrophys. J.* **164**, 131.
Liouville, J.: 1858, *J. Math.* (2) **3**, 1–25.
Lippincott, S. L.: 1966, *Comm. Obs. Roy. Belge Uccle*, (B), **17**, 68.
Lipschitz, R.: 1863, *J. Crelle* **62**, 1.
Lohsen, E.: 1975, *Inform. Bull. Var. Stars*, No. 988.
Lubow, S. H. and Shu, F. H.: 1975, *Astrophys. J.* **198**, 383.
Lubow, S. H. and Shu, F. H.: 1976, *Astrophys. J.* **207**, L 53.
Lucy, L. B.: 1967, *Z. Astrophys.* **65**, 89.
Lucy, L. B.: 1968, *Astrophys. J.* **151**, 1123; **153**, 877.
Lucy, L. B.: 1973, *Astrophys. Space Sci.* **22**, 381.
Luyten, W. J., Struve, O., and Morgan, W. W.: 1939, *Publ. Yerkes Obs.* **7**, pt. 4.
Lyttleton, R. A.: 1953, *The Stability of Rotating Liquid Masses*, Cambridge, Univ. Press.
Lyttleton, R. A.: 1972, *Monthly Notices Roy. Astron. Soc.* **160**, 255.
Lyutyi, V. M.: 1975, in 'Variable Stars and Stellar Evolution' (ed. by V. E. Sherwood and L. Plaut), *IAU Symp.* **67**, 465.
Lyutyi, V. M., Sunyaev, R. A., and Cherepashchuk, A. M.: 1973, *Astron. Zh.* **50**, 3.
Lyutyi, V. M., Sunyaev, R. A., and Cherepashchuk, A. M.: 1974, *Astron. Zh.* **51**, 1150.
Mac Donald, G. J. F.: 1964, *Rev. Geophys.* **2**, 661.
MacMillan, W. D.: 1919, *Amer. Math. Monthly* **26**, 327.
MacMillan, W. D.: 1958, *The Theory of the Potential*, Dover Publ., New York.
Mandelshtam, S. L., Tindo, I. P., Cheremukhin, G. S., Sorokin, L. S., and Dimitriev, A. B.: 1968, *Kosmich. Issled.* **6**, 119.
Mausbach, P.: 1970, *Astrophys. J.* **160**, 135.
Markowitz, W.: 1933, *Astrophys. J.* **77**, 337.
Masters, A. R. and Roberts, D. H.: 1975, *Astrophys. J.* **195**, L 107.
Mathis, J. S.: 1967, *Astrophys. J.* **149**, 619.
Mathis, J. S. and Odell, A. P.: 1973, *Astrophys. J.* **180**, 517.
Mauder, H.: 1972, *Astron. Astrophys.* **17**, 1.
McLaughin, D. B.: 1924, *Astrophys. J.* **60**, 22.
McLaughlin, D. B.: 1934, *Publ. Univ. Obs. Michigan* **6**, 3.
Melchior, P.: 1966, *The Earth Tides*, Pergamon Press, London and New York.
Meltzer, A.: 1957, *Astrophys. J.* **125**, 359.
Meščerskii, F.: 1893, *Astron. Nachr.* **132**, 129.
Meščerskii, F.: 1902, *Astron. Nachr.* **159**, 229.
Mestel, L.: 1965, in *Stellar Structure* (ed. by L. H. Aller and D. B. McLaughlin), Univ. of Chicago Press, pp. 465–497.
Michell, J.: 1767, *Phil. Trans. Roy. Soc. London* **57**, 234.
Michell, J.: 1784, *Phil. Trans. Roy. Soc. London* **74**, 35.
Milgrom, M. and Salpeter, E. E.: 1975, *Astrophys. J.* **196**, 589.
Milne, E. A.: 1929, *Monthly Notices Roy. Astron. Soc.* **89**, 518.
Milne, E. A.: 1930, *Quart. J. Math. (Oxford)* **1**, 1.
Mitchell, R. I.: 1954, *Astrophys. J.* **120**, 274.
Mochnacki, S. W. and Doughty, N. A.: 1972, *Monthly Notices Roy. Astron. Soc.* **156**, 51 and 243.
Montanari, G.: 1671, 'Sopra la sparizione d'alcune stelle e altre novità celesti' in *Prose di Signori Academici Gelati di Bologna*.

Morton, D. C.: 1960, *Astrophys. J.* **132**, 146.
Moss, D. L.: 1971, *Monthly Notices Roy. Astron. Soc.* **153**, 41.
Moss, D. L.: 1972, *Monthly Notices Roy. Astron. Soc.* **157**, 433.
Moss, D. L. and Whelan, J. A J.: 1970, *Monthly Notices Roy. Astron. Soc.* 149, 147.
Motz, L.: 1952, *Astrophys. J.* **115**, 562.
Müller, P.: 1956, *Comptes Rend. Acad. Sci. Paris* **242**, 460.
Müller, G. and Kempf, P.: 1903, *Astrophys. J.* **17**, 201.
Nairai, K.: 1971, *Publ. Astron. Soc. Japan* **23**, 529.
Nelson, M. R.: 1976, *Astrophys. J.* **209**, 168.
Nelson, B. and Young, A.: 1970, *Publ. Astron. Soc. Pacific* **82**, 699.
Nishimura, H. and Mori, H.: 1961, *Progress Theor. Phys. Japan* **26**, 967.
Novikov, I. D. and Thorne, K. S.: 1973, in *Black Holes* (Proc. Les Houches Conf., 1972; ed. by C. and B. S. De Witt), Gordon and Breach, New York and London, pp. 347–450 (cf. p. 448).
Novotny, E.: 1973, *Introduction to Stellar Atmospheres and Interiors*, Oxford Univ. Press; Part III.
Odell, A. P.: 1974, *Astrophys. J.* **192**, 417.
Okamoto, I. and Sato, K.: 1970, *Publ. Astron. Soc. Japan* **22**, 317.
Olson, E. C.: 1968, *Publ. Astron. Soc. Pacific* **80**, 185.
Oosterhoff, P. Th.: 1937, *Leiden Obs. Ann.* **17**, No. 1.
Opolski, A.: 1952, *Arkiv Astron.* **1**, 249.
Oppenheimer, S.: 1885, *Astron. Nachr.* **113**, 209.
Orlov, A. A.: 1960, *Astron. Zh.* **37**, 902.
Osaki, Y.: 1965, *Publ. Astron. Soc. Japan* **17**, 97.
Oster, L.: 1957, *Z. Astrophys.* **42**, 228.
Ostriker, J.: 1970, in *Stellar Rotation* (ed. by A. Slettebak), D. Reidel Publ. Co., pp. 147–155.
Owen, F. N., Jones, T. W., and Gibson, D. M.: 1976, *Astrophys. J.* **210**, L 27.
Paczynski, B.: 1965, *Acta Astron.* **15**, 197.
Paczynski, B.: 1966, *Acta Astron.* **16**, 231.
Paczynski, B.: 1967, *Acta Astron.* **17**, 1, 193, 287, 355.
Paczynski, B.: 1971, *Ann. Rev. Astron. Astrophys.* **9**, 183–208.
Paczynski, B. and Ziolkowski, J.: 1967, *Acta Astron.* **17**, 7.
Payne-Gaposchkin, C. H.: 1957, *The Galactic Novae*, North-Holland Publ. Co., Amsterdam; pp. 250–251.
Pearce, J. A.: 1939, *Monthly Notices Roy. Astron. Soc.* **99**, 354.
Pekeris, C. L.: 1938, *Astrophys. J.* **88**, 189.
Penston, M. V.: 1973, *Astrophys. J.* **183**, 505.
Peters, P. C. and Mathews, J.: 1963, *Phys. Rev.* **131**, 435.
Petrie, R. M.: 1960, *Ann. Astrophys.* **23**, 744.
Petrie, R. M. and Heard, J. F.: 1970, *Publ. Dom. Astrophys. Obs.* **13**, 329.
Petro, L. D. : 1975, *Astrophys. J.* **195**, 709.
Petro, L. D. and Hiltner, W. A.: 1973, *Astrophys. J.* **181**, L39.
Petro, L. D. and Hiltner, W. A.: 1974, *Astrophys. J.* **190**, 661.
Petty, A. F.: 1973, *Astrophys. Space Sci.* **21**, 189.
Pickering, E. C.: 1880, *Proc. Amer. Acad. Sci.* **16**, 1.
Piotrowski, S. L.: 1947, *Astrophys. J.* **106**, 472.
Piotrowski, S. L.: 1948, *Astrophys. J.* **108**, 36 and 510.
Piotrowski, S. L.: 1964a, *Bull. Acad. Sci. Polonaise (ser. math. astr. phys.)* **12**, 323.
Piotrowski, S. L.: 1964b, *Acta Astron.* **14**, 251.
Piotrowski, S. L.: 1967, *Comm. Obs. Roy. Belg. Uccle*, (B), No. 17, p. 133 ff.
Plaskett, J. S. and Pearce, J. A.: 1931, *Publ. Dominion Astrophys. Obs.* 5, 99.
Plavec, M.: 1958, *Mem. Soc. Roy. Sci. Liège* (4) **20**, 11.
Plavec, M.: 1959, *Bull. Astron. Inst. Czech.* **10**, 185.
Plavec, M.: 1960, *Bull. Astron. Inst. Czech.* **11**, 148.
Plavec, M.: 1964, *Bull. Astron. Inst. Czech.* **15**, 156.
Plavec, M.: 1967a, *Comm. Obs. Roy. Belg. Uccle*, B, No. 17, pp. 87–97.
Plavec, M.: 1967b, *Bull. Astron. Inst. Czech.* **18**, 253 and 334.
Plavec, M.: 1968, in *Adv. Astron. Astrophys.* (ed. by Z. Kopal), Academic Press, New York **6**, 202–268.
Plavec, M.: 1968a, *Astrophys. Sapce Sci.* **1**, 239.
Plavec, M.: 1970, *Publ. Astron. Soc. Pacific* **82**, 957.
Plavec, M. and Kratochvîl, P.: 1964, *Bull. Astron. Inst. Czech.* **15**, 165.

Plavec, M. and Křĩž, S.: 1965, *Bull. Astron. Inst. Czech.* **16**, 297.
Plavec, M. and Smetanová, M.: 1959, *Bull. Inst. Astr. Czech.* **10**, 192.
Plavec, M. and Křĩž, S.: 1965, *Bull. Astron. Inst. Czech.* **16**, 297.
Plavec, M., Pěkny, Z., and Smetanová, M.: 1960, *Bull. Astron. Inst. Czech.* **11**, 180.
Plavec, M., Sehnal, L., and Mikuláš, J.: 1964, *Bull. Astron. Inst. Czech.* **15**, 171.
Plavec, M., Ulrich, R. K., and Polidan, R. S.: 1973, *Publ. Astron. Soc. Pacific* **85**, 769.
Poincaré, H.: 1885, *Acta Math.* 7, 259.
Poincaré, H.: 1892, *Les méthodes nouvelles de la mécanique céleste* (Gauthier-Villars, Paris), vol. 1, Chap. V.
Poincaré, H.: 1902, *Phil. Trans. Roy. Soc. London* **198** A.
Poincaré, H.: 1903, *Leçons sur les figures d'equilibre d'une masse fluide* (Gauthier-Villars, Paris).
Poincaré, H.: 1910, *Bull. Astron.* **27**, 321.
Pooley, G. C. and Ryle, M.: 1973, *Nature (phys. Sci.)* **244**, 270.
Popov, M. V.: 1964, *Per. Zvjozdy* **15**, 115.
Popper, D. M. and Plavec, M.: 1976, *Astrophys. J.* **205**, 462.
Prendergast, K. H.: 1960, *Astrophys. J.* **132**, 162.
Prendergast, K. H. and Taam, R. E.: 1974, *Astrophys. J.* **189**, 125.
Radau, R.: 1885a, *Comptes Rend. Acad. Paris* **100**, 972.
Radau, R.: 1885b, *Bull. Astron.* **2**, 157.
Radzievsky, V. V. and Surkova, L. P.: 1973, *Astron. Zh.* **50**, 1200.
Rappaport, S. and Joss, P. C.: 1977, *Nature* **266**, 123.
Refsdahl, S. and Weigert, A.: 1969, *Astrophys. Space Sci.* **3**, 175.
Rigal, J. L.: 1955, *Comptes Rend. Acad. Paris* **240**, 50.
Roach, G. F.: 1968, *Astrophys. Space Sci.* **1**, 32.
Roach, G. F.: 1975, *Astrophys. Space Sci.* **36**, 159.
Roberts, A. W.: 1902, *Proc. Roy. Soc. Edinburgh* **24**, 73.
Robinson, E. L.: 1972, *Publ. Astron. Soc. Pacific* **84**, 51.
Roche, E. A.: 1849, *Mém. Acad. Sci. Montpellier* **1**, 243 and 333.
Roche, E. A.: 1851, *Mém. Acad. Sci. Montpellier* **2**, 21.
Roche, E. A.: 1873, *Ann. de l'Acad. Sci. Montpellier* **8**, 235.
Rosenthal, J. E.: 1931, *Astron. Nachr.* **244**, 169.
Rosseland, S.: 1926, *Astrophys. J.* **63**, 342.
Rosseland, S.: 1932, *Astron. Publ. Oslo* No. 2.
Rossi, B.: 1973, in 'X-Ray and γ-Ray Astronomy' (ed. by H. Bradt and R. Giacconi), *IAU Symp.* **55**, 1.
Roxburgh, I. W.: 1966, *Astrophys. J.* **143**, 111.
Rucinski, S. M.: 1973, *Acta Astron.* **23**, 79.
Rucinski, S. M.: 1974, *Acta Astron.* **24**, 119.
Russell, H. N.: 1912, *Astrophys. J.* **35**, 315; **36**, 54.
Russell, H. N.: 1939, *Astrophys. J.* **90**, 641.
Russell, H. N. and Shapley, H.: 1912, *Astrophys. J.* **36**, 239 and 385.
Samter, H.: 1922, *Astron. Nachr.* **217**, 129.
Sandage, A., Osmer, P., Giacconi, R., Gorenstein, P., Gursky, H., Waters, J., Bradt, H., Garmire, G., Sreekantan, B., Oda, M., Osawa, K., and Jugaku, J.: 1966, *Astrophys. J.* **146**, 316.
Sanford, R. F.: 1949, *Astrophys. J.* **109**, 81.
Sargent, W. L. W.: 1957, M.Sc. Thesis (Manchester), unpublished.
Sato, K.: 1971, *Publ. Astron. Soc. Japan* **23**, 335.
Sawyer Hogg, H.: 1975, *David Dunlap Obs. Publ.* **3**, No. 6.
Scarfe, C. D., Barlow, D. J., and Niehaus, R. J.: 1976, *Astrophys. Space Sci.* **39**, 129.
Schatzman, E.: 1958, *White Dwarfs* (North-Holland Publ. Co., Amsterdam), p. 106.
Schatzman, E. and Rigal, J. L.: 1954, *Comptes Rend. Acad. Paris* **238**, 2392.
Schoenberg, M. and Chandrasekhar, S.: 1942, *Astrophys. J.* **96**, 161.
Schuerman, D. W.: 1972, *Astrophys. Space Sci.* **19**, 351.
Schwarzschild, K.: 1897, *Neue Ann. Sternw. München* **3**, 275.
Schwarzschild, M.: 1958, *Structure and Evolution of the Stars*, Princeton Univ. Press.
Seguin, F. H.: 1976, *Astrophys. J.* **207**, 848.
Semeniuk, I. and Paczynski, B.: 1968, *Acta Astron.* **18**, 33.
Shakura, N. I. and Sunyaev, R. A.: 1973, *Astron. Astrophys.* **24**, 337.
Shapley, H.: 1948, in *Harvard Centennial Symposia* (Harv. Obs. Mono. No. 7), p. 249 ff.
Shaw, G. B.: 1912, *Pygmalion* (cf. Collected Works, vol. 14; Constable and Co., Ltd., London), p. 253.
Siscoe, G. L. and Heinemann, M. A.: 1974, *Astrophys. Space Sci.* **31**, 363.

Sistero, R. F.: 1968, *Inf. Bull. Var. Stars*, No. 316.
Slettebak, A.: 1968, *Astrophys. J.* **115**, 1043.
Smak, J.: 1961, *Acta Astron.* **11**, 171.
Smak, J.: 1962, *Acta Astron.* **12**, 28.
Smak, J.: 1965, *Acta Astron.* **15**, 326.
Smak, J.: 1969, *Acta Astron.* **19**, 155.
Smak, J.: 1970, *Acta Astron.* **20**, 311.
Smak, J.: 1971a, *Acta Astron.* **21**, 15.
Smak, J.: 1971b, *Veröffentl. Remeis Sternw. Bamberg* **9**, 248–267.
Smak, J.: 1976, *Acta Astron.* **26**, 277.
Smarr, L. L. and Blandford, R.: 1976, *Astrophys. J.* **207**, 574.
Smart, W. M.: 1925, *Monthly Notices Roy. Astron. Soc.* **85**, 423.
Smith, R. C. and Worley, R.: 1974, *Monthly Notices Roy. Astron. Soc.* **167**, 199.
Söderhjelm, S.: 1974 *Astron. Astrophys.* **34**, 59.
Sorensen, S. A., Matsuda, T., and Sakurai, T.: 1974, *Progress Theor. Phys. Japan* **52**, 333.
Sorensen, S. A., Matsuda, T., and Sakurai, T.: 1975, *Astrophys. Space Sci.* **33**, 465.
Sterne, T. E.: 1937, *Monthly Notices Roy. Astron. Soc.* **97**, 582.
Sterne, T. E.: 1939, *Monthly Notices Roy. Astron. Soc.* **99**, 451.
Sterne, T. E.: 1941, *Proc. U.S. Nat. Acad. Sci.* **27**, 99.
Stothers, R.: 1974, *Astrophys. J.* **194**, 651.
Strong, I. B. and Klebesadel, R. W.: 1976, *Sci. Amer.* **235**, 66.
Struve, O.: 1944, *Astrophys. J.* **99**, 222.
Struve, O.: 1952, *Publ. Astron. Soc. Pacific* **64**, 180.
Struve, O. and Sahade, J.: 1957, *Publ. Astron. Soc. Pacific* **6**, 41.
Süer, H. B.: 1970, *J. Pure Appl. Sci. (Ankara)* **3**, 59.
Surkova, L. P.: 1973, *Per. Zvjozdy* **18**, 589.
Swings, P.: 1936, *Z. Astrophys.* **12**, 40.
Synge, J. L.: 1957, *The Relativistic Gas* (North-Holland Publ. Co., Amsterdam), Appendix (pp. 96–104).
Szebehely, V. G.: 1967, *Theory of Orbits: the Restricted Problem of Three Bodies*, Academic Press, New York.
Szebehely, V. G.: 1974, *Astron. J.* **79**, 1449.
Taam, R. E. and Faulkner, J.: 1975, *Astrophys.* **J. 198**, 435.
Takase, B.: 1953, *Ann. Tokyo Obs.* **3**, 192.
Takeda, S.: 1934, *Mem. Coll. Sci. Kyoto Univ., Ser. A.* **17**, 197.
Takeda, S.: 1937, *Mem. Coll. Sci. Kyoto Univ., Ser. A.* **20**, 47.
Thackeray, A. D.: 1964, in 'The Galaxy and Magellanic Clouds' *IAU-URSI Symp.* (ed. by F. J. Kerr and A. W. Rodgers) **20**, pp. 18–22.
Theokas, A.: 1977, *Astrophys. Space Sci.* **52**, 213.
Thomas, H. C.: 1974, *Astrophys. J.* **191**, L 25.
Thomas, L. H.: 1930, *Quart. J. Math. (Oxford)*, **1**, 239.
Thompson, W. (Lord Kelvin): 1863, *Phil. Trans. Roy. Soc. London* **153**, 612.
Thompson, W. (Lord Kelvin): 1886, *Proc. Roy. Soc. Edinburgh* **5**, 512.
Tisserand, F.: 1884, *Comptes Rend. Acad. Paris* **99**, 579.
Tisserand, F.: 1889, *Traité de la mécanique céleste* (Gauthier-Villars, Paris), vol. I.
Tisserand, F.: 1891, *Traité de la mécanique céleste* (Gauthier-Villars, Paris), vol. II.
Tokis, J. N.: 1973, *Astrophys. Space Sci.* **26**, 447 and 477.
Tokis, J. N.: 1974, *Astrophys. Space Sci.* **31**, 349.
Treanor, P. J.: 1960, *Monthly Notices Roy. Astron. Soc.* 121, 503.
Trimble, V. and Thorne, K. S.: 1969, *Astrophys. J.* **156**, 1013.
Underhill, A.: 1969, *Astrophys. Space Sci.* **3**, 109.
Underhill, A.: 1973, in 'Wolf-Rayet and High-Temperature Stars' (ed. by M. K. V. Bappu and J. Sahade), *IAU Symp.* 49, 237–252.
Van Horn, H. M., Sofia, S. Savedoff, M. P., Guthie, J. G. and Berg, R. A.: 1975, *Science* **188**, 930.
Véronnet, A.: 1912, *J. Math.* **(6) 8**, 331.
Vogel, H. C.: 1890, *Astron. Nachr.* **123**, 289.
Vogt, H.: 1928, *Astron. Nachr.* **232**, 1.
Walker, M.: 1956, *Astrophys. J.* **123**, 68.
Walker, M.: 1971, *Kl. Veröffentl. Remeis Sternw. Bamberg* **9**, 243.
Walter, K.: 1931, *Königsberg Obs. Veröff.*, No. 2.

Warner, B.: 1971, *Kl. Veröffentl. Remeis Sternw. Bamberg* **9**, 144.
Warner, B. and Peters, W. L.: 1972, *Monthly Notices Roy. Astron. Soc.* **160**, 15.
Weigert, A.: 1967, *Comm. Obs. Royal Belg. Uccle* (B) No. 17, p. 79.
Whelan, J. A. J.: 1972, *Monthly Notices Roy. Astron, Soc.* **156**, 115.
Whelan, J. A. J., Worden, S. P., and Mochnacki, S. W.: 1973. *Astrophys. J.* **183**, 133.
Whelan, J. A. J., Mochnacki, S. W., and Worden, S. P.: 1974, *Monthly Notices Roy. Astron. Soc.* **168**, 31.
Will, C. M.: 1975, *Astrophys. J.* **196**, L 3.
Wilson, D. C.: 1966, *Astrophys. J.* **144**, 695.
Wilson, R. E.: 1971a, *Astrophys. J.* **170**, 529.
Wilson, R. E.: 1971b, *Nature (Phys. Sci.)* **234**, 406.
Wilson, R. E.: 1972, *Astrophys. Space Sci.* **19**, 165.
Wilson, R. E. and Biermann, P.: 1976, *Astron. Astrophys.* **48**, 349.
Woolley, R. v. d. R., Epps, E. A., Penston, M. J., and Pococks, S. B.: 1970, *Roy. Obs. Ann.*, **No. 5**.
Worden, S. P. and Whelan, J. A. J.: 1973, *Monthly Notices Roy. Astron. Soc.* **163**, 391.
Yabushita, S.: 1966, *Monthly Notices Roy. Astron. Soc.* **133**, 133.
Yamasaki, A.: 1975, *Astrophys. Space Sci.* **34**, 413.
Young, A. and Lanning, H. H.: 1975, *Publ. Astron. Soc. Pacific* **87**, 461.
Zahn, J. P.: 1966, *Ann. Astrophys.* **29**, 313, 489 and 565.
Zahn, J. P.: 1967, *Comm. Obs. Roy. Belg. Uccle* (B), No. 17, p. 124 ff.
Zahn, J. P.: 1970, *Astron. Astrophys.* **4**, 452.
Zeipel, H. v.: 1924a, in *Festschrift für Hugo v. Seeliger,* Leipzig; p. 144 ff.
Zeipel, H. v.: 1924b, *Monthly Notices Roy. Astron. Soc.*, **84**, 702.
Zeldovich, Ya. B. and Novikov, I. T.: 1967, *Relativistic Astrophysics,* Izd. Nauka, Moscow.
Ziolkowski, J.: 1969, *Astrophys. Space Sci.* **3**, 14.
Zwicky, F.: 1957, *Morphological Astronomy*, Julius Springer, Berlin.

INDEX OF NAMES

INDEX OF SUBJECTS

ASTROPHYSICS AND SPACE SCIENCE LIBRARY

Edited by

J. E. Blamont, R. L. F. Boyd, L. Goldberg, C. de Jager, Z. Kopal, G. H. Ludwig, R. Lüst, B. M. McCormac, H. E. Newell, L. I. Sedov, Z. Švestka, and W. de Graaff

1. C. de Jager (ed.), *The Solar Spectrum, Proceedings of the Symposium held at the University of Utrecht, 26–31 August, 1963.* 1965, XIV + 417 pp.
2. J. Ortner and H. Maseland (eds.), *Introduction to Solar Terrestrial Relations, Proceedings of the Summer School in Space Physics held in Alpbach, Austria, July 15–August 10, 1963 and Organized by the European Preparatory Commission for Space Research.* 1965, IX + 506 pp.
3. C. C. Chang and S. S. Huang (eds.), *Proceedings of the Plasma Space Science Symposium, held at the Catholic University of America, Washington, D.C., June 11–14, 1963.* 1965, IX + 377 pp.
4. Zdeněk Kopal, *An Introduction to the Study of the Moon.* 1966, XII + 464 pp.
5. B. M. McCormac (ed.), *Radiation Trapped in the Earth's Magnetic Field. Proceedings of the Advanced Study Institute, held at the Chr. Michelsen Institute, Bergen, Norway, August 16–September 3, 1965.* 1966, XII + 901 pp.
6. A. B. Underhill, *The Early Type Stars.* 1966, XII + 282 pp.
7. Jean Kovalevsky, *Introduction to Celestial Mechanics.* 1967, VIII + 427 pp.
8. Zdeněk Kopal and Constantine L. Goudas (eds.), *Measure of the Moon. Proceedings of the 2nd International Conference on Selenodesy and Lunar Topography, held in the University of Manchester, England, May 30–June 4, 1966.* 1967, XVIII + 479 pp.
9. J. G. Emming (ed.), *Electromagnetic Radiation in Space. Proceedings of the 3rd ESRO Summer School in Space Physics, held in Alpbach, Austria, from 19 July to 13 August, 1965.* 1968, VIII + 307 pp.
10. R. L. Carovillano, John F. McClay, and Henry R. Radoski (eds.), *Physics of the Magnetosphere, Based upon the Proceedings of the Conference held at Boston College, June 19–28, 1967.* 1968, X + 686 pp.
11. Syun-Ichi Akasofu, *Polar and Magnetospheric Substorms.* 1968, XVIII + 280 pp.
12. Peter M. Millman (ed.), *Meteorite Research. Proceedings of a Symposium on Meteorite Research, held in Vienna, Austria, 7–13 August, 1968.* 1969, XV + 941 pp.
13. Margherita Hack (ed.), *Mass Loss from Stars. Proceedings of the 2nd Trieste Colloquium on Astrophysics, 12–17 September, 1968.* 1969, XII + 345 pp.
14. N. D'Angelo (ed.), *Low-Frequency Waves and Irregularities in the Ionosphere. Proceedings of the 2nd ESRIN-ESLAB Symposium, held in Frascati, Italy, 23–27 September, 1968.* 1969, VII + 218 pp.
15. G. A. Partel (ed.), *Space Engineering. Proceedings of the 2nd International Conference on Space Engineering, held at the Fondazione Giorgio Cini, Isola di San Giorgio, Venice, Italy, May 7–10, 1969.* 1970, XI + 728 pp.
16. S. Fred Singer (ed.), *Manned Laboratories in Space. Second International Orbital Laboratory Symposium.* 1969, XIII + 133 pp.
17. B. M. McCormac (ed.), *Particles and Fields in the Magnetosphere. Symposium Organized by the Summer Advanced Study Institute, held at the University of California, Santa Barbara, Calif., August 4–15, 1969.* 1970, XI + 450 pp.
18. Jean-Claude Pecker, *Experimental Astronomy.* 1970, X + 105 pp.
19. V. Manno and D. E. Page (eds.), *Intercorrelated Satellite Observations related to Solar Events. Proceedings of the 3rd ESLAB/ESRIN Symposium held in Noordwijk, The Netherlands, September 16–19, 1969.* 1970, XVI + 627 pp.
20. L. Mansinha, D. E. Smylie, and A. E. Beck, *Earthquake Displacement Fields and the Rotation of the Earth, A NATO Advanced Study Institute Conference Organized by the Department of Geophysics, University of Western Ontario, London, Canada, June 22–28, 1969.* 1970, XI + 308 pp.
21. Jean-Claude Pecker, *Space Observatories.* 1970, XI + 120 pp.
22. L. N. Mavridis (ed.), *Structure and Evolution of the Galaxy. Proceedings of the NATO Advanced Study Institute, held in Athens, September 8–19, 1969.* 1971, VII + 312 pp.
23. A. Muller (ed.), *The Magellanic Clouds. A European Southern Observatory Presentation: Principal Prospects, Current Observational and Theoretical Approaches, and Prospects for Future Research, Based on the Symposium on the Magellanic Clouds, held in Santiago de Chile, March 1969, on the Occasion of the Dedication of the European Southern Observatory.* 1971, XII + 189 pp.

24. B. M. McCormac (ed.), *The Radiating Atmosphere. Proceedings of a Symposium Organized by the Summer Advanced Study Institute, held at Queen's University, Kingston, Ontario, August 3–14, 1970.* 1971, XI + 455 pp.
25. G. Fiocco (ed.), *Mesospheric Models and Related Experiments. Proceedings of the 4th ESRIN-ESLAB Symposium, held at Frascati, Italy, July 6–10, 1970.* 1971, VIII + 298 pp.
26. I. Atanasijević, *Selected Exercises in Galactic Astronomy.* 1971, XII + 144 pp.
27. C. J. Macris (ed.), *Physics of the Solar Corona. Proceedings of the NATO Advanced Study Institute on Physics of the Solar Corona, held at Cavouri-Vouliagmeni, Athens, Greece, 6–17 September 1970.* 1971, XII + 345 pp.
28. F. Delobeau, *The Environment of the Earth.* 1971, IX + 113 pp.
29. E. R. Dyer (general ed.), *Solar-Terrestrial Physics/1970. Proceedings of the International Symposium on Solar-Terrestrial Physics, held in Leningrad, U.S.S.R., 12–19 May 1970.* 1972, VIII + 938 pp.
30. V. Manno and J. Ring (eds.), *Infrared Detection Techniques for Space Research. Proceedings of the 5th ESLAB-ESRIN Symposium, held in Noordwijk, The Netherlands, June 8–11, 1971.* 1972, XII + 344 pp.
31. M. Lecar (ed.), *Gravitational N-Body Problem. Proceedings of IAU Colloquium No. 10, held in Cambridge, England, August 12–15, 1970.* 1972, XI + 441 pp.
32. B. M. McCormac (ed.), *Earth's Magnetospheric Processes. Proceedings of a Symposium Organized by the Summer Advanced Study Institute and Ninth ESRO Summer School, held in Cortina, Italy, August 30–September 10, 1971.* 1972, VIII + 417 pp.
33. Antonin Rükl, *Maps of Lunar Hemispheres.* 1972, V + 24 pp.
34. V. Kourganoff, *Introduction to the Physics of Stellar Interiors.* 1973, XI + 115 pp.
35. B. M. McCormac (ed.), *Physics and Chemistry of Upper Atmospheres. Proceedings of a Symposium Organized by the Summer Advanced Study Institute, held at the University of Orléans, France, July 31–August 11, 1972.* 1973, VIII + 389 pp.
36. J. D. Fernie (ed.), *Variable Stars in Globular Clusters and in Related Systems. Proceedings of the IAU Colloquium No. 21, held at the University of Toronto, Toronto, Canada, August 29–31, 1972.* 1973, IX + 234 pp.
37. R. J. L. Grard (ed.), *Photon and Particle Interaction with Surfaces in Space. Proceedings of the 6th ESLAB Symposium, held at Noordwijk, The Netherlands, 26–29 September, 1972.* 1973, XV + 577 pp.
38. Werner Israel (ed.), *Relativity, Astrophysics and Cosmology. Proceedings of the Summer School, held 14–26 August, 1972, at the BANFF Centre, BANFF, Alberta, Canada.* 1973, IX + 323 pp.
39. B. D. Tapley and V. Szebehely (eds.), *Recent Advances in Dynamical Astronomy. Proceedings of the NATO Advanced Study Institute in Dynamical Astronomy, held in Cortina d'Ampezzo, Italy, August 9–12, 1972.* 1973, XIII + 468 pp.
40. A. G. W. Cameron (ed.), *Cosmochemistry. Proceedings of the Symposium on Cosmochemistry, held at the Smithsonian Astrophysical Observatory, Cambridge, Mass., August 14–16, 1972.* 1973, X + 173 pp.
41. M. Golay, *Introduction to Astronomical Photometry.* 1974, IX + 364 pp.
42. D. E. Page (ed.), *Correlated Interplanetary and Magnetospheric Observations. Proceedings of the 7th ESLAB Symposium, held at Saulgau, W. Germany, 22–25 May, 1973.* 1974, XIV + 662 pp.
43. Riccardo Giacconi and Herbert Gursky (eds.), *X-Ray Astronomy.* 1974, X + 450 pp.
44. B. M. McCormac (ed.), *Magnetospheric Physics. Proceedings of the Advanced Summer Institute, held in Sheffield, U.K., August 1973.* 1974, VII + 399 pp.
45. C. B. Cosmovici (ed.), *Supernovae and Supernova Remnants. Proceedings of the International Conference on Supernovae, held in Lecce, Italy, May 7–11, 1973.* 1974, XVII + 387 pp.
46. A. P. Mitra, *Ionospheric Effects of Solar Flares.* 1974, XI + 294 pp.
47. S.-I. Akasofu, *Physics of Magnetospheric Substorms.* 1977, XVIII + 599 pp.
48. H. Gursky and R. Ruffini (eds.), *Neutron Stars, Black Holes and Binary X-Ray Sources.* 1975, XII + 441 pp.
49. Z. Švestka and P. Simon (eds.), *Catalog of Solar Particle Events 1955–1969. Prepared under the Auspices of Working Group 2 of the Inter-Union Commission on Solar-Terrestrial Physics.* 1975, IX + 428 pp.
50. Zdeněk Kopal and Robert W. Carder, *Mapping of the Moon.* 1974, VIII + 237 pp.
51. B. M. McCormac (ed.), *Atmospheres of Earth and the Planets. Proceedings of the Summer Advanced Study Institute, held at the University of Liège, Belgium, July 29--August 8, 1974.* 1975, VII + 454 pp.
52. V. Formisano (ed.), *The Magnetospheres of the Earth and Jupiter. Proceedings of the Neil Brice Memorial Symposium, held in Frascati, May 28–June 1, 1974.* 1975, XI + 485 pp.

53. R. Grant Athay, *The Solar Chromosphere and Corona: Quiet Sun.* 1976, XI + 504 pp.
54. C. de Jager and H. Nieuwenhuijzen (eds.), *Image Processing Techniques in Astronomy. Proceedings of a Conference, held in Utrecht on March 25–27, 1975*, XI + 418 pp.
55. N. C. Wickramasinghe and D. J. Morgan (eds.), *Solid State Astrophysics. Proceedings of a Symposium, held at the University College, Cardiff, Wales, 9–12 July 1974.* 1976, XII + 314 pp.
56. John Meaburn, *Detection and Spectrometry of Faint Light.* 1976, IX + 270 pp.
57. K. Knott and B. Battrick (eds.), *The Scientific Satellite Programme during the International Magnetospheric Study. Proceedings of the 10th ESLAB Symposium, held at Vienna, Austria, 10–13 June 1975.* 1976, XV + 464 pp.
58. B. M. McCormac (ed.), *Magnetospheric Particles and Fields. Proceedings of the Summer Advanced Study School, held in Graz, Austria, August 4–15, 1975.* 1976, VII + 331 pp.
59. B. S. P. Shen and M. Merker (eds.), *Spallation Nuclear Reactions and Their Applications.* 1976, VIII + 235 pp.
60. Walter S. Fitch (ed.), *Multiple Periodic Variable Stars. Proceedings of the International Astronomical Union Colloquium No. 29, Held at Budapest, Hungary, 1–5 September 1975.* 1976, XIV + 348 pp.
61. J. J. Burger, A. Pedersen, and B. Battrick (eds.), *Atmospheric Physics from Spacelab. Proceedings of the 11th ESLAB Symposium, Organized by the Space Science Department of the European Space Agency, held at Frascati, Italy, 11–14 May 1976.* 1976, XX + 409 pp.
62. J. Derral Mulholland (ed.), *Scientific Applications of Lunar Laser Ranging. Proceedings of a Symposium held in Austin, Tex., U.S.A., 8–10 June, 1976.* 1977, XVII + 302 pp.
63. Giovanni G. Fazio (ed.), *Infrared and Submillimeter Astronomy. Proceedings of a Symposium held in Philadelphia, Penn., U.S.A., 8-10 June, 1976.* 1977, X+226 pp.
64. C. Jaschek and G. A. Wilkins (eds.), *Compilation, Critical Evaluation and Distribution of Stellar Data. Proceedings of the International Astronomical Union Colloquium No. 35, held at Strasbourg, France, 19-21 August, 1976.* 1977, XIV+316 pp.
65. M. Friedjung (ed.), *Novae and Related Stars. Proceedings of an International Conference held by the Institut d'Astrophysique, Paris, France, 7-9 September, 1976.* 1977, XIV+228 pp.
66. David N. Schramm (ed.), *Supernovae. Proceedings of a Special IAU Session on Supernovae held in Grenoble, France, 1 September, 1976.* 1977, X+192 pp.
67. Jean Audouze (ed.), *CNO Isotopes in Astrophysics. Proceedings of a Special IAU Session held in Grenoble, France, 30 August, 1976.* 1977, XIII+195 pp.
68. Z. Kopal, *Dynamics of Close Binary Systems*, forthcoming.
69. A. Bruzek and C. J. Durrant (eds.), *Illustrated Glossary for Solar and Solar-Terrestrial Physics.* 1977, approx. 216 pp.
70. H. van Woerden (ed.), *Topics in Interstellar Matter.* 1977, VIII + 295 pp.
71. M. A. Shea, D. F. Smart, and T. S. Wu (eds.), *Study of Travelling Interplanetary Phenomena.* 1977, XII+439 pp.